ANNUAL REVIEW OF
NEUROSCIENCE

EDITORIAL COMMITTEE (1995)

ANNUAL REVIEW OF NEUROSCIENCE

VOLUME 18, 1995

W. MAXWELL COWAN, *Editor*
Howard Hughes Medical Institute

ERIC M. SHOOTER, *Associate Editor*
Stanford University School of Medicine

CHARLES F. STEVENS, *Associate Editor*
Salk Institute for Biological Studies

RICHARD F. THOMPSON, *Associate Editor*
University of Southern California

ANNUAL REVIEWS INC.   4139 EL CAMINO WAY   P.O. 10139   PALO ALTO, CALIFORNIA 94303-0139

ANNUAL REVIEWS INC.
Palo Alto, California, USA

International Standard Serial Number: 0147–006X
International Standard Book Number: 0–8243–2418-8

Annual Review and publication titles are registered trademarks of Annual Reviews Inc.

⊗ The paper used in this publication meets the minimum requirements of American National Standard for Information Sciences—Permanence of Paper for Printed Library Materials, ANSI Z39.48-1984.

Typesetting by Kachina Typesetting Inc., Tempe, Arizona; John Olson, President; Jeannie Kaarle, Typesetting Coordinator; and by the Annual Reviews Inc. Editorial Staff

PRINTED AND BOUND IN THE UNITED STATES OF AMERICA

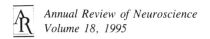

Annual Review of Neuroscience
Volume 18, 1995

CONTENTS

(*Continued*)   v

vi   CONTENTS   *(continued)*

SOME RELATED ARTICLES IN OTHER *ANNUAL REVIEWS*

From the *Annual Review of Cell Biology*, Volume 10 (1994):

Receptor Protein-Tyrosine Kinases and Their Signal Transduction Pathways, Peter van der Geer, Tony Hunter, and Richard A. Lindberg
Dyneins: Molecular Structure and Cellular Function, E. L. F. Holzbaur and Richard Vallee
Cell Biology of the Amyloid β-Protein Precursor and the Mechanism of Alzheimer's Disease, Dennis J. Selkoe

From the *Annual Review of Pharmacology and Toxicology*, Volume 35 (1995):

P_2-*Purinergic Receptors: Subtype-Associated Signaling Responses and Structure*, T. Kendall Harden, José L. Boyer, and Robert A. Nicholas
Pharmacology of Cannabinoid Receptors, Allyn C. Howlett
Adenosine-Receptor Subtypes: Characterization and Therapeutic Regulation, Mark E. Olah and Gary L. Stiles
Inhibition of Constitutive and Inducible Nitric Oxide Synthase: Potential Selective Inhibition, J. M. Fukuto and G Chaudhuri
Nitric Oxide in the Nervous System, J. Zhang and S. H. Snyder
Polyamine Regulation of N-*Methyl-D-Aspartate Receptor Channels*, Robert L. Macdonald and David M. Rock
Relationships Between Lead-Induced Learning Impairments and Changes in Dopaminergic, Cholinergic, and Glutaminergic Neurotransmitter System Functions, D. A. Cory-Slechta

From the *Annual Review of Physiology*, Volume 57 (1995):

Molecular Pathway of the Skeletal Muscle Sodium Channel, Robert L. Barchi
Review of Mechanosensitive Channels, Henry Sackin
Mechanisms of Activation of Glutamate Receptors and the Time Course of Excitatory Synaptic Currents, B. Edmonds, A. J. Gibbs, and D. Colquhoun
Mechanisms of Activation of Muscle Nicotinic Acetylcholine Receptors and the Time Course of Endplate Currents, B. Edmonds, A. J. Gibbs, and D. Colquhoun
Physiological Diversity of Nicotinic Acetylcholine Receptors Expressed by Vertebrate Neurons, Daniel S. McGehee and Lorna W. Role
The 5-HT_3 Receptor Channel, Meyer B. Jackson and Jerrell L. Yakel
Nitric Oxide Signaling in the Central Nervous System, J. Garthwaite and C. L. Boulton
Nitric Oxide as a Neurotransmitter in Peripheral Nerves: Nature of Transmitter and Mechanism of Transmission, M. J. Rand and C. G. Li

From the *Annual Review of Psychology,* Volume 46 (1995):

ANNUAL REVIEWS INC. is a nonprofit scientific publisher established to promote the advancement of the sciences. Beginning in 1932 with the *Annual Review of Biochemistry*, the Company has pursued as its principal function the publication of high-quality, reasonably priced *Annual Review* volumes. The volumes are organized by Editors and Editorial Committees who invite qualified authors to contribute critical articles reviewing significant developments within each major discipline. The Editor-in-Chief invites those interested in serving as future Editorial Committee members to communicate directly with him. Annual Reviews Inc. is administered by a Board of Directors, whose members serve without compensation.

For the convenience of readers, a detachable order form/envelope is bound into the back of this volume.

Annu. Rev. Neurosci. 1995. 18:1–18

GENE TARGETING IN ES CELLS

Philippe Soriano

Program in Molecular Medicine, Fred Hutchinson Cancer Research Center, 1124 Columbia Street, Seattle, Washington 98104

KEY WORDS: homologous recombination, gene disruption, mouse genetics

---

HISTORY OF TARGETED GENETICS IN THE MOUSE

Mutant mice have been collected for many years. At the start, this collection, referred to as the mouse fancy, included mice selected on the basis of morphological criteria, such as tail length or coat color. In the early 1900s, mouse genetics began in earnest with the derivation of inbred strains, and later, irradiation or exposure to a variety of chemical agents was used to derive novel mutations. From there on, however, establishing a link between the mutated locus and an individual gene proved to be difficult and time consuming. One of the first neurological mutants analyzed in this manner for which the gene could be identified was the shiverer mutation, which contains a deletion of part of the myelin basic protein gene (Roach et al 1985).

The development of transgenic technologies provided an alternative method to derive mouse mutants. Transgenic mice were first derived by infecting embryos with viruses (Jaenisch & Mintz 1974, Jaenisch 1976) and later derived by using DNA microinjection into zygotes (Gordon et al 1980; Brinster et al 1981; Costantini & Lacy 1981; Wagner et al 1981a,b). In some instances, insertion of a transgene within a cellular gene might be expected to cause a mutation, in which case the transgene could serve as a tag to identify the mutated gene. The first example of recessive insertional mutagenesis in a transgenic mouse line was observed in 1983 when proviral insertion of Moloney murine leukemia virus (Mo-MuLV) into the Mov 13 strain was shown to cause embryonic lethality (Jaenisch et al 1983). Schnieke et al (1983) subsequently demonstrated that the provirus had inserted into the first intron of the α1(I) collagen gene, thus leading to embryonic lethality at midgestation. Many

1

other cases of mutations by transgene insertion have followed, including several cases of insertion in previously identified neurological loci, such as purkinje cell degeneration (Krulewski et al 1989) or *dystonia musculorum* (Kothary et al 1988). Several reviews cover insertional mutagenesis (Gridley et al 1987, Gridley 1991). In general, researchers have observed that whereas DNA microinjection often leads to rearrangements in the genome, which complicates if not prevents the identification of the mutant loci, retroviral infection leads to a predictable, clean insertion, which allows the mutated locus to be cloned with relative ease.

Insertional mutagenesis has one serious limitation, however: identifying an insertional mutant requires considerable effort. Because most phenotypes are recessive, transgenic mice have to be bred to homozygosity, and on the average, only one out of twenty strains exhibits a mutant phenotype. Cloning the mutated locus can be time consuming, especially if the mutant strain was generated by DNA microinjection. Therefore, alternative methods for producing mutations in transgenic mice were developed. Following the pioneering work of Tarkowski, Mintz, and Gardner in the 1960s, in which chimeras were formed between embryos with different genotypes (Tarkowski 1961, Mintz 1962, Gardner 1968), researchers suspected that there might be a stem cell that could colonize all lineages of a chimeric mouse, including the germ line. At first, several groups of investigators used embryonal carcinoma cells for this purpose (reviewed in Robertson & Bradley 1986). Unfortunately, these cells never reproducibly contributed to the germ line of chimeric animals. Subsequent efforts were made to derive stem cells from the early embryo proper. In 1981, two groups reported the derivation of embryonic stem (ES) cells from the inner cell mass of preimplantation embryos (Evans & Kaufman 1981, Martin 1981). In 1984, Bradley et al (1984) demonstrated that ES cells could contribute to the germ line of chimeric mice with high efficiency.

In many respects, the limitations of insertional mutagenesis have been resolved by the use of gene traps in ES cells (Gossler et al 1989, Friedrich & Soriano 1991, von Melchner et al 1992). In this approach, a promoterless reporter gene (typically βgalactosidase) is introduced into ES cells. Expression of the reporter only occurs if the reporter has inserted in the correct orientation and reading frame within another gene, and βgalactosidase expression can be used to monitor the normal expression of the mutated gene. In one study performed in our laboratory, 20 out of 42 such strains exhibited recessive mutant phenotypes (Friedrich & Soriano 1991; G Friedrich, Z Chen & P Soriano, unpublished observations). However, in these studies, as in the analysis of any classical mutation, it is nearly impossible to predict the nature of the mutated gene based on mutant phenotype. The reverse genetic approach offered by gene targeting offers an attractive way out of this impasse.

With the advent of ES cells, it became possible to envision deriving mice from ES cells in which a specific genetic modification had been selected. This method was first used in 1987 when two groups reported the derivation of hypoxanthine-phosphoribosyl transferase (HPRT) deficient mice (Hooper et al 1987, Kuehn et al 1987). To derive mutations in nonselectable genes, however, targeting strategies relied heavily on the ability of exogenous DNA to undergo homologous recombination with a target locus. At first, the efficiency of homologous recombination was very low (Smithies et al 1985, Thomas et al 1986), but better vector design soon made it possible to target genes in ES cells (Thomas & Capecchi 1987). Thus, it became possible to derive mouse mutants in any gene by using gene targeting in ES cells. The first reports on the phenotypic consequences of deriving mutant mice using this approach date only from 1990, but they have been followed by an explosion of activity in many laboratories. Because the field is advancing very rapidly, I present an outline of the methodologies used and examples of results that highlight some of the contributions of this approach in neurobiology.

HOMOLOGOUS RECOMBINATION IN ES CELLS

Two types of vectors can be used for targeting mutations to individual loci: replacement (Ω) vectors or insertion vectors (Thomas & Capecchi 1987). Both types of vectors are represented schematically in Figure 1. A replacement vector (Figure 1A) is linearized in such a way that the vector sequences remain colinear with the target sequences. The vector can be linearized either within the region of homology or in the plasmid backbone. This feature allows a screening procedure, known as the positive-negative selection, to select against nonhomologous recombination events (Mansour et al 1988). In this approach, a thymidine kinase gene from herpes simplex virus (HSVTK) is inserted at either end of the linearized vector. Cells in which the construct has integrated randomly can be selected against using a toxic nucleotide analogue, such as gancyclovir or 1-(2-deoxy-2-fluoro-β-D-arabinofuranosyl)-5-iodouracil (FIAU), which are substrates for the HSVTK gene, but not cellular thymidine kinases. An insertion vector (Figure 1B) is linearized within the region of homology, and homologous recombination will lead to a duplication of genomic sequences. Each type of vector has its own distinct advantages, as described below, but replacement vectors have been used more broadly.

Several parameters govern the rate at which homologous recombination can occur. First, higher recombination rates are observed with increases in the length of vector homology, mostly up to 6 kb (Thomas & Capecchi 1987, Hasty et al 1991b, Deng & Capecchi 1992). Second, te Riele et al (1992) observed that using DNA isogenic with the strain of mice from which the ES cells are derived (typically 129Sv) promotes high recombination rates, presumably because mismatches between vector and target sequences prevent

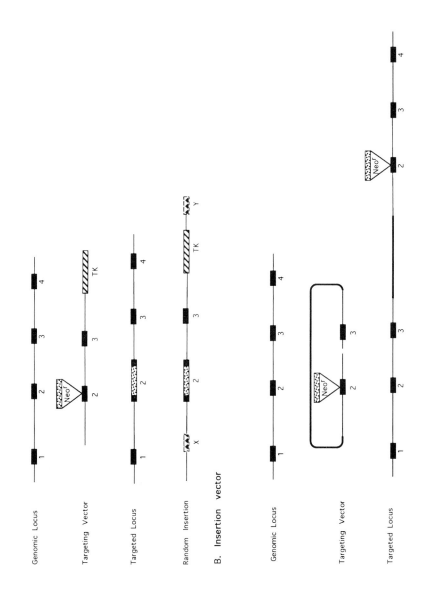

A. Replacement vector

B. Insertion vector

sequence recognition. Last, Hasty et al (1991c, 1992) have observed that higher rates of recombination are obtained by using insertion instead of replacement vectors, although Deng & Capecchi (1992) have suggested that this result might have been due to the use of nonisogenic DNA. Although the mechanism and the enzymology of homologous recombination remain important issues to resolve, the frequency of homologous recombination is often very high, thus making it simple to isolate several targeted clones among a few hundred colonies. For practical approaches, the reader is referred to comprehensive reviews (e.g. Ramírez-Solis et al 1993).

Most of the work done so far using gene targeting has been directed towards the generation of null mutations. This can be done either by inserting the drug resistance marker into an exon critical for the gene's function or by making a deletion encompassing one or several exons of the gene. However, the collection of hypomorphic alleles at individual murine loci, such as Steel or White Spotting (Besmer 1991), and genetic studies in a variety of other species underline the importance of creating an allelic series for each gene. This can be accomplished with several different procedures.

First, small mutations can be created, such as making an amino acid substitution, by using either the hit-and-run/in-and-out method (Hasty et al 1991a, Valancius & Smithies 1991) or the tag-and-exchange method (Askew et al 1993). In the hit-and-run approach, small changes are introduced into the locus by a two-step recombination procedure (Figure 2A). In the first step, an insertion vector with the small mutation is inserted into the target locus along with selectable markers that allow both forward and reverse selection. Only forward selection is used in this step; homologous recombination that uses such a vector will create a duplication of genomic sequences. In the second step, an intrachromosomal recombination event that excises the integrated vector sequences can be isolated by negative selection. In the tag-and-exchange method, the mutation is created by using two successive replacement vectors (Figure 2B). The first vector contains markers that can be selected both forward and backward, but only forward selection is used to isolate targeted events (this can be used to generate a null allele, for instance). Recombination of the second vector, which only carries the small mutation, can be selected in the second step by using negative selection.

Mutations can also be created that will act at specific times and/or locations during development or adulthood by using site-specific recombinases, such as *Cre* from bacteriophage P1 and FLP from yeast (for a review, see Kilby et al

----

Figure 1 Targeting vectors: (*A*) replacement vector and (*B*) insertion vector. Shaded boxes (numbered 1 to 4) represent exons; thin lines, introns; thick lines, the plasmid backbone; stippled box, the neomycin resistance (*Neo^r*) gene; hatched box, the HSV TK (TK) gene; and X and Y, exons of a nontargeted gene.

A. Hit and Run

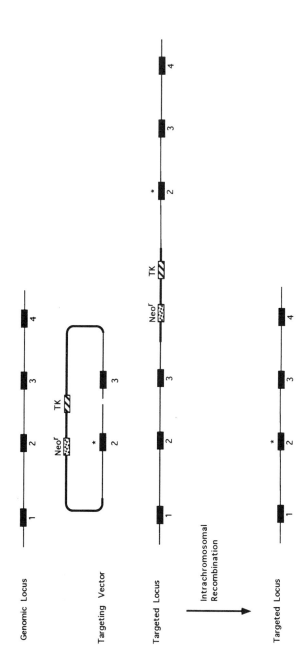

Genomic Locus

Targeting Vector

Targeted Locus

Intrachromosomal
Recombination

Targeted Locus

B. Tag and Exchange

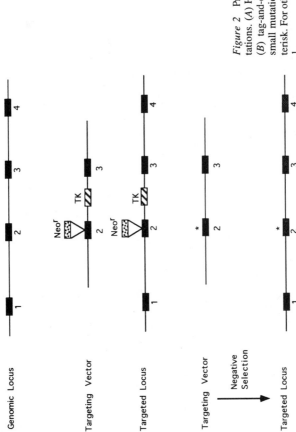
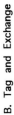

Figure 2 Production of small mutations. (*A*) Hit-and-run method and (*B*) tag-and-exchange method. The small mutation is denoted by an asterisk. For other symbols, see Figure 1.

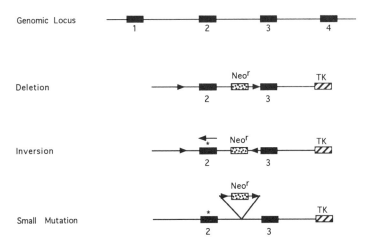

Figure 3 Potential uses of site-specific recombinases. *Cre* recombinase will catalyze excision between two *lox* sites (*arrowheads*) if they are in the same orientation, and inversion if the *lox* sites are in the opposite orientation. Three replacement vectors are represented. In a deletion of exon 2 (to produce a null allele), the *lox* sites are in the same orientation in the introns; in an inversion procedure (e.g. to introduce the activating mutation * into a null allele, exon 2 is in the opposite orientation to the gene), the *lox* sites in the introns are inverted. To create a small mutation, and delete the *Neo* gene leaving a single *lox* site behind, *Cre* recombinase can be introduced transiently in the targeted ES cell clones. For other symbols, see Figure 1.

1993). The *Cre* and FLP recombinases can catalyze recombination between two *lox*P or two FRT sites, respectively (Figure 3). If these sites are in the same orientation, the intervening sequences are deleted. If they are in opposite orientation, the intervening sequences are inverted. This system has been used successfully in mammalian cells, including ES cells, and in transgenic mice. In one study (Orban et al 1992), tissue-specific DNA recombination in thymocytes was observed by mating mice containing a *Cre* expression cassette under the control of the *lck* distal promoter with mice containing βgalactosidase flanked by *lox*P sites. X-gal positive thymocytes were observed in the progeny of these mice. A number of possible experiments can be considered using this system, such as tissue-specific gene inactivation, transgene insertion at *lox*P sites targeted to a specific locus, and the creation of small mutations.

TARGETED GENE MUTANTS

At the time of this writing, over one hundred reports have been published on targeted gene disruptions, and perhaps several hundred are being analyzed. Because of this rapid rate of progress, I focus on only some of these mutations, particularly mutations of genes expressed in the nervous system, to illustrate the contributions made by these studies and the limitations of this approach.

Developmental Mutants

One of the first genes to be disrupted by gene targeting which led to a neural phenotype was *Wnt1*. *Wnt1*, originally known as *Int1*, was first identified as a preferred site of integration of mouse mammary tumor virus (MMTV). It was later shown to be homologous to the *wingless* (*wg*) gene of *Drosophila,* and to encode a secreted glycoprotein. *Wnt1* may act like a growth factors by signaling through a yet unidentified receptor (for a review, see Nusse & Varmus 1992). In the mouse, there are at least 10 members in the *Wnt* gene family, which have restricted but occasionally overlapping patterns of expression. Mutant mice carrying a targeted disruption of the *Wnt1* gene usually die within the first 24 h after birth (McMahon & Bradley 1990, Thomas & Capecchi 1990). Further analysis of the affected mutants reveal that they are lacking the midbrain, one of the tissues of normal *Wnt1* expression, as early as embryonic day 14 (E14). This phenotype might arise from a defect in pattern formation leading to the lack of appearance of the hindbrain during development. Alternatively, the hindbrain may be formed and then lost. This result could be expected if *Wnt1* were required for normal cell growth, although extensive areas of cell death have not been observed in the mutant embryos. *Wnt1* is also expressed in the neural tube, but no defect is observed at this location, perhaps because of compensation by other family members. Interestingly, Thomas et al (1991) have found another mutation in the *Wnt1* gene in *swaying* mice, a spontaneous mutant strain. These mice survive longer than the targeted mutants, but they but suffer from ataxia and lack a normal anterior cerebellum. Variation in this phenotype indicates either that *swaying* is not a null mutation or that certain genetic modifiers of the phenotype may segregate according to the genetic background.

As stated previously, *Wnt1* is the vertebrate homologue of the *Drosophila wg* gene, which plays a central role in establishing segment polarity during fly development. One of the targets of *wg* in flies is *engrailed* (*en*), for which there are two homologues in mice, *en1* and *en2* (Joyner & Martin 1987). *Engrailed* encodes a homeodomain transcription factor, and in flies, *wg* is required to maintain the expression of *en* in an adjacent row of cells. Using in situ hybridization and antibody staining, McMahon et al (1992) demonstrated that *Wnt1* is required for expression of *en1* and *en2*. This raises the possibility that *en* mutants might exhibit similar defects to *Wnt* mutants in mice. However, a targeted mutation in the *en2* gene does not lead to an embryonic phenotype but to a surprisingly mild alteration in cerebellar foliation (Joyner et al 1991). As *en1* is also expressed in a similar domain as *en2,* the two genes may be functionally compensating for each other. It will be of considerable interest to examine the consequences of disrupting the *en1* gene.

A number of other genes are also expressed in a restricted pattern during neural development. One of these, *Krox20,* was initially identified in fibro-blasts as a serum-inducible immediate early gene encoding a zinc finger protein. *Krox20* is expressed during development in the third and fifth rhombo-meres, which are transient structures that may represent compartments in the hindbrain (Wilkinson et al 1989). The targets of *Krox20* remain largely un-known, but several homeobox (*Hox*) genes are expressed in a similarly regu-lated pattern, suggesting that *Krox20* might be involved in the regulation of some of these genes. Indeed, there are several *Krox20* binding sites in the enhancer region upstream of the *Hoxb-2* (*Hox2.8*) gene (Sham et al 1993). The disruption of the *Krox20* gene was accomplished recently by two groups (Schneider-Maunoury et al 1993, Swiatek & Gridley 1993). In these mutants, rhombomeres 3 and 5 are reduced or eliminated, as judged by screening for a variety of rhombomere markers, including *Hox* genes. The absence of the two rhombomeres results in fusion of the trigeminal ganglia with the facial and vestibular ganglia, as well as fusion of the superior ganglia of the glossopha-ryngeal and vagus nerves. The majority of the mutant mice die shortly after birth, most likely because the defects in cranial nerves lead to an inability to feed, but there is some variation in penetrance of this phenotype.

As indicated above, potential targets of *Krox20* include *Hox* genes. Targeted disruption of *Hox* genes is being carried out in a number of laboratories, and in general, mutations in these genes have been observed to lead to pattern formation defects as evidenced by alterations in bone structure. However, at least one of these mutations, in the *Hoxa-1* (*Hox1.6*) gene, leads to defects in cranial nerve and hindbrain formation (Lufkin et al 1991, Chisaka et al 1992) and to death at birth. The phenotype is variably penetrant, and the mutations reported by two groups (Lufkin et al 1991, Chisaka et al 1992) differ with respect to neural tube closure and rhombomere formation. As one of the mutations only eliminates the homeodomain, and the other presumably repre-sents a null mutation, a homeodomain-less protein might still have some function.

Another gene involved in neural development is the *mouse achaete scute homologue 1* (*Mash1*). In the fly, *Achaete-Scute* is involved in neuroblast development in both the central and peripheral nervous systems, particularly in the decision to become neuroblasts or epidermis. In mice, *Mash1* is broadly expressed in the CNS during development but is restricted in the adult to regions undergoing neurogenesis; *Mash1* is also transiently expressed in the peripheral nervous system (PNS). Mutant mice die at birth, and further analysis reveals that they are lacking olfactory, sympathetic, parasympathetic, and part of the enteric neuron lineages (Guillemot et al 1993). However, for the most part, development of the central nervous system is normal. Because *Mash1* deficiency leads to defects in several classes of neurons, it must have a generalized, although not universal, function rather than be involved in the specification of a specific lineage.

Signaling Mutants

In the past several years, the structure of the neurotrophin receptors has been under considerable investigation. From these studies, it was shown that there are several receptors with varying affinities for different neurotrophins (for a review, see Glass & Yancopoulos 1993). The first of these is the so-called low affinity nerve growth factor (NGF) receptor, known as p75. This receptor lacks any recognizable catalytic motif, is more broadly expressed than the high affinity receptors, and binds all neurotrophins. The high affinity receptors, in contrast, belong to the Trk family of receptor tyrosine kinases. There are at least three known *Trk* genes that will bind neurotrophins: TrkA, TrkB, and TrkC. TrkA binds NGF, TrkB will bind BDNF as well as NT4/5 and NT3, and *TrkC* only binds NT3.

The first of the receptor genes to be disrupted was the p75 gene (Lee et al 1992). Mice carrying a mutation in this gene are viable, but they exhibit defects in innervation in the distal extremities. Despite these sensory defects, other sensory neurons, or sympathetic neurons that are NGF dependent, are unaffected in p75 mutants. These results add to the controversy about the role of this low-affinity receptor (see Glass & Yancopoulos 1993 for discussion).

In contrast, mutations in the genes that encode Trk receptors lead to much more severe defects. Mice carrying a mutation in the *TrkB* gene die shortly after birth because of lack of feeding (Klein et al 1993). This symptom is due to neuronal deficiencies in both the central and peripheral nervous system. There is a significant loss of motor neurons in the facial motor nucleus as well as in the spinal cord, and the increased number of pyknotic cells suggests that this is a result of extensive cell death. These mice also show significant loss of neurons in the trigeminal ganglia. Preliminary results indicate that *TrkA* mutants show extensive sensory and sympathetic neuropathies, and *TrkC* mutants show some loss of proprioception (M Barbacid, personal communication). In addition, crossing the various *Trk* mutants with the p75 mutants might help elucidate the role of the low-affinity receptor in neurotrophin response.

It will be interesting to examine the phenotypic consequences of disrupting the various neurotrophins, to see if their phenotype is similar to that of their predicted receptors. At least two of these, BDNF and NT3, have been disrupted in ES cells (P Ernfors, K-F Lee & R Jaenisch, personal communication). The phenotype of the *BDNF* mutants might be expected to be less severe than that of the *TrkB* mutants, as *TrkB* binds both BDNF and another neurotrophin, NT4. Preliminary results indicate that *BDNF*-deficient mice indeed survive birth and that sympathetic and some dopaminergic and motor neurons are unaffected. However, the mutant mice have severe defects in coordination and balance and show periods of hyperactivity followed by immobility. Excessive neurodegeneration was observed in the dorsal root, trigeminal and nodose ganglia, trigeminal mesencephalic nucleus, and axons innervating the inner ear. Preliminary evidence indicates that *NT3*-deficient mice have a loss of proprioception, similar to the *TrkC* mutant mice.

Other molecules involved in signaling include the Src family of nonreceptor tyrosine kinases, of which at least three, Src, Fyn, and Yes, are highly expressed in the developing and adult CNS. The genes encoding all three kinases have been disrupted (Soriano et al 1991; Grant et al 1992; Stein et al 1992; P Soriano, unpublished data), but only one of these mutations, in the *Fyn* gene, leads to a demonstratable neural phenotype (see below). These kinases are negatively regulated by a kinase, Csk, that phosphorylates a carboxy-terminal tyrosine. In contrast to the mild defects observed in the *Src* family kinase mutants, mice carrying a mutation in the *Csk* gene die at embryonic day 9 (E9) (Imamoto & Soriano 1993, Nada et al 1993). These mutants exhibit defects in notochord development, as well as a folding of the neural tube, perhaps because of altered adhesive properties of neurectoderm cells (Imamoto & Soriano 1993). In the *Csk* mutants, the activity of Src family kinases is elevated because of reduced phosphorylation of a carboxy-terminal tyrosine. The *Csk* phenotype is at least partially due to increased activity of Src family kinases, as it can be alleviated by crossing with *Src* or *Yes* mutants, which are not critical for neural development (A Imamoto & P Soriano, unpublished results).

Memory Mutants

Whereas silent mutations created by random mutagenesis may be missed if they do not lead to an overt phenotype, an inherent advantage of targeted mutagenesis is the knowledge that a particular gene has been mutated. Mutant mice can thus be screened for discreet defects that may not be immediately apparent, including behavioral abnormalities. Long-term potentiation (LTP), the continuing increase in synaptic response following brief synaptic stimulation, has been broadly implicated in the processes of learning and memory. Several protein kinases, including the calcium-calmodulin-dependent protein kinase type II (αCamKinaseII) and a neuronal isoform of the calcium-phospholipid-dependent protein kinase (PKCγ), are thought to play an important role in this process. To establish a direct link between LTP and learning, mice carrying targeted disruptions in each of the kinase genes were produced. αCamKinaseII is neural specific and is present both presynaptically and postsynaptically at sites that express LTP. Mutant mice were shown to be impaired both in LTP (Silva et al 1992b), and in the ability to perform spatial learning tasks in a Morris maze test (Silva et al 1992a). However, LTP in the mutant animals was not completely eliminated, suggesting that another kinase could substitute. *PKCγ* mutant mice exhibit greatly diminished LTP, which can be induced, however, under specific experimental conditions (Abeliovich et al 1993a). The mutant mice only exhibit mild defects in spatial learning (Abeliovich et al 1993b). It was proposed that PKCγ plays a regulatory rather than a determining role in LTP.

Another kinase that plays a significant role in LTP is the Src family non-receptor tyrosine kinase, Fyn. It had been previously shown that drugs that can inhibit nonreceptor tyrosine kinases could impair LTP (O'Dell et al 1991). However, these studies did not distinguish which kinase was responsible for this feature. When mice mutant for *Src, Fyn, Yes,* and *Abl* were tested, only mice lacking Fyn exhibited a significant and reproducible defect in LTP, although two other measures of synaptic plasticity, paired pulse facilitation and posttetanic potentiation, were normal (Grant et al 1992). When challenged in a Morris maze test, the *Fyn* mutants also showed restricted spatial learning. However, these mutants also display abnormal hippocampal anatomy, with an increased number of granule cells in the dentate gyrus, and of pyramidal cells in the CA3 region. Therefore, whether the defect in LTP observed in these animals is due to defective signaling, as could be predicted from the inhibitor studies, or to a defect in the development of the hippocampus is still unclear. A conditional mutation (for instance using a site specific recombinase) that would eliminate Fyn activity only in the adult hippocampus would help resolve this issue.

As shown above, with the exception of Fyn, many genes implicated in LTP do not lead to anatomical defects in the brain. Nitric oxide (NO) is believed to play an important role as a retrograde neurotransmitter, and LTP is sensitive to inhibitors of nitric oxide synthase (NOS). Mice carrying a targeted disruption of the neuronal NOS gene also do not show histopathological defects in the central nervous system, although they exhibit defects in the digestive system that may be due to defective innervation (Huang et al 1993). It will be interesting to study LTP in these mutant animals.

Other Mutants

Several targeted mutations lead to phenotypes affecting the peripheral nervous system of the adult. One of these is in the P0 gene (Giese et al 1992). P0 is a major glycoprotein highly expressed in Schwann cells, which has structural motifs similar to other cell adhesion molecules. P0 mutant mice are deficient in normal motor coordination and exhibit tremors. Axons in the peripheral nerves are severely hypomyelinated and in some cases degenerated. Interestingly, P0 has recently been implicated in a subset of human patients affected by Charcot-Marie-Tooth disease, a peripheral neuropathy (Hayasaka et al 1993).

Another targeted mutation that leads to motor neuron degeneration is in the ciliatory neurotrophic factor (CNTF) gene (Masu et al 1993). CNTF has many activities including promoting survival of neurons in culture. CNTF interacts with a heterodimeric receptor that also binds oncostatin M, IL6, or leukemia inhibitory factor (LIF). In vivo, expression of CNTF is restricted to myelinating Schwann cells and to a subset of astrocytes. No morphological changes are observed in CNTF mutant mice at birth, but progressive degeneration of spinal motor neurons was observed in animals several months old.

FUTURE PERSPECTIVES

These examples illustrate the extraordinary power of gene targeting to decipher the physiological role of individual genes. This approach has several limitations, however. First and foremost is the fact that genetic analysis only tells us if the gene is essential, but the lack of a phenotype does not necessarily indicate that the gene has no function. The case of Src family kinases is a good example. Src is highly expressed in the nervous system, and it was widely believed that a disruption of the *Src* gene would lead to a neural phenotype. Instead, mutant mice develop osteopetrosis, a bone remodeling disease due to defective osteoclast function (Soriano et al 1991). Functional redundancy can also be documented by investigating the phenotype of double mutants (for a discussion, see Thomas 1993). However, care must be taken in the interpretation of double mutant phenotypes, which might be additive rather than synergistic.

Second, mutant phenotypes that act early in development can lead to a lethal phenotype, thereby precluding functional analysis of the mutation at later stages. Alternatively, if the phenotype is not lethal early, some of the defects observed later may be due to the accumulative load of defects earlier in development. In both cases, it may be useful to create conditional mutations by using site-specific recombinases.

Third, some of the functions of the mutated gene may be better addressed in tissue culture, and the availability of targeted mutants provides a convenient source for the derivation of specific cell lines that lack a given gene product. Another example again involves Src. Several studies link a variety of signaling genes to molecules involved in cell adhesion (for a review, see Gumbiner 1993). Although *Src* mutant mice fail to exhibit a neural phenotype, Src⁻ cerebellar neurons fail to extend dendrites efficiently by using the neural cell adhesion molecule L1 as a substrate, whereas neurons deficient for the related kinases Fyn and Yes do not exhibit such a defect (Ignelzi et al 1994). These experiments might uncover synergism between unrelated genes, which may be interesting to test genetically.

With the increased scrutiny of various classical mutants (see the chapter by

Hatten & Heinz in this volume) and the generation of targeted mutations using homologous recombination in ES cells, the physiological role of many molecules that play important roles in neurobiology may likely become better understood. Therefore, using such approaches, we should expect an avalanche of information in the next few years that uses the mouse as a genetic system.

ACKNOWLEDGMENTS

I thank Mariano Barbacid, Rudolf Jaenisch, and Susumu Tonegawa for communicating results prior to publication, and my laboratory colleagues for comments on the manuscript. I apologize to colleagues whose work I may have missed in this survey. Work from my laboratory has been supported by grants from the National Institute of Child Health and Human Development.

> Any *Annual Review* chapter, as well as any article cited in an *Annual Review* chapter,
> may be purchased from the Annual Reviews Preprints and Reprints service.
> 1-800-347-8007; 415-259-5017; email: arpr@class.org

Literature Cited

Abeliovich A, Chen C, Goda Y, Silva A, Stevens C, Tonegawa S. 1993a. Modified hippocampal long-term potentiation in PKCγ-mutant mice. *Cell* 75:1253–62

Abeliovich A, Paylor R, Chen C, Kim J, Wehner J, Tonegawa S. 1993b. PKCγ mutant mice exhibit mild defects in spatial and contextual learning. *Cell* 75:1263–71

Askew G, Doetschman T, Lingrel J. 1993. Site-directed point mutations in embryonic stem cells: a gene-targeting tag-and-exchange strategy. *Mol. Cell Biol.* 13:4115–24

Besmer P. 1991. The kit ligand encoded at the murine Steel locus: a pleiotropic growth and differentiation factor. *Curr. Opin. Cell Biol.* 3:939–46

Bradley A, Evans M, Kaufman MH, Robertson E. 1984. Formation of germ-line chimaeras from embryo-derived teratocarcinoma cell lines. *Nature* 309:255–56

Brinster RL, Chen HY, Trumbauer ME, Senear AW, Warren R, Palmiter RD. 1981. Somatic expression of herpes thymidine kinase in mice following injection of a fusion gene into eggs. *Cell* 27:223–31

Chisaka O, Musci T, Capecchi M. 1992. Developmental defects of the ear, cranial nerves and hindbrain resulting from targeted disruption of the mouse homeobox gene *Hox-1.6*. *Nature* 355:516–20

Costantini F, Lacy E. 1981. Introduction of a rabbit β-globin gene into the mouse germ line. *Nature* 294:92–94

Deng C, Capecchi M. 1992. Reexamination of gene targeting frequency as a function of the extent of homology between the targeting vector and the target locus. *Mol. Cell Biol.* 12:3365–71

Evans MJ, Kaufman MH. 1981. Establishment in culture of pluripotential cells from mouse embryos. *Nature* 292:154–56

Friedrich G, Soriano P. 1991. Promoter traps in embryonic stem cells: a genetic screen to identify and mutate developmental genes in mice. *Genes Dev.* 5(9):1513–23

Gardner RL. 1968. Mouse chimaeras obtained by the injection of cells into the blastocyst. *Nature* 220:596–97

Giese K, Martini R, Lemke G, Soriano P, Schachner M. 1992. Mouse P0 gene disruption leads to hypomyelination, abnormal expression of recognition molecules, and degeneration of myelin and axons. *Cell* 71: 565–76

Glass D, Yancopoulos G. 1993. The neurotrophins and their receptors. *Trends Cell Biol.* 3:262–68

Gordon JW, Scangos GA, Plotkin DJ, Barnosa JA, Ruddle FH. 1980. Genetic transformation of mouse embryos by microinjection of purified DNA. *Proc. Natl. Acad. Sci. USA* 77:7380–84

Gossler A, Joyner AL, Rossant J, Skarnes WC. 1989. Mouse embryonic stem cells and reporter constructs to detect developmentally regulated genes. *Science* 244:463–65

Grant SGN, O'Dell TJ, Karl KA, Stein PL, Soriano P, Kandel ER. 1992. Impaired long-term potentiation, spatial learning, and hippocampal development in *fyn* mutant mice. *Science* 258:1903–10

Gridley T. 1991. Insertional versus targeted mutagenesis in mice. *New Biol.* 3:1025–34

Gridley T, Soriano P, Jaeniscch R. 1987. Insertional mutagenesis in mice. *Trends Genet.* 3:162–66

Guillemot F, Lo L-C, Johnson J, Auerbach A, Anderson D, Joyner A. 1993. Mammalian achaete-scute homolog 1 is required for the early development of olfactory and autonomic neurons. *Cell* 75:463–76

Gumbiner B. 1993. Proteins associated with the cytoplasmic surface of adhesion molecules. *Neuron* 11:551–64

Hasty P, Ramírez-Solis R, Krumlauf R, Bradley A. 1991a. Introduction of a subtle mutation into the *Hox-2.6* locus in embryonic stem cells. *Nature* 350:243–46

Hasty P, Rivera-Perez J, Bradley A. 1991b. The length of homology required for gene targeting in embryonic stem cells. *Mol. Cell Biol.* 11:5586–91

Hasty P, Rivera-Perez J, Bradley A. 1992. The role and fate of DNA ends for homologous recombination in embryonic stem cells. *Mol. Cell Biol.* 12:2464–74

Hasty P, Rivera-Perez J, Chang C, Bradley A. 1991c. Target frequency and integration pattern for insertion and replacement vectors in embryonic stem cells. *Mol. Cell Biol.* 11: 4509–17

Hatten ME, Heintz N. 1995. Mechanisms of neural patterning and specification in the developing cerebellum. *Annu. Rev. Neurosci.* 18:385–408

Hayasaka K, Ohnishi A, Takada G, Fukushima Y, Murai Y. 1993. Mutation of the mouse P0 gene in Charcot-Marie-Tooth neuropathy type 1. *Biochem. Biophys. Res. Commun.* 194:1317–22

Hooper M, Hardy K, Handyside A, Hunter S, Monk M. 1987. HPRT-deficient (Lesch-Nyhan) mouse embryos derived from germline colonization by cultured cells. *Nature* 326:292–95

Huang P, Dawson T, Bredt D, Snyder S, Fishman M. 1993. Targeted disruption of the neuronal nitric oxide synthase gene. *Cell* 75: 1273–86

Ignelzi M, Miller D, Soriano P, Maness P. 1994. Impaired neurite outgrowth of Src-minus neurons on the cell adhesion molecule L1. *Neuron* 12:873–84

Imamoto A, Soriano P. 1993. Disruption of the *csk* gene, encoding a negative regulator of Src family tyrosine kinases, leads to neural tube defects and embryonic lethality in mice. *Cell* 73:1117–24

Jaenisch R. 1976. Germ line integration and mendelian transmission of the exogenous Moloney Leukemia virus. *Proc. Natl. Acad. Sci. USA* 73:1260–64

Jaenisch R, Harbers K, Schnieke A, Löhler J, Chumakov I, et al. 1983. Germline integration of Moloney murine leukemia virus at the Mov 13 locus leads to recessive lethal mutation and early embryonic death. *Cell* 32:209–16

Jaenisch R, Mintz B. 1974. Simian virus 40 DNA sequences in DNA of healthy adult mice derived from preimplantation embryos injected with viral DNA. *Proc. Natl. Acad. Sci. USA* 71:1250–54

Joyner AL, Herrup K, Auerbach BA, Davis CA, Rossant J. 1991. Subtle cerebellar phenotype in mice homozygous for a targeted deletion of the En-2 homeobox. *Science* 251(4998): 1239–43

Joyner AL, Martin GR. 1987. En-1 and En-2, two mouse genes with sequence homology to the *Drosophila* engrailed gene: expression during embryogenesis. *Genes Dev.* 1(1):29–38

Kilby N, Snaith M, Murray J. 1993. Site-specific recombinases: tools for genome engineering. *Trends Genet.* 9:413–21

Klein R, Smeyne RJ, Wurst W, Long LK, Auerbach BA, et al. 1993. Targeted disruption of the *trkB* neurotrophin receptor gene results in nervous system lesion and neonatal death. *Cell* 75:113–22

Kothary R, Clapoff S, Brown A, Campbell R, Peterson A, Rossant J. 1988. A transgene containing lacZ inserted into the dystonia locus is expressed in neural tube. *Nature* 335:435–37

Krulewski R, Neumann P, Gordon J. 1989. Insertional mutation in a transgenic mouse allelic with Purkinje cell degeneration. *Proc. Natl. Acad. Sci. USA* 86:3709–12

Kuehn MR, Bradley A, Robertson EJ, Evans MJ. 1987. A potential model for Lesch-Nyhan syndrome through introduction of HPRT mutations in mice. *Nature* 326:295–98

Lee K-F, Li E, Huber J, Landis SC, Sharpe AH, et al. 1992. Targeted mutation of the gene encoding the low affinity NGF receptor p75 leads to deficits in the peripheral sensory nervous system. *Cell* 69:737–49

Lufkin T, Dierich A, LeMeur M, Mark M, Chambon P. 1991. Disruption of *Hox-1.6* homeobox gene results in defects in a region corresponding to its rostral domain of expression. *Cell* 66:1105–19

Mansour SL, Thomas KR, Capecchi MR. 1988. Disruption of the proto-oncogene int-2 in mouse embryo-derived stem cells: a general strategy for targeting mutations to nonselectable genes. *Nature* 336:348–52

Martin GR. 1981. Isolation of a pluripotent cell line from early mouse embryos cultured in medium conditioned by teratocarcinoma stem cells. *Proc. Natl. Acad. Sci. USA* 78: 7634–38

Masu Y, Wolf E, Holtmann B, Sendtner M, Brem G, Thoenen H. 1993. Disruption of the CNTF gene results in motor neuron degeneration. *Nature* 365:27–32

McMahon A, Joyner A, Bradley A, McMahon J. 1992. The midbrain-hindbrain phenotype of *Wnt-1/Wnt-1* mice results from stepwise deletion of engrailed-expressing cells by 9.5 days postcoitum. *Cell* 69:581–95

McMahon AP, Bradley A. 1990. The *Wnt-1 (int-1)* proto-oncogene is required for development of a large region of the mouse brain. *Cell* 62:1073–85

Mintz B. 1962. Formation of genotypically mosaic mouse embryos. *Am. Zool.* 2:432

Nada S, Yagi T, Takeda H, Tokunaga T, Nakagawa H, et al. 1993. Constitutive activation of Src family kinases in mouse embryos that lack csk. *Cell* 73:1125–35

Nusse R, Varmus H. 1992. *Wnt* genes. *Cell* 69:1073–87

O'Dell T, Kandel E, Grant S. 1991. Long term potentiation in the hippocampus is blocked by tyrosine kinase inhibitors. *Nature* 353:558–60

Orban PC, Chui D, Marth JD. 1992. Tissue- and site-specific DNA recombination in transgenic mice. *Proc. Natl. Acad. Sci. USA* 89:6861–65

Ramírez-Solis R, Davis A, Bradley A. 1993. Gene targeting in embryonic stem cells. In *Guides to Techniques in Mouse Development*, pp. 855–78. San Diego, CA: Academic

Roach A, Takahashi N, Pravtcheva D, Ruddle F, Hood L. 1985. Chromosomal mapping of mouse myelin basic protein gene and structure and transcription of the partially deleted gene in Shiverer mutant mice. *Cell* 42:149–55

Robertson E, Bradley A. 1986. Production of permanent cell lines from early embryos and their use in studying developmental problems. In *Experimental Approaches to Mammalian Embryonic Development*, ed. J Rossant, RA Pedersen, pp. 475–508. Cambridge, England: Cambridge Univ. Press

Schneider-Maunoury S, Topilko P, Seitanidou T, Levi G, Cohen-Tannoudji M, et al. 1993. Disruption of *Krox-20* results in alteration of rhombomeres 3 and 5 in the developing hindbrain. *Cell* 75:1199–14

Schnieke A, Harbers K, Jaenisch R. 1983. Embryonic lethal mutation in mice introduced by retroviral insertion into the α1(I) collagen gene. *Nature* 304:315–20

Sham MH, Vesque C, Nonchev S, Marshall H, Frain M, et al. 1993. The zinc finger gene *Krox20* regulates *HoxB2 (Hox2.8)* during hindbrain segmentation. *Cell* 72:183–96

Silva A, Paylor R, Wehner J, Tonegawa S. 1992a. Impaired spatial learning in α-calcium-calmodulin kinase II mutant mice. *Science* 257:206–11

Silva A, Stevens C, Tonegawa S, Wang Y. 1992b. Deficient hippocampal long-term potentiation in α-calcium-calmodulin kinase II mutant mice. *Science* 257:201–6

Smithies O, Gregg R, Boggs S, Koralewski M,

Kucherlapati R. 1985. Insertion of DNA sequences into the human chromosome β-globin locus by homologous recombination. *Nature* 317:230–34

Soriano P, Montogomery C, Geske R, Bradley A. 1991. Targeted disruption of the *c-src* proto-oncogene leads to osteopetrosis in mice. *Cell* 64:693–702

Stein P, Lee H-M, Rich S, Soriano P. 1992. pp59fyn mutant mice display differential signaling in thymocytes and peripheral T cells. *Cell* 70:741–50.

Swiatek P, Gridley T. 1993. Perinatal lethality and defects in hindbrain development in mice homozygous for a targeted mutation of the zinc finger gene *Krox20*. *Genes Dev.* 7:2071–84

Tarkowski AK. 1961. Mouse chimaeras developed from fused eggs. *Nature* 190:857–60

te Riele H, Maandag ER, Berns A. 1992. Highly efficient gene targeting in embryonic stem cells through homologous recombination with isogenic DNA constructs. *Proc. Natl. Acad. Sci. USA* 89:5128–32

Thomas J. 1993. Thinking about genetic redundancy. *Trends Genet.* 9:395–99

Thomas K, Capecchi M. 1990. Targeted disruption of the murine *int-1* proto-oncogene resulting in severe abnormalities in midbrain and cerebellar development. *Nature* (346):847–50

Thomas K, Folger K, Capecchi M. 1986. High frequency targeting of genes to specific sites in the mammalian genome. *Cell* 44:419–28

Thomas K, Musci T, Neumann P, Capecchi M. 1991. *Swaying* is a mutant allele of the proto-oncogene *Wnt-1*. *Cell* 67:969–76

Thomas KR, Capecchi MR. 1987. Site-directed mutagenesis by gene targeting in mouse embryo-derived stem cells. *Cell* 51:503–12

Valancius V, Smithies O. 1991. Testing an "in-out" targeting procedure for making subtle genomic modifications in mouse embryonic stem cells. *Mol. Cell Biol.* 11:1402–8

von Melchner H, DeGregori JV, Rayburn H, Reddy S, Friedel C, Ruley HE. 1992. Selective disruption of genes expressed in totipotent embryonal stem cells. *Genes Dev.* 6:919–27

Wagner EF, Stewart TA, Mintz B. 1981a. The human β-globin gene and a functional viral thymidine kinase gene in developing mice. *Proc. Natl. Acad. Sci. USA* 78:5016–20

Wagner TE, Hoppe PC, Jollick JD, Scholl DR, Hodinka RL, Gault JB. 1981b. Microinjection of a rabbit β-globin gene into zygotes and its subsequent expression in adult mice and their offspring. *Proc. Natl. Acad. Sci. USA* 78:6376–80

Wilkinson D, Bhatt S, Chavrier P, Bravo R, Charnay P. 1989. Segment-specific expression of a zinc finger gene in the developing nervous system of the mouse. *Nature* 337:461–64

NOTE ADDED IN PROOF

The following articles have appeared in the past few months and highlight exciting new developments in mouse genetics:

Crowley C, Spencer SD, Nishimura MC, Chen KS, Pitts-Mek S, et al. 1994. Mice lacking nerve growth factor display perinatal loss of sensory and sympathetic neurons yet develop basal forebrain cholinergic neurons. *Cell* 76: 1001–12

Ernfots P, Lee K-F, Jaenisch R. 1994. Mice lacking brain-derived neurotrophic factor develop with sensory deficits. *Nature* 368:147–50

Ernfots P, Lee K-F, Kucera J, Jaenisch R. 1994. Lack of neurotrophin-3 leads to deficiencies in the peripheral nervous system and loss of limb proprioceptive afferents. *Cell* 77:503–12

Farinas I, Jones KR, Backus C, Wang X-Y, Reichardt LF. 1994. Severe sensory and sympathetic deficits in mice lacking neurotrophin-3. *Nature* 369:658–61

Jones KR, Farinas I, Backus C, Reichardt LF. 1994. Targeted disruption of the brain-derived neurotrophic factor gene perturbs brain and sensory neuron but not motor neuron development. *Cell* 76:989–1000

Klein R, Silos-Santiago I, Smeyne RJ, Lira SA, Brambilla R, et al. 1994. Disruption of the neurotrophin-3 receptor gene *trkC* eliminates Ia muscle afferents and results in abnormal movements. *Nature* 368:249–51

Li Y, Erzurumlu RS, Chen C, Jhaveri S, Tonegawa S. 1994. Whisker-related neuronal patterns fail to develop in the trigeminal brainstem nuclei of NMDAR1 knockout mice. *Cell* 76:427–37

Millen KJ, Wurst W, Herup K, Joyner AL. 1994. Abnormal embryonic development and patterning of postnatal foliation in two mouse Engrailed-2 mutants. *Development* 120:695–706

Smeyne RJ, Klein R, Schnapp A, Long LK, Bryant S, et al. 1994. Severe sensory and sympathetic neuropathies in mice carrying a disrupted *trk*/NGF receptor gene. *Nature* 368:246–49

Snider WD. 1994. Functions of the neurotrophins during nervous system development: what the knockouts are teaching us. *Cell* 77:627–38

Takada S, Stark KL, Shea MJ, Vassileva G, McMahon JA, McMahon AP. 1994. *Wnt-3a* regulates somite and tailbud formation in the mouse embryo. *Genes Dev.* 8:174–89

Wurst W, Auerbach AB, Joyner AL. 1994. Multiple developmental defects in Engrailed-1 mutant mice: an early mid-hindbrain deletion and patterning defects in forelimbs and sternum. *Development* 120:2065–75

Annu. Rev. Neurosci. 1995. 18:19–43

CREATING A UNIFIED REPRESENTATION OF VISUAL AND AUDITORY SPACE IN THE BRAIN

E. I. Knudsen and M. S. Brainard

Department of Neurobiology, Stanford University School of Medicine, Stanford, California 94305-5401

KEY WORDS: superior colliculus, experience-dependent plasticity, auditory plasticity, sound localization, space perception

INTRODUCTION

The accurate and reliable perception of complex stimuli requires the integration of information provided by a variety of sensory cues. For example, the perception of a face involves the integration of shape, depth, color, and texture. Within the nervous system, the representation of complex stimuli results from both analytic and synthetic processes. First, complex stimuli are analyzed into their constituent components by low-order sensory neurons that are selective for simple stimulus features. Then, increasingly complex stimulus selectivities are synthesized by combining the selectivities of appropriate lower-order neurons (Knudsen et al 1987, Van Essen et al 1992). At the highest levels in such sensory hierarchies, neurons may be selective for stimuli that have unique significance for the individual (Margoliash 1986, Miyashita 1988), indicating that experience may play a critical role in establishing the response properties of such high-order neurons. Although much progress has been made in understanding how sensory systems analyze stimuli into constituent features, much less is known about how information is recombined across features to create selectivity for complex stimuli. This article discusses the integration of visual and auditory spatial information that underlies the localization of stimulus sources as an example of integrative processes that lead to complex stimulus selectivity of high-order neurons.

The brain derives great advantage from combining visual and auditory

19

0147-006X/95/0301-0019$05.00

spatial information because vision and audition offer different and complementary information for stimulus localization. Vision provides extremely reliable information, because of the direct projection of stimulus location onto the retinae, and high spatial resolution near the line of sight. Audition, on the other hand, provides information about the location of stimuli in any direction, behind obstacles or camouflage, and in the dark. Consequently, a bimodal representation of stimulus location is more versatile and reliable than are the representations based on either modality alone.

In this article, we first describe the nature of the information that is conveyed by visual and auditory localization cues, and present behavioral and neurophysiological data that illustrate how this information interacts in the determination of stimulus location. We then describe experiments that reveal the importance of experience in creating circuitry that enables the integration of visual and auditory information to take place. Finally, we discuss the strategies used by the brain to maintain the congruence of visual and auditory representations of space in spite of the capacity for independent movements of the eyes and ears. The review focuses specifically on the problem of creating a unified bimodal representation of stimulus location, but the underlying principles may apply more generally to the integration of information from independent sensory cues.

General Considerations for the Integration of Visual and Auditory Spatial Information

In order to combine information from different sensory cues, the nervous system must be able to integrate information that is encoded in coordinate frames that are different from one another. This problem is particularly apparent for the integration of visual and auditory information for stimulus localization. The location of an object is encoded topographically on the retinae. In contrast, the location of a sound source is computed from cues that result from the interaction of the head and external ears (pinnae) with the incoming sound (Blauert 1983, Middlebrooks & Green 1991): differences in the timing and amplitude of sound at the two ears and the distribution of energy across frequencies at each ear (monaural spectrum). Visual-auditory interactions that are based on stimulus location depend, therefore, on the brain forming valid associations between topographic spatial information in the visual system and localization cue values in the auditory system.

The problem of integrating visual and auditory localization cues is compounded by the fact that there is not a fixed correspondence between auditory localization cues and locations in the visual field. Rather, the specific values of localization cues that correspond with each spatial location depend on the geometry of the individual's head and external ears. Accurate determination of sound-source location requires that individuals learn unique associations of localization cues with spatial locations. Thus, the processes that underlie bi-modal localization, like those that underlie other high-order functions, must be calibrated by experience. Moreover, as the eyes or pinnae move with respect to the head, the correspondence between auditory localization cues and loca-tions in the visual field will change. Consequently, integration of auditory and visual spatial cues also requires that these cues be interpreted in a dynamically varying context.

BEHAVIORAL EXAMPLES OF VISUAL-AUDITORY INTERACTIONS

Cooperative Influences

The processes that mediate visual and auditory localization interact in a coop-erative fashion (Stein et al 1989; Perrott et al 1990, 1991). When visual and auditory stimuli are present at low intensity levels, bimodal localization is far superior to unimodal localization: The probability of detecting a bimodal stimulus greatly exceeds the probability of detecting either unimodal stimulus, and bimodal stimuli are localized with greater precision and with a much shorter latency than are unimodal stimuli (Stein et al 1989, Perrott et al 1991). An improvement in the precision of bimodal over unimodal localization is also observed for high intensity stimuli, although the magnitude of improvement is relatively less (Auerbach & Sperling 1974). This improvement probably reflects the fact that visual spatial information and auditory spatial information are derived independently. Consequently, the variance of the spatial estimate from each modality is reduced by combining information across modalities (Howard 1982).

Visual Dominance in Stimulus Localization

Visual spatial information dominates over auditory spatial information in the perception of stimulus location. When visual and auditory spatial information is discordant, humans perceive the auditory stimulus as originating from near the location of the visual stimulus, a phenomenon referred to as visual capture or ventriloquism (Figure 1). When the plausibility that a sound originates from a particular visual object is high, based on the temporal correlation of the stimuli (Jack & Thurlow 1973) and on cognitive factors (Warren et al 1981), visual localization can influence auditory localization even when visual and auditory stimuli are separated by angular distances of as much as 30° (Jackson 1953, Thurlow & Rosenthal 1976).

The dominance of visual localization over auditory localization is also manifested in the long-term effects of sustained visual-auditory discordance. Visual-auditory discordance can be induced by optically shifting the projection of the visual field onto the retinae with displacing prisms. Humans adapt quickly to displacing prisms by shifting the perceived zero positions of the

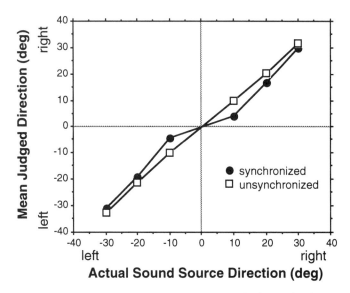

Figure 1 Visual capture of auditory localization in humans. Subjects pointed to the apparent location of an auditory stimulus with a hand that was out of view. Visual stimuli consisted of two different 30-s film clips of a woman talking projected onto a screen located directly in front (0° azimuth) of the subject. Auditory stimuli were the sound tracks for the film clips, presented from a hidden loudspeaker located 10, 20, or 30° to the left or right of the screen. The sound track was either for the film clip being viewed (synchronized) or for the other film clip (unsynchronized). The subjects were told that the sound may or may not come from the direction of the visual image and that they were to point at the sound source with their unseen hand. The plotted points are group averages, but every subject showed a similar effect. Note that even with instructions that minimized the belief that the visual and auditory stimuli originated from the same source, auditory localization was biased strongly by the location of a plausible, synchronized visual image. From Nachmias & Spelke (J Nachmias & E Spelka, unpublished data).

eyes in the head to compensate for the optical displacement of the prisms (Howard 1982). Barn owls, on the other hand, cannot move the eyes in the head by more than a few degrees (Knudsen 1989). Consequently, for barn owls, displacing prisms cause a sustained discordance between visual and auditory spatial information. Barn owls that are raised with displacing prisms alter auditory localization systematically in the direction of the optical displacement of the visual field (Figure 2) (Knudsen & Knudsen 1989).

NEURAL INTEGRATION OF VISUAL AND AUDITORY SPATIAL INFORMATION

The behavioral data presented in the previous section illustrate that strong interactions can occur between the visual and auditory modalities during stim-

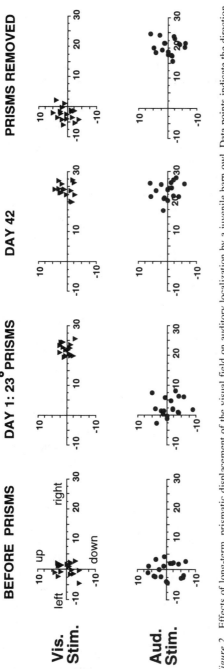

Figure 2 Effects of long-term, prismatic displacement of the visual field on auditory localization by a juvenile barn owl. Data points indicate the direction of head-orientation response to a visual (*upper panels*) or an auditory stimulus (*lower panels*) measured with a search coils. Stimuli were presented in a darkened, sound-attenuated chamber. Responses are plotted relative to the actual location of the stimulus. Visual localization was tested with a 2-s glow of a light-emitting diode presented from random locations; visual localizations demonstrate the optical effects of the prisms. Auditory localization was tested by using a 2-s noise burst presented from a movable loudspeaker positioned at random locations. Before the prisms were attached, visual and auditory responses were similar. After one day of experience with 23° right–displacing prisms, auditory responses were still accurate. After 42 days of continuous prism-experience, however, auditory orientation responses were altered to match the optical displacement of the visual field. The data on the right were collected immediately after the prisms were removed. Adapted from Knudsen & Knudsen (1990).

ulus localization. It follows that, at some level within the nervous system, similar interactions must occur between the neural activity elicited by visual and auditory stimuli.

The region of the brain where the representation and integration of auditory and visual spatial information is understood best is the optic tectum, also referred to as the superior colliculus in mammals. The optic tectum is a layered midbrain nucleus involved both in the localization of sensory stimuli and in the orientation of gaze and attention towards stimuli (Stein & Meredith 1993). The optic tectum receives convergent spatial information appropriate to these functions from the visual and auditory systems.

The organization and function of the optic tectum have been studied in a wide variety of species from all vertebrate classes. Principles derived from the study of one species usually obtain for others. In the discussion that follows, the term tectum refers to both the superior colliculus of mammalian species and the optic tectum of other classes of vertebrates. Where no explicit mention of species is made, statements refer to findings that generalize across species.

Visual and Auditory Space Maps in the Tectum

The tectum contains a map of visual space. The superficial layers of the tectum receive a direct topographic projection from the contralateral retina. In addition, both superficial and deep layers of the tectum receive indirect retinotopic maps from descending cortical projections and possibly from intrinsic tectal connections (Berson 1988, Stein & Meredith 1993). As a result of these connections, tectal units are excited by visual stimuli in a restricted region of the visual field (receptive field) and are inhibited by stimuli located outside of this region (Sterling & Wickelgren 1969, Cynader & Berman 1972, Stein & Meredith 1993). The receptive fields are organized systematically across the surface of the tectum to form a visual map that represents the entire visual field of the contralateral retina (Figure 3), except in primates where the map represents contralateral space only and stops at the representation of the vertical meridian (Lane et al 1973).

In the deep layers of the tectum (and also in the superficial layers in the barn owl), units respond to auditory stimuli as well. Auditory units also exhibit excitatory receptive fields that are surrounded by inhibitory regions (Meredith & Stein 1986a, Brainard et al 1992). Auditory receptive fields are larger than visual receptive fields. In the barn owl, a nocturnal predator that relies heavily on sound localization to capture prey, auditory receptive fields average approximately 30° in azimuth and 50° in elevation compared with visual receptive fields, which average only about 12° in diameter (Knudsen 1982). In other species, such as cats and monkeys, auditory receptive fields are generally even larger and for many neurons responses can be elicited by sounds located within an entire hemifield (Middlebrooks & Knudsen 1984, Jay & Sparks 1987).

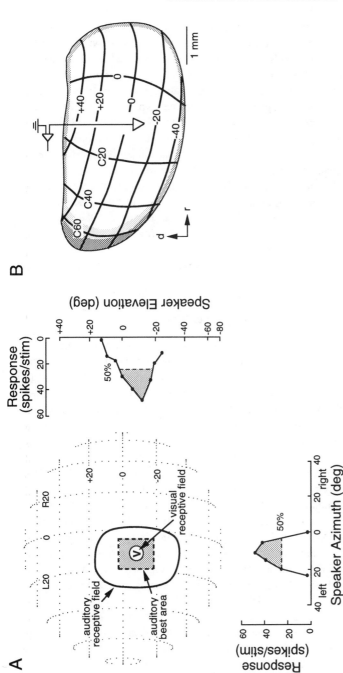

Figure 3 Auditory and visual maps of space in the optic tectum of the barn owl. (*A*) Auditory and visual receptive fields of a single unit are plotted on a dotted globe of space, calibrated in double pole coordinates. The auditory stimulus was a 100-ms noise burst presented at 20 dB above threshold; the visual stimulus was a bar of light swept across the visual field. The solid, vertical oblong indicates the extent of the auditory excitatory receptive field. The response of the unit to noise bursts presented from various locations within the receptive field are shown below and to the right. The auditory best area (the area from which sounds elicited responses greater than 50% of maximum) is shaded. The visual excitatory receptive field is indicated by the circled V. The data are adapted from Knudsen (1985). (*B*) Contour lines, indicating the locations of neurons with receptive fields centered at equivalent azimuths or elevations, are projected onto a lateral view of the right optic tectum. The contours are calibrated in double pole coordinates, as shown in *A*. The approximate location of the recording site for the data plotted in *A* is indicated. Scale = 1 mm. Abbreviations: c, contralateral; d, dorsal; r, rostral. Adapted from Knudsen (1982).

However, even for neurons with large receptive fields, the strength of response usually varies with the location of auditory stimuli so that strong responses are only elicited when sounds arise from a much more restricted region of space, the auditory best area (shaded box in Figure 3A). Just as with visual receptive fields, auditory best areas are systematically organized across the tectum, thereby creating a map of auditory space (Knudsen 1982, King & Palmer 1983, Middlebrooks & Knudsen 1984, King & Hutchings 1987).

The map of auditory space, which is based on the tuning of neurons to sound localization cues (Wise & Irvine 1985, Olsen et al 1989), is approximately aligned with the map of visual space in the tectum. A close correspondence exists between the centers of auditory best areas and visual receptive fields of neurons encountered in penetrations perpendicular to the tectal surface (Knudsen 1982, King & Palmer 1983, Middlebrooks & Knudsen 1984, King & Hutchings 1987). Moreover, many neurons in the intermediate and deep layers of the tectum receive convergent auditory and visual information. These neurons have closely aligned auditory and visual receptive fields and thus form a bimodal space map (Figure 3B).

Visual-Auditory Integration in the Tectum

The representations of visual and auditory space are not merely superimposed in the tectum; they are integrated in a nonlinear manner by bimodal neurons to yield a unified representation of stimulus location. Single-unit studies in cats and guinea pigs have revealed some of the principles that underlie this integration (Meredith & Stein 1983, 1986a,b; King & Palmer 1985; Meredith et al 1987). In general, responses to combinations of auditory and visual stimuli are not simple sums of the responses elicited by the same stimuli presented in isolation. Instead, depending on the neuron and on the conditions under which the stimuli are combined, responses to bimodal stimuli may be greatly enhanced or reduced relative to the sum of the responses elicited by the unimodal components presented alone. The most extreme examples of such nonlinear interactions are observed when responses to the unimodal stimulus components are weak or absent (Meredith & Stein 1986b). In many of these cases, responses to the combined presentation of the same stimuli are vigorous and reliable. Such response enhancement has the potential to increase the sensitivity of tectal neurons to bimodal stimuli and, thus, is analogous to the dramatic increases in sensitivity to weak bimodal stimuli that are observed behaviorally.

The nature of the interactions between auditory and visual stimuli depends both on the relative timing and on the relative spatial locations of the stimuli. Responses to bimodal stimuli are typically enhanced relative to the responses to the component stimuli, if the components occur at about the same time (Figure 4) and arise from a similar region of space (Figure 5). When these

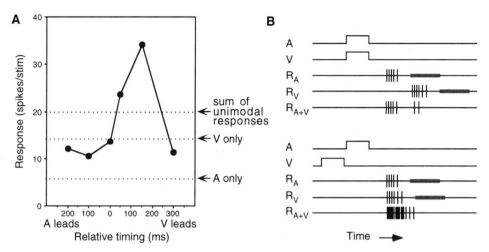

Figure 4 Dependence of auditory-visual interaction on stimulus timing. (*A*) The response of a bimodal neuron in the cat tectum is plotted as a function of the relative timing of brief visual and auditory stimuli. Unimodal stimuli presented individually elicited only moderate responses. When these stimuli were combined, the response could be enhanced or depressed relative to the sum of the responses to the individual stimuli. Maximal enhancement occurred when the visual stimulus preceded the auditory stimulus by approximately 150 ms, and this enhancement gradually declined and reversed in sign as the relative timing of the stimuli was shifted from this optimal value. The data are adapted from Meredith et al (1987). (*B*) Schematic illustration of the principles governing temporal integration of bimodal stimuli. At the top is shown an example in which auditory (A) and visual (V) stimuli are presented simultaneously. The response to the auditory stimulus alone (RA) or to the visual stimulus alone (RV) consists of a period of excitation (indicated by tick marks) followed by a period of inhibition (indicated by shading). In this case, the latency of the visual response exceeds the latency of the auditory response so that visual excitation coincides with auditory inhibition and the response to the combined stimuli ($R_{A + V}$) is reduced relative to the sum of the unimodal responses. The situation for the same neuron when the visual stimulus precedes the auditory stimulus by the difference between the latencies of the visual and auditory responses is shown at the bottom. In this case, the periods of maximal excitation for auditory and visual stimuli coincide. The response to the combined stimuli may then equal or exceed the sum of the unimodal responses.

conditions are not met, responses to the bimodal stimulus may be reduced relative to the responses to the unimodal components.

The optimal time interval between visual and auditory stimulation can be predicted for any given neuron from the relative latencies of the neuron's responses to visual and auditory stimuli presented alone (Figure 4B). Maximal response enhancement occurs when the excitatory inputs resulting from visual and auditory stimulation converge simultaneously on the unit (King & Palmer 1985, Meredith et al 1987). Because response latencies are typically much longer for visual than for auditory stimuli, maximal response enhancement for

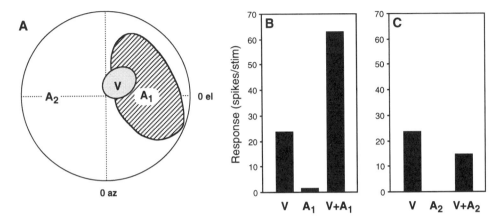

Figure 5   Dependence of auditory-visual interaction on stimulus location. (*A*) The visual excitatory receptive field (*shaded region*) and auditory excitatory receptive field (*striped region*) of a bimodal tectal neuron in the cat tectum are illustrated on a circle representing the frontal hemifield. V, A_1, and A_2 indicate locations from which visual and auditory stimuli were presented. (*B*) When a visual stimulus located within the visual receptive field was combined with an auditory stimulus located within the auditory receptive field (A_1), the response was greatly enhanced relative to the sum of the unimodal responses. (*C*) The response to the same visual stimulus was inhibited by an auditory stimulus located outside the auditory excitatory receptive field (A_2). Adapted from Meredith & Stein (1986a).

most units requires a delay in the auditory relative to the visual stimulus (King & Palmer 1985, Meredith et al 1987).

The nature of the spatial dependence of bimodal interactions can be understood, at least qualitatively, in terms of the structure of visual and auditory receptive fields (Figure 5). Stimuli that are located outside of a neuron's excitatory receptive field for that stimulus modality inhibit the neuron. Hence, the response of a bimodal neuron will be enhanced only when both auditory and visual stimuli are located within their respective excitatory receptive fields, and will be reduced if either stimulus falls into an inhibitory surround.

The angular distance that can separate visual and auditory stimuli and still result in facilitatory interactions in tectal neurons depends on the sizes of their visual and auditory receptive fields. Because visual receptive fields are consistently smaller than auditory receptive fields (Knudsen 1982, King & Palmer 1983, Middlebrooks & Knudsen 1984, King & Hutchings 1987), bimodal tectal neurons are more sensitive to displacements of a visual stimulus from its optimal location than to displacements of an auditory stimulus. As a consequence, the site in the bimodal tectal map that is activated by visual and auditory stimuli should be more sensitive to the location of the visual stimulus than to the location of the auditory stimulus. For example, consider a bimodal

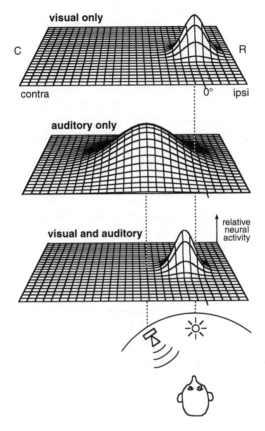

Figure 6 Hypothetical neural representations of spatially separate visual and auditory stimuli (*bottom*), schematically illustrated on a plane representing the tectal surface. The relative activity of different tectal loci is indicated by the relative height above the plane. Neurons located outside of the zones of excited neurons are inhibited (not shown) by the stimulus. Top: A frontal visual stimulus results in a sharp peak of activity centered in the rostral (R) tectum. Middle: An auditory stimulus located more peripherally results in a peak of activity centered further caudal (C) in the tectum. The peak is broader because auditory receptive fields are much larger than visual receptive fields. Bottom: The combination of visual and auditory stimuli results in a single peak of activity located between the peaks for the unimodal stimuli but biased towards the location at which the visual stimulus was represented.

stimulus consisting of a frontally located visual stimulus and a peripherally located auditory stimulus (Figure 6). The visual stimulus will activate neurons at a site in the rostral tectum where neurons have frontal visual receptive fields (Figure 6, *top plot*). Because auditory receptive fields are relatively large, the auditory stimulus will excite neurons over a large region, including the rostral tectum (Figure 6, *middle plot*). When the auditory and visual stimuli are presented concurrently, neurons in the rostral tectum will receive convergent excitation and respond strongly. In contrast, neurons at more caudal sites in the tectum will be inhibited (not shown) by the frontally located visual stimulus because it is located in the inhibitory surround of their visual receptive fields. As a result, the bimodal stimulus should elicit strong neuronal responses in the rostral tectum, near the location corresponding to the visual stimulus but biased toward the location that corresponds to the auditory stimulus (Figure 6, *bottom plot*). Thus, just as vision dominates in the perception of the location of misaligned auditory and visual stimuli (Figure 1), so too it dominates in the representation of stimulus location in the tectum.

Benefits of Cross-Modal Integration for Information Processing

The mechanisms of cross-modal integration may enhance the tectal representation of bimodal stimuli. For example, the location of a weak auditory stimulus, as represented by a peak of activity in the tectal map, may be poorly defined against a background of spontaneous activity or activity elicited by other stimuli. However, if the stimulus also contains a visual component, the data from the previous section suggest that the corresponding peak of tectal activity will be greatly increased while activity elsewhere will be reduced. The more sharply defined peak of activity that would result from this nonlinear integration potentially enables the nervous system to detect bimodal stimuli for which the unimodal components are below behavioral threshold and to determine the location of the stimulus with increased precision. Thus, the interaction of visual and auditory signals at the level of single neurons in the tectum parallels, and therefore potentially contributes to, the improvement in localization of bimodal stimuli that has been reported in behavioral studies (Auerbach & Sperling 1974, Stein et al 1988, Stein & Meredith 1990).

Another benefit of integrating visual and auditory spatial information into a common neurophysiological code is that it enables the information to be used subsequently without regard to the modality of the stimulus. The tectum participates in the execution of a variety of orienting movements, including coordinated changes in the position of eyes, pinnae, head, and body (Sparks 1990, Stein & Meredith 1993). Additionally, the tectum is involved in shifting the direction of attention to different spatial locations (Desimone et al 1990, Colby 1991). These functions are primarily dependent on the locations of salient environmental stimuli and can be elicited by either visual or auditory stimuli. Because each function requires its own complex and specialized circuitry, there is an obvious efficiency to allowing both auditory and visual inputs access to as much common circuitry as possible. The alignment of visual and auditory maps in the tectum and their convergence into a bimodal map is therefore a convenient organizational scheme for information processing, in that it allows either modality to supply information to the same motor and attentional circuitry.

Visual-Auditory Integration in the Forebrain

The principles that govern the integration of visual and auditory information in the tectum may apply to other brain regions. Nowhere has bimodal integration been investigated as extensively as in the tectum. However, convergence of auditory and visual information has been documented in a variety of forebrain areas, at least some of which are likely to participate in spatial analysis based on the presence of spatial tuning for visual and auditory stimuli. These areas include the superior temporal polysensory (STP) area and hippocampus of primates and the anterior ectosylvian sulcus (AES), lateral suprasylvian cortex, and visual cortex in cats (Stein & Meredith 1993). Where examined, the principles critical to bimodal integration of spatial information in the tectum also apply in these cortical areas. First, many individual neurons [including a surprising 40% in striate and extrastriate visual cortex (Morrell 1972, Fishman & Michael 1973)] are sensitive to both visual and auditory stimuli (Hikosaka et al 1988, Watanabe & Iwai 1991, Tamura et al 1992, Wallace et al 1992, Stein et al 1993). Second, the auditory and visual receptive fields of bimodal neurons tend to be overlapping or aligned, even in areas STP and AES where there is no topographic representation of space (Morrell 1972, Hikosaka et al 1988, Wallace et al 1992, Stein et al 1993). Finally, responses to bimodal stimuli may be nonlinear functions of the responses to the unimodal components and, where examined, the nature of the bimodal interaction is governed by the same principles of temporal interaction that operate in the tectum (Fishman & Michael 1973, Wallace et al 1992, Stein et al 1993).

DEVELOPMENT OF ALIGNED VISUAL AND AUDITORY SPACE MAPS

The interactions between the visual and auditory modalities described in the preceding sections require that the brain establish appropriate associations between visual spatial locations and the values of auditory cues produced by sound sources at those locations. The alignment of auditory receptive fields with visual receptive fields of bimodal neurons provides a neurophysiological assay of the auditory cue values that are associated with specific locations in (visual) space in an individual animal. The alignment of visual and auditory space maps in the tectum has, therefore, been used extensively to elucidate the role of experience in establishing appropriate visual-auditory associations for stimulus localization.

The visual map of space is created by reproducing centrally the topographic representations of space that exist in the retinae. This is accomplished in large part by molecular markers that guide retinotectal and, possibly, corticotectal fibers to the correct regions of the tectum (Sanes 1993). Visual receptive fields are then sharpened by activity-dependent mechanisms that reshape visual axonal arbors on the basis of temporal correlations between their activity and that of neighboring arbors (Fawcett & O'Leary 1985, Udin & Fawcett 1988). Temporal correlations in the spontaneous activity of neighboring neurons are strong even before birth (Meister et al 1991). Thus, even without visual experience, self-organizational mechanisms can sharpen receptive fields and refine the map of visual space that is specified by genetic instructions (Miller 1989).

In contrast to the visual map, the auditory map of space bears little relation to the topographic representation of frequency that exists in the cochleae. In order to create neurons with spatial receptive fields, the auditory system must combine information from arrays of neurons that are tuned to the appropriate cue values (Knudsen 1984, Wagner et al 1987, Brainard et al 1992). Selecting the complicated patterns of connectivity that are necessary to enable a neuron to respond to a particular constellation of cue values could be accomplished by activity-dependent mechanisms that strengthen the connections of concurrently active afferents (Miller 1989). Thus, the same mechanisms that refine the visual space map could create restricted auditory spatial receptive fields on the basis of the temporally correlated activity that is evoked across arrays of cue-sensitive neurons by acoustic sources.

In order to create a map of space, the tuning of neighboring neurons to auditory cue values must change in accordance with the changes in cue values across space. The patterns of cue values across space (spatial patterns) are, to some extent, systematic and predictable (Knudsen et al 1991, Middlebrooks & Green 1991). For example, left-ear leading interaural time differences (ITDs) always correspond with sound sources located to the left, and the larger the ITD, the further the sound source is located to the left. The maximum left-ear leading ITD corresponds with a source directly to the left, and an ITD at the center of the experienced range, which normally is 0 µs, corresponds with a source on the midline. Similarly, in species with symmetrical ears, left-ear greater interaural level differences (ILDs) always correspond with sound sources to the left, and ILDs at the center of the experienced range correspond with a sound source on the midline. These basic relationships do not vary significantly with frequency, across individuals, or during development. Moreover, the central auditory system represents frequency-specific cue values topographically in certain brainstem nuclei (Wenstrup 1986, Wagner et al 1987, Manley et al 1988, Carr & Konishi 1990). Because of the predictability of the basic relationships between cue values and source locations, genetic mechanisms could guide systematic projections from these nuclei that could yield a crude map of space at higher levels.

The consistency of the cue-location relationships discussed above suggests that auditory receptive fields and an auditory space map could develop to some degree without visual instruction. This has been shown to be the case by raising animals with eyelids sutured closed. In the tecta of blind-reared guinea pigs, ferrets, and barn owls, auditory receptive fields are organized into a map of space that is normal in several respects (Knudsen et al 1991, Withington 1992, King & Carlile 1993). The region of space represented in the map and the orientation of the map in the tectum are normal [with the notable exception of the elevation map in some blind-reared barn owls (Knudsen et al 1991)]. Moreover, the region of space directly in front of the animal is always represented at the correct site in the rostral tectum. These aspects of the map are based on predictable associations between cue values and spatial locations, and this predictability may provide the basis for positioning and orienting the auditory space map independently of visual experience.

Auditory spatial tuning is not entirely normal, however, in blind-reared animals. In guinea pigs, azimuth tuning is abnormally broad (Withington 1992); in barn owls, elevation tuning is abnormally broad, and the map of elevation is oriented upside-down in some animals (Knudsen et al 1991); in ferrets, the proportion of units exhibiting multiple receptive fields increases (King & Carlile 1993); and in all species, there is a degradation in the auditory map relative to normal, as evidenced by increased misalignment between auditory and visual receptive field centers. These aspects of the auditory map may depend on the correct associations of frequency-dependent ILD or monaural spectrum cues with spatial locations. Because the correspondences between these cues and spatial locations vary greatly with the size, shape, and orientation of the individual's pinnae (Carlile & Pettigrew 1987, Carlile 1990, Knudsen et al 1991), bimodal sensory experience may be required to learn them.

Displacement of the Visual Field

The influence of vision on the topography of the auditory space map has been demonstrated by raising barn owls with prisms that displace the visual field laterally (Knudsen & Brainard 1991). As discussed earlier, this manipulation causes systematic changes in sound localization behavior (Figure 2). The prisms optically shift the correspondence between spatial locations and retinal loci. Consequently, the region of visual space represented in the retinotopic tectal map is also shifted (Figure 7A). Because the auditory cues reaching the ears are not altered, the immediate effect of prisms is to create a misalignment between the locations of auditory and visual receptive fields in the tectum. Prolonged exposure to prisms causes auditory receptive fields to shift by an amount matching the prismatic displacement (Figure 7B and C). In young ferrets, surgical deviation of one eye results in similar changes in auditory

Auditory Space Map

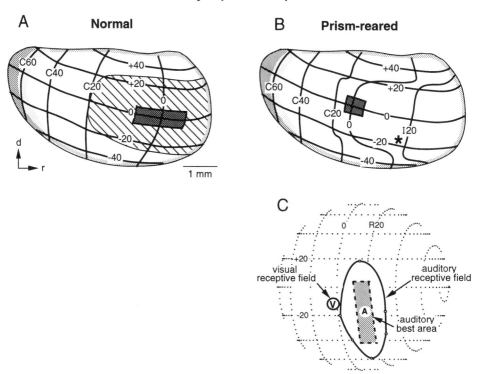

Figure 7 Adaptive adjustment of the auditory space map in the barn owl's optic tectum in response to long-term, prismatic displacement of the visual field during development. Isoazimuth and isoelevation contours are projected onto a lateral view of the right optic tectum, as described in Figure 3. (*A*) The normal auditory space map. The hatched region is the part of the tectum that represents the portion of the visual field that was optically displaced in the prism-reared animal. Most of the rest of the tectum represents portions of the visual field that were obstructed by the prism frames. The shaded rectangle indicates the representation of the frontal 10°. (*B*) The auditory space map in an owl raised with prisms that displaced the visual field 23° leftward. The map is altered substantially only in the part of the tectum that represents the optically displaced portion of the visual field, shown in *A*. The shaded rectangle indicates the representation of the frontal 10°, which is decreased by about 50% relative to normal. (*C*) The visual receptive field (*circled V*), measured with the prisms removed, and the auditory best area (*shaded rectangle*) of a single unit recorded in the prism-reared owl at the site in the tectum shown by the asterisk in *B*. The data were gathered and plotted as described in Figure 3. Abbreviations: c, contralateral; i, ipsilateral; d, dorsal; r, rostral. Adapted from Knudsen & Brainard (1991).

spatial tuning (King et al 1988). Thus, visual information, when available, overrides all other sources of spatial information in controlling the topography of the auditory space map in the tectum.

In prism-reared owls, the auditory space map in the rostral end of the tectum, which represents the optically displaced portion of the visual field, is highly abnormal (Figure 7B). The extent of space that is represented is altered, with the rostral edge of the map representing locations that differ by as much as 30° from the normally represented locations (Knudsen & Brainard 1991). The magnification of different regions of space (quantified as micrometers of tissue per degree of space) is also altered dramatically (Brainard & Knudsen 1993). In normal owls, the region of space directly in front is magnified maximally and about twice as much as the region located 20° contralateral (Figure 7A) (Knudsen 1982). This differential magnification of frontal space is unrelated to the rate at which auditory cues vary across space because auditory cue values vary almost linearly with the azimuth of a sound source (Knudsen et al 1994). Instead, it results from a visually controlled expansion in the amount of tissue that comes to represent particular cue values. In owls raised with 23° displacing prisms, the region of space that is magnified maximally is shifted to the side, and the magnification of frontal space (0° azimuth) is decreased by approximately 50% (shaded box in Figure 7B). In addition, abrupt changes in magnification occur at locations in the map that represent borders between the optically displaced and obstructed regions of the visual field. Thus, vision exerts a powerful influence on the topography of the auditory space map in the tectum.

Experience-dependent changes have been reported in the topography of many other sensory maps throughout the brain by using other manipulations (Devor & Wall 1978, Jenkins et al 1990, Kaas 1991, Pons et al 1991, Gilbert & Wiesel 1992, Recanzone et al 1993). However, in nearly all of these cases, changes in map topography have resulted from changes in the balance of activity between the afferents that convey the mapped information. In contrast, in the case of the prism-induced shift in the auditory space map, auditory input is not altered. Instead, the changes in topography result from visually based instructive signals that control the mapping of auditory spatial information in the tectum (Knudsen 1994).

Monaural Occlusion

Visual instruction may also play an important role in calibrating abnormal auditory localization cues that result from monaural hearing impairment. Monaural occlusion, which simulates a conductive hearing loss, alters the ranges of cue values that an animal experiences. As a result, highly abnormal combinations of ILD and ITD are produced by a sound source at any location. Long-term exposure to monaural occlusion leads to adaptive adjustments in the tectal auditory space map in young animals. In barn owls and ferrets raised with one ear occluded, tectal auditory receptive fields are aligned with visual receptive fields as long as the ear remains occluded (Knudsen 1983, 1985; King et al 1988). In barn owls, these adaptive adjustments in auditory spatial tuning result from changes in unit tuning to ITD and ILD toward the abnormal ranges of values imposed by the earplug (Mogdans & Knudsen 1992). Although some adjustment in unit tuning proceeds in birds that are deprived of vision (Knudsen & Mogdans 1992), these adjustments are small compared with those made in sighted birds. This indicates that visual experience plays an important role in the adaptation to monaural occlusion.

Benefit of Merging Visual and Auditory Information for Calibrating Auditory Cues

A potential advantage of converging visual and auditory spatial information onto single neurons is that it provides a substrate for visually guided, auditory learning to take place (Knudsen 1994). In response to bimodal stimulation, visual inputs would depolarize bimodal neurons; concurrently active auditory inputs to those same neurons could be selectively strengthened by a mechanism such as associative long-term potentiation (Kelso & Brown 1986, Brown et al 1991). Thus, auditory cue values that correspond with a particular stimulus location could be identified and associated with the site in the map representing that location. An auditory map, calibrated in this manner, could be used to interpret auditory spatial information in other areas of the brain.

DYNAMIC MAINTENANCE OF MAP ALIGNMENT

In order for the nervous system to benefit from the convergence of visual and auditory spatial information, the representations of space derived from these modalities must remain aligned. However, the maintenance of this alignment is problematic in that the nervous system is provided with visual and auditory spatial information in different coordinate frames. The location of visual stimuli is encoded in retinotopic coordinates, while the location of auditory stimuli is derived from cues that depend on the orientation of the pinnae. Because visual and auditory receptive fields in the tectum are based, respectively, on retinotopic and pinna-dependent localization cues, movements of the eyes or ears relative to the head should disrupt the alignment of receptive fields.

Close alignment of visual and auditory receptive fields in the tectum has been documented typically in anesthetized animals in which the eyes and pinnae were oriented in their primary (normal alert) positions (Knudsen 1982, King & Palmer 1983, Middlebrooks & Knudsen 1984, King & Hutchings 1987). In awake animals, this alignment may be preserved by behavior that minimizes the amount of time that the eyes and pinnae spend oriented toward the side. For example, cats usually maintain the eyes in their primary positions, and shifts in gaze initiated by eye movements are often followed by equivalent movements of the head that restore the eyes to their primary positions (Harris et al 1980, Guitton et al 1990, Guitton 1992). In animals with mobile pinnae, coordination of pinna movements with eye movements could also contribute to maintaining alignment between visual and auditory receptive fields.

In addition to behaviors that tend to maintain the alignment between the visual and auditory space maps, neural signals related to eye position also modify auditory receptive fields in the tectum. Such influences of eye position on auditory spatial tuning have been demonstrated in both monkeys and cats (Jay & Sparks 1987, Peck et al 1993). In both species there is a range of effects of eye position. Some auditory receptive fields shift systematically with the direction of visual fixation so that essentially perfect alignment is maintained between visual and auditory receptive fields (Figure 8A). More common, however, are auditory receptive fields that shift only partially to compensate for changes in eye position or that do not shift at all. Additionally, there are many neurons for which auditory receptive field locations do not shift with eye position, but for which the strength of response to sound stimuli within the receptive field is strongly modulated by eye position (Figure 8B). It is unclear why this range of influence of eye position on auditory responses is found in the tectum. However, one possibility is that a transformation of auditory information from pinna-centered coordinates to retinotopic coordinates is accomplished within the tectum and that neurons with receptive fields that only partially compensate for changes in eye position represent intermediate stages in this transformation. This hypothesis predicts that tectal neurons for which the compensation for eye position is most complete are tectal output neurons, a prediction that has not yet been tested.

The conversion of auditory spatial information into retinotopic coordinates is appropriate for the control of gaze. However, for other tasks, the most useful coordinate system may differ. For example, reaching or moving towards an object may best be served by representing stimulus location in body-centered coordinates. Likewise, navigation through a familiar environment may be facilitated by an allocentric representation in which stimulus location is represented relative to fixed landmarks in extrapersonal space. Recent evidence suggests that visual stimuli may be transformed into such body-centered and allocentric coordinate systems in different areas of the brain (Andersen et al 1993). Presumably, for the auditory system to contribute to these tasks, auditory stimuli would have to be similarly transformed. Preliminary reports suggest that both egocentric and allocentric representations of auditory space may indeed occur in the primate hippocampus (Tamura et al 1990, 1992).

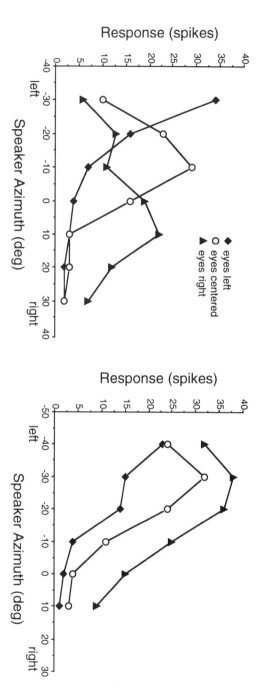

Figure 8 Effect of eye orientation on auditory responses in the optic tectum of the monkey. Noise bursts (90 dB SPL) were presented while the eyes were oriented either 24° to the left, straight ahead, or 24° to the right, as indicated. The orientations of the pinnae remained constant during the measurements. (A) The auditory spatial tuning of this unit shifted with eye orientation so that the unit always responded best when the sound source was located about 10° to the left of the direction of gaze. (B) The auditory responses of this unit increased as the eyes were oriented further toward the left. Adapted from Jay & Sparks (1987).

CONCLUDING REMARKS

The visual and auditory systems interact during development and throughout adult life to create a unified, bimodal percept of space. This percept is more sensitive, reliable, and versatile than the percept of space that is provided by either modality operating in isolation. The relative contributions of vision and audition to the percept of space vary with conditions. Whenever visual information is available, however, it dominates. The dominance of vision is manifested behaviorally by the immediate shift in sound-localization that is induced by the sight of a plausible stimulus source (Warren et al 1981) and by the gradual shift in sound localization behavior that occurs during development in response to a sustained discordance between visual and auditory localization (Knudsen & Knudsen 1990). The dominance of vision is manifested neurophysiologically by the gradual shift of the auditory space map in the optic tectum that is caused by a sustained displacement of the visual field during development (Knudsen & Brainard 1991).

There are advantages to translating unimodal spatial information into a shared neurophysiological code, as illustrated by the cross-modal interactions that occur in the optic tectum. The mutual alignment of the visual and auditory space maps in the tectum enables local, nonlinear integrative mechanisms to operate on the combined information from both modalities. Facilitatory interactions specifically amplify patterns of activity that are coincident both in space and time across the modalities (King & Palmer 1985, Meredith & Stein 1986a, Meredith et al 1987). Low-level signals that otherwise might be lost in noise can thus be detected, ambiguities in unimodal information can be resolved, and stimulus location can be estimated with enhanced precision. Behavioral correlates of these advantages include increased sensitivity and localization acuity when stimuli contain both visual and auditory components (Stein et al 1989, Perrott et al 1991). In addition, a shared code permits efficient, multimodal input to neural networks supporting specialized functions that require spatial information. Again, this is exemplified by the optic tectum, the output of which is used to control the direction of an animal's gaze and attention.

Another advantage of integrating visual and auditory spatial information is that it provides a potential substrate for visual calibration of auditory localization cues (Brainard & Knudsen 1993, Knudsen 1994). Although the nervous system can develop a representation of auditory space in the absence of visual information, the representation is degraded. Vision seems essential for interpreting certain details in the relationship between auditory cue values and locations in space. The convergence of visual and auditory spatial information onto single neurons potentially enables common learning mechanisms, such as associative long-term potentiation, which increases the strength of concurrently active inputs (Kelso & Brown 1986), to calibrate the representation of auditory localization cues.

The advantages of integrating spatial information from different sensory modalities apply as well to the integration of information from different cues analyzed within single sensory modalities. Facilitatory interactions among the representations of cues within a single modality can similarly improve stimulus detection and specification, and can resolve ambiguities in the information conveyed by individual cues (Middlebrooks & Green 1991, Brainard et al 1992, Van Essen et al 1992). In addition, the cues that are the most reliable for certain aspects of stimulus specification could be used to calibrate the interpretation of less reliable cues, just as vision calibrates auditory spatial cues for stimulus localization. Thus, the principles that have been elucidated by the study of visual-auditory integration for stimulus localization may apply generally to integrative processes that underlie a wide variety of brain functions.

Literature Cited

Auerbach C, Sperling P. 1974. A common auditory-visual space: evidence for its reality. *Percept. Psychophys.* 16:129–35

Andersen RA, Snyder LH, Li CS, Stricanne B. 1993. Coordinate transformations in the representation of spatial information. *Curr. Opin. Neurobiol.* 3:171–76

Berson DM. 1988. Retinal and cortical inputs to cat superior colliculus: composition, convergence and laminar specificity. *Prog. Brain Res.* 75:17–26

Blauert J. 1983. *Spatial Hearing: The Psychophysics of Human Sound Localization.* Cambridge, MA: MIT Press

Brainard MS, Knudsen EI. 1993. Experience-dependent plasticity in the inferior colliculus: a site for visual calibration of the neural representation of auditory space in the barn owl. *J. Neurosci.* 13:4589–608

Brainard MS, Knudsen EI, Esterly SD. 1992. Neural derivation of sound source location: resolution of spatial ambiguities in binaural cues. *J. Acoust. Soc. Am.* 91:1015–27

Brown TH, Zador AM, Mainen ZF, Claiborne BJ. 1991. Hebbian modifications in hippocampal neurons. In *LTP: A Debate of Current Issues,* ed. M Baudry, J Davis, pp. 357–89. Cambridge, MA: MIT Press

Carlile S. 1990. The auditory periphery of the ferret. I: directional response properties and the pattern of interaural level differences. *J. Acoust. Soc. Am.* 88:2180–95

Carlile S, Pettigrew AG. 1987. Directional properties of the auditory periphery in the guinea pig. *Hear. Res.* 31:111–22

Carr CE, Konishi M. 1990. A circuit for detection of interaural time differences in the brainstem of the barn owl. *J. Neurosci.* 10:3227–46

Colby CL. 1991. The neuroanatomy and neurophysiology of attention. *J. Child Neurol.* 6:S90–S118

Cynader M, Berman N. 1972. Receptive-field organization of monkey superior colliculus. *J. Neurophysiol.* 35:187–201

Desimone R, Wessinger M, Thomas L, Schneider W. 1990. Attentional control of visual perception: cortical and subcortical mechanisms. *Cold Spring Harbor Symp. Quant. Biol., Vol. LV,* pp. 963–71. Cold Spring Harbor, NY: Cold Spring Harbor Lab. Press

Devor M, Wall PD. 1978. Reorganisation of spinal cord sensory map after peripheral nerve injury. *Nature* 276:75–76

Fawcett JW, O'Leary DDM. 1985. The role of electrical activity in the formation of topographic maps in the nervous system. *Trends Neurosci.* 8:201–6

Fishman MC, Michael CR. 1973. Integration of

auditory information in the cat's visual cortex. *Vision Res.* 13:1415–19

Gilbert CD, Wiesel TN. 1992. Receptive field dynamics in adult primary visual cortex. *Nature* 356:150–52

Guitton D. 1992. Control of eye-head coordination during orienting gaze shifts. *Trends Neurosci.* 15(5):174–79

Guitton D, Munoz DP, Galiana HL. 1990. Gaze control in the cat: studies and modeling of the coupling between orienting eye and head movements in different behavioral tasks. *J. Neurophysiol.* 64:509–31

Harris LR, Blakemore C, Donaghy M. 1980. Integration of visual and auditory space in the mammalian superior colliculus. *Nature* 288:56–59

Hikosaka K, Iwai E, Saito H, Tanaka K. 1988. Polysensory properties of neurons in the anterior bank of the caudal superior temporal sulcus of the macaque monkey. *J. Neurophysiol.* 60:1615–37

Howard IP. 1982. *Human Visual Orientation.* New York: Wiley

Jack CE, Thurlow WR. 1973. Effects of degree of visual association and angle of displacement on the "ventriloquism" effect. *Percept. Mot. Skills* 37:967–79

Jackson CV. 1953. Visual factors in auditory localization. *Q. J. Exp. Psychol.* 5:52–65

Jay MF, Sparks DL. 1987. Sensorimotor integration in the primate superior colliculus. II. Coordinates of auditory signals. *J. Neurophysiol.* 57:35–55

Jenkins WM, Merzenich MM, Ochs MT, Allard T, Guic-Robles E. 1990. Functional reorganization of primary somatosensory cortex in adult owl monkeys after behaviorally controlled tactile stimulation. *J. Neurophysiol.* 63:82–104

Kaas JH. 1991. Plasticity of sensory and motor maps in adult mammals. *Annu. Rev. Neurosci.* 14:137–67

Kelso SR, Brown TH. 1986. Differential conditioning of associative synaptic enhancement in hippocampal brain slices. *Science* 232:85–87

King AJ, Carlile S. 1993. Changes induced in the representation of auditory space in the superior colliculus by rearing ferrets with binocular eyelid suture. *Exp. Brain Res.* 94:444–55

King AJ, Hutchings ME. 1987. Spatial response properties of acoustically responsive neurons in the superior colliculus of the ferret: a map of auditory space. *J. Neurophysiol.* 57:596–624

King AJ, Hutchings ME, Moore DR, Blakemore C. 1988. Developmental plasticity in the visual and auditory representations in the mammalian superior colliculus. *Nature* 332:73–76

King AJ, Palmer AR. 1983. Cells responsive to free-field auditory stimuli in guinea-pig superior colliculus: distribution and response properties. *J. Physiol.* 342:361–81

King AJ, Palmer AR. 1985. Integration of visual and auditory information in bimodal neurons in the guinea-pig superior colliculus. *Exp. Brain Res.* 60:492–500

Knudsen EI. 1982. Auditory and visual maps of space in the optic tectum of the owl. *J. Neurosci.* 2:1177–94

Knudsen EI. 1983. Early auditory experience aligns the auditory map of space in the optic tectum of the barn owl. *Science* 222:939–42

Knudsen EI. 1984. Synthesis of a neural map of auditory space in the owl. In *Dynamic Aspects of Neocortical Function,* ed. GM Edelman, WM Cowan, WE Gall, pp. 375–96. New York: Wiley

Knudsen EI. 1985. Experience alters the spatial tuning of auditory units in the optic tectum during a sensitive period in the barn owl. *J. Neurosci.* 5:3094–109

Knudsen EI. 1989. Fused binocular vision is required for development of proper eye alignment in barn owls. *Vis. Neurosci.* 2:35–40

Knudsen EI. 1994. Supervised learning in the brain. *J. Neurosci.* 14:3985–97

Knudsen EI, Brainard MS. 1991. Visual instruction of the neural map of auditory space in the developing optic tectum. *Science* 253:85–87

Knudsen EI, du Lac S, Esterly SD. 1987. Computational maps in the brain. *Annu. Rev. Neurosci.* 10:41–65

Knudsen EI, Esterly SD, du Lac S. 1991. Stretched and upside-down maps of auditory space in the optic tectum of blind-reared owls; acoustic basis and behavioral correlates. *J. Neurosci.* 11:1727–47

Knudsen EI, Esterly SD, Olsen JF. 1994. Adaptive plasticity of the auditory space map in the optic tectum of adult and baby barn owls in response to external ear modification. *J. Neurophysiol.* 71:79–94

Knudsen EI, Knudsen PF. 1989. Vision calibrates sound localization in developing barn owls. *J. Neurosci.* 9:3306–13

Knudsen EI, Knudsen PF. 1990. Sensitive and critical periods for visual calibration of sound localization by barn owls. *J. Neurosci.* 63:131–49

Knudsen EI, Mogdans J. 1992. Vision-independent adjustment of unit tuning to sound localization cues in response to monaural occlusion in developing owl optic tectum. *J. Neurosci.* 12:3485–93

Lane RH, Allman JM, Kaas JH, Miezin FM. 1973. The visuotopic organization of the superior colliculus of the owl monkey (*Aotus trivirgatus*) and the bush baby (*Galago senegalensis*). *Brain Res.* 60:335–49

Manley GA, Koeppl C, Konishi M. 1988. A neural map of interaural intensity difference in the brainstem of the barn owl *Tyto alba*. *J. Neurosci.* 8:2665–76

Margoliash D. 1986. Preference for autogenous song by auditory neurons in a song system nucleus of the white-crowned sparrow. *J. Neurosci.* 6(6):1643–61

Meister M, Wong ROL, Baylor DA, Shatz CJ. 1991. Synchronous bursts of action potentials in ganglion cells of the developing mammalian retina. *Science* 252:939–43

Meredith MA, Nemitz JW, Stein BE. 1987. Determinants of multisensory integration in superior colliculus neurons. I. Temporal factors. *J. Neurosci.* 7:3215–29

Meredith MA, Stein BE. 1983. Interactions among converging sensory inputs in the superior colliculus. *Science* 221:389–91

Meredith MA, Stein BE. 1986a. Spatial factors determine the activity of multisensory neurons in cat superior colliculus. *Brain Res.* 365:350–54

Meredith MA, Stein BE. 1986b. Visual, auditory, and somatosensory convergence on cells in superior colliculus results in multisensory integration. *J. Neurophysiol.* 56: 640–62

Middlebrooks JC, Green DM. 1991. Sound localization by human listeners. *Annu. Rev. Psychol.* 42:135–59

Middlebrooks JC, Knudsen EI. 1984. A neural code for auditory space in the cat's superior colliculus. *J. Neurosci.* 4:2621–34

Miller KD. 1989. Correlation-based models of neural development. In *Neuroscience and Connectionist Theory*, ed. MA Gluck, DE Rumelhart, pp. 267–352. Hillsdale, NJ: Lawrence Erlbaum Assoc.

Miyashita Y. 1988. Neuronal correlate of visual associative long-term memory in the primate temporal cortex. *Nature* 335:817–20

Mogdans J, Knudsen EI. 1992. Adaptive adjustment of unit tuning to sound localization cues in response to monaural occlusion in developing owl optic tectum. *J. Neurosci.* 12: 3473–84

Morrell F. 1972. Visual system's view of acoustic space. *Nature* 238:44–46

Olsen JF, Knudsen EI, Esterly SD. 1989. Neural maps of interaural time and intensity differences in the optic tectum of the barn owl. *J. Neurosci.* 9:2591–605

Peck CK, Baro JA, Warder SM. 1993. Sensory integration in the deep layers of superior colliculus. *Prog. Brain Res.* 95:91–102

Perrott DR, Saberi K, Brown K, Strybel TZ. 1990. Auditory psychomotor coordination and visual search behavior. *Percept. Psychophys.* 48:214–26

Perrott DR, Sadralodabai T, Saberi K, Strybel T. 1991. Aurally aided visual search in the central visual field: effects of visual load and visual enhancement of the target. *Hum. Factors* 33:389–400

Pons TP, Garraghty PE, Ommaya AK, Kaas JH, Taub E, Mishkin M. 1991. Massive cortical reorganization after sensory deafferentation in adult macaques. *Science* 252:1857–60

Recanzone GH, Schreiner CE, Merzenich MM. 1993. Plasticity in the frequency representation of primary auditory cortex following discrimination training in adult owl monkeys. *J. Neurosci.* 13(1):87–103

Sanes JR. 1993. Topographic maps and molecular gradients. *Curr. Opin. Neurobiol.* 3:67–74

Sparks DL. 1990. Signal transformations required for the generation of saccadic eye movements. *Annu. Rev. Neurosci.* 13:309–36

Stein BE, Huneycutt WS, Meredith MA. 1988. Neurons and behavior: the same rules of multisensory integration apply. *Brain Res.* 448: 355–58

Stein BE, Meredith MA. 1990. Multisensory integration: neural and behavioral solutions for dealing with stimuli from different sensory modalities. *Ann. NY Acad. Sci.* 608:51–70

Stein BE, Meredith MA. 1993. *The Merging of the Senses.* Cambridge, MA: MIT Press

Stein BE, Meredith MA, Huneycutt WS, McDade L. 1989. Behavioral indices of multisensory integration: orientation to visual cues is affected by auditory stimuli. *J. Cogn. Neurosci.* 1:1–12

Stein BE, Meredith MA, Wallace MT. 1993. The visually responsive neuron and beyond: multisensory integration in cat and monkey. *Prog. Brain Res.* 95:79–90

Sterling P, Wickelgren BG. 1969. Visual receptive fields in the superior colliculus of the cat. *J. Neurophysiol.* 32:1–15

Tamura R, Ono T, Fukuda M, Nakamura K. 1990. Recognition of egocentric and allocentric visual and auditory space by neurons in the hippocampus of monkeys. *Neurosci. Lett.* 109:293–98

Tamura R, Ono T, Fukuda M, Nakamura K. 1992. Spatial responsiveness of monkey hippocampal neurons to various visual and auditory stimuli. *Hippocampus* 2:307–22

Thurlow WR, Rosenthal, TM. 1976. Further study of existence regions for the "ventriloquism effect." *J. Am. Audiol. Soc.* 1:280–86

Udin SB, Fawcett JW. 1988. Formation of topographic maps. *Annu. Rev. Neurosci.* 11: 289–327

Van Essen DC, Anderson CH, Felleman DJ. 1992. Information processing in the primate visual system: an integrated systems perspective. *Science* 255:419–23

Wagner H, Takahashi T, Konishi M. 1987. Representation of interaural time difference in the central nucleus of the barn owl's inferior colliculus. *J. Neurosci.* 7:3105–16

Wallace MT, Meredith MA, Stein BE. 1992. The integration of multiple sensory inputs in cat cortex. *Exp. Brain Res.* 91:484–88

Warren DH, Welch RB, McCarthy TJ. 1981. The role of visual-auditory "compellingness" in the ventriloquism effect: implications for transitivity among the spatial senses. *Percept. Psychophys.* 30:557–64

Watanabe J, Iwai E. 1991. Neuronal activity in visual, auditory and polysensory areas in the monkey temporal cortex during visual fixation task. *Brain Res. Bull.* 26:583–92

Wenstrup JJ, Ross LS, Pollak GD. 1986. Binaural response organization within a frequency-band representation of the inferior colliculus: implications for sound localization. *J. Neurosci.* 6:962–73

Wise LZ, Irvine DRF. 1985. Topographic organization of interaural intensity difference sensitivity in deep layers of cat superior colliculus: implications for auditory spatial representation. *J. Neurophysiol.* 54:185–211

Withington DJ. 1992. The effect of binocular lid suture on auditory responses in the guinea-pig superior colliculus. *Neurosci. Lett.* 136:153–56

Annu. Rev. Neurosci. 1995. 18:45–75

BIOLOGY AND GENETICS OF HEREDITARY MOTOR AND SENSORY NEUROPATHIES

Ueli Suter

Department of Cell Biology, Swiss Federal Institute of Technology,
ETH-Hönggerberg, CH-8093 Zürich, Switzerland

G. Jackson Snipes

Departments of Neurobiology and Pathology (Neuropathology), Stanford University
School of Medicine, Stanford, California 94305

KEY WORDS: myelin, *trembler,* Charcot-Marie-Tooth disease, hereditary neuropathy with
liability to pressure palsies, *PMP22,* P0, connexin 32

INTRODUCTION

The molecular bases of several hereditary sensorimotor peripheral neuropathies are now understood. The natures of the individual genes that have been implicated in these diseases are, in many cases, surprising but provide a satisfying framework for our understanding of these well-described, but historically controversial, group of neuropathies. In order to gain an appreciation of these findings, however, a discussion of the intersecting disciplines that have made our present level of understanding of these diseases possible must be developed. Thus, we consider this group of hereditary peripheral neuropathies from historical, clinical, genetic, and biological viewpoints in order to introduce the facts, the ideas, and the controversies that surround them.

The peripheral neuropathies comprise a large and diverse group of disorders in which one or more components of the peripheral autonomic and/or the cranial and spinal nerves and their peripheral components [as opposed to their central nervous system (CNS) projections] are damaged (for a comprehensive overview, see Dyck et al 1993). Only a small subset of peripheral neuropathies are considered to be hereditary; the majority are the result of toxic, metabolic,

45

0147-006X/95/0301-0045$05.00

infectious, or autoimmune processes. The hereditary peripheral neuropathies can further be subcategorized based on whether autonomic nerves, sensory nerves, motor fibers, or a specific combination thereof are primarily affected. This review focuses on the most numerically important members of this group, the hereditary neuropathies involving both motor and sensory nerves. We further restrict this review to those diseases for which a molecular basis has been identified, specifically, Charcot-Marie-Tooth (CMT) disease, Dejerine-Sottas syndrome (DSS), hereditary neuropathy with liability to pressure palsies (HNPP), and appropriate animal models, which include the *trembler* (*Tr*) and *trembler-J* (*Tr-J*) mice as well as genetically engineered mice. Although these disorders represent only a small percentage of all peripheral neuropathies, they are still numerically and clinically important. CMT alone affects up to 1 in 2,500 people (Skre 1974).

HISTORICAL PERSPECTIVE

Description and Classification of Disease Entities

CHARCOT-MARIE-TOOTH DISEASE   Many of the luminaries of the 19th century, including Friedreich, Osler, Marinesco, and Virchow whom we credit for their clinical acumen and their contributions towards a scientific understanding of disease, reported patients and their relatives who had muscle weakness and wasting that was particularly prominent in the distal lower extremities, a process that was collectively termed progressive muscular atrophy (cited in Brust et al 1978). In 1886, two reports, one authored by two French physicians, Jean Martin Charcot and Pierre Marie (Charcot & Marie 1886), and the other by an English physician, Howard Henry Tooth (Tooth 1886), described five additional cases and provided complementary evidence from the literature for a syndrome with a marked hereditary component that Tooth called "the peroneal type of progressive muscular atrophy" to reflect the often severe weakness and atrophy of the muscles of the lower leg and foot innervated by the peroneal nerves. The autosomal dominantly inherited syndrome described by Charcot, Marie, and Tooth was characterized by progressive weakness followed by atrophy of the lower extremities, often originating in the feet, and progressing to the hands and finally the arms. Both reports noted the striking genetic component and described variable degrees of sensory impairment such as sensory loss and paresthesias. Although Charcot, Marie, and Tooth described identical syndromes for which their names are now synonymous, they differed in their interpretations on its pathogenesis: Charcot & Marie (1886) favored a process involving the spinal cord (which contains the cell bodies of the motor neurons) while Tooth (1886) felt that the primary defect resides in the peripheral nerves themselves. This intriguing controversy persists to the present day where it has been recast in terms of the cellular and molecular mechanisms that underlie axon and Schwann cell interactions.

DEJERINE-SOTTAS SYNDROME   In 1893, Dejerine & Sottas (1893) described a more severe form of peroneal muscular atrophy with progressive generalized loss of muscle function, severe sensory loss, and limb ataxia. Their patients, like many CMT cases, showed marked hypertrophy of the peripheral nerves, an enlargement that was attributed to increased production of interstitial tissue, which forms the onion bulb structures that are observed in histologic cross sections of affected nerves. Subsequent electron microscopic studies showed that the onion bulb generally consists of a central thinly myelinated axon surrounded by supernumerary Schwann cell processes that are separated by an increased amount of endoneurial collagen. Thus, DSS and CMT have been referred to as hypertrophic neuropathies. Autosomal dominant and recessive inheritance patterns of DSS can be inferred from early studies (Austin 1956), but most authors have considered DSS to be transmitted by autosomal recessive inheritance. Because forms of DSS have been described that differ with regards to individual features, such as age of onset, duration, and involvement of sensory, cranial, or autonomic nerves, existence of DSS as a distinct clinical entity has been questioned (Austin 1956, Harding & Thomas 1980). As we see later in this review, based on molecular genetic studies, the clinical features of DSS can reasonably be considered as similar to severe forms of CMT with an onset in early childhood, irrespective of the inheritance pattern.

HEREDITARY NEUROPATHY WITH LIABILITY TO PRESSURE PALSIES   A seemingly unrelated hereditary neuropathy, the hereditary neuropathy with liability to pressure palsies (HNPP) (also called tomaculous neuropathy) was described by Davies (1954). He and others have described an autosomal, dominantly inherited, clinically homogeneous syndrome characterized by a tendency toward repeated sensory and motor nerve palsies, which are brought about by minor pressure or trauma to an affected peripheral nerve bundle. Whereas normal individuals frequently experience transient nerve palsies (lasting seconds to minutes) following similar nerve insults, patients with HNPP frequently have residual nerve damage that lasts days to months before it resolves. The Carpal Tunnel syndrome and other entrapment neuropathies, in particular, are common in HNPP (Lupski et al 1993). The pathologic hallmark of this disorder is multiple focal thickenings of myelin that form sausage-shaped, or tomaculous, enlargements along individual nerve fibers. Of particular relevance is the fact that many of these patients will develop an indolent sensorimotor neuropathy in later life that resembles CMT (for details, see Windebank

1993). Interestingly, in several families with HNPP, some members show no evidence of HNPP but instead manifest features of CMT, such as reduced nerve conduction velocities (NCVs) with axon loss and evidence for demyelination and remyelination without significant onion-bulb formation (Windebank 1993).

NOMENCLATURE OF HEREDITARY MOTOR AND SENSORY NEUROPATHIES   Thus far we have used the designations peroneal muscular atrophy and Charcot-Marie-Tooth disease interchangeably as if they were synonyms. Although this is true in a general sense, in that the cardinal clinical feature of Charcot-Marie-Tooth disease is the distal lower extremity weakness and wasting implied by peroneal muscular atrophy, a more precise view may be appropriate. A century of pathologic evaluations, supplemented by electron microscopic studies, and extensive clinical evaluations, supplemented by electrodiagnostic studies, have classified these syndromes into two major groups [see Harding & Thomas (1980) and references therein], designated Charcot-Marie-Tooth disease type 1 and type 2 (CMT1 and CMT2) (Dyck et al 1993). CMT1 is more common than CMT2 and is characterized by demyelination, as revealed electrophysiologically by significantly lowered NCVs in both sensory and motor nerves (for review, see Lupski et al 1991b) and pathologically by segmental demyelination and remyelination and onion bulb formation. The progressive nature of CMT1 does not appear to arise from the demyelination and remyelination processes revealed by decreased NCVs but rather from a progressive loss of axons as revealed by denervation changes by electromyography and decreased amplitudes of motor action potentials (Lupski et al 1991b). CMT2, on the other hand, is characterized by normal or near normal NCVs, and the major pathologic features are a decreased number of large myelinated axons without significant evidence for demyelination and remyelination and an increased number of tomaculae in nerve tease preparations (Dyck et al 1993). Thus, CMT1 has been termed the demyelinating form and CMT2 the axonal form of CMT. These classifications, however, appear to be oversimplifications. As early as 1927, Dawidenkow (1927) maintained that there were at least 12 types of hereditary neuropathies. Interestingly, one of these neuropathies was unique in that only males were affected through four generations, thus suggesting an X-linked mode of inheritance. Part of the difficulty in classification undoubtedly arises from the inherently heterogeneous nature of hereditary sensorimotor neuropathies. There is marked clinical heterogeneity in the severity of the disease, even within a single family, among affected siblings, and between identical twins (Lupski et al 1991b). This heterogeneity is also reflected in the chromosomal localization and identity of the genes that, when perturbed, give rise to remarkably similar neuropathies (see below). Dyck's widely accepted classification of the hereditary neuropathies, which takes into account these and many additional considerations, maintains the major entities as outlined above (Dyck 1975). Specifically, CMT1 is equivalent to hereditary motor and sensory neuropathy I (HMSN I); CMT2 is equivalent to HMSN II; DSS is equivalent to HMSN III; and HNPP is considered an inherited recurrent focal neuropathy. The Roussy-Levy syndrome, an inherited peripheral neuropathy that is associated with the features of CMT1 and intention tremors, is considered to be a variant of HMSN I (Dyck et al 1993). Nevertheless, we adhere to the CMT designation because the remainder of this review emphasizes the genetic aspects of hereditary sensorimotor neuropathies, and the genetics literature employs the CMT terminology.

GENETICS OF CMT   The complicated genetics of hereditary sensorimotor neuropathies has begun to yield to the persistence of the geneticists and the power of their techniques. In 1989, the defect in the majority of patients with CMT1 was mapped to chromosome 17 (termed subtype CMT1A) (Raeymaekers et al 1989, Vance et al 1989), and in 1991, an intrachromosomal duplication on the short arm of chromosome 17 was found to be associated with CMT1A (Raeymaekers et al 1991, Lupski et al 1991a). At about the same time, Suter et al (1992a,b) showed that mutations in the gene for the peripheral myelin protein PMP22 were responsible for the severe peripheral neuropathies in *Tr* and *Tr-J* mice. The human *PMP22* gene was implicated as a candidate gene for CMT1A based on these results, its mapping to within the CMT1A duplication (Matsunami et al 1992, Patel et al 1992, Timmerman et al 1992, Valentijn et al 1992b), and the finding that a subset of CMT1A patients carry mutations in the *PMP22* gene (Valentijn et al 1992a; Roa et al 1993b,c). Soon thereafter, Chance et al (1993) showed that the reciprocal deletion of the same 1.5-Mb region that is duplicated in CMT1A is associated with HNPP. In addition, mutations in the major myelin protein of the peripheral nervous system (PNS), protein zero (P0), have been found in rare CMT1 families (subgroup CMT1B) (Hayasaka et al 1993a,c; Kulkens et al 1993), and additional mutations in the *PMP22* (Roa et al 1993a) and P0 (Hayasaka et al 1993b) genes have been associated with DSS. Bergoffen et al (1993a) have identified mutations in the gap junction protein gene that encodes connexin 32 in families with an X-linked dominant inheritance pattern (subtype CMTX).

   Additional rare genetic variants of CMT1 include another autosomal dominant form (CMT1C) (Chance et al 1992) and an autosomal recessive form of CMT, characterized by slow NCVs and hypomyelination, which has been linked to chromosome 8 (termed CMT4A) (Ben Othmane et al 1993a). More autosomal loci have been suggested (CMT4B and CMT4C) (Ben Othmane et al 1993a). Furthermore, up to three additional X-linked CMT loci, including X-linked recessive forms, have been identified (Ionasescu et al 1992). One disease locus for CMT2 has been identified on chromosome 1 (CMT2A) (Ben Othmane et al 1993b), and evidence suggests additional loci exist.

COMPARISON OF THE CLINICAL FEATURES OF CMT1A, CMT1B, AND DOMINANT CMTX   CMT1A is the most common form of CMT [accounts for probably more than 50% of all CMT cases (Wise et al 1993)] and clinically shows all of the features historically associated with peroneal muscular atrophy (reviewed in Lupski et al 1993). Onset of disease symptoms typically occurs in the first two decades. Progressive distal muscle weakness is the predominant symptom in CMT1A; sensory problems are relatively minor. Symptoms associated with the distal weakness include abnormalities of gait (often a steppage gait), diminished tendon reflexes, foot deformities (including pes cavus), distal muscle atrophy often associated with cramps, and difficulty with fine muscle movements in the hands. Patients with CMT1A have decreased NCVs (<40m/s, 100% penetrance) (Nicholson 1991), but as mentioned previously, the severity of the weakness varies from patient to patient.

Most reports suggest that CMT1B and CMT1A are clinically indistinguishable (Dyck et al 1993), although early reports suggested that patients with CMT1B had lower NCVs and more extensive onion bulb formation and a more severe neuropathy than CMT1A patients (Bird et al 1983, Dyck et al 1989). Given the emerging genetic heterogeneity of both CMT1A and CMT1B, the severity of the disease appears to depend, in large part, on the nature of the individual mutations in *PMP22* or P0.

Clinical genetic studies indicate that up to 10% of CMT cases are linked to the X chromosome (prevalence ~1/100,000) (Skre 1974). The hallmark of X-linked inheritance is that primarily male family members are affected via maternal, but not paternal, transmission. Generally, affected males have normal mental status and cranial nerve function, but as in other forms of CMT, the clinical features vary in severity among affected patients in the same family (Rozear et al 1987, Hahn et al 1990). The onset of CMTX is in childhood. Clinically, patients show progressive distal weakness and sensory loss that is, overall, more severe than in CMT1. Sensory and motor NCVs of affected males are significantly reduced but less so than in CMT1 patients. In addition, the amplitude of compound motor action potentials may be severely reduced. Significant findings in biopsies of dominant CMTX patients include the loss of large myelinated axons with onion bulb formation and evidence of demyelination and remyelination and axon regeneration. Carrier females seem to have a variable, mild disease as assessed by clinical criteria, and reduced NCVs that are intermediate in severity between affected males with CMTX and normal individuals.

GENES INVOLVED IN HEREDITARY MOTOR AND SENSORY NEUROPATHIES

The Role of the Peripheral Myelin Protein 22 Gene in Inherited Peripheral Neuropathies

Peripheral Myelin Protein 22 (PMP22) is a recently described component of PNS myelin (for review, see Snipes et al 1993a, Suter et al 1993). The events that led to the discovery of PMP22 and its role in CMT1A provide a classic example of how basic research on general biological processes (in this case, the mechanisms of nerve injury and repair) can synergize with the clinical descriptions of a disease and the emerging findings from modern human genetics to reveal the molecular basis of a fundamental disease process. Because *PMP22* was the first of the CMT genes to be elucidated, the biology of *PMP22* is discussed in detail.

IDENTIFICATION AND CLONING OF PMP22   The peripheral myelin protein PMP22 (SR13, CD25, PASII, gas-3) was cloned by a differential hybridization strategy designed to isolate cDNA sequences that are regulated after a focal crush injury to the rat sciatic nerve (De Leon et al 1991, Spreyer et al 1991, Welcher et al 1991). DNA sequence analysis revealed a single open reading frame that predicted a protein of 160 amino acids and a core molecular weight of 18 kDa. Computer-assisted hydrophobicity plots and in vitro translation experiments characterized the predicted PMP22 protein as an extremely hydrophobic integral membrane protein, containing four membrane-associated, potentially membrane-spanning domains (Manfioletti et al 1990, Suter et al 1992b) (Figure 1). In addition, a conserved consensus site for N-linked glycosylation was found at asparagine 41. Indeed, PMP22 has been shown to contain an N-linked carbohydrate chain in vitro and in vivo (Kitamura et al 1976, Manfioletti et al 1990, Welcher et al 1991, Pareek et al 1993), which is consistent with the apparent molecular weight of 22 kDa of the peripheral nerve-derived PMP22 protein on SDS-PAGE (Snipes et al 1992, Pareek et al 1993).

EXPRESSION AND REGULATION OF PMP22   Northern blot, RNase protection, and in situ hybridization analysis revealed that PMP22 transcripts are found mainly in myelinating Schwann cells in the PNS and that PMP22 mRNA expression is regulated by axonal contact (Spreyer et al 1991, Welcher et al 1991, Snipes et al 1992). Minor expression of PMP22 mRNA can also be found in the central nervous system, particularly in the motor nuclei of most of the cranial nerves and in the motoneurons in the ventral horn of the spinal cord (Parmantier et al 1993).

The striking correlation between myelin formation and PMP22 expression during development and after injury suggested that this gene might encode a

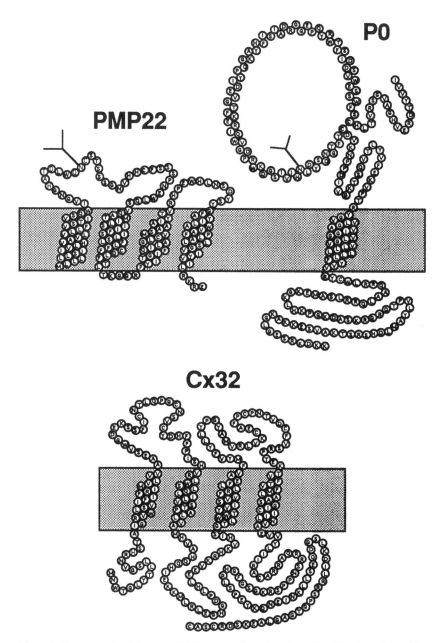

Figure 1   Topological models of PMP22, P0 and Cx32. Each amino acid residue is symbolized by an open circle. Amino acids are abbreviated with the one-letter code. The Y-shaped symbols indicate N-linked glycosylation sites.

novel myelin protein. In support of this hypothesis, immunohistochemical analysis at the light microscopic level revealed intense PMP22 immunoreactivity in the PNS myelin sheath (Snipes et al 1992), and subsequent electron microscopic studies have further confirmed the localization of PMP22 to the compact portion of the PNS myelin (C Haney, B Trapp, GJ Snipes & U Suter, unpublished data). Complementary evidence that PMP22 is a myelin protein was obtained by comparing the partial amino acid sequence of the previously described bovine myelin protein PASII (Kitamura et al 1976) to the cDNA-predicted rat PMP22 amino acid sequence. The nearly identical N-terminal sequences indicated that these two proteins are most probably species homologues and that the putative N-terminal signal sequence of the PMP22 protein is not cleaved during biosynthesis in Schwann cells (Welcher et al 1991). Quantitative estimations by Western blot analysis suggest that PMP22 comprises approximately 2–5% of total protein in the rat PNS myelin (Pareek et al 1993).

PMP22 expression is, however, not restricted to the nervous system. Northern blot and RNase protection analysis have revealed the presence of PMP22 mRNA in a variety of tissues, most notably in lung, gut, and heart (Spreyer et al 1991, Welcher et al 1991). Computer-aided database searches suggested that the PMP22 mRNA is the rat homologue of the growth-arrest-specific mouse mRNA, gas-3 (Manfioletti et al 1990). The gas family of mRNAs was isolated from nonproliferating NIH3T3 fibroblasts, based on the observation that the steady-state levels of these mRNAs are elevated in resting cells (by serum starvation or contact inhibition) (Schneider et al 1988). Recent work suggests that the upregulation of PMP22 mRNA expression in NIH3T3 cells is also reflected at the protein level, which supports the idea that nonneural PMP22 expression may be of functional significance (S Pareek, personal communication). Furthermore, this type of PMP22 regulation in vitro appears to be a widespread phenomenon because it is also observed in the C6 glioma cell line and human and rat fibroblasts (GJ Snipes & U Suter, unpublished data). Interestingly, PMP22 mRNA also appears to be upregulated in PC12 cells treated with nerve growth factor (De Leon et al 1994). Whether any of this observed regulation reflects a direct role for PMP22 in growth-arrest phenomena or whether it arises as a consequence of cellular differentiation remains to be determined.

THE HUMAN *PMP22* GENE AND ITS REGULATORY REGIONS   The human *PMP22* gene spans approximately 40 kilobases and contains four coding exons and two untranslated exons in its 5′-flanking region (Patel et al 1992, Suter et al 1994) (Figure 2). Mapping of PMP22 mRNAs and functional analysis by using reporter constructs suggest that each of the 5′ untranslated exons is preceded by a tissue-specific promoter. Both PMP22 promoters appear to be highly

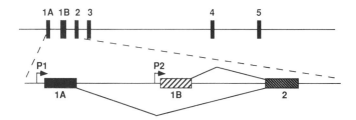

Figure 2 Structure of the human *PMP22* gene. Each exon is symbolized with a filled rectangular box and consecutively numbered. Either exon 1A or 1B are alternatively spliced to exon 2. P1 and P2 indicate alternatively used tissue-specific promoters.

active in human peripheral nerve with the expression of the specific transcript driven by promoter 1 predominating. In contrast, the PMP22 transcripts originating from promoter 2 are more abundant in all examined nonneural tissues, although total PMP22 mRNA is much reduced compared to the expression in the PNS. Only minor levels of the alternative transcript derived from promoter 1 could be detected in nonneural tissues, which supports the view that the observed PMP22 expression is not solely due to innervation.

The structure of the regulatory 5′-flanking region in the rat PMP22 gene is identical to its human counterpart. RNase protection analysis during development and regeneration of the rat peripheral nerve suggested a very tight correlation of the myelination process with the activity of the putative PMP22 promoter 1 while a similar but less pronounced association was observed with the activity of promoter 2.

PMP22 MUTATIONS IN *TREMBLER* AND *TREMBLER-J*   Mutations in several major myelin proteins lead to spontaneous neurological mouse mutants like *shiverer* (deletion in MBP) or *jimpy* [point mutation and splice defect in proteolipid protein (PLP)] (for review see Lemke 1988). Such mouse strains are powerful tools in developmental neurobiology as they yield important insights into the function of the affected genes in vivo. Furthermore, neurological mouse mutants provide excellent model systems to study related diseases in animals and humans. Obviously, it is crucial for the interpretation of these models to identify the primary genetic defects that lead to the specific disease phenotypes.

The characterization of PMP22 as a PNS myelin protein led to the search for possible spontaneous PMP22 mutations in mice. The autosomal dominant *Tr* mouse was singled out as a likely candidate based on its PNS-specific defect in myelination and aberrant Schwann cell growth (Falconer 1951, Henry et al 1983, Henry & Sidman 1988; for an extensive list of references, see Heath et al 1991). In support of this hypothesis, the *pmp22* gene was mapped to mouse chromosome 11 in the immediate vicinity of the *Tr* locus (Suter et al 1992a,b). Subsequent analysis of inbred *Tr* mice identified a point mutation in the *pmp22* gene that replaced a glycine residue at position 150 in the fourth putative transmembrane domain of PMP22 with an aspartic acid residue (G150D) (Suter et al 1992b). In addition, a point mutation in the first transmembrane region of PMP22 (L16P) was detected in the *Tr-J* mouse that had been previously hypothesized to be allelic to *Tr* mice (Henry et al 1983, Suter et al 1992a). Both, *Tr* and *Tr-J* mice show limb quadriparesis, which is worse in the hind limbs. Young animals are also prone to transient convulsions and develop body tremors with advancing age. Histologically, a severe PNS hypomyelination coupled with an overproduction of Schwann cells is observed. Furthermore, the *Tr* and *Tr-J* alleles exhibit a dosage effect, e.g. homozygotes have a more severe PNS myelin deficiency than heterozygotes (Henry & Sidman 1983, Lemke 1993). The main difference between *Tr* and *Tr-J* mice appears to be the limited viability of homozygous *Tr-J* (these animals die at postnatal day 17 and/or 18) whereas homozygous *Tr* live a long time) (Henry & Sidman 1988). However, these findings and additional subtle differences in the pathology of *Tr* and *Tr-J* have to be interpreted cautiously because the two strains have been maintained on different genetic backgrounds.

MUTATIONS AFFECTING THE *PMP22* GENE ASSOCIATED WITH HEREDITARY PE-RIPHERAL NEUROPATHIES IN HUMAN   The majority of CMT patients genetically belong to the autosomal dominant subtype CMT1A and almost all of the CMT1A patients carry a 1.5-Mb tandem chromosomal duplication on the short arm of chromosome 17 [for review see Suter et al (1993)]. Furthermore, the reciprocal deletion of the same chromosomal region has been shown to cause the dominantly inherited HNPP disease (Chance et al 1993). CMT1A and HNPP appear to be the result of a reciprocal, unequal nonsister chromatid exchange (unequal crossing-over) mediated through a misalignment of homologous repeated sequences that are flanking the normal 1.5-Mb monomeric region (Chance et al 1994). This mutational mechanism convincingly explains the high incidence of CMT1A (and probably HNPP) in the population and constitutes the first example of two human disorders with a Mendelian pattern of inheritance that are the reciprocal results of an unequal crossing-over involving large internal chromosomal segments. Interestingly, studies in de novo CMT1A duplication patients suggest that this unequal crossover event is overwhelmingly of paternal origin (Palau et al 1993). The mechanism(s) by which male-specific factors, presumably operating during spermatogenesis, give rise to these diseases remain(s) to be established.

Pmp22 is located on mouse chromosome 11 in a region of conserved synteny with human chromosome 17 near the CMT1A-HNPP locus (Buchberg et al 1989). As predicted from this observation, the human *PMP22* gene maps firmly within the CMT1A duplication without interruption (Matsunami et al 1992, Patel et al 1992, Timmerman et al 1992, Valentijn et al 1992b). These results add further plausibility to the hypothesis that one or more dosage-sensitive gene(s) within the CMT1A duplication and HNPP deletion are likely to be responsible for the respective disease phenotypes. This has been suggested by the finding of a CMT1A-like phenotype (decreased NCVs) in several patients with partial trisomies that encompass the CMT1A-HNPP locus (reviewed by Lupski et al 1993). Several lines of evidence support the hypothesis that *PMP22* is the main dosage-sensitive gene within the CMT1A duplication and HNPP deletion. First, a CMT1A family has been found that carries a considerably smaller duplication of only 460 kilobases. Because this particular duplication still includes the entire *PMP22* gene, it excludes most of the other genes present in the frequent 1.5-Mb CMT1A duplication from further consideration (Valentijn et al 1993). Secondly, several different missense point mutations in *PMP22* have been found in various families diagnosed with CMT1A [L16P (Valentijn et al 1992a), S79C (Roa et al 1993c), and T118M (Roa et al 1993b)] and autosomal dominant DSS [M69K and S72L (Roa et al 1993a)]. Interestingly, one of the mutations in a European CMT1A family is identical to the mutation in *Tr-J* (L16P). This result suggests that the *Tr-J* mouse is an ideal animal model for this specific subtype of CMT1A. Third, a 2-bp frame-shifting deletion within the first translated exon of *PMP22* has been found to be associated with HNPP (Nicholson et al 1994). This mutation would be expected to generate a null allele. The similarity of the phenotypes generated by this mutation and the HNPP deletion directly implicates *PMP22* as the critical deleted gene in HNPP.

MUTATIONS AFFECTING THE *PMP22* GENE PROVIDE HINTS ABOUT IMPORTANT DOMAINS WITHIN THE PMP22 PROTEIN   A common feature of all PMP22 mutations found so far in CMT1A, DSS, *Tr,* and *Tr-J* is their location in putative transmembrane domains. Because the same domains are also the most strongly conserved parts, evolutionarily, of the PMP22 protein (Patel et al 1992), one might speculate that the membrane-associated regions are significant to the proper function of PMP22. This hypothesis is indirectly supported by the finding that the particular mutation, T118M (Roa et al 1993b), which affects a position that is predicted to exist just inside of the third putative transmembrane region, is only manifested as CMT1 in a compound heterozygote with one allele carrying the T118M mutation and the second allele affected by the HNPP deletion. Another family member, with no signs of HNPP or CMT1A, carries the T118M on one allele and is normal on the other allele, while a third group of family members suffers from HNPP by carrying the HNPP deletion in combination with a normal allele. These findings suggest that the T118M mutation is likely to be inherited as a recessive allele and may help to explain some rare recessive CMT1 pedigrees.

The PMP22 mutations observed in *Tr* and *Tr-J* are both nonconservative amino acid changes. They introduce either a helix-breaking amino acid (proline in *Tr-J*) or a charged amino acid (aspartic acid in *Tr*) into the probably highly structured putative transmembrane domains of the PMP22 protein. Thus, these mutations might substantially alter the tertiary structure of the PMP22 protein by affecting its membrane association. The reason both mutations confer a dominantly inherited phenotype, however, remains puzzling. The most plausible hypothesis suggests a strong dominant-negative effect, possibly due to the potential formation of defective interactions of PMP22 with itself or other Schwann cell proteins. In contrast, the *Tr-J* mutation is unlikely to result in a null allele given the quite severe CMT phenotype (strongly reduced NCVs) in human patients carrying the same mutation as *Tr-J* mice (Valentijn et al 1992a, Hoogendijk et al 1993). This is in contrast to the much milder neuropathy observed in HNPP patients who carry a *PMP22* null allele due to a complete deletion of the PMP22 locus.

Several intriguing questions derived from the results of human genetic studies remain unanswered: What is the biological basis of the dominantly inherited dosage-sensitivity of *PMP22* that leads to either CMT1A by an increase of only 50% in PMP22 expression or HNPP by a 50% reduction of PMP22 expression? Why is the disease phenotype of a homozygous CMT1A patient [doubled PMP22 expression (Lupski et al 1991a)] much worse (similar to DSS) than in heterozygous CMT1A patients, and is there a connection to the dosage-sensitivity observed in *Tr* and *Tr-J*? How can the putative over-expression of PMP22 and one class of point mutations in the same gene lead to similar phenotypes? A mechanism whereby the function of PMP22 is affected by mutation or overexpression that alters a critical stoichiometric relationship of PMP22 to itself or another component of myelin could explain many of these questions, but other questions remain: Why is a second class of point mutations in the *PMP22* gene associated with the more severe phenotype DSS? And finally, it remains unclear why a 5′ splice site mutation in the last intron of the *PMP22* gene results in autosomal dominant CMT1A (as opposed to HNPP) (Nelis et al 1994).

THE PARTICULAR GENETICS OF *PMP22* ARE NOT WITHOUT PRECEDENT   The available genetic data strongly suggest that *PMP22* is the dosage-sensitive gene within the CMT1A duplication (or HNPP deletion). How either over-expression of PMP22 or point mutations in the *PMP22* gene can lead to very similar disease phenotypes remains difficult to reconcile. Recent experiments reveal that a similar, albeit artificially created, situation exists after expression of a mutant rhodopsin transgene or overexpression of the wild-type rhodopsin allele in transgenic mice. Both lines developed quite similar photoreceptor degeneration (Olsson et al 1992). However, the detailed biological mechanisms leading to the observed effects in these experiments have yet to be determined.

An interesting parallel in the genetics of CMT1A and HNPP can be found for the major myelin protein in the CNS—PLP. PMP22 and PLP are considered to have similar functions based on their relative abundance and reciprocal expression in PNS and CNS myelin, respectively (Suter et al 1993). Mutations in the X chromosome–linked *PLP* gene have been detected in the *jimpy* and *rumpshaker* mouse mutants as well as in the human CNS myelin disorder, Pelizaeus-Merzbacher disease (PMD) (for review, see Snipes et al 1993a). Interestingly, an intrachromosomal duplication spanning the entire *PLP* locus has been described in a particular patient with a general, PMD-like myelin disorder (Cremers et al 1987). Similarly, a patient carrying a deletion spanning the *PLP* gene also appears to be affected by PMD (Raskind et al 1991). In further support of a postulated PLP dosage effect, transgenic mice with only a twofold overexpression of the *PLP* gene under its natural regulatory elements exhibit severe hypomyelination and astrocytosis in the CNS indicating that the differentiation of oligodendrocytes requires correct levels of *PLP* expression (Readhead et al 1994).

The Role of Mutations in the P0 Gene in CMT1B

The first reported linkage of a CMT1 locus was to the Duffy locus on chromosome 1 (Bird et al 1982). Although this finding has been confirmed, it soon became obvious that the main disease locus was linked to chromosome 17 (CMT1A) and that the CMT phenotype could only be linked to chromosome 1 (CMT1B) in rare cases (Ionasescu et al 1993).

The identification of the myelin protein PMP22 as the culprit in CMT1A prompted the search for mutations in other genes that encode myelin proteins in the PNS in the rarer forms of CMT1. Indeed, mutations in the P0 gene were found in CMT1B.

STRUCTURE AND EXPRESSION OF P0 IN PERIPHERAL MYELIN   P0 is found throughout compact myelin where it comprises 40 to 50% of the total protein content (for review, see Lemke 1988), and it first appears during development when Schwann cells and large calibre axons associate in a 1:1 relationship (Martini et al 1988). The cDNA sequence, isolated from a sciatic nerve library, predicts that P0 is a membrane glycoprotein derived from a precursor protein after cleavage of its signal sequence (Lemke & Axel 1985). This hypothesis has been confirmed by complete amino acid sequencing of mature P0 (Sakamoto et al 1987). Thus, P0 consists of a single extracellular, a transmembrane, and an intracellular domain (Figure 1). The extracellular part contains an immunoglobulin (Ig)-like domain, a motif common to a large family of cell adhesion (cell recognition) molecules that also includes myelin-associated glycoprotein (MAG) and the neural cell adhesion molecule (NCAM). Experimental evidence obtained by using various transfection paradigms in nonneural cell types strongly suggest that P0 exhibits homotypic adhesive properties, presumably mediated by its Ig-like domain (for review, see Lemke 1993). Furthermore, the N-linked carbohydrate chain attached to the extracellular domain of P0 (a subset of which carries the L2-HNK1 epitope) is implicated in this adhesive function (Bollensen & Schachner 1987). It has also been suggested that the highly positively charged intracellular domain of P0 might be involved in an adhesive function via electrostatic interactions with the head groups of acidic lipids in compact myelin (Lemke 1988, Lemke et al 1988). This idea has emerged from considering potential functional compensatory mechanisms to explain the minimal changes in PNS myelin compared to CNS myelin in homozygous *shiverer* mice that lack myelin basic protein (MBP). Whether the intracellular domain of P0 is indeed involved in adhesion processes and/or signal transduction is an intriguing but unanswered question.

THE P0 GENE AND ITS REGULATORY ELEMENTS   The rat, mouse, and human P0 genes each span approximately 7 kb and are comprised of six coding exons that appear to correspond to functional domains of the P0 protein (Lemke et al 1988). The 5'-flanking region of the rat P0 gene has been mapped and the putative P0 promoter was analyzed by using transfections of reporter constructs into various cultured cells (Lemke et al 1988). The results of this analysis suggested that the isolated P0 promoter is sufficient to direct Schwann cell–specific expression, and subsequent experiments in transgenic mice convincingly demonstrated that the rat P0 promoter is indeed capable of directing the expression of various transgenes specifically to Schwann cells (Messing et al 1992).

FUNCTION OF P0 IN PNS MYELIN   Most of the initial information concerning the function of P0 in the process of myelination has been obtained in vitro. In particular, Owens & Boyd (1991) applied a strategy aimed at decreasing P0 expression by using retroviral gene transfer to express high levels of P0 antisense mRNA in cultured Schwann cells. If such P0-reduced Schwann cells were cocultured with sensory neurons under conditions that allow myelination, a severe inhibition of the capability of these Schwann cells to elaborate compact myelin was observed.

The most convincing support for a crucial role of the P0 protein in the compaction of myelin in the PNS has been obtained by generating a targeted disruption of the P0 gene in transgenic mice (Giese et al 1992). Behaviorally, P0 minus (P0⁻) mice are deficient in motor coordination and exhibit tremors and occasional convulsions. Microscopic examination of the peripheral nerves of homozygous P0⁻ mice shows severe but incomplete loss of myelin in the PNS, similar to the *Tr* mouse. Initially, Schwann cells and axons establish the correct 1:1 association in the P0⁻ mouse, but the subsequent process of forming compact myelin is greatly disturbed. Although the pathological phenotype in P0⁻ mice appears quite heterogeneous with regards to myelination, the overall interpretation is consistent with a critical role of P0 in the compaction of myelin. In particular, regular intraperiod lines are virtually absent, which indicates a function for the extracellular domain of P0 in establishing this structure. Evidence for axonal degeneration can be observed in some fibers. Interestingly, the degenerating axons might correspond to the largest caliber fibers in the normal animal because the degenerating axons in the P0⁻ mouse are surrounded by the most extensive myelin-like envelopes. In addition, the P0⁻ mouse exhibits a paradoxical proliferation of nonmyelinating Schwann cells as previously described in the *Tr* and *Tr-J* mice. However, the interpretation of the pathology in the P0⁻ mouse mutant is confounded by the disregulation observed in several other PNS proteins.

MUTATIONS IN THE P0 GENE IN HUMAN   Several mutations in the P0 gene have been found and the resulting phenotypes were clinically classified as CMT1B or DSS. Most of these mutations are missense mutations that are located in the extracellular domain of P0 and might interfere with its proposed adhesive function (D61E, K67E; numbering according to the putative mature P0 protein) (Hayasaka et al 1993a). Mutants of this class are associated with CMT1B. In one case, the very first amino acid of the putative mature P0 protein is altered (I1M) (Hayasaka et al 1993c). This mutation might directly interfere with the maturation of the P0 precursor by preventing the cleavage of the signal sequence. Such a mechanism could potentially explain the observed dominant inheritance pattern and the early onset of disease in this particular family. In another CMT1B family, a 3-bp in-frame deletion that leads to the loss of serine-34 (Kulkens et al 1993) in the extracellular domain has been described. Interestingly, the same serine-34 residue is converted to a cysteine in a sporadic patient diagnosed with DSS (S34C) (Hayasaka et al 1993b). The introduction of an additional cysteine residue might lead to aberrant formation of disulfide bridges, possibly explaining the observed severe phenotype by virtue of a strong dominant-negative effect. A second sporadic case of DSS shows a glycine-to-arginine exchange (G138R) in the single transmembrane region of the P0 protein (Hayasaka et al 1993b). One might speculate that this mutation is likely to affect the membrane anchoring of the P0 protein because a charged amino acid has been introduced into a hydrophobic transmembrane stretch, reminiscent of a similar amino acid exchange in the fourth transmembrane domain of PMP22 in the *Tr* mouse.

The observation of a dominant phenotype of the P0 mutations found so far is somewhat surprising because the heterozygous P0⁻ mouse mutant does not exhibit unusual signs of demyelination. It should be kept in mind, however, that CMT is characterized as a pleotrophic late-onset disease. Thus, a potential phenotype in the heterozygous P0⁻ might be subtle and only manifested in a subgroup of aged animals. Alternatively, the known mutations in the P0 gene that are associated with CMT1B might exert their effect via a dominant-negative mechanism. Such a hypothesis might predict that the P0 protein forms complexes in a *cis-* or *trans-*configuration with itself or other proteins that are required for the maintenance of the PNS myelin sheath. While we already know that P0 can interact homotypically, other proteins that are encoded by CMT1 genes are also candidates for participating in such a multiprotein complex.

One might also anticipate finding (possibly recessive) mutations in the intracellular domain of P0, given the proposed role of this domain in myelin compaction. A single family with a potential splice defect that could possibly lead to alterations in the cytoplasmic P0 domain has been described (Hayasaka et al 1993, Su et al 1993). The mutational mechanism, however, as well as the nature of the mutation in this particular family, is controversial and remains to be clarified (Hayasaka et al 1993, Su et al 1993). Interestingly, two members of this family showed features of an HNPP-like tomaculous neuropathy (Thomas et al 1994).

Connexin 32 and X-Linked Dominant CMT

The gene that carries the mutations responsible for dominant CMTX has been identified as encoding the gap junction protein connexin 32 (Cx32) (Bergoffen et al 1993a). Eight families with dominant CMTX were screened for mutations in the connexin 32 gene (designated *GJB1*), and nonidentical mutations were found in seven of them. This discovery was the result of efforts to increase the resolution of the linkage of the CMTX disease phenotype at band Xq13 (Bergoffen et al 1993b) followed by a mutational analysis of candidate genes that mapped to that region.

STRUCTURE AND FUNCTION OF Cx32   Cx32 is a member of a family of homologous proteins that are assembled into highly ordered hexamers (termed connexons) in the plasma membrane of a cell and that interact with connexons on the surface of an adjacent cell (or different regions of the same cell) (for review, see Bennett et al 1991). The resulting dodecamer is the smallest unit of a gap junction that functions as a potentially regulatable pore through which ions and small molecules can cross from one cell to another. Morphologic techniques have identified gap junctions in many different cell types, and biochemical studies have established that electrically coupled cells express connexins, often more than one that can be localized by immunofluorescent techniques to the region of the gap junction (Bennett et al 1991). Experimentally, it has been shown that functional gap junction dodecamers can be entirely composed of a single connexin (Swenson et al 1989) or, alternatively, by the apposition of nonidentical but individually homogenous connexons (Werner et al 1989). Nicholson et al (1987) have questioned whether or not the hexameric connexons themselves can be composed of more than one type of connexin, but this is still unknown.

THE CONNEXIN FAMILY OF GAP JUNCTION PROTEINS   At least nine different rat connexin cDNAs besides Cx32 have been cloned, and some homologues have been identified in other species (for review, see Dermietzel & Spray 1993). The rat and mouse Cx32 genes carry the entire open reading frame on a single exon, a structural feature that is conserved in all connexin genes that have been isolated so far. Interestingly, the rat Cx32 gene contains at least one (or potentially two) 5'-untranslated exon(s), each of which might be preceded by individual promoters (Miller et al 1988). This structure of the regulatory regions is reminiscent of the *PMP22* gene.

Connexin proteins are relatively small molecules with molecular weights ranging from 20–50 kDa. Structurally, monomeric connexins resemble PMP22 in that they contain four transmembrane domains (TMs) with two extracellular loops that bridge TM1 with TM2 (designated EL1) and TM3 with TM4 (designated EL2) (Milks et al 1988) (Figure 1). The amino and carboxy termini and a loop bridging TM2 and TM3 are intracellular. Sequence comparisons between different connexins have suggested common structural motifs and demonstrate significant sequence conservation, particularly within the extracellular and transmembrane domains (Bennett et al 1991). Specifically, TM3 contains conserved amphipathic amino acids that are thought to align at the central core of the hexameric connexon to form a hydrophilic pore (Milks et al 1988). Furthermore, the TM2 region has been implicated in voltage gating of gap junction channels (Suchyna et al 1993). EL1 and EL2 are hypothesized to mediate the binding between the connexons of adjacent plasma membranes. Interestingly, both EL1 and EL2 contain three cysteines at conserved locations that appear to form intramolecular disulfide bonds and are important for channel function (Rahman et al 1993). The intracellular amino terminus is relatively conserved (approximately 50% amino acid identity among the connexins) and is believed to be important for correct membrane insertion and orientation (Bennett et al 1991). The length and amino acid sequence of the intracellular carboxy termini diverge widely among the different connexins, perhaps contributing to the observed heterogeneity of permeabilities and gating properties of individual gap junctions in different cell types (Saez et al 1990), although this is debated (Werner et al 1991).

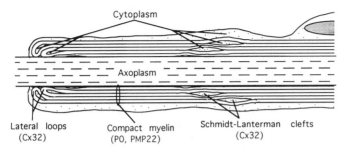

Figure 3 Schematic drawing of the structure of myelin. Particular components of the fine structure are indicated.

MUTATIONS OF Cx32 IN CMTX   Given the number of presumed functional domains of Cx32, it is probably no surprise that many different mutations have
been found. These mutations may be summarized as point mutations that occur
in the intracellular amino-terminal part (G12S), TM3 (V139M and R142W)
and EL2 (L156R, P172S, and E186K), and a frameshift mutation in EL2 (at
position 175) (Bergoffen et al 1993a). It remains to be determined whether
these mutations result in a gain or loss of function of their respective functional
protein domains or whether these mutations interfere with the general intracellular processing, insertion, or assembly of connexons. Given the frequency
of dominant CMTX in the population, there is little doubt that additional
mutations in Cx32 will be identified as causes of CMTX.

Because Cx32 is expressed in a variety of tissues, including liver, testis,
kidney, pancreas, uterus, and brain (Dermietzel & Spray 1993), it is surprising
that mutations in Cx32 result in dysfunction that is apparently limited to the
PNS. In the PNS, previous morphological studies have failed to demonstrate
the existence of gap junctions despite the fact that significant amounts of Cx32
are expressed in peripheral nerves (Bergoffen et al 1993a). This discrepancy
is partially reconciled by the localization of Cx32, by immunofluorescent
techniques, to the Schmidt-Lanterman clefts and the lateral loops of the myelin
sheath formed by Schwann cells of the PNS (Bergoffen et al 1993a) (Figure
3). These myelin subdomains are similar in that they both are regions where
significant amounts of cytoplasm can be found between the leaflets of adjacent
myelin membranes. The fact that some cells can express more than one connexin may serve to explain why mutations in Cx32 specifically interfere with
its function in the PNS and not in other Cx32-expressing tissues. Alternatively,
Cx32 may have a uniquely important role in Schwann cell and myelin biology.
The most obvious function would be to allow the rapid transfer of ions,
nutrients, or signal molecules between the main body of the Schwann cell and

the many layers of the myelin membrane, and perhaps to and from the en-sheathed axon.

CMT1 and CMT2 are Distinct Genetic and Clinical Entities

It is important to distinguish clearly between the two fundamentally different forms of CMT, CMT1 and CMT2. These two CMT subtypes can only be reliably differentiated by electrodiagnostic studies that demonstrate the characteristically reduced NCVs in CMT1. Although most of the features of CMT1 can be explained at the molecular level by mutations in myelin proteins, the nature and consequences of CMT2 mutations remain obscure. CMT2A has been linked to the distal region of the short arm of chromosome 1 to 1p35–p36 (Ben Othmane et al 1993b), and genetic heterogeneity of CMT2 is also likely (the CMT1B gene P0 maps to 1q22–q23 and has been excluded as a potential candidate gene). No candidate genes for CMT2 have been identified so far.

Dejerine-Sottas Syndrome Can Be Caused By Mutations Affecting Different Myelin Proteins

As discussed earlier, DSS is now considered a pronounced congenital hypo-myelinating syndrome with either recessive or dominant inheritance. By these more relaxed criteria, the severe hypomyelinating phenotypes of the *Tr* and *Tr-J* mice would quite possibly qualify for the diagnosis of DSS. Furthermore, the phenotype observed in the human pedigree that carries the identical mutation as *Tr-J* is unusually severe although the disease in this family has been classified as CMT1A (Valentijn et al 1992a). Thus, the finding of various mutations in PMP22 and P0 that result in neuropathies with differing severities suggests that it is conceptually reasonable to consider DSS and classic CMT1 extremes in a continuum of an essentially single disease process, which is caused by various mutations affecting peripheral myelin proteins. The elucidation of the biological functions of the affected domains in the PMP22 and P0 proteins should provide a rational basis to relate the already known and as yet unidentified mutations in PMP22 and P0 to their respective CMT1 and DSS phenotypes. We do not, however, wish to dismiss the usefulness of the DSS and CMT1 classification because a similar situation exists with regards to the prognostically important classification involving Becker's and Duchenne's muscular dystrophies, which essentially represent relatively mild and severe forms, respectively, of a single disease process involving the dystrophin gene.

THE RELATIONSHIP OF HEREDITARY PERIPHERAL NEUROPATHIES TO THE BIOLOGY OF THE PERIPHERAL NERVOUS SYSTEM

Several mutations affecting the *PMP22*, Cx32, and P0 genes are associated with inherited peripheral neuropathies. The functional mechanisms by which these mutations cause the disease phenotype remain elusive but can be considered from several vantages. As outlined before, these mutations may cause functional disturbances that are best appreciated at the protein structure level whereby a single mutation affects a functional domain of a protein. Alternatively, the effects of these mutations can be considered in the context of myelin structure, myelin function, general Schwann cell function, or axon-glia interactions.

PMP22, P0, and Cx32 Are Critical Components of Myelin

The major common feature of the genes mutated in CMT1 (*PMP22*, P0) and, with some limitations, CMTX (Cx32) is that they are produced by Schwann cells and encode proteins that are components of PNS myelin (Figure 3). Consequently, one way of approaching the problem of understanding these mutations is to view them in the context of myelin structure and function. The function of myelin appears to be rather simple. Specifically, myelin acts as an electrical insulator to allow the rapid propagation of electrical impulses along the length of the axon by saltatory conduction. The structure of myelin, on the other hand, is complex and needs to be discussed in detail with particular reference to PMP22, Cx32, and P0.

Peripheral nerve myelin is a highly ordered, repetitive structure comprised of multiple spiral wrapping of the Schwann cell plasma membrane around a centrally placed cylindrical axon (for a comprehensive description, see Peters et al 1991). When viewed by electron microscopy in cross sections, PNS myelin can be resolved into a single spiral of repeated units with a periodicity of approximately 12 nm between each wrap of myelin. These repeating units represent the apposition of the extracellular and intracellular leaflets of the Schwann cell plasma membrane that are brought closely together as the myelin sheath is compacted, thereby forming the intraperiod and major dense lines, respectively, as observed by electron microscopy. X-ray diffraction and high-resolution electron microscopic studies can further resolve the intraperiod line and the major dense line into two components that are separated by approximately 2–5 nm. Thus, while there is apparently a small space between the apposing membranes, the majority of myelin is largely inaccessible to the diffusion of relatively large intracellular and extracellular molecules, such as proteins. The major dense line houses the intracellular domains of PMP22 and P0, and the P2 and MBP molecules. The 2–5 nm gap separating the intraperiod lines in compact myelin contains the extracellular loops of PMP22 and the immunoglobulin-like adhesion domain of P0 and defines the maximum distance required for the interactions that are believed to occur between proteins located on the apposing surfaces of the plasma membrane.

A single Schwann cell produces myelin that ensheaths only part of an axon. Normally, myelin-forming Schwann cells are evenly spaced along the length of the axon, and the individual Schwann cells are physically separated from each other by nodes of Ranvier into internodal segments delineating a single Schwann cell and its myelin. Immediately adjacent to the node of Ranvier lies the paranodal region where the edges of the myelin leaflets are separated by Schwann cell cytoplasm, thereby forming a continuous spiral of Schwann cell cytoplasm that extends from the main body of the Schwann cell to the innermost portion of the myelin spiral, termed the inner mesaxon (for review, see Raine 1984). Junctional complexes with features of tight junctions are observed at the interface between the lateral loops of myelin at the internode (Suzuki & Zagoren 1978) and desmosome-like structures are observed at the interface between the internode and the paranodal regions. Additional intercellular junctional complexes with features of zonula occludens are formed in the region of apposition between the paranodal myelin loops and the axonal membrane (Peters et al 1991, Raine 1984). Another subdomain of myelin, the Schmidt-Lanterman clefts, resemble the paranodal region in that the myelin membranes are separated by a spiraling channel of Schwann cell cytoplasm that extends through the full thickness of the myelin sheath. The Schmidt-Lanterman clefts are distributed at semiregular intervals along the internode and are bounded from the compact internodal myelin on either side by junctional complexes that resemble desmosomes (Raine 1984). Cx32 has been localized to the Schmidt-Lanterman clefts and the lateral loop of myelin at the paranode with immunohistochemistry by using light microscopy (Bergoffen et al 1993a). So far, freeze-fracture studies have failed to reveal the hexagonal arrays of intramembranous particles characteristic of gap junctions either within the compact myelin (as might be expected if PMP22 were a gap junction protein), or at the Schmidt-Lanterman clefts and paranodal regions where Cx32 is localized. Linear arrays of intramembranous particles are observed, however, in the freeze-fracture replicas of peripheral nerve myelin that are thought to correspond to the juctional complexes at the paranodes visualized by transmission electron microscopy (EM) (Peters et al 1991). Although these junctional complexes may contain Cx32, its precise localization and the nature of any complexes that Cx32 might form in Schwann cells requires further investigation.

The simplicity of myelin structure implied by its periodicity when viewed by static techniques such as X-ray diffraction and electron microscopy betrays the underlying complexity of myelin assembly and maintenance that is necessary to generate the highly ordered morphologically and functionally identified subdomains that define mature myelin (because a thorough description of the formation of myelin is beyond the scope of this review, the reader is referred to Raine 1984 for details). Descriptions of the CMT1 pathology have been interpreted as consistent with demyelination, thus suggesting that normally formed myelin is progressively lost. After comparing the CMT1 mutations and phenotypes with the similar mutations and severe hypomyelinating phenotypes of *Tr* and *Tr-J* and patients with DSS, the alternative hypothesis can be proposed that not all of the myelin originally formed in CMT1 patients may be normal to begin with. Because many hereditary neuropathies in humans have their onset in late childhood [although decreased NCVs can normally be measured earlier (Lupski et al 1991b)] and are progressive, one must also consider the possible effects of the underlying mutations on the mechanisms of myelin formation, maintenance, and renewal. Unfortunately, however, the dynamics of these processes are still poorly understood.

Are PMP22, P0, and Cx32 Associated With Inter- and Intracellular Adhesion Processes?

We have already discussed the compelling evidence that P0 is involved in homotypic adhesive processes, and it is evident from connexin expression studies in *Xenopus* oocytes that a gap junction protein must be able to interact homotypically in order to form communicating junctions between adjacent cells. Indirect evidence suggests that PMP22 might be involved in adhesion. Computer-aided secondary structure predictions of the PMP22 protein reveal a striking structural parallel to a class of polypeptides that includes PLP, the connexins, synaptophysin, the retinitis pigmentosa disease gene rds-peripherin, M6, and the tight junction protein occludin, all of which have four putative membrane-spanning domains (Kitagawa et al 1993). PLP and its splice variant DM20 may function as components of adhesive pores based on evolutionary considerations and a low amino acid identity to channel proteins. Similarly, the connexins can be considered adhesive pores and, although no direct evidence that occludin forms pores exists, it is believed to directly mediate the adhesion between membranes on adjacent cells. Although PMP22 is not related to PLP-DM20 or any other known channel proteins based on primary amino acid sequence comparisons, it fits the model of a component of an adhesive pore in PNS myelin that might be required to maintain the homeostasis of the nerve by mediating the physiological communication between the outer surface of myelin with the periaxonal space and/or the axonal membrane.

In support of the hypothesis that PMP22 might be involved in adhesion, PMP22 has been shown to carry the L2-HNK1 epitope, a carbohydrate configuration that has been implicated in mediating adhesive processes (Hammer et al 1993, Snipes et al 1993b). In this context, it is interesting to note that a subset of patients with autoimmune IgM neuritis (monoclonal IgM gammopathy) have autoantibodies that are directed against HNK or HNK-like epitopes present on a subset of myelin proteins and glycolipids (Nobile-Orazio et al 1984, Burger et al 1992). However, the exact identity of the disease-causing, epitope-carrying molecule(s) remains unclear. These patients show features of PNS demyelination, including a characteristic widening of the myelin lamellae at the intraperiod line. Recent studies have suggested that this widening is due to binding of autoimmune IgM antibodies to an antigen present on the extracellular surface of the widened myelin lamellae (Lach et al 1993). The demyelination may be due to perturbation of the adhesive function of HNK-bearing glycolipids and proteins like P0, MAG, PMP22, and possibly other, still undescribed HNK-positive PNS myelin components.

In summary, one can speculate that P0, PMP22, and Cx32 could form homotypic or heterotypic protein complexes in myelin (and possibly build up adhesive pores), especially since such a model would explain the similar (dominant) phenotypes of mutants affecting each of these proteins. However, direct evidence for such complex(es) in myelin has not been forthcoming, with the exception of the involvement of P0 in homotypic interactions.

Can Schwann Cells Influence Axon Properties?

The strong expression of the genes mutated in CMT1 in Schwann cells suggests that the disease is primarily due to a defect in this glial cell type. Yet, as we have discussed, both dominant CMTX and CMT1 are associated with an axonopathy as revealed by lowered amplitudes of compound action potentials and by detailed morphologic studies of the affected peripheral nerves. Thus, the question arises as to whether there are mechanisms to explain how Schwann cell abnormalities involving myelin can cause axonal loss. A first hint that such mechanisms exist comes from grafting studies between the peripheral nerves of wild-type and *Tr* mice. These experiments suggest that Schwann cells have a profound influence on axons by affecting neurofilament density, neurofilament phosphorylation, slow axonal transport and axonal calibers (de Waegh et al 1992). Interestingly, a similar increase in neurofilament density as seen in *Tr* mice and in *Tr* grafts into wild-type mice has been observed in the poorly myelinated fibers of CMT1 patients (Gabreel-Festen et al 1992). Although these findings demonstrate that Schwann cells profoundly affect the cytoarchitecture of axons, it remains to be determined whether these changes can explain the long-term axon atrophy and degeneration observed in CMT patients. However, the fact that axonal degeneration is seen in P0⁻ mice clearly demonstrates that a mutation in a Schwann cell–specific myelin protein can be responsible for axon atrophy.

Can Axons Influence Schwann Cell Properties?

Alternatively, a converse mechanism in which primary axonopathies or neuronopathies affect Schwann cell and myelin properties could be envisaged as a disease mechanism in CMT1. In fact, CMT1 has been widely considered to be primarily an axonopathy with secondary loss of myelin (for a more detailed discussion, see Dyck 1984). Ample evidence demonstrates that axons can directly modulate myelin production by Schwann cells. For example, in Wallerian degeneration, focal axon damage leads to a rapid and profound destruction of myelin and repression of myelin gene expression by Schwann cells surrounding the degenerating axons (see Snipes et al 1992 and references therein). Furthermore, the contact of axons with Schwann cells during nerve regeneration can be temporally correlated with the upregulation of myelin gene expression (Lemke & Chao 1988). More subtle examples of axon effects on Schwann cells include the direct relationship between the caliber of the axon and the thickness of the surrounding myelin sheath. Furthermore, axons can have a mitogenic effect on Schwann cells (Pellegrino & Spencer 1985). Thus, because axons and neuron-derived factors profoundly influence Schwann cell functions such as proliferation and myelin formation, it seems premature to rule out the possibility that neurons are directly affected in CMT1 until it is conclusively determined whether or not neurons express Cx32, PMP22, or P0 and whether such an (hypothetical) expression affects neuronal function.

CMT1 Genes May Affect Cell Growth and Differentiation

Interestingly, PMP22 cDNA was first isolated as a growth arrest–specific gene, gas-3, from resting fibroblasts in culture and it has been suggested that PMP22 might convey a survival signal that is required to maintain the viability of a differentiated cell that has been removed from the cell cycle (Schneider et al 1992b). Alternatively, PMP22 could act as an active regulator of cell growth. With respect to Schwann cells, one might hypothesize that PMP22 could play a role in the development and differentiation of the Schwann cell and is later recruited as a structural myelin protein. Indeed, the striking Schwann cell–proliferation defect observed in Tr mice was one of the features, in addition to the hypomyelination, that suggested that PMP22 mutations might underlie the Tr defect. A similar view has been proposed for PLP based on speculations concerning the jimpy and the rumpshaker mouse mutant. Although both strains are PLP mutants, they vary considerably in their respective phenotypes. While jimpy shows hypomyelination and extensive glial cell death, rumpshaker exhibits only a myelin deficiency, and the oligodendrocytes appear normal without signs of premature cell death (Schneider et al 1992a).

There are indications that other CMT1 genes may, perhaps indirectly, also influence cell proliferation. Like the Tr and Tr-J, the P0− and the P0-diphtheria toxin A chain (see below) transgenic mice show an increased number of Schwann cells in their hypomyelinated peripheral nerves. In addition, the connexin family has also been implicated in the regulation of cell growth. Like PMP22 mRNA levels, connexin mRNA levels are inversely correlated with cell growth, which suggests that quiescent cells (perhaps related to differentiation) might be more likely to form intercellular junctions. Alternatively, the formation of intercellular junctions may directly or indirectly regulate cell growth. This hypothesis is supported by gene transfer experiments into cultured cells (Zhu et al 1992). Whether wild-type or mutated Cx32 expression has any direct effect on Schwann cell growth is still unknown.

Experimental Schwann Cell Death Can Mimic the CMT1 and Tr Phenotypes

The relationship between Schwann cell proliferation and myelin deficiency can best be discussed based on an interesting transgenic mouse in which the P0 promoter was utilized to direct Schwann cell–specific expression of diphtheria toxin A chain (Messing et al 1992). As a consequence, Schwann cells that try to activate the P0 gene are killed. These mice show a profound hypomyelination and behaviorally resemble the *Tr* and *Tr-J* mice. More interestingly, however, older mice have significantly increased numbers of Schwann cells in spite of the cell death that must have occurred during the abortive myelination process. This finding has been interpreted as evidence for a dynamic equilibrium between myelinating and nonmyelinating Schwann cells in the normal nerve, and it raises the fundamentally important question of whether hypomyelination generally leads to Schwann cell proliferation because of a prolonged exposure of the Schwann cells to axon-derived or autocrine factors that are normally masked after myelination has been completed. By analogy, one might further speculate that *Tr, Tr-J,* and some CMT1 mutations may be deleterious to Schwann cells, and the resulting Schwann cell death could contribute to the observed hypomyelination.

OUTLOOK AND CONCLUSION

We are indebted to the early clinician-scientists who identified and recognized the significance of the particular constellation of signs, symptoms, and heritability that constitute CMT disease in humans and the *Tr* phenotype in mice. Their successors further refined the clinical, electrophysiological, pathological, and genetic similarities and differences between the different disease entities. The subsequent studies of disease-causing mutations affecting peripheral nerve function have provided an important framework for our current understanding of PNS biology. Conversely, the studies on peripheral nerve biology enabled us to put the disease-causing genes into perspective. Many additional CMT1 mutations will likely be identified within *PMP22,* Cx32, and P0 genes or their regulatory elements. Furthermore, additional CMT genes have been linked to other loci that will encode additional CMT disease proteins. While we anxiously await the identification of additional CMT genes, we can anticipate based on current knowledge that these genes will cause CMT or DSS phenotypes through a variety of mechanisms affecting myelination, Schwann cell regulation, and/or axon-glia interactions. The elucidation of the roles of CMT-causing genes in the structure and function of the PNS and the correlation of the findings with the different mutations observed in CMT disease should continue to expand our ability to understand the functions of the peripheral nervous system.

ACKNOWLEDGMENTS

US is supported by the Swiss National Science Foundation and GJS by the NIH.

Literature Cited

Austin JH. 1956. Observations on the syndrome of hypertrophic neuritis (the hypertrophic interstitial radiculoneuropathies). *Medicine* 35: 187–37

Ben Othmane K, Hentati F, Lennon F, Hamida CB, Biel S, et al. 1993a. Linkage of a locus (CMT4A) for autosomal recessive Charcot-Marie-Tooth disease to chromosome 8q. *Hum. Mol. Genet.* 2:1625–28

Ben Othmane K, Middleton LT, Loprest LJ, Wilkinson KM, Lennon F, et al. 1993b. Localization of a gene (CMT2A) for autosomal dominant Charcot-Marie-Tooth disease type 2 to chromosome 1p and evidence for genetic heterogeneity. *Genomics* 17:370–75

Bennett MVL, Barrio LC, Bargiello TA, Spray DC, Hertzberg E, Saez JC. 1991. Gap junctions: new tools, new answers, new questions. *Neuron* 6:305–20

Bergoffen J, Scherer SS, Wang S, Oronzi Scott M, Bone LJ, et al. 1993a. Connexin mutations in X-linked Charcot-Marie-Tooth disease. *Science* 262:2039–42

Bergoffen J, Trofatter J, Pericak-Vance MA, Haines JL, Chance PF, Fischbeck KH. 1993b. Linkage localization of X-linked Charcot-Marie-Tooth disease. *Am. J. Hum. Genet.* 52:312–18

Bird TD, Ott J, Giblett ER. 1982. Evidence for linkage of Charcot-Marie-Tooth disease to the duffi locus on chromosome 1. *Am. J. Hum. Genet.* 34:388–94

Bird TD, Ott J, Giblett ER, Chance PF, Sumi SM, Kraft GH. 1983. Genetic linkage evidence for heterogeneity in Charcot-Marie-Tooth Neuropathy (HMSN type I). *Ann. Neurol.* 14:679–84

Bollensen E, Schachner M. 1987. The peripheral myelin glycoprotein P0 expresses the L2/HNK-1 and the L3 carbohydrate structures shared by neural adhesion molecules. *Neurosci. Lett.* 82:77–82

Brust JC, Lovelace RE, Devi S. 1978. Clinical and electrodiagnostic features of Charcot-Marie-Tooth syndrome. *Acta. Neurol. Scand.* 58(Suppl. 68):3–140

Buchberg AM, Brownell E, Nagata S, Jenkins NA, Copeland NG. 1989. A comprehensive linkage map of murine chromosome 11 reveals extensive linkage conservation between mouse and human. *Genetics* 122: 153–61

Burger D, Perruisseau G, Simon M, Steck AJ. 1992. Comparison of the N-linked oligosaccharide structures of the two major human myelin glycoproteins MAG and P0: assessment of the structures bearing the epitope for HNK-1 and human monoclonal immunoglobulin M found in demyelinating neuropathy. *J. Neurochem.* 58:854–61

Chance PF, Abbas N, Lensch MW, Pentao L, Roa BB, et al. 1994. Two autosomal dominant neuropathies result from reciprocal DNA duplication/deletion of a region on chromosome 17. *Hum. Mol. Genet.* 3:223–28

Chance PF, Alderson MK, Leppig KA, Lensch MW, Matsunami N, et al. 1993. DNA deletion associated with hereditary neuropathy with liability to pressure palsies. *Cell* 72: 143–51

Chance PF, Matsunami N, Lensch W, Smith B, Bird TD. 1992. Analysis of the DNA duplication 17p11.2 in Charcot-Marie-Tooth neuropathy type 1 pedigrees: additional evidence for a third autosomal CMT1 locus. *Neurology* 42:2037–41

Charcot J-M, Marie P. 1886. Sur une forme particulière d'atrophie musculaire progressive souvent familiale debutant par les pieds et les jambes et atteignant plus tard les mains. *Rev. Méd.* 6:97–138

Cremers FPM, Pfeiffer RA, Van de Pol TJR, Hofker MH, Kruse TA, et al. 1987. An interstitial duplication of the X-chromosome in a male allows physical fine mapping of probes from the Xq13-q22 region. *Hum. Genet.* 77:23–27

Davies DM. 1954. Recurrent peripheral nerve palsies in a family. *Lancet* 2:266–68

Dawidenkow S. 1927. Uber die neurotische Muskelatrophie Charcot-Marie: klinische-genetische Studien. *Z. Neurol. fr Neurologie* 107:259–320

De Leon M, Nahin RL, Mendoza ME, Ruda MA. 1994. SR13/PMP-22 expression in rat nervous system, in PC12 cells, and C6 glial cell lines. *J. Neurosci. Res.* 38:167–81

De Leon M, Welcher AA, Suter U, Shooter EM. 1991. Identification of transcriptionally regulated genes after sciatic nerve injury. *J. Neurosci. Res.* 29:437–48

de Waegh SM, Lee VM-Y, Brady ST. 1992. Local modulation of neurofilament phosphorylation, axonal caliber, and slow axonal transport by myelinating Schwann cells. *Cell* 68:451–63

Dejerine J, Sottas J. 1893. Sur la névrite interstitielle, hypertrophique et progressive de l'enfance. *CR Soc. Biol.* 45:63–96

Dermietzel R, Spray DC. 1993. Gap junctions in the brain: where, what type, how many and why? *Trends Neurosci.* 16:186–92

Dyck PJ. 1975. Inherited neuronal degeneration and atrophy affecting periperal motor, sensory and autonomic neurons. In *Peripheral Neuropathy,* ed. PJ Dyck, PK Thomas, EH Lambert, pp. 825–67. Philadelphia: Saunders. 1st ed.

Dyck PJ. 1984. Inherited neuronal degeneration and atrophy affecting periperal motor, sensory and autonomic neurons. In *Peripheral Neuropathy.* ed. PJ Dyck, PK Thomas, EH Lambert, R Bunge, pp. 1600–55. Philadelphia: Saunders. 2st ed.

Dyck PJ, Karnes JL, Lambert EH. 1989. Longitudinal study of neuropathic deficits and nerve conduction abnormalities in hereditary motor and sensory neuropathy type 1. *Neurology* 39:1302–8

Dyck PJ, Thomas PK, Griffin JW, Low PA, Poduslo JF, eds. 1993. *Peripheral Neuropathy.* Philadelphia: Saunders. 1728 pp. 3rd ed.

Falconer DS. 1951. Two new mutants, "Trember" and "Reeler" with neurological actions in the house mouse (*Mus musculus L.*) *J. Genet.* 50:192–201

Gabreëls-Festen AAWM, Joosten EMG, Gabreëls FJM, Jennekens FGI, Janssen-van Kempen TW. 1992. Early morphological features in dominantly inherited demyelinating motor and sensory neuropathy (HMSN type I). *J. Neurol. Sci.* 107:145–54

Giese KP, Martini R, Lemke G, Soriano P, Schachner M. 1992. Mouse P0 gene disruption leads to hypomyelination, abnormal expression of recognition molecules, and degeneration of myelin and axons. *Cell* 71: 565–76

Hahn AF, Brown WF, Koopman WJ, Feasby TE. 1990. X-linked dominant hereditary motor and sensory neuropathy. *Brain* 113: 1511–25

Hammer JA, O'Shannessy DJ, De Leon M, Gould R, Zand D, et al. 1993. Immunoreactivity of PMP-22, P0, and other 19 to 28 kDa glycoproteins in peripheral nerve myelin of mammals and fish with HNK1 and related antibodies. *J. Neurochem.* 35:546–58

Harding AE, Thomas PK. 1980. The clinical features of hereditary motor and sensory neuropathy types I and II. *Brain* 103:259–80

Hayasaka K, Himoro M, Sato W, Takada G, Uyemura K, et al. 1993a. Charcot-Marie-Tooth neuropathy type 1B is associated with mutations of the myelin P0 gene. *Nat. Genet.* 5:31–34

Hayasaka K, Himoro M, Sawaishi Y, Nanao K, Takahashi T, et al. 1993b. De novo mutation of the myelin P0 gene in Dejerine-Sottas disease (hereditary motor and sensory neuropathy type III). *Nat. Genet.* 5:266–69

Hayasaka K, Takada G, Ionasescu VV. 1993c. Mutation of the myelin Po gene in Charcot-Marie-Tooth neuropathy type 1B. *Hum. Mol. Genet.* 2:1369–72

Heath JW, Inuzuka T, Quarles RH, Trapp BD. 1991. Distribution of P0 protein and the myelin-associated glycoprotein in peripheral nerves from Trembler mice. *J. Neurocytol.* 20:439–49

Henry EW, Cowen JS, Sidman RL. 1983. Comparison of trembler and trembler-J phenotypes: varying severity of peripheral hypomyelination. *J. Neuropathol. Exp. Neurol.* 42:688–706

Henry EW, Sidman RL. 1988. Long lives for homozygous trembler mutant mice despite virtual absence of peripheral myelin. *Science* 241:344–46

Hoogendijk JE, Janssen EAM, Gabreëls-Festen AAWM, Ongerboer de Visser BW, Visser M, Bolhuis PA. 1993. Allelic heterogeneity in hereditary motor and sensory neuropathy type Charcot-Marie-Tooth disease type 1a. *Neurology* 43:1010–15

Ionasescu VV, Ionasescu R, Searby C. 1993. Screening of dominantly inherited Charcot-Marie-Tooth neuropathies. *Muscle Nerv.* 16: 1232–38

Ionasescu VV, Trofatter J, Haines JL, Summers AM, Ionasescu R, Searby C. 1992. X-linked recessive Charcot-Marie-Tooth neuropathy: clinical and genetic study. *Muscle Nerv.* 15: 368–73

Kitagawa K, Sinoway MP, Yang C, Gould RM, Colman DR. 1993. A proteolipid protein gene family: expression in sharks and rays and possible evolution from ancestral gene encoding a pore-forming polypeptide. *Neuron* 11:433–48

Kitamura K, Suzuki M, Uyemura K. 1976. Purification and partial characterization of two glycoproteins in bovine peripheral nerve myelin membrane. *Biochim. Biophys. Acta* 455: 806–16

Kulkens T, Bolhuis PA, Wolterman RA, Kemp S, te Nijenhuis S, et al. 1993. Deletion of the serine 34 codon from the major peripheral myelin protein P0 gene in Charcot-Marie-Tooth disease type 1B. *Nat. Genet.* 5:35–39

Lach B, Rippstein P, Atack D, Afar DEH, Gregor A. 1993. Immunoelectron microscopic localization of monoclonal IgM antibodies in gammopathy associated with peripheral demyelinative neuropathy. *Acta Neuropathol.* 85:298–307

Lemke G. 1988. Unwrapping the genes of myelin. *Neuron* 1:535–43

Lemke G. 1993. The molecular genetics of myelination: an update. *Glia* 7:263–71

Lemke G, Axel R. 1985. Isolation and sequence of a cDNA encoding the major structural protein of peripheral myelin. *Cell* 40:501–8

Lemke G, Chao M. 1988. Axons regulate Schwann cell expression of the major myelin and NGF receptor genes. *Development* 102: 499–504

Lemke G, Lamar E, Patterson J. 1988. Isolation and analysis of the gene encoding peripheral myelin protein zero. *Neuron* 1:73–83

Lupski JR, Chance PF, Garcia CA. 1993. Inherited primary peripheral neuropathies: molecular genetics and clinical implications of CMT1A and HNPP. *JAMA* 270:2326–30

Lupski JR, de Oca-Luna LM, Slaugenhaupt S, Pentao L, Guzzetta V, et al. 1991a. DNA duplication associated with Charcot-Marie-Tooth disease type 1A. *Cell* 66:219–32

Lupski JR, Garcia CA, Parry GJ, Patel PI. 1991b. Charcot-Marie-Tooth polyneuropathy syndrome: clinical, elecrophysiological, and genetic aspects. In *Current Neurology,* ed. S Appel, pp. 1–25. Chicago: Mosby-Yearbook

Manfioletti G, Ruaro ME, Del Sal G, Philipson L, Schneider C. 1990. A growth arrest-specific (gas) gene codes for a membrane protein. *Mol. Cell. Biol.* 10:2924–30

Martini R, Bollensen E, Schachner M. 1988. Immunocytological localization of the major peripheral nervous system glycoprotein P0 and the L2/HNK-1 and L3 carbohydrate structures in the developing and adult mouse sciatic nerve. *Dev. Biol.* 129:330–38

Matsunami N, Smith B, Ballard L, Lensch MW, Robertson M, et al. 1992. Peripheral myelin protein-22 gene maps in the duplication in chromosome 17p11.2 associated with Charcot-Marie-Tooth 1A. *Nat. Genet.* 1:176–79

Messing A, Behringer RR, Hammang JP, Palmiter RD, Brinster RL, Lemke G. 1992. P0 promoter directs expression of reporter and toxin genes to Schwann cells of transgenic mice. *Neuron* 8:507–20

Milks LC, Kumar NM, Houghton R, Unwin N, Gilula NB. 1988. Topology of the 32 kD liver gap junction protein determined by site-directed antibody localization. *EMBO J.* 7: 2967–75

Miller T, Dahl G, Werner R. 1988. Structure of a gap junction gene: rat connexin-32. *Biosci. Rep.* 8:455–64

Nelis E, Timmerman V, De Jonghe P, Van Broeckhoven C. 1994. Identification of a 5′ splice site mutation in the PMP-22 gene in the autosomal dominant Charcot-Marie-Tooth disease type 1. *Hum. Mol. Genet.* In press

Nicholson B, Dermietzel R, Teplow D, Traub O, Willecke K, Revel J-P. 1987. Two homologous protein components of hepatic gap junctions. *Nature* 329:732–34

Nicholson GA. 1991. Penetrance of the hereditary motor and sensory neuropathy Ia mutation: assessment by nerve conduction studies. *Neurology* 41:547–52

Nicholson GA, Valentijn LJ, Cherryson AK, Kennerson ML, Bragg TL, et al. 1994. A frame shift mutation in the PMP22 gene in hereditary neuropathy with liability to pressure palsies. *Nat. Genet.* 6:263–66

Nobile-Orazio E, Hays AP, Latov N, Perman G, Golier J, et al. 1984. Specificity of mouse and human monoclonal antibodies to myelin-associated glycoprotein. *Neurology* 34: 1336–42

Olsson JE, Gordon JW, Pawlyk BS, Roof D, Hayes A, et al. 1992. Transgenic mice with a rhodopsin mutation (Pro23His): a mouse model of autosomal dominant retinitis pigmentosa. *Neuron* 9:815–30

Owens GC, Boyd CJ. 1991. Expressing antisense P0 RNA in Schwann cells perturbs myelination. *Development* 112:639–49

Palau F, Löfgren A, De Jonghe PD, Bort S, Nelis E, et al. 1993. Origin of the de novo duplication in Charcot-Marie-Tooth disease type 1A: unequal nonsister chromatid exchange during spermatogenesis. *Hum. Mol. Genet.* 2:2031–35

Pareek S, Suter U, Snipes GJ, Welcher AA, Shooter EM, Murphy RA. 1993. Detection and processing of peripheral myelin protein PMP22 in cultured Schwann cells. *J. Biol. Chem.* 268:10372–9

Parmantier E, Cabon F, Zalc B. 1993. *In the CNS PMP22 RNA is expressed in Motoneurons.* Presented at ISN Satellite Mtg., La Londe Les Maures, France

Patel PI, Roa BB, Welcher AA, Schoener-Scott R, Trask BJ, et al. 1992. The gene for the peripheral myelin protein PMP-22 is a candidate for Charcot-Marie-Tooth disease type 1A. *Nat. Genet.* 1:159–65

Pellegrino R, Spencer P. 1985. Schwann cell mitosis in response to regenerating peripheral axons in vivo. *Brain. Res.* 341:16–25

Peters A, Palay SL, Webster H de F. 1991. *The Fine Structure of the Nervous System. Neurons and Their Supporting Cells.* New York: Oxford Univ. Press. 491 pp. 3rd ed.

Raeymaekers P, Timmerman V, De Jonghe P, Swerts L, Gheuens J, et al. 1989. Localization of the mutation in an extended family with Charcot-Marie-Tooth neuropathy (HMSN I). *Am. J. Hum. Genet.* 45:953–58

Raeymaekers P, Timmerman V, Nelis E, De Jonghe P, Hoogendijk JE, et al. 1991. Duplication in chromosome 17p11.2 in Charcot-Marie-Tooth neuropathy type 1a (CMT 1a). *Neuromuscular Disord.* 1:93–97

Rahman S, Carlile G, Evans WH. 1993. Assembly of hepatic gap junctions. Topography and distribution of connexin 32 in intracellular and plasma membranes determined using sequence-specific antibodies. *J. Biol. Chem.* 268:1260–65

Raine CS. 1984. Morphology of myelin and myelination. In *Myelin*, ed. P Morell, pp. 1–50. New York: Plenum

Raskind WH, Williams CA, Hudson LD, Bird TD. 1991. Complete deletion of the proteolipid protein gene (PLP) in a family with X-linked Pelizaeus-Merzbacher disease. *Am. J. Hum. Genet.* 49:1355–60

Readhead C, Schneider A, Griffiths I, Nave KA. 1994. Premature arrest of myelin formation in transgenic mice with increased proteolipid gene dosage. *Neuron* 12:583–95

Roa BB, Dyck PJ, Marks HG, Chance PF, Lupski JR. 1993a. Dejerine-Sottas syndrome associated with point mutation in the peripheral myelin protein 22 (*PMP22*) gene. *Nat. Genet.* 5:269–72

Roa BB, Garcia CA, Pentao L, Killian JM, Trask BJ, et al. 1993b. Evidence for a recessive *PMP22* point mutation in Charcot-Marie-Tooth disease type 1A. *Nat. Genet.* 5:189–94

Roa BB, Garcia CA, Suter U, Kulpa DA, Wise CA, et al. 1993c. Charcot-Marie-Tooth disease type 1A: association with a spontaneous point mutation in the PMP22 gene. *N. Engl. J. Med.* 329:96–101

Rozear MP, Pericak-Vance MA, Fischbeck K, Stajich JM, Gaskell PC, et al. 1987. Hereditary motor and sensory neuropathy, X-linked: a half century follow-up. *Neurology* 37:1460–65

Saez JC, Nairn AC, Czernik AJ, Spray DC, Hertzberg EL, et al. 1990. Phosphorylation of connexin 32, a hepatocyte gap-junction protein, by cAMP-dependent protein kinase, protein kinase C and Ca^{2+}/calmodulin protein kinase II. *Eur. J. Biochem.* 192:263–73

Sakamoto Y, Kitamura K, Yosjimura K, Nishijima T, Uyemura K. 1987. Complete amino acid sequence of P0 protein from bovine peripheral nerve. *J. Biol. Chem.* 262:4208–14

Schneider A, Montague P, Griffiths I, Fanarraga M, Kennedy P, et al. 1992a. Uncoupling of hypomyelination and glial cell death by a mutation in the proteolipid protein gene. *Nature* 358:758–61

Schneider C, Del Sal G, Brancolini C, Gustincich S, Manfioletti G, Ruaro ME. 1992b. *43. Colloquium Mosbach; DNA Replication and the Cell Cycle.* Berlin: Springer-Verlag

Schneider C, King RM, Philipson L. 1988. Genes specifically expressed at growth arrest of mammalian cells. *Cell* 54:787–92

Skre H. 1974. Genetic and clinical aspects of Charcot-Marie-Tooth disease. *Clin. Genet.* 6:98–118

Snipes GJ, Suter U, Shooter EM. 1993a. Genetics of myelin. *Curr. Opin. Neurobiol.* 3:694–702

Snipes GJ, Suter U, Shooter EM. 1993b. Human peripheral myelin protein-22 carries the L2/HNK-1 carbohydrate epitope. *J. Neurochem.* 61:1961–64

Snipes GJ, Suter U, Welcher AA, Shooter EM. 1992. Characterization of a novel peripheral nervous system myelin protein (PMP-22/SR13). *J. Cell Biol.* 117:225–38

Spreyer P, Kuhn G, Hanemann CO, Gillen C, Schaal H, et al. 1991. Axon-regulated expression of a Schwann cell transcript that is homologous to a "growth arrest-specific" gene. *EMBO J.* 10:3661–68

Su Y, Brooks DG, Li L, Lepercq J, Trofatter JA, et al. 1993. Myelin protein zero gene mutated in Charcot-Marie-Tooth type 1B patients. *Proc. Natl. Acad. Sci. USA* 90:10856–60

Suchyna TM, Xu LX, Gao F, Fourtner CR, Nicholson BJ. 1993. Identification of a proline residue as a transduction element involved in voltage gating of gap junctions. *Nature* 365:847–49

Suter U, Moskow JJ, Welcher AA, Snipes GJ, Kosaras B, et al. 1992a. A leucine-to-proline mutation in the putative first transmembrane domain of the 22-kDa peripheral myelin protein in the trembler-J mouse. *Proc. Natl. Acad. Sci. USA* 89:4382–86

Suter U, Welcher AA, Ozcelik T, Snipes GJ, Kosaras B, et al. 1992b. Trembler mouse carries a point mutation in a myelin gene. *Nature* 356:241–44

Suter U, Welcher AA, Snipes GJ. 1993. Progress in the molecular understanding of hereditary peripheral neuropathies reveals new insights into the biology of the peripheral nervous system. *Trends Neurosci.* 16:50–56

Suter U, Snipes GJ, Schoener-Scott R, Welcher AA, Pareek S, et al. 1994. Regulation of tissue-specific expression of alternative Peripheral Myelin Protein-22 (PMP22) gene transcripts by two promoters. *J. Biol. Chem.* In press

Suzuki K, Zagoren JC. 1978. Studies on the copper binding affinity of fibers in the peripheral nervous system. *Neuroscience* 3: 477–55

Swenson KI, Jordan JR, Beyer EC, Paul DL. 1989. Formation of gap junctions by expression of connexins in *Xenopus* oocyte pairs. *Cell* 57:145–55

Thomas FP, Lebo RV, Rosokija GD, Lovelace RE, et al. 1994. Toamculus neuropathy in chromosome 1 Charcot-Marie-Tooth syndrome. *Acta Neuropathol.* 87:91–97

Timmerman V, Nelis E, Van Hul W, Nieuwenhuijsen BW, Chen KL, et al. 1992. The peripheral myelin protein gene PMP-22 is contained within the Charcot-Marie-Tooth disease type 1A duplication. *Nat. Genet.* 1: 171–75

Tooth HH. 1886. *The Peroneal Type of Progressive Muscular Atrophy.* London: Lewis

Valentijn LJ, Baas F, Wolterman RA, Hoogendijk JE, Bosch NHA, et al. 1992a. Identical point mutations of PMP-22 in Trembler-J mouse and Charcot-Marie-Tooth disease type 1a. *Nat. Genet.* 2:288–91

Valentijn LJ, Baas F, Zorn I, Hensels GW, de Visser M, Bolhuis PA. 1993. Alternatively sized duplication in Charcot-Marie-Tooth disease type 1A. *Hum. Mol. Genet.* 2:2143–46

Valentijn LJ, Bolhuis PA, Zorn I, Hoogendijk JE, van den Bosch N, et al. 1992b. The peripheral myelin gene PMP-22/GAS-3 is duplicated in Charcot-Marie-Tooth disease type 1A. *Nat. Genet.* 1:166–70

Vance JM, Nicholson GA, Yamaoka LH, Stajich J, Stewart CS, et al. 1989. Linkage of Charcot-Marie-Tooth neuropathy type 1a to chromosome 17. *Exp. Neurol.* 104:186–89

Welcher AA, Suter U, De Leon M, Snipes GJ, Shooter EM. 1991. A myelin protein is encoded by the homologue of a growth arrest-specific gene. *Proc. Natl. Acad. Sci. USA* 88:7195–99

Werner R, Levine E, Rabadan-Diehl C, Dahl G. 1989. Formation of hybrid cell-cell channels. *Proc. Natl. Acad. Sci. USA* 86:5380–84

Werner R, Levine E, Rabadan-Diehl C, Dahl G. 1991. Gating properties of connexin32 cell-cell channels and their mutants expressed in Xenopus oocytes. *Proc. R. Soc. Lond. Ser. B* 243:5–11

Windebank AJ. 1993. See Dyck et al 1993, pp. 1137–48

Wise CA, Garcia CA, Davis SN, Heju Z, Pentao L, et al. 1993. Molecular analysis of unrelated Charcot-Marie-Tooth (CMT) disease patients suggest a high frequency of the CMT1A duplication. *Am. J. Hum. Genet.* 53:853–63

Zhu D, Kidder GM, Caveney S, Naus CCG. 1992. Growth retardation in glioma cells cocultured with cells overexpressing a gap junction protein. *Proc. Natl. Acad. Sci. USA* 89:10218–21

Annu. Rev. Neurosci. 1995. 18:77–99
Copyright © 1995 by Annual Reviews Inc. All rights reserved

TRIPLET REPEAT EXPANSION MUTATIONS: The Example of Fragile X Syndrome

Stephen T. Warren and Claude T. Ashley, Jr.

Howard Hughes Medical Institute and Departments of Biochemistry and Pediatrics, Emory University School of Medicine, Atlanta, Georgia 30322

KEY WORDS: *FMR1* gene, mental retardation, X chromosome, unstable DNA, DNA
             methylation

INTRODUCTION

Mental retardation represents a deficiency in intelligence, as measured by IQ, with limited adaptive behavior that is normally reflected in maturation, learning, or social adjustment (American Psychiatric Association 1987). Approximately 1 to 3% of the population, depending upon definitions of adaptive behavior, is mentally retarded (Popper 1988). The etiologies and determinants of mental retardation are diverse and include socioeconomic influences leading to extreme malnutrition and/or inadequate prenatal care; toxic insults, such as that leading to fetal alcohol syndrome; trauma and infection; and genetic factors (Popper 1988). At least 300 genetic disorders include mental retardation as part of the phenotype, and genetic components are considered important influences on related disorders such as attention deficit disorder and learning disability (Smith et al 1983, Biederman et al 1987, McKusick et al 1992).

Since the early twentieth century a male predominance at all levels of mental retardation, ranging from 1.5 to 3 times the incidence in females, has been acknowledged (Penrose 1938). Although a variety of explanations have been put forth, including the probable ascertainment bias of mentally retarded males being more frequently institutionalized because of uncontrollable or violent behavior, there is good reason to believe X-linked loci contribute significantly to this gender inequity (Opitz 1986). Perhaps as many as 95 genes have been tentatively assigned to the X chromosome where mutations lead to mental retardation as at least part of the phenotype (Schwartz 1993). As the vast

77

majority of these mutations appear recessive, females, protected by an additional X chromosome, are rarely affected, and hence this leads to the excess male mental retardates. However, most of these disorders are exceedingly rare, and as single loci, none would be expected to appear frequently in the population. Thus, an aggregate of all X-linked mutations involving mental retardation appeared responsible for gender differences until a single locus, that responsible for fragile X syndrome, became recognized as a clinical entity that eventually would account for nearly half of the male predominance of mental retardation (Opitz & Sutherland 1984) .

HISTORICAL AND CLINICAL ASPECTS

Fragile X syndrome, an X-linked dominant disorder with reduced penetrance, is the most frequent form of familial mental retardation, with a prevalence of 1 in 1500 males and 1 in 2500 females (Gustavson et al 1986, Webb et al 1986). It is the single most common etiology of X-linked mental retardation (Neri et al 1992, Brown & Jenkins 1992), and ranks second only to Down's syndrome, which is most often sporadic, as the most frequent genetic cause of mental retardation overall. First described by Martin & Bell (Martin & Bell 1943), the syndrome cosegregates with a marker X chromosome that is observed cytogenetically as a nonstaining constriction in the long arm of

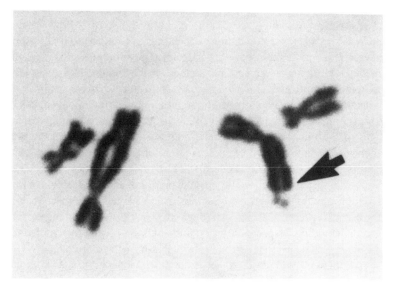

Figure 1 Partial metaphase spread from a patient with fragile X syndrome showing the isochromatid gap (*arrow*) near the distal end of the X chromosome long arm, which is indicative of the fragile X site (From Warren & Nelson 1994. Used by permission).

the X chromosome near the distal tip (Lubs 1969). Sutherland (1977) later identified this constriction as an inducible fragile site at Xq27.3 (Figure 1); it is induced in tissue culture by perturbation of folic acid or thymidine pools, which results in limited amounts of dTTP for DNA synthesis (Sutherland 1977, Glover 1981). Upon analysis by electron microscopy, the fragile site was described as an isochromatid gap at Xq27.3 (Harrison et al 1983). In fragile X males, levels of fragile site expression in metaphase spreads from peripheral lymphocytes range from 5–50% of X chromosomes analyzed, whereas the extent of expression in female heterozygotes is on the order of 1–30% (Sutherland 1977, Sutherland et al 1985). Levels of induced fragile site expression in metaphase spreads of normal controls have been shown to be less than 1% (Jenkins et al 1986). For many years, prenatal diagnosis and carrier testing relied upon scoring the fragile X marker in metaphase spreads, but recent advances in the field have rendered cytogenetic tests supplementary (Warren & Nelson 1994).

The major clinical manifestations of fragile X syndrome in adult males most notably include: mental retardation, ranging from mild to severe; mild facial dysmorphia, characterized by long, narrow facies, large dysmorphic ears, and prominence of the jaw and forehead; and macroorchidism with testicular volumes greater than 25 ml (Hagerman 1991, Hagerman et al 1991). With the obvious exception of macroorchidism, all of the phenotypic features of fragile X described above may be exhibited by adult carrier females, although they tend to be less severe (Hagerman et al 1992). Additional clinical signs often contributing to the fragile X phenotype include connective tissue abnormalities, such as joint hyperextensibility, pes planus, pectus excavatum, and mitral valve prolapse (Opitz et al 1984, Hagerman & Synhorst 1984, Hagerman et al 1984, Loehr et al 1986). The presence of autistic-like behaviors is frequently reported in fragile X males (and occasionally in female carriers). These behaviors include hand-flapping, rocking, or hand-biting (repetitive motor behaviors); gaze aversion; tactile defensiveness (decreased social and nonverbal communication skills); and repetitive speech patterns, perseveration, and echolalia (dysfunctional verbal communication) (Brown et al 1982, Reiss et al 1986, Reiss & Freund 1992). Psychiatric disorders, such as chronic depression, schizotypal personality disorder, social avoidance, social anxiety, and withdrawal, are reportedly more prevalent in fragile X carrier females than in multifeature-matched controls (Reiss et al 1988, Hagerman & Sobensky 1989, Freund et al 1993). Hyperactivity and attention deficit disorder (ADHD) appear to be common in fragile X males before puberty (Hagerman & Sobensky 1989), although it is unclear if these features are more frequent in fragile X syndrome than is normally encountered in general mental retardation (Fisch 1993). However, because of the subtlety of the fragile X phenotype prior to puberty, hyperactivity represents the presenting illness in many fragile X males (Hagerman & Sobensky 1989). Additionally, because the phenotype of fragile X syndrome is quite pleiotropic and difficult to diagnose de novo prior to puberty, detailed checklists involving precise anthropometric measurements have been devised (Butler et al 1991a, Butler et al 1991b, Butler et al 1992).

INHERITANCE PATTERN

The inheritance pattern of fragile X syndrome is both complex and unusual for a sex-linked Mendelian trait. Although the inclusion of affected heterozygous females as well as hemizygous males would suggest an X-linked dominant mode of inheritance, the occurrence of apparently unaffected males who carry the fragile X chromosome has been well documented (Martin & Bell 1943, Dunn et al 1963, Howard-Peebles & Friedman 1985, Froster-Iskenius et al 1986). In analyses of a total of 206 fragile X pedigrees by both classical and complex segregation analyses, Sherman et al (1984, 1985) reported a 20% deficit in affected males in the absence of evidence of any sporadic cases, a finding best explained by reduced penetrance in males of 80%. Given the fact that affected males rarely reproduced, these unusual nonpenetrant carrier males were termed transmitting males because they passed the fragile X chromosome to all of their obligate carrier daughters. Interestingly, nonpenetrant transmitting males were also cytogenetically negative for fragile site expression in lymphocytes (Howard-Peebles & Friedman 1985, Froster-Iskenius et al 1986). Penetrance of mental impairment in females was also found to be reduced, with an overall penetrance of 35%, in the study by Sherman et al (1984), and disease expression was again found to be in direct correlation with cytogenetic expression of the fragile site. Thus, a model of an X-linked dominant gene with reduced penetrance was imposed (Sherman et al 1984, 1985).

Sherman et al (1984, 1985) made a most peculiar observation that the risk of mental impairment associated with fragile X syndrome was a function of one's position within the pedigree. As mentioned, carrier males are both nonpenetrant and cytogenetically negative for fragile X syndrome. Interestingly, Sherman et al (1984, 1985) found that the risk of mental impairment in obligate carrier daughters of carrier males approached 0%. Siblings of these males were also at relatively low risks of 9% and 5% in males and females, respectively. Offspring of normal carrier females, however, incurred risks of 38% (males) and 16% (females) for mental impairment, which corresponded to penetrances of 76% and 32%, while offspring of mentally impaired carrier females displayed penetrances of 100% (males) and 56% (females). Thus, disease expression was dependent upon passage of the fragile X through a female, which led some to hypothesize that a premutation in carrier males causing the fragile X chromosome to be more susceptible to alteration in female gametes might account for this observation (Pembrey et al 1985). This odd pattern of increasing risk with vertical transcention through a pedigree, referred to as the Sherman paradox, was unique in human genetics, and it was only after the molecular basis of fragile X syndrome was elucidated that this peculiar pattern of inheritance was fully understood.

IDENTIFICATION OF THE FRAGILE X MUTATION

In 1991, the combined approaches of genetic mapping, physical mapping, and somatic cell genetics culminated in the positional cloning of the fragile X locus and the identification of the mutation responsible for fragile X syndrome. Prior to this time, the fragile X disorder had been mapped genetically through linkage analysis to the same interval as the Xq27 fragile site between the factor VIII (12 cM distal) and factor IX (5 cM proximal) loci, which narrowed the region of interest to approximately 20 megabases (Goodfellow et al 1985, Oberlé et al 1986). Also localized to the region surrounding the fragile site were the enzymes HPRT (Xq26) (Pai et al 1980) and G6PD (Xq28) (Filippi et al 1983). By using rodent cells deficient for both of these enzymes, Warren and coworkers (Warren et al 1987, Warren et al 1990) successfully constructed somatic cell hybrids containing either exclusively proximal or exclusively distal portions of the human fragile X chromosome. They did this by selection either for or against the human HPRT enzyme activity followed by histochemical staining for G6PD activity. Because it had previously been shown that the fragile X site was cytogenetically expressed in hybrid cells, chromosomal breakage at or near the fragile site was expected for most of the hybrids (Warren & Davidson 1984). This was confirmed through hybridization of known proximal and distal X chromosome markers to the obtained hybrids (Warren et al 1990). The availability of proximal and distal hybrids, as well as other X-breakpoint hybrids, allowed a number of new polymorphic loci to be mapped relative to the fragile site, which narrowed the gap to within 3 megabases (Suthers et al 1990, Hirst et al 1991, Suthers et al 1991, Rousseau et al 1991b).

The newly identified battery of DNA markers bracketing the fragile X locus allowed isolation of yeast artificial chromosomes (YACs) with inserts spanning the fragile X hybrid breakpoints (Dietrich et al 1991, Heitz et al 1991, Verkerk et al 1991, Kremer et al 1991b). Two groups independently identified an aberrantly methylated CpG island within this region of DNA in fragile X patients. Because of the absence of normal cleavage by methylation-sensitive enzymes in fragile X cells, a band of anomalous mobility on pulsed-field gel electrophoresis were used to identify the abnormal CpG island (Bell et al 1991, Vincent et al 1991). Oberlé et al (1991) identified a 9-kb DNA segment from the YAC 209G4 that contained the CpG island hypermethylated in fragile X patients and that detected an apparent insertion in fragile X patients on Southern blots. Yu et al (1991) and Verkerk et al (1991) concurrently yet independently identified a single restriction fragment of approximately 5 kb that contained both the proximal and distal fragile X hybrid breakpoints and also detected length variation in fragile X patients on Southern blots. Kremer et al (1991a) further localized this instability to a single 1.0-kb Pst I fragment, pfxa2, that was sequenced and found to contain an unusual CGG trinucleotide repeat of 43 copies. By using the polymerase chain reaction (PCR) and primers flanking the repeat, Kremer et al (1991a) mapped the region of instability in pfxa2 to within the CGG trinucleotide repeat itself, thereby suggesting that the length variation observed in fragile X patients could result from length variation within the trinucleotide repeat. The translocation breakpoints of various hybrids were also mapped to within the CGG repeat, thus colocalizing the fragile site and the fragile X mutation to this novel trinucleotide repeat (Kremer et al 1991a).

In order to identify expressed sequences in the fragile X breakpoint cluster region, four cosmid subclones of the YAC 209G4, which spanned the region encompassing the erroneously methylated CpG island, were used as probes against a cDNA library from human fetal brain (Verkerk et al 1991). Two cDNA clones, BC22 and BC72, comprising a contig of 3765 nucleotides, were identified and the DNA sequence of each was obtained. Analysis of this sequence revealed a single major open reading frame that remained open at the 5′ end and encoded a predicted polypeptide of 657 amino acids. The CGG repeat potentially displaying length variation in fragile X patients was contained within the 5′ portion of the open reading frame where it was initially thought to encode an uninterrupted stretch of 30 arginine residues. However, later studies demonstrate that the CGG repeat is confined to the 5′ untranslated region and is, therefore, not translated (Ashley et al 1993a) (see below). Upon northern analysis, a 4.8-kb transcript was detected in RNA from human brain and placenta, which suggested that approximately 1 kb of sequence remained to be determined. A zoo blot containing genomic DNA from a number of eukaryotes, including lower organisms such as nematode and yeast, displayed band(s) in lanes of all organisms except *Drosophila* when probed with BC22 under high stringency, thus indicating strong evolutionary conservation of this gene, referred to as *FMR1* (fragile X mental retardation 1), and implying functional importance of its protein product, FMRP. Initial data base searches at both the nucleotide and amino acid levels, however, produced no significant homologies with any previously reported sequences and provided no immediate clues as to the normal function of this predicted protein (see below).

FRAGILE X TRINUCLEOTIDE REPEAT EXPANSION

The delimitation of the region of DNA instability to within the CGG repeat suggested the possibility that length variation observed in fragile X carriers might result from the inherent instability of this reiterated trinucleotide repeat, which would be reminiscent of other tandemly repeated sequences such as the highly polymorphic dinucleotide repeats (Weber 1990). Indeed, analysis of the CGG repeat of *FMR1* across normal populations revealed that it was highly polymorphic, with repeat sizes ranging from 6–54 triplets and a mode of 30 (Fu et al 1991, Brown et al 1993, Jacobs et al 1993, Snow et al 1993). In analysis of repeat sizes in fragile X pedigrees, Fu et al (1991) discovered that individuals nonpenetrant for fragile X syndrome, carrier males and their obligate carrier daughters, displayed CGG repeat lengths ranging from 54–200 repeats, whereas affected individuals exhibited repeat sizes in excess of 200 and often greater than 1000 repeats. This finding was in agreement with the previous reports of apparently smaller insertions in carriers and larger insertions in affected individuals (Oberlé et al 1991, Yu et al 1991, Kremer et al 1991a), which Oberlé et al (1991) referred to as premutations and full mutations, respectively, in order to relate the size of the insertion with penetrance of the disease. This correlation between repeat size and penetrance not only allowed resolution of the Sherman paradox (Fu et al 1991), but it also paved the way toward identification of other trinucleotide repeat expansion mutations by substantiating earlier claims (Harper 1989) that diseases displaying genetic anticipation, increasing severity, or decreasing age of onset of disease from one generation to the next might share a common form of mutation.

In addition to establishing an association between repeat size and penetrance for fragile X syndrome, studies of the repeat length of *FMR1* also aided in defining the boundaries of meiotic instability. In the study by Fu et al (1991), all *FMR1* CGG repeats of 54 and greater displayed meiotic instability, and smearing of the full mutation bands on Southern blots suggested mitotic instability as well. In a recent analysis of 116 families referred for fragile X testing, a *FMR1* CGG repeat of 51 was shown to be meiotically stable through five generations, whereas a repeat of 57 displayed meiotic instability in the next generation (Snow et al 1993). Also in this study, a CGG repeat of only 61 in a carrier female expanded into the full mutation range in the following generation. However, another premutation allele of 90 in a different carrier female only expanded to 115 upon transmission to her offspring. Therefore, repeats of 54 and above clearly display meiotic instability and should be considered premutations; however, the expansion of premutation alleles into the full mutation range must be dependent upon other factors yet to be identified.

METHYLATION OF THE *FMR1* GENE

In addition to trinucleotide repeat expansion in fragile X patients, abnormal methylation of a CpG island located 250 base pairs upstream of the CGG repeat of *FMR1* was also found in individuals expressing the fragile X phenotype (Bell et al 1991, Heitz et al 1991, Oberlé et al 1991, Vincent et al 1991). Like expansion of the repeat, the extent of CpG-island methylation is directly correlated with disease expression because carrier males, which are unaffected, display unmethylated CpG islands. Individuals with the full mutation, however, have fully methylated CpG islands (Oberlé et al 1991). Unaffected carrier females display methylation only on the inactive X chromosome, while affected females tend to be skewed toward excess methylation (> 50%), although this is not always the case (Oberlé et al 1991). The CGG repeat, which is itself rich in CpG dinucleotides, is also fully methylated in affected individuals with full mutations or at normal repeat copy number on the inactive X (Hansen et al 1992, Hornstra et al 1993). The identification of abnormal methylation associated with fragile X syndrome is in agreement with Laird's (1987) hypothesis, which states that fragile X syndrome results from persistence of an imprint that is due to failure of erasure of X inactivation in a carrier female.

Methylation of the CpG island adjacent to the *FMR1* gene correlates with transcriptional silencing of *FMR1* and expression of the fragile X phenotype. In a study by Pieretti et al (1991), 16 of 20 affected males with expanded repeats in the full mutation range displayed fully methylated CpG islands and produced no *FMR1* transcript. Of the four affected males with *FMR1* expression, three were identified as mosaics for fully methylated full mutations and unmethylated premutations, and one was a methylation mosaic with a partially unmethylated full mutation. In a prenatal diagnosis of an affected fetus with a full mutation, Sutcliffe et al (1992) reported an absence of methylation in chorionic villus samples that correlated with normal levels of *FMR1* expression, while in fetal tissue a fully methylated CpG island was found but not *FMR1* message. Recently, four carrier males with repeat sizes between 200 and 400 but with unmethylated CpG islands and repeats were found to be intellectually and physically normal by observation (Loesch et al 1993). Conversely, two males exhibiting partially methylated CpG islands and CGG repeats between 200 and 300 copies also displayed mild clinical and physical manifestations of fragile X syndrome (McConkie-Rosell et al 1993).

Although many carrier females have premutation alleles, a significant proportion of the normal carrier females carry full mutations with expansions well beyond 200 repeats as well as the abnormal methylation of the *FMR1* gene. Rousseau et al (1991a) demonstrated that approximately half (53%) of women with full mutations are penetrant with IQ levels in the borderline and mentally retarded range. The other half presumably escape the effect of the full mutation because of X inactivation (lyonization) patterns. However, Taylor et al (1994) showed that the pattern of X inactivation, i.e the proportion of fragile X chromosomes active versus inactive, in peripheral lymphocytes of full mutation females was not a predictor of intellectual function. Thus, it is most reasonable to conclude that the protein encoded by *FMR1* is cell autonomous and that the proportion of cells in key regions of the brain with the fragile X chromosome in the active state (but still not expressing *FMR1* because of the abnormal methylation) are responsible for penetrance in females.

Thus, the absence or reduction of *FMR1* expression as a function of methylation status of the upstream CpG island and the CGG repeat is thought to be the basis for fragile X syndrome. The fact that at least four fragile X patients lacking repeat expansion but possessing other mutations of the *FMR1* gene itself, three large deletions and one point mutation, have been identified supports this hypothesis (Gideon et al 1992, Wöhrle et al 1992, De Boulle et al 1993, Tarleton et al 1993). The role of repeat expansion in abnormal methylation, or vice versa, remains to be determined.

DETECTION OF THE FRAGILE X MUTATION

Both the expansion of the CGG repeat and the abnormal methylation of the *FMR1* gene are important diagnostic indicators. Methods have been developed to assess these mutational parameters in patients (reviewed by Warren & Nelson 1994). Detection can employ either Southern blotting (Figure 2*a*) or the polymerase chain reaction (Figure 2*b*). Southern blotting most commonly utilizes the digestion of genomic DNA with EcoR I and the methylation-sensitive enzyme BssH II or Eag I. EcoR I digestion liberates a 5.2-kb fragment containing the promoter and first exon of *FMR1*, which includes the CGG repeat. Double digestion with BssH II or Eag I cleaves normal male DNA into 2.4-kb and 2.8-kb fragments; the latter fragment contains the trinucleotide repeat. The active X chromosome of normal females also displays this pattern, but the inactive normal X in a female is methylated, as is common for many X-linked loci, and is resistant to BssH II or Eag I digestion. Thus, normal females exhibit three bands as shown in lane two of Figure 2*a*.

Premutation alleles exhibit a shift in the 2.8-kb band to a slightly higher molecular weight that reflects the increase in the CGG repeat to the 54–230 repeat range. Full mutations display a large, somewhat diffuse band that is generally not cleaved by BssH II or Eag I. These bands reveal the expansion of the CGG repeat into the range of many hundreds of triplets as well as the abnormal methylation of the *FMR1* gene. Females with full mutations display a similar pattern superimposed upon that of the normal X (in both the active and inactive states).

Analysis by PCR has allowed a much more precise estimation of repeat length, particularly in normal and premutation alleles (Fu et al 1991). However, amplification of full mutations is difficult, though not impossible (Pergolizzi et al 1992). As shown in Figure 2*b*, PCR amplification of the CGG repeat by using flanking primers of unique sequence demonstrates the remarkable insta-

(A)

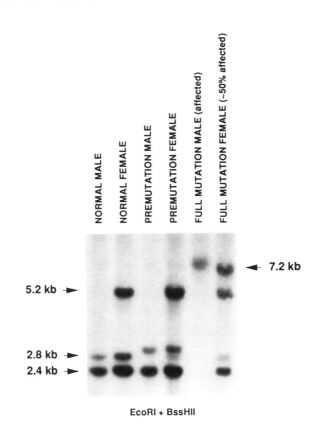

Figure 2   Molecular detection of the *FMR1* gene. (*A*) Southern blot of genomic DNA digested with EcoRI and BssHII, resolved through agarose, transferred to membrane, and probed with radiolabeled *FMR1* gene fragment pE5.1 (Verkerk et al 1991). (*B*) PCR analysis of the *FMR1* repeat by using flanking primers as described by Pergolizzi et al (1993). Numbers next to pedigree symbols reflect the number of *FMR1* triplet repeats (Both figures from Warren & Nelson 1994. Used by permission).

bility of the premutation alleles, where siblings often have alleles unique from one another as well as distinct from the parent. Analysis by PCR is useful for defining accurate premutation repeat lengths and for demonstrating instability, which may be of some utility in genetic counseling (Warren & Nelson 1994), and is useful in situations where limited DNA is a constraint.

OTHER TRINUCLEOTIDE REPEAT EXPANSION DISEASES

Since the emergence of the trinucleotide repeat expansion mutation responsible for fragile X syndrome, six other diseases caused by trinucleotide expansions

(B)

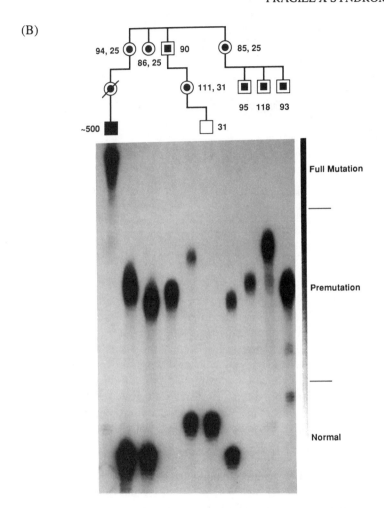

have been identified (Table 1) (reviewed by Warren & Nelson 1993, 1994): spinal and bulbar muscular atrophy (SBMA) (La Spada et al 1991); myotonic dystrophy (DM) (Aslanidis et al 1992, Brook et al 1992, Fu et al 1992, Harley et al 1992, Mahadevan et al 1992); Huntington's disease (HD) (Group 1993); spinal and cerebellar ataxia type 1 (SCA1) (Orr et al 1993); FraX E mental retardation (Knight et al 1993); and dentatorubral pallidoluysian atrophy (DPA) (Koide et al 1994, Nagafuchi et al 1994). Interestingly, all trinucleotide repeat diseases described thus far are neurologic diseases, although there is currently no good hypothesis to account for this pattern. Similarities between the trinucleotide repeats of these diseases include that they are all normally

Table 1   Genetic diseases associated with unstable trinucleotide repeats

| Disease | Trinucleotide repeat[a] | Repeat number | | | Change in gene function |
		Normal	Carrier	Affected	
Fragile X syndrome	CGG 5'UTR	6–52	50–200	230 to >1000	Loss
Myotonic dytrophy	CTG 3'UTR	5–37	—	50 to >1000	mRNA stability
Spinal & bulbar muscular atrophy	CAG coding	12–34	—	40–62	Gain
Huntington's disease	CAG coding	11–36	—	42–100	Gain[b]
Spinocerebellar ataxia type 1	CAG coding	19–36	—	43–81	Gain[b]
Dentatorubral pallidoluysian atrophy	CAG coding	7–23	—	49–75	Gain[b]
FRAXE mental retardation	CCG —	6–25	116–133	200 to >800	Loss[b]

[a] All repeat are exonic with the position shown; UTR is untranslated region.
[b] Functional change is not established and is shown based upon similarity to other examples.

polymorphic, GC-rich, and expressed sequences, while differences include the exact nucleotide content of the repeat and the location of the repeats within the various messages (Richards & Sutherland 1992). In the majority of these diseases (HD, SBMA, DPA, and probably SCA1), an unstable CAG repeat within the coding region of the respective genes undergoes expansion, which produces an amplified polyglutamine tract in the nascent protein. In the case of X-linked SBMA, Warren & Nelson (1993) have suggested that the expansion of the polyglutamine tract in the androgen receptor must constitute a gain of function, perhaps through allowing the receptor to interact with promoters that it does not normally recognize, because the absence of manifestations of testicular feminization make it unlikely that the receptor is nonfunctional. Although the function of the gene products in HD and SCA1 are currently unknown, the dominant inheritance pattern of these autosomally inherited disorders argues that expansion of the polyglutamine tract in these two diseases will constitute a gain-of-function as well. In myotonic dystrophy, a CTG repeat in the 3' untranslated region (UTR) of the myotonin protein kinase gene is enlarged, which results in altered message stability and aberrant intracellular signaling due to abnormal dosage of this protein kinase (Fu et al 1993, Sabourin et al 1993). As previously discussed, expansion of a CGG repeat in the 5' UTR of the *FMR1* gene is associated with abnormal methylation of the repeat and an upstream CpG island leading to transcriptional silencing of *FMR1*. No information concerning the mechanism of disease in FraX E mental retardation is currently available because the expanded CGG repeat has not yet been ascribed to a gene.

Of all of these repeat expansion mutations, only the repeats of DM and fragile X syndrome, which are within untranslated regions, have full mutations with repeat expansions exceeding several hundred copies. Repeat expansions of the other diseases are usually 2–4 times normal, and maximum repeat sizes rarely exceed 100 copies. This implies that the mechanism of expansion of fragile X syndrome and myotonic dystrophy could be similar, although it is puzzling that DM does not display the parent-of-origin requirement for expansion observed in fragile X syndrome. However, reductions in trinucleotide repeat sizes, which are almost never seen in fragile X syndrome, are often seen in paternal DM alleles (Reyniers et al 1993). Nevertheless, given the fact that the full mutations of DM or fragile X are not amenable to cloning, the other trinucleotide diseases with more subtle trinucleotide expansions will most likely be more informative with regard to the mechanism of trinucleotide instability.

MECHANISM OF THE FRAGILE X REPEAT EXPANSION

The molecular mechanism of trinucleotide repeat expansion still remains an enigma, although recent advances in the fragile X field have been enlightening. In previous segregation analyses of over 200 fragile X pedigrees, no spontaneous mutations in female gametes causing fragile X syndrome were identified (Sherman et al 1984, 1985). Given the fact that fragile X alleles are constantly being removed from the gene pool because of reduced fitness in affected individuals, Sherman et al (1984, 1985) predicted an extraordinarily high mutation rate in male gametes, on the order of 7.2×10^{-4}, to account for the high prevalence of fragile X syndrome within the population. However, subsequent analyses of fragile X pedigrees have failed to detect any new mutations producing fragile X syndrome (Jacobs et al 1986, Rousseau et al 1991a, Yu et al 1992, Smits et al 1993). Consistent with the absence of new mutations, Smits et al (1993) recently found that five fragile X males related through common ancestors from the eighteenth century were all concordant for the same allele at the DXS548 locus, a polymorphic marker locus located 150 kb proximal to *FMR1* (Riggins et al 1992). This finding attests to the very old nature of the fragile X mutation in this family and suggests the possibility of linkage disequilibrium in the region encompassing the fragile X locus. Haplotype analyses of markers flanking the fragile X locus have, in fact, revealed significant linkage disequilibrium between fragile X syndrome and certain haplotypes, which indicates that a small group of founder chromosomes, perhaps as few as six, might account for the vast majority of fragile X mutations observed today (Richards et al 1992, Oudet et al 1993). The observation of linkage disequilibrium in the region of the fragile X locus is reminiscent of the situation in individuals with myotonic dystrophy (DM), the gene of which is in complete linkage disequilibrium with a single insertion/deletion polymorphism located 5 kb from the unstable CTG repeat (Imbert et al 1993). However, unlike the myotonic dystrophy gene, the fragile X locus appears to be in linkage equilibrium with a discrete subset of haplotypes, which suggests that multiple mutations of the primordial *FMR1* repeat are responsible for the fragile X chromosomes in today's population (Smits et al 1993).

A low prevalence of new mutations and marked linkage disequilibrium suggesting highly ancient mutations are unusual findings for X-linked mutations (Chakravarti 1992), and this scenario is hard to reconcile in light of the high prevalence of fragile X syndrome in the population. One possible explanation for this phenomena is an unexpectedly high number of premutations in the population that go undetected in the absence of any obvious phenotype. In favor of this hypothesis, Snow et al (1993) found the frequency of premutation alleles in the normal population to be 0.8%. A multistep model has been proposed that invokes a discrete set of mutations of a normal primordial allele to a pre-premutation, or a predisposed allele, that is relatively stable for many years (thus accounting for the linkage disequilibrium observed) but has the potential to convert to unstable premutations that rapidly expand to full mutations with concurrent disease expression (Morton & Macpherson 1992). In light of this model and the above linkage disequilibrium data, Oudet et al (1993) asserted that the most common 30 allele of the CGG repeat of *FMR1* (Brown et al 1993, Snow et al 1993) was probably the primordial allele and that larger alleles in the range of 38–40 repeats represent the predisposed alleles that ultimately account for all of the fragile X chromosomes. In addition, it has been suggested that a mutation of one of the interspersed AGG triplets to CGG within the normally cryptic repeat of *FMR1* (Pieretti et al 1991, Verkerk et al 1991) could account for the primordial mutation event (Oudet et al 1993) because repeats that are purer in content are less stable (Weber 1990). Finally, superimposed upon this model is the likelihood of polymerase slippage during replication of these reiterated sequences (Imbert et al 1993), which accounts for the smaller, inert variation in the normal CGG repeat length contributing to the normally polymorphic nature of the repeats. If this model is valid, then it follows that the primary new mutations responsible for the cases of fragile X syndrome today probably occurred many generations ago, and any new mutations identified today may not manifest themselves phenotypically for many years to come.

The apparent dearth of expansion of premutation alleles into the full mutation range upon transmission from carrier males to their obligate carrier daughters has also recently been addressed. Upon discovery of a rare mating event by an affected male, Willems et al (1992) followed the segregation of the fragile X mutation from an affected male through his normal carrier daughter to an affected grandson. At the molecular level, the daughter received only a premutation allele, whereas the grandfather was mosaic for full and premutation-size alleles, which suggested a priori that the grandfather displayed germ-line mosaicism as well as somatic mosaicism. However, subsequent analyses of affected males with full mutations in somatic tissues revealed only premutation alleles in their sperm (Reyniers et al 1993). One interpretation of these findings is that partial trinucleotide repeat expansion occurs within the male gamete, with subsequent selection against sperm with fully expanded repeats. Another interpretation is that trinulceotide repeat expansion occurs postzygotically following the sequestering of primordial germ cells that are spared the massive expansion. However, a maternal imprint must be invoked that leads to somatic expansion of maternal premutations. With regard to this model of postzygotic expansion, it will be crucial to determine whether oocytes with full mutations are observed in ovaries of carrier females. To date, no such analysis of repeat sizes in oocytes has been reported.

A prediction of such a postzygotic expansion model is that a variable degree of repeat length mosaicism should be present in the somatic tissues of affected patients. Indeed, somatic mosaicism in cells of fragile X patients has been well documented (Fu et al 1991, Pieretti et al 1991, Rousseau et al 1991a). In a study of 511 individuals from 63 fragile X families, Rousseau et al (1991b) reported that approximately 15% of individuals with mutations within the fully expanded range also displayed premutation alleles in a subset of cells. In individuals such as these, the premutation allele is predominately unmethylated and transcriptionally active, whereas the full mutation allele is most often fully methylated and transcriptionally silent (Pieretti et al 1991). In analysis of a 13-week male fetus, nearly identical mosaic patterns were observed across all tissues analyzed, and subsequent prolonged cell culture provided no evidence of mitotic instability of the alleles initially present (Wöhrle et al 1993). Interestingly, chorionic villi obtained at 10 weeks gestation for prenatal diagnoses displayed fully expanded, unmethylated *FMR1* repeats, thereby delimiting the window for trinucleotide repeat expansion to before this time and methylation of expanded repeats to sometime afterwards. Thus, many data support the idea that expansion occurs postconceptionally, and of the two models described, it is the most widely accepted at the present time.

One major shortcoming of the postzygotic model of repeat expansion is its inability to satisfactorily account for the strict parental influence observed in fragile X pedigrees. If this model is valid, then an imprinting phenomenon clearly must also occur to account for the absence of repeat expansion in daughters of carrier males. As previously mentioned, Laird (1987) has proposed that fragile X syndrome results from the abnormal persistence of an imprint due to incomplete erasure of X-inactivation on the fragile X chromosome in female gametes. Although the observation of abnormal methylation associated with repeat expansion is consistent with this hypothesis, the occurrence of mosaicism for full and premutation alleles and the fact that premutation alleles are unmethylated (Pieretti et al 1991) are difficult to reconcile with regard to Laird's hypothesis. Furthermore, corroborating data from two labs indicate that expansion precedes methylation in the embryo (Sutherland et al 1991, Sutcliffe et al 1992), which makes it questionable whether methylation in any form contributes to the lack of expansion in daughters of carrier males. So, validation of the postzygotic model of repeat expansion is incumbent upon identification of the factor responsible for the strict maternal origin of expanded *FMR1* alleles.

CHARACTERIZATION OF *FMR1* AND THE GENE PRODUCT

Since the identification of the *FMR1* gene, much has been gleaned with regard to its physical structure and pattern of expression. The *FMR1* gene includes 17 exons encompassing 38 kb of the human genome (Eichler et al 1993). The CGG repeat resides within the first exon of the gene and is positioned 250 basepairs downstream of the CpG island that is aberrantly methylated in fragile X patients (Fu et al 1991, Eichler et al 1993) and may serve as the promoter of the *FMR1* gene (Hwu et al 1993). On Northern analysis, expression in a number of human tissues both related and unrelated to the fragile X phenotype is observed; these include brain, testes, placenta, lung, kidney, and heart (Hinds et al 1993). In situ RNA expression studies in adult mouse tissues expanded the list of tissues by adding esophageal epithelium, thymus, spleen, ovary, eye, colon, uterus, thyroid, and liver; expression in heart, aorta, or muscle was not observed in the adult mouse (Hinds et al 1993). In cross sections of mouse brain, Hinds et al (1993) reported expression predominately in neurons as opposed to white matter or glial cells, with highest levels of expression in the cerebellum, habenula, and the granular layers of the hippocampus and cerebral cortex. In a recent in situ RNA expression analysis of cross sections of brain from a 25-week-old human fetus (Abitbol et al 1993), the authors report *FMR1* expression mainly in neurons; the nucleus basalis magnocellularis and areas of the hippocampus are most intense. These results are consistent with the observations by Hinds et al (1993) in that predominant labeling is observed in areas critical to limbic circuitry, which is a pathway in the brain involved in cognition, memory, and behavior.

Immunohistochemical analysis with antibodies against FMRP confirmed the predominant neuronal and sparse glial distribution of FMRP previously indicated by in situ RNA studies (Devys et al 1993). In addition, two reports of intracellular immunolocalization of FMRP concur that the protein is localized predominately to the cytosol, although nuclear localization in the absence of cytosolic expression was observed in a small proportion of cells (Devys et al 1993, Verheij et al 1993). Interestingly, the amino terminal half of the protein displayed exclusive nuclear localization according to Devys et al (1993), although this half of the protein displays no nuclear localization signal and lacks the putative nuclear localization signal near the carboxy terminus described by Verkerk et al (1991). Thus, it remains unclear whether FMRP might, under certain conditions, have a functional role in the nucleus.

The mouse *FMR1* homologue has been identified and shown to be remarkably conserved, with nucleotide and amino acid identity values of 95% and 97%, respectively, within the coding region (Ashley et al 1993a). The CGG repeat responsible for the fragile X phenotype in humans is also conserved in mouse both in its position and nucleotide sequence, although the repeat number is reduced to nine copies (8 CGGs and 1 CGA) (Ashley et al 1993a). An in-frame stop codon identified upstream of the human repeat and a conserved ATG located 66 and 69 nucleotides downstream of the CGG repeat in mouse and human, respectively, delimit the CGG repeat to the 5′ untranslated region (5′ UTR) of the *FMR1* message and predict a putative protein product of 69 kD (Ashley et al 1993a). Additionally, alternative splicing of the *FMR1* gene has been documented. The results of this splicing predict at least twelve potential isoforms of the *FMR1* predicted protein, FMRP, in all tissues analyzed and suggest the potential for intracellular functional diversity of the *FMR1* gene product (Verkerk et al 1993, Ashley et al 1993a). In favor of the notion of functional heterogeneity is the fact that a subset of the alternative splicing events in *FMR1* alter the normal reading frame and produce isoforms with novel carboxy termini and quite distinct hydropathy profiles (Ashley et al 1993a).

Two copies of a 30–amino acid domain conserved across evolution from bacteria to humans have been identified within the amino terminal half of the FMRP predicted sequence (Ashley et al 1993b, Siomi et al 1993b). This repeated domain, termed a KH domain (Siomi et al 1993a), is also present in a number of proteins that have an interaction with RNA in common, including one member of the hnRNP class of ribonucleoproteins, hnRNP K. Also identified within the carboxy terminal portion FMRP amino acid sequence are two copies of the RGG box (arginine-glycine-glycine) (Ashley et al 1993b), an amino acid motif implicated in both RNA as well as DNA binding (Christensen & Fuxa 1988). Nucleic acid binding assays of in vitro translated FMRP reveal that it is an RNA binding protein that binds to its own RNA transcript with high affinity ($Kd = 5.7$ nM) and to a lesser extent to a pool of RNAs from human fetal brain; weak binding of FMRP to both single-strand and double-strand DNA is observed as well (Ashley et al 1993b). Selective binding of FMRP to individual human fetal brain transcripts occurs, and it is suggested that approximately 4% of the messages present in human fetal brain might be bound by FMRP, which is consistent with the highly pleiotropic nature of the fragile X phenotype (Ashley et al 1993b). Finally, a 2:1 stoichiometry of binding of RNA:protein is reported, which implies that each protein molecule has the potential to interact with two RNA molecules (Ashley et al 1993b).

CONCLUSIONS

Fragile X syndrome accounts for a major proportion of mental retardation in humans. In addition to the intellectual deficit, these patients also exhibit a wide range of behavioral problems, including autistic-like features. Thus, the molecular understanding of this syndrome should lead to further understanding of the higher cognitive functioning in humans. Indeed, much has already been learned from the cloning of the fragile X gene, *FMR1*. The mutational basis of the syndrome is the extraordinary expansion of a trinucleotide repeat within the 5' untranslated region of the gene. Since this discovery, six other genetic diseases, all neurological in nature, have now been attributed to unstable triplet repeats within exons of different genes. This novel mutational mechanism will likely be found responsible for even more genetic disorders. The mechanism of repeat expansion is poorly understood and is a fertile and important area of inquiry.

When the CGG repeat in the *FMR1* gene exceeds 230 repeats, the gene is heavily methylated and transcriptionally suppressed. The absence of the encoded protein, FMRP, is therefore believed responsible for the mental retardation and associated phenotype. This cytosolic protein has been demonstrated to be a selective RNA-binding protein, interacting with as much as 4% of human brain mRNA. Thus, the further understanding of this interaction and the identification of those genes whose messages interact could lead to a great deal of insight into the mechanisms of cognitive function in humans.

ACKNOWLEDGMENTS

STW is an investigator with the Howard Hughes Medical Institute and CTA is a predoctoral fellow with the March of Dimes Birth Defects Foundation.

Literature Cited

Abitbol M, Menini C, Delezoide A-L, Rhyner T, Vekemans M, Mallet J. 1993. Nucleus basalis magnolaris and hippocampus are the major sites of FMR-1 expression in the human fetal brain. *Nature Genet.* 4:147–52

American Psychiatric Association. 1987. *Diagnostic and Statistical Manual of Mental Dis-* *orders,* pp. 28–33. Washington DC: Am. Psychiatr. Assoc. 567 pp. 3rd ed.

Ashley CT, Sutcliffe JS, Kunst CB, Leiner HA, Eichler EE, et al. 1993a. Human and murine FMR-1: alternative splicing and translational initiation downstream of the CGG-repeat. *Nature Genet.* 4:244–51

Ashley CT, Wilkinson KD, Reines D, Warren ST. 1993b. FMR1 protein: conserved RNP family domains and selective RNA binding. *Science* 262:563–66

Aslanidis C, Jansen G, Amemiya C, Shutler G, Mahadevan M, et al. 1992. Cloning of the essential myotonic dystrophy region and mapping of the putative defect. *Nature* 355:548–51

Bell MV, Hirst MC, Nakahori Y, MacKinnon RN, Roche A, et al. 1991. Physical mapping across the fragile X: hypermethylation and clinical expression of the fragile X syndrome. *Cell* 64:861–66

Biederman J, Munir K, Knee D, Armentano M, Autor S, et al. 1987. High rate of affective disorders in probands with attention deficit disorders and their relatives—a controlled family study. *Am. J. Psychiatry* 144:330–33

Brook JD, McCurrach ME, Harley HG, Buckler AJ, Church D, et al. 1992. Molecular basis of myotonic dystrophy: expansion of a trinucleotide (CTG) repeat at the 3' end of a transcript encoding a protein kinase family member. *Cell* 68:799–808

Brown WT, Houck GE, Jeziorowska A, Levinson FN, Ding X, et al. 1993. Rapid fragile X carrier screening and prenatal diagnosis using a nonradioactive PCR test. *JAMA.* 270(13):1569–75

Brown WT, Jenkins EC. 1992. The fragile X syndrome. *Mol. Genet. Med.* 2:39–65

Brown WT, Jenkins EC, Friedman E, Brooks J, Wisniewski K, et al. 1982. Autism is associated with fragile X syndrome. *J. Autism Dev. Dis.* 12:303–8

Butler MG, Allen A, Haynes JL, Singh DN, Watson MS, Breg WR. 1991a. Anthropometric comparison of mentally retarded males with and without the fragile X syndrome. *Am. J. Med. Genet.* 38:260–68

Butler MG, Brunschwig A, Miller LK, Hagerman RJ. 1992. Standards for selected anthropometric measurements in males with the fragile X syndrome. *Pediatrics* 89:1059–62

Butler MG, Mangrum T, Gupta R, Singh DN. 1991b. A 15-item checklist for screening mentally retarded males for the fragile X syndrome. *Clin. Genet.* 39:347–54

Chakravarti A. 1992. Fragile X founder effect. *Nature Genet.* 1:237–38

Christensen ME, Fuxa KP. 1988. The nucleolar protein, B-36, contains a glycine and dimethylarginine-rich sequence conserved in several other nuclear RNA-binding proteins. *Biochem. Biophys. Res. Comm.* 155(3): 1278–83

De Boulle K, Verkerk AJMH, Reyniers E, Vits L, Hendrickx J, et al. 1993. A point mutation in the FMR-1 gene associated with fragile X mental retardation. *Nature Genet.* 3(1): 31–35

Devys D, Lutz Y, Rouyer N, Bellocq J-P, Mandel J-L. 1993. The FMR-1 protein is cytoplasmic, most abundant in neurons, and appears normal in carriers of the fragile X premutation. *Nature Genet.* 4:335–40

Dietrich A, Kioschis P, Monaco AP, Gross B, Korn B, et al. 1991. Molecular cloning and analysis of the fragile X region in man. *Nucl. Acids Res.* 19:2567–72

Dunn HG, Renpenning H, Gerrard JW, Miller JR, Tabata T, Federoff S. 1963. Mental retardation as a sex-linked defect. *Am. J. Mental Def.* 67:827–48

Eichler EE, Richards S, Gibbs RA, Nelson DL. 1993. Fine structure of the human FMR1 gene. *Hum. Mol. Genet.* 2:1147–53

Filippi G, Rinaldi A, Archidiacono N, Rocchi M, Balasz I, Sinas-Calco M. 1983. Brief report: linkage between G6PD and fragile X syndrome. *Am. J. Med. Genet.* 15:113–19

Fisch GS. 1993. What is associated with the fragile X syndrome? *Am. J. Med. Genet.* 48(2):112–21

Freund LS, Reiss AL, Abrams MT. 1993. Psychiatric disorders associated with fragile X in the young female. *Pediatrics* 91:321–29

Froster-Iskenius U, McGillivray BC, Dill FJ, Hall JG, Herbst DS. 1986. Normal male carriers in the fra(X) form of X-linked mental retardation (Martin-Bell syndrome). *Am. J. Med. Genet* 232:619–32

Fu Y-H, Friedman DI, Richards S, Pearlman JA, Gibbs RA, et al. 1993. Decreased expression of myotonin-protein kinase messenger RNA and protein in adult form of myotonic dystrophy. *Science* 260:235–38

Fu Y-H, Kuhl DP, Pizzuti A, Pieretti M, Sutcliffe JS, et al. 1991. Variation of the CGG repeat at the fragile X site results in genetic instability: resolution of the Sherman paradox. *Cell* 67:1047–58

Fu Y-H, Pizzuti A, Fenwock RG, King J, Rajnarayan S, et al. 1992. An unstable triplet repeat in a gene related to myotonic muscular dystrophy. *Science* 255:1256–58

Gideon AK, Baker E, Robinson H, Partington MW, Gross B, et al. 1992. Fragile X syndrome without CCG amplification has an FMR1 deletion. *Nature Genet.* 1:341–44

Glover TW. 1981. FUdr induction of the X chromosome fragile site: evidence for the mechanism of folic acid and thymidine deprivation. *Am. J. Hum. Genet.* 33:234–42

Goodfellow PN, Davies KE, Ropers HH. 1985. Report of the committee on the genetic constitution of the X and Y chromosomes. *Cytogenet. Cell. Genet.* 40:296–352

Group THDCR. 1993. A novel gene containing a trinucleotide repeat that is expanded and unstable on Huntington's disease chromosomes. *Cell* 72:971–83

Gustavson K-H, Blomquist H, Holmgren G. 1986. Prevalence of fragile-X syndrome in mentally retarded children in a Swedish county. *Am. J. Med. Genet* 23:581–88

Hagerman RJ. 1991. Physical and behavioral phenotype. In *Fragile X Syndrome, Diagnosis, Treatment and Research,* ed. RJ Hagerman, AC Silverman, pp. 3–68. Baltimore: Johns Hopkins Univ. Press

Hagerman RJ, Amiri K, Cronister A. 1991. Fragile X checklist. *Am. J. Med. Genet* 38: 283–87

Hagerman RJ, Jackson C, Amiri K, Silverman AC, OConner R, Sobensky W. 1992. Girls with fragile X syndrome: physical and neurocognative status and outcome. *Pediatrics* 89(3):395–400

Hagerman RJ, Sobensky WE. 1989. Psychopathology in fragile X syndrome. *Am. J. Orthopsychiatry* 59:142–52

Hagerman RJ, Synhorst DP. 1984. Mitral valve prolapse and aortic dilitation in fragile X syndrome. *Am. J. Med. Genet* 17:123–31

Hagerman RJ, Van Housen K, Smith ACM, McGavran L. 1984. Consideration of connective tissue dysfunction in fragile X syndrome. *Am. J. Med. Genet* 17:111–21

Hansen RS, Gartler SM, Scott CR, Chen S-H, Laird CD. 1992. Methylation analysis of CGG sites in the CpG island of the human FMR1 gene. *Hum. Mol. Genet.* 1(8):571–78

Harley HG, Brook JD, Rundle SA, Crow S, Reardon W, et al. 1992. Expansion of an unstable DNA region and phenotypic variation in myotonic dystrophy. *Nature* 355:545–46

Harper PS. 1989. Myotonic dystrophy and related disorders. In *Principles and Practice of Medical Genetics,* ed. AEH Emery, DL Rimoin, pp. 579–97. Philadelphia: Saunders. 894 pp.

Harrison CJ, Jack EM, Allen TD, Harris R. 1983. The fragile X: a scanning electron microscope study. *J. Med. Genet.* 20:280–85

Heitz D, Rousseau F, Devys D, Saccone S, Abderrahim H, et al. 1991. Isolation of sequences that span the fragile X and identification of a fragile X-related CpG island. *Science* 251:1236–39

Hinds HL, Ashley CT, Nelson DL, Warren ST, Housman DE, Schalling M. 1993. Tissue specific expression of FMR1 provides evidence for a functional role in fragile X syndrome. *Nature Genet.* 3:36–43

Hirst MC, Roche A, Flint TJ, MacKinnon RN, Bassett JHD, et al. 1991. Linear order of new and established markers around the fragile site at Xq27.3. *Genomics* 10:243–49

Hornstra IK, Nelson DL, Warren ST, Yang TP. 1993. High resolution methylation analysis of the FMR1 gene trinucleotide repeat region in fragile X syndrome. *Hum. Mol. Genet.* 2(10):1659–65

Howard-Peebles PN, Friedman JM. 1985. Un-

affected carrier males in families with fragile X syndrome. *Am. J. Hum. Genet.* 37:956–64

Hwu W-L, Lee Y-M, Lee S-C, Wang T-R. 1993. In vitro DNA methylation inhibits FMR1 promoter. *Biochem. Biophys. Res. Commun.* 193:324–29

Imbert G, Kretz C, Johnson K, Mandel J-L. 1993. Origin of the expansion mutation in myotonic dystrophy. *Nature Genet.* 4:72–76

Jacobs PA, Bullman H, Macpherson J, Youings S, Rooney V, et al. 1993. Population studies of the fragile X: a molecular approach. *J. Med. Genet.* 30:454–59

Jacobs PA, Sherman S, Turner G, Webb T. 1986. The fragile (X) syndrome: the mutation problem. *Am. J. Med. Genet* 23:611–17

Jenkins ED, Brooks J, Duncan CJ, Sanz MM, Silverman WP, et al. 1986. Low frequencies of apparently fragile X chromosomes in normal control cultures: a possible explanation. *Exp. Cell Biol.* 54:40–48

Knight SJL, Flannery AV, Hirst MC, Campbell L, Christodoulou Z, et al. 1993. Trinucleotide repeat amplification and hypermethylation of a CpG island in FRAXE mental retardation. *Cell* 74:127–34

Koide R, Ikeuchi T, Onodera O, Tanaka H, Igarashi S, et al. 1994. Unstable expansion of CAG repeat in hereditary dentatorubral-pallidoluysian atrophy (DRPLA). *Nature Genet.* 1:9–13

Kremer EJ, Pritchard M, Lynch M, Yu S, Holman K, et al. 1991a. Mapping of DNA instability at the fragile X to a trinucleotide repeat sequence p(CCG)n. *Science* 252:1711–14

Kremer EJ, Yu S, Pritchard M, Nagaraja R, Heitz D, et al. 1991b. Isolation of a human DNA sequence which spans the fragile X. *Am. J. Hum. Genet.* 49:656–61

La Spada AR, Wilson EM, Lubahn DB, Harding AE, Fischbeck KH. 1991. Androgen receptor gene mutations in X-linked spinal and bulbar muscular atrophy. *Nature* 352:77–79

Laird CD. 1987. Proposed mechanism of inheritance and expression of the human fragile X syndrome of mental retardation. *Genetics* 117:587–99

Loehr JP, Synhorst DP, Wolfe RR, Hagerman RJ. 1986. Aortic root dilatation and mitral valve prolapse in the fragile-X syndrome. *Am. J. Med. Genet.* 23:189–94

Loesch DZ, Huggins R, Hay DA, Gedeon AK, Mulley JC, Sutherland GR. 1993. Genotype-phenotype relationships in fragile X syndrome: a family study. *Am. J. Hum. Genet.* 53:1064–73

Lubs JA Jr. 1969. A marker X chromosome. *Am. J. Hum. Genet.* 21:231–44

Mahadevan M, Tsilfidis C, Sabourin L, Shutler G, Amemiya C, et al. 1992. Myotonic dystrophy mutation: an unstable CTG repeat in the 3' untranslated region of the gene. *Science* 255:1253–55

Martin JP, Bell J. 1943. A pedigree of mental defect showing sex linkage. *J. Neurol. Neurosurg. Psychol.* 6:154–57

McConkie-Rosell A, Lachiewicz AM, Spiridigliozzi GA, Tarleton J, Schoenwald S, et al. 1993. Evidence that methylation of the FMR-1 locus is responsible for variable phenotype expression of the fragile X syndrome. *Am. J. Hum. Genet.* 53:800–9

McKusick VA, Francomano CA, Antonarakis SE, eds. 1992. *Mendelian Inheritance in Man: Catalogs of Autosomal Dominant, Autosomal Recessive, and X-linked Phenotypes.* Baltimore: John Hopkins Univ. Press. 2028 pp. 10th ed.

Morton NE, Macpherso, JN. 1992. Population genetics of the fragile X syndrome: multiallelic model for the FMR1 locus. *Proc. Natl. Acad. Sci. USA* 89:4215–17

Nagafuchi S, Yanagisawa H, Sato K, Shirayama T, Ohsaki E, et al. 1994. Dentatorubral pallidoluysian atrophy expansion of an unstable CAG trinucleotide on chromosome 12p. *Nature Genet.* 1:14–18

Neri G, Chiurazzi P, Arena F, Lubs HA, Glass IA. 1992. XLMR genes: update 1992. *Am. J. Med. Genet.* 43:373–82

Oberlé I, Heilig R, Moisan JP, Kloepfer C. 1986. Fragile-X mental retardation syndrome with two flanking polymorphic DNA markers. *Proc. Natl. Acad. Sci. USA* 83:1016–20

Oberlé I, Rousseau F, Heitz D, Kretz C, Devys D, et al. 1991. Instability of a 550-base pair DNA segment and abnormal methylation in fragile X syndrome. *Science* 252:1097–102

Opitz JM. 1986. On the gates of hell and a most unusual gene. *Am. J. Med. Genet.* 23:1–10 (erratum 1987. 26:37)

Opitz JM, Sutherland GR. 1984. International workshop on the fragile X and X-linked mental retardation. *Am. J. Med. Genet.* 17:5–94

Opitz JM, Westphal J, Daniel A. 1984. Discovery of a connective tissue dysplasia in the Martin-Bell syndrome. *Am. J. Med. Genet.* 17:101–9

Orr HT, Chung M, Banfi S, Kwiatkowski TJ, Servadio A, et al. 1993. Expansion of an unstable trinucleotide (CAG) repeat in spinocerebellar ataxia type 1. *Nature Genet.* 4: 221–26

Oudet C, Mornet E, Serre JL, Thomas F, Lentes-Zengerling SL, et al. 1993. Linkage disequilibrium between the fragile X mutation and two closely linked CA repeats suggests that fragile X chromosomes are derived from a small number of founder chromosomes. *Am. J. Hum. Genet.* 52:297–304

Pai GS, Sprenkle JA, Do TT, Mareni CE, Migeon BR. 1980. Localization of loci for hypoxanthine phosphoribosyltransferase and glucose-6-phosphate dehydrogenase and biochemical evidence of nonrandom X chromosome expression from studies of a human X-autosome translocation. *Proc. Natl. Acad. Sci. USA* 77(5):2810–13

Pembrey ME, Winter RM, Davies KE. 1985. A premutation that generates a defect at crossing over explains the inheritance pattern of fragile X mental retardation. *Am. J. Med. Genet.* 21:709–17

Penrose LS. 1938. *A Clinical and Genetic Study of 1,280 Cases of Mental Defect.* London: MRC

Pergolizzi RG, Erster SH, Goonewardena P, Brown WT. 1992. Detection of full fragile X mutations by polymerase chain reaction. *Lancet* 339:271–72

Pieretti M, Zhang F, Fu YH, Warren ST, Oostra BA, et al. 1991. Absence of expression of the *FMR-1* gene in fragile X syndrome. *Cell* 66: 817–22

Popper CW. 1988. Disorders usually first evident in infancy, childhood, or adolescence. In *Textbook of Psychiatry,* ed. JA Talbot, RE Hales, SC Yudofsky, pp. 649–735. Washington, D.C: Am. Psychiatr.

Reiss AL, Feinstein C, Toomey KE, Goldsmith B, Rosenbaum K, Caruso MA. 1986. Psychiatric disability associated with the fragile X syndrome. *Am. J. Med. Genet.* 23:393–401

Reiss AL, Freund L. 1992. Behavioral phenotype of fragile X syndrome. *Am. J. Med. Genet.* 43:35–46

Reiss AL, Hagerman RJ, Vingradov S, Abrams M, King RJ. 1988. Psychiatric disability in carrier females of the fragile X syndrome. *Arch. Gen. Psychiatr.* 45:25–30

Reyniers E, Vits L, De Boulle K, Van Roy B, Van Velzen D, et al. 1993. The full mutation in the FMR-1 gene of male fragile X patients is absent in their sperm. *Nature Genet.* 4: 143–46

Richards RI, Holman K, Friend K, Kremer E, Hillen D, et al. 1992. Evidence of founder chromosomes in fragile X syndrome. *Nature Genet.* 1:257–60

Richards RI, Sutherland GR. 1992. Heritable unstable DNA sequences. *Nature Genet.* 1: 7–9

Riggins GJ, Sherman SL, Oostra BA, Sutcliffe JS, Feitell D, et al. 1992. Characterization of a highly polymorphic dinucleotide repeat 150 kb proximal to the fragile X site. *Am. J. Med. Genet.* 43:237–43

Rousseau F, Heitz D, Biancalana V, Blumenfeld S, Kretz C, et al. 1991a. Direct diagnosis by DNA analysis of the fragile X syndrome of mental retardation. *N. Engl. J. Med.* 325:1673–81

Rousseau F, Vincent A, Rivella S, Heitz D, Triboli C, et al. 1991b. Four chromosomal breakpoints and four new probes mark out a 10-cM region encompassing the fragile-X locus (FRAXA). *Am. J. Hum. Genet.* 48: 108–16

Sabourin LA, Mahadevan MS, Narang M, Lee

DSC, Surh LC, Korneluk RG. 1993. Effect of the myotonic dystrophy (DM) mutation on mRNA levels of the DM gene. *Nature Genet.* 4(3):233–38

Schwartz CE. 1993. Invited editorial: X-linked mental retardation: in pursuit of a gene map. *Am. J. Hum. Genet.* 52:1025–31

Sherman SL, Jacobs PA, Morton NE, Froster-Iskenius U, Howard-Peebles PN, et al. 1985. Further segregation analysis of the fragile X syndrome with special reference to transmitting males. *Hum. Genet.* 69:289–99

Sherman SL, Morton NE, Jacobs PA, Turner G. 1984. The marker (X) syndrome: a cytogenetic and genetic analysis. *Ann. Hum. Genet.* 48:21–37

Siomi H, Matunis MJ, Michael WM, Dreyfuss G. 1993a. The pre-mRNA binding K protein contains a novel evolutionarily conserved motif. *Nucl. Acids Res.* 21(5):1193–98

Siomi H, Siomi MC, Nussbaum RL, Dreyfuss G. 1993b. The protein product of the fragile X gene, FMR1, has characteristics of an RNA binding protein. *Cell* 74:291–98

Smith SD, Kimberling WJ, Pennington BF, Lubs HA. 1983. Specific reading disability-identification of an inherited form through linkage analysis. *Science* 219:1345–47

Smits APT, Dreesen JCFM, Post JG, Smeets DFCM, de Die-Smulders C, et al. 1993. The fragile X syndrome: no evidence for any recent mutations. *J. Med. Genet.* 30:94–96

Snow K, Doud LK, Hagerman R, Pergolizzi RG, Erster SH, Thibodeau SN. 1993. Analysis of a CGG sequence at the FMR-1 locus in fragile X families and in the general population. *Am. J. Hum. Genet.* 53:1217–28

Sutcliffe JS, Nelson DL, Zhang F, Pieretti M, Caskey CT, et al. 1992. DNA methylation represses FMR-1 transcription in fragile X syndrome. *Hum. Mol. Genet.* 1:397–400

Sutherland GR. 1977. Fragile sites on human chromosomes: demonstration of their dependence on the type of tissue culture medium. *Science* 197:265–66

Sutherland GR, Gedeon A, Kornman L, Donnelly A, Byard RW, et al. 1991. Prenatal diagnosis of fragile X syndrome by direct detection of the unstable DNA sequence. *N. Engl. J. Med.* 325:1720–22

Sutherland GR, Hecht F, Mulley JC, Glover TW, Hecht BK. 1985. *Fragile Sites on Human Chromosomes.* New York: Oxford Univ. Press. 280 pp.

Suthers GK, Hyland VJ, Callen DF, Oberlé I, Rocchi M, et al. 1990. Physical mapping of new DNA probes near the fragile X mutation (FRAXA) by using a panel of cell lines. *Am. J. Hum. Genet.* 47:187–95

Suthers GK, Mulley JC, Voelckel MA, Dahl N, Vaisanen ML, et al. 1991. Linkage heterogeneity near the fragile X locus in normal and fragile X families. *Genomics* 10:576–82

Tarleton J, Richie RX, Schwartz C, Rao K, Aylsworth AS, Lachiewicz A. 1993. An extensive de novo deletion removing FMR1 in a patient with mental retardation and the fragile X syndrome phenotype. *Hum. Mol. Genet.* 2(11):1973–74

Taylor A, Safanda JF, Fall MZ, Quince C, Lang KA, et al. 1994. Molecular predictors of cognitive involvement in female carriers of fragile X syndrome. *JAMA* 271(7):507–14

Verheij C, Bakker CE, deGraff E, Keulemans J, Willemson R, et al. 1993. Characterization and localization of the FMR-1 gene product associated with fragile X syndrome. *Nature* 363:722–24

Verkerk AJMH, de Graff E, De Boulle K, Eichler EE, Konecki DS, et al. 1993. Alternative splicing in the fragile X gene FMR1. *Hum. Mol. Genet.* 2(4):399–404

Verkerk AJMH, Pieretti M, Sutcliffe JS, Fu YH, Kuhl DPA, et al. 1991. Identification of a gene (FMR-1) containing a CGG repeat coincident with a breakpoint cluster region exhibiting length variation in fragile X syndrome. *Cell* 65:905–14

Vincent A, Heitz D, Petit C, Kretz C, Oberl I, Mandel JL. 1991. Abnormal pattern detected in fragile-X patients by pulsed-field gel electrophoresis. *Nature* 349:624–26

Warren ST, Davidson RL. 1984. Expression of fragile X chromosome in human-rodent somatic cell hybrids. *Somat. Cell Mol. Genet.* 10:409–13

Warren ST, Knight SJL, Peters JF, Stayton CL, Consalez GG, Zhang F. 1990. Isolation of the human chromosomal band Xq28 within somatic cell hybrids by fragile X site cleavage. *Proc. Natl. Acad. Sci. USA* 87:3856–60

Warren ST, Nelson DL. 1993. Trinucleotide repeat expansions in neurologic disease. *Curr. Opin. Neurobiol.* 3:752–59

Warren ST, Nelson DL. 1994. Advances in molecular analysis of fragile X syndrome. *JAMA* 271:536–42

Warren ST, Zhang F, Licameli GR, Peters JF. 1987. The fragile X site in somatic cell hybrids: an approach for molecular cloning of fragile sites. *Science* 237:420–23

Webb TP, Bundey SE, Thake AI, Todd J. 1986. Population incidence and segregation ratios in Martin-Bell syndrome. *Am. J. Med. Genet.* 23:573–80

Weber JL. 1990. Informativeness of human (dC-dA)n • (dG-dT) polymorphisms. *Genomics* 7:524–30

Willems PJ, Van Roy B, De Boulle K, Vits L, Reyniers E, et al. 1992. Segregation of the fragile X mutation from an affected male to his normal daughter. *Hum. Mol. Genet.* 1(7):511–17

Wöhrle D, Hennig I, Vogel W, Steinbach P. 1993. Mitotic stability of fragile X mutations in differentiated cells indicates early postconceptional trinucleotide repeat expansion. *Nature Genet.* 4:140–42

Wöhrle D, Kotzot D, Hirst MC, Manca A, Korn B, et al. 1992. A microdeletion of less than 250 kb, including the proximal part of the FMR-1 gene and the fragile-X site, in a male with the clinical phenotype of fragile-X syndrome. *Am. J. Hum. Genet.* 51:299–306

Yu S, Mulley J, Loesch D, Turner G, Donnelly A, et al. 1992. Fragile-X syndrome: unique genetics of the heritable unstable element. *Am. J. Hum. Genet.* 50:968–80

Yu S, Pritchard M, Kremer E, Lynch M, Nancarrow J, et al. 1991. Fragile X genotype characterized by an unstable region of DNA. *Science* 252:1179–81

Annu. Rev. Neurosci. 1995. 18:101–28

THE NEUROBIOLOGY OF INFANTILE AUTISM

Andrea L. Ciaranello

Harvard University, Cambridge, Massachusetts 02138

Roland D. Ciaranello

Nancy Pritzker Laboratory of Developmental and Molecular Neurobiology, Department of Psychiatry, Stanford University School of Medicine, Stanford, California 94305-5485

KEY WORDS: neurodevelopment, genetic

FEATURES OF INFANTILE AUTISM

Clinical Characteristics

Early infantile autism constitutes the most devastating of a group of neurodevelopmental syndromes called the pervasive developmental disorders. Autism is not a single disease entity but a syndrome with diverse causes. The clinical features of autism vary greatly but, by definition, include deficits in social relatedness, communication, and interests or routines. Onset of autistic signs and behaviors typically occurs in infancy, and the syndrome is usually fully present by the fourth year. The presence of mental retardation may affect the clinical picture greatly. Severely autistic children may be retarded, mute, and preoccupied with repetitive activities; they often exhibit motor stereotypies such as rocking or hand-flapping. They are profoundly withdrawn and may show extreme aversion to social or physical contact. More mildly affected children may have normal or even superior intelligence, with well-developed language skills. Their deficits in social relatedness and preoccupation with rituals and routines may set them apart as very odd but not necessarily autistic. Although there are no localizing neurologic signs in autism, mild, or soft, neurologic signs are common, and grand mal seizures are frequently present

101

after puberty (reviewed in Gillberg & Coleman 1992, Lotspeich & Ciaranello 1993).

Diagnosis

Autism was first described by Kanner (1943). The early psychiatric classification schemes mingled autism and a childhood form of schizophrenia; both were so vaguely defined as to be useless for research purposes. It was not until the 1970s that formal diagnostic criteria for autism were developed (Ritvo & Freeman 1978, Rutter & Hersov 1977). Autism was included in the *Diagnostic and Statistical Manual* (DSM) of the American Psychiatric Association for the first time in 1980 and is now a well-recognized clinical entity. With the refinement in diagnostic methodology has come a greater awareness of the remarkable phenotypic diversity of the disorder. Currently the Autism Diagnostic Interview (ADI) (LeCouteur et al 1989) and the Autism Diagnostic Observation Scale (ADOS) (Lord et al 1989), designed to be used concurrently, are widely accepted as the diagnostic gold standard for autism. The ADI is an investigator-based structured interview that captures symptoms of autism in four areas: social interaction and behavior, language and communication, interests and daily routines, and age of onset. The interview is administered to the parent and takes 2–3 hours to complete. The ADOS is a structured behavior-observation instrument administered to the child; it takes 20 to 45 min. After completing tasks designed to elicit social, language, and communication behaviors, the examiner rates the child on several rating scales. In contrast to autism, the diagnostic criteria for the other pervasive developmental disorders remain somewhat more ill defined, and their distinction as separate entities, rather than forms of autism, remains controversial (Rutter & Schopler 1992, Volkmar & Cohen 1991).

Autism occurs in 1 out of 2000 live births; boys outnumber girls about 4:1. However, the development of new diagnostic instruments and a growing realization of the phenotypic diversity of autism may alter these estimates. Two major questions have emerged: Is there a mild form of autism, and what is the relation of autism to other pervasive developmental disorders?

In recent years, numerous investigators have described social, cognitive, and psychiatric deficits in the nonautistic relatives of autistic probands that suggest the syndrome described by Kanner (1943) may be the core presentation of a broader phenotype (Bolton & Rutter 1990; Folstein & Rutter 1987b, 1988; Freeman et al 1989). Asperger (1944) defined a syndrome in which mental retardation was not prominent but which otherwise resembled a milder, or high-functioning, form of autism. Taken together, these studies suggest that a broader autism phenotype with a greater prevalence than initially recognized exists (Gillberg et al 1991a).

Relation of Autism to Other Disorders

PERVASIVE DEVELOPMENTAL DISORDERS   In the current diagnostic nomenclature (American Psychiatric Association 1987), autism is one of the two pervasive developmental disorders; the other is pervasive developmental disorder not otherwise specified (PDD-NOS). Autism and PDD-NOS share a childhood onset and a common array of clinical features. An earlier version of this nomenclature (American Psychiatric Association 1980) described five disturbances that made up the PDD spectrum: autism, residual-state autism, child-onset PDD, residual state PDD, and atypical autism. These disorders share considerable clinical similarity with autism, usually varying in age of onset or clinical severity.

ASPERGER'S SYNDROME   Within a year after Kanner's first paper describing autism (Kanner 1943), Asperger (1944) published a report describing a group of children with milder autistic behaviors. The concept of high-functioning autism, or Asperger's syndrome, languished until Wing (1981) revived it. The major differences between autism and Asperger's syndrome are in severity: Both groups of children suffer from social interaction deficits, impaired communication skills, and unusual or bizarre behaviors, but children with Asperger's syndrome are less impaired and ususally do not show signs of mental retardation (Frith 1992). Szatmari and colleagues have proposed formal diagnostic criteria for Asperger's syndrome (Szatmari et al 1989a,b,c). Although there is general acceptance that autism, Asperger's syndrome, and high-functioning autism are fundamentally the same disorder (Szatmari 1991; Szatmari et al 1989b, 1990), this view is not universally held. Ozonoff et al (1991) argue that high-functioning autism and Asperger's syndrome are separate disorders and are empirically distinguishable without diagnostic criteria because "theory of mind" measures in neuropsychological profiles are impaired only in high-functioning autistic patients and not in subjects with Asperger's. This assertion has been challenged by Klin et al (1992).

Schizoid personality disorder, which is one of the Axis II diagnoses in the adult psychiatric diagnostic nomenclature, is now being closely evaluated as a form of autism in adults. Studies by Wolff and colleagues (Wolff & Barlow 1979) and Cull et al (1984) noted the similarity in clinical features between children diagnosed with schizoid personality disorder followed into adulthood and those with Asperger's syndrome followed into adulthood; both research groups concluded the two disorders were identical. In studies comparing normal children, children with schizoid personality disorder, and children with Asperger's syndrome, Wolff & Chick (1980) observed considerable overlap between the schizoid and autistic groups. The schizoid group was more distractable, less perseverative, and less likely to use emotional terms in describing people.

The concept of milder forms of autism has been extended in family studies to include the siblings and parents of autistic children. Wolff et al (1988) observed that 16 out of 35 parents of autistic children met criteria for schizoid personality disorder, whereas none of 39 parents of normal children met these criteria. DeLong & Dwyer (1988) studied a total of 929 relatives of 44 autistic probands and found 17 with Asperger's syndrome. Piven et al (1990) studied 67 siblings of 37 autistic probands and found 3% with autism, 4.4% with severe social dysfunction and isolation, and 15% with cognitive disorders. All groups reported an increased incidence of affective disorders—bipolar illness and recurrent depression—in the relatives of autistic probands.

The importance of these studies lies in the hypothesis they generate: that classical or typical autism is the most severe expression of a broader phenotype that includes Asperger's syndrome, schizoid personality disorder, and perhaps even cognitive deficits occurring in conjuction with deficits in reciprocal social interaction. This hypothesis, if verified, becomes an extremely valuable tool in the exploration of the genetics of autism.

MENTAL RETARDATION SYNDROMES   A number of investigators have noted the direct relationship between the clinical severity of autism and IQ (reviewed in Lotspeich & Ciaranello 1993). Most autistic children score in the retarded range (IQ < 70), and many fall into severely or profoundly retarded categories (IQ < 30). Many retarded children exhibit some autistic-like behaviors such as rocking, hand-flapping, or head-banging; these children are usually differentiated from autistic children with similar IQs on the basis of deficits in social relatedness, which are more marked in autistic children. The most common cause of mental retardation, Down's Syndrome, is rarely associated with autism, while the second most common cause, fragile X syndrome, is associated infrequently. Nonetheless, considerable ambiguity and diagnostic uncertainty remain over whether a child with very low IQ (<30) and features of autism suffers from autism or a primary mental retardation syndrome. Given that all of the defining signs and behaviors of autism persist even in children with superior intelligence, it may be reasonable to hypothesize that the primary neurologic insult causing autism does so with or without mental retardation, and the presence of the latter is neither a constant nor necessary accompaniment to the disorder.

Autism and fragile X syndrome   Fragile X syndrome [Fra(X)] is the form of mental retardation most frequently associated with infantile autism. Fragile X syndrome is an X-linked inherited disorder. Affected children have a thin and elongated face, prominent mandible, large ears, dysmorphic facies, enlarged testicles, and behavioral disorders, including hyperactivity, attention deficits, and autistic behaviors. Fragile X can be identified cytogenetically by a constriction on the long arm of the X chromosome at Xq27.3, which is visible only in metaphase chromosomes of lymphyocytes grown in folate-deficient media (Lubs 1968, Sutherland 1977). Yu et al (1991) and Verkerk et al (1991) have isolated a gene responsible for Fra(X), which they have called *FMR1* (fragile X mental retardation 1). The gene contains an unstable (CGG)n repeat located in the 5'-untranslated region of the transcription unit. The presence of an amplified (CGG)n repeat results in abnormal DNA methylation and transcriptional suppression of *FMR1* (Richards et al 1993). Normal individuals have 5–50 such repeats, while affected persons have 200 or more (Snow et al 1993). Females with 50–200 repeats carry a premutation that is unstable and may amplify during meiosis, thus transmitting the full mutation to their offspring. The full mutation is also unstable and amplifies, so the disease may become increasingly severe with progressive generations. The FMR1 gene product is an RNA binding protein (Ashley et al 1993, Siomi et al 1993) that is expressed in the brain, testis, and numerous other organs (Abitol et al 1993).

Fragile X syndrome affects 1 in 1250 males and 1 in 2500 females (Tsongalis & Silverman 1993), and it exhibits marked phenotypic and genotypic diversity. It is possible to carry a mutant form of *FMR1* without showing the clinical symptoms or chromosomal marker of the syndrome, but Hagerman et al (1992) argue that 35% of heterozygous carriers of the gene demonstrate some level of cognitive impairment.

The degree of association of fragile X with autism has been widely estimated, at values ranging from 0 to 53% (for reviews see Bolton & Rutter 1990, Payton et al 1989). Smalley et al (1988) pooled data from 12 studies to produce an estimate of 8%. Hagerman et al (1986) found 16% of 50 fragile X males met the DSMIII criteria for autism, but as many as 98% met at least one criterion. Payton et al (1989) studied 85 autistic males and found a frequency of fragile X syndrome of 2.4%. Piven et al (1991) used the ADI to identify autistic subjects; their estimated association of 2.7% (in 75 autistic subjects) has become widely accepted. Ascertainment bias and varying laboratory methods, however, become significant in the comparison of the studies that generate these estimates (Piven et al 1991).

There is considerable controversy about whether the occurrence of autism with fragile X represents a phenotypic variant of fragile X without a change in the fundamental molecular defect, a variant of fragile X caused by a distinct molecular event in *FMR1*, or the comorbidity of two unrelated diseases. One way to address this is to determine the frequency with which both disorders associate with mental retardation. Fisch (1992) reviewed 21 studies of autistic males and 19 studies of mentally retarded males. Of 5601 mentally retarded males, 5.5% expressed fragile X syndrome; of 1006 autistic males, 5.4% expressed fragile X. These figures produced a proportionate excess risk of autism associated with exposure to fragile X of 0.0. Using 45 pairs of subjects matched for age, sex, and IQ, Einfeld et al (1989) studied whether individuals with fragile X are more autistic than mentally retarded controls and found no significant differences. Both Fisch (1992) and Einfeld et al (1989) suggest that clinicians may have been misled into associating autism and fragile X by the appearance of mental retardation in both disorders and by the hand-flapping and gaze avoidance, which are more common in fragile X subjects than in mentally retarded controls. Hallmayer et al (1994) studied 81 children from multiplex autistic families (2 or more affected members) and found none with expanded (CGG)n repeats, supporting the hypothesis that autism is not an occult form of Fra(X). In the same study they tested for linkage of autism to *FMR1* and excluded linkage with a high degree of certainty (lod score < -30). In an ongoing study, our group (JF Hallmayer, E Pintado, L Lotspeich, D Spiker, et al, unpublished data) has examined fragile X–positive autistic children from multiplex families and found only the expanded (CGG)n repeat characteristic of fragile X. Taken together, all these studies favor the hypothesis that the autism that occurs with fragile X represents a form of the fragile X syndrome, although the debate remains lively (Cohen et al 1991). The recent discovery of two additional fragile sites in the Xq27–q28 region complicates the picture; these sites have been named FRAXE (Knight et al 1993) and FRAXF (Hirst et al 1993). Mentally retarded FRAXE-positive individuals have more than 200 copies of a CGG repeat located in an uncharacterized gene at Xq28 (Knight et al 1993). With the discovery of these additional sites, the fragile site around *FMR1* has been designated FRAXA. The relationship, if any, of these new fragile sites to autism has not been tested.

Phenylketonuria   Phenylketonuria (PKU) is one of a handful of metabolic diseases that appear infrequently with autism. PKU is an autosomal recessive disorder occurring in 1 in 14,000 births; 1 in 60 people carry the mutant gene. The disorder is a defect in phenylalanine hydroxylase, the liver enzyme that converts phenylalanine to tyrosine. An affected fetus develops normally because the maternal enzyme is functional, but after birth, toxic levels of phenylalanine accumulate and disrupt normal brain development. In untreated cases, mental retardation develops within one year after birth, often accompanied by light hair and skin, eczematoid dermatitis, seizures, and self-injurious and autistic behaviors. A low-phenylalanine diet during the first eight years of life permits normal intellectual development. Phenylketonuric females return to this diet during pregnancy to prevent the occurrence of a particularly severe form of PKU in their offspring (Mabry et al 1963).

The association of PKU with autism was first reported by Friedman (1969) in a review of studies presenting 50 children with PKU and autism (as well as mental retardation). The association of autism and PKU has been confirmed by Knobloch & Pasamanick (1975) and by Lowe et al (1980), who reported 3 PKU patients among a group of 65 autistic subjects. This association has been useful in studying the relationship of autism to mental retardation. Friedman (1969) examined children who were treated for PKU with a low-phenylalinine diet late in childhood, when the treatment was first developed. These children had already exhibited mental retardation and autism; once they were placed on phenylalanine-restricted diets, their mental retardation remained, but their autistic behaviors disappeared. When they went off the diets, their autistic behaviors returned.

Tuberous sclerosis   The neurocutaneous disease tuberous sclerosis is another genetic disease with a reported association to autism. Tuberous sclerosis is a rare autosomal dominant disorder, occurring in 1 in 30,000 births (Berberich & Hall 1979, Bundy & Evans 1969), that leads to mental deficiency, seizures, skin lesions, and intracranial lesions. Genetic heterogeneity is indicated by linkage of tuberous sclerosis to DNA markers at 9q34 and at 11q22 (Sampson et al 1989, Smith et al 1990).

Most studies of the association of tuberous sclerosis and autism have reported only one or two cases (reviewed in Lotspeich & Ciaranello 1993), but several large-scale systematic studies have been performed. Smalley estimates that tuberous sclerosis and fragile X together account for 8 to 11% of all cases of autism (Smalley 1991, Smalley et al 1991). Hunt & Dennis (1987) found autism in 40 of 90 subjects with tuberous sclerosis, and Curatolo et al (1991) found autism in 6 of 34 tuberous sclerosis subjects; both studies used the same diagnostic instruments. Smalley et al (1991) found 5 autistic subjects in a group of 24 with tuberous sclerosis. Later, Smalley et al (1992) reviewed several studies and obtained pooled estimates that 17 to 58% of tuberous sclerosis subjects showed autism, while 0.4 to 3% of autistic probands expressed tuberous sclerosis. Hunt & Shepard (1993) estimated that between 25 and 50% of children with tuberous sclerosis were autistic. Smalley et al (1991) performed family studies on 13 tuberous sclerosis and 14 autistic probands and found seven cases of coincidence of the two disorders. These seven subjects received scores on the ADI similar to autistic subjects without tuberous sclerosis, although they displayed fewer repetitive rituals. This study found a significantly higher frequency of seizures and mental retardation among autistic than among nonautistic tuberous sclerosis subjects, and more male tuberous sclerosis subjects with autism than female, despite an equal sex ratio among the tuberous sclerosis probands.

Rett syndrome   Rett syndrome is a neurodegenerative disease that is also associated with autism. It affects only females between the ages of six and eighteen months and involves the loss of purposeful hand use; development of stereotypic hand movements such as wringing; motor and cognitive impairment; and, in some cases, spasticity and loss of ambulation, mental retardation, autism, hyperventilation, or seizures (Hagberg 1985, Hagberg et al 1983, Trevathan & Naidu 1988). Rett syndrome affects between 1 in 10,000 and 1 in 15,000 girls (Hagberg 1985, Hagberg et al 1983, Kerr & Stephenson 1985) and is considered a sex-linked dominant trait, lethal in males, with sporadic new mutations (Zoghbi 1988, Zoghbi et al 1985). Rett syndrome is reportedly associated with autism, and the two disorders display strong overlap of symptoms. Gillberg et al (1985b) and Gillberg (1989) found a break in the X chromosome at Xq22 in 6 of 15 subjects with Rett syndrome; they found the same break in 4 of 46 autistic subjects. These data are inconclusive because the frequency of the break at Xq22 in the general population is not known.

CAUSES OF AUTISM

To understand the etiology of autism, it is convenient to divide the syndrome into nongenetic and genetic causes. Clinically, there is no distinction between these classifications; from all outward appearances, they are indistinguishable. The nongenetic causes are all associated with disruption, usually prenatal, to the normal pattern of brain development. Given this and the overlapping clinical picture, it seems reasonable to conclude that genetic forms of autism arise from mutations in genes controlling brain development and that both genetic and nongenetic etiologies cause damage to the same brain centers and regions (Ciaranello et al 1982, Ciaranello & Wong 1987).

Nongenetic Causes of Autism

The principal nongenetic cause of autism is prenatal viral infection. Desmond et al (1967) and Chess (1977) reported that 8–13% of children born during the 1964 rubella pandemic developed autism along with the other birth defects associated with congenital rubella syndrome. Infection with the rubella virus causes congenital malformations, skeletal deformities, cardiac dysgenesis, and blindness. CNS defects vary from mild cognitive disabilities to severe mental retardation, and are associated with widespread focal neuronal death, gliosis, and vasculitis (Desmond et al 1967).

   Although other infectious agents have also been associated with autism, these are mostly single cases (reviewed in Lotspeich & Ciaranello 1993). Taken together, they constitute additional evidence that prenatal infection can disrupt brain development in such a way that autism ensues. Prenatal toxoplasmosis, syphilis (Rutter & Bartak 1971), varicella (Knobloch & Pasamanick 1975), and rubeola (Deykin & MacMahon 1979) have been linked to single cases of autism. There are six reported cases of prenatal cytomegalovirus infection and autism (Ivarsson et al 1990, Markowitz 1983, Stubbs 1978, Stubbs et al 1984). Interestingly, eight cases of postnatal infection with mumps virus leading to autism have been reported (Deykin & MacMahon 1979), as well as a single case of herpes simplex infection (DeLong et al 1981).

Although a popular notion exists that autism is associated with prenatal, perinatal, or neonatal trauma, there is in fact relatively little evidence to support this view. Nelson (1991) has examined the literature on autism and a variety of birth complications and reports no consistent or specific link between maternal history, pregnancy, delivery, or neonatal events and autism. Mason-Brothers et al (1990) compared a survey of the records of 233 autistic subjects in Utah with those of 62 nonautistic siblings, four previous surveys, and normative data. Of 36 potentially pathologic prenatal, perinatal, or postnatal factors, no factor or group of factors appeared more frequently among the autistic group. Lord et al (1991) studied 23 males and 23 females with high-functioning autism and similar age and IQ against a control group of 54 of their normally developing siblings. She found a higher frequency of pregnancy and delivery complications in the autistic subjects; also, in the autistic group, the frequency of first-born and fourth-or-later-born children was higher. Of several risk factors examined, gestation length of more than 42 weeks was the only factor occurring more frequently in the autistic group. Lord found slight support for the role of pre- and perinatal factors in the development of autism, but these factors seemed to play a larger role in cases of autism associated with severe mental retardation than in cases of high-functioning autism.

Other nongenetic factors leading to autism may include hypothyroidism and maternal cocaine or alcohol use during pregnancy. Gillberg et al (1992) report five children with autism or autistic-like conditions. Three of the five had congenital hypothyroidism; two had mothers who were likely to have been hypothyroid during pregnancy. Davis et al (1992) studied 70 cocaine- or polydrug- (including cocaine) using mothers in an inner-city hospital: 11.4% gave birth to autistic children, and 94% of the children born showed language delay. Nanson (1992) reported six children with fetal alcohol syndrome, histories of maternal alcohol abuse during pregancy, and autism. Comparing these children to eight nonautistic children with fetal alcohol syndrome, Nanson found the autistic subjects all to be moderately or severely retarded.

Maternal antibody formation and rejection by embryonic lymphocytes may also play a role in infantile autism. Warren et al (1990) studied eleven mothers of autistic children aged six or younger; six of the mothers showed antibodies that reacted with lymphocytes of their autistic child. Five of these six mothers had histories of complications during pregnancy. Because antigens expressed on lymphocytes are also expressed on cells of the CNS in the developing embryo, Warren and colleagues (Warren et al 1990) suggest that aberrant maternal immunity may lead to autism. Further exploring this suggestion, they found that 25 autistic subjects showed significantly lower numbers of T (CD2+) cells, numbers of B (CD20+) cells, numbers and percentages of total lymphocytes, and numbers and percentages of helper T (CD4+) cells when compared to siblings and normal children. The differences were greater for autistic girls than boys when compared to normal children of the same sex (Yonk et al 1990). Warren et al (1990) confirm these data, finding lower levels of not only CD4+ and CD2+ cells but also of CD4+CD45RA cells in 36 autistic subjects when compared with controls. No significant differences in the levels of B (CD20+), suppressor T (CD8+), inducers of helper function (CD+ CDw29+), or natural killer (CD56+) cells were observed.

Genetic Causes of Autism

There is considerable evidence that genetic factors play a major role in the pathogenesis of autism. The discussion that follows examines the literature on the genetics of autism, dividing it into epidemiological studies, twin studies, family studies, and linkage and association studies.

EPIDEMIOLOGY STUDIES   Evidence of a genetic etiology for autism has been provided by numerous epidemiologic studies, many of which have estimated the frequency with which siblings of autistic probands are affected with autism. Estimates for sibling frequency have ranged between 2 and 6%, or 50 to 150 times the frequency in the general population (Rutter & Bartak 1971). Other studies have examined the recurrence risk of autism in families, which is calculated by dividing the number of autistic siblings after the proband by the total number of siblings after the proband. Estimates have included 2% (Minton et al 1982, Rutter 1967), 2.8% (August et al 1981), 4.5% (Jorde et al 1991, Ritvo et al 1989), and 5.9% (Baird & August 1985). Jorde et al (1991) report that recurrence risk does not vary with proband IQ. Smalley et al (1988) combined the recurrence risk values from six studies involving 886 relatives of 24 probands in 285 families to obtain a pooled recurrence risk estimate of 2.7%. In a second study of 1698 siblings, Smalley et al (1991) determined a recurrence risk of 3.3%. Ritvo et al (1989) sampled a large autistic population from Utah and estimated the overall recurrence risk to be 8.6%; if the first autistic child was male, the recurrence risk was 7%, and if the first autistic child was female, the recurrence risk was 14.5%. They computed the relative risk, the proportional increase in risk due to having an affected family member, to be 215. Jorde et al (1991) determined that the kinship coefficient, a measure of familial aggregation and an indication of the likelihood of a genetic basis, was 0.001 in autism, 22 times higher than that for a control group. The data produced by all of these studies are likely to be skewed by stoppage rules, the tendency of a family to stop bearing children after the birth of an autistic child. The extent to which stoppage rules are a factor in autism is unknown, so the actual risk of transmitting the disorder genetically may be greater than these studies indicate.

Epidemiologic data have been used to perform segregation analyses in autism, but the results have been inconsistent. Segregation analyses that attempt to specify a mode of inheritance for autism are complicated by the sex-influenced inheritance, reduced penetrance, and variable expression of the disorder itself, as well as by ascertainment biases, diagnostic ambiguities, and stoppage rules. Smalley et al (1988, 1991) have proposed a combination of multifactorial inheritance and genetic heterogeneity. These conclusions have been supported as possible models by Jorde et al (1991), who found evidence of major-gene, polygenic, sibling effect, and mixed (shared sibling effect and major-gene) models in the cases of 209 autistic individuals in 185 families. These same investigators also found that a multifactorial model was suggested when sporadic cases of autism were included. They did not, however, find evidence to support major-locus inheritance in the entire sample or in families in which the autistic proband had an IQ of less than 50. The results of this study contrast with earlier data from this same group that produced a segregation ratio of 0.19 ± 0.07 in an analysis of 46 families, compatible with autosomal recessive inheritance, (Ritvo et al 1985b). Hallmayer et al (JF Hallmayer, J Mountain & LL Cavalli-Sforza, unpublished data) analyzed the data from Jorde et al (1991) and observed evidence for a single major dominant locus when some parents of autistic subjects were treated as affected. The dependence of the segregation analysis on parental affected status points to the importance of resolving the question of a parental phenotype in autism before a mode of inheritance can be established.

TWIN STUDIES   Twin studies also suggest a genetic basis for autism. Folstein & Rutter (1977, 1978) studied pairs of same-sex twins, with diagnosis of the second twin made blind to the affected status of the first. They found concordance in four of eleven (36%) monozygotic (MZ) twins and in zero of ten (0%) dizygotic (DZ) twins. Steffenburg et al (1989) found even greater differences between monozygotic (91%) and dizygotic twins (0%) in a population-based study in Northern Europe. Ritvo et al (1985a) reported 24% dizygotic and 96% monozygotic twin concordance. This study has been criticized, however, for its ascertainment bias (national newsletter recruitment of subjects) and its inclusion of opposite-sex twin pairs (Folstein & Rutter 1987a). Smalley et al (1988) pooled the data from these studies as well as from eleven single case studies to produce adjusted concordance estimates (assuming one proband per twin pair) of 64% for monozygotic twins and 9% for dizygotic twins. These studies all reveal a much greater degree of concordance in monozygotic than in dizygotic twins and, therefore, suggest a substantial genetic component in autism. Adoptive twin studies would allow an even clearer differentiation of genetic and environmental factors. To date, however, adoptive twin studies have not been performed, probably because it has not been possible to identify a large sample of twin pairs, with at least one autistic proband, that have been reared apart.

FAMILY STUDIES   Family studies provide further evidence for a genetic basis for autism. These studies are also important because they could help clarify a mode of inheritance. Two issues are prominent. First, is there a broader autistic phenotype than the tightly defined syndrome in general use that would identify more relatives of autistic probands as affected? Second, does this or a related phenotype also include parents? Folstein & Rutter (1977) reported a significant increase in the concordance rates for MZ (from 36 to 82%) and DZ (from 0 to 10%) twins and in sibling risk rates if cognitive dysfunction was included in the definition of the autism phenotype. They were the first to propose that this broader phenotype may represent a genetic liability for autism. This hypothesis is also suggested by Bolton & Rutter (1990), recent work from Folstein's group (Folstein & Piven 1991; Folstein & Rutter 1987b, 1988), and Smalley (1991). Smalley & Asarnow (1990) and Minton et al (1982) also report an atypical cognitive profile in siblings of autistic probands, in which performance scores are higher than verbal scores and higher than those of controls. The atypical cognitive profiles examined in these studies are reminiscent of the "pockets of intelligence" described by Kanner (1943) in autistic subjects. These pockets of intelligence refer to uneven cognitive abilities, such as rote memory skills, that distinguish autistic patients from mentally retarded subjects with uniformly depressed intelligence. Smalley et al (1988) report that familial clustering of these uneven cognitive abilities is more pronounced than clustering of mental retardation (uniform intelligence deficits) in siblings of autistic probands.

These data support the hypothesis that familial transmission of autism involves a mild form of the disorder. Such a view accounts for some cases in which autistic probands do not have a close family member who fits the diagnostic criteria for autism; if the phenotype can be shown to include a mild form of autism, then the sibling frequencies may be found to approach those predicted by simple genetic models. A high-functioning form of autism also explains how the gene or genes for the disorder can be maintained in the gene pool although autistic individuals rarely reproduce.

Many studies have explored mild social, psychiatric, and cognitive defects in parents and other relatives of autistic subjects. Many of these studies are reviewed by Smalley et al (1988, 1991) and by Folstein & Piven (1991). Folstein's group has focused on awkward social interactions, poor pragmatic language skills (including abnormal prosody of speech and poor nonverbal language), Asperger's syndrome, schizoid personality disorder, and high-functioning autism. They find substantial support for the concept of an expanded autistic phenotype that occurs in the siblings, parents, and extended relatives of autistic probands (Folstein & Piven 1991; Folstein & Rutter 1987b, 1988; Garber et al 1989; Landa et al 1991, 1992). Dykens et al (1991) have observed poverty of speech, poor reality testing, and perceptual distortions in high-functioning autistic adults. Landa et al (1991, 1992) report higher numbers of "abnormal social discourse behaviors" and "skeletal or rambling stories" in parents of autistic subjects who performed narrative discourse (storytelling) tasks, compared to parents of control children. Bowman (1988) report the case of a family with two autistic children in which a third child and the father had Asperger's syndrome. Wolff et al (1988) found schizoid personality disorder in 16 of 35 parents of autistic children and in 0 of 39 parents of children with deafness, mental retardation, or seizure disorder. DeLong & Dwyer (1988) examined 929 first- and second-degree relatives of 44 autistic probands; 17% of the relatives were diagnosed with Asperger's syndrome. Piven et al (1990), in a study of 67 siblings of 37 autistic subjects, found 3% were autistic themselves, 4.4% fit the criteria for Asperger's syndrome, and 15% showed cognitive disorders. Gillberg et al (1992) compared speech, language, and reading problems; social deficits; and psychiatric disorders in siblings and parents of autistic children with siblings and parents of normal children and siblings and parents of children with deficits in attention, motor control, and perception (DAMP) with relatives of normal children. Learning disorders were equally common among relatives of autistic and normal subjects and more common in relatives of subjects with the DAMP tetrad. Asperger's syndrome was more common in relatives of autistic subjects than in the relatives of controls, while mothers of autistic children showed a tendency for schizoaffective disorder. The results of these studies do not conclusively support the existence of mild autism, in part because of the lack of a standardized instrument for the description and quantification of clinical observations in adults. The development of the Pragmatic Rating Scale by Folstein's group (Landa et al 1991) may meet the need for such an instrument.

Support for complex modes of inheritance in autism also comes from cognitive family studies. Uneven cognitive functioning clusters more heavily in families with autistic fragile X probands than in families with nonautistic fragile X probands. Smalley (1988) suggests that autism may occur with fragile X because of the involvement of a gene for affective disorder separate from and loosely linked to the fragile X gene. Mental retardation also affects clustering of cognitive disorders in famililes of autistic probands. Smalley (1991) reports different recurrence risks of neuropsychiatric disorders when autism and mental retardation are both present compared to autism alone. The effects of sex, mental retardation, and fragile X on the familial clustering associated with autism suggest either genetic heterogeneity, in which more than one gene leads to autism, or a multifactorial model of inheritance with varying thresholds for expression, although the data are not sufficient to support one hypothesis over the other.

Greater clustering of cognitive dysfunctions has also been observed in autism multiplex families (August et al 1981, Baird & August 1985). This observation is consistent with the hypothesis that two forms of autism exist: a familial, genetic form and a sporadic form due to single environmental insults or nontransmitted genetic events such as chromosome deletions (Smalley et al 1988). The genetic form of autism may find its etiology in genetic heterogeneity, in multifactorial inheritance, or in the maintenance of the gene in the gene pool through a carrier trait of cognitive impairment. The existence of a familial and a sporadic form of autism is supported by the studies of Links et al (1980), in which autistic subjects with low frequencies of minor physical anomalies, high IQs, and more relatives with psychiatric disorders appear to express the familial form of autism, while autistic probands with more frequent minor physical anomalies, lower IQs, and fewer affected relatives represent the sporadic form.

LINKAGE AND ASSOCIATION STUDIES   In recent years, linkage analysis has become an extremely powerful strategy for identifying genes responsible for diseases. Linkage studies seek to identify cosegregation of a trait or disorder with a known DNA marker. Linkage analysis does not require knowledge of the gene product or of the pathophysiology of the disease under investigation. Indeed, isolation of a gene through linkage studies can often lead to an understanding of the abnormal proteins that cause a genetic disease, as has been the case with Duchenne's muscular dystrophy (Gusella 1989, Hoffman et al 1987, Hoffman & Kunkel 1989) and cystic fibrosis (Kerem et al 1989, Riordan et al 1989, Rommens et al 1989). In order to perform linkage analysis, however, the definition of *affected* for the trait under consideration must be clearly stated. This is especially true for psychiatric diseases in which no biological markers exist and phenotypic expression varies greatly (Price 1993). Although conventional linkage strategies work best with monogenic disorders, the lod scores produced are informative even if the trait actually displays polygenic or oligogenic inheritance (Greenberg & Hodge 1989, Konigsburg et al 1990, Vieland et al 1992). In the absence of a known mode of inheritance, linkage can be performed by using different models of presumed inheritance (Risch et al 1989). To offset some of the statistical limitations of this, Risch has proposed a method of analyzing only affected siblings (Risch 1990a,b, 1991; Risch et al 1989), a strategy that may lend itself particularly well to psychiatric diseases in which large, densely affected pedigrees are rare. Based on these considerations, autism is well suited to linkage analysis because of its high relative risk (100–200). The only published linkage study in autism reported absence of linkage to autism to blood group polymorphisms (Spence et al 1985). Several groups are now assembling multiple-incidence families and conducting linkage analyses using DNA markers (Hallmayer et al 1991, Spiker et al 1994).

Association studies can also be helpful in identifying disease genes. An association is the increased frequency of occurrence of a specific marker with a particular genetic trait or disorder. Several studies testing the association between autism and specific markers have been reported. Spence et al (1985) compared the number of shared HLA haplotypes among autistic siblings and found no evidence for association of autism with HLA haplotype. Warren et al (1990) noted that autism has several autoimmune features and that C4 deficiency had previously been reported in association with autoimmune disorders. Warren et al (1991) then examined the frequency of the null allele at C4A and C4B in 19 autistic subjects, and found that the autistic subjects and their mothers presented a 58% increase in the frequency of the null allele at C4B (compared to a 27% increase in the control group). No significant increase was found in the C4B of the fathers or siblings of the autistic subjects or in the C4A of any participants in the study. A follow-up study (Warren et al 1992) noted that the possibly associated C4B allele lies on the extended or ancestral haplotype [B44-SC30-DR4]. This extended haplotype increased by almost sixfold in 21 autistic children compared to healthy controls. The total number of extended haplotypes expressed on chromosomes of autistic subjects was significantly increased when compared to the chromosomes of healthy controls. This may suggest that a gene related to or included in the major histocompatability complex may be associated with autism.

Herault et al (1993) examined the association between restriction fragment length polymorphisms (RFLPs) in the regions of the genes encoding the biogenic amine biosynthetic enzymes tyrosine hydroxylase, dopamine β-hydroxylase, and tryptophan hydroxylase, and found no association with autism. They reported a significant association, however, between autism and the locus containing the gene for hras-1 on chromosome 11. No linkage results were reported in this study. Comings et al (1991) found that the prevalence of the A1 allele of the TaqI polymorphism in the dopamine D2 receptor gene is significantly increased in autism (54.5% of 33 autistic subjects compared to 24.5% of 314 controls), as well as in Tourette's syndrome, attention deficit hyperactivity disorder (ADHD), and alcoholism. They suggest that the A1 allele is associated with behavioral disorders in which it acts as a modifying gene but not as the primary etiological agent.

CYTOGENETIC ABNORMALITIES AND AUTISM   Cytogenetic abnormalities may be invaluable in identifying a genetic locus for autism. There have been numerous reports of cytogenetic defects in autism, but none are frequent or consistent. Fragile X is the most common cytogenetic abnormality reported in association with autism. Gillberg et al (1991b) described the cases of six boys with autism, moderate to severe mental retardation, and mild to moderate physical stigmata; some of the boys also displayed muscular hyptonia, epilepsy, kyphoscoliosis, short stature, or low weight, and all six boys also possessed a supernumerary (marker) chromosome 15. In the same study, Gillberg et al (1991b) reported a unique and separate syndrome associated with partial trisomy 15. Seshadri et al (1992) described the case of a two-and-one-half-year-old boy with autism and an 18q abnormality in the absence of the phenotypic features of 18q⁻ syndrome, and Lopreiato & Wulfsberg (1992) reported a six-and-one-half-year-old boy with typical infantile autism and a complex chromosome rearrangement of chromosomes 1, 7, and 21. Murayama et al (1991) described three children with de novo terminal deletion of the long arm of chromosome 1 [46, XX, del(1)(q43)]. All three children displayed autistic-like behavior: lack of eye contact and interest in people, little emotion, stereotypic repeated movements, and repeated unusual sounds. A three-year-old boy with microcephaly but an otherwise normal autistic phenotype was described by Blackman et al (1991); he also had partial duplication of the short arm of the Y chromosome. This is the sixth reported case of Y chromosome aneuploidy associated with autism, but the first reported case involving an isodicentric Y chromosome. Other chromosomal abnormalities associated with autism are described in Gillberg & Coleman (1992) and Lotspeich & Ciaranello (1993).

NEUROPATHOLOGY OF AUTISM

The literature on the neuropathology of autism is large and often contradictory. Much of it has been examined in several reviews from our group (Ciaranello et al 1990, Lotspeich 1994, Lotspeich & Ciaranello 1993, Rubenstein et al 1990). We divided the literature by methodologic approach into studies on histopathologic findings, neuroradiologic findings, and neurophysiologic and neurochemical findings. Rather than restate those reviews here, we try to examine this literature from a different perspective.

The basic hypothesis linking all this work, which spans four decades of inquiry, is that autism is a developmental disorder of the central nervous system, and thus there should be some characteristic signature of neuroanatomic or neurochemical deficits. What is now well recognized but was not appreciated much in earlier studies is the enormous clinical heterogeneity that autism presents. As described earlier, autistic children may exhibit intellectual functioning ranging from severely retarded to intellectually superior, they may be mute or have highly developed language skills, and their social handicaps may be mild or severe, as may their stereotypic rituals and behaviors. In light of this, it is not too surprising that the literature seeking to answer the question of what the underlying neurobiologic deficit(s) in autism is/are has produced confusing and often contradictory results. In the sections that follow, we examine the literature seeking to find sites of neuronal dysgenesis in autism.

Cerebellum

Evidence implicating the cerebellum in the pathophysiology of autism was originally put forward by Ornitz and colleagues in the 1960s and 1970s (Ornitz 1985, Ornitz & Ritvo 1968; reviewed in Todd & Ciaranello 1985). Their studies showed that autistic children exhibited clinical and neurophysiologic deficits indicative of cerebellar dysfunction. Some years later, the role of the cerebellum was pursued at the anatomic level in studies by Bauman & Kemper (1985) and by Ritvo et al (1986). Bauman & Kemper (1985) conducted autopsy studies on the brains of six autistic subjects varying in age from 9–29 years. They noted the loss of granule and Purkinje cells from the posterolateral neocerebellum and the archicerebellar cortex. The anterolateral cerebellum and the vermis were largely spared. Coincident with this was the observation that neurons in the deep cerebellar nuclei and in the inferior olivary nuclei were abnormally large, suggesting the retention of an immature neuronal phenotype. This was more pronounced in younger subjects. The hypothesis that emerged from these studies was that early loss of Purkinje cells disrupts the ontogeny of the cerebellar circuitry. Immature neurons perisist, and a nonfunctional fetal pattern of circuitry is retained. With maturation, the fetal neuronal pattern is lost but is not replaced by an adult pattern, so the normal circuitry of the cerebellum does not develop (Bauman & Kemper 1989).

The question of cerebellar deficits in autism has been pursued by other investigators using different tools. The question whether there is a difference in cerebellar size in autism was studied by Rumsey et al (1988), who did not observe cerebellar atrophy in autistic men. Courchesne et al (1988) reported a 25% decrease in vermal lobules VI and VII in 14 autistic subjects, a result confirmed by Murakami et al (1989). However, studies by Garber et al (1989) and Piven et al (1990) failed to find evidence supporting this observation. Holttum et al (1992) have written a recent review of this subject, concluding that there is no evidence to support the notion of cerebellar size abnormalities in autism.

Brainstem

Early neurophysiologic studies (Ornitz 1985, 1989; Ornitz & Ritvo 1968) implicated auditory and vestibular system pathways in autism. Support for this contention, however, has not been forthcoming by any of the anatomic, imaging, or neurophysiological methods now available. Reviews by Minshew

(1991) and Klin (1993) identify several neurophysiological abnormalities in autistic subjects, but these are not specific for autism and are found in non-autistic matched controls.

Limbic and Forebrain Structures

Kemper & Bauman (1993) reported small, densely packed cells in the hippocampus and amygdala of four autistic subjects and suggested this reflected an immature pattern of neuronal development in these structures. They observed immature-looking neurons in the diagonal band of Broca, which projects cholinergic afferents to the hippocampus and amygdala. To date, the only abnormality noted in limbic and forebrain structures has been observed at the microscopic level. Magnetic resonance imaging studies have not found any alterations in volume or area of any of the structures in these regions (Gaffney et al 1989).

Cerebral Cortex

Because of the role of the cerebral cortex in information processing, cognition, language, and learning, it would seem intuitive that the cerebrum would be the focus of investigation in autism. At the microscopic level only a handful of studies have been performed, however, and these have only examined limited cortical regions. Coleman et al (1985) found no differences in cell density in the auditory cortical structures, while Williams et al (1980) observed normal pyramidal cell architecture in the cases they studied. Abnormalities in gross cortical anatomy have been reported by Piven et al (1990), who observed malformations in the gyri of 13 autistic subjects by using magnetic resonance imaging. Seven of the 13 autistic subjects had at least one gyral malformation; matched controls showed none. These malformations arise from defects in neuronal migration during the first six months of gestation.

There have been several magnetic resonance imaging studies of cortical size in autism (Gaffney et al 1987, 1989; Gaffney & Tsai 1987), but none of these has demonstrated significant differences. In early neurophysiological studies, Ornitz & Ritvo (1968) postulated a reversal of the normal cerebral asymmetry (L > R) in autistic subjects. This has since been examined by magnetic resonance imaging. Hashimoto et al (1988) examined several brain structures and found a reversed L-R cortical asymmetry in autistic and retarded subjects (R > L) compared to controls (L > R). Although the autistic subjects showed greater reversed asymmetry, this study would be more informative if the confound of mental retardation had been eliminated.

Cortical metabolism has been studied in autism without conclusive findings. Rumsey et al (1985) showed an increase in cortical glucose metabolism by positron emission tomography in autistic subjects, but there was substantial overlap with the control group. De Volder et al (1987) found no differences in glucose utilization between autistic and control subjects, as did Buchsbaum et al (1992) and Siegel et al (1992). However, in the latter two groups, several *outliers,* regions of increased or diminished metabolic activity, were found which raised speculation that specific subjects might have glucose utilization abnormalities. Minshew et al (1993) have studied phosphorous metabolism in autism using magnetic resonance spectroscopy, and found evidence of diminished adenosine triphosphate (ATP) production and increased phosphodiester production. They postulated that there might be a hypermetabolic state in autism coupled with diminished synthesis or increased destruction of membrane phospholipid constituents.

Neurochemical Studies

Many studies have sought to find neurochemical deficits in autism. The fundamental hypothesis underlying all these studies is that autism is a metabolic disease, like phenylketonuria, arising from a defect in some biochemical pathway. The association of autism with certain metabolic diseases, reviewed earlier, has provided some basis for this view. Nonetheless, the outcome of these many years of investigation has been extremely disappointing, so much so that the fundamental hypothesis must be called into question. With the exception of the occasional association of autism with a known metabolic disorder, there have been no consistent or conclusive findings that could unravel a biochemical basis for autism. Virtually every chemical that can be measured in autism has been, and isolated abnormalities have been reported in many; however, no consistent defect has been observed. As a consequence, there is dawning acceptance that other avenues of inquiry may be more productive.

There have been several reviews of the literature on neurochemical deficits in autism, including three from our own group (Elliott & Ciaranello 1987, Todd & Ciaranello 1985, Wong & Ciaranello 1987). Gillberg & Coleman (1992) have summarized this literature in a review as well. Most biochemical studies of autism have focused on neurotransmitters, metabolites, and hormones. Of these, the neurotransmitter serotonin (5HT) has been the most widely studied. Schain & Freedman (1961) originally reported elevated blood serotonin levels in autistic (~35%) and retarded (~50%) subjects compared to controls. This finding has been extensively repeated, and although its significance remains obscure, it is one of the few enduring biochemical findings in all of psychiatric research (reviewed in Anderson et al 1990; Cook 1990). Piven et al (1991) found that 5HT levels in five autistic subjects with affected relatives (autism or PDD) were significantly higher than the 5HT levels in 23 autistic subjects without affected relatives; these levels were again significantly higher than in 10 controls. These data were still significant after adjustment for sex, age, and IQ. The authors concluded that serotonin level may be associated with a "genetic liability" for autism. Yuwiler & Freedman (1987) extensively studied hyperserotonemia in autism but were unable to account for elevated blood serotonin by any biochemical alteration in neurotransmitter synthesis or degradation. Because cerebrospinal fluid serotonin or serotonin metabolite levels do not correlate with blood levels (Elliott & Ciaranello 1987), the biological significance of hyperserotonemia is unclear. Lauder (1983, 1990) and Lauder & Krebs (1978) have proposed that serotonin may play a trophic role in brain development. It is intriguing to speculate that some early deficit in brain serotonin may disrupt the maturation of CNS neurons, thus altering neuritic differentiation and synaptogenesis. A similar role has been proposed for dopamine in brain ontogeny (Todd 1992). Deficits in norepinephrine or its metabolites (Launay et al 1987, Young et al 1978), dopamine (Cohen et al 1974, 1977; Gillberg & Hamilton-Hellberg 1983), dopamine β-hydroxylase (Lake et al 1977), $5HT_2$ receptor binding (Cook et al 1993), and β-endorphin (Gillberg et al 1985a) have been reported in some studies, but no consistent findings have emerged.

TREATMENT

As is the case for most chronic neurologic disorders, there is no cure for autism. Treatment strategies have focused on alleviating specific behaviors, improving social skills, and developing cognitive and language abilities. In conjuction with these, treatment of more disruptive and disabling behaviors with specific medications, such as haloperidol and pimozide, has been moderately successful (reviewed in Campbell et al 1987, 1990). It was not so many years ago that autistic children were relegated to the back wards of state institutions, where, not surprisingly, they regressed and became progressively more retarded. More targeted intervention strategies, particularly those aimed at developing language and communication skills, have combined with the advent of federal and state legislation mandating educational programs to serve autistic children and their families to create a more favorable environment in which autistic children can learn and function more effectively. A detailed discussion of treatment strategies for autistic children is beyond the scope of this review, but the topic has been treated thoroughly and well in recent books and reviews by Gillberg & Coleman (1992) and Lotspeich (1994).

SUMMARY AND FUTURE DIRECTIONS

Autism is a severe neurodevelopmental disorder characterized by deficits in social interaction and communication, with unusual or bizarre routines and rituals. Onset is typically in infancy, and the full syndrome is almost always present by the fourth year of life. The clinical presentation is highly variable, ranging from severe to mild, and it can be greatly affected by the presence of mental retardation. There is growing evidence that the classical clinical picture of autism represents the core of a much broader phenotype that includes cognitive deficits, language and communication impairments, and deficits in reciprocal social interaction. Schizoid personality disorder may be one manifestation of this expanded autistic phenotype in adults.

Autism has been associated with numerous genetic and metabolic diseases, although the vast majority of cases are ideopathic. Prenatal infection with rubella virus has been an important nongenetic cause of autism, but other prenatal and perinatal complications do not appear causal. Epidemiologic, twin, and family studies all suggest a genetic etiology in autism, particularly in multiplex families. Segregation analyses have not identified a clear mode of inheritance, however, and several models of inheritance ranging from simple to complex have been invoked.

Although there is no consistent neuroanatomic defect in autism, a growing body of evidence implicates neuronal maturation defects, particularly in the cerebellum and limbic structures. These deficits do not appear to be reflected in the size or metabolic activity of these structures, however. Neurophysiologic deficits are widespread but not specific for autism; the same may be said for neurochemical abnormalities.

Where does that leave us? Much of the current work on autism seeks to clarify the site of neurologic deficit, its structural characteristics, and its molecular basis. Neuropsychologic studies aimed at localizing and clarifying functional deficits have contributed important information and should continue to do so as research instruments become more refined. The weight of neuroanatomic and neuropsychologic evidence implicates the cerebellum and the limbic forebrain and, at the cellular level, suggests that examining deficits in neuronal migration, maturation, or synaptic connectivity may be productive areas of research. Carrying out studies in more homogeneous subsets of the autistic population, such as multiplex families, may help reduce some of the background noise created by the phenotypic and etiologic diversity in autism. Genetic studies hold promise for localizing mutant genes, but the problems of phenotypic classification, uncertain mode of inheritance, and an almost certain genetic heterogeneity remain significant challenges. Taking the diverse array of neurobiologic findings in autism together, we might speculate that the underlying genetic defect is in a gene or genes that regulate neuronal maturation in cortical or subcortical information processing centers, leading to disruption of normal synaptogenesis and circuit formation. Testing this hypothesis will require isolation of the responsible gene(s), mapping the encoded transcript(s), and characterizing the cellular localization, temporal course of expression, and function of the protein product(s).

ACKNOWLEDGMENTS

Work from the authors' laboratory was supported by a grant from the National Institute of Mental Health (MH39437) and by the Solomon and Rebecca Baker Fund, the Spunk Fund, Inc., the Scottish Rite Schizophrenia Research Foundation, and the National Alliance for Research in Schizophrenia and Depression (NARSAD). RDC is the recipient of a Research Career Scientist Award from the National Institute of Mental Health (MH 00219).

Literature Cited

Abitol M, Menini C, Delezoide AL, Rhyner T, Bekemans M, et al. 1993. Nucleus basalis magnocellularis and hippocampus are the major sites of FMR-1 expression in the human fetal brain. *Nat. Genet.* 4(2):147–53

Anderson GM, Horne WC, Chtterjee D, Cohen DJ. 1990. The hyperserotonemia of autism. *Ann. NY Acad. Sci.* 600:331–40

Ashley CT, Wilkinson KD, Reines D, Warren ST. 1993. FMR-1 protein: conserved RNP family domains and selective RNA binding. *Science* 262:563–66

Asperger H. 1944. Die "autistischen Psychopathen" im Kindesalter. *Arch. Psychiatr. Nervenkr.* 177:76–137

American Psychiatric Association. 1980. *Diagnostic and Statistical Manual of Mental Disorders. III.* Washington DC: APA. 494 pp.

American Psychiatric Association. 1987. *Diagnostic and Statistical Manual of Mental Disorders III-R.* Washington DC: APA. 567 pp.

August GJ, Steware MA, Tsai L. 1981. The incidence of cognitive disabilities in the siblings of autistic children. *Br. J. Psychiatry* 138:416–22

Baird TD, August GJ. 1985. Familial heterogeneity in infantile autism. *J. Autism Dev. Disord.* 15(3):315–21

Bauman M, Kemper TL. 1985. Histoanatomic observations of the brain in early infantile autism. *Neurology* 35:866–74

Bauman ML, Kemper TL. 1989. Abnormal cerebellar circuitry in autism? *Neurology* 39(Suppl 1):186

Berberich MS, Hall BD. 1979. Penetrability and variability in tuberous sclerosis. *Birth Defects* 15:297–304

Blackman JA, Selzer SC, Patil S, Van Dyke DC. 1991. Autistic disorder associated with an iso-dicentric chromosome. *Dev. Med. Child Neurol.* 33(2):162–66

Bolton P, Rutter M. 1990. Genetic influences in autism. *Int. Rev. Psychiatry* 2:67–80

Bowman EP. 1988. Asperger's syndrome and autism: the case for a connection. *Br. J. Psychiatry* 152:377–82

Buchsbaum MS, Siegel BV, Wu JC, Hazlett E, Sicotte N, et al. 1992. Attention performance in autism and regional brain metabolic rate assessed by positron emission tomography. *J Autism Dev. Disord.* 22(1):115–25

Bundy S, Evans K. 1969. Tuberous sclerosis: a genetic study. *J. Neurol. Neurosurg. Psychiatry* 32:591–603

Campbell M, Anderson LT, Small AM. 1990. Pharmacotherapy in autism: a summary of research at Bellevue/New York University Hospital. *Brain Dysfunct.* 3:299–307

Campbell M, Perry R, Small AR, Green WH. 1987. Overview of drug treatment in autism. In *Neurobiological Issues in Autism,* ed. E Schopler, GB Mesibov, pp. 341–52. New York: Plenum

Chess S. 1977. Report on autism in congenital rubella. *J. Autism Child. Schizophr.* 7:68–81

Ciaranello RD, VandenBerg SR, Anders TF. 1982. Intrinsic and extrinsic determinants of neuronal development: relations to infantile autism. *J. Autism Devel. Disord.* 12(2):115–46

Ciaranello RD, Wong DL, Rubenstein JLR. 1990. Molecular neurobiology and disorders of brain development. In *Application of Basic Neuroscience to Child Psychiatry,* ed. SI Deutsch, A Weizman, R Weizman, pp. 9–31. New York: Plenum

Cohen DJ, Caparulo BK, Shaywitz BJ, Bowers MB. 1977. Dopamine and serotonin metabolism in neuropsychiatrically disturbed children. *Arch. Gen. Psych.* 34:545–50

Cohen DJ, Shaywitz BA, Johnson WT, Bowers M. 1974. Cerebrospinal fluid measures of homovanillic acid and 5-hydroxyindole-acetic acid. *Arch. Gen. Psychiatry* 34: 545–50

Cohen IL, Sudhalter V, Pfadt A, Jenkins EC, Brown WT, et al. 1991. Why are autism and the fragile-X syndrome associated? Conceptual and methodological issues. *Am. J. Hum. Genet.* 48(2):195–202

Coleman PD, Romano J, Laphan L, Simon W. 1985. Cell counts in cerebral cortex of an autistic patient. *J. Autism. Devel. Disord.* 15: 245–55

Comings DE, Comings BG, Muhleman D, Dietz G, Shahbahrami B, et al. 1991. The dopamine D2 receptor locus as a modifying gene in neuropsychiatric disorders. *JAMA* 266(13):1793–800

Cook DH. 1990. Autism: review of neurochemical investigation. *Synapse* 6:292–308

Cook EH Jr, Arora RC, Anderson GM, Berry-Kravis EM, Yan SY, et al. 1993. Platelet serotonin studies in hyperserotonemic relatives of children with autistic disorder. *Life Sci.* 52(25):2005–15

Courchesne E, Yeung-Courchesne R, Press RA, Hesselink JR, Jernigan TL. 1988. Hypoplasia of cerebellar vermal lobules VI and VII in autism. *New Engl. J. Med.* 318(21):1349–53

Cull J, Chick J, Wolff S. 1984. A consensual validation of schizoid personality in childhood and adult life. *Br. J. Psychiatry* 144: 646–48

Curatolo P, Cusmai R, Cortesi F. 1991. Neuropsychiatric aspects of tuberous sclerosis. *Ann. NY Acad. Sci.* 615:8–16

Davis E, Fennoy I, Laraque D, Kanem N, Brown G, et al. 1992. Autism and developmental abnormalities in children with perinatal cocaine exposure. *J. Nat. Med. Assoc.* 84(4):315–19

DeLong GR, Bean SC, Brown FR. 1981. Acquired reversible autistic syndrome in acute encephalopathic illness in children. *Arch. Neurol.* 38:191–94

DeLong GR, Dwyer JT. 1988. Correlation of family history with specific autistic subgroups: asperger's syndrome and bipolar affective disease. *J. Autism Dev. Disord.* 18(4): 593–600

Desmond MM, Wilson GS, Melnick JL, Singer DB, Zion TE, et al. 1967. Congenital rubella encephalitis: course and early sequelae. *J. Pediatr.* 71(3):311–31

De Volder A, Bol A, Michel C, Congneau M, Goffinet AM. 1987. Brain glucose metabolism in children with the autistic syndrome: positron tomography analysis. *Brain Dev.* 9(6):581–87

Deykin EF, MacMahon B. 1979. Viral exposure and autism. *Am. J. Epidemiol.* 109:628–38

Dykens E, Volkmar FR, Glick M. 1991. Thought disorder in high-functioning autistic adults. *J. Autism Dev. Disord.* 21(3):291–301

Einfeld S, Molony H, Hall W. 1989. Autism is not associated with the fragile X syndrome. *Am. J. Med. Genet.* 34(2):187–93

Elliott GR, Ciaranello RD. 1987. Neurochemical hypotheses of childhood psychoses. In *Neurobiological Issues in Autism,* ed. E Schopler, G Mesibov, pp. 245–61. New York: Plenum

Fisch GS. 1992. Is autism associated with the fragile X syndrome? *Am. J. Med. Genet.* 43: 47–55

Folstein S, Rutter M. 1977. Infantile autism: a genetic study of 21 twin pairs. *J. Child Psychol. Psychiatry* 18:297–321

Folstein S, Rutter M. 1978. A twin study of individuals with infantile autism. In *Autism: A Reappraisal of Concepts and Treatment,* ed. M Rutter, E Schopler, pp. 219–42. New York: Plenum

Folstein SE, Piven J. 1991. Etiology of autism: genetic influences. *Pediatrics* 87(5):767–73

Folstein SE, Rutter ML. 1987a. See Elliot & Ciaranello 1987, pp. 83–105

Folstein SE, Rutter ML. 1987b. See Elliot & Ciaranello 1987, p. 83–106

Folstein SE, Rutter ML. 1988. Autism: familial aggregation and genetic implications. *J. Autism Dev. Disord.* 18(1):3–30

Freeman BJ, Ritvo E, Mason-Brothers A, Pingree C, Yokata A, et al. 1989. Psychometric assessment of first-degree relatives of 62 autistic probands in Utah. *Am. J. Psychiatry* 146(3):361–64

Friedman E. 1969. The "autistic syndrome" and phenylketonuria. *Schizophrenia* 1:249–61

Frith U. 1992. Asperger and his syndrome. In *Autism and Asperger Syndrome,* ed. U Frith, pp. 1–36. Cambridge, England: Press Syndicate

Gaffney GR, Kuperman S, Tsai LY, Minchin S. 1989. Forebrain structure in infantile autism. *J. Am. Acad. Child Adolesc. Psychiatry* 28:534–37

Gaffney GR, Kuperman S, Tsai LY, Minchin S, Hassanein KM. 1987. Midsagittal magnetic resonance imaging of autism. *Br. J. Psychiatry* 151:831–33

Gaffney GR, Tsai LY. 1987. Brief report: magnetic resonance imaging of high level autism. *J. Autism Dev. Disord.* 17(3):433–38

Garber HJ, Ritvo ER, Chiu LC, Griswold VJ, Kashanian A, et al. 1989. A magnetic resonance imaging study of autism: normal fourth ventricle size and absence of pathology. *Am. J. Psychiatry* 146(4):532–34

Gillberg C, Coleman M. 1992. *The Biology of the Autistic Syndrome.* New York: Cambridge Univ. Press. 317 pp.

Gillberg C, Hamilton-Hellberg C. 1983. Childhood psychosis and monoamine metabolites in spinal fluid. *J. Autism Dev. Disord.* 13: 383–96

Gillberg C, Steffenburg S, Schaumann H.

1991a. Autism: epidemiology: is autism more common now than 10 years ago? *Br. J. Psychiatry.* 158:403–9

Gillberg C, Steffenburg S, Wahlstrom J, Gillberg IC, Sjostedt A, et al. 1991b. Autism associated with marker chromosome. *J. Am. Acad. Child Adolesc. Psychiatry* 30(3):489–94

Gillberg C, Terenius L, Lonnerholm G. 1985a. Endorphin activity in childhood psychosis. *Arch. Gen. Psychiatry.* 42:780–83

Gillberg C, Wahlstrom JBH. 1985b. A "new" chromosome marker common to the Rett syndrome and infantile autism? The frequency of fragile sites at X P22 in 81 children with infantile autism, childhood psychosis and the Rett syndrome. *Brain Dev.* 7:365–67

Gillberg IC, Gillberg C, Kopp S. 1992. Hypothyroidism and autism spectrum disorders. *J. Child Psychol. Psychiatry Allied Discipl.* 33(3):531–42

Gillberg S. 1989. The borderland of autism and Rett syndrome: five case histories to highlight diagnostic difficulties. *J. Autism Dev. Disord.* 19(4):545–59

Greenberg DA, Hodge SE. 1989. Linkage analysis under "random" and "genetic" reduced penetrance. *Genet. Epidemiol.* 6:259–64

Gusella J. 1989. Location cloning strategy for characterizing genetic defects in Huntington's disease and Alzheimer's disease. *FASEB J.* 3(9):2036–41

Hagberg B. 1985. Rett's syndrome: prevalence and impact on progressive severe mental retardation in girls. *Acta Paediatr. Scand.* 74:405–8

Hagberg G, Aicardi J, Dias K, Ramos O. 1983. A progressive syndrome of autism, dementia, ataxia, and loss of purposeful hand use in girls: Rett's syndrome: report of 35 cases. *Ann. Neurol.* 14:470–71

Hagerman RJ, Jackson AW, Levitas A, Rimland B, Braden M. 1986. An analysis of autism in 50 males with the fragile X syndrome. *Am. J. Med. Genet.* 23:393–401

Hagerman RJ, Jackson C, Amiri K, Silverman AC, O'Connor R, et al. 1992. Girls with fragile X syndrome: physical and neurocognitive status and outcome. *Pediatrics* 89(3):395–400

Hallmayer JF, Pintado E, Lotspeich L, Spiker D, McMahon W, et al. 1994. Molecular analysis and test of linkage between the FMR-1 gene and familial infantile autism in multiplex families. *Am. J. Hum. Genet.* In press

Hashimoto T, Tayama M, Mori K, Fujino K, Miyazaki M, et al. 1988. Magnetic resonance imaging in autism: preliminary report. *Neuropediatrics* 20:142–46

Herault J, Perrot A, Barthelemy C, Buchler M, Cherpi C, et al. 1993. Possible association of c-Harvey-Ras-1 (HRAS-1) marker with autism. *Psychiatry Res.* 46(3):261–67

Hirst MC, Barnicoat A, Flynn G, Wang Q, Daker M, et al. 1993. The identification of a third fragile site, FRAXF, in Xq27-q28 distal to both FRAXA and FRAXE. *Hum. Mol. Genet.* 2(2):197–200

Hoffman EP, Brown RH, Kunkel LM. 1987. Dystrophin: the protein product of the Duchenne muscular dystrophy locus. *Cell* 51:919–28

Hoffman EP, Kunkel LM. 1989. Dystrophon abnormalities in Duchenne/Becker muscular dystrophy. *Neuron* 2:1019–29

Holttum JR, Minshew NJ, Sanders RS, Phillips NE. 1992. Magnetic resonance imaging of the posterior fossa in autism. *Biol. Psychiatry* 32(12):1091–101

Hunt A, Dennis J. 1987. Psychiatric disorder among children with tuberous sclerosis. *Dev. Med. Child Neurol.* 29:190–98

Hunt A, Shepherd C. 1993. A prevalence study of autism in tuberous sclerosis. *J. Autism Dev. Disord.* 23(2):323–39

Ivarsson SA, Bjerre I, Vegfors P, Ahlfors K. 1990. Autism as one of several disabilities in two children with congenital cytomegalovirus infection. *Neuropediatrics* 21:102–3

Jorde LB, Hasstedt SJ, Ritvo ER, Mason-Brothers A, Freeman BJ, et al. 1991. Complex segregation analysis of autism. *Am. J. Hum. Genet.* 49:932–38

Kanner L. 1943. Autistic disturbances of affective contact. *Nerv. Child.* 2:217–50

Kemper TL, Bauman ML. 1993. The contribution of neuropathologic studies to the understanding of autism. *Behav. Neurol.* 11:175–87

Kerem B-S, Rommens JM, Buchanan JA, Markiewica D, Cox TK, et al. 1989. Identification of the cystic fibrosis gene: genetic analysis. *Science* 245:1073–80

Kerr AM, Stephenson JBP. 1985. Rett's syndrome in the west of Scotland. *Br. Med. J.* 291:579–82

Klin A. 1993. Auditory brainstem responses in autism: brainstem dysfunction or peripheral hearing loss. *J. Autism Dev. Disord.* 23:15–33

Klin A, Volkmar FR, Sparrow SS. 1992. Autistic social dysfunction: some limitations of the theory of mind hypothesis. *J. Child Psychol. Psychiatry Allied Discipl.* 33(5):861–76

Knight SJ, Flannery AV, Hirst MC, Campbell L, Christodoulou Z, et al. 1993. Trinucleotide repeat amplification and hypermethylation of a CpG island in FRAXE mental retardation. *Cell* 74(1):127–34

Knobloch H, Pasamanick B. 1975. Some etiologic and prognostic factors in early infantile autism and psychosis. *J. Pediatr.* 55:182–91

Konigsburg LW, Blangero W, Mitchell BD, Kammerer CM. 1990. A simulation study of quantitative linkage trait analysis under a mixed polygenic and major gene model. *Am. J. Hum. Genet.* 47:A139

Lake CR, Ziegler MG, Murphy DL. 1977. Increased norepinephrine levels and decreased dopamine b-hydroxylase activity in primary autism. *Arch. Gen. Psychiatry* 34:553–66

Landa R, Folstein SE, Isaacs C. 1991. Spontaneous narrative discourse performance of parents of autistic individuals. *J. Speech Hear. Res.* 34:1339–45

Landa R, Piven J, Wzorek MM, Gayle JO, Chase GA, et al. 1992. Social language use in parents of autistic individuals. *Psychol. Med.* 22:1339–45

Lauder JM. 1983. Hormonal and humoral influences on brain development. *Psychoneuroendocrinology* 8:121–55

Lauder JM. 1990. Ontogeny of the serotonergic system in the rat: serotonin as a developmental signal. In *The Neuropharmacology of Serotonin,* ed. PM Whitaker-Azmitia SJ Peroutka, pp. 297–314. New York: NY Acad. Sci.

Lauder JM, Krebs H. 1978. Serotonin as a differentiation signal in early neurogenesis. *Dev. Neurosci.* 1:15–30

Launay JM, Bursztein C, Ferrari P. 1987. Catecholamine metabolism in infantile autism: a controlled study of 22 autistic children. *J. Autism Dev. Disord.* 17:333–47

LeCouteur A, Rutter M, Lord C, Rios P, Robertson S, et al. 1989. Autism diagnostic interview: a standardized investigator-based instrument. *J. Autism. Dev. Disord.* 19:363–87

Links PS, Stockwell M, Abichandandi R, Simeon J. 1980. Minor physical anomalies in childhood autism. I. Their relationship to pre- and perinatal conditions. *J. Autism Dev. Disord.* 10:273–92

Lopreiato JO, Wulfsberg EA. 1992. A complex chromosome rearrangement in a boy with autism. *J. Dev. Behav. Pediatr.* 13(4):281–83

Lord C, Mulloy C, Wendelboe M, Schopler E. 1991. Pre- and perinatal factors in high-functioning females and males with autism. *J. Autism Dev. Disord.* 21(2):197–209

Lord C, Rutter M, Good S, Heemsbergen J, Jordan H, et al. 1989. Autism diagnostic observation scale: a standardized observation of communicative and social behavior. *J. Autism Dev. Disord.* 19:185–212

Lotspeich L. 1994. Autism and pervasive developmental disorders. In *Psychopharmacology: The Fourth Generation of Progress,* ed. FE Bloom, DJ Kupfer. New York: Raven. In press

Lotspeich LJ, Ciaranello RD. 1993. The neurobiology and genetics of infantile autism. In *International Review of Neurobiology,* ed. R Bradley, pp. 87–129. San Diego: Academic

Lowe TL, Tanaka K, Seashore MR, Young JG,

Cohen DJ. 1980. Detection of phenylketonuria in autistic and psychotic children. *JAMA* 243(2):126–28

Lubs HA. 1968. A marker X chromosome. *Am. J. Hum. Genet.* 21:231–44

Mabry CC, Denniston JC, Nelson TL, Son CD. 1963. Maternal phenylketonuria. *New Engl. J. Med.* 296:1404–8

Markowitz PI. 1983. Autism in a child with congenital cytomegalovirus infection. *J. Autism Dev. Disord.* 13(3):249–53

Mason-Brothers A, Ritvo ER, Pingree C, Petersen PB, Jenson WR, et al. 1990. The UCLA-University of Utah epidemiologic survey of autism: prenatal, perinatal and postnatal factors. *Pediatrics* 86(4):514–19

Mayes L, Volkmar F, Hooks M, Cicchetti D. 1993. Differentiating pervasive developmental disorder not otherwise specified from autism and language disorders. *J. Autism Dev. Disord.* 23(1):79–90

Minshew NJ. 1991. Indices of neural function in autism: clinical and biologic implications. *Pediatrics* 87(5):774–80 (Suppl.)

Minshew NJ, Goldstein G, Dombrowski SM, Panchalingan K, Pettigrew JW. 1993. A preliminary 31P MRS study of autism: evidence for undersynthesis and increased degradation of brain membranes. *Biol. Psychiatry* 33:762–73

Minton J, Campbell M, Green WH, Jennings S, Samit C. 1982. Cognitive assessment of siblings of autistic children. *J. Am. Acad. Child Psychiatry* 21(3):256–61

Murakami JW, Courchesne E, Press GA, Yeung-Courchesne R, Hesselink R. 1989. Reduced cerebellar hemisphere size and its relationship to vermal hypolasia in autism. *Arch. Neurol.* 46:689–94

Murayama K, Greenwood RS, Rao KW, Aylsworth AS. 1991. Neurological aspects of del (1q) syndrome. *Am. J. Med. Genet.* 40(4):488–92

Nanson JL. 1992. Autism in fetal alcohol syndrome: a report of six cases. *Alcohol. Clin. Exp. Res.* 16(3):558–65

Nelson KB. 1991. Prenatal and perinatal factors in the etiology of autism. *Pediatrics* 83(5):761–66 (Suppl.)

Ornitz EM. 1985. Neurophysiology of infantile autism. *J. Am. Acad. Child Psychiatry* 24(3):251–62

Ornitz EM. 1989. Autism: at the interface between sensory and information processing. In *Autism: Nature, Diagnosis and Treatment,* ed. G Dawson, pp. 174–207. New York: Guilford

Ornitz EM, Ritvo ER. 1968. Perceptual inconstancy in early infantile autism. *Arch. Gen. Psychiatry* 18:76–98

Ozonoff S, Rogers SJ, Pennington BF. 1991. Asperger's syndrome: evidence of an empirical distinction from high-functioning autism. *J. Child Psychol. Psychiatry Allied Discipl.* 32(7):1107–22

Payton JB, Steele MW, Wenger SL, Minshew MJ. 1989. The fragile X marker and autism in perspective. *J. Am. Acad. Child Adolesc. Psychiatry* 28(3):417–21

Piven J, Berthier ML, Starkstein SE, Nehme E, Pearlson G, et al. 1990. Magnetic resonance imaging evidence for a defect of cerebral cortical development in autism. *Am. J. Psychiatry* 147(6):734–39

Piven J, Gayle J, Landa R, Wzorek M, Folstein S. 1991. The prevalence of fragile X in a sample of autistic individuals diagnosed using a standardized interview. *J. Am. Acad. Child Adolesc. Psychiatry* 30(5):825–30

Price RA. 1993. Genetic approaches to mental illness. In *Biological Basis of Brain Function and Disease,* ed. A Frazer, PB Molinoff, A Winokur, pp. 281–99. New York: Raven

Richards RI, Holman K, Yu S, Sutherland GR. 1993. Fragile X syndrome unstable element, p(CGG)n, and other simple tandem repeat sequences are binding sites for specific nuclear proteins. *Hum. Mol. Genet.* 2(9):1429–35

Riordan JR, Rommens JM, Kerem, B-S, Alon N, Rozmahel R, et al. 1989. Identification of the cystic fibrosis gene: cloning and characterization of complementary DNA. *Science* 245:1066–73

Risch N. 1990a. Linkage strategies for genetically complex traits. (III). The effects of marker polymorphism on analysis of affected relative pairs. *Am. J. Hum. Genet.* 46:242–53

Risch N. 1990b. Linkage strategies for genetically complex traits. I. Multilocus models. *Am. J. Hum. Genet.* 46:222–28

Risch N. 1991. A note on multiple testing procedures in linkage analysis. *Am. J. Hum. Genet.* 48:1058–64

Risch N, Claus E, Giuffra L. 1989. *Linkage and Mode of Inheritance in Complex Traits.* New York: Liss

Ritvo ER, Freeman BJ. 1978. National society for autistic children definition of the syndrome of autism. *J. Autism Child Schizophr.* 8:162–67

Ritvo ER, Freeman BJ, Mason-Brothers A, Mo A, Ritvo AM. 1985a. Concordance for the syndrome of autism in 46 pairs of afflicted twins. *Am. J. Psychiatry* 142:74–77

Ritvo ER, Freeman BJ, Schiebel AB, Duong T, Robinson H, et al. 1986. Lower Purkinje cell counts in the cerebellum of four autistic subjects: initial findings of the UCLA-NSAC autopsy research report. *Am. J. Psychiatry* 143(7):862–66

Ritvo ER, Jorde LB, Mason-Brothers A, Freeman BJ, Pingree C, et al. 1989. The UCLA-University of Utah Epidemiologic Survey of Autism: recurrence risk estimates and genetic counseling. *Am. J. Psychiatry* 146(8):1032–36

Ritvo ER, Spence MA, Freeman BJ, Mason-Brothers AAM, et al. 1985b. Evidence for autosomal recessive inheritance in 46 families with multiple incidence of autism. *Am. J. Psychiatry* 142:187–92

Rommens JM, Iannuzzi MC, Kerem B, Drumm ML, Melmer G, et al. 1989. Identification of the cystic fibrosis gene: chromosome walking and chromosome jumping. *Science* 245:1059–65

Rubenstein JLR, Lotspeich L, Ciaranello RD. 1990. The neurobiology of developmental disorders. In *Advances in Clinical Child Psychology,* ed. BJ Lahey, AE Kazdin, pp. 1–52. New York: Plenum

Rumsey JM, Creasey H, Stepanek, JSRD, Patronas N, et al. 1988. Hemispheric, asymmetries, fourth venticular size, and cerebellar morphology in autism. *J. Autism Dev. Disord.* 18(1):127–37

Rumsey JM, Duara R, Grady C, Rapoport JL, Margolin RA, et al. 1985. Brain metabolism in autism, resting cerebral glucose utilization rates as measured with positron emission tomography. *Arch. Gen. Psychiatry* 42:448–55

Rutter M. 1967. Psychotic disorders in early childhood. In *Recent Developments in Schizophrenia, British Journal of Psychiatry Special Publication,* ed. AJ Coppen, A Walk, pp. 133–58

Rutter M, Bartak L. 1971. Causes of infantile autism: some considerations from recent research. *J. Autism Child. Schizophrenia* 1(1):20–32

Rutter M, Hersov R. 1977. *Child Psychiatry: Modern Approaches.* Oxford: Blackwell. 960 pp.

Rutter M, Schopler E. 1992. Classification of pervasive developmental disorders: some concepts and practical considerations. *J. Autism Dev. Disord.* 22(4):459–82

Sampson JR, Yates JRW, Pirrit LA. 1989. Evidence for genetic heterogeneity in tuberous sclerosis. *J. Med. Genet.* 26:511–16

Schain RJ, Freedman DX. 1961. Studies on 5-hydroxyindole metabolism in autistic and other mentally retarded children. *J. Pediatr.* 58:315–18

Seshadri K, Wallerstein R, Burack G. 1992. 18q$^-$ chromosomal abnormality in a phenotypically normal 2 1/2 year old male with autism. *Dev. Med. Child Neurology.* 34(11):1005–9

Siegel BV, Asarnow R, Tanguay P. 1992. Regional cerebral glucose metabolism and attention in adults with a history of childhood autism. *J. Neuropsychiatry* 4:406–14

Siomi H, Siomi MC, Nussbaum RL, Dreyfuss G. 1993. The protein product of the fragile X gene, FMR-1, has characteristics of an RNA-binding protein. *Cell* 74(2):291–98

Smalley SL. 1991. Genetic influences in autism. *Psychiatr. Clin. N. Am.* 14(1):125–39

Smalley SL, Asarnow RF. 1990. Brief report: cognitive subclinical markers in autism. *J. Autism Dev. Disord.* 20(2):271–78

Smalley SL, Asarnow RF, Spence MA. 1988. Autism and genetics. *Arch. Gen. Psychiatry* 45:953–61

Smalley SL, Smith M, Tanguay P. 1991. Autism and psychiatric disorders in tuberous sclerosis. *Ann NY Acad Sci.* 615:382–83

Smalley SL, Tanguay PE, Smith M, Gutierrez G. 1992. Autism and tuberous sclerosis. *J. Autism Dev. Disord.* 22(3):339–55

Smith M, Smalley S, Cantor R. 1990. Mapping of a gene determining tuberous sclerosis to human chromosome 11q14–11q23. *Genomics* 6:105–14

Snow K, Doud LK, Hagerman R, Pergolizzi RG, Erster SH, et al. 1993. Analysis of a CGG sequence at the FMR-1 locus in Fragile X families and in the general population. *Am. J. Hum. Genet.* 53:1217–28

Spence MA, Ritvo ER, Marazita ML, Funderburk SL, Sparkes RS, et al. 1985. Gene mapping studies with the syndrome of autism. *Behav. Genet.* 15:1–13

Spiker DK, Lotspeich L, Kraemer HC, Hallmayer J, McMahon W, et al. 1994. The genetics of autism: characteristics of affected and unaffected children from 37 multiplex families. *Am. J. Med. Genet.* 54:27–35

Steffenburg S, Gillberg C, Hellgren L, Andersson LC, et al. 1989. A twin study of autism in Denmark, Finland, Iceland, Norway and Sweden. *J. Child Psychol. Psychiatry* 30(3): 405–16

Stubbs EG. 1978. Autistic symptoms in a child with congenital cytomegalovirus infection. *J. Autism Child Schizophr.* 8:37–43

Stubbs EG, Ash ES, WCP. 1984. Autism and congenital cytomegalovirus. *J. Autism Dev. Disord.* 14:183–89

Sutherland GR. 1977. Fragile sites on human chromosomes: demonstration of their dependence on the type of tissue culture medium. *Science* 197:265–66

Szatmari P. 1991. Asperger's syndrome: diagnosis, treatment, and outcome. *Psychiatr. Clin. N. Am.* 14(1):81–93

Szatmari P, Bartolucci G, Bremner R. 1989a. A follow-up study of high-functioning autistic children. *J. Autism Dev. Disord.* 19:213–25

Szatmari P, Bartolucci G, Bremner R. 1989b. Asperger's syndrome and autism: comparison of early history and outcome. *Dev. Med. Child Neurol.* 31(6):709–20

Szatmari P, Bremmer R, Nagy J. 1989c. Asperger's syndrome: A review of clinical features. *Can. J. Psychiatry* 34:554–60

Szatmari P, Tuff L, Finlayson MA, Bartolucci G. 1990. Asperger's syndrome and autism:

neurocognitive assessments. *J. Am. Acad. Child Adolesc. Psychiatry* 29:130–36

Todd RD. 1992. Neural development is regulated by classical neurotransmitters: dopoamine D2 receptor stimulation enhances neurite outgrowth. *Biol. Psychiatry* 31(8):794–807

Todd RD, Ciaranello RD. 1985. Early infantile autism and the childhood psychoses. In *Handbook of Clinical Neurology,* ed. PJ Vinken, GW Bruyn, HL Klawans, pp. 189–97. Amsterdam: Elsevier

Trevathan E, Naidu S. 1988. The clinical recognition and differential diagnosis of Rett syndrome. *J. Child Neurol.* 3(Suppl):6–16

Tsongalis GJ, Silverman LM. 1993. Molecular pathology of the Fragile X syndrome. *Arch. Pathol. Lab. Med.* 117(11):1121–25

Verkerk AJ, Pieretti M, Sutcliffe JS, Fu YH, Kuhl DP, et al. 1991. Identification of a gene (FMR-1) containing a CGG repeat coincident with a breakpoint cluster region exhibiting length variation in fragile X syndrome. *Cell* 65:905–14

Vieland V, Greenberg D, Hodge SE, Ott J. 1992. Linkage analysis of two-locus diseases under single-locus and two-locus analysis models. *Cytogenet. Cell Genet.* 59:145–46

Volkmar FR, Cohen DJ. 1991. Debate and argument: the utility of the term pervasive developmental disorders. *J. Child Psychol. Psychiatry Allied Discipl.* 32(7):1171–72

Warren RP, Cole P, Odell JD, Pingree CB, Warren WL, et al. 1990. Detection of maternal antibodies in infantile autism. *J. Am. Acad. Child Adolesc. Psychiatry* 29(6):873–77

Warren RP, Singh VK, Cole P, Odell JD, Pingree CB, et al. 1991. Increased frequency of the null allele at the complement C4B locus in autism. *Clin. Exp. Immunol.* 83(3): 438–40

Warren RP, Singh VK, Cole P, Odell JD, Pingree CB, et al. 1992. Possible association of the extended MHC haplotype B44-SC30-DR4 with autism. *Immunogenetics* 36(4): 203–7

Warren RP, Yonk LJ, Burger RA, Cole P, Odell JD, et al. 1990. Deficiency of suppressor-inducer (CD4+CD45RA+) T cells in autism. *Immunol. Invest.* 19(3):245–51

Williams RS, Hauser SL, Purpura DP, DeLong FR, Swisher CN. 1980. Autism and mental retardation. Neuropathologic studies performed in four retarded persons with autistic behavior. *Arch. Neurol.* 37:749–53

Wing L. 1981. Language, social and cognitive impairments in autism and severe mental retardation. *J. Autism Dev. Disord.* 11:31–44

Wolff S, Barlow A. 1979. Schizoid personality in childhood: a comparative study of schizoid, autistic and normal children. *J. Child Psychol. Psychiatry* 20:29–46

Wolff S, Chick J. 1980. Schizoid personality in

childhood: a controlled follow-up study. *Psychol. Med.* 10:85–100

Wolff S, Narayan S, Moyes B. 1988. Personality characteristics of parents of autistic children: a controlled study. *J. Child Psychol. Psychiatry* 29(2):143–53

Wong DL, Ciaranello RD. 1987. Molecular biological approaches to mental retardation. In *Psychopharmacology: The Third Generation of Progress,* ed. H Meltzer, WF Bunney, JT Coyle, KL Davis, IJ Kopin, R Shuster, R Shader, pp. 861–66. New York: Raven

Yonk LJ, Warrne RP, Burger RA, Cole P, Odell JD, et al. 1990. CD4+ helper T-cell depression in autism. *Immunol. Lett.* 25(4):341–45

Young JG, Cohen DJ, Brown S, Caparulo BK.

1978. Decreased urinary free catecholamines in childhood autism. *J. Am. Acad. Child Psychiatry* 17:671–79

Yu S, Pritchard M, Kremer E, Lynch M, Nancarrow J, et al. 1991. Fragile X ge ype characterized by an unstable region of DNA. *Science* 252:1179–82

Yuwiler A, Freedman DX. 1987. Neurotransmitter research in autism. In *Neurobiological Research in Autism,* ed. E Schopler, M Rutter, pp. 263–84. New York: Plenum

Zoghbi H. 1988. Genetic aspects of Rett syndrome. *J. Child Neurol.* 3(Suppl):76–77

Zoghbi HY, Percy AK, Glaze DG. 1985. Reduction of biogenic amine levels in the Rett syndrome. *N. Engl. J. Med.* 313:921–24

Annu. Rev. Neurosci. 1995. 18:129–58

DYNAMIC REGULATION OF RECEPTIVE FIELDS AND MAPS IN THE ADULT SENSORY CORTEX

Norman M. Weinberger

Department of Psychobiology and Center for the Neurobiology of Learning and Memory, University of California, Irvine, California 92717-3800

KEY WORDS: deafferentation, learning, auditory, somatosensory, visual

INTRODUCTION

A dominant belief in neuroscience is that sensory systems in the adult are stable, in contrast to the extensive and pervasive plasticity that characterizes development of the nervous system. Empirical bases for this dogma of sensory immutability include the usually precise and stable responses of sensory neurons in anesthetized animals and the reduction of sensory cortical plasticity beyond critical periods of development. The subjective experience of neuroscientists also supports the dogma. Perception of the outside world appears to be clear, immediate and effortless. To most workers, this implies that the sensory systems, once having developed, must be stable in order to provide accurate information about the environment. However, a rapidly growing literature attests to a very large degree of short- and long-term modification of receptive fields (RF) and reorganization of representational maps under a variety of circumstances: learning, sensory stimulation, and sensory deafferentation.

This chapter reviews contemporary findings concerning the dynamic regulation of receptive fields and maps in the primary auditory, somatosensory, and visual cortices of the adult brain. In contrast to previous reviews that have been confined largely to the perspective of sensory physiology, this article also emphasizes behavioral considerations. This seems to be appropriate, if not

129

mandatory, because behavior is normally dynamic and adaptive, and sensory cortex is notable for its evolutionary development and implication in higher functions.

The present coverage is highly selective, necessitated by severe constraints of space. Detailed analyses of publications were not possible and coverage of the effects of sensory deafferentation had to be limited to a scant summary of major effects and their possible relevance to sensory stimulation and learning; fortunately the effects of deafferentation have been reviewed in detail (Kaas 1991). Within the literature on sensory stimulation and learning, studies limited to standard learning paradigms were not included, in favor of studies of receptive fields and representational maps. These limitations should not unduly compromise this review because its intention is mainly conceptual. Specifically, the goal is to provide a framework that will be useful for thinking about both current and future research on adult sensory cortical plasticity and reorganization.

This framework is based on an empirical law that is not yet widely appreciated in neuroscience. It may be summarized as follows: Behaving (i.e. waking) animals can continually acquire and retain information about (a) individual sensory stimuli, (b) relationships between various sensory stimuli, and (c) relationships between their own behavior and its sensory consequences. An implication of this law is that the attainment of an adequate understanding of how sensory cortex in the adult subserves perception and behavior also requires achieving an adequate account of the role of learning in sensory cortex. In theory, this role could have been nil. In fact it is not, as attested by the results of explicit learning experiments and other studies that can reasonably be considered to involve learning. The role of learning in denervation-induced plasticity and reorganization is currently largely conjectural but cannot be discounted.

The following topics are discussed in turn: the relationship between sensory physiology and learning, basic forms of learning that are particularly relevant to adult cortical plasticity, methodological considerations, major issues and emerging principles in learning and sensory cortex, examples of these principles from the literature on learning, brief comments on the effects of sensory deafferentation, and conclusions.

SENSORY PHYSIOLOGY AND LEARNING

Research reports generally stress the surprising and sometimes disquieting nature and extent of sensory cortical plasticity. Yet fifty years of research shows that learning produces physiological plasticity in relevant sensory neocortices. Strangely, not only has such learning-related sensory plasticity been largely ignored within the field of sensory physiology, it has been little noticed within the neurobiology of learning and memory. The reasons for this lack of interest probably include the widely held assumption that the acquisition and storage of information occurs only in higher nonsensory cortical structures, such as association cortex, and in nonsensory subcortical structures, such as the hippocampus. As a result, adult sensory cortical plasticity has been disclaimed or ignored by the two disciplines whose subject matter concerns the processing of environmental stimuli.

This situation is changing, in part because the basic experimental paradigms of the two fields may be viewed individually as incomplete but together as complementary aspects of a more comprehensive approach to the role of sensory system function in adaptive behavior. This approach is based on the fact that sensory stimuli simultaneously have two types of parameters, physical and psychological (Weinberger & Diamond 1988). The physical parameters are specified by standard units of measure, e.g. wavelength, hertz, and decibels. Psychological parameters concern the learned behavioral significance of stimuli but are not expressed in comparable universal units of measure. Rather, the acquired significance of a stimulus is determined by objective measurement of some aspect of overt behavior. However, this asymmetry in the degree of standardized units of measurement does not diminish the fact that stimuli have both physical and psychological parameters.

The basic paradigm of sensory physiology is to vary the physical parameters of stimuli while keeping their psychological parameters constant; this is generally accomplished by anesthetizing the subjects so that they cannot acquire information about the stimuli. The basic paradigm of the field of learning is to vary the psychological parameter (e.g. by altering the relationships and significance among stimuli) while keeping the physical parameters constant. The complementary nature of these paradigms is summarized in Table 1.

The two paradigms can be combined within the same subject or between groups by, first, running the sensory design, second, performing an explicit learning experiment, and third, repeating the sensory physiology assessment. This yields the effects of the learning treatment on receptive fields or representational maps. Of course, other treatments can be used in step two. Sensory deafferentation has been employed extensively. However, note that sensory deafferentation that involves a period of recovery in a waking animal, before

Table 1 Complementary nature of experimental paradigms of sensory physiology and learning

Discipline	Stimulus parameters	
	Physical	Psychological
Sensory physiology	Vary	Constant
Learning	Constant	Vary

a final sensory physiology assessment is done, also provides animals with the opportunity to learn.

I do not claim that learning enters into all cases of receptive field or map plasticity in the adult. As I discuss later, there are ample instances of immediate changes in the functional properties of sensory cortices following sensory deafferentation under anesthesia that cannot involve learning, and the role of learning in most cases of chronic denervation has not yet been studied. It would be as misleading to overemphasize learning and behavioral aspects of cortical plasticity as it would be to ignore them. Rather, the present goal is to increase awareness of behavioral aspects of adult sensory cortical plasticity that are relevant, sometimes crucial, to the issues at hand.

A BRIEF RESUME OF RELEVANT TYPES OF LEARNING

The reader who is knowledgeable about basic forms of learning can skip this section. For other readers, a brief explanation should be helpful. Standard reference works can be consulted for more details (e.g. Lieberman 1990; Mackintosh 1974, 1983).

Habituation

Presentation of a single stimulus of weak or moderate intensity generally elicits an attention, or orienting, response on its initial occurrence, as a potential index of food, predator, prey, or simply information about the changed sensory environment. Repeated stimulus presentation without any other consequences results in loss of attentional interest and behavioral response, and usually, a reduction of neural response within relevant sensory cortex. This process of learning not to attend to such a stimulus is termed habituation. It differs from sensory adaptation and fatigue as habituation can occur at long interstimulus intervals, develops more rapidly with weaker stimulus intensity, and is highly specific to the parameters of the repeated stimulus. Repeated sensory stimulation is widely used in studies of sensory cortex; both response decrements and increments, with modification of receptive field properties, have been reported (e.g. Lee & Whitsel 1992).

Sensory Preconditioning

Sensory preconditioning refers to the learning that occurs when two sensory stimuli (S1 and S2) of weak or moderate intensity (e.g. sound and touch) are presented sequentially (within a few tenths to several seconds). The animals learn that S2 follows S1 and may treat S1 as a signal for S2. For example, paired stimulation of two whiskers has been used to study cortical plasticity (e.g. Delacour et al 1987).

Classical Conditioning

Classical conditioning also involves two sequential stimuli, but the second stimulus is strong and biologically significant, i.e. food or a noxious stimulus. The first stimulus is referred to as the conditioned stimulus (CS), and the second stimulus as the unconditioned stimulus (US). Importantly, contemporary research has shown that classical conditioning far transcends the popular notion of simple stimulus-response learning, e.g. conditioned salivation to a bell in Pavlov's dogs. Rather, conditioning is more accurately characterized in terms of the acquisition and retention of a large amount of information, including the detailed physical parameters of the CS, the US, and other contextual stimuli and their relationships (Rescorla 1988).

Instrumental Conditioning

In instrumental conditioning, the presentation of a sensory stimulus is contingent upon a behavioral response. For example, an animal might be required to press a bar (response) to receive food (stimulus), but the stimulus may be any sensory event, not merely food or water. In general, most behavior alters the sensory environment, placing subjects in a feedback loop with their environment. Instrumental conditioning can occur after sensory deafferentation (e.g. amputation of a digit or lesion of the retina), when a subject's attempts to behaviorally compensate for its sensory deficit produce new relationships between behavior and its sensory consequences.

METHODOLOGICAL CONSIDERATIONS

The field of adult sensory cortical plasticity is no more plagued with methodological problems than any other field. In fact, despite such problems, a replicable and consistent account of cortical plasticity is emerging. However, the following topics should be kept in mind, particularly when consulting the primary source literature.

Sensory Stimulus

The nature of sensory stimulation is of paramount importance. The size of a receptive field and the delineation of a representational map are affected by stimulus variables, especially by stimulus intensity.

Receptive Fields

The accepted definition of a receptive field (RF) common to the literature reviewed is that portion of receptor epithelia that when stimulated affects the discharges of sensory neurons. This definition allows inclusion of inhibitory as well as excitatory effects. Recent studies indicate that RFs have pronounced temporal dimensions, such that over the course of a few hundred milliseconds or less, the RF of a neuron can undergo marked changes (e.g. Dinse et al 1990). As temporally dynamic RFs have not yet been widely studied, one should bear in mind that our understanding of cortical plasticity, and other aspects of sensory physiology, may be subject to considerable revision and new perspectives in the future.

Neurophysiological and Metabolic Measures

The recording of neuronal discharges and the determination of the utilization and area of activation of 2-[^{14}C]-deoxy-D-glucose (2DG) have both been extensively used in studies of adult cortical plasticity. The former provides good temporal resolution but restricted spatial coverage, unless used repeatedly to delineate a map. The latter provides poor temporal resolution but excellent spatial coverage and, therefore, yields valuable information on the reorganization of representational maps. Because neurophysiological and metabolic methods do not provide the same information, caution should be exercised comparing results. Excellent spatial resolution of sensory cortex can be achieved with the noninvasive optical imaging of intrinsic signals, and the optically detected regions of activation have been precisely validated electrophysiologically (Masino et al 1993). The human sensory cortex is being studied with magnetoencephalography (MEG), which provides better spatial resolution than electroencephalography.

State of Arousal

Receptive field size can vary as a function of the state of arousal of the subject. In particular, RFs are often reduced with increasing depth of anesthesia (Armstrong-James & George 1988) and can differ greatly within the same subject in the waking vs the anesthetized state (Simons et al 1992). In contrast, the representation of the hand in the monkey seems not to be so affected (Stryker et al 1987).

Cortical Layer

The differential anatomical and physiological characteristics of the lamina within the neocortex strongly suggest that dynamic aspects of receptive fields and maps are probably not the same from the cortical surface to its depths. Reports do not always specify the lamina from which recordings are obtained. Maps are usually obtained from the sites of termination of leminscal thalamocortical projections, i.e. deep layer III and layer IV.

Terminology: Plasticity, Reorganization, and Regulation

Terms intended only as descriptions often acquire functional overtones or are linked to particular mechanisms. The term plasticity is generally used in this paper in a purely descriptive sense as a property of nervous tissue to change its responsivity, derived from Konorski (1967). Other terms should be used to refer to interpretations or mechanisms of such change. Reorganization is a descriptive term used here to refer to plasticity in the central representation of a receptor epithelium, i.e. of a representational map. Dynamic regulation is a broad term that refers both to plasticity and reorganization.

MAJOR ISSUES AND EMERGING PRINCIPLES

Several issues permeate the literature on the dynamic regulation of adult sensory cortex. They are not all explicitly discussed for every topic because they have been explored to highly varying degrees across the subject, but they are revisited at the end of this paper.

1. What treatments produce changes in receptive fields and maps?
2. What are the detailed characteristics of such changes, including their time course, i.e. onset, development, and duration?
3. What are the contributions of subcortical and other structures to the changes observed in primary sensory cortical fields?
4. What are the cellular mechanisms involved with or underlying the receptive field or map changes?
5. What are the behavioral roles of plasticity in receptive fields and reorganization in representational maps?

The following principles seem to be emerging from explicit studies of learning and experiments in which sensory stimulation is presented to waking, often behaving, animals.

L1. Learning about a stimulus systematically modifies the functional organization of primary sensory cortex.
L2. Learning effects are observed at the level of single neurons, as studied by receptive fields, and at a much larger spatial scale as evident by changes in representational maps.
L3. Stimuli that become signals for food or noxious stimulation receive increased response within receptive fields and increased representation within maps, while stimuli that are of no significance receive decreased response and representation.
L4. Plasticity in sensory cortex caused by learning develops rapidly and can be maintained indefinitely.

THE EFFECTS OF LEARNING ON SENSORY CORTEX

Under this heading, I first review receptive field plasticity and then review the reorganization of cortical maps. Studies of sensory stimulation that may not fall strictly within the domain of learning are also included here because they are few in number and are most closely related to this topic. The findings throughout these sections support Principles L1 and L2, respectively. The basic characteristics of plasticity (L3 and L4) are considered separately for receptive fields and maps. A final section regarding possible mechanisms concerns both RFs and maps.

Receptive Field Plasticity

AUDITORY CORTEX   The logic of applying RF analysis to learning is of interest and illustrates the advantages of using concepts and findings from the field of sensory physiology to solve problems in the neurobiology of learning. For decades it had been known that classical conditioning (e.g. tone followed by food or shock) produces facilitated responses in the auditory cortex to acoustic conditioned stimuli (reviewed by Weinberger & Diamond 1987). Such facilitation could be caused either by a general increase in neuronal excitability or by a specific enhancement in the processing of the frequency of the conditioned stimulus.[1] Determination of RFs for frequency (frequency tuning) before and after training can resolve this issue. A general increase in responsivity should produce increased responses to all frequencies, i.e. both the CS and other (non-CS) frequencies that were not included in training. In contrast, a specific change in the processing of the CS should produce facilitation to the CS frequency and little or no change, perhaps even depression, of responses to other frequencies.

Classical fear conditioning (tone-shock) is learned very quickly (5–10 pairings) (Lennartz & Weinberger 1992a). This training produces CS-specific RF plasticity in the primary auditory cortex of the guinea pig (Bakin & Weinberger 1990). Responses to the CS frequency are increased while responses to the pretraining best frequency (BF) and many other frequencies are decreased or show little change. This results in a shifting of tuning toward or even to the CS frequency, which can become the new BF (Figure 1A1,2,3 and B1).[2] This plasticity requires CS-US pairing (Bakin et al 1992) (Figure 1B2). The same type of retuning is also found in the rat (Taylor & Rucker 1993). Thus, classical conditioning to a behaviorally important tone retunes frequency RFs, thereby

[1]It is important to distinguish between nonassociative effects, such as sensitization and pseudoconditioning, which are not at issue here, and associative effects that produce a genuine learning-based increase to all sounds. As a rough metaphor, this would be akin to increasing the volume of a radio, in contrast to frequency-specific changes in RFs, which would be like tuning to another station.

[2]Receptive field analysis was first applied to classical conditioning for secondary auditory cortex of the cat and also showed CS-specific plasticity (Diamond & Weinberger 1986, 1989). The emphasis in this article is on primary sensory cortical fields, which constitute all of the other relevant literature.

increasing responses to the CS and reducing response to other (non-CS) frequencies.

Repeated presentation of a behaviorally unimportant stimulus is an instance of habituation and produces the opposite effects—decreased response to the repeated frequency (Figure 1*B3*). This RF plasticity is due to learned inattention rather then sensory adaptation or fatigue because it develops at rates of stimulus presentation that are too slow to produce these effects (Condon & Weinberger 1991). Further evidence in support of Principle L3, facilitation or depression based on learned stimulus importance, is provided by the results of discrimination training (one tone reinforced, a second tone not reinforced). This type of training also produces frequency-specific RF plasticity with increased response to the reinforced frequency but decreased response to the nonreinforced frequency (and also decreased response to other frequencies) (Edeline & Weinberger 1993; see also Edeline et al 1990a). Frequency-specific effects have also been obtained using a novel discrimination paradigm in the gerbil, but responses to frequencies adjacent to the CS frequency are facilitated, resulting in "lateral contrast enhancement" (Ohl et al 1992; F Ohl & H Scheich, unpublished data). Reconciliation of the gerbil findings with those of the guinea pig remains to be done.

Receptive field plasticity develops as rapidly as behavioral signs of conditioned fear, after only five training trials (Edeline, Pham & Weinberger 1993). Receptive field plasticity is enduring, as seen by within-subject recordings obtained before and at weekly intervals for up to eight weeks following conditioning (Weinberger et al 1993). In this study, subjects were trained while awake but RFs were obtained while they were anesthetized. Therefore, RF plasticity in classical conditioning is sufficiently robust to be expressed under anesthesia; this plasticity cannot be due to arousal during RF determination; and RFs obtained under anesthesia can reflect the results of prior learning experiences.

SOMATOSENSORY CORTEX   Learning-induced receptive plasticity is also found in the cortical representation of the whiskers of rodents. Anatomically this representation consists of discrete aggregates of neurons in layer IV of primary somatosensory cortex (SI). Termed barrels (Woolsey & Van der Loos 1970) for their cylindrical shape, they can be observed in rows and columns that match the matrix of mystacial vibrissae; the latter are designated by letters and numbers (for review, see Kossut 1992).

Learning has been studied by using a sensory preconditioning paradigm, i.e. sequential pairing of two weak or moderate stimuli. In waking rats, Delacour et al (1987) recorded discharges from barrel neurons for which deflection of one whisker produced a consistent response (S2 stimulation), and deflection of another whisker produced a weak or no response (S1 stimulation); S1 preceded S2 by 500 ms. This sensory-preconditioning training regimen produced significant increased responses to S1 in as few as 30–100 pairings. Freely behaving animals can produce similar effects. All whiskers except two neighbors on one side of the face were repeatedly trimmed (Diamond et al 1993). The assumption was that the rats would costimulate the two neighboring whiskers during their ambient behavior. Within 3 days intact whiskers elicited increased discharges in their cortical barrels while previously unstimulated (trimmed) whiskers elicited fewer discharges than controls (Figure 2). The actual monitoring of whisker use by the subjects is necessary to fully understand this plasticity.

VISUAL CORTEX   There are not yet studies of the effects of learning on RFs in the primary visual cortex. An interesting related case, not strictly a case of learning, concerns contextual effects of stimulation outside of the RF on responses to stimuli within the RF. The orientation of lines surrounding the RF of neurons in area 17 of the cat alters the cells' RF properties as a function of the orientation of the surround bars. Effects include increases or decreases in the strength of response, shifts in the preference of orientation, and modification of the bandwidth of orientation tuning (Gilbert & Weisel 1990). These findings are part of a generally expanding literature showing that the responses of cells in the adult visual cortex are dynamic rather than static.

Representational Maps

AUDITORY SYSTEM   Receptive field plasticity extrapolated over the frequency map should increase the representation of a learned behaviorally significant

---

Figure 1   The effects of learning upon receptive fields in the primary auditory cortex of the waking guinea pig. (*A*) An example of CS-specific receptive field modification produced by classical conditioning. The illustrated case is one in which the CS frequency became the best frequency. *1.* Preconditioning the best frequency was 9.5 kHz (*open arrowhead*) and the CS was selected to be 9.0 kHz (*closed arrowhead*) for conditioning, which produced behavioral conditioned responses to this frequency (not shown). *2.* One hour postconditioning, the CS frequency became the best frequency because of increased response to this frequency and decreased response to the preconditioning best frequency and other frequencies. *3.* The receptive field difference function (post-minus pre-RFs) shows that conditioning produces the maximal increase at the CS frequency and maximal decrease at the pretraining best frequency. Open circles show no systematic effect on spontaneous activity. (*B*) Group receptive field mean (± s.e.) difference functions (treatment minus control) for three types of training. *1.* Conditioning produces increased response at the frequency of the conditioned stimulus and decreases at most other frequencies starting at 0.25 octaves from the CS frequency (side-band suppression). *2.* Sensitization training produces a broad, nonspecific increase in response across the auditory receptive field, both for auditory and visual sensitization training. *3.* Habituation produces a frequency-specific decrease for a frequency which developed a decrement in response due to repeated presentation alone. Note the high degree of specificity; frequencies 0.125 octaves from the repeated frequency were little affected. From Weinberger 1994.

A.

1. Pre-Conditioning RF

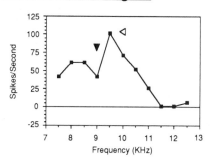

2. One Hour Post-Condit. RF

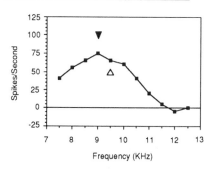

3. One Hour Post minus Pre-RF

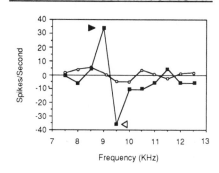

B.

1. Conditioning Modification of RF

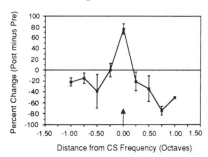

2. Auditory & Visual Sensitization

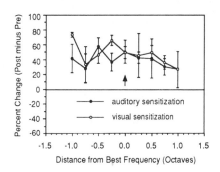

3. Habituation Modification of RF

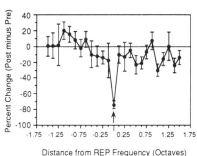

frequency (Weinberger et al 1990). This general conception is supported by findings in owl monkeys that were trained over several months in a difficult frequency discrimination task for a food reward. An increase in the number of cortical sites for which the range of discriminated frequencies were the best frequencies revealed that this type of instrumental conditioning produced an increased representation for the frequency band within which discriminations were made (Figure 3). Subjects that received similar stimulation but were not engaged in the discrimination task did not develop an enlarged frequency representation (Recanzone et al 1993). In classical conditioning, reviewed above, there are decreased responses to a CS–, nonreinforced, frequency. For instrumental conditioning, the frequency band within which difficult discriminations were made exhibited increased representation. These differences may reflect the fact that the instrumental task was a difficult discrimination in which the frequency of the CS– has to be continually attended and processed.

CS-specific reorganization of the frequency representation of the auditory cortex has also been documented in metabolic studies (reviewed in Gonzalez-Lima 1992, Scheich 1991, Scheich et al 1993). Classical conditioning increases response to the CS frequency (Gonzalez-Lima & Scheich 1986) or shifts the frequency representation toward the CS frequency; instrumental conditioned avoidance training increases the size and intensity of response to the CS frequency while decreasing 2DG uptake to discriminated nonreinforced frequencies (Simonis & Scheich as described in Scheich et al 1993).

SOMATOSENSORY SYSTEM   A type of sensory preconditioning paradigm has been used to determine the effects of paired digit stimulation in primates on cortical representation of the digits. (Receptive field findings obtained simultaneously with maps are also presented here.) Paired digit stimulation was induced by experimental syndactyly; adjacent digits were surgically fused for 3–7 months. This treatment abolished the cortical border between the digits. Within the wide common zone of representation, RFs extended across the normal border (Allard et al 1991, Clark et al 1988). The authors believe that correlated stimulation, i.e. presumed temporal coincidence of stimulation during ambient behavior of the monkeys, is largely responsible for the novel RFs and the resultant increases in cortical representation of the digits. A complementary finding is reported for surgery to correct congenital syndactyly in

---

→

Figure 2 Respresentative responses of barrel D2 cells in rats with differing sensory experience. Whisker deflection in each PSTH was at 0 ms. Cell P5U3 (*left*) was recorded in a rat (WP17) with all whiskers intact. Note the vigorous response to the CRF whisker and the symmetry in the response to whiskers D1 and D3 and whiskers C2 and E2. This is in contrast to cell P6U2 (*right*), which was recorded in a rat (WP21) with whiskers D2 and D3 paired during the preceding 64 h. Here, movement of whisker D3 yielded a stronger response (*arrow*) than did the SRF whiskers that had been cut. From Diamond et al 1993.

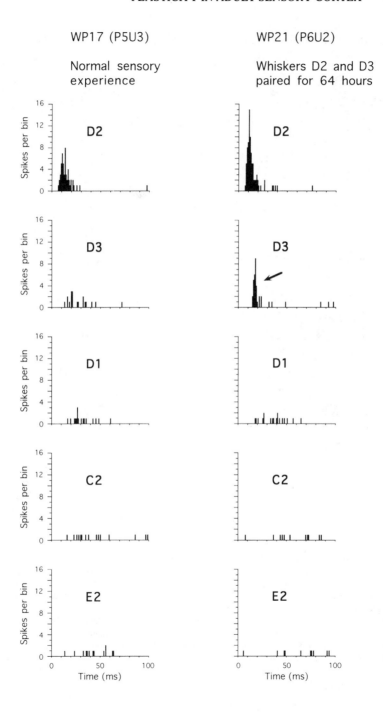

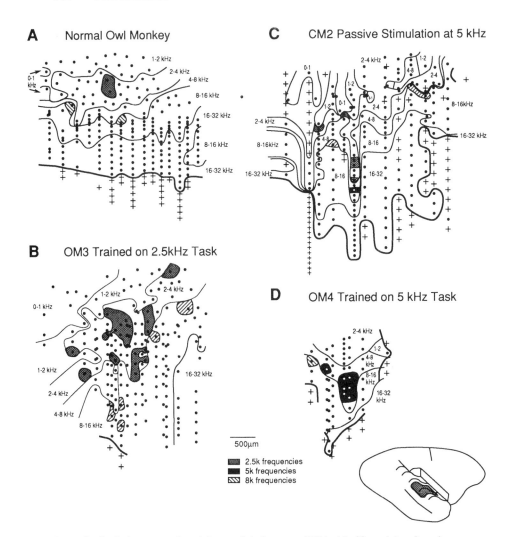

Figure 3 Cortical representation of characteristic frequency (CF) in A1 of four adult owl monkeys. Thin lines define boundaries of cortical locations with CFs within one octave. Stippled regions encircle cortical locations where neurons were recorded with CFs in the frequency range used in the 2.5 kHz task; solid regions represent frequencies used in the 5 kHz task; and hatched regions represent the frequency range used in the 8 kHz task. The cortical areas representing a given frequency range were approximated by connecting the 50% distance values to the neighboring recording sites with CFs outside the given frequency range. Pluses denote recording sites with neuronal response not consistent with properties of A1 neurons. *A* is from a representative normal owl monkey (N2); *B* is from a monkey trained at 2.5 kHz (OM3); *C* shows the monkey passively stimulated with the frequencies used in the 5 kHz task (CM2); and *D* shows the representation of the monkey trained at the 5 kHz task (OM4). From Recanzone et al 1993.

humans. Within-subject determination of digit representation using MEG showed an improved representation of the digits in the somatosensory cortex after surgery (Mogliner et al 1993).

Paired stimulation within digits also yields reorganization. Innervated skin flaps were exchanged between adjacent digits in monkeys. After several months, the cortical maps revealed representations of the skin "islands" within the normal representation of the recipient digits (reported in Merzenich & Jenkins 1993). Temporally correlated stimulation would seem to be very strong in this situation, although the extent to which such peripheral stimulation is responsible for this reorganization is unknown.

Instrumental conditioning alters the maps of digits in the somatosensory cortex. Owl monkeys were trained for several months to maintain contact with the edge of a rotating disc with the tips of one or more digits to obtain a food reward (Jenkins & Merzenich 1987; Jenkins et al 1990). Within-subject maps showed an expanded area of representation of the tips of the stimulated digits; RF size within these representations was decreased. The authors also reported novel cutaneous responses in area 3a, interpreted as expanded representation into another cytoarchitectonic area that normally represents only noncutaneous afferentation. However, Killackey (1989) has argued that these findings probably represent the engagement of muscle spindles or the unmasking of weak cutaneous inputs to area 3a rather than an expansion of area 3b.

Recanzone et al (1992a,c,d,e) determined the effects of instrumental conditioning of owl monkeys to discriminate between two frequencies of tactile stimulation applied to a small region of a single phalange of a digit. After prolonged training, the authors found (*i*) progressive improvement in behavioral discrimination that was specific to the trained site of stimulation; (*ii*) the appearance of a representation of cutaneous input from the glabrous and hairy skin of the trained hand in area 3a, with a loss of much of the normal representation of deep receptors; (*iii*) a greatly enlarged representation of the small skin locus used for tactile discrimination; (*iv*) increased temporal resolution of neuronal responses to the stimulated compared to control skin; and (*v*) lack of any of these effects in control subjects that received the same stimulation, which was behaviorally irrelevant because they were simultaneously performing an auditory discrimination task. Receptive fields were smaller than normal in the rotating disk experiment but larger than normal in the tactile frequency experiment. For a discussion concerning reconciliation of these findings, see Merzenich & Jenkins 1993.

A possibly related finding in humans is that learning to read Braille increases the cortical representation of the relevant finger tip, but some of this effect may occur in childhood (Pascual-Leone & Torres 1993). A similar effect might be obtained in the whisker barrel cortex. When adult rats have only a single whisker available during 90 days of ambient behavior, there is a pronounced metabolic increase in the dimensions of the relevant barrel column (Kossut et al 1988).

Habituation (repeated nonreinforced stroking by the experimenters) produces a significant decrease in 2DG utilization and in the area of the activated cortical barrel in lamina IIIb and IV within 10 days in adults (CL Hand & PJ Hand, submitted). In contrast, classical conditioning produces a significant increase in both 2DG utilization and in the area of activation of these same cortical layers (CL Hand & PJ Hand, submitted). Similar conditioning effects develop in aversive training; whisker-shock pairing in mice increases the cortical 2DG representation of CS vs control vibrissae (Kossut & Siucinska 1993). Thus, acquired stimulus significance increases cortical representation of a stimulus irrespective of its positive or negative reinforcement.

A more literal but less controlled whisker pairing also increases cortical representation. Gluing together the tips of a pair of adjacent whiskers for 4–8 days produces a 5- to 6-fold increase of the cortical representation of the fused whiskers (Yun 1991). However, stimulation imposed by experimenters over a similar time period can produce opposite results. Welker et al (1992) continually stimulated whiskers of mice for 1–4 days. Labeling of 2DG in the barrel field contralateral to the stimulation sites was decreased compared to ipsilateral controls. This experiment appears to be an instance of habituation. It demonstrates that stimulation per se is not sufficient to control representational plasticity; rather the behavioral significance of the stimulation determines the sign of change.

Possible Mechanisms of RF Plasticity and Map Reorganization in Learning

At present, the mechanisms of learning-related plasticity and reorganization are unknown but under increasingly active investigation. Current evidence concerns thalamic involvement, cortical mechanisms, cholinergic effects, and correlated neural activity.

THALAMIC INVOLVEMENT   Thalamic processes probably cannot account for plasticity in the auditory cortex during learning. The ventral medial geniculate body is the lemniscal source of frequency-specific input to granular layers of the auditory cortex, but it develops no plasticity to the CS during training (reviewed in Weinberger & Diamond 1987) and only very weak and highly transient RF plasticity after conditioning (Edeline & Weinberger 1991). The magnocellular medial geniculate body provides nonlemniscal input to upper layers of the auditory cortex. Its cells do develop increased responses to the CS during training, and their RFs are retuned to favor the CS frequency (Edeline & Weinberger 1992, Edeline et al 1990b, Lennartz & Weinberger 1992b). However, their RFs are much more complex and broadly tuned than those of auditory cortical cells, so it seems unlikely that the highly frequency-specific cortical RF plasticity is simply projected from this nucleus, although this cannot yet be discounted. Metabolic studies report effects of classical conditioning in the medial geniculate body, presumably in both the ventral and magnocellular nuclei, during acquisition but not during extinction trials. Furthermore, metabolic changes due to classical conditioning during acquisition and extinction trials have been found at lower levels of the auditory system (Gonzalez-Lima & Scheich 1984, Gonzalez-Lima & Agudo 1990). The differential effect in the thalamus might indicate that the auditory system should not be considered as a simple series circuit. Alternatively, or in addition, different findings in the ventral medial geniculate could reflect differences in neurophysiological vs 2-DG methods or differences in training parameters or both.

In the whisker barrel system, facilitated discharges of spared whiskers in behaving animals appear to be due to a cortical mechanism within the first 10 days of plasticity, but thalamocortical transmission seems to be facilitated thereafter (M Armstrong-James, M Diamond & F Ebner, submitted). Diamond and colleagues have hypothesized thalamic gating of interbarrel interactions (Diamond et al 1992, Diamond & Armstrong-James 1992).

CORTICAL MECHANISMS   In the barrel cortex, laminar recordings indicate that pairing-induced response plasticity occurs primarily in supra and to a lesser extent infragranular layers before appearing in layer IV (Diamond et al 1993). The authors argue that this sequence of changes indicates an intracortical, specifically an interbarrel, basis of plasticity rather than the projection of changes from the ventrobasal complex to the somatosensory cortex.

The raccoon has been used extensively in studies of cortical reorganization because of the large and distinct representation of its forepaw digits (Welker & Seidenstein 1959). Neurons in a zone of representation of the glabrous surface of the digits in SI respond only to stimulation of a single digit, while an adjacent heterogeneous field has convergence of responses from 2–5 digits (Doetsch et al 1988, Rasmusson et al 1991). Experimental syndactyly increases the incidence of excitatory postsynaptic potentials (EPSPs) in the denervated glabrous representation that are elicited by microstimulation of the heterogeneous zone, supporting the hypothesis that representational reorganization in the digit glabrous map results from increased cortico-cortical input from the heterogeneous zone (Zarzecki et al 1993). As expected, syndactyly increased the incidence of EPSPs from digit 3; surprisingly, the same effect was found for digit 5, which had not been joined to digit 4. This finding raises some question about the hypothesis that covariant digit stimulation is necessary for reorganization and the assumption that syndactyly produces increased correlated stimulation of the joined digits vs other digits. Quantification of actual digit use and stimulation (covariant or otherwise) should help resolve this issue.

CHOLINERGIC INVOLVEMENT   There is considerable evidence implicating the neuromodulatory actions of acetylcholine (ACh) (Woody et al 1978) in learning-induced sensory cortical plasticity. For example, ACh enables long-lasting facilitation of responses to cutaneous stimulation, applied cortically (Metherate et al 1988) or released by stimulation of the nucleus basalis (NB) (Rasmusson & Dykes 1988, Tremblay et al 1990, Webster et al 1991). Of particular relevance, ACh appears to be necessary for RF plasticity based on sensory preconditioning. Plasticity caused by whisker pairing (Delacour et al 1987, reviewed above) is impaired by microiontophoretic application of atropine to the barrel cortex, indicating a dependence of this process on muscarinic receptors in the cortex (Delacour et al 1990). Acetylcholine has also been implicated in RF plasticity caused by classical conditioning. In the auditory cortex, iontophoretic application of muscarinic agonists (McKenna et al 1989) or anticholinesterases (Ashe et al 1989) modifies frequency tuning that endures after drug application. Stimulation of the nucleus basalis produces atropine-sensitive long-lasting modification of evoked responses in the auditory cortex, including facilitation of field potentials, cellular discharges and EPSPs elicited by medial geniculate stimulation (Metherate & Ashe 1991, 1993), and facilitation of neuronal discharges to paired tones (Hars et al 1993, Hennevin et al 1992). Further, pairing one tone with iontophoretic application of muscarinic agonists produces pairing-specific, atropine-sensitive modification of RFs (Metherate & Weinberger 1990).

CORRELATED ACTIVITY   Extended Hebbian rules (Hebb 1949, Stent 1973) have been applied to learning-induced plasticity in the adult sensory cortex. As broadly interpreted, correlated pre- and postsynaptic activity would increase synaptic strength, and uncorrelated activity would decrease synaptic strength. Increased responses to paired stimuli and decreased responses to unpaired stimuli have generally been assumed to result from Hebbian processes (e.g. Allard et al 1991, Diamond et al 1993; see also Merzenich & Jenkins 1993). The apparent importance of a behaviorally engaged subject, rather than merely paired sensory stimulation per se, suggests gating of plasticity by one or more neuromodulators (e.g. Greuel et al 1988). For example, the retuning of RFs in the auditory cortex during classical conditioning is hypothesized to depend on the simultaneous strengthening of CS synapses and weakening of non-CS synapses in the cortex if the US produces widespread postsynaptic excitation, perhaps by the release of acetylcholine (Weinberger et al 1990; see also Dykes et al 1988). Presynaptic input should be active during presentation of the CS frequency but inactive for other frequencies not presented during conditioning;

Hebbian rules applied to receptive field plasticity during conditioning and habituation

		Pre-synaptic input active?	
		Yes	No
Postsynaptic cell depolarized?	Yes	1 Strengthen (CS. condition.)	2 Weaken (non CS. condition.)
	No	3 Weaken (CS. habituation)	4 No change (non CS. habituation)

Various frequency inputs converge on cell

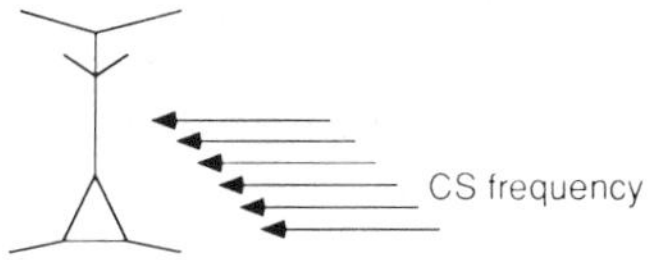

CS frequency

Conditioning : During each trial (CS-US), the CS frequency is the only active frequency input, and the US depolarizes the cell. Therefore, CS synapses are strengthened (1) and non-CS synapses are weakened (2).

Habituation : During each trial ("CS" alone), the CS frequency is the only active frequency input, but the absence of other stimuli results in a lack of postsynaptic depolarization. Therefore, CS synapses are weakened (3) but non-CS synapses are unchanged (4).

Figure 4 Application of extended Hebbian rules to receptive field plasticity for classical conditioning and habituation. Simple combinations of pre- and postsynaptic elements, each of which can be in an active or nonactive state, might account for the effects of conditioning on CS and non-CS synaptic strengths and for the effects of habituation on the repeated and nonrepeated stimuli. From Weinberger 1993.

this would result in an increased pre- and postsynaptic correlation for the former and the inverse for the latter. Hebbian rules might also explain the selective reduction of cortical responses to a habituated stimulus (Figure 4).

Fregnac and his colleagues (Fregnac et al 1988, 1992; Shulz & Fregnac 1992) have obtained more evidence for Hebbian processes in the visual cortex of anesthetized cats. They were able to reverse the orientation and ocular dominance preferences of cells in juvenile and adult cats by postsynaptically increasing neuronal response to an unpreferred stimulus and decreasing neuronal response to the preferred stimulus. It would be interesting to extend this type of approach to behaving subjects to determine if differential reward contingencies also change visual cortical RF preferences.

Additional evidence of Hebbian processes has been found in the auditory cortex of behaving monkeys by altering the discharge contingencies between pairs of neurons (Ahissar et al 1992). When one neuron fired spontaneously, a sound that activated a second cell was delivered immediately. The cross-correlation between these cells increased if the sound was behaviorally significant, i.e. if the monkey had to produce a behavioral response for reward, but was hardly changed if the sound was behaviorally irrelevant. Although these findings do not include RF data, they do show that functional relations between sensory cortical neurons in the adult can be controlled by the degree of correlated activity.

The role of correlated activity in the expansion of RFs and representational maps also is supported by recent studies of intracortical microstimulation (ICMS). This approach assumes that stimulation will excite most local afferents and postsynaptic cells to provide temporally coincident activity, which is hypothesized to be necessary for changing synaptic weights and recruiting cells into a new representation. Recanzone et al (1992b) found that ICMS applied to the somatosensory cortex in both rats and monkeys produced enlarged cortical representations of the restricted skin region that was represented at the site of stimulation. A hypothesized increase in neuronal group cooperativity within the affected zone also has been reported. Dinse et al (1993) found that there was increased correlated activity between pairs of neurons within the cortical sector that had been representationally reorganized but not between cells outside of this affected zone. Application of such cross-correlational analysis to cortical zones that have undergone learning-induced reorganization in behaving animals is feasible and should be undertaken.

EFFECTS OF SENSORY DEAFFERENTATION

As explained earlier, space constraints limit this section to a resume of the salient characteristics of the effects of sensory denervation. For a general review of auditory, somatosensory, and visual deafferentation, see Kaas (1991). For the somatosensory system, relevant reviews and discussions of mechanisms include Calford & Tweedale (1991b), Juliano & Jacobs (1993), Kano et al (1991), Pons et al (1988), and Whitsel & Kelly (1988). For the visual cortex see Gilbert (1992), Hendry & Carder (1992), Jones (1990), and Kasamatsu (1994).

Partial sensory deafferentation by destruction or reversible anesthesia of sensory receptors (e.g. amputation, denervation, or lesion of the cochlea or retina) generally results in responses to other stimuli, usually those that engage sensory epithelium that is adjacent to the site of lesion. Novel responses are often observed immediately, i.e. while subjects are still under general anesthesia. In the somatosensory system, immediate novel, often expanded, receptive fields have been found in primary somatosensory cortex [flying fox (Calford & Tweedale 1988), monkey (Calford & Tweedale 1991a), raccoon (Kehlahan & Doetsch 1984), and rat (Byrne & Calford 1991)], the ventroposteriomedial complex (VPM) of the thalamus [rat (Garraghty & Kass 1991b; Nicoleilis et al 1993a,b)], and the dorsal column nucleus cuneatus [cat (Pettit & Schwark 1993; see also Rhoades et al 1987)]. In the visual system, immediate new receptive fields have been observed in the visual cortex [cat (Chino et al 1992, Gilbert & Wiesel 1992, Pettet & Gilbert 1992; LM Schmid, MGP Rosa, JS Ambler & MB Calford, submitted) and monkey (Fiorani et al 1992, Gilbert & Wiesel 1992)] but not in the lateral geniculate nucleus (Gilbert & Wiesel 1992).

These immediate effects cannot be caused by postdeafferentation sensory-behavioral interactions, i.e. by learning processes. Immediate effects presumably do form the basis for long-term chronic modifications of receptive fields and representational maps that develop over weeks or months following deafferentation. Such long-term changes have been found for all cases studied. Novel responses observed can readily be interpreted as expansions of the representations of the receptor epithelia that are adjacent to the site of receptor damage. These map reorganizations have been found in the auditory cortex [cat (Rajan et al 1993), guinea pig (Robertson & Irvine 1989), monkey (Schwaber et al 1993), and mouse (Willot et al 1993)], somatosensory cortex [monkey (Garraghty & Kaas 1991a; Merzenich et al 1983a,b, 1984; Pons et al 1991; see also Ramachandran et al 1992 for relevant human behavioral findings), raccoon (Rasmusson 1982, Rasmusson & Turnbull 1983, Turnbull & Rasmusson 1991, Rasmusson et al 1992)], and visual cortex [cat (Kaas et al 1990, Schmid et al 1993) and monkey (Heinen & Skavenski 1991)]. Merzenich et al (1983b) and Calford & Tweedale (1988) have observed progressive refinement of new RFs and overall representational maps.

Such findings, and other considerations, have led workers to propose that these dynamic aspects of functional cortical organization following deafferentation are due to use or experience, i.e. sensory stimulation that engages intact sensory receptors of the affected sensory system (e.g. Merzenich 1986). The preceding section that reviews learning and sensory stimulation effects on receptive fields and representational maps supports this hypothesis. It also makes explicit the difference between stimulation per se and stimulation that is behaviorally relevant. Behavioral significance seems to be important, if not necessary, for facilitated cortical responses. Stimulation without behavioral significance appears to produce habituation and decreased cortical responses. In any event, systematic objective measurement of sensory stimulation and behavior following peripheral sensory deafferentation have not yet been reported. It follows that the behavioral significance of such stimulation has not been reported either. Therefore, the role of learning in cortical RF plasticity and reorganization subsequent to deafferentation remains conjectural.

CONCLUSIONS

This section (*a*) attempts to summarize findings, (*b*) presents a capsule over-view of efforts in the field, (*c*) makes recommendations for future inquiry, and (*d*) presents a suggestion.

Major Issues: What Has Been Found?

TREATMENTS THAT MODIFY RFs AND REPRESENTATIONAL MAPS   All of the types of learning studied alter both RFs and maps: habituation, sensory pre-conditioning, classical conditioning, and instrumental conditioning. Sensory deafferentation also changes RFs and maps. Thus, sensory cortical plasticity is produced both by events that animals normally encounter and by traumas that, while less prevalent in life, reveal an apparent life-long capacity for plasticity and altered representations of sensory receptors.

CHARACTERISTICS OF DYNAMIC REGULATION   Learning generally increases responses to and representations of stimuli that acquire behavioral significance and conversely decreases response to and representations of behaviorally un-important stimuli. RF plasticity in classical conditioning can be rapid and last indefinitely. Sensory deafferentation produces expanded representations of adjacent parts of sensory epithelia. New responses may be present immediately or require time to develop, and representations exhibit progressive refinement and precision over weeks and months. Thus, a neural economics appears to be at work: Sensory cortex has a constant number of neurons, but their priorities for being engaged by stimuli are reallocated on the basis of behavioral signif-icance in learning and some sort of fill-in rule in deafferentation. These two rules are not mutually exclusive.

SUBCORTICAL AND OTHER CORTICAL STRUCTURES INVOLVED   Learning in the auditory system does not seem to be evident for classical conditioning in the lemniscal thalamus but rather in nonlemniscal thalamus; however, sensory preconditioning in the whisker barrel system may be present in lemniscal thalamocortical projections in a late stage of plasticity. Denervation produces plasticity in the thalamus and dorsal columns of the somatosensory system. Thus, current findings implicate subcortical structures in denervation more than in learning; however, the relative paucity of data render this conclusion tenuous.

The ability of primary sensory cortex to be a likely site of plasticity in

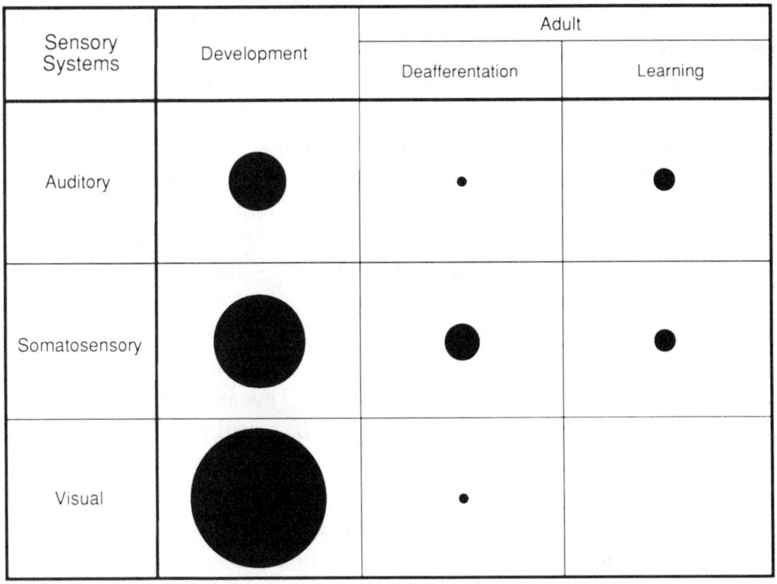

Sensory Systems	Development	Adult	
		Deafferentation	Learning
Auditory	●	•	●
Somatosensory	●	●	●
Visual	●	•	

Figure 5 A subjective estimate of the extent of research in developmental sensory plasticity and adult sensory cortical plasticity. The relative sizes of the dots indicate the relative estimates of research; the dots should be interpreted as occuring on a greatly expanded nonlinear scale.

learning and sensory stimuli is evident in invasive treatments that produce RF and map changes, e.g. direct application of cholinergic agents, release of cortical ACh by stimulation of the nucleus basalis, local modification of postsynaptic excitability, and intracortical microstimulation.

CELLULAR MECHANISMS   Cholinergic actions at the level of sensory cortex appear to be important, perhaps necessary, for learning, but other transmitters have not been studied to the same extent in learning context. Changes in the degree of correlated activity, particularly as formulated in terms of extended Hebbian rules, find support for the effects of learning.

What Has Been Investigated?

Inquiry into RF plasticity and representational reorganization in primary sensory cortex is relatively new. As in other fields, research to date has been uneven across sensory systems and treatments. Figure 5 presents a summary of a subjective impression of the efforts to date. Included is an estimate of efforts in developmental plasticity, as a sort of calibration. The relative sizes of the dots are estimates of the magnitude of research, but they should be viewed as occurring on at least a log scale: The page would be too small for the visual developmental plasticity dot! Clearly research on adult plasticity has room to grow, and several areas are fertile ground for foundational studies. Adult plasticity is distinguished from developmental plasticity not merely in the age of its subjects and certainly not in the use of sensory deafferentation or stimulation, but rather by the study of learning. Of both scientific and sociological interest, and clear benefit, studies of plasticity and learning in adult sensory cortex are originating mainly from sensory physiology laboratories rather than laboratories concerned with learning and memory. The latter are hereby encouraged to bring their particular expertise to this problem area.

What Should Be Studied?

CORTICOTHALAMIC SYSTEMS AND PLASTICITY   This is perhaps the greatest area of ignorance in sensory cortical function. It is well known that thalamocortical relations are in the form of loops rather than an ascending chain. Plasticity in the thalamus may well cause cortical plasticity. However, there are at least two other possibilities. First, tonic influences from the cortex could be involved in thalamic plasticity via descending projections. Second, there can be plasticity at both levels; the thalamic could be of a different nature than the cortical plasticity that it promotes or enables (e.g. Weinberger et al 1990). Plasticity and reorganizations in sensory systems undoubtedly have functional consequences wherever they are observed, regardless of the levels or sites that are causative.

SENSORY STIMULATION AND BEHAVIOR IN CHRONIC DEAFFERENTATION   The effects of denervation may not be understandable unless investigators assess the nature and amount of sensory stimulation and the behavioral significance of such stimulation to the subjects between the time of deafferentation and the time of assessment of sensory cortex. It makes a great deal of difference whether or not a stimulus is behaviorally relevant, and there exist a multitude of standard applicable behavioral techniques that can be used to make this type of assessment.

THE WAKING BRAIN WHENEVER POSSIBLE   The study of waking brains is more difficult in sensory physiology than the study of anesthetized brains. Without question, data obtained under anesthesia are of enormous importance. Nonetheless, we should bear in mind that the use of anesthesia without comparing findings to the waking brain constitutes a sort of drug experiment without a control, that the anesthetized brain did not evolve, and that the use of a general anesthetic, while essential for some types of experiments, may well obscure some fundamental features of brain function. The fact that learning is prevented by anesthesia is not merely an inconvenience but a warning.

BEHAVIORAL FUNCTIONS OF RF PLASTICITY AND MAP REORGANIZATIONS   The last of the five issues listed at the start of this article concerned the behavioral functions of physiological changes in sensory cortices. There are insufficient findings on this central issue to have warranted a summary. Informed speculations concern basic perceptual functions such as perceptual continuities in the absence of complete sensory receptor stimulation; more complex perceptual processes, including constancies and expectation-based effects; increased acuity and discriminative capacities; and selective attention or inattention. Learning, in the broad sense of acquiring information, could subserve several of these functions as well as provide the basis for storing information about the behavioral significance of stimuli where it is analyzed, i.e. in sensory cortex and other levels of sensory systems. Research is needed.

A Suggestion

Discovery of the dynamic regulation of primary sensory cortex in the adult constitutes a severe blow to the hypothesis that cortical perceptual functions are based on static properties of individual cells. The case is by no means closed, however, because there may still be populations of cortical cells that have static response properties so that both static and dynamic processes work together within sensory cortex. In any event, it appears that new conceptions of sensory cortical function are needed to incorporate dynamic regulation. The field currently seems to be in a state of conceptual flux. A broad view of sensory system function within the framework of animal adaptation and behavior should be thoroughly considered during this period of rapid change, large challenges, and great opportunities.

ACKNOWLEDGMENTS

Preparation of this review was supported by NIDCD research grant #DC 02346 and by an unrestricted grant from the Monsanto Company. I am pleased to acknowledge the critical comments of Ron Frostig who reviewed an earlier version of this article, and I thank Jacquie Weinberger for bibliographic assistance and preparation of this manuscript. I am also grateful to the many neuroscientists who provided reprints and preprints of their research and apologize for my inability to include all relevant work.

Literature Cited

Allard T, Clark SA, Jenkins WM, Merzenich MM. 1991. Reorganization of somatosensory area 3b representations in adult owl monkeys after digital syndactyly. *J. Neurophysiol.* 66: 1048–58

Ahissar E, Vaadia E, Ahissar M, Bergman H, Arieli A, Abeles M. 1992. Dependence of cortical plasticity on correlated activity of single neurons and on behavioral context. *Science* 257:1412–15

Armstrong-James M, George MJ. 1988. Influence of anesthesia on spontaneous activity and receptive field size of single units in rat Sm1 neocortex. *Exp. Neurol.* 99:369–87

Ashe JH, McKenna TM, Weinberger NM. 1989. Cholinergic modulation of frequency receptive fields in auditory cortex: II. Frequency-specific effects of anticholinesterases provide evidence for a modulator action of endogenous ACh. *Synapse* 4:44–54

Bakin JS, Lepan B, Weinberger NM. 1992. Sensitization induced receptive field plasticity in the auditory cortex is independent of CS-modality. *Brain Res.* 577:226–35

Bakin JS, Weinberger NM. 1990. Classical conditioning induces CS-specific receptive field plasticity in the auditory cortex of the guinea pig. *Brain Res.* 536:271–86

Byrne JA, Calford MB. 1991. Short-term expansion of receptive fields in rat primary somatosensory cortex after hindpaw digit denervation. *Brain Res.* 565:218–24

Calford MB, Tweedale R. 1988. Immediate and chronic changes in responses of somatosensory cortex in adult flying-fox after digit amputation. *Nature* 332:446–48

Calford MB, Tweedale R. 1991a. Immediate expansion of receptive fields of neurons in area 3b of macaque monkeys after digit denervation. *Somatosens. Mot. Res.* 8:249–60

Calford MB, Tweedale R. 1991b. C-fibres provide a source of masking inhibition to primary somatosensory cortex. *Proc. R. Soc. London Ser. B* 243:269–75

Chino YM, Kaas JH, Smith E, Langston AL, Cheng H. 1992. Rapid reorganization of cortical maps in adult cats following restricted deafferentation in retina. *Vis. Res.* 32:789–96

Clark SA, Allard T, Jenkins WM, Merzenich MM. 1988. Receptive fields in the body-surface map in adult cortex defined by temporally correlated inputs. *Nature* 332: 444–45

Condon CD, Weinberger NM. 1991. Habituation produces frequency-specific plasticity of receptive fields in the auditory cortex. *Behav. Neurosci.* 105:416–30

Delacour J, Houcine O, Costa JC. 1990. Evidence for a cholinergic mechanism of "learned" changes in the responses of barrel field neurons of the awake and undrugged rat. *Neuroscience* 34:1–8

Delacour J, Houcine O, Talbi B. 1987. "Learned" changes in the responses of the rat barrel field neurons. *Neuroscience* 23:63–71

Diamond ME, Armstrong-James M. 1992. Role of parallel sensory pathways and cortical columns. *Concepts Neurosci.* 3:55–78

Diamond ME, Armstrong-James M, Ebner FF. 1992. Somatic sensory responses in the rostral sector of the posterior group (POm) and in the ventral posterior medial nucleus (VPM) of the rat thalamus. *J. Comp. Neurol.* 318:462–76

Diamond ME, Armstrong-James M, Ebner FF. 1993. Experience-dependent plasticity in adult rat barrel cortex. *Proc. Natl. Acad. Sci. USA* 90:2082–86

Diamond ME, Huang W, Ebner FF. 1993. Laminar comparison of plasticity in adult barrel cortex. *Soc. Neurosci. Abstr.* 19:1569

Diamond DM, Weinberger NM. 1986. Classical conditioning rapidly induces specific changes in frequency receptive fields of single neurons in secondary and ventral ectosylvian auditory cortical fields. *Brain Res.* 372:357–60

Diamond DM, Weinberger NM. 1989. Role of context in the expression of learning-induced plasticity of single neurons in auditory cortex. *Behav. Neurosci.* 103:471–94

Dinse HR, Kruger K, Best J. 1990. A temporal structure of cortical information processing. *Concepts Neurosci.* 1:199–238

Dinse HR, Recanzone GH, Merzenich MM. 1993. Alterations in correlated activity parallel ICMS-induced representational plasticity. *NeuroReport* 5:173–76

Doetsch GS, Standage GP, Johnston KW, Lin CS. 1988. Intracortical connections of two functional subdivisions of the somatosensory forepaw cerebral cortex of the raccoon. *J. Neurosci.* 8:1887–900

Dykes RW, Metherate R, Tremblay N. 1988. Cholinergic modulation of the neuronal excitability in cat somatosensory cortex. In *Neurotransmitters and Function: From Molecules to Mind,* ed. M Avoli, T Reader, R Dykes, P Gloor. New York: Plenum

Edeline J-M, Neuenschwander-El Massioui N, Dutrieux G. 1990a. Frequency-specific cellular changes in the auditory system during acquisition and reversal of discriminative conditioning. *Psychobiology* 18:382–93

Edeline J-M, Neuenschwander-el Massioui N, Dutrieux G. 1990b. Discriminative long-term retention of rapidly induced multiunit changes in the hippocampus, medial geniculate and auditory cortex. *Behav. Brain Res.* 39:145–55

Edeline J-M, Pham P, Weinberger NM. 1993. Rapid development of learning-induced receptive field plasticity in the auditory cortex. *Behav. Neurosci.* 107:539–57

Edeline J-M, Weinberger NM. 1991. Thalamic short-term plasticity in the auditory system: associative returning of receptive fields in the ventral medial geniculate body. *Behav. Neurosci.* 105:618–39

Edeline J-M, Weinberger NM. 1992. Associative retuning in the thalamic source of input to the amygdala and auditory cortex: receptive field plasticity in the medial division of the medial geniculate body. *Behav. Neurosci.* 106:81–105

Edeline J-M, Weinberger NM. 1993. Receptive field plasticity in the auditory cortex during frequency discrimination training: selective retuning independent of task difficulty. *Behav. Neurosci.* 107:82–103

Fiorani M Jr, Rosa MGP, Gattass R, Rocha-Miranda CE. 1992. Dynamic surrounds of receptive fields in primate striate cortex: a physiological basis for perceptual completion? *Proc. Natl. Acad. Sci. USA* 89:8547–51

Fregnac Y, Shulz D, Thorpe S, Bienenstock E. 1988. A cellular analogue of visual cortical plasticity. *Nature* 333:367–70

Fregnac Y, Shulz D, Thorpe S, Bienenstock E. 1992. Cellular analogs of visual cortical epigenesis. I. Plasticity of orientation selectivity. *J. Neurosci.* 12:1280–300

Garraghty PE, Kaas JH. 1991a. Large-scale functional reorganization in adult monkey cortex after peripheral nerve injury. *Proc. Natl. Acad. Sci. USA* 88:6976–80

Garraghty PE, Kaas JH. 1991b. Functional reorganization in adult monkey thalamus after peripheral nerve injury. *NeuroReport* 2:747–50

Gilbert CD. 1992. Horizontal integration and cortical dynamics. *Neuron* 9:1–13

Gilbert CD, Wiesel TN. 1990. The influence of contextual stimuli on the orientation selectivity of cells in primary visual cortex of the cat. *Vis. Res.* 30:1689–701

Gilbert CD, Wiesel TN. 1992. Receptive field dynamics in adult primary visual cortex. *Nature* 356:150–52

Gonzalez-Lima F. 1992. Brain imaging of auditory learning functions in rats: studies with fluorodeoxyglucose autoradiography and cytochrome oxidase histochemistry. In *Advances in Metabolic Mapping Techniques for Brain Imaging of Behavioral and Learning Functions*, ed. F Gonzalez-Lima, Th Findenstadt, H Scheich, 68:39–109. NATO ASI Ser. D, Boston/London: Kluwer Academic. 527 pp.

Gonzalez-Lima F, Agudo J. 1990. Functional reorganization of neural auditory maps by differential learning. *NeuroReport* 1:161–64

Gonzalez-Lima F, Scheich H. 1984. Neural substrates for tone-conditioned bradycardia demonstrated with 2-deoxyglucose I activation of auditory nuclei. *Behav. Brain Res.* 14:213–33

Gonzalez-Lima F, Scheich H. 1986. Neural substrates for tone-conditioned bradycardia demonstration with 2 deoxyglucose, II. Auditory cortex plasticity. *Behav. Brain Res.* 20:281–93

Greuel JM, Lunmann HJ, Singer W. 1988. Pharmacological induction of use-dependent receptive field modifications in the visual cortex. *Science* 242:74–77

Hars B, Maho C, Edeline JM, Hennevin E. 1993. Basal forebrain stimulation facilitates tone-evoked responses in the auditory cortex of awake rat. *Neuroscience* 56:61–74

Hebb DO. 1949. *The Organization of Behavior: A Neuropsychological Theory.* New York: Wiley

Heinen SJ, Skavenski AA. 1991. Recovery of visual responses in foveal V1 neurons following bilateral foveal lesions in adult monkey. *Exp. Brain Res.* 83:670–74

Hendry SH, Carder RK. 1992. Organization and plasticity of GABA neurons and receptors in monkey visual cortex. *Prog. Brain Res.* 90:477–502

Hennevin E, Maho C, Hars B. 1992. Learning-induced increase of tone-evoked response in the auditory thalamus during paradoxical sleep. *Soc. Neurosci. Abstr.* 18:1064

Jenkins WM, Merzenich MM. 1987. Reorganization of neocortical representations after brain injury: a neurophysiological model of the bases of recovery from stroke. *Prog. Brain Res.* 71:249–66

Jenkins WM, Merzenich MM, Ochs MT, Allard T, Guic-Robles E. 1990. Functional reorganization of primary somatosensory cortex in adult owl monkeys after behaviorally controlled tactile stimulation. *J. Neurophysiol.* 63:82–104

Jones EG. 1990. The role of afferent activity in the maintenance of primate neocortical function. *J. Exp. Biol.* 153:155–76

Juliano SL, Jacobs SE. 1993. The role of acetylcholine in barrel cortex. In *Cerebral Cortex (Barrel Cortex)*, ed. EG Jones, A Peters, I Diamond. New York: Plenum. In press

Kaas JH. 1991. Plasticity of sensory and motor maps in adult mammals. *Annu. Rev. Neurosci.* 14:137–67

Kaas JH, Krubitzer LA, Chino YM, Langston AL, Polley EH, Blair N. 1990. Reorganization of retinotopic cortical maps in adult mammals after lesions of the retina. *Science* 248:229–31

Kano M, Lino K, Kano M. 1991. Functional reorganization of adult cat somatosensory cortex is dependent on NMDA receptors. *NeuroReport* 2:77–80

Kasamatsu T. 1994. Studies on regulation of

ocular dominance plasticity: strategies and findings. In *Structural and Functional Organization of the Neocorte. A Symposium in the Memory of Otto D. Creutzfeldt,* ed. B Albowitz, K Albus, U Kuhnt, HC Nothdurft, P Wahle, pp. 68–80. Berlin: Springer-Verlag

Kelahan AM, Doetsch GS. 1984. Time-dependent changes in the functional organization of somatosensory cerebral cortex following digit amputation in adult raccoons. *Somatosens. Res.* 2:49–81

Killackey HP. 1989. Static and dynamic aspects of cortical somatotopy: a critical evaluation. *Cogn. Neurosci.* 1:3–11

Konorski J. 1967. *Integrative Activity of the Brain: An Interdisciplinary Approach.* Chicago: Univ. Chicago Press

Kossut M. 1992. Plasticity of the barrel cortex neurons. *Prog. Neurobiol.* 39:389–422

Kossut M, Hand PJ, Greenberg J, Hand CL. 1988. Single vibrissal cortical column in SI cortex of rat and its alterations in neonatal and adult vibrissa-deafferented animals: a quantitative 2DG study. *J. Neurophysiol.* 60:829–52

Kossut M, Siucinska E. 1993. Short-lasting classical conditioning and extinction produce changes of body maps in somatosensory cortex of mice. *Soc. Neurosci. Abstr.* 19:162

Lee C-J, Whitsel BL. 1992. Mechanisms underlying somatosensory cortical dynamics: I. In vivo studies. *Cereb. Cortex* 2:81–106

Lennartz RC, Weinberger NM. 1992a. Analysis of response systems in Pavlovian conditioning reveals rapidly vs slowly acquired conditioned responses: support for two-factors and implications for neurobiology. *Psychobiology* 20:93–119

Lennartz RC, Weinberger NM. 1992b. Frequency-specific receptive field plasticity in the medial geniculate body induced by Pavlovian fear conditioning is expressed in the anesthetized brain. *Behav. Neurosci.* 106:484–97

Lieberman DA. 1990. *Learning Behavior and Cognition.* Belmont, CA: Wadsworth. 500 pp.

Mackintosh NJ. 1974. *The Psychology of Animal Learning.* New York: Academic. 730 pp.

Mackintosh NJ. 1983. *Conditioning and Associative Learning.* New York: Oxford Univ. Press. 316 pp.

Masino SA, Kwon MC, Dory Y, Frostin RD. 1993. Characterization of functional organizations within rat barrel cortex using intrinsic signal optical imaging through a thinned skull. *Proc Natl. Acad. Sci. USA* 90:9998–10002

McKenna TM, Ashe JH, Weinberger NM. 1989. Cholinergic modulation of frequency receptive fields in auditory cortex: I. Frequency-specific effects of muscarinic agonists. *Synapse* 4:30–43

Merzenich MM. 1986. Sources of intraspecies and interspecies cortical map variability in mammals: conclusions and hypotheses. In *Comparative Neurobiology: Modes of Communication in the Nervous System,* ed. MJ Cohen, F Strumwasser, pp. 105–16. New York: Wiley

Merzenich MM, Jenkins WM. 1993. Reorganization of cortical representations of the hand following alterations of skin inputs induced by nerve injury, skin island transfers, and experience. *J. Hand Ther.* 6:89–104

Merzenich MM, Kaas JH, Wall J, Nelson RJ, Sur M, Felleman D. 1983a. Topographic reorganization of somatosensory cortical areas 3b and 1 in adult monkeys following restricted deafferentation. *Neuroscience* 8:33–55

Merzenich MM, Kaas JH, Wall JT, Sur M, Nelson RJ, Felleman DJ. 1983b. Progression of change following median nerve section in the cortical representation of the hand in areas 3b and 1 in adult owl and squirrel monkeys. *Neuroscience* 10:639–65

Merzenich MM, Nelson RJ, Stryker MP, Cynader MS, Schoppmann A, Zook JM. 1984. Somatosensory cortical map changes following digit amputation in adult monkeys. *J. Comp. Neurol.* 224:591–605

Metherate R, Ashe JH. 1991. Basal forebrain stimulation modifies auditory cortex responsiveness by an action at muscarinic receptors. *Brain Res.* 559:163–67

Metherate R, Ashe JH. 1993. Nucleus basalis stimulation facilitates thalamocortical synaptic transmission in the rat auditory cortex. *Synapse* 14:132–43

Metherate R, Tremblay N, Dykes RW. 1988. Transient and prolonged effects of acetylcholine on responsiveness of cat somatosensory cortical neurons. *J. Neurophysiol.* 59:1253–76

Metherate R, Weinberger NM. 1990. Cholinergic modulation of responses to single tones produces tone-specific receptive field alterations in cat auditory cortex. *Synapse* 6:133–45

Mogilner A, Grossman JA, Ribary U, Joliot M, Volkmann J, et al. 1993. Somatosensory cortical plasticity in adult humans revealed by magnetoencephalography. *Proc. Natl. Acad. Sci. USA* 90:3593–97

Nicolelis MA, Lin RC, Woodward DJ, Chapin JK. 1993a. Dynamic and distributed properties of many-neuron ensembles in the ventral posterior medial thalamus of awake rats. *Proc. Natl. Acad. Sci. USA* 90:2212–16

Nicolelis MA, Lin RC, Woodward DJ, Chapin JK. 1993b. Induction of immediate spatiotemporal changes in thalamic networks by peripheral block of ascending cutaneous information. *Nature* 361:533–36

Ohl F, Simonis C, Scheich H. 1992. Coding

associative auditory information by spectral gradient enhancement. *Soc. Neurosci. Abstr.* 18:841

Pascual-Leone A, Torres F. 1993. Plasticity of the sensorimotor cortex representation of the reading finger in Braille readers. *Brain* 116:39–52

Pettet MW, Gilbert CD. 1992. Dynamic changes in receptive-field size in cat primary visual cortex. *Proc. Natl. Acad. Sci. USA* 89:8366–70

Pettit MJ, Schwark HD. 1993. Receptive field organization in dorsal column nuclei during temporary denervation. *Science* 262:2054–56

Pons TP, Garraghty PE, Mishkin M. 1988. Lesion-induced plasticity in the second somatosensory cortex of adult macaques. *Proc. Natl. Acad. Sci. USA* 85:5279–81

Pons TP, Garraghty PE, Ommaya AK, Kaas JH, Taub E, Mishkin M. 1991. Massive cortical reorganization after sensory deafferentation in adult macaques. *Science* 252:1857–60

Rajan R, Irvine DRF, Wise LZ, Heil P. 1993. Effect of unilateral partial cochlear lesions in adult cats on the representation of lesioned and unlesioned cochleas in primary auditory cortex. *J. Comp. Neurol.* 338:17–49

Ramachandran VS, Stewart M, Rogers-Ramachandran DC. 1992. Perceptual correlates of massive cortical reorganization. *NeuroReport* 3:583–86

Rasmusson DD. 1982. Reorganization of raccoon somatosensory cortex following removal of the fifth digit. *J. Comp. Neurol.* 205:313–26

Rasmusson DD, Dykes RW. 1988. Long-term enhancement of evoked potentials in cat somatosensory cortex produced by co-activation of the basal forebrain and cutaneous receptors. *Exp. Brain Res.* 70:276–86

Rasmusson DD, Turnbull BG. 1983. Immediate effects of digit amputation on SI cortex in the raccoon: unmasking of inhibitory fields. *Brain Res.* 288:368–70

Rasmusson DD, Webster HH, Dykes RW. 1992. Neuronal response properties within subregions of raccoon somatosensory cortex 1 week after digit amputation. *Somatosens. Mot. Res.* 9:279–89

Rasmusson DD, Webster HH, Dykes RW, Biesold D. 1991. Functional regions within the map of a single digit in raccoon primary somatosensory cortex. *J. Comp. Neurol.* 313:151–61

Recanzone GH, Jenkins WM, Hradek GT, Merzenich MM. 1992a. Progressive improvement in discriminative abilities in adult owl monkeys performing a tactile frequency discrimination task. *J. Neurophysiol.* 67:1015–30

Recanzone GH, Merzenich MM, Dinse HR.

1992b. Expansion of the cortical representation of a specific skin field in primary somatosensory cortex by intracortical microstimulation. *Cereb. Cortex* 2:181–96

Recanzone GH, Merzenich MM, Jenkins WM, Grajski KA, Dinse HR. 1992c. Topographic reorganization of the hand representation in cortical area 3b owl monkeys trained in a frequency-discrimination task. *J. Neurophysiol.* 67:1031–56

Recanzone GH, Merzenich, MM, Jenkins WM. 1992d. Frequency discrimination training engaging a restricted skin surface results in an emergence of a cutaneous response in cortical area 3a. *J. Neurophysiol.* 67:1057–70

Recanzone GH, Merzenich MM, Schreiner CE. 1992e. Changes in the distributed temporal response properties of SI cortical neurons reflect improvements in performance on a temporally based tactile discrimination task. *J. Neurophysiol.* 67:1071–91

Recanzone GH, Schreiner CE, Merzenich MM. 1993. Plasticity in the frequency representation of primary auditory cortex following discrimination training in adult owl monkeys. *J. Neurosci.* 13:87–103

Rescorla RA. 1988. Behavioral studies of Pavlovian conditioning. *Annu. Rev. Neurosci.* 11:329–52

Rhoades RW, Belford GR, Killackey HP. 1987. Receptive-field properties of rat ventral posterior medial neurons before and after selective kainic acid lesions of the trigeminal brain stem complex. *J. Neurophysiol.* 57:1577–600

Robertson D, Irvine DR. 1989. Plasticity of frequency organization in auditory cortex of guinea pigs with partial unilateral deafness. *J. Comp. Neurol.* 282:456–71

Scheich H. 1991. Auditory cortex: comparative aspects of maps and plasticity. *Curr. Opin. Neurobiol.* 1:236–47

Scheich H, Simonis C, Ohl F, Tillein J, Thomas H. 1993. Functional organization and learning-related plasticity in auditory cortex of the Mongolian gerbil. *Prog. Brain Res.* 97:135–43

Schmid LM, Rosa MGP, Ambler JS, Calford MB. 1993. Changes in responsiveness and ocular dominance of striate cortex neurons following retinal lesions. *Proc. Aust. Neurosci. Soc.* 4:126

Shulz D, Fregnac Y. 1992. Cellular analogs of visual cortical epigenesis. II. Plasticity of binocular integration. *J. Neurosci.* 12:1301–18

Schwaber MK, Garraghty PE, Kaas JH. 1993. Neuroplasticity of the adult primate auditory cortex following cochlear hearing loss. *Am. J. Otol.* 14:252–58

Simons DJ, Carvell GE, Hershey AE, Bryant DP. 1992. Responses of barrel cortex neurons in awake rats and effects of urethane anesthesia. *Exp. Brain Res.* 91:29–72

Stent GS. 1973. A physiological mechanism for Hebb's postulate of learning. *Proc. Natl. Acad. Sci. USA.* 70:997–1001

Stryker MP, Jenkins WM, Merzenich MM. 1987. Anesthetic state does not affect the map of the hand representation within area 3b somatosensory cortex in owl monkey. *J. Comp. Neurol.* 258:297–303

Taylor ME, Rucker HK. 1993. Patterns of response plasticity in the receptive field of the rat auditory cortex after conditioning. *Soc. Neurosci. Abstr.* 19:1688

Tremblay N, Warren RA, Dykes RW. 1990. Electrophysiological studies of acetylcholine and the role of the basal forebrain in the somatosensory cortex of the cat (II) cortical neurons excited by somatic stimuli. *J. Neurophysiol.* 64:1212–22

Turnbull BG, Rasmusson DD. 1991. Chronic effects of total or partial digit denervation on raccoon somatosensory cortex. *Somatosens. Mot. Res.* 8:201–13

Webster HH, Rasmusson DD, Dykes RW, Schliebs R, Schober W, et al. 1991. Longterm enhancement of evoked potentials in raccoon somatosensory cortex following coactivation of the nucleus basalis of Meynert complex and cutaneous receptors. *Brain Res.* 545:292–96

Weinberger NM. 1994. Retuning the brain by fear conditioning. In *The Cognitive Neurosciences,* ed. M Gazzaniga. Cambridge: MIT Press. In press

Weinberger NM. 1993. Learning-induced changes of auditory receptive fields. *Curr. Opin. Neurobiol.* 3:570–77

Weinberger NM, Ashe JH, Metherate R, McKenna TM, Diamond DM, Bakin JS. 1990. Retuning auditory cortex by learning: a preliminary model of receptive field plasticity. *Concepts Neurosci.* 1:91–131

Weinberger NM, Diamond DM. 1987. Physiological plasticity in auditory cortex: rapid induction by learning. *Prog. Neurobiol.* 29:1–55

Weinberger NM, Diamond DM. 1988. Dynamic modulation of the auditory system by associative learning. In *Auditory Function: The Neurobiological Bases of Behavior,* ed. GM Edelman, WE Gall, WM Cowan, pp. 485–512. New York: Wiley. 817 pp.

Weinberger NM, Javid R, Lepan B. 1993. Long-term retention of learning-induced receptive-field plasticity in the auditory cortex. *Proc. Natl. Acad. Sci. USA* 90:2394–98

Welker WI, Seidenstein JI. 1959. Somatic sensory representation in the cerebral cortex of the raccoon (Procyon Lotor). *J. Comp. Neurol.* 111:469–501

Welker E, Rao SB, Dorfl J, Melzer P, Van der Loos H. 1992. Plasticity in the barrel cortex of the adult mouse: effects of chronic stimulation upon deoxyglucose uptake in the behaving animal. *J. Neurosci.* 12:153–70

Whitsel BL, Kelly DG. 1988. Learning acquisition ("learning") by the somatosensory cortex. In *Brain Structure, Learning, and Memory,* ed. JL Davis, RW Newburgh, EJ Wegman, pp. 93–131. Washington, DC: Am. Assoc. Adv. Sci.

Willott JF, Aitkin LM, McFadden SL. 1993. Plasticity of auditory cortex associated with sensorineural hearing loss in adult C57BL/6J mice. *J. Comp. Neurol.* 329:402–11

Woody CD, Swartz BE, Gruen E. 1978. Effects of acetylcholine and cyclic GMP on input resistance of cortical neurons in awake cats. *Brain Res.* 158:373–95

Woolsey TA, Van der Loos H. 1970. The structural organization of layer IV in the somatosensory region (SI) of mouse cerebral cortex. *Brain Res.* 17:205–42

Yun J. 1991. [Functional reorganization of rat somatosensory cortex induced by synchronous afferent stimulation]. *Chung Kuo I Hsueh Ko Hsueh Yuan Hsueh Pao* 13:278–82 (In Chinese)

Zarzecki P, Witte S, Smits E, Gordon DC, Kirchberger P, Rasmusson DD. 1993. Synaptic mechanisms of cortical representational plasticity: somatosensory and corticocortical EPSPs in reorganized raccoon SI cortex. *J. Neurophysiol.* 69:1422–32

Annu. Rev. Neurosci. 1995. 18:159–92

ISOLATION, CHARACTERIZATION, AND USE OF STEM CELLS FROM THE CNS

Fred H. Gage, Jasodhara Ray, and Lisa J. Fisher

Department of Neurosciences, School of Medicine, University of California, San Diego, La Jolla, California 92093-0627

KEY WORDS: progenitor, precursor, neuroblast, CNS transplantation, gene transfer, growth factors

INTRODUCTION

Most of the differentiated cells in the mammalian body are not permanent. In most organs, cells are dying and being replaced at varying rates. The body has developed two strategies for tissue renewal or replacement. The first and most straightforward is duplication, wherein a differentiated cell divides to give rise to a daughter cell of the same geno- and phenotype. Hepatocytes and endothelial cells fall into this category. For hepatocytes, cell number in the adult is regulated in a predictable manner such that when cell damage occurs, the cells divide to generate the number of cells needed to replace those that have been lost. Endothelial cells are somewhat more creative in that they can divide when new blood vessels are formed. The second way that differentiated cells are replaced is by arising from undifferentiated cells through a process analogous to cell genesis. The prototypical cell of this category is the blood cell, since there is evidence that all cells of the hematopoietic lineage can be derived from a single, multipotent stem cell. These multipotent cells are capable of self-replication and the generation of various committed progenitor cells that, under the influence of appropriate cytokine and growth factors, ultimately differentiate into mature blood cells. These mature blood cells then die within a matter of weeks. Although the existence of the multipotent cell is acknowledged, it is an elusive cell whose isolation, characterization, and utilization are subjects of great interest.

159

Anderson (1989) has suggested that the stem cells of the hematopoietic system are a useful analogy to apply to the nervous system and predicts that similar cells exist in both the central nervous system (CNS) and the peripheral nervous system (PNS). Although the presence and nature of multipotent cells within the CNS remain a point of discussion, the nomenclature from the hematopoeitic system has been adapted to describe neural precursors and their progeny. Stem cells thus refer to a population that is capable of exte..Jed self-renewal and the ability to generate multilineage (neurons and glia) cell types (see Figure 1). Progeny of the stem cells are progenitor cells, which are also capable of self-replication but show a limited life span. The progenitors show further lineage restriction upon division by giving rise to progeny that generate either neurons (neuroblasts) or glia (glioblasts) but not both. Although these terms are used often, there is some discrepancy in how they are defined in various studies. The problem is particularly complex in the brain because of the large number of neural phenotypes that exist. Therefore, for a cell to be called a neural stem cell, it would have to be able to differentiate not only into neurons and glia but into the vast diversity of neural cell types that are present within the CNS (e.g. interneurons, projecting neurons, type I astyrocytes). An alternative set of definitions may be less value laden and more consistent. A stem cell can be defined as a cell that (*a*) is not in the final stage of differentiation or terminally differentiated, (*b*) can continue to divide throughout the life of an animal, and (*c*) has progeny that can either continue as stem cells or terminally differentiate. For simplicity, since it is often difficult to determine if a cell within the CNS displays all of these features, the term precursor is used in this review to refer to any cell types that do not appear to be terminally differentiated. In some of the work summarized below, however, the terms used by different investigators are mentioned to provide a point of comparison between studies.

This review summarizes work that has sought to locate, isolate, and characterize putative stem cells within the CNS. The review is divided into several general areas. First, we discuss evidence that precursor populations normally reside within the postnatal mammalian brain and that some of these cells may continue to contribute neuronal progeny to the mature brain through adulthood. Next, we describe the characterization of neural precursor cells in vitro. Much of this work has been facilitated by the identification of factors that increase the survival and growth of primary precursor cells in vitro. Recent work that has revealed the developmental potential of cultured neural precursors when implanted back into the CNS is then discussed. We conclude by presenting work that has begun to explore the potential of neural precursors for treating CNS damage or disease. Because of the scope of this review, there are issues that will not be addressed. In particular, the controversy over the in vivo multipotentiality of progenitors and the migration patterns of neuronal progen-

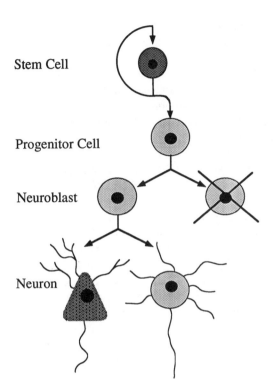

Stem Cell

Progenitor Cell

Neuroblast

Neuron

1. How many divisions per stem cell?
2. What is the potentiality of stem cells?
3. Can stem cells persist quiescently?

4. How long is the cell cycle of a progenitor?
5. Can progenitors persist quiescently?

6. What percentage of progenitors or neuroblasts die?
7. What is the mechanism of cell death?
8. How many times can a neuroblast divide?
9. Can a neuroblast survive indefinitely in a quiescent state?
10. Can neuroblasts be activated to divide in vivo?

11. How long do terminally differentiated cells survive?
12. Can a differentiated cell de-differentiate?

Figure 1 Schematic of neural stem cell lineage and questions that remain to be addressed concerning these cells and their progeny. The stem cell is capable of extended self-replication, as indicated by the arrow that loops around to the top of the stem cell. In this example, the progenitor population that arises from the stem cell is shown to generate two cells: one that dies without further commitment (circle with X) and another that is committed to the neuronal lineage (neuroblast). The neuroblast should be able to generate all of the different types of neurons within the CNS (only two types are shown). Figure adapted from figures 17–34 in Alberts et al (1989).

itors in different regions of the brain are summarized elsewhere (Hatten 1993). Rather, we have focused on the use of neural precursors to understand brain development and adult plasticity and as potential sources of cellular replacement in the damaged adult brain.

PRECURSOR POPULATIONS IN THE POSTNATAL BRAIN

The epithelium of the ectoderm gives rise to the neurons and glia that make up the adult central nervous system. The mechanisms that lead to this diversity are difficult to ascertain because cells migrate from their birth place and because of the large number of different cell types that make up the brain. Nevertheless, enormous headway is being made in determining how cells decide what they will become and how they find their site of terminal differentiation. Most of the mechanisms that are responsible for cell proliferation and migration are shut off in the adult mammalian brain, and thus the capacity for self repair and cellular replacement is greatly diminished. Although controversial, there is evidence that cell division continues in the adult brain and that some of the resulting cells become neurons. This controversy has been at least partially resolved over the last 30 years with the introduction of [^{3}H]thymidine labeling techniques (Altman 1962, Sidman 1970, Rakic 1985). Quantification of those cells that incorporate [^{3}H]thymidine into their DNA during mitosis can reveal patterns of new cell formation and the length of cell cycles. The route of administration of the label, the duration after injection to time of sacrifice, and the concentration of the isotope can all influence the interpretation of the results, but careful and methodical studies have provided a reliable set of conclusions. More recently, retroviral-mediated gene transfer has increased the arsenal of techniques that can be used to identify dividing populations within the CNS. The most common retroviral vectors contain the *Escherichia coli lacZ* (β-galactosidase) gene. This marker gene is integrated into the DNA of proliferating cells and then inherited by all the subsequent progeny (Sanes et al 1986, Price et al 1987). The expression of this gene is easily detected either immunocytochemically or through enzyme histochemistry with the light and electron microscope. The four areas of the adult brain that have been studied most and provide the strongest case for postnatal neurogenesis in vivo—the dentate gyrus of the hippocampus, the olfactory epithelium, the ependymal zone, and the cerebellum—are reviewed here.

Dentate Gyrus of the Hippocampus

The number of neurons in the hippocampus of the rat increases 43% between postnatal day 30 and postnatal day 365 (Bayer 1982). Two morphological changes occur in the hippocampus during this time: The overall number of cells increases, and the volume of cells in the ventral leaf of the dentate gyrus (DG) decreases. Even though pyknotic cells are rarely seen in the DG, some cell death is expected to occur. Neural precursors appear to arise from a stem cell population that lies in the basal region of the granule cell layer. This has been suggested from labeling studies that reveal marked cells in this region shortly after [^{3}H]thymidine pulsing and few cells in the granule cell layer when the area is examined at longer intervals after labeling. Convincing evidence from several studies indicates that this dividing cell population develops into mature neurons. Using electron microscopy and [^{3}H]thymidine autoradiography, Kaplan (1981) observed that in the postnatal dentate gyrus there are mitotic neuroblasts in prophase and telophase with synapses on their cell bodies

Table 1 Immunocytochemical markers used to identify neuronal and glial cells

Antigenic markers	Cell types	Nature of antigens	Reference
Nestin	Stem	Intermediate filament	Fredriksen & McKay 1988
Vimentin	Precursor	Intermediate filament	Houle & Fedoroff 1983
A2B5	Precursor	Ganglioside	Eisenbarth et al 1979
L1	Premigratory neuron	Cell adhesion molecule	Rathjen & Schachner 1984
NF (L,M,H)	Neuron	Intermediate filament	Carden et al 1987
NSE	Neuron	Enolase (isoenzyme)	Schmechel et al 1980
MAP-2	Neuron	Microtubule-associated protein	Dotti et al 1987
Calbindin-D	Neuron	Ca^{2+} binding protein	Sloviter 1989
Substance P	Neuron	Peptide	Cooper et al 1982
GABA	Neuron and glia	Amino acid	Cooper et al 1982
GFAP	Astrocyte	Intermediate filament	Debus et al 1983
Gal C	Oligodendrocyte	Galactosphingolipid	Ranscht et al 1982
O4	Oligodendrocyte	Sulfatide	Sommer & Schachner 1981

and with axonal-like processes. Stanfield & Trice (1988) labeled cells with [^{3}H]thymidine, allowed them to mature for longer than a month, and then injected fluorescent dye into CA3, the terminal field of dentate granule neurons. They observed that some of the thymidine-labeled cells also contained the fluorescent dye, indicating that the newborn cells displayed the characteristic neuronal feature of an axonal process. More recently, Cameron et al (1993) examined the timecourse of the maturation of the proliferative population within the hippocampus by pulsing the adult brain with [^{3}H]thymidine and then using combined autoradiography and immunolabeling for neuron-specific enolase (NSE) and the glial marker glial fibrillary acidic protein (GFAP) (see Table 1). Their results showed that the number of cells double-labeled for NSE and [^{3}H]thymidine progressively increased after the thymidine injection; 85% of the labeled cells were NSE positive four weeks postlabeling. In contrast, the number of dual-labeled GFAP cells remained constant across the same four-week time period. The authors concluded that the majority of newborn cells in the dentate gyrus differentiate into neurons, not glia, and that some of the cells are born in the granule cell layer while others migrate from the hilus to the granule cell layer.

The significance of neurogenesis in the adult hippocampus is presently unknown. It is perhaps too easy to speculate that this neuronal plasticity may be important in a structure that is involved in the formation of new memories. Perhaps, unlike areas of the brain that are involved in long-term memory, there may be an advantage to having cell replacement and addition occur in areas involved in cognitive plasticity. Clearly, the functional ramifications of neurogenesis within the hippocampus remain to be elucidated.

Olfactory Epithelium

In the olfactory epithelium, the sensory neurons undergo continual degeneration and replacement into adulthood (Graziadei & Monti Graziadei 1978, 1979). The neurons are generated from two basal unipotent stem cell populations—the horizontal basal cell and the globose basal cell, which exists in the base of the epithelium (Kratzing 1978, Graziadei et al 1980). In most mammals, the lifetime of an olfactory cell lasts about one month; this includes the birth, migration, synaptogenesis, and degeneration of these neurons. The factors that regulate this cell cycle are unknown, but it has been hypothesized that the rate of cell birth is regulated by death of the mature neurons. Furthermore, the final differentiation and survival of olfactory neurons is predicted to be dependent on synaptic contact with target cells in the olfactory bulb. Studies showing that damage to mature cells of the epithelium by a variety of manipulations can result in an increase in olfactory neuronal production have provided evidence that supports the role of cell death in olfactory neurogenesis (Matulonius 1975, Graziadei et al 1979, Camara & Harding 1984, Verhaagen et al 1990). Schwartz et al (1991) showed that only the globose basal cells contribute to the neurogenesis that occurs following experimental damage (bulbectomy). Although bulbectomy can increase the rate of cell neurogenesis in the olfactory epithelium, a parallel increase in the rate of cell death also occurs (Carr & Farbman 1992). This latter observation is consistent with a role of target cells in maintaining the survival of differentiated neurons within the olfactory bulb.

Subependymal Zone

In the embryo, proliferation occurs in the ventricular zone, a cellular lining of the ventricles that generates postmitotic cells that migrate from the ventricular zone to form the forebrain. The subventricular zone expands greatly during late gestation and early postnatal periods and is a germinal matrix of the immature dividing neuroectodermal cells. The subventricular zone persists in the adult as a mitotically active area called the subependymal zone (Boulder Committee 1970).

The fate of the subependymal cells is not fully understood. Many recent retroviral lineage-tracer studies are not conclusive, but several intriguing issues important to the isolation of stem cells have been revealed. Levison & Goldman (1993) gave postnatal day 2 (P2) rats injections of a retrovirus harboring either β-galactosidase (β-gal) or alkaline phosphatase, or both. Their results demonstrate that both astroctyes and oligodendrocytes are born in the subependymal zone postnatally. Furthermore, when two different markers were used, evidence supported the conclusion that a single progenitor from the subependymal zone produces both types of glia. Other work by Luskin (1993) presents evidence that the postnatal subependymal zone is heterogeneous. In this study, a retroviral tracer was injected into different parts of the subependymal zone of postnatal rats to reveal the presence of marked cells and cell clusters at various times after the injection. She found that a discrete region of the anterior part of the subependymal zone that lies at the anterior pole generated a large number of neurons that differentiated into granule cells and periglomerular cells of the olfactory bulb. These results are in agreement with a recent report that neurogenesis of olfactory bulb interneurons occurs postnatally, and for granule cells, it extends for many months (Corotto et al 1993).

The ability of the subventricular zone to generate glia and neurons has been suggested to decrease with age. Morshead & van der Kooy (1992) used a retroviral tracer to label mitotically active cells in the adult mouse brain and then looked for migration of labeled progeny out of the subependymal zone and into the surrounding brain tissue. They concluded that the proliferating cells of the subependymal zone self-replicate and also generate progeny that die soon after birth without apparent migration from their birth site. However, Lois & Alvarez-Buylla (1994) recently provided convincing evidence that the subventricular zone is capable of generating neuronal progeny for extended periods in the adult brain. Their discrete microinjections of [^{3}H]thymidine into the lateral ventricles of adult mice revealed clusters of proliferative cells in the subventricular zone that migrated to the olfactory bulb over distances as great as 5 mm before differentiating into either granule neurons or periglomerular cells. Thus even in the mature brain, the subventricular zone harbors an active precursor population that contributes some of the neuronal cell types that reside within the olfactory bulb.

Cerebellum

The five major neuronal cell types of the cerebellum have a distinct spatial and temporal developmental pattern. There are two separate waves of development originating from two separate germinal zones. The deep cerebellar neurons, the Purkinje cells, the Golgi II, and the Golgi epithelial cells all emerge from a zone in the roof of the fourth ventricle called the rhombic lip. A second wave of cell birth occurs later in development and extends into the postnatal period in most vertebrate species. This second germinal zone, termed the external germinal layer (EGL), is formed immediately beneath the pia covering the cerebellar plate and gives rise to granule cells, stellate cells, basket cells, and some glial cells (Addison 1911, Raaf & Kernohan 1944, Phemister & Young 1968). The ease of localizing, dissecting, and manipulating the EGL has made it a focus of developmental studies. In all vertebrates, the EGL expands from a single layer to six to eight cell layers during cell mitogenesis. The EGL persists for varying periods in different species and eventually reduces in size and disappears. For example, the EGL is seen up through P25 in rodents (Addison 1911), to approximately P70 in the dog (Phemister &

Young 1968), and for almost two years in the human (Raaf & Kernohan 1944). Although the proliferative cells in the EGL do not appear to survive as a self-replicating population throughout the life of an organism, potential stem cells derived from the EGL may become quiescent in the maturing CNS but retain the capacity for proliferation if exposed to appropriate environmental signals.

Other

Other areas of the adult brain have been demonstrated to undergo cell division; however, less information and fewer published reports are available about them. Kaplan (1981) reported that 30 days after a three-month-old rat was injected with [^{3}H]thymidine, labeled cells were observed in layer IV of the visual cortex. Ultrastructurally, these labeled cells had the morphology of neurons, including synapses along their dendrites and axons. Although this labeling accounted for only .011% of the cells per section through the entire thickness of the visual cortex, it represents an important population whose neurogenesis in the adult could be significant. There are probably many areas of the adult brain undergoing neurogenesis at a slow rate or low frequency that, when examined in detail, could represent the presence of quiescent stem cells activated for reasons that are presently unknown.

IN VITRO CHARACTERIZATIONS OF PRECURSOR CELLS

One of the first studies that appeared to isolate a multipotent cell in vitro was conducted by Price et al (1987), who explored the characteristics of cells cultured from the embryonic cortex of rats that were transduced to express the β-gal gene. Using morphological criteria, Price et al (1987) demonstrated that discrete clusters of labeled cells that presumably arose from a single β-gal-ex-pressing cell appeared to contain multiple cell types. This observation was subsequently confirmed and extended by Williams and colleagues (1991), who have reported that 18% of the cells within the embryonic cortex generate both neurons and oligodendrocytes in vitro.

An early characterization of multipotent cells isolated from the CNS revealed an influence of exogenous factors on these precursor populations (Temple 1989). Cells derived from the embryonic septum of rats were placed into culture either alone or in combination with fetal striatal tissue. Nearly 50% of the septal cells in the cocultures were found to proliferate for extended periods, whereas none of those cultured alone was capable of more than one division. Some isolated clones of proliferating cells were clearly multipotent, as indicated by the emergence of mixed progeny that expressed either the glial marker GFAP or the neuronal marker neurofilament (NF). This finding indicated that a soluble factor produced within the developing CNS strongly influences the mitogenesis of neural precursors. Similar results obtained with single cells isolated from the embryonic cortex and hippocampus suggested that this response was not unique to precursors derived from a particular region of the CNS. In the following section, studies that have identified multipotent precursor cells and sought to identify some of the exogenous factors that influence these precursor populations in vitro are discussed. Although the focus of this review is on CNS-derived cells, we have included some work on peripheral precursors that has contributed to an understanding of trophic and substrate influences on stem cells.

Immortalized Neural Populations

Historically, it has been difficult to maintain the survival and/or proliferation of primary neural precursors in vitro for periods sufficiently long enough to characterize them and examine cellular responses to exogenous factors. Therefore, many groups have developed neural cell lines by using gene transfer techniques to model precursor responsiveness to experimental manipulations. In addition to being able to readily expand these cell lines for study, the cells can be cloned to obtain homogeneous populations of precursors and enhance reproducibility.

Neural cell lines have typically been generated by the retroviral transduction of oncogenes into cells derived from the developing brain (for review see Cepko 1988, 1989; Lendhal & McKay 1990). The immortalization process arrests cells at defined stages of development and generally halts terminal differentiation. Thus, cells at intermediate stages of differentiation can be propagated for a long period of time. The most commonly used oncogenes for immortalizing cells are members of the *myc* oncogene family (Cepko 1988, 1989) or a temperature-sensitive (ts) mutant of SV40 large T antigen (Jat & Sharp 1989). In contrast to the *myc* oncogene, the large T antigen gene induces proliferation when cells are maintained at 33°C (permissive temperature) but not at 39°C (nonpermissive temperature). Interestingly, not all cells expressing an oncogene become immortal (Ryder et al 1990). It is unclear why some oncogene-expressing cells display a limited life span, but this may reflect an inability of the oncogene to override a differentiation program that was initiated prior to the immortalization.

Several groups have reported the presence of multipotent cells in cultures of immortalized neural populations. Ryder and colleagues (1990) described *myc*-immortalized clones derived from the cerebral cortex or olfactory bulbs of early postnatal rats that generate neuronal and glial progeny. Similarly, immortalized populations derived from the mesencephalon of embryonic day 10 (E10) mice generate multiple cell types in vitro (Bernard et al 1989). That the diverse progeny arose from a single precursor was confirmed by showing a unique viral integration site for each clonal line (Bernard et al 1989, Ryder et al 1990). These multipotent populations did not appear to be an abnormal consequence of the immortalization procedure because many other *myc* clones failed to spontaneously generate multiple cell types (Bartlett et al 1988, Cepko 1989, Ryder et al 1990). At least some of the *myc*-immortalized clones have been found to be responsive to exogenous stimuli. One of the precursor lines that remained in an undifferentiated state in vitro could be induced to differentiate within 24 hours of the addition of acidic fibroblast growth factor (aFGF) or basic FGF (bFGF) to the culture medium (Bartlett et al 1988). Some of the progeny that arose in the FGF-enriched environment displayed NF protein, whereas others expressed GFAP, indicating the generation of both neurons and glia from the precursor population. In some of the lines that showed spontaneous multipotency in vitro, members of the FGF family could induce a rapid increase in NF expression (Bernard et al 1989). In addition, substances secreted from some of the *myc*-clones that did not appear to be in the FGF family have been found to be mitogenic for immortalized precursors (Bernard et al 1989). All of these results must be interpreted with caution, however, since diverse culture conditions are used in different studies (see Table 2).

In contrast to the *myc*-expressing clones, cells immortalized with the tsSV40 large T antigen often do not show spontaneous multipotency in vitro. T antigen–expressing cells derived from the cerebellum, striatum, or medullary raphe of rodents typically generate progeny that express either glial markers (Fredriksen et al 1988, Evard et al 1990, Redies et al 1991) or neuronal phenotypes (Redies et al 1991, White & Whittemore 1992) but not both. In many of these clonal lines, alterations in the morphology and/or phenotype of T antigen–expressing clones can be induced by manipulating the extracellular environment of the cells in vitro. The cerebellar-derived clonal line ST15A, characterized by a neuronal morphology and expression of NF, shows conversion to a GFAP phenotype when the cells are cocultured with primary cells derived from the P3 cerebellum (Redies et al 1991). A cell line derived from the P1 striatum of mice that solely showed gliogenesis in a serum-enriched medium generated both NF-expressing cells (neurons) and glia when switched to a chemically defined medium and a polyornithine substrate (Evard et al 1990). Also, both the survival and differentiation of a clonal medullary raphe population (RN33B) are enhanced when the cells are cocultured with embryonic hippocampus or cortex tissues (Whittemore & White 1993). Furthermore, changes in substratum affect the differentiation of RN33B and the expression of neuronal antigens (Whittemore & White 1993). Cells grown on poly-D-lysine show reduced levels of neuron-specific enolase (NSE) and a complete abolition of NF immunoreactivity. In contrast, laminin and fibronectin (FN) both slightly enhance the NSE expression of RN33B cells without affecting NF. The influence of exogenous factors on the T antigen–expressing cells is

Table 2 Culture conditions for immortalized stem and precursor cells

Species	Region/Age	Cell type[a]	Immortalizing oncogene	Substratum	Medium	Supplement	Exogenous factor	Reference
Mouse	Cerebellum/P4	Multipotent progenitor	v-myc	Poly-L-lysine	DME[b] + 20% FBS	No	No	Ryder et al 1990
Mouse	Olfactory bulb/newborn	Multipotent progenitor	v-myc	Poly-L-lysine	DME + 20% FBS	No	No	Ryder et al 1990
Mouse	Mesencephalon/E10	Precursor	c-myc and N-myc	Uncoated plastic	DME + 10% FBS	No	No	Bartlett et al 1988, Bernard et al 1989
Mouse	Striatum/newborn	Bipotential precursor	SV 40 large T antigen	Uncoated plastic	DME + 10% FBS	No	No	Evard et al 1990
Rat	Adrenal gland/E14.5	SA progenitor MAH	v-myc	Uncoated plastic	L-15 + 10% FBS	Dexamethasone + additives[c]	No	Birren & Anderson 1990
Rat	Cerebellum/P2	Oligopotent precursor	tsSV40 A58	PORN[d]	DME + 10% FBS	No	No	Fredriksen et al 1988
Rat	Cerebellum/P2	Oligopotent precursor	tsSV40 A58	PORN and Laminin	DME + 10% FBS	No	No	Redies et al 1991
Rat	Hippocampus/E16	Stem	tsSV40 A58	PORN	DME + 10% FBS	No	No	Refranz et al 1991
Rat	Medullary raphe/E13	Neuroblast RN33B	tsSV40 A58	Collagen and Poly-L-lysine	DME:F-12 + 10% FBS	No	No	Whittemore & White 1993
Rat	Neural tube/E10.5	Neural crest progenitor NCM-1	v-myc	Fibronectin	L-15 + 10% FBS	No	No	Lo et al 1991

[a] Cell types as defined by authors.
[b] DME = Dulbecco's modified eagles medium.
[c] Doupe et al 1985.
[d] PORN = Polyornithine.

consistent with results obtained with *myc* clones, thus supporting the use of immortalized precursors as a model system to explore CNS development. In general, these studies have revealed a role for both cell surface cues and soluble molecules in the proliferation and differentiation of neural precursors.

Immortalized Cells Derived from the PNS

Sympathoadrenal (SA) precursor cells of the peripheral nervous system are bipotential and differentiate into adrenal chromaffin cells or sympathetic neurons depending on environmental influences. To elucidate the exogenous factors that influence the fate choice and the differentiation of SA progenitor cells in vitro, a *v-myc* immortalized cell line, termed MAH (*myc* immortalized, adrenal derived, HNK-1 positive), was generated from the E14.5 adrenal gland (Birren & Anderson 1990). This cell line displays many of the properties of the SA precursors in situ and has thus been used to explore the development of nerve growth factor (NGF) responsiveness in the sympathoadrenal lineage. In vitro, NGF induces chromaffin cells to show the morphological and antigenic characteristics typical of sympathetic neurons (Doupe et al 1985). However, the SA progenitors are unresponsive to NGF (Anderson & Axel 1986). Like their in situ counterparts, MAH cells do not respond to NGF in vitro, a property that appears to reflect the absence of both the low- (p75) and high-affinity (trkA) forms of the NGF receptor in this SA cell line (Birren & Anderson 1990, Birren et al 1992). Birren & Anderson (1990) have found that bFGF is mitogenic for MAH cells and induces neurite outgrowth and expression of p75. Once the NGF receptor is expressed on these cells, they then become dependent on NGF for continued survival. These results indicate that NGF is not the initial determinant of neuronal differentiation in the SA lineage, but rather bFGF plays this role (see also Stemple et al 1988).

Lo et al (1991) have suggested that another clonal cell line generated by immortalization of neural crest cells with *v-myc,* termed NCM-1 cells, is a glial progenitor population. These cells spontaneously differentiate into Schwann cells in vitro, a process that can be blocked by culturing the progenitors in the mitogenic factor TGF-β (transforming growth factor-β). Some subclones of NCM-1 cells generate progeny that do not express glial markers. Rather, the subclones give rise to cells that show features characteristic of SA progenitors, such as the induction of tyrosine hydroxylase and NF in response to bFGF and dexamethasone. Thus, NCM-1 cells may represent multipotent neural crest stem cells that can generate both glial precursors and SA progenitor-like cells depending on environmental signals.

Primary Precursor Cells

Although the cellular immortalization technique offers many advantages, this process may alter some of the fundamental properties of the parent cells.

Several studies have reported that differences in these two cell types exist. Immortalized cells have shown altered expression of some proteins that are characteristic of the original parent population (Birren & Anderson 1990, Renfranz et al 1991, Vandenberg et al 1991, Whittemore & White 1993). Also, the growth rate of immortalized cells is often faster than that of their primary counterpart (Ryder et al 1990), a property that can be reflected in abnormal karyotypes (Bianchi et al 1993). Thus, many recent studies have focused on characterizing the properties of primary precursor cells in vitro. Much of this work has been aided by the identification of growth factors that enhance the survival and/or proliferation of primary precursors in culture. For comparative purposes, the divergent culture conditions used to examine primary cells in the studies discussed below are listed in Table 3.

Primary Precursor Cells in the Developing Neural Crest

Stemple & Anderson (1992) examined the developmental potentials of serially propagated mammalian neural crest cells in vitro. They found that single neural crest cells are multipotent and able to produce multipotent progeny, indicating that these cells are capable of self renewal and are thus likely to be stem cells. Assessments of the cells in different plating conditions have revealed that substratum plays a role in the fate choice of these stem cells. This was first suggested by observations that neural crest clones that were first established on FN and then overlaid with poly-D-lysine (pDL) at various times postplating gave rise to neuron-only clones more often than glia-only clones (Stemple & Anderson 1992). To demonstrate that the substratum was influencing the lineage decision of the stem cells, clones established on FN were picked and replated at clonal density on either FN or pDL-FN substrates. Subclones of the FN-derived founder were seen to generate neurons on the pDL-FN substratum, but sister cultures plated onto FN generated only glial cells. Although exposure to pDL appeared to drive the precursor population toward a neuronal lineage, this was not always the case. For example, clones founded on pDL-FN and then subcultured were only capable of generating neurons when the FN substratum was enriched with pDL. These results suggest that neural crest cells exposed to FN retain the capacity for neurogenesis and that pDL influences, but does not dictate, neuronal differentiation of these stem cells. Growth factors can also influence or bias the lineage choice of uncommitted neural crest stem cells. Glial growth factor (GGF), also known as Schwann cell–mitogen, suppresses neuronal differentiation while allowing or promoting glial differentiation (Shah et al 1994).

Glial Precursor Cells in the Optic Nerve

The rat optic nerve contains three different types of glial cells: type 1 astrocytes, type 2 astrocytes, and oligodendrocytes. Two of these populations, the

Table 3 Culture conditions for primary stem and precursor cells

Species	Region/Age	Cell type[a]	Substratum	Medium	Supplement	Exogenous factors	References
Mouse	Mesencephalon and Telencephalon/E10	Precursor	Uncoated plastic	DME + 1% FBS	N2[b]	bFGF (50 ng/ml) and heparin (8 µg/ml)	Murphy et al 1990, Drago et al 1991a,b
Mouse	Mesencephalon and Telencephalon/E10	Precursor	Uncoated plastic	Monomed + 10% FBS	No	bFGF (20 ng/ml) and heparin (8 µg/ml)	Kilpatrick & Bartlett 1993
Mouse	Striatum/E14	Multipotent progenitor	PORN	DME:F-12	N2	EGF (20 ng/ml)	Reynolds et al 1992, Vescovi et al 1993
Rat	Cerebral hemispheres or spinal cord E13-14	Neuroblast	Poly-L-lysine	DME	N2	bFGF (5 ng/ml)	Gensburger et al 1987, Deloulme et al 1991
Rat	Cortex/E15	Multipotent precursor	Astrocyte monolayer	DME + 10% FBS	No	No	Price et al 1987
Rat	Cortex/E12-18	Multipotent precursor	Astrocyte monolayer	DME + 0.5% FBS	N2	No	Williams et al 1991
Rat	Hippocampus or spinal cord/E16-18	Neuroblast	PORN and Laminin	DME:F-12	N2	bFGF (20 ng/ml)	Ray et al 1993, Ray & Gage 1994
Rat	Neural crest/E10.5	Stem	Fibronectin	L-15	N2 + additives[c]	EGF (100 ng/ml), bFGF (4 ng/ml), and NGF (20 ng/ml)	Stemple & Anderson 1992
Rat	Optic nerve/PO-adult	O-2A progenitor	Poly-L-lysine	DME + 0.5%	N2	No	Wilswijk et al 1990, Wren et al 1992
Rat	Striatum/E13.5-14.5	Stem	PORN	DME:F-12	N2	bFGF (5 ng/ml), NGF (150 ng/ml), or bFGF → NGF	Cattaneo & McKay 1990
Mouse	Brain/Adult	Precursor	Uncoated plastic	DME + 10% FBS	No	bFGF (20 ng/ml), EGF (20 ng/ml), or Ast-1 CM[d] → NGF	Richards et al 1992
Mouse	Striatum/Adult	Stem	Uncoated plastic	DME:F-12	N2	EGF (20 ng/ml)	Reynolds & Weiss 1992
Rat	Hippocampus/Adult	Neuroblast	Uncoated plastic	DME:F-12	N2	bFGF (20 nl)	Gage et al 1994

[a] Cell types as defined by authors.
[b] N2 = Bottenstein & Sato 1980.
[c] Sieber-Blum & Chokshi 1985.
[d] CM = Conditioned medium.

type 2 astrocytes and oligodendrocytes, have a common precursor that has been named the O-2A progenitor (Raff et al 1983). Both the proliferation and differentiation of O-2A progenitors isolated from the neonatal optic nerve are influenced by exogenous factors. In an early in vitro study, the progenitor population showed strong mitogenic activity when cocultured with type 1 astrocytes (Noble & Murray 1984). It was subsequently determined that these astrocytes secrete platelet-derived growth factor (PDGF), which induces the proliferation of O-2A cells in culture (Noble et al 1988). However, even in the presence of PDGF, the progenitor cells eventually differentiate spontaneously and give rise to the two glial cell types. This terminal differentiation can be blocked by further exposing the O-2A cells to bFGF, which traps them into a self-renewal loop for prolonged periods (Bögler et al 1990). The type of glial cell generated from the progenitor varies depending on the presence of serum in the culture media; type 2 astrocytes arise in serum-rich environments, whereas oligodendrocytes are generated in serum-poor environments. Another factor that can promote O-2A differentiation into type 2 astrocytes is ciliary neurotrophic factor (Hughes et al 1988). These in vitro studies thus support a role for endogenous factors in glial cell development.

Interestingly, bipotent O-2A progenitors are evident within the optic nerve into adulthood (Wolswijk & Noble 1989). However, these adult progenitors display a different morphology, cell cycle length, and migration rate than O-2A cells found within the perinatal optic nerve. A number of observations suggest that the neonatal O-2A cells in the adult optic nerve are completely replaced with a self-renewing, adult form of the progenitor. For example, perinatal O-2A cells serially passaged for 3 months gradually convert to cells that express a marker (O4 antigen) characteristic of adult precursors (Wren et al 1992), and the number of O4-positive cells within the optic nerve in vivo increases from <2% at P7 to >95% in 8-month-old or older rats (Wolswijk et al 1990). These observations further confirm the presence of precursor cells within the adult CNS and also indicate that precursors can show altered properties as the nervous system develops.

Effects of Epidermal Growth Factor on Embryonic Neural Precursors

The mitogenic growth factor epidermal growth factor (EGF) has proliferative effects on neural cells isolated from the embryonic (E14) mouse striatum (Reynolds et al 1992). These EGF-responsive cells, which have been suggested to be a progenitor population within the CNS, also proliferate in the presence of TGF-α but are unresponsive to NGF, bFGF, PDGF, or TGF-β. EGF can induce the proliferation of a single progenitor cell plated on polyornithine (PORN) substratum in serum-free culture conditions and can generate a cluster of undifferentiated progeny that solely express nestin (Reynolds et al 1992), a marker of intermediate filaments within neuroepithelial cells during development (Fredricksen & McKay 1988, Lendahl et al 1990). After multiple cell divisions, cells migrate from the clusters and give rise to NSE-immunoreactive (IR) or GFAP-IR cells. In addition, new proliferating clusters of cells are also seen. The secondary clusters that form in the EGF-enriched culture contain cells with the same undifferentiated morphology as the original progenitor, suggesting the presence of a self-renewing population of cells. Many of the NSE-IR cells also express NF protein, and a small number of cells with neuronal morphologies express markers characteristic of striatal neurons, such as GABA, substance P, and methionine-enkephalin. Since multiple types of progeny are generated in EGF-enriched conditions, these results support a role for EGF in mitosis but not in differentiation of the precursors (see below).

Effects of bFGF on Embryonic Neural Precursors

Catteneo & McKay (1990) isolated cells from E13.5–14.5 rat striatum that expressed nestin, which were termed a stem cell population. In serum-free conditions, bFGF promoted both the survival and proliferation of these precursors in vitro, an effect that was potentiated by NGF. Removal of growth factors halted the proliferation of the precursors, which then differentiated into cells that displayed neuronal morphologies and expression of NF. The proliferating precursors may thus have represented a neuroblast population, rather than stem cells, because they appeared to give rise solely to neurons. The mitogenic action of bFGF on neuroblasts is not restricted to cells isolated from the striatum; similar results have also been observed in cultures derived from embryonic cerebral hemispheres (E13), hippocampus (E16), and spinal cord (E14–16) of rats (Gensburger et al 1987, Deloulme et al 1991, Ray et al 1993, Ray & Gage 1994). Cells in all of these regions that proliferate in response to bFGF generate progeny that predominantly express neuronal markers such as NF, NSE, and microtubule-associated protein-2 (MAP-2) (Gensburger et al 1987, Deloulme et al 1991, Ray et al 1993, Ray & Gage 1994).

The mitogenic action of bFGF is not restricted to neuroblasts; multipotent precursors isolated from the telencephalon and mesencephalon of embryonic (E10) mice also proliferate in response to the growth factor (Kilpatrick & Bartlett 1993). In serum-enriched conditions supplemented with bFGF, two types of clones are generated by these precursors: type A cells (37% of the clones) with a large, amorphous morphology, and type B cells (54% of the clones) that display a cubodial, epithelial morphology. Among the proliferative type B clones, approximately 12% generated only neurons, whereas 39% generated a mixture of undifferentiated cells, neurons, and astrocytes. The undifferentiated progeny of this latter population were also capable of generating neuronal and glial cell types, which suggests that the proliferating parent cell may be a neural stem cell. Unlike the striatal cells characterized by

Cattaneo & McKay (1990), the differentiation of multipotent precursors did not appear to be influenced by the removal of bFGF (Murphy et al 1990, Kilpatrick & Bartlett 1993). These disparate results may reflect phylogenic or ontogenic differences or differing culture conditions between the studies. The study of rat striatal cells used defined serum-free medium (Catteneo & McKay 1990), whereas the studies with neuroepithelial cells used 10% FBS. Alternatively, these differences may reflect the properties of two different precursor populations: stem cells (Kilpatrick & Bartlett 1993) and neuroblasts (Catteneo & McKay 1990). The multipotent bFGF-responsive cells were also completely unaffected by EGF (Kilpatrick & Bartlett 1993). This contrasts with the results of Reynolds and colleagues (Reynolds et al 1992) and may reflect a difference in age between the two putative neural stem cell populations or the fact that multipotent cells in different brain regions respond differently to epigenetic signals.

Evidence that a cascade of growth factors may differentially influence the proliferation and differentiation of multipotent cells has been suggested from work with the EGF-responsive striatal precursors. Although bFGF was not mitogenic for this population (Reynolds et al 1992), it did appear to strongly influence the differentiation of the cells (Vescovi et al 1993). Transient exposure of EGF-generated cell clusters to bFGF in a serum-enriched environment generated two different types of secondary progenitor cells. One was a bipotential cell (5% of the total population) that gave rise to cells with morphological characteristics and antigenic properties of neurons and astrocytes. The other cell type (4% of the total cell population) appeared to be neuroblasts that generated only neurons. Thus, bFGF appeared to be involved in the generation of secondary progenitor populations from the original precursors that were initially unresponsive to bFGF (Reynolds et al 1992).

Interestingly, the presence of serum is essential for achieving both the proliferative and the differentiation effects of bFGF on multipotent precursors (Kilpatrick & Bartlett 1993, Vescovi et al 1993). This finding indicates that there are other factors that supplement and/or potentiate the effects of bFGF on neural precursors. Indeed, bFGF-induced proliferation of neuroepithelial cells has been reported to be dependent on the presence of insulin-like growth factor I (IGF I) (Drago et al 1991a). IGF I appears to act primarily as a survival factor that increases the number of healthy cells that can respond to the mitogenic action of bFGF. The synthesis of laminin, an extracellular matrix protein, is also upregulated by bFGF in a subpopulation of glial precursors (Drago et al 1991b). Laminin appears to act in a paracrine manner to stimulate the differentiation of neuroepithelial cells. Drago et al (1991b) have demonstrated this by showing that antibodies to laminin block the effects of exogenously added laminin on neuroepithelial cell proliferation and neuronal differentiation. These results indicate that bFGF is mitogenic for neural precursors but that the survival and differentiation of cells in bFGF-enriched environments are secondary and are most likely regulated by cellular factors produced in response to bFGF.

Primary Precursor Cells from the Adult Brain

The isolation of precursor cells from the adult brain has been reported by a number of groups (Reynolds & Weiss 1992, Richards et al 1992, Lois & Alvarez-Buylla 1993). This finding is not completely unexpected, since neurogenesis does continue through adulthood in some parts of the brain (see above). Perhaps more striking are observations that these adult precursors respond to many of the same exogenous substances that influence multipotent precursors isolated from the developing CNS. Such results support the possibility that a resident stem cell population within the CNS persists through the lifetime of an organism.

Effects of EGF on Adult Neural Precursors

Reynolds & Weiss (1992) have identified a multipotent population of cells derived from the striatum of adult mice (3–18 months old) in EGF-enriched cultures. These cells proliferate in response to EGF and are unresponsive to bFGF, NGF, or PDGF. The EGF-responsive cells form spheres of proliferating cells after 6–8 days in vitro (DIV) on uncoated substratum. These cells express nestin but no neural markers. Cells within these primary spheres can continue to proliferate in secondary cultures when the primary spheres are dissociated and replated as single cells. Cells that migrate out from the spheres can develop morphologically and express antigenic markers for neurons and astrocytes. The similarities between the EGF-responsive precursors isolated from the adult striatum and the population derived from the fetal striatum (Reynolds et al 1992) suggest that these cells may represent a persistent population of neural stem cells. Although the precise location of these cells within the brain is unknown, they may arise from the subependymal zone, a region that has been reported to harbor stem cells in the adult mouse brain (Lois & Alvarez-Buylla 1993; see above).

Effects of bFGF on Adult Neural Precursors

Another growth factor that stimulates the proliferation of precursor cells from the adult (>60 days) mouse brain in vitro is bFGF (Richards et al 1992). Cells have been cultured in serum-enriched or serum-free conditions supplemented with bFGF, EGF, or both. In some experiments, conditioned medium collected from the astrocyte cell line Ast-1 has also been used in conjunction with the growth factors. In all of the culture conditions, astocytes were the most prevalent cell type that appeared to be generated. Neurons were most frequently observed in cultures containing Ast-1 medium and bFGF. The origin of these growth factor–responsive cells remains to be clarified.

Cells isolated from the hippocampus of adult (>3 months) rats also show a proliferative response to bFGF (Gage et al 1994). Using bromodeoxyuridine (BrdU) incorporation as an index of cellular proliferation, both neurons and astrocytes are generated in these cultures for extended periods (>200 DIV). However, there is an increasing proportion of neuronal-like cells in long-term cultures, suggesting that bFGF is predominantly stimulating the proliferation of neuroblasts. Such a possibility is consistent with the effect of bFGF on neural precursors isolated from the embryonic brain (see above). Alternatively, bFGF may act as a differentiation factor that drives undifferentiated precursors toward a neuronal lineage, as has been observed in the sympathoadrenal system (see above).

Taken together, the in vitro studies of neural precursors from the embryonic and adult CNS have indicated that although genetic factors may play a role in the fate choice and development of a cell, environmental factors are also important components in these processes. A number of exogenous signals, such as growth factors, have clearly been implicated in the development and differentiation of progeny arising from precursor cells. In addition, recent studies examining the effects of growth factors on cells cultured from the adult brain have demonstrated that a population of embryonic stem cells survives through adulthood and that factors that influence precursors early in development appear to remain important into adulthood.

IN VIVO CHARACTERIZATIONS OF NEURAL PRECURSORS

Characterizations of neural precursors in vitro have identified growth factors and substrates that influence cellular proliferation and differentiation. More recently, precursor populations have been implanted back into the brain to assess the developmental plasticity of these cells in different environmental conditions in vivo. Although there is an extensive literature that has explored the properties of heterogeneous mixtures of fetal neural tissues implanted into various regions of the CNS (see Fisher & Gage 1993), this portion of the review focuses on studies that have assessed the developmental potential of isolated and well-characterized populations of neural precursors.

Immortalized Hippocampal Precursors

Cells obtained from the embryonic rat hippocampus during active neurogenesis (E16) have been immortalized using the temperature-sensitive oncogene described above (Renfranz et al 1991). A clonal population selected from this immortalized population, named HiB5 cells, shows some evidence of differentiation at 39°C by the formation of multiple processes but fails to express proteins that are characteristic of either glial (GFAP) or neuronal (NF) populations. However, some similarities between the immortal cells and the primary hippocampal precursors in vitro suggested that the HiB5 cells would provide a model population for exploring the growth and plasticity of neural precursors in the developing CNS. It was of particular interest to assess the fate decisions of the immortalized hippocampal cells when exposed to regions of the postnatal brain that show active neurogenesis. The HiB5 cells were therefore labeled with [³H]thymidine or fluorescent markers, implanted into the hippocampus or cerebellum of rats on postnatal day 2, and assessed 1–6 weeks postgrafting. Regardless of the transplantation site, the vast majority of labeled HiB5 cells were associated with the proliferative subregions of these areas. Specifically, HiB5 cells were predominantly incorporated into the granule cell layer of the dentate gyrus or into the granular layer of the cerebellum. Labeled HiB5 cells in the dentate gyrus displayed ovoid somata, a rich dendritic outgrowth into the molecular layer and an axon that was often found to extend through the mossy fiber pathway to the CA3 region. These engrafted HiB5 cells also expressed the marker calbindin and upregulated *c-fos* in response to kainic acid injections (McKay 1992). These combined properties are characteristic of endogenous granule neurons within the dentate gyrus (Feldmann & Christakos 1983, Smeyne et al 1992), suggesting that the immortalized precursors were driven by cues in the extracellular environment to differentiate into the newly emerging granule cell phenotype.

   The HiB5 cells implanted into the cerebellum displayed markedly different properties from those transplanted into the hippocampus (Renfranz et al 1991). The immortalized cells placed into the cerebellum showed over a twofold increase in cell number after grafting, whereas less than 25% of the HiB5 cells implanted into the hippocampus survived. These findings may suggest that there are factors within the postnatal cerebellum that promote the proliferation of neural precursors. Morphological assessments of cells engrafted within the cerebellum indicated that the HiB5 cells predominantly localized to the granular layer. In this site, labeled cells extended short processes that remained confined to the granular layer and a longer, finer process that bifurcated within the molecular layer. Both of these features are characteristic of cerebellar granule neurons in situ (Palay & Chan-Palay 1974). A smaller population of HiB5 cells was found within the molecular layer and displayed processes with a branching pattern reminiscent of Bergmann glia. Although many of the immortal cells were in or near the Purkinje cell layer, none of these cells appeared to differentiate into Purkinje cells. These results, combined with those from the hippocampal transplants, suggest that the survival and/or differentiation of immortalized embryonic precursors is enhanced in regions of active histogenesis. Site-specific factors within the mitotic areas appear to strongly influence the differentiated phenotype of precursors. This was particularly striking in cases in which the hippocampal-derived HiB5 cells engrafted into the cerebellum differentiated into cerebellar-like cell types. Most importantly, the clonal population differentiated into multiple lineages (neurons and glia) in vivo, indicating that exogenous signals play a role in the fate choice of neural precursors.

Immortalized Cerebellar Precursors

The properties of the HiB5 cells after grafting may have been unique to a precursor population isolated from the embryonic hippocampus. However, this possibility is discounted by studies of immortalized precursors derived from the neonatal cerebellum. In this work, mitotic cells obtained from the cerebellar external germinal layer (EGL) of four-day-old mice were immortalized with a retroviral vector carrying the v-myc oncogene (Ryder et al 1990; see above). In contrast to the HiB5 cells, several cerebellar lines established from this infection were multipotent in vitro and generated progeny with neuronal, astrocyte, and oligodendrocyte morphologies and/or antigenic phenotypes. Two of the multipotent clonal populations generated from the mouse cerebellum were infected a second time with a retrovirus containing the lacZ gene to easily identify the lines for in vivo characterizations (Snyder et al 1992). The clones were then implanted into the EGL of the cerebellum of newborn mice and analyzed both during periods of active histogenesis in the cerebellum (≤postnatal day 7) and after the development and differentiation of intrinsic cerebellar cells had stopped (1–22 months of age).

Early after grafting (≤1 week), labeled cells assumed a spindle-shaped morphology indicative of cells migrating from the EGL. Subsequently, the grafted cells were distributed in many cerebellar structures and showed site-specific differentiation into several different cell types. For example, cells engrafted within the glial-rich molecular layer displayed an astrocyte morphology and labeling for the astrocyte marker GFAP, whereas the cells that localized within white matter tracks resembled oligodendrocytes. In addition to these glial subtypes, two distinct neuronal populations were observed. Cells engrafted within the internal granular layer possessed a morphology suggestive of granule neurons. This identification was confirmed by electron microscopic analyses of the cells, which revealed small somata, few cytoplasmic organelles, and large nuclei containing condensed chromatin blocks (Figure 2A). At very long periods postimplantation (22 months), some of the engrafted granule neurons established synaptic interactions with endogenous fibers in the host brain (Figure 2B–D). A second population of immortalized cells within the lower molecular layer resembled basket neurons with electron microscopic features such as large somata and indented nuclei containing dispersed, non-aggregated chromatin. The clonal origin of these diverse cell types was con-

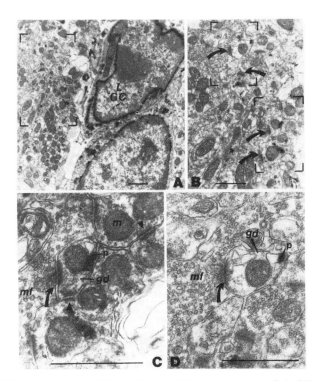

Figure 2   Electron micrographs of immortalized cerebellar precursors engrafted within the internal granular layer (IGL) of the cerebellum 22 months postimplantation. (*A*) Cells expressing the *lacZ* gene are recognized by a dense histochemical precipitate that rings the nucleus. The grafted cell indicated in this view [LGC (labeled granule cell)] displays ultrastructural features, such as condensed chromatin blocks within the nucleus and meager cytoplasm, that are characteristic of endogenous granule cells. (*B*) Synaptic interactions between engrafted cells and host cells (boxed area in *A*) shown at higher power. The two blocked regions in *B* are shown at higher power in *C* and *D* to indicate mossy fiber terminals (*mf*) on precipitate filled (*p*) dendrites of engrafted granule cells (*gd*) (*arrows*) and puncta adherentia (*arrowheads*) between engrafted and endogenous granule cell dendrites. Reprinted with permission from Snyder et al 1992. Copyright *Cell Press*.

firmed by inverse polymerase chain reaction performed on tissue dissected from engrafted regions of the cerebellum. These in vivo results were consistent with those obtained with the cerebellar clones in vitro and indicated that a single cerebellar precursor has the potential for generating multilineage progeny within the brain. The cytoarchitecturally appropriate manner in which the grafted cells integrated into the host cerebellum suggests that local environmental signals play a key role in the differentiation of the precursor population. Finally, the multipotency displayed both by the cerebellar cells derived from the neonatal mouse and the HiB5 cells derived from the embryonic rat strongly suggests that such lability is not unique to a particular subset of neural precursors within the CNS.

The marked plasticity of immortalized precursor cells in regions of active histogenesis suggests that the developmental potential of the precursors may be extremely broad if exposed to appropriate environmental signals. This has been confirmed in a study that assessed the fate of immortalized cerebellar precursors implanted into the embryonic CNS (Snyder et al 1993). The cell line was injected into the ventricular system of fetal mice, where cells would have access to multiple regions of the developing CNS, and then examined at adulthood. As observed for immortalized cells grafted into localized areas of the neonatal brain, cerebellar cells implanted into the embryos showed highly site-specific differentiation. For example, cells distributed in the corpus callosum displayed a glial morphology, whereas those localized to the striatum were predominantly neuronal. Most striking were observations that the immortal cells showed a wider range of cellular morphologies than those implanted into neonates that included cell types only born during embryogenesis, such as pyramidal neurons in the hippocampus and Purkinje cells in the cerebellum. The possibility that signals present during mitosis strongly influence the differentiation of immortal precursors has been further supported by observations that the cerebellar cells are restricted to a glial lineage when implanted into adult mice.

The plasticity of immortalized cells following transplantation to the brain may reflect an unusual property of an immortalized population that may not be characteristic of the primary parent cells. This possibility is discounted by recent preliminary work that explored the developmental features of primary neural precursors exposed to a novel environment. In this study, proliferating cells from the neonatal cerebellum of rats were labeled with [^{3}H]thymidine and implanted into the hippocampus of neonatal rats (Vicario et al 1993). As observed for the HiB5 cells implanted into the developing hippocampus, the primary cerebellar cells localized within the granule cell layer of the dentate gyrus. As observed for granule neurons of the host hippocampus, the engrafted cerebellar-derived cells expressed calbindin (a marker that is not typically expressed by cerebellar cells) and showed upregulation of c-fos in response to kainic acid. These results indicate that some primary cells within the newborn brain can differentiate into cell types that are normally beyond their developmental potential when exposed to changes in environmental signals.

Immortalized Medullary Raphe Precursors

Work with immortal precursors suggests that quiescent regions of the CNS may lack or produce few of the signals that promote cellular development. However, there is evidence that the adult CNS expresses environmental signals that can guide the differentiation of neural precursors. This has been shown in studies of RN33B cells derived from the medullary raphe. As described above, these cells appeared to be neuroblasts because they displayed solely neuronal morphologies upon differentiation in vitro and expressed neuronal markers such as NSE and NF (Whittemore & White 1993).

The RN33B cells were used to explore two issues: first, to assess whether the precursor population would respond to signals in the mature CNS, and second, to determine if diverse target regions of the raphe cells in situ would differentially influence the morphology of the immortal cells as had been observed in vitro (Whittemore & White 1993; see above). For these studies, the RN33B cells were infected a second time with a *lacZ*-containing retrovirus to facilitate in vivo identification (Onifer et al 1993a) and were implanted into the spinal cord and hippocampus of rats (Onifer et al 1993b). Histological examination of grafted regions indicated that the adult CNS supported the survival of the RN33B cells for at least two weeks postimplantation. In both implantation sites, the immortal precursors exhibited a range of morphologies that was often reflective of endogenous cells in the local environment. For example, immortal cells engrafted within the pyramidal layer of the hippocampus displayed oval or pyramidal somata with elaborate apical and basal processes (Figure 3), which are features characteristic of pyramidal neurons in situ. And some cells implanted within the spinal cord resembled the multipolar morphology of spinal cord interneurons.

There were clear differences in the extent to which RN33B cells differentiated in different target regions, however. Most marked was the observation that the majority of the cells engrafted within the spinal cord exhibited a bipolar morphology with very small somata that did not resemble the morphology of any endogenous neurons surrounding the implanted cells. Rather, these RN33B cells were very similar to the undifferentiated morphology of the line in vitro, suggesting that the spinal cord environment did not adequately supply the factors necessary for complete precursor differentiation. In contrast, immortal cells implanted into the hippocampus generally displayed larger somata with more complex process formation. The location of these cells typically correlated with regions of the hippocampus that produce high levels of BDNF, NGF, and NT-3, suggesting a potential role for these neurotrophins in cellular differentiation in vivo. This possibility is supported by the finding of less precursor differentiation within the adult spinal cord, a region that does not synthesize these neurotrophic molecules. Differentiation cannot be solely linked to soluble factors, however, since the RN33B cells often displayed diverse cell types in close proximity to one another. Substrate cues most likely provide an additional guide for the developing neuroblasts. Although the factors that influence precursor development remain to be fully elucidated, these results demonstrate that the adult CNS retains the capacity to direct the differentiation of neural precursors. Furthermore, the microenvironment in

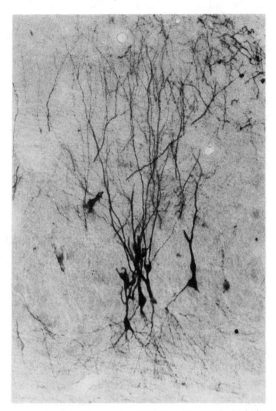

Figure 3 Photomicrograph of *lacZ*-expressing RN33B cells engrafted within the CA3 region of the adult hippocampus 8 weeks postgrafting. Implanted cells displaying β-gal immunoreactivity show extensive apical and distal processes that are characteristic of endogenous cells located in the pyramidal layer of CA3. Micrograph courtesy of LS Shihabuddin, SR Whittemore & VR Holets.

which the precursors develop clearly plays a role in cellular identity, which is consistent with results obtained with immortalized precursors implanted into the immature CNS.

THERAPEUTIC POTENTIAL OF NEURAL PRECURSORS

There has been a great deal of interest in restoring neural function within the damaged CNS through the use of neuronal transplantation (see Fisher & Gage 1993). Most of this work has focused on the use of freshly dissected embryonic neural tissues for grafting, since fetal cells survive and function very well when introduced into a host CNS. The cultured precursor populations provide a powerful alternative cellular source for transplantation because these cells have the capacity for generating site-specific neural cell types. In the following section, some recent studies that have explored the therapeutic potential of cultured precursor populations are discussed.

Immortalized Neural Cells

The immortalized cerebellar precursors that successfully expressed the *lacZ* transgene for as long as 22 months (Snyder et al 1992; see above) have been explored for their ability to transport a therapeutic gene into the CNS of mice with a neurovisceral lysosomal storage disease (Snyder et al 1994). The model used to assess gene transfer with neural precursors is the mucopolysaccharidosis type VII mouse, which has a deficiency in the enzyme β-glucuronidase (GUS). This enzyme deficiency results in early death of the animals from lysosomal storage of undegraded glycosaminoglycans in many areas, including the liver, kidney, and CNS (Birkenmeier et al 1989). Two approaches that have been used to try and correct this enzyme deficiency are bone marrow transplantation and somatic gene therapy directed toward bone marrow cells (Birkenmeier et al 1991, Wolfe et al 1992). However, even when substantial improvements in GUS levels are obtained in peripheral organs, there are only minor corrections in the brain (Birkenmeier et al 1991). Thus, the development of cells that can function as enzyme carriers to the brain would provide an important adjunct therapy to peripheral treatments.

The immortalized cerebellar precursors that expressed the *lacZ* transgene (Snyder et al 1992; see above) were further modified to express human GUS (Snyder et al 1994). These cells were then implanted into newborn mucopolysaccharidosis type VII mice by using a technique that resulted in diffuse engraftment throughout the neural axis. Grafted mice were analyzed as adults for transgene expression (*lacZ* and GUS), GUS enzymatic activity, and anatomical changes in lysosomal storage. The modified precursors showed robust engraftment throughout the brain, which was associated with a diffuse elevation of GUS activity to 1–4% of normal levels. Levels of GUS as high as 20% of normal, which is sufficient to have a neuropathologic impact, have been observed in some regions of the grafted brain. Indeed, some areas of the brain have shown a reduction in abnormal lysosomal storage. To date, the modified precursors have been seen to survive and express GUS for at least 8 months postgrafting, suggesting that such a strategy may provide an effective long-term treatment for enzyme deficiencies.

Primary Neural Cells

The bipotent O-2A precursors that generate oligodendrocytes and type 2 astrocytes in vitro (see above) have been explored as a potential source of grafting material to replace myelin in animal models of glial loss (e.g. spinal cord damage) or degeneration (e.g. multiple sclerosis). Demyelinating lesions in adult rats have been achieved by injecting 0.1% ethidium bromide into an area of the spinal cord that was previously subjected to localized irradiation by X rays. Such treatments produce a region of demyelinated axons within a glial-poor environment that prohibits endogenous repair (Crang et al 1992). This damage thus provides a system for assessing the ability of transplanted glial precursors to reconstruct appropriate myelination in the injured zone (for review see Blakemore & Franklin 1991).

The O-2A cells used for grafting are typically derived from the optic nerve of P7 rats. These cells can be expanded and enriched for the precursor population by culturing the cells in the presence of PDGF and bFGF (Bögler et al 1990). When implanted into the damaged spinal cord, O-2A precursors predominantly differentiate into oligodendrocytes while a smaller percentage (1%) appear to generate astrocytes (Crang et al 1992). The graft-derived oligodendrocytes have been seen to remyelinate as much as 90% of the denuded axons in the damaged region. These newly formed myelin sheaths are generally thinner than those found in the intact spinal cord but are otherwise indistinguishable from CNS myelin. The graft-derived origin of the replacement myelin has been confirmed in some cases by premarking the O-2A cells with the *lacZ* gene prior to transplantation (Crang et al 1991). These results indicate that cultured O-2A cells retain their bipotent nature when reintroduced into the CNS. Furthermore, oligodendrocytes that arise from these implanted progenitors function appropriately, as indicated by the formation of myelin sheaths around axons. Finally, in addition to their ability to repair regions of glial loss, the ability of the O-2A cells to express a transgene after grafting indicates that these precursors can provide a cellular platform for transporting therapeutic gene products into the CNS.

Primary neural populations that are capable of generating both neurons and glia for extended periods in vitro have also been explored as a potential source of neural tissue for CNS transplantation (Gage et al 1994, Ray 1994). Cells obtained from the embryonic hippocampus of rats show proliferative activity in vitro when exposed to high concentrations of bFGF (Ray et al 1993; see above). Differentiated cells that arise in these cultures express several neuronal markers, including NSE, NF, calbindin-D, MAP- 2, MAP-5, and glutamic acid decarboxylase (GAD). These cells, marked with ^{3}H-thymidine for in vivo identification, have been implanted into the hippocampus of adult rats and assessed for cell survival and differentiation for up to 6 weeks post-grafting (Ray 1994). Upon histological analyses, many of the implanted cells were found distributed throughout the dentate gyrus of the hippocampus, where some expressed the granule cell marker calbindin. Other engrafted cells displayed the markers gamma-aminobutyric acid (GABA) or GFAP. These results indicate that bFGF-expanded populations of cultured embryonic neural cells provide a viable source of neurons in vivo that may be useful for repopulating areas of the CNS that have been damaged. Recently, similar results have been obtained with cells isolated from the hippocampus of adult rats (Gage et al 1994). Moreover, since bFGF enhances the proliferation of neural precursors in vitro, transgenes have been successfully incorporated into the adult hippo-campus-derived cells in vitro. These genetically modified cells have been implanted into the adult rat brain where a subgroup of engrafted cells has been seen to express the transgene for at least 6 weeks postimplantation. Thus, in addition to providing a replacement source of neurons and glia for the CNS, cultured neural precursors can be manipulated to carry genes into the brain or spinal cord that may encode for factors that supplement or replace vital mol-ecules.

Finally, the subventricular zone may also provide a source of neuronal cells for transplantation. Lois & Alvarez-Buylla (1994) recently isolated cells from the subventricular zone of transgenic mice that expressed the β-gal gene from a neuron-specific promoter. When implanted into the lateral ventricles of adult hosts, the grafted cells generated progeny that selectively migrated through the adult brain to the olfactory bulb, where they differentiated into two types of olfactory interneurons. Although the engrafted subventricular cells are solely committed to olfactory cell types, exposing the cells to factors such as bFGF may be useful for driving subventricular-derived precursors toward alternative neural fates (see Vescovi et al 1993). The successful manipulation of the migration path and/or lineage decisions of subventricular precursors may offer an additional therapeutic strategy for replacing neuronal populations in the damaged brain.

Neural Precursors In Situ

Overwhelming evidence for a role of growth factors in precursor proliferation and/or differentiation suggests a potential route for activating putative stem cells within the CNS to replace neurons that are lost through damage or disease. Such an intriguing notion was recently pursued by Tao and colleagues (1993); they injected bFGF systemically into neonatal rats to determine if the growth factor could induce neurogenesis in vivo. Mitotic activity was assessed using [³H]thymidine. In comparison to vehicle-injected controls, the bFGF-treated animals showed as much as a 50% increase in [³H]thymidine incorporation in cerebellar tissues. Granule cells isolated from the cerebellum of the bFGF-treated animals by centrifugation appeared to be at least one of the populations that contributed to the proliferative pool. Although quite preliminary, these results highlight the potential for manipulating resident precursor populations within the CNS. Developing methods for successfully provoking selective neurogenesis in the brain will have wide ranging ramifications for repairing neural dysfunction.

SUMMARY

The nervous system of adult mammals, unlike the rest of the organs in the body, has been considered unique in its apparent inability to replace neurons following injury. However, in certain regions of the brain, neurogenesis occurs postnatally and continues through adulthood. The nature, fate, and longevity of cells undergoing proliferation within the CNS are unknown. These cells are increasingly becoming the focus of intense scrutiny; this is a recent development that has led to considerable controversy over the appropriate terminology to describe neural cells as they pass through different stages of proliferation, migration, and differentiation. Continuing studies detailing the properties of mitotic populations in the adult CNS will provide a better understanding of the nature of these cells during their development and should lead to a more consistent nomenclature.

Studies of neural precursors isolated from the embryonic brain have indicated that many subgroups of cells undergo mitosis and subsequent differentiation into neurons and glia in vitro. A number of substances, such as growth factors and substrate molecules, are essential for these processes and also for lineage restriction and fate determination of these cells. Recent studies have shown that cells with proliferative capabilities can also be isolated from the adult brain. The nature of these cells is unknown, but there is evidence that both multipotent cells (stem cells) and lineage-restricted cells (neuroblasts or glioblasts) are resident within the mature CNS and that they can be maintained and induced to divide and differentiate in response to many of the same factors that influence their embryonic counterparts. Presently, it is unclear how many potentially quiescent precursor cells exist in the adult brain or what combination of growth factors and substrate molecules is involved in the proliferation and differentiation of these cells. Some of these questions are currently being addressed by using immortalized neural precursors or growth factor–expanded populations of primary precursors to model precursor responsiveness to environmental manipulations.

Because in vitro culture conditions are unlikely to provide all of the factors necessary for inducing the proliferation and differentiation of neural precursors, recent studies have explored the properties of well-characterized precursor populations after implantation back into specific regions of the developing or adult CNS. These studies have highlighted the importance of the microenvironment in precursor differentiation and further suggested that precursor plasticity is a characteristic that is probably common to neural precursors throughout the CNS. In a therapeutic context, the identification of factors that drive neural precursors toward a desired phenotype will have far-ranging implications for using such cells to repair select regions of the damaged brain. Ongoing work may show that many areas of the adult CNS that are not known to undergo neurogenesis do so, albeit slowly and with spatial diversity. Furthermore, as the brain matures, many quiescent populations may retain a proliferative capability that can be reactivated. As the factors that control the proliferation and differentiation of these cells in situ are clarified, it may become possible to achieve site-specific cellular replacement in vivo in the diseased or injured CNS.

Literature Cited

Altman J. 1962. Are new neurons formed in the brains of adult mammals? *Science* 135:1127–29

Addison WHF. 1911. The development of the Purkinje cells and of the cortical layers in the cerebellum of the albino rat. *J. Comp. Neurol.* 21:459–85

Alberts B, Bray D, Lewis J, Raff M, Roberts K, Watson JD. 1989. *Molecular Biology of the Cell.* New York: Garland. 1219 pp. 2nd ed.

Anderson DJ. 1989. The neural crest cell lineage problem: neuropoiesis? *Neuron* 3:1–12

Anderson DJ, Axel R. 1986. A bipotential neuroendocrine precursor whose choice of cell fate is determined by NGF and glucocorticoids. *Cell* 47:1079–90

Bayer SA. 1982. Changes in the total number of dentate granuale cells in juvenile and adult rats: a correlated volumetric and ^{3}H-thymidine autoradiographic study. *Exp. Brain Res.* 46:315–23

Bartlett PF, Reid HH, Bailey KA, Bernard O. 1988. Immortalization of mouse neural precursor cells by the *c-myc* oncogene. *Proc. Natl. Acad. Sci. USA* 85:3255–59

Bernard O, Reid HH, Bartlett PF. 1989. Role of the c-myc and the N-myc proto-oncogenes in the immortalization of neural precursors. *J. Neurosci. Res.* 24:9–20

Bianchi DW, Wilkins-Haug LE, Enders AC, Hay ED. 1993. Origin of extraembryonic mesoderm in experimental animals: relevance to chorionic mosaicism in humans. *Am. J. Med. Genet.* 46:542–50

Birkenmeier EH, Barker JE, Vogler CA, Kyle JW, Sly WS, et al. 1991. Increased life span and correction of metabolic defects in murine mucopolysaccharidosis type VII after syngeneic bone marrow transplantation. *Blood* 78:3081–92

Birkenmeier EH, Davisson MT, Beamer WG, Ganschow RE, Vogler CA, et al. 1989. Murine mucopolysaccharidosis type VII. Characterization of a mouse with β-glucuronidase deficiency. *J. Clin. Invest.* 83:1258–66

Birren SJ, Anderson DJ. 1990. A v-myc-immortalized sympatho-adrenal progenitor cell line in which neuronal differentiation is initiated by FGF but not NGF. *Neuron* 4:189–201

Birren SJ, Verdi J, Anderson DJ. 1992. Membrane depolarization induces $p140^{trk}$ and NGF-responsiveness, but not $p75^{LNGFR}$ in MAH cells. *Science* 257:395–97

Blakemore WF, Franklin RJM. 1991. Transplantation of glial cells into the CNS. *Trends Neurosci.* 14:323–27

Bögler O, Wren D, Barnett SC, Land H. Noble M. 1990. Cooperation between two growth factors promotes extended self-renewal and inhibits differentiation of oligodendrocyte-type-2 astrocyte (O-2A) progenitor cells. *Proc. Natl. Acad. Sci. USA* 87:6368–72

Bottenstein JE, Sato G. 1980. Growth of a rat neuroblastoma cell line in serum-free supplemented medium. *Proc. Natl. Acad. Sci. USA* 76:514–17

Boulder Committee. 1970. Embryonic vertebrate central nervous system: revised terminology. *Anat. Rec.* 166:257–62

Camara CG, Harding JW. 1984. Thymidine incorporation in the olfactory epithelium of mice: early exponential response induced by olfactory neurectomy. *Brain Res.* 308:63–68

Cameron HA, Woolley CS, McEwen BS, Gould E. 1993. Differentiation of newly born neurons and glia in the dentate gyrus of the adult rat. *Neuroscience* 56:337–44

Carden MJ, Trojanowski JQ, Schlaepfer WW, Lee VM. 1987. Two-stage expression of neurofilament polypeptides during rat neurogenesis with early establishment of adult phosphorylation patterns. *J. Neurosci.* 7: 3789–804

Carr VM, Farbman AI. 1992. Ablation of the olfactory bulb up-regulates the rate of neurogenesis and induces precocious cell death in olfactory epithelium. *Exp. Neurol.* 115: 55–59

Cattaneo E, McKay R. 1990. Proliferation and differentiation of neuronal stem cells regulated by nerve growth factor. *Nature* 347: 762–65

Cepko CL. 1989. Immortalization of neural cells via retrovirus-mediated oncogene transduction. *Annu. Rev. Neurosci.* 12:47–65

Cepko CL. 1988. Immortalization of neuronal cells via oncogene transduction. *Trends Neurosci.* 11:6–8

Cooper JR, Bloom FE, Roth RH. 1982. *The Biochemical Basis of Neuropharmacology,* New York: Oxford Univ. Press. 367 pp. 4th ed.

Corotto FS, Henegar JA, Maruniak JA. 1993. Neurogenesis persists in the subependymal layer of the adult mouse brain. *Neurosci. Lett.* 149:111–14

Crang AJ, Franklin RJM, Blakemore WF, Trotter J, Schachner M, et al. 1991. Transplantation of normal and genetically-engineered glia into areas of demyelination. *Ann. NY Acad. Sci.* 633:563–66

Crang AJ, Franklin RJM, Blakemore WF, Noble M, Barnett SC, et al. 1992. The differentiation of glial cell progenitor populations following transplantation into non-repairing central nervous system glial lesions in adult animals. *J. Neuroimmunol.* 40:243–54

Debus E, Weber K, Osborn M. 1983. Monoclonal antibody specific for glial fibrillary acidic (GFA) protein and for each of the neurofilament triplet polypeptides. *Differentiation* 25: 193–203

Deloulme JC, Baudier J, Sensenbrenner M. 1991. Establishment of pure neuronal cultures from fetal rat spinal cord and proliferation of the neuronal precursor cells in the presence of fibroblast growth factor. *J. Neurosci. Res.* 29:499–509

Dotti CG, Banker GA, Binder LI. 1987. The expression of and distribution of the microtubule-associated proteins tau and microtubule-associated protein 2 in hippocampal neurons in the rat in situ and in cell culture. *Neuroscience* 23:121–30

Doupe AJ, Landis SC, Patterson PH. 1985. Environmental influences in the development of neural crest derivatives: glucocorticoids,

growth factors, and chromaffin cell plasticity. *J. Neurosci.* 5:2119–42

Drago J, Murphy M, Carrol SM, Harvey RP, Bartlett PF. 1991a. Fibroblast growth factor-mediated proliferation of central nervous system precursors depends on endogenous production of insulin-like growth factor I. *Proc. Natl. Acad. Sci. USA* 88:2199–203

Drago J, Nurcombe V, Pearse MJ, Murphy M, Bartlett PF. 1991b. Basic fibroblast growth factor upregulates steady-state levels of laminin B1 and B2 chain mRNA in cultred neuroepithelial cells. *Exp. Cell Res.* 196: 246–54

Eisenbarth GS, Walsh FS, Nirenberg M. 1979. Monoclonal antibody to a plasma membrane antigen of neurons. *Proc. Natl. Acad. Sci. USA* 77:4165–69

Evard C, Borde I, Marin P, Galiana E, Premont J, et al. 1990. Immortalization of bipotential and plastic glio-neuronal precursor cells. *Proc. Natl. Acad. Sci. USA* 87:3062–66

Feldmann SC, Christakos S. 1983. Vitamin D-dependent calcium-binding protein in rat brain: biochemical and immunocytochemical characterization. *Endocrinology* 112: 290–302

Fisher LJ, Gage FH. 1993. Grafting in the mammalian central nervous system. *Physiol. Rev.* 73:583–616

Frederiksen K, Jat PS, Valtz N, Levy D, McKay R. 1988. Immortalization of precursor cells from the mammalian CNS. *Neuron* 1:439–48

Frederiksen K, McKay RDG. 1988. Proliferation and differentiation of rat neuroepithelial precursor cells in vivo. *J. Neurosci.* 8:1144–51

Gage FH, Coates PW, Ray J, Peterson DA, Suhr ST, et al. 1994. Adult hippocampal neural cells survive in vitro and following grafting to the adult hippocampus. *5th Int. Symp. Neural Transplant.* In press

Gensburger C, Labourdette G, Sensenbrenner M. 1987. Brain basic fibroblast growth factor stimulates the proliferation of rat neuronal precursor cells in vitro. *FEBS Lett.* 217:1–5

Graziadei PPC, Kaplan MS, Monti Graziadei GA, Bernstein JJ. 1980. Neurogenesis of sensory neurons in the primate olfactory system after section of fila olfactoria. *Brain Res.* 186:289–300

Graziadei PPC, Levine RR, Monti Graziadei GA. 1979. Plasticity of connections of the olfactory sensory neuron: regeneration into the forebrain following bulbectomy in the neonatal mouse. *Neuroscience* 4:713–27

Graziadei PPC, Monti Graziadei GA. 1978. Continuous nerve cell renewal in the olfactory system. In *Development of Sensory Systems,* ed. M Jacobsen, pp. 55–83. Berlin: Springer-Verlag

Graziadei PPC, Monti Graziadei GA. 1979.

Neurogenesis and neuron regeneration in the olfactory system of mammals. I. Morphological aspects of differentiation and structural organization of the olfactory neurons. *J. Neurocytol.* 8:1–18

Hatten ME. 1993. The role of migration in central nervous system neuronal development. *Curr. Opin. Neurobiol.* 3:38–44

Houle J, Federoff S. 1983. Temporal relationship between the appearance of vimentin and neural tube development. *Dev. Brain Res.* 9:189–95

Hughes SM, Lillien LE, Raff MC, Rohrer H, Sendtner M. 1988. Ciliary neurotrophic factor induced type-2 astrocyte differentiation in culture. *Nature* 335:70–73

Jat PS, Sharp PA. 1989. Cell lines established by a temperature-sensitive simian virus 40 large-T-antigen gene are growth restricted at the nonpermissive temperature. *Mol. Cell. Biol.* 9:1672–81

Kaplan MS. 1981. Neurogenesis in the 3-month-old rat visual cortex. *J. Comp. Neurol.* 195:323–38

Kilpatrick TJ, Bartlett PF. 1993. Cloning and growth of multipotential neural precursors: requirements for proliferation and differentiation. *Neuron* 10:255–65

Kratzing JE. 1978. The olfactory apparatus of the bandicoot (Isoodon macrourus): fine structure and presence of a septal olfactory organ. *J. Anat.* 125:601–13

Lendhal U, McKay RDG. 1990. The use of cell lines in neurobiology. *Trend Neurosci.* 13: 132–37

Lendahl U, Zimmerman LB, McKay RDG. 1990. CNS stem cells express a new class of intermediate filament protein. *Cell* 60:585–95

Levison SW, Goldman JE. 1993. Both oligodendrocytes and astrocytes develop from progenitors in the subventricular zone of postnatal rat forebrain. *Neuron* 10:201–12

Lo LC, Birren SJ, Anderson DJ. 1991. V-myc immortalization of early rat neural crest cells yields a clonal cell line which generates both glial and adrenergic progenitor cells. *Dev. Biol.* 145:139–53

Lois C, Alvarez-Buylla A. 1993. Proliferating subventricular zone cells in the adult mammalian forebrain can differentiate into neurons and glia. *Proc. Natl. Acad. Sci. USA* 90:2074–77

Lois C, Alvarez-Buylla A. 1994. Long-distance neuronal migration in the adult mammalian brain. *Science* 264:1145–48

Luskin MB. 1993. Restricted proliferation and migration of postnatally generated neurons derived from the forebrain subventricular zone. *Neuron* 11:173–89

Matulionis DH. 1975. Ultrastructural study of mouse olfactory epithelium following destruction by ZnSO4 and its subsequent regeneration. *Am. J. Anat.* 142:67–90

McKay RDG. 1992. Immortal mammalian neuronal stem cells differentiate after implantation into the developing brain. In *Gene Transfer and Therapy in the Nervous System,* ed. F Gage, Y Christen, pp. 76–85. Berlin: Springer-Verlag

Morshead CM, van der Kooy D. 1992. Postmitotic death is the fate of constitutively proliferating cells in the subependymal layer of the adult mouse brain. *J. Neurosci.* 12: 249–56

Murphy M, Drago J, Bartlett PF. 1990. Fibroblast growth factor stimulates the proliferation and differentiation of neural precursor cells in vitro. *J. Neurosci. Res.* 25:463–75

Noble M, Murray K. 1984. Purified astrocytes promote the in vitro division of a bipotential glial progenitor cell. *EMBO J.* 3:2243–47

Noble M, Murray K, Stroobant P, Waterfield MD, Riddle P. 1988. Platelet-derived growth factor promotes division and motility and inhibits premature differentiation of the oligodendrocyte/type-2 astrocyte progenitor cell. *Nature* 333:560–62

Onifer SM, White LA, Whittemore SR, Holets VR. 1993a. In vitro labeling strategies for identifying primary neural tissue and a neuronal cell line after transplantation in the CNS. *Cell Transplant.* 2:131–49

Onifer SM, Whittemore SR, Holets VR. 1993b. Variable morphological differentiation of a raphe-derived neuronal cell line following transplantation into the adult rat CNS. *Exp. Neurol.* 122:130–42

Palay SL, Chan-Palay V. 1974. *Cerebellar Cortex Cytology and Organization.* Berlin: Springer-Verlag

Phemister RD, Young S. 1968. The postnatal development of the canine cerebellar cortex. *J. Comp. Neurol.* 134:243–54

Price J, Turner D, Cepko C. 1987. Lineage analysis in the verebrate nervous system by retrovirus-mediated gene transfer. *Proc. Natl. Acad. Sci. USA* 84:156–60

Raaf J, Kernohan JW. 1944. A study of the external granular layer in the cerebellum. *Am. J. Anat.* 75:151–72

Raff MC, Miller RH, Noble M. 1983. A glial progenitor cell that develops in vitro into an astrocyte or an oligodendrocyte depending on the culture medium. *Nature* 303:390–96

Rakic P. 1985. Limits of neurogenesis in primates. *Science* 227:154–56

Ranscht B, Clapshaw PA, Price J, Noble M, Seifert W. 1982. Development of oligodendrocytes and Schwann cells studies with a monoclonal antibody against galactocerebrosidase. *Proc. Natl. Acad. Sci. USA* 79:2709–13

Rathjen FG, Schachner M. 1984. Immunocytological and biochemical characterization of a new neuronal cell surface component (L1 antigen) which is involved in cell adhesion. *EMBO J.* 3:1–10

Ray J. 1994. Basic fibroblast growth factor responsive primary fetal CNS cells express neural markers in vitro and after grafting into the adult animals. *Neuropsychopharmacology* 10:908S

Ray J, Gage FH. 1994. Spinal cord neuroblasts proliferate in response to basic fibroblast growth factor. *J. Neurosci.* 14:3548–64

Ray J, Peterson DA, Schinstine M, Gage FH. 1993. Proliferation, differentiation, and long-term culture of primary hippocampal neurons. *Proc. Natl. Acad. Sci. USA* 90:3602–6

Redies C, Lendahl U, McKay RDG. 1991. Differentiation and heterogeneity in T-antigen immortalized precursor cell lines from mouse cerebellum. *J. Neurosci. Res.* 30:601–15

Renfranz PJ, Cunningham MG, McKay RDG. 1991. Region-specific differentiation of the hippocampal stem cell lines HiB5 upon implantation into the developing mammalian brain. *Cell* 66:713–29

Reynolds BA, Tetzlaff W, Weiss S. 1992. A multipotent EGF-responsive striatal embryonic progenitor cell produces neurons and astrocytes. *J. Neurosci.* 12:4564–74

Reynolds BA, Weiss S. 1992. Generation of neurons and astrocytes from isolated cells of the adult mammalian central nervous system. *Science* 255:1707–10

Richards LJ, Kilpatrick TJ, Bartlett PF. 1992. De novo generation of neuronal cells from the adult mouse brain. *Proc. Natl. Acad. Sci. USA* 89:8591–95

Ryder EF, Snyder EY, Cepko CL. 1990. Establishment and characterization of multipotent neural cell lines using retrovirus vector-mediated oncogene transfer. *J. Neurobiol.* 21:356–75

Sanes JR, Rubenstein JL, Nicolas JF. 1986. Use of recombinant retrovirus to study post-implantation cell lineage in mouse embryos. *EMBO J.* 5:3133–42

Schmechel DE, Brightman MW, Marangos PJ. 1980. Neurons switch from non-neuronal enolase to neuron-specific enolase during differentiation. *Brain Res.* 190:195–214

Schwartz Levey M, Chikaraishi DM, Kauer JS. 1991. Characterization of potential precursor populations in the mouse olfactory epithelium using immunocytochemistry and autoradiography. *J. Neurosci.* 11:3556–64

Shah NM, Marchionni MA, Issacs I, Stroobant P, Anderson DJ. 1994. Glial growth factor restricts mammalian crest stem cells to a glial fate. *Cell* 77:349–60

Sidman RL. 1970. Autoradiographic methods and principles for study of the nervous system with thymidine-H³. In *Contemporary Research Methods in Neuroanatomy*, ed. WJH Nauta, JOE Ebbesson, pp. 25–274. Berlin: Springer-Verlag

Sieber-Blum M, Chokshi HR. 1985. In vitro proliferation and terminal differentiation of quail neural crest cells in a defined culture medium. *Exp. Cell Res.* 158:267–72

Sloviter RS. 1989. Calcium-binding proteins (calbindin-D28k) and parvalbumun immunocytochemistry: localization in the rat hippocampus with specific reference to the selective vulnerability of hippocampal neurons to seizure activity. *J. Comp. Neurol.* 280:477–85

Smeyne RJ, Schilling K, Robertson L, Kuk D, Oberdick J, et al. 1992. Fos-lacZ transgenic mice: mapping sites of gene induction in the central nervous system. *Neuron* 8:13–23

Snyder EY, Deitcher DL, Walsh C, Arnold-Aldea S, Hartweig EA, Cepko CL. 1992. Multipotent neural cell lines can engraft and participate in development of mouse cerebellum. *Cell* 68:33–51

Snyder EY, Taylor RM, Wolfe JH. 1994. Transplantation of β-glucuronidase (GUSB)-expressing immortalized neural progenitors for gene transfer into CNS of the mucopolysaccaridosis (MPS) VII mouse, a model of a neurovisceral lysosomal storage disease. *J. Cell Biochem.* 18A:246 (Abstr.)

Snyder EY, Yandava BD, Pan Z-H, Yoon CH, Macklis JD. 1993. Multipotent neural progenitor cell lines can engraft and participate in development of multiple structures at multiple stages along mouse neuraxis. *Soc. Neurosci.* 19:613 (Abstr.)

Sommer I, Schachner M. 1981. Monoclonal antibodies (O1 to O4) to oligodendrocyte surfaces: an immunocytological study in the central nervous system. *Dev. Biol.* 83:311–27

Stanfield BB, Trice JE. 1988. Evidence that granule cell generated in the dentate gyrus of adult rats extend axonal projections. *Exp. Brain Res.* 72:399–406

Stemple DL, Anderson DJ. 1992. Isolation of a stem cell for neurons and glia from the mammalian neural crest. *Cell* 71:973–85

Stemple DL, Mahanthappa NK, Anderson DJ. 1988. Basic FGF induces neuronal differentiation, cell division, and NGF responsiveness in chromaffin cells: a sequence of events in sympathetic development. *Neuron* 1:517–25

Tao Y, Black IB, DiCicco-Bloom E. 1993. In vivo regulation of cerebellar granule cell neurogenesis by bFGF. *Soc. Neurosci.* 19:30 (Abstr.)

Temple S. 1989. Division and differentiation of isolated CNS blast cells in microculture. *Nature* 340:471–73

Vandenbergh DJ, Mori N, Anderson DJ. 1991. Co-expression of multiple neurotransmitter enzyme genes in normal and immortalized sympathoadrenal progenitor cells. *Dev. Biol.* 148:10–22

Verhaagen J, Oestreicher AB, Grillo M, Khew-Goodall YS, Gispen WH, Margolis FL. 1990. Neuroplasticity in the olfactory system: differential effects of central and peripheral lesions of the primary olfactory pathway on the expression of B-50/GAP43 and the olfactory marker protein. *J. Neurosci. Res.* 26:31–44

Vescovi AL, Reynolds BA, Fraser DD, Weiss S. 1993. bFGF regulates the proliferative fate of unipotent (neuronal) and bipotent (neuronal/astroglial) EGF-generated CNS progenitor cells. *Neuron* 11:951–66

Vicario C, Cunningham MG, McKay RDG. 1993. Fate shifting of cerebellar precursor cells upon transplantation into the developing dentate gyrus. *Soc. Neurosci.* 19:1512 (Abstr.)

White LA, Whittemore SR. 992. Immortalization of raphe neurons: an approach to neuronal function in vitro and in vivo. *J. Chem. Neuroanat.* 5:327–30

Whittemore SR, White LA. 1993. Target regulation of neuronal differentiation in a temperature-sensitive cell line derived from medullary raphe. *Brain Res.* 615:27–40

Williams BP, Read J, Price J. 1991. The generation of neurons and oligodendrocytes from a common precursor cell. *Neuron* 7:685–93

Wolfe JH, Sands MS, Barker JE, Gwynn B, Rowe LB, et al. 1992. Reversal of pathology in murine mucopolysaccharidosis type VII by somatic cell gene transfer. *Nature* 360:749–53

Wolswijk G, Noble M. 1989. Identification of an adult-specific glial progenitor cell. *Development* 105:387–400

Wolswijk G, Riddle PN, Noble M. 1990. Coexistence of perinatal and adult forms of a glial progenitor cell during development of the rat optic nerve. *Development* 109:691–98

Wren D, Wolswijk G, Noble M. 1992. In vitro analysis of the origin and maintenan. of O-2A adult progenitor cells. *J. Cell Biol.* 116:167–76

Annu. Rev. Neurosci. 1995. 18:193–222

NEURAL MECHANISMS OF SELECTIVE VISUAL ATTENTION

Robert Desimone[1] *and John Duncan*[2]

[1]Laboratory of Neuropsychology, NIMH, Building 49, Room 1B80, Bethesda, Maryland 20892 and [2]MRC Applied Psychology Unit, 15 Chaucer Road, Cambridge CB2 2EF, England

KEY WORDS: vision, cortex, primates, visual search, neglect

INTRODUCTION

The two basic phenomena that define the problem of visual attention can be illustrated in a simple example. Consider the arrays shown in each panel of Figure 1. In a typical experiment, before the arrays were presented, subjects would be asked to report letters appearing in one color (targets, here black letters), and to disregard letters in the other color (nontargets, here white letters). The array would then be briefly flashed, and the subjects, without any opportunity for eye movements, would give their report. The display mimics our usual cluttered visual environment: It contains one or more objects that are relevant to current behavior, along with others that are irrelevant.

The first basic phenomenon is limited capacity for processing information. At any given time, only a small amount of the information available on the retina can be processed and used in the control of behavior. Subjectively, giving attention to any one target leaves less available for others. In Figure 1, the probability of reporting the target letter N is much lower with two accompanying targets (Figure 1a) than with none (Figure 1b).

The second basic phenomenon is selectivity—the ability to filter out unwanted information. Subjectively, one is aware of attended stimuli and largely unaware of unattended ones. Correspondingly, accuracy in identifying an attended stimulus may be independent of the number of nontargets in a display (Figure 1a vs 1c) (see Bundesen 1990, Duncan 1980).

193

Figure 1 Displays demonstrating limited processing capacity and selectivity in human vision. Subjects are shown the displays briefly and asked to report only the black letters. Limited capacity is shown by reduced accuracy as the number of targets is increased (compare *b* and *a*). Selectivity is shown by negligible impact of nontargets (compare *a* and *c*).

Taken together, such results suggest the following general model (Broadbent 1958; Neisser 1967; Treisman 1960, 1993). At some point (or several points) between input and response, objects in the visual input compete for representation, analysis, or control. The competition is biased, however, towards information that is currently relevant to behavior. Attended stimuli make demands on processing capacity, while unattended ones often do not.

In the following sections, we first outline the major behavioral characteristics of competition and consider the limitations within the nervous system that make competition necessary. We then describe selectivity, or how the competition may be resolved, at both the behavioral and neural level. To some extent, our account builds on early models of biased competition by Walley & Weiden (1973) and Harter & Aine (1984). The approach we take differs from the standard view of attention, in which attention functions as a mental spotlight enhancing the processing (and perhaps binding together the features) of the illuminated item. Instead, the model we develop is that attention is an emergent property of many neural mechanisms working to resolve competition for visual processing and control of behavior.

COMPETITION

Behavioral Data

In one simple type of experiment, two objects are presented in the visual field. Subjects must identify some property of both objects, with a separate response for each. Such studies reveal several important facts. First, dividing attention between two objects almost always results in poorer performance than focusing attention on one. Identifying simple properties of each object such as size, brightness, orientation, or spatial position gives much the same result as identifying more complex properties such as shape (see Duncan 1984, 1985, 1993).

A possible exception is simple detection of simultaneous energy onsets or offsets (Bonnel et al 1992).

Second, as long as the experiment uses brief stimulus exposures and measures the accuracy of stimulus identification, the major performance limitation appears to occur at stimulus input rather than subsequent short-term storage and response. For example, interference from processing two objects is abolished if they are shown one after the other, with an interval of perhaps a second between them (Duncan 1980), even though the two responses called for must still be remembered and made together at the end of the trial.

Third, interference is independent of eye movements. Even though gaze is always maintained at fixation, it is easier to identify one object in the periphery than two.

Fourth, interference is largely independent of the spatial separation between two objects, at least when the field is otherwise empty (Sagi & Julesz 1985, Vecera & Farah 1994). Though attention is sometimes seen as a mental spotlight illuminating or selecting information from a restricted region of visual space (Eriksen & Hoffman 1973, Posner et al 1980), performance seems not to depend on the absolute spatial distribution of information.

An enduring issue is the underlying reason for between-object competition. It has often been argued that full visual analysis of every object in a scene would be impossibly complex (Broadbent 1958, Tsotsos 1990). Competition reflects a limit on visual identification capacity. Equally strong, however, has been the view that competition concerns control of response systems (Allport 1980, Deutsch & Deutsch 1963). Certainly, some response activation often occurs from objects a person has been told to ignore (Eriksen & Eriksen 1974), which shows that unwanted information is not entirely filtered out in early vision. Very probably, competition between objects occurs at multiple levels between sensory input and motor output (Allport 1993).

Neural Basis for Competition

If the nervous system had unlimited capacity to process information in parallel throughout the visual field, competition between objects would presumably be necessary only at final motor output stages. Before discussing these motor stages, we first consider what limitations in the visual system make competition necessary at the input.

Objects in the visual field compete for processing within a network of 30 or more cortical visual areas (Desimone & Ungerleider 1989, Felleman & Van Essen 1991). These areas appear to be organized within two major cortico-cortical processing pathways, or streams, each of which begins with the primary visual cortex, or V1 (see Figure 2). The first, a ventral stream, is directed into the inferior temporal cortex and is important for object recognition, while the other, a dorsal stream, is directed into the posterior parietal cortex and is

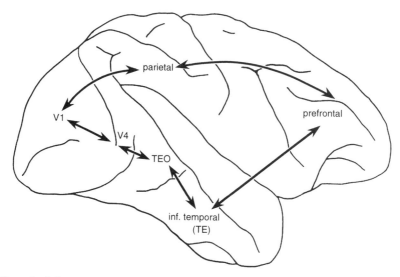

Figure 2 Striate cortex, or V1, is the source of two cortical visual streams. A dorsal stream is directed into the posterior parietal cortex and underlies spatial perception and visuomotor performance. A ventral stream is directed into the inferior temporal cortex and underlies object recognition. Both streams have further projections into prefrontal cortex. Adapted from Mishkin et al (1983) and Wilson et al (1993). For a "wiring diagram" of the areas and connections of the two streams, see Desimone & Ungerleider (1989) and Felleman & Van Essen (1991).

important for spatial perception and visuomotor performance (Ungerleider & Haxby 1994, Ungerleider & Mishkin 1982). Since competition impacts object recognition, we would expect to find one basis for it in the ventral stream.

The ventral stream includes specific anatomical subregions of area V2 (thin and interstripe regions), area V4, and areas TEO and TE in the inferior temporal (IT) cortex (see Desimone & Ungerleider 1989). As one proceeds from one area to the next along this pathway, neuronal properties change in two obvious ways. First, the complexity of visual processing increases. For example, whereas many V1 cells function essentially as local spatiotemporal energy filters, V2 neurons may respond to virtual or illusory contours in certain figures (von der Heydt et al 1984), and IT neurons respond selectively to global or overall object features, such as shape (Desimone et al 1984, Schwartz et al 1983, Tanaka et al 1991). Second, the receptive field size of individual neurons increases at each stage. As one moves from V1 to V4 to TEO to TE, typical receptive fields in the central field representation are on the order of 0.2, 3, 6, and 25° in size, respectively (see Boussaoud et al 1991, Ungerleider & Desimone 1989). Large receptive fields may contribute towards the recognition of objects over retinal translation (Gross & Mishkin 1977, Lueschow et al 1994).

These receptive fields can be viewed as a critical visual processing resource, for which objects in the visual field must compete (Desimone 1992, Olshausen et al 1993, Tsotsos 1990). If one were to add ever more independent objects to a V4 or IT receptive field, the information available about any one of them would certainly decrease. If, for example, a color-sensitive IT neuron were to integrate wavelength over its large receptive field, one might not be able to tell from that cell alone if a given level of response was due to, say, one red object or two yellow ones or three green ones at different locations in the field. Such ambiguity may be responsible for the interference effects found in divided attention.

This ambiguity may be reduced, in part, by linking objects and their features to retinal locations. It is sometimes presumed that location information is absent from the ventral "what" stream altogether and must be supplied by the dorsal "where" stream. In fact, the ventral stream itself contains information about the retinal location of complex object features. V4 and TEO neurons process relatively sophisticated information about object shape (Desimone & Schein 1987, Gallant et al 1993, Tanaka et al 1991) and have retinotopically organized receptive fields (Boussaoud et al 1991, Gattass et al 1988). At any given retinotopic locus in these areas, receptive fields show considerable scatter. One could, in principle, derive information about the relative locations of nearby features from a population of cells with partially overlapping fields the same way one could derive information about a specific color from a population of neurons with broad but different color tuning. Similarly, although receptive fields in IT cortex may span 20–30 degrees or more, they are not homogeneous. Typically, the fields have a hot spot at the center of gaze, which may extend asymmetrically into the upper or lower contralateral visual field. Although the stimulus preferences of IT neurons remain the same over large retinal regions, for a large minority of cells the absolute response to a given stimulus changes significantly with retinal location, i.e. cells are tuned to retinal location the same way they are tuned to other object features (Desimone et al 1984, Lueschow et al 1994, Schwartz et al 1983; also see Chelazzi et al 1993a). Thus, in principle, objects and their locations might be linked to some extent within the ventral stream. Even so, parallel processing across the visual field is likely to be limited.

To sum up, retinal location, as with other object features, is coarsely coded in the ventral stream. Information about more than one object may, to some extent, be processed in parallel, but the information available about any given object will decline as more and more objects are added to receptive fields. Therefore, objects must compete for processing in the ventral stream, and the visual system should use any information it has about relevant objects to bias the competition in their favor. This issue, which we term selectivity, is considered in later sections.

If the dorsal stream receives its visual input in parallel to the ventral stream as the anatomy suggests (Desimone & Ungerleider 1989), then it is presumably faced with competition among objects as well. As in IT cortex, receptive fields in posterior parietal cortex are very large, and it seems likely that increasing the number of independent objects in the visual field will eventually exceed the capacity of parietal cortex to extract the locations of each of them in parallel. Likewise, neural systems for visuomotor control must also deal with competition, to the extent that distractors are not already filtered out of the visual input (e.g. Munoz & Wurtz 1993a,b). Ultimately, for example, it is possible to move the eyes to only one target at a time. A critical issue is how selectivity is coordinated across the different systems so that the same target object is selected for perceptual and spatial analysis as well as for motor control.

SELECTIVITY: SCREENING OUT UNWANTED STIMULI

Behavioral Data

The ability to screen out irrelevant objects (Figure 1) is not absolute. It is easy in some cases and difficult in others, as is well illustrated in visual search. The subject detects or identifies a single target presented in an array of nontargets. Examples are shown in Figure 3. In easy cases, the target appears to "pop out" of the array, as if attention were drawn directly to it (Donderi & Zelnicker 1969, Treisman & Gelade 1980). Under such circumstances, the number of nontargets has little effect on the speed or accuracy of target detection or identification. In hard cases, however, nontargets are not filtered out well. In these instances, the number of nontargets in the display has a large effect on performance. An increase of 50 ms in target detection time for each nontarget added to the array is typical (Treisman & Gelade 1980), though in fact, this

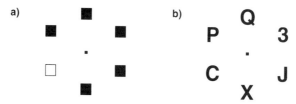

Figure 3   Selectivity in visual search. Target pop-out is revealed when the target is a mismatching element in an otherwise homogeneous field (*panel a*). Search is also extremely easy, however, whenever targets and nontargets are highly discriminable. Pop-out can also be based on more complex properties (*panel b;* search for the single digit).

figure varies widely and continuously from one task to another (Treisman & Gormican 1988).

According to the biased competition model, targets and nontargets compete for processing capacity in visual search. One factor influencing selectivity is bottom-up bias. It is very easy, for example, to find a unique target in an array of homogeneous nontargets (Figure 3a), perhaps reflecting an enduring competitive bias towards local inhomogeneities (Sagi & Julesz 1984). There may be similar biases towards sudden appearances of new objects in the visual field (Jonides & Yantis 1988) and towards objects that are larger, brighter, faster-moving, etc (Treisman & Gormican 1988).

An attentional system, however, would be of little use if it were entirely dominated by bottom-up biases. What is needed is a way to bias competition towards whatever information is relevant to current behavior. That is, one needs top-down control in addition to bottom-up, stimulus-driven biases. Correspondingly, there are many cases of easy search that do not depend on local inhomogeneity or sudden target onset. A colored target in a multicolored display, for example, may show good pop-out if the colors are highly discriminable (Duncan 1989). At least after a little practice, pop-out can be obtained during search for a single digit among letters (Figure 3b) (see Egeth et al 1972, Schneider & Shiffrin 1977).

Even when target selection is guided by top-down control, the ability to find targets is still dependent on bottom-up stimulus factors, especially the visual similarity of targets to nontargets. Provided that targets and nontargets are sufficiently different, however, easy search can be based on many different visual attributes, including simple features, such as size or color, and more complex conjunctions of these features (Duncan & Humphreys 1989, McLeod et al 1988, Wolfe et al 1989). Conjunction search provides a good example of the importance of similarity. In Figures 4a and b, the target is a large, white vertical bar. This target is much harder to find in Figure 4a, where each nontarget shares two properties with the target, than in Figure 4b, where only one property is shared (Quinlan & Humphreys 1987). Indeed, the latter case can give excellent pop-out; a similar result can be produced simply by increasing the discriminability of each conjunction's component features (Wolfe et al 1989).

Such results suggest the following model of biased competition. According to the task, any kind of input—objects of a certain kind, objects with a certain color or motion, objects in a certain location, etc—can be behaviorally relevant. Some kind of short-term description of the information currently needed must be used to control competitive bias in the visual system, such that inputs matching that description are favored in the visual cortex (Bundesen 1990, Duncan & Humphreys 1989). This short-term description has been called the attentional template (Duncan & Humphreys 1989); it may be seen as one aspect

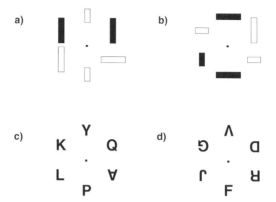

Figure 4   (*a, b*) Discriminability between targets and nontargets in conjunction search. Searching for a large, white vertical bar is harder when nontargets share two (*panel a*) rather than one (*panel b*) property with the target. In the latter case good pop-out can be obtained. (*c, d*) Novelty bias. It is easier to find a single inverted letter among upright nontargets (*panel c*) than the reverse (*panel d*).

of working memory (Baddeley 1986). The template can specify any property of required input—shape, color, location, etc.

Visual search is easy if targets and nontargets are easily discriminable. In this case, nontargets are poor matches to the attentional template and receive a weak competitive bias. Thus, the time it takes to find the target may be independent of the number of nontargets in the display. By contrast, search is difficult if nontargets are similar to the target. In this case, the competitive advantage of the target is reduced because each nontarget shares in the bias provided by the attentional template. Thus, each nontarget added to the display interferes with target detection. Alternative, serial-search accounts are considered below.

A great deal of work has dealt specifically with spatial selection, i.e. selection based on some cue to the location of target information (Eriksen & Hoffman 1973, Posner et al 1980, Sperling 1960). Indeed, spatial selection is often dealt with as a special case. We do not review this work in detail; it was covered earlier by Posner & Petersen (1990), and Colby (1991) has reviewed the neural mechanisms of spatial selection. Certainly, however, space is only one of the many cues that can be used in efficient target selection. A general account of selectivity must deal with both spatial and nonspatial cases. In terms of the biased competition model, prior knowledge of the target's spatial location is just another type of attentional template that can be used to bias competition in favor of the target.

A final consideration is bias derived from long-term memory. One interesting case is bias to novelty. As shown in Figures 4c and d, for example, it is much easier to find an inverted (novel) target among upright (familiar) non-targets (Figure 4c) than the reverse (Figure 4d) (Reicher et al 1976). In fact, the time it takes to find an inverted character may be independent of the number of upright ones in a display (Wang et al 1992), which implies that multiple objects have parallel access to memory and that familiarity is a type of object feature that can be used to bias attentional competition. A second consideration is long-term learned importance. In a busy room, attention can be attracted by the sound of one's own name spoken nearby (Moray 1959). Similarly, long practice with one set of visual targets makes them hard to ignore when they are subsequently made irrelevant (Shiffrin & Schneider 1977). Thus, the top-down selection bias of a current task can sometimes be overturned by information of long-term or general significance acting in a bottom-up fashion. In the next sections we consider both bottom-up and top-down mechanisms for resolving competition.

Bottom-Up Neural Mechanisms for Object Selection

The first neural mechanisms for resolving competition we consider are those that derive from the intrinsic or learned biases of the perceptual systems towards certain types of stimuli. We describe them here as bottom-up processes, not because they do not involve feedback pathways in visual cortex (they may well do so) but because they appear to be largely automatic processes that are not dependent on cognition or task demands.

Stimuli that stand out from their background are processed preferentially at nearly all levels of the visual system. In visual cortex, the responses of many cells to an otherwise optimal stimulus within their classically defined receptive field may be completely suppressed if similar stimuli are within a large surrounding region (for reviews see Allman et al 1985, Desimone et al 1985). The greater the density of stimuli in the surround, the greater the suppression (Knierim & Van Essen 1992). In the middle temporal area (MT), for example, a cell that normally responds to vertically moving stimuli within its receptive field may be unresponsive if the same stimuli are part of a larger moving pattern covering the receptive field and surround (Allman et al 1985, Tanaka et al 1986). These mechanisms almost certainly contribute to the pop-out effects of targets in visual search.

As indicated above, the visual system also seems to be biased towards new objects or objects that have not been recently seen. Thus, the temporal context of a stimulus may contribute as much to its saliency as its spatial context. In the temporal domain, stimuli stored in memory may function as the temporal surround, or context, against which the present stimulus is compared.

Striking examples of such temporal interactions have been found in the anteroventral portion of IT cortex. Most studies in this region recorded cells while monkeys performed delayed matching-to-sample (DMS) tasks with either novel or familiar stimuli. In DMS, a sample stimulus is followed by one or more test stimuli, and the animal signals when a test stimulus matches the sample. For up to a third of the cells in this region, responses to novel sample stimuli become suppressed as the animal acquires familiarity with them (Fahy et al 1993, Li et al 1993, Miller et al 1991, Riches et al 1991). The cells are not novelty detectors, in that they do not respond to any novel stimulus. Rather, they remain stimulus selective both before and after the visual experience.

In fact, this shrinkage in the population of activated neurons as stimuli become familiar may increase the selectivity of the overall neuronal population for those stimuli. As one learns the critical features of a new stimulus, cells activated in a nonspecific fashion drop out of the activated pool of cells (Li et al 1993), leaving those that are most selective. There is also direct evidence that some IT cells selective for faces become more tuned to a familiar face following experience (Rolls et al 1989).

An effect akin to the novelty effect is also found for familiar stimuli that have been seen recently. When a test stimulus matches the previously seen sample in the DMS trial, responses to that stimulus tend to be suppressed (Miller et al 1991, 1993; also see Baylis & Rolls 1987, Eskandar et al 1992, Fahy et al 1993, Riches et al 1991). Although it was originally proposed that this suppressive effect was dependent on active working memory for the sample, recent work has shown it to be an automatic outcome of any stimulus repetition (Miller & Desimone 1994). For many cells, this suppression occurs even if the repeated stimuli differ in size or appear in different retinal locations (Lueschow et al 1994). Thus, the detection of novelty and recency apparently occurs at a high level of stimulus representation.

Taken together, the results indicate that both novel stimuli and stimuli that have not been recently seen will have a larger neural signal in the visual cortex, giving them a competitive advantage in gaining control over attentional and orienting systems. This would explain the bias towards novelty in the human behavioral data described above. The longer the organism attends to the object, the more knowledge about the object is incorporated into the structure of the cortex; this reduces the visual signal. It will also reduce the drive on the orienting system so that the organism is free to orient to the next new object (Li et al 1993, Desimone et al 1994). This view is compatible with Adaptive Resonance Theory (Carpenter & Grossberg 1987), in which novel stimuli activate attentional systems that allow new long-term memories to be formed. Consistent with these neurophysiological results in animals, a reduction in neural activation with stimulus repetition in human subjects has been seen in both event-related potentials of the temporal cortex (Begleiter et al 1993) and in brain-imaging studies (Squire et al 1992).

Top-Down Control of Selection in the Ventral Stream

As we have said, top-down biases on visual processing, or the attentional template, derive from the requirements of the task at hand. Although we consider mechanisms for spatial and object selection separately, they in fact share many features.

SELECTION BASED ON SPATIAL LOCATION   As we described above, one central resource for which stimuli compete in the ventral stream seems to be the receptive field. Not surprisingly then, spatial selection in this stream does not simply enhance processing of the stimulus at the attended location but rather seems to resolve competition between stimuli in the receptive field.

In one study of cells in V4 and IT cortex, monkeys performed a discrimination task on target stimuli at one location in the visual field, ignoring simultaneously presented distractors at a second location (Moran & Desimone 1985). The target location for a given run was indicated to the monkey by special instruction trials at the start of that run, i.e. the spatial bias was purely top down and presumably required spatial working memory. When target and distractor were both within the receptive field of the recorded cell, the neuronal response was determined primarily by the target; responses to the distractor were greatly attenuated. The cells responded as though their receptive fields had shrunk around the target. Consistent with this, Richmond et al (1983) found that the presence of a central fixation target in the receptive field of an IT neuron may block the response to a more peripheral stimulus in the field.

In the Moran & Desimone (1985) study, when one of the two locations was placed outside the receptive field of the recorded cell, attention no longer had any effect on the response. This was consistent with the biased competition model: Target and distractor were no longer competing for the cell's response, and thus, top-down spatial bias no longer had any effect.

Receptive fields and the region of space over which attention operated were much larger in the IT cortex. However, even here attentional effects were larger when target and distractor were located within the same hemifield and, therefore, more likely to be in competition (Sato 1988).

In V1, receptive fields were too small to test the effects of placing both target and distractor within them. However, when one stimulus was located inside, and one outside (at the same spatial separation used in area V4), there was no effect of attention on V1 cells in this paradigm. These results suggest that target selection is a two-stage process: The first stage works over a small spatial range in V4, and the second stage works over a much larger spatial range in IT cortex; both are in line with their receptive field sizes (Moran & Desimone 1985). Studies of event-related potentials in humans have also localized a region modulated by spatial attention in lateral prestriate cortex; this region may correspond to area V4 (Mangun et al 1993).

Recently, Motter (1993) has reported attentional effects on responses of cells in V1, V2, and V4. In contrast to the Moran & Desimone (1985) study, these effects were found when one stimulus was inside the field, and others outside. Most surprisingly, cueing the animal for the target location was almost as likely to suppress responses to the target as to facilitate them. A possible reason for the discrepancy between the two studies is that Motter (1993) found these effects only when there were a large number of distractors in the visual field, whereas Moran & Desimone (1985) used only a single distractor. Increasing the competition among objects in the visual field may have increased the role of attentional biases. Other differences include the fact that Motter used an explicit spatial cue to indicate the target location in the display, and the target (but not any of the distractors) was physically added to the cue, possibly inducing some complex sensory effects. In any event, other recent studies have confirmed that attentional effects in V4 are much larger when target and distractor compete within the same receptive field than in any other configuration (Luck et al 1993; L Chelazzi, unpublished data).

CIRCUITRY UNDERLYING SPATIAL SELECTION   Although the synaptic mechanisms mediating the gating of V4 and IT responses are unknown, anatomy dictates that they fall into either of two classes (Desimone 1992). In the first class, spatial biasing inputs to visual cortex determine which specific subset of a cell's inputs causes the cell to fire, whereas in the second class, the inputs determine which specific cells in a population are allowed to fire. In other words, one can either gate some of the inputs to a cell on or off, or one can gate some of the cells on or off. Theoretical models for both classes of circuitry have been developed (Anderson & Van Essen 1987, Crick & Koch 1990, Desimone 1992, Niebur et al 1993, Olshausen et al 1993, Tsotsos 1994). All of the models resolve competition when there are mulitple stimuli within the receptive field. Presently, there are insufficient data to decide between them.

If the gating of V4 and IT responses occurs as a result of an external input that biases competition in favor of the target, one might expect to see some evidence for it. A possible candidate has been found in a new study of spatial attention in V4 (Luck et al 1993). V4 cells in this study showed a sustained elevation of their baseline (prestimulus) firing rates whenever the animal's attention was directed inside their receptive field. This elevation of activity with attention could be the neural analogue of the attentional template for location. The elevation occurred at the start of each trial before any stimulus had appeared. Since the only information about where to attend was given to the animal minutes earlier at the start of a block of trials, the relevant location must have been stored in working memory. The spatial resolution of this source was very high; when attention was shifted to different regions within the same receptive field, the magnitude of the baseline shift varied according to the distance between the focus of attention and the receptive field center. Thus, whatever spatial bias signal enters the cortex, it apparently has a spatial resolution finer than the receptive field dimensions of V4 cells.

SELECTION BASED ON FEATURES   The mechanism underlying the selection of objects by their features (when their location is not known in advance) requires a means to hold the sought-after object in working memory and to use this memory (or attentional template) to resolve competition among the elements in the scene. Recently, evidence for this selection mechanism has been found in the anteroventral portion of IT cortex, the same portion in which memory-related activity has been found (Chelazzi et al 1993a).

Monkeys were briefly presented with a complex picture (the cue) at the center of gaze to hold in memory. The cue on a given trial was either a good stimulus that elicited a strong response from the cell or a poor stimulus that elicited little or no response when presented by itself. Following a delay, the good and the poor stimuli were both presented simultaneously as choice stimuli, at an extrafoveal location. The monkey made a saccadic eye movement to the target stimulus that matched the cue, ignoring the nonmatching stimulus (the distractor).

As shown in Figure 5, the choice array initially activated IT cells tuned to the properties of either stimulus, in parallel, irrespective of which stimulus was the target. Within 200 ms after array onset, however, the response changed dramatically depending on whether the animal was about to make an eye movement to the good or poor stimulus. When the target was the good stimulus, the response remained high. However, when the target was the poor stimulus for the recorded cell, the response to the good distractor stimulus was suppressed even though it was still within the receptive field. This change in response occurred about 100 ms before the onset of the eye movement. The cells responded as though the target stimulus captured their response, so neuronal activity in IT cortex reflected only the target's properties. Cells selective for the nontargets were suppressed within 200 ms and remained suppressed until well after the eye movement was made. Similar effects were found for choice arrays of larger sizes.

Just as with spatially directed attention, these effects of object selection in the IT cortex were much smaller when target and nontargets were located in opposite hemifields than when they were in the same hemifield, i.e. when they were maximally in competition. Interestingly, similar competitive effects are seen even at high levels of oculomotor control in the frontal eye field. Cells in this region were recorded while monkeys made eye movements to a target in a field of distractors (Schall & Hanes 1993). Responses to distractors in the

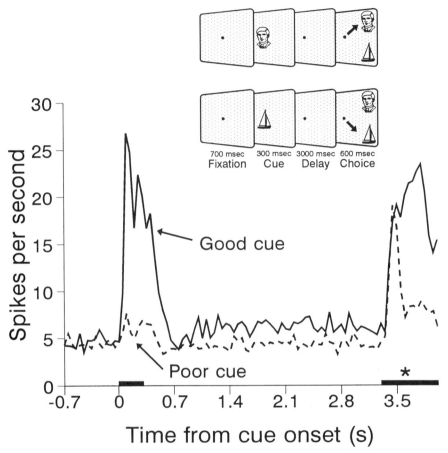

Figure 5 Effects of object selection on responses of cells in the IT cortex. The upper insert illustrates the general visual search task. Graphs show the average response of 22 cells recorded while monkeys performed the task. The cue was chosen to be either a good or a poor stimulus for the recorded cell. When the choice array was presented, the animal made a saccadic eye movement to the stimulus (target) that matched the previous cue. The saccadic latency was 300 ms, indicated by the asterisk. Cells had a higher firing rate in the delay preceding the choice array when their preferred stimulus was the cue. Following the delay, cells were activated (on the average) by their preferred stimulus in the array, regardless of whether it was the target. However, 100 ms before the eye movement was made, responses diverged depending on whether the target was the good or the poor stimulus. The two dark horizontal bars indicate when the cue and the choice were presented. Adapted from Chelazzi et al (1993a).

receptive field were more suppressed when the competing target was located just outside the receptive field, and thus maximally competitive, than when it was further away.

Two findings suggest that the target is selected in the IT cortex as a result of inputs (initiated at the time of the cue) that bias competition in favor of the target. First, IT cells selective for the properties of the cue-target show higher maintained activity in the delay following the cue than do cells selective for the distractor. This could be the neural correlate of the attentional template for the target. Second, a subpopulation of the cells gives enhanced responses to the choice array when their preferred stimulus is the target, even during the first 200 ms in which all cells tend to be active. Together, the results indicate that cells selective for the target are primed to respond to it by an external source before the onset of the choice array; they then give an enhanced response to the target when it appears. Eventually, as a result of competitive interactions during the initial visual activation, cells selective for the distractors are suppressed. At some point in time, mechanisms for spatial selection may also be engaged to facilitate localization of the target for the eye movement.

Cue-, or template-, related activity during delay periods (Fuster & Jervey 1981, Miller et al 1993, Miyashita & Chang 1988) and enhanced responses to (target) stimuli matching a prior cue (Miller & Desimone 1994) have also been found in studies of working memory in IT cortex. Visual search simply appears to be a variant of a working memory task, in which the distractors are distributed in space rather than time. Importantly, the same seems to be true of spatial selection, which shares many features with object selection, including template-related activity during delays followed by response-suppression to competing nontargets. The major difference may simply be the nature of the template. The potential sources of the template that primes IT cells in working memory is considered below.

Somewhat similar mechanisms may be reflected in studies of human brain activation that use positron emission tomography (PET). In one study (Corbetta et al 1991), subjects were asked to compare one feature of two successive displays, each containing a moving field of colored shapes. Different portions of extrastriate cortex were preferentially activated depending on whether the relevant feature was motion, on the one hand, or color or shape, on the other. Physiological studies have also shown a variety of other nonspatial, top-down influences on ventral stream neurons that may influence object selection bias, but they are beyond the scope of this review (e.g. Maunsell et al 1991, Spitzer et al 1988, Spitzer & Richmond 1991).

Neural Sources of Spatial Selection Bias

LESION STUDIES IN HUMANS   We now turn to the neural systems that might be the source of the attentional template for spatial location. The lesion data are readily explained by the biased competition model but, unfortunately, do not by themselves pin down the critical sources.

Following the formation of a lesion on one side of the brain, there is often a disregard or neglect of objects and actions in contralateral space (for review see Bisiach & Vallar 1988). Neglect can be manifested as failure to copy one half of a drawing, to read text on one half of a page, to shave one half of the face, etc.

Neglect of one form or another has been associated with damage to a great variety of brain structures, including the parietal cortex (e.g. Bisiach & Vallar 1988), the frontal cortex (Heilman & Valenstein 1972), the cingulate gyrus (Watson et al 1973), the basal ganglia (Hier et al 1977), the thalamus (Rafal & Posner 1987, Watson & Heilman 1979), the midbrain and superior colliculus (Posner et al 1982), and even the temporal lobe (Shelton et al 1990).

Most importantly, neglect of the contralesional side can be exaggerated by competing events on the other, unimpaired, side—a phenomenon termed extinction. Thus, neglect manifests more as a competitive bias against one side than as an absolute inability to deal with that side (Kinsbourne 1993). If there are critical spatial gating inputs to the ventral stream, they probably arise from more than one structure.

In terms of the biased competition model, damage to the spatially mapped areas of one hemisphere may cause two different types of behavioral impairment. First is the loss of whatever functions are mediated by the damaged areas, which may include perceptual, visuospatial, and oculomotor functions. Second is the loss of competitive weights afforded to objects in the affected portion of the contralesional field, which may be manifested anywhere between sensory input and motor output. In the visual system, this loss could affect visual cortex either directly, through the elimination of structures that contribute to stimulus saliency or that supply top-down spatial selection inputs, or indirectly, through the elimination of structures that supply the critical ones with inputs. The superior colliculus does not project directly to the visual cortex, for example, but ultimately provides inputs to other structures that do. Either unilateral or focal damage to the colliculus could affect competition within these other parts of the system, thus throwing them out of balance (Desimone et al 1990b). A loss of competitive weights would also explain why neglect and extinction most commonly follow unilateral rather than bilateral lesions; with bilateral lesions, neither hemifield has a competitive advantage over the other. It seems likely that competition in multiple brain systems is coordinated so that a loss of competitive weights in one system has general effects in others.

Considering that lesions will typically result in both a loss of function and a loss of competitive weights and that competitive weights may be affected at any level between sensory input and motor output, it is not surprising that there are many reports of dissociation between one form of neglect and another. For example, there are reports of neglect of body vs environmental objects (e.g. Guariglia & Antonucci 1992), neglect of close vs far space (e.g. Halligan &

Marshall 1991), and sensory vs motor neglect (e.g. Tegner & Levander 1991). There are also strong laterality effects, which we do not cover here (see Posner & Peterson 1990).

One interesting possibility raised by lesion studies is that the posterior parietal cortex specifically mediates disengaging attention from its current focus (Posner et al 1984), or in our terms, shifting the balance of competitive weights from one object to another. A disengagement deficit may partly explain rare cases of Balint's syndrome and simultagnosia following extensive bilateral damage to the parietal lobe (e.g. Humphreys & Riddoch 1993). In these patients, attention can become locked onto one object; nonattended objects seem to disappear. According to Posner and colleagues, this disengage function of the parietal cortex differs from that of the superior colliculus and pulvinar, which they propose mediate moving attention and focusing attention, respectively (reviewed in Posner & Petersen 1990). This division is based primarily on reaction time data from patients with large unilateral lesions affecting, but generally not limited to, one of the three structures. However, monkeys with discrete unilateral lesions or deactivation of any one of these structures all show a general slowing of reaction times for targets in the contralesional field as well as a disengagement impairment when attention is switched from the ipsilesional to contralesional field (see below). These impairments may simply follow from a loss of competitive weights in the affected field. Thus, a specific role for parietal cortex in disengagement is still an open question.

LESION STUDIES IN PRIMATES   The general rule for lesion effects in monkeys is the same as in humans: Unilateral lesions of structures with a contralateral field representation result in a loss of whatever functions are mediated by the damaged area as well as neglect and extinction syndromes from a loss of competitive weights in the contralesional field. Bilateral lesions, which do not upset the competitive balance between the fields, tend to have less effect on spatial attention.

In fact, there are at least two instances when adding a lesion in one hemisphere corrects an attentional impairment caused by a lesion in the other. Monkeys with unilateral lesions of the posterior parietal cortex tend to make voluntary eye movements into the ipsilesional field when presented with bilateral stimuli. However, this bias is corrected when an additional lesion is subsequently made in the posterior parietal cortex of the opposite hemisphere (Lynch & McLaren 1989). Similarly, cats with unilateral lesions of striate cortex show a severe contralateral neglect; however, a lesion of the substantia nigra in the opposite hemisphere substantially reduces the neglect (Wallace et al 1990).

PULVINAR   The most frequently proposed source of attentional inputs to the cortex has probably been the pulvinar (e.g. see Crick 1984, Olhausen et al

1993). This large structure contains several different nuclei, each of which contains one or more functionally distinct regions connected anatomically to a specific region of the visual cortex (Bender 1981; Benevento & Rezak 1976; Ungerleider et al 1983, 1984). The pulvinar has been implicated in attentional control based on neuropsychological studies of humans with thalamic brain damage (Rafal & Posner 1987), PET activation studies (LaBerge & Buchsbaum 1990), and physiological recording and chemical deactivation studies in monkeys (Desimone et al 1990b; Petersen et al 1985, 1987; Robinson et al 1986). However, pulvinar lesions raise the same issues of interpretation as lesions in other structures we have considered.

In one study, the portion of the pulvinar termed Pdm, which is anatomically interconnected with the posterior parietal cortex, was reversibly deactivated in one hemisphere (Petersen et al 1987). Following deactivation, reaction times to targets in the contralesional field were slower than normal, especially when attention was first misdirected into the ipsilesional field (i.e. a disengage impairment). Thus, Pdm deactivation seemed to reduce the saliency of con-tralesional stimuli thereby reducing their competitive weights for either visual processing or control over behavior (Robinson & Petersen 1992). This loss of weights may have simply resulted from the loss of Pdm inputs to the posterior parietal cortex of the same hemisphere, as the latter structure is implicated in attentional control in its own right (see below). Both unilateral deactivation of the superior colliculus and unilateral lesions of the posterior parietal cortex had effects similar to those of Pdm deactivation (see Colby 1991).

Analogous results were found with unilateral chemical deactivation of the lateral pulvinar (PL), the part connected with areas V4 and IT cortex. Monkeys discriminated the color of a target in the (contralesional) field opposite the deactivated pulvinar, with or without a distractor in the unaffected (ipsilesio-nal) field (Desimone et al 1990b). The deactivation had no effect on the monkey's ability to discriminate the target unless it was paired with a distractor, a result reminiscent of extinction. If PL was the source of critical gating inputs to extrastriate cortex, moving the distractor closer to the target should have had a devastating effect on performance. However, when the distractor was moved into the same hemifield as the target, the impairment was substantially diminished, presumably because neither stimulus then had a competitive ad-vantage. As with Pdm, deactivation of PL most likely deprived visual cortex in the same hemisphere of excitatory inputs and reduced target saliency. Bilateral pulvinar lesions have no effect on the ability of monkeys to find a target embedded in distractors, which further suggests that PL does not have a necessary role in attentional gating (Bender & Butter 1987).

In fact, the biased competition model predicts results similar to those of pulvinar deactivation from partial lesions in any spatially mapped visual struc-ture that makes a contribution to saliency and hence competitive weight. Such an outcome is observed in monkeys with lesions affecting one quadrant of the visual field representation in area V4. In one study, animals were trained to make eye movements to an odd-man-out target in an array of stimuli presented around an imaginary ring (Schiller & Lee 1991). If the target was located in the lesion quadrant and if it was dimmer than the other stimuli in the unaffected parts of the field, the animals were impaired. However, there was no impairment if the target was brighter than the other stimuli, suggesting that the V4 lesion reduced target saliency. In another study, animals were especially impaired in discriminating the shape of a target located in the lesion quadrant when a distractor was located in an unaffected part of the field; however, there was little impairment when both the target and distractor were located within the lesion quadrant (Desimone et al 1990a). In the latter configuration, neither stimulus had a competitive advantage from the lesion.

In summary, the biased competition model affords a ready explanation for the effects of unilateral or partial lesions on attention. At this time, the pulvinar is no more likely than other structures to be a critical source of gating inputs to the ventral stream. To pin down these sources will likely require converging evidence from lesion and physiological studies.

PHYSIOLOGICAL STUDIES   The classic paradigm for studying cells within the presumed control system for spatial attention has been the saccadic enhancement paradigm (Goldberg & Wurtz 1972, Wurtz & Goldberg 1972). In this task, the monkey fixates a central stimulus while a second stimulus is presented within a cell's receptive field in the periphery. In the experimental condition, the fixation stimulus is turned off and the animal saccades to the receptive field stimulus when it appears. In a control condition, the fixation stimulus stays on and the monkey is rewarded for signaling when it dims, ignoring the receptive field stimulus. The control over eye movements is largely top down in this task, although the experimental condition has some automatic, or reflexive, components.

Some of the cells in virtually all structures implicated in spatial attention give larger responses to the receptive field stimulus in the experimental condition (the target) than in the control (the distractor), a result usually termed the enhancement effect (although, in fact, it is often unclear whether the target response is enhanced or the distractor response is suppressed). This effect is found in the superior colliculus, the substantia nigra, the Pdm nucleus of the pulvinar, the posterior parietal cortex, the frontal eye fields (Goldberg & Wurtz 1972, Hikosaka & Wurtz 1983, Lynch et al 1977, Robinson et al 1978, Petersen et al 1985, Wurtz & Mohler 1976; also see Colby 1991), and the dorsolateral prefrontal cortex (di Pelligrino & Wise 1993b). However, in both the superficial layers of the colliculus and the frontal eye fields, the effect is known to be specific for saccadic eye movements; no enhancement is found when the animal simply attends to the peripheral stimulus and signals when it dims by releasing a bar (Colby et al 1993, Goldberg & Bushnell 1981, Wurtz & Mohler 1976). Thus, these cells appear to be involved in the selection of targets for eye movements rather than in selection for visual processing. A remarkable implication of the fact that these visuomotor cells respond equally to targets and distractors in the absence of eye movements is that visual input to these parts of the oculomotor system does not derive from cells in the dorsal and ventral streams whose responses are gated by spatial attention. Competition between stimuli must take place independently within the oculomotor system and yet be coordinated with competition within visual processing systems.

Although cells in the substantia nigra and intermediate layers of the colliculus have not yet been tested in this condition of attention without eye movements, cells in Pdm, dorsolateral prefrontal cortex, and posterior parietal cortex all show the enhancement effect in this purely attentional condition (Bushnell et al 1981, Colby et al 1993, di Pelligrino & Wise 1993b, Petersen et al 1985). Of these three regions, the posterior parietal and prefrontal cortices may be the most critical for spatial attention, as the enhancement in Pdm may simply reflect the input it receives from the posterior parietal cortex. Furthermore, studies with PET have shown activation of posterior parietal cortex in a task involving shifting attention (Corbetta et al 1993). Thus, based on presence of the enhancement effect, both posterior parietal and prefrontal cortex are possible sources of a spatial-biasing signal to visual cortex.

If the top-down selection of spatial locations for attention typically involves working memory, as we have suggested, an important clue to the identity of the relevant cells would be response activation in working memory tasks. In fact, in such tasks cells in the dorsolateral prefrontal and posterior parietal cortexes are tonically active whenever the animal holds "in mind" a location within a cell's receptive field (in the absence of any stimulus) (Chelazzi et al 1993b, Colby et al 1993, di Pelligrino & Wise 1993a, Funahashi et al 1989, Fuster 1973, Gnadt & Andersen 1988, Quintana & Fuster 1992, Wilson et al 1993). Furthermore, these two regions are heavily interconnected anatomically and appear to form part of a distributed system for spatial cognition (for a review, see Goldman-Rakic 1988). These physiological data, in conjunction with data showing neglect and extinction effects following both prefrontal and posterior parietal lesions, argue that both structures may work together in generating top-down spatial selection biases.

Sources of Object Selection Bias

As with spatial selection, the attentional templates for objects and their features may derive from mechanisms underlying working memory. If so, then the prefrontal cortex most likely plays an important role. Just as lesions of the dorsolateral prefrontal cortex impair working memory for space (see Funahashi et al 1993), lesions of the ventral prefrontal cortex impair working memory for objects (Mishkin & Manning 1978). Furthermore, Wilson et al (1993) report that cells in the dorsolateral cortex have maintained activity for object location whereas cells in the ventral cortex have maintained activity for object identity. Indeed, the dorsal and ventral prefrontal cortices appear to be the frontal extensions of the dorsal and ventral processing streams, respectively (Mishkin et al 1983, Wilson et al 1993).

Just as the posterior parietal cortex may work together with the dorsolateral prefrontal cortex in generating spatial templates, the anterior IT cortex may play an analogous role with the inferior prefrontal cortex in generating object and feature templates (Desimone et al 1994, Fuster et al 1985). Both are heavily interconnected anatomically (Ungerleider et al 1989), and neurons in both structures are activated during identical working memory tasks for objects (Chelazzi et al 1993b).

OBJECTS, GROUPING, AND THE BINDING PROBLEM

So far we have not dealt specifically with the representation of objects in the cortex. Although this is a key issue for understanding attention, little is actually known about the neural representations of objects. We review just a few of the relevant behavioral and neurophysiological facts.

As described above, when human subjects divide attention between two objects, the decrement in performance is rather insensitive to spatial separation. What does matter in divided attention is whether two properties to be identified belong to the same or different objects. It is far easier to identify two properties (e.g. orientation and contrast) of one object than properties of two different objects (Duncan 1993, Lappin 1967), even when the two objects overlap (Duncan 1984). Indeed, under simple conditions subjects can identify two properties of a single object just as easily as they can identify one (Duncan 1984, 1993).

The operations that segment and group visual input into discrete objects or chunks are beyond the scope of this review. Many factors combine to determine which parts of the visual input belong together, including spatial proximity, shared motion or color, contour features such as local concavities and T-junctions, and long-term familiarity with the object (see e.g. Beck et al 1983, Hummel & Biederman 1992, Grossberg et al 1994, Palmer 1977). The data suggest, however, that the objects so constructed behave as wholes when they compete for visual representation and/or control of behavior.

Strengthening the perceived grouping between irrelevant and relevant display items by, for example, giving them a common motion makes the irrelevant items harder to ignore (Driver & Baylis 1989, Kramer & Jacobson 1991). The ease of visual search in homogenous arrays (Figure 3a) partly reflects the tendency of identical nontargets to group together and apart from the target. Visual grouping determines which parts of the input belong together; subsequent competitive operations tend to respect, or preserve, these groupings (Duncan & Humphreys 1989).

Other than a general tendency for neuronal responses in the visual cortex to be influenced by the overall distribution of items within the receptive field, the neural mechanisms underlying object grouping are unknown. Grossberg et al (1994) have attempted to model grouping as a product of mechanisms for image segmentation. Given the importance of grouping for attentional control, this is a ripe area for future research.

Closely related to grouping is the binding problem, or the problem in a distributed representation of keeping together parts or attributes of the same object or entity (Hinton & Lang 1985). It is often presumed that there are separate representations for different features, such as color and orientation, in the cortex. If so, then an obvious question is how the color red, say, becomes bound to the bar of the appropriate orientation when there are multiple colored bars of different orientation in the visual field. A common view is that attention helps solve the binding problem by linking together different features at the attended location (Treisman & Gelade 1980, Treisman & Schmidt 1982). One problem with this view is that a complex object such as a face has many different features that would need to be bound together, one at a time. Multipart objects, such as the human body, may have hierarchical part-whole relationships that would require comparable binding hierarchies (e.g. a finger may be seen as part of a hand, a limb, or the entire body). Additionally, as we have said, targets may pop out of a visual search display before they are the focus of attention, even when they are defined by the conjunction of elementary attributes (Duncan & Humphreys 1989, McLeod et al 1988, Wolfe et al 1989). This implies some type of solution to the binding problem that works in parallel across the visual field.

At the neural level, the necessity to bind together the output of cells specialized for different elemental features may be overstated. To our knowledge, no cortical cell has ever been reported that is influenced by only one stimulus feature. Neurons may convey more information about some features than others, but their responses often vary along many different feature dimensions, particularly in area V4 and IT cortex (Desimone et al 1984, Tanaka 1993). Some cells in temporal cortex respond specifically to objects with highly complex conjunctions of features, such as faces, even under anesthesia when selective attention is presumably absent (Desimone et al 1984). A possible role for correlated activity of neurons in binding is considered elsewhere in this volume (Singer & Gray 1995). As with grouping, much more needs to be known about object representations in the cortex before we understand the role of attention in binding.

SERIAL AND PARALLEL MODELS

When targets are selected by spatial location, all models of the underlying mechanism posit some type of spatial gating mechanism. It is when the target's location is unknown and it must be found on the basis of its identity (e.g. searching for a face in a crowd) that different classes of models diverge significantly. According to serial search accounts, scenes are searched element by element by a spotlight of attention (Olhausen et al 1993, Schneider & Shiffrin 1977, Treisman & Gelade 1980), unless the target pops out from the background on the basis of an elemental feature difference. As each element is selected in turn by attention, it is evaluated by a recognition memory process, and the scan of the array is terminated when the target is found. As more and more nontarget elements are added to the scene, it takes longer and longer to scan the array to find the target; 50 ms per item is typical (although as we have said, this time varies continuously over a large range of values).

In the other major class of models, all elements of the visual input compete in parallel for visual processing (Atkinson et al 1969, Bundesen 1990, Duncan & Humphreys 1989, Sperling 1967). This class includes the biased competition account that has been the theme of this review so far.

The difficulties of distinguishing between parallel and serial models on the basis of reaction time data are well known (Townsend 1971), particularly because recent serial models have become hybrids with both serial and parallel component processes. To explain pop-out effects with targets defined by the conjunction of several features, for example, both Guided Search (Wolfe et al 1989) and Feature Integration Theory (Treisman & Sato 1990) incorporate parallel top-down processes to identify all regions in the visual field that share target features. Another interesting hybrid is the spatial and object search model of Grossberg et al (1994), which explains both easy and difficult search on the basis of grouping and recognition operations recursively applied in parallel across the visual field.

Unfortunately, the physiological data on object search in IT cortex (Chelazzi et al 1993) described above do not allow us to distinguish conclusively between serial and parallel mechanisms. The fact that search arrays initially activate cells selective for any of the component elements, targets or nontargets, is consistent with the biased competition model in which all objects are prcessed in parallel. However, it is possible that what seems to be an initial parallel activation lasting 200 ms is actually a serial activation, with the serial scanner switching between elements at a rate too rapid to discern in the neural data. The strongest argument against the serial model is that known memory mechanisms in IT cortex are sufficient to explain the results without invoking a hidden serial process.

The time it takes to recognize one object and release processing capacity for another, or the attentional dwell time, is a critical issue in comparing the different models. To make such a measurement, a brief temporal interval is introduced between two targets, e.g. two letters to be identified. If the time between presentation of the targets is shorter than the attentional dwell time, there should be interference, as both targets will compete for processing capacity. According to typical serial models, which posit rapid attentional scanning of objects in a scene, each object consumes processing capacity for only a few dozen milliseconds; thus, the attentional dwell time is short and interference should be eliminated with correspondingly short interstimulus intervals. Attentional dwell times can be much greater in parallel models because more than one object in a scene is processed at once (with increasing interference as the number of objects is increased). Thus, interference may last for far longer periods of time. A recent study using this method of sequential target presentation found interference lasting for several hundred milliseconds, consistent with parallel models (Duncan et al 1994; see also Pashler & Badgio 1987).

CONCLUSIONS

By way of contrast, it would be useful to consider again the standard model of selective visual attention widely accepted in neuroscience. According to this view, attention focuses on one region of the visual field at a time. It is mediated by a system of spatially mapped structures that enhance processing in visual cortex at attended locations and reduce it at unattended ones. The components of this system are revealed by neglect and extinction syndromes following lesions. Attention is unnecessary for simple feature discriminations but resolves the binding problem by linking together the output of cells coding different elemental features of the attended object. It is a serial, high-speed scanning mechanism moving from one location to the next in around 50 ms.

The data we have reviewed cast doubt over many of the postulates of the standard view. Instead, they suggest the following conclusions:

1. At several points between input and response, objects in the visual field compete for limited processing capacity and control of behavior.
2. This competition is biased in part by bottom-up neural mechanisms that separate figures from their background (in both space and time) and in part by top-down mechanisms that select objects of relevance to current behavior. Such bias can be controlled by many stimulus attributes, including selection by spatial location, by simple object features, and by complex conjunctions of features.
3. Within the ventral stream, which underlies object recognition, top-down biasing inputs resolve competition mainly between objects located within

the same receptive field. These mechanisms may work in a similar fashion for both object and spatial selection. In some cases, these inputs are directly revealed through elevation of the maintained activity of cells coding the location or feature of the expected item. The critical difference between spatial and feature selection may be the source and nature of the selection template.

4. Because many spatially mapped structures contribute to competition, unilateral lesions will often cause neglect and extinction syndromes that do not necessarily imply a specific role in attentional control.

5. The top-down selection templates for both locations and objects are probably derived from neural circuits mediating working memory, perhaps especially in prefrontal cortex.

6. Objects act as wholes in neural competition. The construction of object representations from the conjunction of many different features appears, in many cases, to occur in parallel across the visual field before individual objects are selected and, hence, prior to any attentional binding.

7. Though the matter remains controversial, according to our analysis attention is not a high-speed mental spotlight that scans each item in the visual field. Rather, attention is an emergent property of slow, competitive interactions that work in parallel across the visual field.

ACKNOWLEDGMENTS

We are grateful to L Chelazzi, C Colby, P DeWeerd, EK Miller, M Mishkin, DL Robinson, and LG Ungerleider for helpful comments on the manuscript; to R Hoag for help with references; and to M Adams for help with figures. The work was supported in part by the Human Frontiers Science Program Organization.

Literature Cited

Allman J, Miezin F, McGuinness E. 1985. Stimulus specific responses from beyond the classical receptive field: neurophysiological mechanisms for local-global comparisons in visual neurons. *Annu. Rev. Neurosci.* 8:407–30

Allport DA. 1980. Attention and performance. In *Cognitive Psychology: New Directions,* ed. G. Claxton, pp. 112–53. London: Routledge & Kegan Paul

Allport DA. 1993. Attention and control: have we been asking the wrong questions? A critical review of twenty-five years. In *Attention and Performance XIV,* ed. DE Meyer, S Kornblum, pp. 183–218. Cambridge, MA: MIT Press

Anderson CH, Van Essen DC. 1987. Shifter circuits: a computational strategy for dynamic aspects of visual processing. *Proc. Natl. Acad. Sci. USA* 84:6297–6301

Atkinson RC, Holmgren JE, Juola JF. 1969. Processing time as influenced by the number of elements in a visual display. *Percept. Psychophys.* 6:321–26

Baddeley AD. 1986. *Working Memory.* Oxford: Oxford Univ. Press

Baylis GC, Rolls ET. 1987. Responses of neurons in the inferior temporal cortex in short term and serial recognition memory tasks. *Exp. Brain Res.* 65:614–22

Beck J, Prazdny K, Rosenfeld A. 1983. A theory of texture segmentation. In *Human and Machine Vision,* ed. J Beck, B Hope, A Rosenfeld, pp. 1–38. London: Academic

Begleiter H, Porjesz B, Wang W. 1993. A neurophysiologic correlate of visual short-term memory in humans. *Electroenceph. Clin. Neurophysiol.* 87:46–53

Bender DB. 1981. Retinotopic organization of macaque pulvinar. *J. Neurophysiol.* 46:672–93

Bender DB, Butter CM. 1987. Comparison of the effects of superior colliculus and pulvinar lesions on visual search and tachistoscopic pattern discrimination in monkeys. *Exp. Brain Res.* 69:140–54

Benevento LA, Rezak M. 1976. The cortical projections of the inferior pulvinar and adjacent lateral pulvinar in the rhesus monkey (macaca mulatta): an autoradiographic study. *Brain Res.* 108:1–24

Bisiach E, Vallar G. 1988. Hemineglect in humans. In *Handbook of Neuropsychology,* ed. F Boller, J Grafman, 1:195–222. Amsterdam: Elsevier

Bonnel A-M, Stein J-F, Bertucci P. 1992. Does attention modulate the perception of luminance changes? *Q. J. Exp. Psychol.* 44A:601–26

Boussaoud D, Desimone R, Ungerleider LG. 1991. Visual topography of area TEO in the macaque. *J. Comp. Neurol.* 306:554–75

Broadbent DE. 1958. *Perception and Communication.* London: Pergamon

Bundesen C. 1990. A theory of visual attention. *Psychol. Rev.* 97:523–47

Bushell MC, Goldberg ME, Robinson DL. 1981. Behavioral enhancement of visual responses in monkey cerebral cortex. I. Modulation in posterior parietal cortex related to selective visual attention. *J. Neurophysiol.* 46:755–72

Carpenter GA, Grossberg S. 1987. A massively parallel architecture for a self-organizing neural pattern recognition machine. *Comp. Vis. Graph. Image Process.* 37:54–115

Chelazzi L, Miller EK, Duncan J, Desimone R. 1993a. A neural basis for visual search in inferior temporal cortex. *Nature* 363:345–47

Chelazzi L, Miller EK, Lueschow A, Desimone R. 1993b. Dual mechanisms of short-term memory: ventral prefrontal cortex. *Soc. Neurosci. Abstr.* 19:975

Colby CL. 1991. The neuroanatomy and neurophysiology of attention. *J. Child Neurol.* 6:S90–S118

Colby CL, Duhamel J, Goldberg ME. 1993. The analysis of visual space by the lateral intraparietal area of the monkey: the role of extraretinal signals. In *Progress in Brain Research,* ed. TP Hicks, S Molotchnikoff, T Ono, pp. 307–16. Amsterdam: Elsevier

Corbetta M, Miezin FM, Dobmeyer S, Shulman GL, Petersen SE. 1991. Selective and divided attention during visual discriminations of shape, color, and speed: functional anatomy by positron emission tomography. *J. Neurosci.* 11:2383–2402

Corbetta M, Miezin FM, Shulman GL, Petersen SE. 1993. A PET study of visuospatial attention. *J. Neurosci.* 13:1202–26

Crick F. 1984. The function of the thalamic reticular complex: the searchlight hypothesis. *Proc. Natl. Acad. Sci. USA* 81:4586–90

Crick F, Koch C. 1990. Some reflections on visual awareness. *Cold Spring Harbor Symp. Quant. Biol.* 55:953–62

Desimone R. 1992. Neural circuits for visual attention in the primate brain. In *Neural Networks for Vision and Image Processing,* ed. GA Carpenter, S Grossberg, pp. 343–64. Cambridge, MA: MIT Press

Desimone R, Albright TD, Gross CG, Bruce C. 1984. Stimulus-selective properties of inferior temporal neurons in the macaque. *J. Neurosci.* 4:2051–62

Desimone R, Li L, Lehky S, Ungerleider L, Mishkin M. 1990a. Effects of V4 lesions on visual discrimination performance and on responses of neurons in inferior temporal cortex. *Soc. Neurosci. Abstr.* 16:621

Desimone R, Miller EK, Chelazzi L, Lueschow A. 1994. Multiple memory systems in the visual cortex. In *The Cognitive and Neural Sciences,* ed. M. Gazzaniga. Cambridge, MA: MIT Press. In press

Desimone R, Schein SJ. 1987. Visual properties of neurons in area V4 of the macaque: sensitivity to stimulus form. *J. Neurophysiol.* 57:835–68

Desimone R, Schein SJ, Moran J, Ungerleider LG. 1985. Contour, color and shape analysis beyond the striate cortex. *Vis. Res.* 25:441–52

Desimone R, Ungerleider LG. 1989. Neural mechanisms of visual processing in monkeys. In *Handbook of Neuropsychology, Vol. 2,* ed. F Boller, J Grafman, pp. 267–99. New York: Elsevier

Desimone R, Wessinger M, Thomas L, Schneider W. 1990b. Attentional control of visual perception: cortical and subcortical mechanisms. *Cold Spring Harbor Symp. Quant. Biol.* 55:963–71

Deutsch JA, Deutsch D. 1963. Attention: some theoretical considerations. *Psychol. Rev.* 70:80–90 di Pellegrino G, Wise SP. 1993a. Primate frontal cortex: visuospatial vs. visuomotor activity, premotor vs. prefrontal cortex. *J. Neurosci.* 13:1227–43

di Pellegrino G, Wise SP. 1993b. Effects of attention on visuomotor activity in the premotor and prefrontal cortex of a primate. *Somatosens. Motor Res.* 10:245–62

Donderi DC, Zelnicker D. 1969. Parallel processing in visual same-different decisions. *Percept. Psychophys.* 5:197–200

Driver J, Baylis GC. 1989. Movement and visual attention: the spotlight metaphor breaks down. *J. Exp. Psychol.* 15:448–56

Duncan J. 1980. The locus of interference in the perception of simultaneous stimuli. *Psychol. Rev.* 87: 272–300

Duncan J. 1984. Selective attention and the organization of visual information. *J. Exp. Psychol.* 13:501–17

Duncan J. 1985. Visual search and visual attention. In *Attention and Performance XI,* ed. MI Posner, OSM Marin, pp. 85–104. Hillsdale, NJ: Erlbaum

Duncan J. 1989. Boundary conditions on parallel processing in human vision. *Perception* 18:457–69

Duncan J. 1993. Similarity between concurrent visual discriminations: dimensions and objects. *Percept. Psychophys.* 54:425–30

Duncan J, Humphreys GW. 1989. Visual search and stimulus similarity. *Psychol. Rev.* 96: 433–58

Duncan J, Ward R, Shapiro K. 1994. Direct measurement of attentional dwell time in human vision. *Nature* 369:313–15

Egeth H, Jonides J, Wall S. 1972. Parallel processing of multielement displays. *Cogn. Psychol.* 3:674–98

Eriksen BA, Eriksen CW. 1974. Effects of noise letters upon the identification of a target letter in a non-search task. *Percept. Psychophys.* 16:143–49

Eriksen CW, Hoffman JE. 1973. The extent of processing of noise elements during selective encoding from visual displays. *Percept. Psychophys.* 14:155–60

Eskandar EN, Richmond BJ, Optican LM. 1992. Role of inferior temporal neurons in visual memory: I. Temporal encoding of information about visual images, recalled images, and behavioral context. *J. Neurophysiol.* 68:1277–95

Fahy FL, Riches IP, Brown MW. 1993. Neuronal activity related to visual recognition memory: long-term memory and the encoding of recency and familiarity information in the primate anterior and medial inferior and rhinal cortex. *Exp. Brain Res.* 96:457–72

Felleman DJ, Van Essen DC. 1991. Distributed hierarchical processing in the primate cortex. *Cereb. Cortex* 1:1–47

Funahashi S, Bruce CJ, Goldman-Rakic PS.

1989. Mnemonic coding of visual space in the monkey's dorsolateral prefrontal cortex. *J. Neurophysiol.* 61:331–49

Funahashi S, Bruce CJ, Goldman-Rakic PS. 1993. Dorsolateral prefrontal lesions and oculomotor delayed-response performance: evidence of mnemonic "scotomas." *J. Neurosci.* 13(4):1479–97

Fuster JM. 1973. Unit activity in prefrontal cortex during elayed-response performance: neuronal correlates of transient memory. *J. Neurophysiol.* 36:61–78

Fuster JM, Bauer RH, Jervey JP. 1985. Functional interactions between inferotemporal and prefrontal cortex in a cognitive task. *Brain Res.* 330:299–307

Fuster JM, Jervey JP. 1981. Inferotemporal neurons distinguish and retain behaviorally relevant features of visual stimuli. *Science* 212:952–55

Gallant JL, Braun J, Vanessen DC. 1993. Selectivity for polar, hyperbolic, and cartesian gratings in macaque visual cortex. *Science* 259:100–3

Gattass R, Sousa AP, Gross CG. 1988. Visuotopic organization and extent of V3 and V4 of the macaque. *J. Neurosci.* 8:1831–45

Gnadt JW, Andersen RA. 1988. Memory related motor planning activity in posterior parietal cortex of macaque. *Exp. Brain Res.* 70:216–20

Goldberg ME, Bushnell MC. 1981. Behavioral enhancement of visual responses in monkey cerebral cortex. II. Modulation in frontal eye fields specifically related to saccades. *J. Neurophysiol.* 46:773–87

Goldberg ME, Wurtz RH. 1972. Activity of superior colliculus in behaving monkey. II. Effect of attention on neuronal responses. *J. Neurophysiol.* 35:560–74

Goldman-Rakic PS. 1988. Topography of coginition: paralleled distribution networks in primate association cortex. *Annu. Rev. Neurosci.* 11:137–56

Gross CG, Mishkin M. 1977. The neural basis of stimulus equivalence across retinal translation. In *Lateralization in the Nervous System,* ed. S Harned, R Doty, J Jaynes, L Goldberg, G Krauthamer, pp. 109–22. New York: Academic

Grossberg S, Mingolla E, Ross WD. 1994. A neural theory of attentive visual search: interactions of boundary, surface, spatial, and object representations. *Psychol. Rev.* 101:In press

Guariglia C, Antonucci G. 1992. Personal and extrapersonal space: a case of neglect dissociation. *Neuropsychologia* 30:1001–9

Harter MR, Aine CJ. 1984. Brain mechanisms of visual selective attention. In *Varieties of Attention,* ed. R Parasuraman, DR Davies, pp. 293–321. Orlando, FL: Academic

Halligan PW, Marshall JC. 1991. Left neglect

for near but not far space in man. *Nature* 350:498–500

Heilman KM, Valenstein E. 1972. Frontal lobe neglect in man. *Neurology* 22:660–64

Hier DB, Davis KR, Richardson EP, Mohr JP. 1977. Hypertensive putaminal hemorrhage. *Ann. Neurol.* 1:152–159

Hinton GE, Lang KJ. 1985. Shape recognition and illusory conjunctions. *Proc. Int. Jt. Conf. Artif. Intell.* 9:252–59

Hikosaka O, Wurtz RH. 1983. Visual and oculomotor functions of monkey substantia nigra pars reticulata. I. Relation of visual and auditory responses to saccades. *J. Neurophysiol.* 49:1230–53

Hummel JE, Biederman I. 1992. Dynamic binding in a neural network for shape recognition. *Psychol. Rev.* 99:480–517

Humphreys GW, Riddoch MJ. 1993. Interactions between object and space systems revealed through neuropsychology. In *Attention and Performance XIV*, ed. DE Meyer, S Kornblum, pp. 183–218. Cambridge, MA: MIT Press

Jonides J, Yantis S. 1988. Uniqueness of abrupt visual onset in capturing attention. *Percept. Psychophys.* 43:346–54

Kinsbourne M. 1993. Orientational bias model of unilateral neglect: evidence from attentional gradients within hemispace. In *Unilateral Neglect: Clinical and Experimental Studies*, ed. IH Robertson, JC Marshall, pp. 63–86. Hillsdale, NJ: Erlbaum

Kniermin JJ, Van Essen DC. 1992. Neuronal responses to static texture patterns in area V1 of the alert macaque monkey. *J. Neurophysiol.* 67:961–80

Kramer AF, Jacobson A. 1991. Perceptual organization and focused attention: the role of objects and proximity in visual processing. *Percept. Psychophys.* 50:267–84

LaBerge D, Buchsbaum MS. 1990. Positron emission tomographic measurements of pulvinar activity during an attention task. *J. Neurosci.* 10:613–19

Lappin JS. 1967. Attention in the identification of stimuli in complex displays. *J. Exp. Psychol.* 75:321–28

Li L, Miller EK, Desimone R. 1993. The representation of stimulus familiarity in anterior inferior temporal cortex. *J. Neurophysiol.* 69:1918–29

Luck S, Chelazzi L, Hillyard S, Desimone R. 1993. Effects of spatial attention on responses of V4 neurons in the macaque. *Soc. Neurosci. Abstr.* 19:27

Lueschow A, Miller EK, Desimone R. 1994. Inferior temporal mechanisms for invariant object recognition. *Cereb. Cortex* In press

Lynch JC, McLaren JW. 1989. Deficits of visual attention and saccadic eye movements after lesions of parietooccipital cortex in monkeys. *J. Neurophysiol.* 61:74–90

Lynch JC, Mountcastle VB, Talbot WH, Yin TC. 1977. Parietal lobe mechanisms for directed visual attention. *J. Neurophysiol.* 40:362–89

Mangun GR, Hillyard SA, Luck SJ. 1993. Electrocortical substrates of visual selective attention. In *Attention and Performance XIV*, ed. DE Meyer, S Kornblum, pp. 183–218. Cambridge, MA: MIT Press

Maunsell JHR, Sclar G, Nealey TA, DePriest DD. 1991. Extraretinal representations in area V4 in the macaque monkey. *Vis. Neurosci.* 7:561–73

McLeod P, Driver J, Crisp J. 1988. Visual search for a conjunction of movement and form is parallel. *Nature* 332:154–55

Miller EK, Li L, Desimone R. 1991. A neural mechanism for working and recognition memory in inferior temporal cortex. *Science* 254:1377–79

Miller EK, Li L, Desimone R. 1993. Activity of neurons in anterior inferior temporal cortex during a short-term memory task. *J. Neurosci.* 13:1460–78

Miller EK, Desimone, R. 1994. Parallel neuronal mechanisms for short-term memory. *Science* 263:520–22

Mishkin M, Manning FJ. 1978. Non-spatial memory after selective prefrontal lesions in monkeys. *Brain Res.* 143:313–23

Mishkin M, Ungerleider LG, Macko KA. 1983. Object vision and spatial vision: two cortical pathways. *Trends Neurosci.* 6:414–17

Miyashita Y, Chang HS. 1988. Neuronal correlate of pictorial short-term memory in the primate temporal cortex. *Nature* 331:68–70

Moran J, Desimone R. 1985. Selective attention gates visual processing in the extrastriate cortex. *Science* 229:782–84

Moray N. 1959. Attention in dichotic listening: affective cues and the influence of instructions. *Q. J. Exp. Psychol.* 11:56–60

Motter BC. 1993. Focal attention produces spatially selective processing in visual cortical areas V1, V2 and V4 in the presence of competing stimuli. *J. Neurophysiol.* 70:909–19

Munoz DP, Wurtz RH. 1993a. Fixation cells in monkey superior colliculus. I. Characteristics of cell discharge. *J. Neurophysiol.* 70:559–75

Munoz DP, Wurtz RH. 1993b. Fixation cells in monkey superior colliculus. II. Reversible activation and deactivation. *J. Neurophysiol.* 70:576–89

Neisser U. 1967. *Cognitive Psychology*. New York: Appleton-Century-Crofts

Niebur E, Koch C, Rosin C. 1993. An oscillation-based model for the neuronal basis of attention. *Vis. Res.* 33:2789–2802

Olshausen BA, Anderson CH, Van Essen DC. 1993. A neurobiological model of visual attention and invariant pattern recognition based on dynamic routing of information. *J. Neurosci.* 13(11):4700–19

Palmer SE. 1977. Hierarchical structure in perceptual representation. *Cogn. Psychol.* 9: 441–74

Pashler H, Badgio PC. 1987. Attentional issues in the identification of alphanumeric characters. In *Attention and Performance, Vol. 2,* ed. M. Coltheart, 2:63–81. Hillsdale, NJ: Erlbaum

Petersen SE, Robinson DL, Keys W. 1985. Pulvinar nuclei of the behaving rhesus monkey: visual responses and their modulation. *J. Neurophysiol.* 54:867–86

Petersen SE, Robinson DL, Morris JD. 1987. Contributions of the pulvinar to visual spatial attention. *Neuropsychologia* 25:97–105

Posner MI, Cohen Y, Rafal RD. 1982. Neural systems control of spatial orienting. *Philos. Trans. R. Soc. London Ser. B* 298:187–98

Posner MI, Petersen SE. 1990. The attention system of the human brain. *Annu. Rev. Neurosci.* 13:25–42

Posner MI, Snyder CRR, Davidson BJ. 1980. Attention and the detection of signals. *J. Exp. Psychol.* 109:160–74

Posner MI, Walker JA, Friedrich FJ, Rafal RD. 1984. Effects of parietal injury on covert orienting of attention. *J. Neurosci.* 4:1863–1974

Quinlan PT, Humphreys GW. 1987. Visual search for targets defined by combinations of color, shape and size: an examination of the task constraints on feature and conjunction searches. *Percept. Psychophys.* 41:455–72

Quintana J, Fuster JM. 1992. Mnemonic and predictive functions of cortical neurons in a memory task. *Neuroreport* 3:721–24

Rafal RD, Posner MI. 1987. Deficits in human visual spatial attention following thalamic lesions. *Proc. Natl. Acad. Sci. USA* 84:7349–53

Reicher GM, Snyder CRR, Richards JT. 1976. Familiarity of background characters in visual scanning. *J. Exp. Psychol.* 2:522–30

Riches IP, Wilson FA, Brown MW. 1991. The effects of visual stimulation and memory on neurons of the hippocampal formation and the neighboring parahippocampal gyrus and inferior temporal cortex of the primate. *J. Neurosci.* 11:1763–79

Richmond BJ, Wurtz RH, Sato T. 1983. Visual responses of inferior temporal neurons in awake rhesus monkey. *J. Neurophysiol.* 50: 1415–32

Robinson DL, Goldberg ME, Stanton GB. 1978. Parietal association cortex in the primate: sensory mechanisms and behavioral modulations. *J. Neurophysiol.* 41:910–32

Robinson DL, Petersen SE. 1992. The pulvinar and visual salience. *Trends Neurosci.* 15: 127–32

Robinson DL, Petersen SE, Keys W. 1986. Saccade-related and visual activities in the pulvinar nuclei of the behaving rhesus monkey. *Exp. Brain Res.* 62:625–34

Rolls ET, Baylis GC, Hasselmo ME, Nalwa V. 1989. The effect of learning on the face selective responses of neurons in the cortex in the superior temporal sulcus of the monkey. *Exp. Brain Res.* 76:153–64

Sagi D, Julesz B. 1984. Detection versus discrimination of visual orientation. *Perception* 13:619–28

Sagi D, Julesz B. 1985. Fast noninertial shifts of attention. *Spat. Vis.* 2:141–49

Sato T. 1988. Effects of attention and stimulus interaction on visual responses of inferior temporal neurons in macaque. *J. Neurophysiol.* 60:344–64

Schall JD, Hanes DP. 1993. Neural basis of saccade target selection in frontal eye field during visual search. *Nature* 366:467–69

Schiller PH, Lee K. 1991. The role of the primate extrastriate area V4 in vision. *Science* 251:1251–53

Schneider W, Shiffrin RM. 1977. Controlled and automatic human information processing: I. Detection, search, and attention. *Psychol. Rev.* 84:1–66

Schwartz EL, Desimone R, Albright TD, Gross CG. 1983. Shape recognition and inferior temporal neurons. *Proc. Natl. Acad. Sci. USA* 80:5776–78

Shelton PA, Bowers D, Heilman KM. 1990. Peripersonal and vertical neglect. *Brain* 113: 191–205

Shiffrin RM, Schneider W. 1977. Controlled and automatic human information processing: II. Perceptual learning, automatic attending, and a general theory. *Psychol. Rev.* 84:127–90

Singer W, Gray CM. 1995. Visual feature integration and the temporal correlation hypothesis. *Annu. Rev. Neurosci.* 18:555–86

Sperling G. 1960. The information available in brief visual presentations. *Psychol. Monogr.* 74 (11, Whole No. 498)

Sperling G. 1967. Successive approximations to a model for short-term memory. In *Attention and Performance,* ed. AF Sanders, 1: 285–292. Amsterdam: North-Holland

Spitzer H, Desimone R, Moran J. 1988. Increased attention enhances both behavioral and neuronal performance. *Science* 240:338–40

Spitzer H, Richmond BJ. 1991. Task difficulty: ignoring, attending to, and discriminating a visual stimulus yield progressively more activity in inferior temporal cortex. *Exp. Brain Res.* 83:340–48

Squire LR, Ojemann JG, Miezin FM, Petersen SE, Videen TO, Raichle ME. 1992. Activation of the hippocampus in normal humans: a functional anatomical study of memory. *Proc. Natl. Acad. Sci. USA* 89:1837–41

Tanaka K. 1993. Neuronal mechanisms of object recognition. *Science* 262:685–88

Tanaka K, Hikosaka K, Saito H, Yukie M,

Fukada Y, Iwai E. 1986. Analysis of local and wide-field movements in the superior temporal visual areas of the macaque monkey. *J. Neurosci.* 6:134–44

Tanaka K, Saito H, Fukada Y, Moriya M. 1991. Coding visual images of objects in the inferotemporal cortex of the macaque monkey. *J. Neurophysiol.* 66:170–89

Tegnér R, Levander M. 1991. Through a looking glass: a new technique to demonstrate directional hypokinesia in unilateral neglect. *Brain* 114:1943–51

Townsend JT. 1971. A note on the identifiability of parallel and serial processes. *Percept. Psychophys.* 10:161–63

Treisman AM. 1960. Contextual cues in selective listening. *Q. J. Exp. Psychol.* 12:242–48

Treisman AM, Gelade G. 1980. A feature integration theory of attention. *Cogn. Psychol.* 12:97–136

Treisman AM, Gormican S. 1988. Feature analysis in early vision: evidence from search asymmetries. *Psychol. Rev.* 95:15–48

Treisman, A, Sato S. 1990. Conjunction search revisited. *J. Exp. Psychol.* 16:459–78

Treisman AM, Schmidt H. 1982. Illusory conjunctions in the perception of objects. *Cogn. Psychol.* 14:107–41

Tsotsos JK. 1990. Analyzing vision at the complexity level. *Behav. Brain Sci.* 13:423–69

Tsotsos JK. 1994. Towards a computational model of visual attention. In *Early Vision and Beyond,* ed. T Papathomas. Cambridge, MA: MIT Press. In press

Ungerleider LG, Desimone R, Galkin TW, Mishkin M. 1984. Subcortical projections of area MT in the macaque. *J. Comp. Neurol.* 223:368–86

Ungerleider LG, Gaffan D, Pelak VS. 1989. Projections from inferior temporal cortex to prefrontal cortex via the uncinate fascicle in rhesus monkeys. *Exp. Brain Res.* 76:473–84

Ungerleider LG, Galkin TW, Mishkin M. 1983. Visuotopic organization of projections from striate cortex to inferior and lateral pulvinar in rhesus monkey. *J. Comp. Neurol.* 217:137–57

Ungerleider LG, Haxby J. 1994. What and where in the human brain. *Curr. Opin. Neurobiol.* 4:157–65

Ungerleider LG, Mishkin M. 1982. Two cortical visual systems. In *Analysis of Visual Behavior,* ed. J Ingle, MA Goodale, RJW Mansfield, pp. 549–86. Cambridge, MA: MIT Press

Vecera SP, Farrah MJ. 1994. Does visual attention select objects or locations? *J. Exp. Psychol.* In press

von der Heydt R, Peterhans E, Baumgartner G. 1984. Illusory contours and cortical neuron responses. *Science* 224:1260–62

Watson RT, Heilman KM. 1979. Thalamic neglect. *Neurology* 29:690–94

Watson RT, Heilman KM, Cauthen JC, King FA. 1973. Neglect after cingulectomy. *Neurology* 23:1003–7

Wallace SF, Rosenquist AC, Sprague JM. 1990. Ibotenic acid lesions of the lateral substantia nigra restore visual orientation behavior in the hemianopic cat. *J. Comp. Neurol.* 296: 222–52

Walley RE, Weiden TD. 1973. Lateral inhibition and cognitive masking: a neuropsychological theory of attention. *Psychol. Rev.* 80:284–302

Wang Q, Cavanagh P, Green M. 1992. Familiarity and pop-out in visual search. *Assoc. Res. Vis. Opthamol. Abs.* 33:1262

Wilson FAW, O Scalaidhe SP, Goldman-Rakic PS. 1993. Dissociation of object and spatial processing domains in primate prefrontal cortex. *Science* 260:1955–58

Wolfe JM, Cave KR, Franzel SL. 1989. Guided search: an alternative to the feature integration model for visual search. *J. Exp. Psychol.* 15:419–33

Wurtz RH, Goldberg ME. 1972. The primate superior colliculus and the shift of visual attention. *Invest. Ophthalmol.* 11:441–50

Wurtz RH, Mohler CW. 1976. Enhancement of visual responses in monkey striate cortex and frontal eye fields. *J. Neurophysiol.* 39:766–72

Annu. Rev. Neurosci. 1995. 18:223–53

FUNCTIONAL INTERACTIONS OF NEUROTROPHINS AND NEUROTROPHIN RECEPTORS

Mark Bothwell

Department of Physiology and Biophysics, University of Washington, Seattle, Washington 98195

KEY WORDS: nerve growth factor, brain-derived neurotrophic factor, neurotrophin-3, trk receptor, retrograde signaling

INTRODUCTION

The Neurotrophic Hypothesis

Programmed neuronal cell death is a prominent feature of vertebrate embryonic development. Timing and extent of cell death are dependent on the influence of the targets those neurons innervate. The number of neurons in the adult nervous system and the density of innervation of their targets are determined by a competitive process in which tissues produce, in limiting quantities, trophic factors that support neuronal survival. Neurons actively compete to establish optimum contact with the target, and those making the least effective contacts with the target are eliminated by cell death. If each target of innervation elaborated a unique trophic factor, the resulting mechanism would provide a means of enforcing the correct patterns of connectivity between specific neuronal populations and their intended terminal targets.

Since the discovery of nerve growth factor (NGF) (Levi-Montalcini 1951, Cohen 1960, Levi-Montalcini 1987), over four decades of research have provided abundant evidence that the function of NGF closely obeys the neurotrophic hypothesis. NGF acts on a relatively limited variety of neuronal populations, however, including sympathetic and subpopulations of sensory neurons of the peripheral nervous system, as observed in the studies cited above, and striatal and septal cholinergic neurons in the brain (Gnahn et al 1983, Hefti et al 1984, Mobley et al 1985). The initial presumption that many neurotrophic factors must account for the complexity of connectivity of the nervous system

223

may be incorrect. The number of neurotrophic factors presently appears to be relatively small. Remarkably, NGF and several recently discovered neurotrophic factors constitute a family of structurally and functionally similar proteins known as neurotrophins.

Although the number of neurotrophins is small, their actions are enormously subtle and complex. Thus, neurotrophins may contribute substantially to establishing the complex pattern of connectivity of the nervous system, not by the action of a large variety of neurotrophic factors as first imagined, but by complex interactions of a smaller number of neurotrc̲ ic factors. The growing list of neuronal populations regulated by neurotrophins includes sensory and sympathetic neurons, hippocampal pyramidal and dentate neurons, motoneurons, retinal ganglion cells, septal and striatal cholinergic neurons, cerebellar Purkinje cells, and neurons of substantia nigra and of the locus ceruleus (see Korsching 1993 for a review).

Additional Functions of Neurotrophins

Although many of the actions of NGF are consistent with the neurotrophic hypothesis stated above, NGF and other neurotrophins also regulate other aspects of neuronal function and have important activities on nonneuronal cells as well. NGF promotes branching of both axons and dendrites of responsive neurons (Snider 1988, Doubleday & Robinson 1992) and promotes neurotransmitter synthesis, for example, by enhancing expression of choline acetyltransferase in cholinergic neurons (Gnahn et al 1983), enhancing tyrosine hydroxylase in adrenergic neurons at both the transcriptional and post-translational level (Greene et al 1984, Gizang-Ginzberg & Ziff 1990), and enhancing substance P expression in sensory neurons (Lindsay & Harmar 1989). NGF action on neurons is mediated both by direct local action on axonal morphology that does not require nuclear function (Nichols et al 1989) and by effects mediated by regulating nuclear gene expression (e.g. Wood et al 1992). In the nervous system, neurotrophin action is not limited to neurons but is also directed against neuronal stem cells and glia, for which neurotrophins are mitogens and/or differentiating agents (Kalcheim & Gendreau, Confort et al 1991, Sieber-Blum 1991, Birren et al 1992, Kalcheim et al 1992, McNulty et al 1993, Spoerri et al 1993, Barres et al 1994). Indeed, some of the postulated functions of neurotrophins fall quite outside the range of actions generally ascribed to growth factors and more closely resemble those of neuropeptides, for neurotrophins have been found to have acute affects on the electrophysiological properties of neurons (Ueyama et al 1991, Ritter & Mendell 1992, Berninger et al 1993, Lohof et al 1993, Palmer et al 1993, Tancredi et al 1993).

In addition, neurotrophins have manifold activities on nonneural cell types. These activities, which are generally less studied, may act in functional coordination of neural and nonneural tissues. NGF has mitogenic, survival-promoting, chemotactic, or differentiating activities on a wide variety of blood cells, including mast cells (Matsuda et al 1991, Horigome et al 1993), lymphocytes (Thorpe & Perez-Polo 1987, Otten et al 1989, Kimata et al 1991, Brodie & Gelfand 1992), monocytes (Ehrhard et al 1993), polymorphonuclear leukocytes (Gee et al 1983), neutrophils (Kannan et al 1991), and basophils (Matsuda et al 1988, Bischoff & Dahinden 1992). Other cell types responsive to NGF or other neurotrophins include skeletal and smooth muscle cells (Brodie & Sampson 1990, Hempstead et al 1993), keratinocytes (Di Marco et al 1993), corneal epithelial cells (Woost et al 1992), and chromaffin cells (Lillien & Claude 1985).

Scope of This Review

Excellent recent reviews have discussed the nature of neurotrophin receptors, their signal transduction mechanisms, and the functions of neurotrophins for particular neuronal systems (Chao 1992, Meakin & Shooter 1992, Barbacid 1993, Bradshaw et al 1993, Korsching 1993). The intent of the present review is neither to thoroughly discuss the cellular mechanism of action of neurotrophins nor to completely catalogue the range of biological activities of neurotrophins. Instead, the discussion focuses on the interface of mechanistic and biological investigations, in order to explore the manner in which the mechanistic complexity of neurotrophin signaling serves to accomplish complex regulatory functions.

NEUROTROPHINS

NGF

NGF exists as a 26-kDa homodimer of 13-kDa polypeptides, referred to as βNGF. Although the association of the chains of the dimer is noncovalent, the interaction is stable even at physiologically low NGF concentrations (Bothwell & Shooter 1977).

Discovery of BDNF

While NGF provides trophic support to sympathetic neurons, a subset of sensory neurons derived from the neural crest and cholinergic neurons of the striatal and septal regions of the brain, other peripheral nervous system (PNS) neurons and the majority of CNS neurons are nonresponsive. An example of this specificity is the selective action of NGF on discrete populations of sensory neurons. NGF supports the survival of a subpopulation of (mainly nociceptive) sensory neurons derived from the neural crest, while the remaining neural crest–derived sensory neurons, and all of the sensory neurons of cranial ganglia (e.g. nodose ganglia) derived from epithelial placodes, are unresponsive. Yet, these NGF-insensitive sensory neurons are as fully dependent for survival on target-derived factors as are the NGF-sensitive ones. These observations led to the discovery and molecular cloning of a neurotrophic factor for nodose ganglion neurons: brain-derived neurotrophic factor (BDNF) (Leibrock et al 1989, Ernfors et al 1990). Remarkably, BDNF showed substantial structural similarity to NGF; there is 50% identity of amino acid sequence.

Discovery of NT-3 and NT-4/5

Noting the existence of segments of near-identity of NGF and BDNF sequence, several groups initiated attempts to clone other related factors by employing polymerase chain reaction with primers corresponding to the conserved sequences. Neurotrophin-3 (NT-3) was cloned from several species (Hohn et al 1990, Jones & Reichardt 1990, Kaisho et al 1990, Maisonpierre et al 1990, Rosenthal et al 1990), and NT-4 was cloned from *Xenopus laevis* (Halböök et al 1991). Another mammalian neurotrophin was cloned by two groups, one of which concluded that this factor was a homologue of the frog NT-4, while the other group concluded that it was structurally too divergent to be considered a mammalian homologue of NT-4. Consequently, the factor was named NT-4 or NT-5, and is commonly referred to now as NT-4/5. (Berkemeier et al 1991, Ip et al 1992).

RECEPTORS

Discovery of p75

Early studies revealed that NGF receptors on various neuronal populations and on tumor cell lines were heterogeneous with regard to binding affinity and binding kinetics; the distinct receptor populations were referred to as low and high affinity or as kinetically fast and slow (Frazier et al 1974, Sutter et al 1979, Schechter & Bothwell 1981, Godfrey & Shooter 1986). Purification attempts in different laboratories yielded proteins with divergent properties: Receptor proteins of 130–140 and 75–80 kDa were described (Puma et al 1983, Kouchalakos & Bradshaw 1986, Marano et al 1987). Chemical cross-linking of ^{125}I-NGF to receptors also suggested the existence of two receptor proteins of about 130–140 and 80 kDa (Massagué et al 1981, Buxser et al 1983b, Grob et al 1983).

The first receptor to be cloned was the 75- to 80-kDa (hereafter referred to simply as p75). When expressed in fibroblasts, this receptor had properties similar to the low-affinity and kinetically fast receptors previously described (Chao et al 1986, Johnson et al 1986, Radeke et al 1987). This receptor is commonly referred to as the low-affinity receptor, but this term is misleading, for reasons that become clear in this review. p75 binds all neurotrophins with similar affinity (Rodriguez-Tébar et al 1990, 1992; Squinto et al 1991). Transfection of fibroblasts with p75 fails to yield detectable cellular responses. This, coupled with the absence of obvious structural motifs that might mediate intracellular signal transduction and the inability of p75 expressed in fibroblasts to generate the higher affinity class of binding sites suggested that another receptor protein must exist, although it was not initially clear whether this putative protein was itself capable of binding NGF or whether it simply enhanced the binding affinity of p75. The former possibility turned out to be correct.

Discovery of trk Receptors

TRKA   The observations that NGF stimulates tyrosine kinase activity (Maher 1988) and that the tissue distribution of transcripts of the receptor tyrosine kinase protooncogene *trk* (Martin-Zanca et al 1990) was similar to the distribution of NGF-responsive neurons led to the discovery that the encoded protein p140^{trk} is an NGF receptor (Kaplan et al 1991a,b; Klein et al 1991). These studies demonstrated that NGF stimulates tyrosine phosphorylation of the trk protein as expressed endogenously in rat PC12 cells and after transfection of fibroblastic cells. Expression of trk in mouse 3T3 cell lines confers on these cells a range of NGF responses, including enhanced survival in low serum, enhanced cell proliferation, and enhanced anchorage-independent growth. Similar but weaker activities are observed with NT-3 and NT-4/5, while BDNF is essentially inactive.

TRKB AND TRKC   Low-stringency cDNA library screens led several groups to isolate clones that encoded structural homologues of trk. These were named *trkB* (Klein et al 1990, Middlemas et al 1991) and *trkC* (Cordon-Cardo et al 1991, Lamballe et al 1991). Because the term trk is now frequently used to refer to the family of trk-related receptors, I henceforth refer to the original trk receptor as trkA. When trkB and trkC proteins are expressed in 3T3 fibroblasts, the appropriate neurotrophins yield functional responses that are similar to those obtained with trkA. trkB is activated equally by BDNF and NT-4/5, and to a lesser degree by NT-3, while only NT-3 is an effective activator of trkC (Berkemeier et al 1991, Lamballe et al 1991, Ip et al 1992). Neurotrophin binding affinity for trkB and trkC and the ability of neurotrophins to stimulate tyrosine autophosphorylation of trkB and trkC have similar patterns of specificity (Berkemeier et al 1991, Glass et al 1991, Klein et al 1991b, Lamballe et al 1991, Soppet et al 1991, Squinto et al 1991, Cordon-Cardo et al 1992).

EVIDENCE THAT TRK RECEPTORS ARE THE FUNCTIONAL RECEPTORS IN NEURONS
Although these studies made it clear that the trk receptors are functional

neurotrophin receptors that yield mitogenic responses in 3T3 cells, other studies revealed that the same receptors mediated the differentiating and survival responses of neurotrophins in neurons. All three of the trk receptors are expressed in highest levels in the nervous system (Klein et al 1989, Martin-Zanca et al 1990, Lamballe et al 1991). Neuronal cell lines and neurons in primary culture express various trk receptors, and these show neurotrophin-induced tyrosine autophosphorylation (Kaplan et al 1991; Klein et al 1991a,b; Soppet et al 1991; Squinto et al 1991; Marsh et al 1993). Transfection of PC12 cells with trkB confers the ability to differentiate in the presence of BDNF (Squinto et al 1991). A PC12 cell variant deficient in trkA does not respond to NGF, but NGF induces differentiation of these cells following transfection with trkA (Loeb et al 1992). Also following transfection with trkA, enhanced expression of trkA in PC12 cells increases the rapidity of differentiation (Hempstead et al 1992). K252a and K252b, relatively specific inhibitors of the trk receptor tyrosine kinase activity, effectively block neurotrophin responses of both PC12 cells and cultured CNS neurons (reviewed by Knüsel & Hefti 1992). Finally, targeted mutation of *trkA, trkB,* and *trkC* genes in mice causes the anticipated disruption of neural development (Klein et al 1993; M Barbacid, personal communication). Thus, the functional role of trkA, trkB, and trkC as the neurotrophin receptors in neurons has been clearly established.

OTHER TRK RECEPTOR HOMOLOGUES   Several other receptor tyrosine kinases have been described that resemble trkA, trkB, and trkC (Pulido et al 1992, Di Marco et al 1993a, Jennings et al 1993, Wilson et al 1993). However, none of these has significant sequence similarity to the trkA, trkB, or trkC in the extracellular domain, and none has been shown to bind neurotrophins. Di Marco et al (1993a,b) have suggested that one of these, trkE, is an NGF receptor, but this conclusion is based on indirect evidence and requires confirmation. Dtrk (Pulido et al 1992), a *Drosophila* trk-related species, does not appear to function as a neurotrophin receptor. Although suggestive evidence for functional neurotrophins and receptors in invertebrates exists (Ridgway et al 1991, Hayashi et al 1992), no functional homologue of p75, trk receptors, or neurotrophins has been cloned from an invertebrate.

NGF BINDING PROPERTIES OF TRKA   The distinction between low- and high-affinity NGF binding sites found on cells has received considerable attention because of the common belief, perhaps poorly founded, that biological response is mediated exclusively by the high-affinity population of receptors. Quotations of dissociation equilibrium binding constants (Kd) vary widely; values of approximately 10 to 100 pM and 100 pM to 2 nM for the high- and low-affinity sites are typically reported. Although some studies have assumed that the equilibrium high-affinity and kinetically slowly dissociating binding sites represent equivalent receptor pools, in PC12 cells this view is too simplistic; a substantial proportion of 2-nM–Kd binding sites release bound NGF slowly. For these sites, slower rates for dissociation are balanced by correspondingly slower rates of association, which yield a relatively low Kd (Schechter & Bothwell 1981). Conflicting studies concluded that binding to p75 contributed to both high- and low-affinity sites (Green & Greene 1986) or that the higher molecular weight receptor (now known to be trkA) was exclusively of higher affinity, while the p75 receptor was exclusively of lower affinity (Hosang & Shooter 1985). That the truth may lie somewhere between these extremes is suggested by Weskamp & Reichardt (1991) who found that antibodies that blocked binding of NGF to p75 eliminated one half of the high-affinity binding.

Published findings concerning binding of NGF to trkA receptors expressed in the absence of p75 are also controversial. trkA receptors expressed in transfected fibroblasts have binding properties resembling those of the slow, low-affinity receptor population in PC12 cells (Schechter & Bothwell 1981). They have a dissociation equilibrium constant in the nonmolar range similar to that of p75, but they bind and release NGF much more slowly than does p75 (Kaplan et al 1991, Hartman et al 1992, Meakin et al 1992). However, although some investigators find that high-affinity (low-Kd) receptors are formed only by the simultaneous expression of p75 and trk (Hempstead et al 1991), others find that a small proportion of high-affinity sites are generated by trk receptors in the absence of p75 and that introduction of p75 does not affect this proportion (Jing et al 1992).

p75 Structural Details

TNF RECEPTOR HOMOLOGY   Although p75 represented a novel type of receptor when it was first cloned (Johnson et al 1986, Radeke et al 1987), a large number of structurally related cytokine receptors have since been discovered, the best characterized of which are the two tumor necrosis factor (TNF) receptors (reviewed by Banner et al 1993, Suda et al 1993). The proteins encoded by the gene superfamily all have a single transmembrane domain; a small, poorly conserved cytoplasmic domain; and an extracellular domain containing a variable number (typically 4) of repeats of a sequence motif defined by the characteristic spacing of 6 cysteinyl residues within a span of about 40 amino acid residues. The effector cytokines for these receptors are structurally related but share no sequence similarity with the neurotrophins (Suda et al 1993).

The three-dimensional structure of the extracellular domain of one member of this gene superfamily, the TNF-R55 receptor, as a complex with TNF, has been characterized by X-ray crystallography (Banner et al 1993). The configuration of the polypeptide backbone and the disulfide pairing of the extracel-

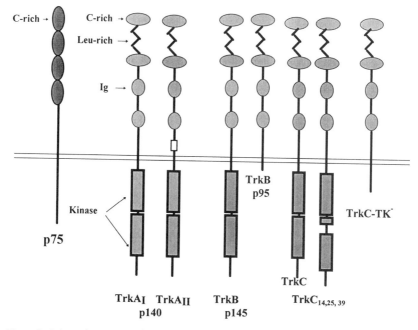

Figure 1   Schematic representation of neurotrophin receptors.

lular domains of the other members of the gene family, including the p75 neurotrophin receptor, will probably be similar. Thus, by analogy to TNF-R55, the four cysteine repeat motifs of p75 are probably arranged in a collinear manner, as depicted in Figure 1, and the 6 cysteines within each repeat form disulfides paired 1:2, 3:5, and 4:6. This provides a useful framework for future studies examining mutations of p75. From deletional analysis of the cysteine repeat motifs of p75, it has been concluded that all four repeats contribute to formation of the NGF binding site (Welcher et al 1991, Yan & Chao 1991, Baldwin et al 1992).

SEQUENCE CONSERVATION   Comparison of the sequence of mammalian and avian p75 reveals that the amino acid sequence has been highly conserved during evolution; the most highly conserved region is the membrane-spanning and juxtamembrane region. The high degree of conservation and the high hydrophilic moment of this region suggest that it represents a region of protein-protein interaction (Large et al 1989, Heuer et al 1990).

DIMERIC STRUCTURE   A variable fraction of p75 exists as a disulfide-linked dimer (Grob et al 1985). p75 is heavily phosphorylated on serine residues in every cell type that has been examined, but the extent of phosphorylation is independent of NGF (Grob et al 1985, Taniuchi et al 1986b). In cells that contain both monomeric and dimeric p75, phosphorylation is present almost exclusively on the monomeric component (Grob et al 1985). This suggests that phosphorylation inhibits dimerization and/or that dimerization prevents phosphorylation.

trk Receptor Structural Details

SEQUENCE SIMILARITY AND STRUCTURAL MOTIFS OF TRK RECEPTORS   Pairwise comparisons of trkA, trkB, and trkC reveal 66–68% amino acid–sequence identity (Lamballe et al 1991). In addition to the presence of highly similar tyrosine kinase catalytic domains, the three trk neurotrophin receptors have similar extracellular domains containing three repeats of an immunoglobulin-like domain (Schneider & Schweiger 1991). The tyrosine kinase domain of the trk receptors is interrupted by a short insert sequence, in a position similar to the insert present in several other tyrosine kinases, including the PDGF receptor (Tsoulfas et al 1993).

SPLICE VARIANTS OF TRK RECEPTORS   The trk receptors exhibit a large variety of splice variants, and many of these variations produce significant changes in structure of the encoded proteins. A summary of the structure of the trans-lation products is shown in Figure 1. Of the trk receptors, *trkA* transcripts show the least complexity. Only two splice variants have been described (Barker et al 1993). The two encoded proteins differ in the presence or absence of a six–amino acid insertion in the extracellular domain. The insert-containing form is the only isoform in neuronal tissues, while the insert minus form is mainly expressed in nonneural tissues.

TRUNCATED VARIENTS OF TRKB   The alternate receptor forms encoded by the splice variants of *trkB* and *trkC* are clearly different in function. Two splice variants of *trkB* have been identified that encode receptors in which the tyrosine kinase domain has been replaced with novel shorter cytoplasmic domains only 21 or 23 amino acids in length (Klein et al 1990, Middlemas et al 1991). The truncated isoform is the most abundant form in a variety of nonneuronal tissues, and in the nervous system it increases in abundance relative to the full-length receptor as embryonic development progresses (Allendoerfer et al 1994). In adult brain the truncated receptor is heavily expressed on ependymal cells and epithelial cells of the choroid plexus (Klein et al 1990, Middlemas et al 1991). Several functions of the truncated form have been proposed. The truncated variants could be negative regulators of trkB signal transduction. Artificially truncated trkA receptors inhibit signaling of the full-length trkA receptors (Jing et al 1992), and truncated trkB receptors inhibit signaling of the full-length trkB receptors (S Skirboll & M Bothwell, unpublished), presumably by formation of functionally inactive receptor heterodimers. Alternatively, these receptors might restrict the diffusion of BDNF from sites of release. It also has been suggested that the truncated receptors might be positive regulators of BDNF action by presenting BDNF to full-length receptors; however, they may interact with src-like cytoplasmic tyrosine kinases, in the manner of many cytokine receptors (Klein et al 1990, Middlemas et al 1991).

TRUNCATED VARIANTS OF TRKC   Like *trkB, trkC* transcripts have been identified that encode a receptor lacking the tyrosine kinase domain (Tsoulfas et al 1993, Okazawa et al 1993) In addition, three splice variants generate inserts of varying length within the tyrosine kinase domain of trkC. Isoforms with inserts of 14, 25, and 39 amino acids have been described. These isoforms have normal NT-3-induced activation of receptor autophosphorylation but are inactive with regard to mitogenic effects on 3T3 cells (Tsoulfas et al 1993, Valenzuela et al 1993). Although it is possible that these isoforms are incapable of signal transduction, it seems more likely that they couple to cellular responses that have not yet been identified.

Signal Transduction by the trk Receptors

The signal transduction mechanisms employed by the trk receptors are similar to those employed by other receptor tyrosine kinases (Ullrich & Schlessinger 1991). The mechanisms have been most extensively studied for trkA but appear to be generally similar for trkB and trkC (Marsh et al 1993, Widmer et al 1993). NGF binding to trkA leads to homodimerization, followed by autophosphorylation of tyrosyl residues (Jing et al 1992). Phosphotyrosine residues serve as sites for interaction of intracellular proteins such as grb2 (Suen et al 1993), which binds the ras-activator protein mSos1 (Rosakis-Adcock et al 1993). Activated ras binds and activates raf kinase, which in turn activates the MAP kinases ERK1 and ERK2 (reviewed by Barbacid 1993). ERK1 is physically associated with trk (Loeb et al 1992), while both ERK1 and ERK2 are associated with p75 in PC12 cells (Volonté et al 1993a). NGF signal transduction through trkA receptors involves several parallel pathways (D'Arcangelo & Halegoua 1993). One pathway activates phospholipase C-γ1, which leads to hydrolysis of phosphatidylinositol and phosphatidylglycans, generating diacylglycerol and inositol trisphosphate (Bellini et al 1988, Altin & Bradshaw 1990, Belia et al 1991, Ohmichi et al 1991, Vetter et al 1991). Not surprisingly, therefore, trk activation promotes mobilization of intracellular Ca^{2+} ion (Pandiella-Alonso et al 1986, van-Calker et al 1989, Grohovaz et al 1991, Yamashita & Kawana 1991, Berninger et al 1993).

NEUROTROPHIN STRUCTURE AND DETERMINANTS OF SPECIFICITY

Crystal Structure of NGF

The three-dimensional structure of NGF obtained from crystallographic data (McDonald et al 1991) provides useful information for characterizing the nature of neurotrophin-receptor interactions. The NGF folded structure is defined by three antiparallel pairs of β strands with the β sheets arranged predominantly along the long axis of the protomer, linked by hairpin loops at the top and bottom. The NGF protomers associate along their long axes, forming a dimer with twofold symmetry. Viewed with the axis of symmetry aligned vertically, the amino-terminal and carboxy-terminal ends of both chains lie at the same end (the bottom of the structure, as conventionally depicted) and are not well resolved in the X-ray structure, apparently because they have considerable flexibility within the crystal. The residues that are of greatest importance in defining the backbone structure, including all six of the cyteines involved in disulfide bond formation, are well conserved among the neurotrophins, suggesting that BDNF and NT-3 probably have a similar folded structure. The region of interface of the two NGF protomers is highly conserved among the neurotrophins, suggesting that BDNF and NT-3 should form similar dimers, and even raising the possibility that BDNF and NT-3 form heterodimers. The dimeric structure of BDNF and NT-3 and the ability of neurotrophins to form heterodimers have been confirmed biochemically (Rad- ziejewski et al 1992, Narhi et al 1993).

Sites of Interaction With Receptors

The sequence segments that differ most substantially among the neurotrophins fall on three hairpin loops at the top of the NGF molecule, on the amino-terminal and carboxy-terminal segments, on one β strand, and on a reverse turn at the bottom of the molecule. These regions are probably important for selective binding to trkA, trkB, and trkC (McDonald et al 1991, Ibáñez et al 1993). This conclusion is largely consistent with the results of studies examining the effects of in vitro mutagenesis on neurotrophin receptor selectivity. Mutants comprised of segments derived from different neurotrophins, and point mutations have been examined (Ibáñez et al 1990, 1991, 1992, 1993; Suter et al 1992). Chimeric neurotrophins, with segments of varying regions from different neurotrophins, combine the biological activities of NGF, BDNF, and NT-3 (Ibáñez et al 1993). This implies that the folded structure of the various neurotrophins is similar and that the trk receptor selectivity resides in the neurotrophin variable regions. The NGF regions conferring trkA specificity

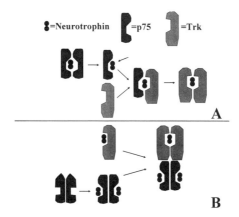

Figure 2   Models for interaction of p75 and trk receptors. (*A*) Binding of NGF to p75 or trk monomers causes formation of homodimeric receptors. NGF induces formation of p75-trk receptor heterodimers, which serve as intermediates in the formation of trk homodimers. (*B*) p75 exists primarily in a dimeric form that undergoes a conformational change upon binding two molecules of dimeric NGF. The conformationally altered p75 dimer binds trk and promotes formation of trk homodimers.

all lie along one face of the NGF dimer, with residues from both protomers contributing to the putative trkA binding site. The twofold symmetry of the NGF dimer allows the dimer to potentially bind two trkA molecules, possibly promoting dimerization and, therefore, activation of trkA receptors. Similar regions of BDNF confer trkB specificity. The importance of the amino terminus (Kahle et al 1992) and carboxy terminus (Drinkwater et al 1993) of NGF for trkA interactions has also been demonstrated by deletion.

Neurotrophin sequences responsible for NGF binding to the p75 receptor are located primarily at the top of the dimer, partly overlapping with the trkA binding sites (Ibáñez et al 1992). These results are generally in agreement with the results of immunochemical mapping of p75 and trkA binding sites for NGF (Nanduri et al 1994) except that this latter study found that p75 and trkA bound to the top and bottom of the dimer, respectively, without evidence for overlapping binding sites. Either arrangement allows a single NGF dimer to bind p75 and trkA simultaneously, stabilizing the formation of a p75-trk heterodimer. Ibáñez et al (1992) noted that such a molecule might be an intermediate in the formation of an NGF-associated trk-trk homodimer, as shown in Figure 2*A*; this is consistent with the proposed role of p75 in ligand recruitment (Glass et al 1991, Jing et al 1992).

CONTROVERSY CONCERNING FUNCTIONALITY OF p75

Introduction

Several investigators have suggested that the p75 receptor may not play a direct role in neurotrophin action on cells. Alternative functions have been proposed for p75, such as buffering the concentration of available extracellular NGF and maintaining a high concentration of NGF (or another neurotrophin) near the site of release, or allowing NGF bound to p75 on one cell to be presented to trk receptors on another cell (Taniuchi et al 1986a, Barbacid 1993). Others have concluded, however, that p75 serves as an accessory subunit to trk receptors, facilitating trk signaling.

p75 Is Not Essential for Response to Neurotrophins

Clearly, p75 is not essential for at least some trk receptor–mediated responses to neurotrophins. Mouse 3T3 cells do not express p75, yet when transfected with trk receptors they display a number of different responses to the appropriate neurotrophin, including enhanced proliferation rate, enhanced survival in low serum, and enhanced anchorage-independent growth (Barbacid 1993). Although there is extensive coexpression of p75 with trk receptors in both neural and nonneural tissue, some normal cell types express trk receptors in the absence of p75 receptors, and these cells are functionally regulated by neurotrophins. Examples include striatal cholinergic neurons (Knüsel et al 1994), sympathetic neuronal precursors (Birren et al 1992), and mast cells (Horigone et al 1993), all of which express trkA and respond to NGF, as well as hippocampal neurons, which express trkB and trkC and respond to BDNF and NT-3 (Collazo et al 1992, Marsh et al 1992).

The PC12 neuronal cell line expresses p75 and trkA but yields NGF responses (including induction of c-fos transcription) after the NGF binding of virtually all fast (p75) receptors has been destroyed by proteolysis (Schechter & Bothwell 1981). Inhibition of NGF binding to p75 by p75 antibodies also fails to inhibit NGF response (Weskamp & Reichardt 1991). Furthermore, NGF mutants with minimal affinity for binding to p75 but intact affinity for trkA are capable of yielding normal responses from PC12 cells as well as from primary neuronal cultures (Ibáñez et al 1992). These results indicate that NGF binding to p75 is not essential for response of PC12 cells.

Evidence that p75 Enhances NGF Activation of trkA Receptors

In contrast to these findings, a number of reports suggest that p75 enhances NGF responsiveness of cells expressing trkA. Transfection of an NGF nonresponsive neuroblastoma cell line with p75 yielded a cell line that initiated neurites in response to NGF (Matsushima & Bogenmann 1990), and transfection of p75 into a PC12 mutant cell line lacking p75 reconstituted the ability of NGF to induce c-fos transcription (Hempstead et al 1989) and to induce tyrosine kinase activity (Berg et al 1991). Furthermore, transfection of PC12 cells with a chimeric receptor comprised of the EGF receptor–ligand binding domain attached to the juxtamembrane and cytoplasmic domains of p75 yielded EGF-dependent neuronal differentiation of PC12 cells (Yan et al 1991). Also, transfection of PC12 cells with truncated p75 mutants lacking the ligand binding domain caused loss of endogenous expression of p75 protein and diminished the sensitivity of these cells to NGF. Interestingly, these cells also showed a relative enhancement of sensitivity to NT-3, leading the authors to suggest that p75 might influence the neurotrophin selectivity of trk receptors (Benedetti et al 1993), as had been speculated previously (Bothwell 1991). Expression of p75 following transfection of a trkA-expressing rat sympathetic precursor neuronal cell line enhances the sensitivity of these cells to NGF-induced neurite outgrowth (Verdi et al 1994). However, mice lacking p75 by virtue of gene-targeted mutation have spinal sensory neurons for which NGF-enhanced survival is shifted to significantly higher concentrations (Davies et al 1993).

Contributions of p75 and trkA to High-Affinity Neurotrophin Binding

An equivalent controversy exists concerning the functional role of p75 in generating high-affinity receptors. Chao and coworkers have found that p75 and trkA individually have equilibrium constants for binding NGF comparable to the low-affinity class of receptors found in neurons and PC12 cells but that the combined presence of p75 and trkA generates high-affinity sites (Hempstead et al 1991, Kaplan et al 1991a, Battleman et al 1993). On the other hand, Barbacid and coworkers found that trk receptors generate high-affinity binding sites by themselves (as a small percentage of the total), apparently via formation of trkA homodimers, and p75 has no influence on NGF binding affinity of trk receptors (Klein et al 1991, Lamballe et al 1991, Jing et al 1992). It has not yet been established whether the analysis of one group is in error or, alternatively, whether the discrepancy represents real differences between the different cell lines employed.

Although the origin of high-affinity binding, as defined by equilibrium constant, is still contested, the origin of the heterogeneity of NGF binding sites with regard to kinetic rate constants is more consistently defined by the above studies. p75 receptors (in the absence of trk receptors) exhibit the rapid rate constants for binding and releasing NGF, as well as the sensitivity to trypsin, characteristic of the fast class of receptors on neurons and PC12 cells, while trkA receptors uniformly exhibit the much slower rate constants for binding and releasing NGF, and the trypsin insensitivity, characteristic of the slow class of receptors on neurons and PC12 cells. Even the trkA receptor sub-population exhibiting low-affinity equilibrium constants has a slow rate constant for NGF dissociation—the low affinity at equilibrium results from the exceptionally slow rate constant for NGF association.

p75 Knockout Mouse

Mice with a targeted mutation of the *p75* gene exhibit a substantial loss of neurons within sensory ganglia, and a deficit of nocioceptive function (Lee et al 1992). Subsequent studies have revealed that there are also substantial sympathetic abnormalities in these mice (Lee et al 1994). Thus, p75 has an important function in the two best-characterized sites of NGF action in vivo.

Alternative Modes of p75-trk Interaction

If one accepts that there is a functional interaction of p75 with trkA (and perhaps of trkB and trkC), the nature of that interaction remains to be clearly defined. It has been proposed that the twofold symmetry of receptor binding sites on a neurotrophin might facilitate formation of p75 or trk homodimers, or of p75-trk heterodimers, as illustrated in Figure 2A (Hempstead et al 1991, Jing et al 1992, Ibáñez et al 1993). The high-affinity receptor has been variously ascribed to such p75-trk heterodimers (Hempstead et al 1991) or to trk homo-dimers (Jing et al 1992). As discussed above, p75 and trk binding sites on NGF may overlap to a significant extent. Given the apparent distribution of receptor binding sites on NGF (and other neurotrophins), it does not seem likely that NGF bound to a p75 homodimer would be capable of associating simultaneously with monomeric or dimeric trk receptors, and conversely, neurotrophin bound to a trk receptor is unlikely to be able to associate with p75. However, Ibáñez et al (1993) suggested that NGF associated with a p75-trk heterodimer might exist as a transient intermediate in the formation of an NGF-associated trk-trk homodimer, consistent with the proposed role of p75 in ligand recruitment as proposed by Glass et al 1991 and Jing et al 1992. This model is depicted in Figure 2A.

It seems unlikely that there is a tight physical association between trk and p75 receptors in the plasma membrane because chemical cross-linking studies have failed to detect such complexes (e.g. Jing et al 1992), and immuno-precipitation of p75, with antibodies against a variety of p75 epitopes, failed to coprecipitate trkA protein (Meakin & Shooter 1991). However, cross-linking may fail simply because reactive amino acid side chains are not suitably juxtaposed in protein complexes, and interactions that are strong enough to maintain protein-protein association in the plasma membrane are not always strong enough to maintain the association during detergent extraction and immunoprecipitation. Thus such negative results are not conclusive.

Fluorescence recovery after photobleaching (FRAP) has been employed to examine the lateral mobility of neurotrophin receptors expressed in insect cells employing bacculovirus vectors (Ross et al 1993). These studies revealed that p75 was freely mobile when expressed alone but became restricted in mobility when trk receptors were coexpressed and that this interaction required an intact trkA tyrosine kinase domain. A similar interaction between p75 and trkB was not observed. The simplest explanation for this behavior is that p75 associates directly with trkA. Less direct mechanisms of interaction are possible, however. Both ERK1 and protein kinase N are physically associated with both trkA and p75 (Volonté et al 1993a,b,c). Thus p75 and trkA association might involve bridging molecules of ERK1 or protein kinase N.

An Alternative Model for p75-trk Interactions

INTRODUCTION   Most models that postulate p75-trk interactions propose that a jointly bound neurotrophin bridges the two receptors, as depicted in Figure 2A. Substantial data support an alternative model, however, in which neurotrophins bind to p75 and trk independently, with functional interactions communicated between the two receptors via neurotrophin-induced conformational changes, as illustrated in Figure 2B.

EVIDENCE FOR ALLOSTERIC PROPERTIES OF p75   Rodriguez-Tébar et al (1990, 1992) demonstrated that the binding of BDNF and NT-3 to rat p75 receptors expressed in mouse L cells is positively cooperative. In contrast, positive cooperativity was not apparent for NGF binding. However, a limited degree of positive cooperativity for NGF binding to p75 has been observed by other investigators (Venkatakrishnan et al 1990). Positively cooperative binding necessarily requires a minimum of two interacting binding sites. Because there is no evidence for two neurotrophin binding sites on a p75 monomer, the cooperative binding must reflect the dimeric form of p75. Evidence was presented above suggesting that phosphorylation of p75 prevents dimerization. Thus, the quantity of dimeric p75 and, consequently, the degree to which neurotrophin binding is cooperative may be regulated by the extent of phosphorylation of p75.

A powerful model for positively cooperative binding was proposed by Monod et al (1965). The Monod/Wyman/Changeaux (MWC) model postulates that positively cooperative binding results from binding ligands to a protein that exists in an equilibrium between two conformations having different binding affinity for the ligand in question. The only assumption made by the model is that individual subunits within an oligomeric protein undergo the

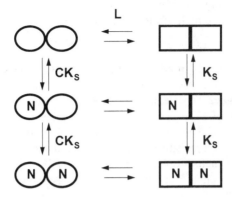

Figure 3   MWC model for conformational transitions of p75. In the MWC model, L represents the equilibrium constant for the interconversion of two conformations of p75, in the absence of bound neurotrophin (N). K_s represents the equilibrium constant for binding of the ligand to the higher affinity (*square*) conformation, and c represents the ratio of the equilibrium constants for neurotrophin binding to high- and low-affinity states. Consequently, the equilibrium constant for neurotrophin binding to the low-affinity (*circle*) conformation is the product of c and K_s.

conformational change in a concerted manner. The model is illustrated in Figure 3 for the case of a dimeric protein (appropriate for p75) with subunits existing in two conformational states represented by circles and squares.

The essence of the model is that ligand binding preferentially (i.e. with higher affinity) to the square conformation will, by mass action, shift the equilibrium, thereby yielding an increased concentration of the square conformation. Thus, the second ligand molecule will bind with greater affinity than the first, because a greater proportion of available sites will be in the higher affinity conformation. The MWC model describes this behavior quantitatively, in terms of three constants of the system. C represents the equilibrium constant for the interconversion of the two conformations in the absence of bound ligand. L represents the ratio of the equilibrium constants for binding of a ligand to the circle and square conformation. K_s represents the equilibrium constant for ligand binding to the square conformation. K_c, the equilibrium constant for binding to the circle conformation, may be calculated as K_s/L. Maximum cooperativity results from large values of L coupled with values of c much smaller than 1. The MWC model describes two equations of state. If N represents the concentration of unbound neurotrophin, Y_N—the fractional saturation of binding as a function of neurotrophin concentration—may be calculated as

$$Y_N = [LcN/K_s(1+cN/K_s) + N/K_s(1 + N/K_s)]/[L(1 + cN/K_s)^2 + (1 + N/K_s)^2].$$

The function describing the dependence on ligand concentration of the fraction of protein in the square high-affinity conformation is given as

$$R_N = [(1 + N/K_s)^2]/[L(1 + cN/K_s)^2 + (1 + N/K_s)^2].$$

I have taken the liberty of adjusting the MWC model constants to achieve an optimum fit for Y_N to the data for NGF, BDNF, and NT-3 binding to p75 obtained by Rodriguez-Tébar et al (1991, 1992), as shown in Figure 4. Even though the binding of NGF does not appear to be positively cooperative on subjective inspection (i.e. the binding curve is not obviously sigmoid), a value of c consistent with slight positive cooperativity fits the data well. Remarkably, K_s for p75 binding to neurotrophins (particularly BDNF and NT-3) approaches the value of the so-called high-affinity receptor, as shown in the legend to Figure 4.

The function R_N is plotted with these constants in Figure 5, which reveals that BDNF and NT-3 induce the p75 conformational change more effectively than NGF. This finding is consistent with the results of Tim et al (1992), who employed spectroscopic techniques to examine conformational changes associated with binding of neurotrophins to the extracellular domain of p75, and observed greater changes for BDNF and NT-3 than for NGF.

EFFECTS OF ANTIBODIES AND LECTINS ON p75 CONFORMATION   The MWC model also accounts for the allosteric interactions between two heterologous ligands that bind at distinct sites. A ligand that binds preferentially to the square conformation will shift equilibrium towards production of that conformation and decrease the effective value of L, while a ligand binding preferentially to the circle conformation will increase the value of L. Tanaka et al (1989) described a hybridoma, 7902, that produces a monoclonal antibody to an avian motoneuron protein subsequently identified as p75. Antibody 7902 has the remarkable property of substantially enhancing the binding of ^{125}I-labeled neurotrophins to chicken p75 receptors expressed in mouse L cells. For NGF, the affinity is increased 30-fold (M Bothwell, unpublished data). The results obtained are consistent with the hypothesis that 7902 binds preferentially to the p75 square state with a value of c of 0.01 or smaller. This effect of 7902 is not entirely novel or unique because a similar but smaller effect has been reported for the antibody 192 when interacting with rat p75 (Chandler et al 1984). Antibody 192 increases NGF binding affinity to rat p75 by about 3- to 4-fold, which is consistent with 192 possessing a value of c of about 0.1. Furthermore, the lectin wheat germ agglutinin causes conversion of low-affinity (p75) receptors to a high-affinity state (Buxser et al 1983a, Grob et al 1983, Vale et al 1983). This effect also may represent the preferential association of the lectin with the higher affinity (square) conformation of p75.

The MWC model also provides a means of accounting for the otherwise

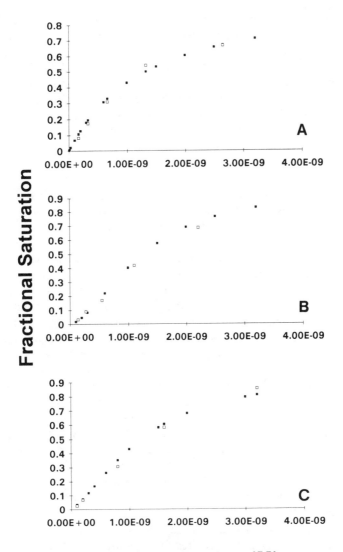

Figure 4 Fit of data for neurotrophin-p75 binding interactions fit to the MWC model. Published data for neurotrophin binding to p75 (Rodriguez-Tébar et al 1990, 1992) are fit to the MWC model, using a value of L of 1000. Fraction of full saturation is plotted as a function of neurotrophin concentration. Experimental data points are represented by open symbols, while theoretical points are represented by filled symbols. (*A*) NGF: C = 0.06, K_s = 120 pM. (*B*) BDNF: C = 0.004, K_s = 40 pM. (*C*) NT-3: C = 0.01, K_s = 40 pM.

puzzling behavior of other p75 monoclonal antibodies. Several monoclonal antibodies against the human p75 that dramatically inhibit NGF binding are nonetheless capable of efficiently immunoprecipitating p75 with bound NGF chemically cross-linked (Marano et al 1987). This seems paradoxical because if the antibody inhibits binding by interacting with the NGF binding site, then the antibody binding site should be blocked when NGF is cross-linked to that site. However, in the MWC model, if these antibodies bind with higher affinity to the circle conformation of p75, then they may inhibit NGF binding by increasing the value of L and need not interact directly with the NGF binding site.

TRK INTERACTIONS   The allosteric model provides a framework for a novel mechanism whereby neurotrophin-p75 interactions may modulate trk receptor function. If only the square conformation of p75 binds to trkA (and possibly also trkB and trkC), then the dimeric nature of p75 will promote trk dimerization and activation, and neurotrophin binding to p75 will enhance this facilitory interaction. The essential elements of this model are depicted in Figure 2*B*. In this model, neurotrophins promote trk dimerization and activation in two independent ways: (*a*) by directly binding to and dimerizing trk and (*b*) by binding to p75 and inducing a p75 conformation that binds to and dimerizes trk.

The thermodynamic principles that apply to linked chemical equilibria such as these indicate that the binding of a neurotrophin to p75, by favoring trk dimerization, increases the effective affinity for neurotrophin binding to trk receptors; conversely, neurotrophin binding to trk receptors enhances trk interactions with a form of p75 with greater affinity for neurotrophin and thereby increases the effective affinity of p75 for neurotrophin. In other words, both p75 and trk exist in forms with higher and lower affinity for binding neurotrophins, and allosteric interactions may allow neurotrophin binding at p75 sites to enhance neurotrophin binding to trk sites, and vice versa. If this model is correct, it becomes clear why efforts to distinguish between the contributions of p75 and of trk to low- and high-affinity binding have proved to be confusing and controversial!

A model such as this may account for conflicting findings with regard to whether the presence of p75 is required for high-affinity binding and cellular response to neurotrophins. In cells with a relatively high concentration of trk receptors in the plasma membrane, dimerization leading to formation of high-affinity binding sites will occur readily and in the absence of p75. In cells with lower concentrations of trk receptors, however, trk dimerization may not occur to a significant extent unless p75 is present to promote trk dimerization. p75 is normally expressed within the sensory nervous system, predominantly on trkA-expressing neurons (e.g. Verge et al 1992). Consequently, it is not surprising that the p75 knockout mouse primarily affects these cells. Presumably,

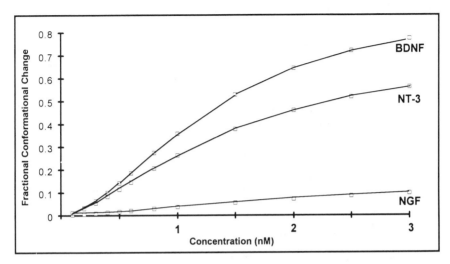

Figure 5   Effect of neurotrophins on conformational transition of p75, as predicted by MWC model. The fractional conformational change, as predicted by the MWC model-state function, is plotted as a function of neurotrophin concentration for NGF, BDNF, and NT-3. MWC constants are taken from Figure 4.

trkB- and trkC-positive sensory neurons express these receptors at concentrations sufficient to allow effective dimerization without p75.

Another surprising prediction of the model requires consideration of the important differences in the manner of action of the different neurotrophins on p75 depicted in Figure 5. BDNF activates p75 much more effectively than NGF. For example, 2-nM BDNF would place 65% of p75 in the active state, while 2-nM NGF would yield only about 7%, and even 100-nM NGF would yield only about 15% activation. In effect, NGF acts as a partial agonist and, therefore, in the presence of BDNF (a strong agonist), would act antagonistically. Addition of 100-nM NGF to 2-nM BDNF would competitively displace all bound BDNF and would cause p75 to drop from 65% active to 15% active. As is discussed in the next section, NGF may act as a functional antagonist of BDNF action in some systems, which is consistent with the above prediction (von Bartheld et al 1994). Although the model in Figure 2B depicts a direct association of p75 with trk, the interaction need not be direct. Similar behavior would result if p75 and trk were associated via a bridging molecule such as ERK1 or protein kinase N.

trk-Independent Action of p75

Although most studies have suggested that p75 is incapable of signaling in the absence of trk receptors, a few studies have suggested that p75 receptors may have intrinsic activity. Feinstein & Larhammar (1990) noted that p75 contains, within its cytoplasmic domain, a sequence predicted to form an amphiphilic alpha-helix similar to G-protein binding sites present in other receptors. Clear evidence that p75 signals via a G protein–mediated mechanism is still lacking. Several protein kinases are physically associated with p75, including MAP kinase and protein kinase N, but binding of NGF to p75 does not appear to enhance activity of these protein kinases (Volonté et al 1993a,c). Many studies suggest, however, that p75 can signal in the absence of trk receptors. Expression of p75 in *Xenopus* oocytes enhances progesterone-induced maturation (Sehgal et al 1988). Although it is possible that this effect reflects p75 cooperation with an endogenously expressed trk receptor, expression of such receptors cannot be detected (S Skirboll & M Bothwell, unpublished data). Rabiadeh et al (1993) have reported that expression of p75 (in the absence of trkA) following transfection of several neuronal cell lines leads to enhanced apoptotic death, which is prevented by NGF, and Anton et al (1994) found that p75 mediates migratory responses of Schwann cells to NGF.

RETROGRADE AXONAL SIGNALING

What Is the Retrograde Signal?

In most investigations of trk receptor–mediated signal transduction and of p75-trk functional interactions, a key feature of normal neurotrophin action has been largely ignored. In vivo, neurotrophins act upon receptors at the axon terminus. Ultimately, the trophic signal must be transmitted retrogradely along the axon to the somatic nucleus, which in a large mammal may be many centimeters or even meters distant. The retrograde signal apparently is associated with retrogradely transported vesicular components because the signal is blocked by manipulations that block vesicular transport.

Although the identity of the retrograde signal for neurotrophin action has not been demonstrated, several good candidates can be considered. Following binding to receptors at the nerve terminus, neurotrophins are transported retrogradely to the cell body (DiStefano et al 1992). Both p75 and trkA receptors are also retrogradely transported, in response to neurotrophin binding (Johnson et al 1987, Ehlers et al 1993, Loy et al 1993). Ehlers et al (1993) demonstrated that the trkA receptor is transported in a tyrosine-phosphorylated state, suggesting that this species may constitute the retrograde signal. Whether the neurotrophins remain bound to the trk receptor during transport or, alternatively, whether the trk receptor is capable of remaining in the activated state for many hours after a transient association with neurotrophin remains unknown. Preliminary studies indicate, however, that both p75 and trkA complexes with NGF have significant stability at pH 5.5 (M Bothwell, unpublished data), suggesting that neurotrophin-receptor complexes may remain intact within acidified endosomal compartments.

Vacuum Cleaner Model for Retrograde Transport

In addition to its role in signal transduction, retrograde transport has another important function. The transport process actively removes neurotrophin from the vicinity of the neurotrophin receptor–bearing axon terminus. The diminution of local neurotrophin supply requires neurons to actively compete for a dwindling supply of trophic factors, and this competition represents the essence of trophic function (Davies et al 1987, Kessler 1985). The ability of neurotrophins to upregulate neurotrophin receptors (e.g. Verge et al 1992) represents a feed-forward regulatory system that may sharpen trophic competition, according to a "rich-get-richer" scheme. Neurons that make good target contacts get a good supply of neurotrophin, upregulate receptors in order to become more competent to detect dwindling supplies of neurotrophin, and remove remaining neurotrophin all the more effectively because of their more efficient receptor-mediated retrograde transport, while neurons that fail to find a good supply of neurotrophin initially are unable to upregulate neurotrophin receptors and consequently die as neurotrophin supply becomes depleted.

A Specialized Role for p75 Function in Retrograde Signaling of Projection Neurons

From numerous published accounts, as well as unpublished studies from my laboratory, it is clear that most cells that express p75 also express trkA, trkB, or trkC. This generalization is consistent with the concept that p75 functions primarily in the context of modulating trk activation. Conversely, however, many cell populations express trk receptors without p75. This is consistent with the view that trk receptors function effectively in the absence of p75. p75 receptors are expressed almost exclusively in projection neurons, rather than in local circuit neurons, whereas trk receptors are expressed in both types of neurons. Examples of projection neurons expressing p75 (permanently or transiently during development) include retinal ganglion cells, motoneurons, basal forebrain magnocellular cholinergic neurons, cerebellar Purkinje cells, as well as primary sensory and postganglionic sympathetic neurons, and all of these neurons are functionally responsive to neurotrophins. Various local circuit neurons that express trk receptors in the absence of p75 (such as hippocampal neurons) are responsive to neurotrophins, nonetheless (Marsh et al 1993). A cogent example is provided by brain cholinergic neurons. The cholinergic neurons of rat brain striatum and septum both express trkA and respond to NGF, but the septal projection neurons express p75, and the striatal local circuit neurons do not (Steininger et al 1993). Such correlations suggest that projection neurons may have a special need for p75 function.

A difference between projection neurons and local circuit neurons is that the former must obtain their trophic support by neurotrophins signaling from a distant source by retrograde transport, while the latter, with only short axonal projections, are presumably not as dependent on retrograde signaling processes. Thus, p75 may have a specialized role in facilitating retrograde signaling. Several lines of evidence support this notion. The rat p75 antibody 192 has little effect on NGF responses of sensory or sympathetic neurons in tissue culture but devastates these neurons in vivo (Johnson et al 1989). Neurons exposed to NGF in tissue culture receive trophic signals from receptors on the cell body, so retrograde axonal signaling is not essential. However, sensory and sympathetic neurons in vivo receive their NGF exclusively at the axon termini and are thus entirely dependent on retrograde axonal signaling. The toxic effect of p75 antibody only in vivo is therefore consistent with a role of p75 specific to retrograde axonal signaling, as previously pointed out by Korsching (1993).

von Bartheld et al (1994) have provided additional evidence in support of this hypothesis. Neurons of the avian isthmo-optic (IO) nucleus, which innervate the retina, express p75, trkB, and trkC, but not trkA. BDNF, which is presumably derived from the retinal target in vivo, is trophic for IO neurons, both in vivo and in vitro. In contrast, NGF enhances developmental death of IO neurons in vivo but has no effect in vitro. NGF appears to cause its negative effect by competing with BDNF for association with p75, since NGF and antibody to p75 have equivalent ability to prevent retrograde transport of ^{125}I-BDNF and to enhance developmental death of IO neurons (von Bartheld et al 1994; CS von Bartheld, unpublished results). Again, the restriction of the toxic effect of NGF to the in vivo situation can be explained if p75 function is essential for effective retrograde axonal signaling but is not essential for signaling via receptors on the neuronal soma. If p75 function merely enhances the efficiency of retrograde signaling by trk receptors, then p75 function may be necessary only in neurons with relatively low levels of expression of trk receptors or with exceptionally nonabundant sources of neurotrophin.

CONCLUSIONS

We have much to learn about the complex interactions among the neurotrophins and their receptors. In this system, basic questions concerning cell biology have direct and immediate relevance for understanding biological function in the living animal. Although the complexity of the system is daunting from an experimental point of view, this same complexity suits the neurotrophin system to play a major role in regulating the development and adult function of the nervous system.

Literature Cited

Allendoerfer KL, Cabelli RJ, Escandon E, Kaplan DR, Nikolics K, Shatz CJ. 1994. Regulation of neurotrophin receptors during the maturation of the mammalian visual system. *J. Neurosci.* In press

Altin JG, Bradshaw RA. 1990. Production of 1,2-diacylglycerol in PC12 cells by nerve growth factor and basic fibroblast growth factor. *J. Neurochem.* 54:1666–76

Anton ES, Weskamp G, Reichardt LF, Matthew WD. 1994. Nerve growth factor and its low-affinity receptor promote Schwann cell migration. *Proc. Natl. Acad. Sci. USA* 91: 2795–99

Baldwin AN, Bitler CM, Welcher AA, Shooter EM. 1992. Studies on the structure and binding properties of the cysteine-rich domain of rat low affinity nerve growth factor receptor (p75NGFR). *J. Biol. Chem.* 267: 8352–59

Banner DW, D'Arcy A, Janes W, Gentz R, Schoenfeld H-J, et al. 1993. Crystal structure of the soluble human 55 kd TNF receptor-human TNFβ complex: implications for TNF receptor activation. *Cell* 73:431–45

Barbacid M. 1993. Nerve growth factor: a tale of two receptors. *Oncogene* 8:2033–42

Barker PA, Lomen-Hoerth C, Gensch EM, Meakin SO, Glass DJ, Shooter EM. 1993. Tissue-specific alternative splicing generates two isoforms of the trkA receptor. *J. Biol. Chem.* 268:15150–57

Barres BA, Raff MC, Gaese F, Bartke I, Dechant G, Barde Y-A. 1994. A crucial role for neurotrophin-3 in oligodendrocyte development. *Nature* 367:371–75

Battleman DS, Geller AI, Chao MV. 1993. HSV-1 vector-mediated gene transfer of the human nerve growth factor receptor p75hNGFR defines high-affinity NGF binding. *J. Neurosci.* 13:941–51

Belia S, Fulle S, Antonioli S, Salvatore AM, Fano G. 1991. NGF induces activation of phospholipase C (membrane bound) in PC12 cells. *Physiol. Chem. Phys. Med. NMR* 23: 35–41

Bellini F, Toffano G, Bruni A. 1988. Activation of phosphoinositide hydrolysis by nerve growth factor and lysophosphatidylserine in rat peritoneal mast cells. *Biochim. Biophys. Acta* 970:187–93

Benedetti M, Levi A, Chao MV. 1993. Differential expression of nerve growth factor receptors leads to altered binding affinity and neurotrophin responsiveness. *Proc. Natl. Acad. Sci. USA* 90:7859–63

Berg MM, Sternberg DW, Hempstead BL, Chao MV. 1991. The low-affinity p75 nerve growth factor (NGF) receptor mediates NGF-induced tyrosine phosphorylation. *Proc. Natl. Acad. Sci.* 88:7106–10

Berkemeier LR, Winslow JW, Kaplan DR, Nikolics K, Goeddel DV, Rosenthal A. 1991. Neurotrophin-5: a novel neurotrophic factor that activates trk and trkB. *Neuron* 7:857–66

Berninger B, Garcia DE, Inagaki N, Hahnel C, Lindholm D. 1993. BDNF and NT-3 induce intracellular Ca^{2+} elevation in hippocampal neurones. *Neuroreport* 4:1303–6

Birren SJ, Verdi JM, Anderson DJ. 1992. Membrane depolarization induces p140trk and NGF responsiveness, but not p75LNGFR, in MAH cells. *Science* 257:395–97

Bischoff SC, Dahinden CA. 1992. Effect of nerve growth factor on the release of inflammatory mediators by mature human basophils. *Blood* 79:2662–69

Bothwell M. 1991. Keeping track of neurotrophin receptors. *Cell* 65:915–18

Bothwell M, Shooter EM. 1977. Dissociation equilibrium constant of β-nerve growth factor. *J. Biol. Chem.* 252:8532–36

Bradshaw RA, Blundell TL, Lapatto R, McDonald NQ, Murray-Rust J. 1993. Nerve growth factor revisited. *Trends Biochem. Sci.* 18:48–52

Brodie C, Gelfand EW. 1992. Functional nerve growth factor receptors on human B lymphocytes. Interaction with IL-2. *J. Immunol.* 148: 3492–97

Brodie C, Sampson SR. 1990. Nerve growth factor and fibroblast growth factor influence post-fusion expression of Na-channels in cultured rat skeletal muscle. *J. Cell. Physiol.* 144:492–97

Buxser SE, Kelleher DJ, Watson L, Puma P, Johnson GL. 1983b. Change in state of nerve growth factor receptor. Modulation of receptor affinity by wheat germ agglutinin. *J. Biol. Chem.* 258:3741–49

Buxser SE, Watson L, Johnson GL. 1983a. A comparison of binding properties and structure of NGF receptor on PC12 pheochromocytoma and A875 melanoma cells. *J. Cell. Biochem.* 22:219–33

Chao MV. 1992. Growth factor signaling: Where is the specificity? *Cell* 68:995–97

Chao MV, Bothwell MA, Ross AH, Koprowski

H, Lanahan AA, et al. 1986. Gene transfer and molecular cloning of the human NGF receptor. *Science* 232:518–21

Chandler CE, Parsons LM, Hosang M, Shooter EM. 1984. A monoclonal antibody modulates the interaction of nerve growth factor with PC12 cells. *J. Biol. Chem.* 259:6882–89

Cohen S. 1960. Purification of a nerve-growth promoting protein from the mouse salivary gland and its neurocytotoxic antiserum. *Proc. Natl. Acad. Sci. USA* 46:302–11

Collazo D, Takahashi H, McKay RD. 1992. Cellular targets and trophic functions of neurotrophin-3 in the developing rat hippocampus. *Neuron* 9:643–56

Confort C, Charrasse S, Clos J. 1991. Nerve growth factor enhances DNA synthesis in cultured cerebellar neuroblasts. *Neuroreport* 2:566–68

Cordon-Cardo C, Tapley P, Jing SQ, Nanduri V, O'Rourke E, et al. 1991. The trk tyrosine protein kinase mediates the mitogenic properties of nerve growth factor and neurotrophin-3. *Cell* 66:173–83

D'Arcangelo G, Halegoua S. 1993. A branched signaling pathway for nerve growth fadctor is revealed by src-, ras-, and raf-mediated gene inductions. *Mol. Cell. Biol.* 13:3146–55

Davies AM, Bandtlow C, Heumann R, Korsching S, Rohrer H, Thoenen H. 1987. Timing and site of nerve growth factor synthesis in developing skin in relation to innervation and expression of the receptor. *Nature* 326:353–58

Davies AM, Lee K-F, Jaenisch R. 1993. p75-deficient trigeminal sensory neurons have an altered response to NGF but not to other neurotrophins. *Neuron* 11:565–74

Di Marco E, Cutuli N, Guerra L, Cancedda R, De Luca M. 1993a. Molecular cloning of trkE, a novel trk-related putative tyrosine kinase receptor isolated from normal human keratinocytes and widely expressed by normal human tissues. *J. Biol. Chem.* 268: 24290–95

Di Marco E, Mathor M, Bondanza S, Cutuli N, Marchisio PC, et al. 1993b. Nerve growth factor binds to normal human keratinocytes through high and low affinity receptors and stimulates their growth by a novel autocrine loop. *J. Biol. Chem.* 268:22838–46

DiStefano PS, Friedman B, Radziejewski C, Alexander C, Boland P, et al. 1992. The neurotrophins BDNF, NT-3, and NGF display distinct patterns of retrograde axonal transport in peripheral and central neurons. *Neuron* 8:983–93

Doubleday B, Robinson PP. 1992. The role of nerve growth factor in collateral reinnervation by cutaneous C-fibres in the rat. *Brain Res.* 593:179–84

Drinkwater CC, Barker PA, Suter U, Shooter EM. 1993. The carboxyl terminus of nerve growth factor is required for biological activity. *J. Biol. Chem.* 268:23202–7

Ehlers MD, Kaplan DR, Koliatsos VE. 1993. Retrograde transport of neurotrophin receptors. *Soc. Neurosci. Abstr.* 19:1299

Ehrhard PB, Ganter U, Bauer J, Otten U. 1993. Expression of functional trk protooncogene in human monocytes. *Proc. Natl. Acad. Sci. USA* 90:5423–27

Ernfors P, Ibáñez CF, Ebendal T, Olson L, Persson H. 1990. Molecular cloning and neurotrophic activities of a protein with structural similarities to nerve growth factor: developmental and topographical expression in the brain. *Proc. Natl. Acad. Sci. USA* 87:5454–58

Feinstein DL, Larhammar D. 1990. Identification of a conserved protein motif in a group of growth factor receptors. *FEBS Lett.* 272: 7–11

Frazier WA, Boyd LF, Bradshaw RA. 1974. Properties of the specific binding of ^{125}I-nerve growth factor to responsive peripheral neurons. *J. Biol. Chem.* 249:5513–19

Gee AP, Boyle MD, Munger KL, Lawman MJ, Young M. 1983. Nerve growth factor: stimulation of polymorphonuclear leukocyte chemotaxis in vitro. *Proc. Natl. Acad. Sci. USA* 80:7215–18

Gizang-Ginsberg E, Ziff EB. 1990. Nerve growth factor regulates tyrosine hydroxylase gene transcription through a nucleoprotein complex that contains c-Fos. *Genes Dev.* 4: 477–91

Glass DJ, Nye SH, Hantzopoulos P, Macchi MJ, Squinto SP, et al. 1991. TrkB mediates BDNF/NT-3-dependent survival and proliferation in fibroblasts lacking the low affinity NGF receptor. *Cell* 66:405–13

Gnahn H, Hefti F, Heumann R, Schwab ME, Thoenen H. 1983. NGF-mediated increase of choline acetyltransferase (ChAT) in the neonatal rat forebrain: evidence for a physiological role of NGF in the brain? *Brain Res.* 285:45–52

Godfrey EW, Shooter EM. 1986. Nerve growth factor receptors on chick embryo sympathetic ganglion cells: binding characteristics and development. *J. Neurosci.* 6:2543–50

Green SH, Greene LA. 1986. A single Mr approximately 103,000 ^{125}I-beta-nerve growth factor-affinity-labeled species represents both the low and high affinity forms of the nerve growth factor receptor. *J. Biol. Chem.* 261:15316–26

Greene LA, Seeley PJ, Rukenstein A, DiPiazza M, Howard A. 1984. Rapid activation of tyrosine hydroxylase in response to nerve growth factor. *J. Neurochem.* 42:1728–34

Grob PM, Berlot CH, Bothwell MA. 1983. Affinity labeling and partial purification of nerve growth factor receptors from rat pheochromocytoma and human melanoma cells. *Proc. Natl. Acad. Sci. USA* 80:6819–23

Grob PM, Bothwell MA. 1983. Modification of nerve growth factor receptor properties by wheat germ agglutinin. *J. Biol. Chem.* 258: 14136–43

Grob PM, Ross AH, Koprowski H, Bothwell M. 1985. Characterization of the human melanoma nerve growth factor receptor. *J. Biol. Chem.* 260:8044–49

Grohovaz F, Zacchetti D, Clementi E, Lorenzon P, Meldolesi J, Fumagalli G. 1991. [Ca^{2+}]i imaging in PC12 cells: multiple response patterns to receptor activation reveal new aspects of transmembrane signaling. *J. Cell Biol.* 113:1341–50

Hallböök F, Ibáñez CF, Persson H. 1991. Evolutionary studies of the nerve growth factor family reveal a novel member abundantly expressed in *Xenopus* ovary. *Neuron* 6:845–58

Hartman DS, McCormack M, Schubenel R, Hertel C. 1992. Multiple trkA proteins in PC12 cells bind NGF with a slow association rate. *J. Biol. Chem.* 267:24516–22

Hayashi I, Perez-Magallanes M, Rossi JM. 1992. Neurotrophic factor-like activity in *Drosophila. Biochem. Biophys. Res. Commun.* 184:73–79

Hefti F, Dravid A, Hartikka J. 1984. Chronic intraventricular injections of nerve growth factor elevate hippocampal choline acetyltransferase activity in adult rats with partial septo-hippocampal lesions. *Brain Res.* 293: 305–11

Hempstead BL, Martin-Zanca D, Kaplan DR, Parada LF, Chao MV. 1991. High-affinity NGF binding requires coexpression of the trk proto-oncogene and the low-affinity NGF receptor. *Nature* 350:678–83

Hempstead BL, Rabin SJ, Kaplan L, Reid S, Parada LF, Kaplan DR. 1992. Overexpression of the trk tyrosine kinase rapidly accelerates nerve growth factor-induced differentiation. *Neuron* 9:883–96

Hempstead BL, Schleifer LS, Chao MV. 1989. Expression of functional nerve growth factor receptors after gene transfer. *Science* 243: 373–75

Heuer JG, Fatemie-Nainie S, Wheeler EF, Bothwell M. 1990. Structure and developmental expression of the chicken NGF receptor. *Dev. Biol.* 137:287–304

Hohn A, Leibrock J, Bailey K, Barde YA. 1990. Identification and characterization of a novel member of the nerve growth factor/brain-derived neurotrophic factor family. *Nature* 344: 339–41

Horigome K, Pryor JC, Bullock ED, Johnson EM Jr. 1993. Mediator release from mast cells by nerve growth factor. Neurotrophin specificity and receptor mediation. *J. Biol. Chem.* 268:14881–87

Hosang M, Shooter EM. 1985. Molecular characteristics of nerve growth factor receptors on PC12 cells. *J. Biol. Chem.* 260:655–62

Ibáñez CF, Ebendal T, Barbany G, Murray-Rust J, Blundell TL, Persson H. 1992. Disruption of the low affinity receptor-binding site in NGF allows neuronal survival and differentiation by binding to the trk gene product. *Cell* 69:329–41

Ibáñez CF, Ebendal T, Persson H. 1991. Chimeric molecules with multiple neurotrophic activities reveal structural elements determining the specificities of NGF and BDNF. *EMBO J.* 10:2105–10

Ibáñez CF, Hallböök F, Ebendal T, Persson H. 1990. Structure-function studies of nerve growth factor: functional importance of highly conserved amino acid residues. *EMBO J.* 9:1477–83

Ibáñez CF, Ilag LL, Murray-Rust J, Persson H. 1993. An extended surface of binding of Trk tyrosine kinase receptors in NGF and BDNF allows the engineering of a multifunctional pan-neurotrophin. *EMBO J.* 12:2281–93

Ip NY, Ibáñez CF, Nye SH, McClain J, Jones PF, et al. 1992. Mammalian neurotrophin-4: structure, chromosomal localization, tissue distribution, and receptor specificity. *Proc. Natl. Acad. Sci. USA* 89:3060–64

Jennings CG, Dyer SM, Burden SJ. 1993. Muscle-specific trk-related receptor with a kringle domain defines a distinct class of receptor tyrosine kinases. *Proc. Natl. Acad. Sci. USA* 90:2895–99

Jing S, Tapley P, Barbacid M. 1992. Nerve growth factor mediates signal transduction through trk homodimer receptors. *Neuron* 9: 1067–79

Johnson D, Lanahan A, Buck CR, Sehgal A, Morgan C, et al. 1986. Expression and structure of the human NGF receptor. *Cell* 47:545–54

Johnson EM Jr, Osborne PA, Taniuchi M. 1989. Destruction of sympathetic and sensory neurons in the developing rat by a monoclonal antibody against the nerve growth factor (NGF) receptor. *Brain Res.* 478:166–70

Johnson EM Jr, Taniuchi M, Clark HB, Springer JE, Koh S, et al. 1987. Demonstration of the retrograde transport of nerve growth factor receptor in the peripheral and central nervous system. *J. Neurosci.* 7:923–29

Jones KR, Reichardt LF. 1990. Molecular cloning of a human gene that is a member of the nerve growth factor family. *Proc. Natl. Acad. Sci. USA* 87:8060–64

Kahle P, Burton LE, Schmelzer CH, Hertel C. 1992. The amino terminus of nerve growth factor is involved in the interaction with the receptor tyrosine kinase p140trkA. *J. Biol. Chem.* 267:22707–10

Kaisho Y, Yoshimura K, Nakahama K. 1990.

Cloning and expression of a cDNA encoding a novel human neurotrophic factor. *FEBS Lett.* 266:187–91

Kalcheim C, Carmeli C, Rosenthal A. 1992. Neurotrophin 3 is a mitogen for cultured neural crest cells. *Proc. Natl. Acad. Sci. USA* 89:1661–65

Kalcheim C, Gendreau M. 1988. Brain-derived neurotrophic factor stimulates survival and neuronal differentiation in cultured avian neural crest. *Brain Res.* 469:79–86

Kannan Y, Ushio H, Koyama H, Okada M, Oikawa M, et al. 1991. 2.5S nerve growth factor enhances survival, phagocytosis, and superoxide production of murine neutrophils. *Blood* 77:1320–25

Kaplan DR, Hempstead BL, Martin-Zanca D, Chao MV, Parada LF. 1991a. The trk proto-oncogene product: a signal transducing receptor for nerve growth factor. *Science* 252: 554–56

Kaplan DR, Martin-Zanca D, Parada LF. 1991b. Tyrosine phosphorylation and tyrosine kinase activity of the trk proto-oncogene product induced by NGF. *Nature* 350:158–60

Kessler JA. 1985. Parasympathetic, sympathetic, and sensory interactions in the iris: nerve growth factor regulates cholinergic ciliary ganglion innervation in vivo. *J. Neurosci.* 5:2719–25

Kimata H, Yoshida A, Ishioka C, Mikawa H. 1991. Stimulation of Ig production and growth of human lymphoblastoid B-cell lines by nerve growth factor. *Immunology* 72:451–52

Klein R, Conway D, Parada LF, Barbacid M. 1990. The trkB tyrosine protein kinase gene codes for a second neurogenic receptor that lacks the catalytic kinase domain. *Cell* 61: 647–56

Klein R, Jing SQ, Nanduri V, O'Rourke E, Barbacid M. 1991a. The trk proto-oncogene encodes a receptor for nerve growth factor. *Cell* 65:189–97

Klein R, Nanduri V, Jing SA, Lamballe F, Tapley P, Bryant S, Cordon-Cardo C, Jones KR, Reichardt LF, Barbacid M. 1991b. The trkB tyrosine protein kinase is a receptor for brain-derived neurotrophic factor and neurotrophin-3. *Cell* 66:395–403

Klein R, Parada LF, Coulier F, Barbacid M. 1989. trkB, a novel tyrosine protein kinase receptor expressed during mouse neural development. *EMBO J.* 8:3701–9

Klein R, Smeyne RJ, Wurst W, Long LK, Auerbach BA, et al. 1993. Targeted disruption of the trkB neurotrophin receptor gene results in nervous system lesions and neonatal death. *Cell* 75:113–22

Knüsel B, Hefti F. 1992. K-252 compounds: modulators of neurotrophin signal transduction. *J. Neurochem.* 59:1987–96

Knüsel B, Rabin S, Hefti F, Kaplan DR. 1994. Regulated neurotrophin receptor responsiveness during neuronal migration and early differentiation. *J. Neurosci.* In press

Korsching S. 1993. The neurotrophic factor concept: a reexamination. *J. Neurosci.* 13: 2739–48

Kouchalakos RN, Bradshaw RA. 1986. Nerve growth factor receptor from rabbit sympathetic ganglia membranes. Relationship between subforms. *J. Biol. Chem.* 261: 16054–59

Lamballe F, Klein R, Barbacid M. 1991. trkC, a new member of the trk family of tyrosine protein kinases, is a receptor for neurotrophin-3. *Cell* 66:967–79

Large TH, Weskamp G, Helder JC, Radeke MJ, Misko TP, et al. 1989. Structure and developmental expression of the nerve growth factor receptor in the chicken central nervous system. *Neuron* 2:1123–34

Lee KF, Bachman K, Landis S, Jaenisch R. 1994. Dependence on p75 for innervation of some sympathetic targets. *Science* 263: 1447–49

Lee KF, Li E, Huber LJ, Landis SC, Sharpe AH, et al. 1992. Targeted mutation of the gene encoding the low affinity NGF receptor p75 leads to deficits in the peripheral sensory nervous system. *Cell* 69:737–49

Leibrock J, Lottspeich F, Hohn A, Hofer M, Hengerer B, et al. 1989. Molecular cloning and expression of brain-derived neurotrophic factor. *Nature* 341:149–52

Levi-Montalcini L. 1951. Selective growth-stimulating effects of mouse sarcomas on the sensory and sympathetic nervous system of chick embryos. *J. Exp. Zool.* 116:321–62

Levi-Montalcini R. 1987. The nerve growth factor: thirty-five years later. *EMBO J* 6: 2856–67

Lillien LE, Claude P. 1985. Nerve growth factor is a mitogen for cultured chromaffin cells. *Nature* 317:632–34

Lindsay RM, Harmar AJ. 1989. Nerve growth factor regulates expression of neuropeptide genes in adult sensory neurons. *Nature* 337: 362–64

Loeb DM, Tsao H, Cobb MH, Greene LA. 1992. NGF and other growth factors induce an association between ERK1 and the NGF receptor, gp140prototrk. *Neuron* 9:1053–65

Lohof AM, Ip NY, Poo MM. 1993. Potentiation of developing neuromuscular synapses by the neurotrophins NT-3 and BDNF. *Nature* 363: 350–53

Loy R, Poluha DK, Ross AH. 1993. Bidirectional axonal transport of p140trk high affinity NGF receptor. *Soc. Neurosci. Abstr.* 19: 1295

Maisonpierre PC, Belluscio L, Squinto S, Ip NY, Furth ME, et al. 1990. Neurotrophin-3:

a neurotrophic factor related to NGF and BDNF. *Science* 247:1446–51

Marano N, Dietzschold B, Earley JJ Jr, Schatteman G, Thompson S, et al. 1987. Purification and amino terminal sequencing of human melanoma nerve growth factor receptor. *J. Neurochem.* 48:225–32

Marsh HN, Scholz WK, Lamballe F, Klein R, Nanduri V, 1993. Signal transduction events mediated by the BDNF receptor gp145trkB in primary hippocampal pyramidal cultures. *J. Neurosci.* 13:4281–92

Martin-Zanca D, Barbacid M, Parada LF. 1990. Expression of the trk proto-oncogene is restricted to the sensory cranial and spinal ganglia of neural crest origin in mouse development. *Genes Dev.* 4:683–94

Massagué J, Guillette BJ, Czech MP, Morgan CJ, Bradshaw RA. 1981. Identification of a nerve growth factor receptor protein in sympathetic ganglia membranes by affinity labeling. *J. Biol. Chem.* 256:9419–24

Matsuda H, Kannan Y, Ushio H, Kiso Y, Kanemoto T, et al. 1991. Nerve growth factor induces development of connective tissue-type mast cells in vitro from murine bone marrow cells. *J. Exp. Med.* 174:7–14

Matsuda H, Switzer J, Coughlin MD, Bienenstock J, Denburg JA. 1988. Human basophilic cell differentiation promoted by 2.5S nerve growth factor. *Int. Arch. Allergy Appl. Immunol.* 86:453–57

Matsushima H, Bogenmann E. 1990. Nerve growth factor (NGF) induces neuronal differentiation in neuroblastoma cells transfected with the NGF receptor cDNA. *Mol. Cell. Biol.* 10:5015–20

McDonald NQ, Lapatto R, Murray-Rust J, Gunning J, Wlodawer A, Blundell TL. 1991. New protein fold revealed by a 2.3-Å resolution crystal structure of nerve growth factor. *Nature* 354:411–14

McNulty JA, Fox LM, Silberman S. 1993. Immunocytochemical demonstration of nerve growth factor (NGF) receptor in the pineal gland: effect of NGF on pinealocyte neurite formation. *Brain Res.* 610:108–14

Meakin SO, Shooter EM. 1991. Molecular investigations on the high-affinity nerve growth factor receptor. *Neuron* 6:153–63

Meakin SO, Shooter EM. 1992. The nerve growth factor family of receptors. *Trends Neurosci.* 15:323–31

Meakin SO, Suter U, Drinkwater CC, Welcher AA, Shooter EM. 1992. The rat trk protooncogene product exhibits properties characteristic of the slow nerve growth factor receptor. *Proc. Natl. Acad. Sci. USA* 89:2374–78

Middlemas DS, Lindberg RA, Hunter T. 1991. trkB, a neural receptor protein-tyrosine kinase evidence for a full-length and two truncated receptors. *Mol. Cell. Biol.* 11:143–53

Mobley WC, Rutkowski JL, Tennekoon GI, Buchanan K, Johnston MV. 1985. Choline acetyltransferase activity in striatum of neonatal rats increased by nerve growth factor. *Science* 229:284–87

Monod J, Wyman J, Changeaux J-P. 1965. On the nature of allosteric transitions: a plausible model. *J. Mol. Biol.* 12:88–118

Nanduri J, Vroegtop SM, Buxser SE, Neet KE. 1994. Immunological determinants of nerve growth factor (NGF) involved in p140trk (Trk) receptor binding. *J. Neurosci. Res.* 37: 420–31

Narhi LO, Rosenfeld R, Talvenheimo J, Prestrelski SJ, Arakawa T, et al. 1993. Comparison of the biophysical characteristics of human brain-derived neurotrophic factor, neurotrophin-3, and nerve growth factor. *J. Biol. Chem.* 268:13309–17

Nichols RA, Chandler CE, Shooter EM. 1989. Enucleation of the rat pheochromocytoma clonal cell line, PC12: effect on neurite outgrowth. *J. Cell. Physiol.* 141:301–9

Ohmichi M, Decker SJ, Pang L, Saltiel AR. 1991. Nerve growth factor binds to the 140 kd trk proto-oncogene product and stimulates its association with the src homology domain of phospholipase Cg 1. *Biochem. Biophys. Res. Commun.* 179:217–23

Okazawa H, Kamei M, Kanazawa I. 1993. Molecular cloning and expression of a novel truncated form of chicken trkC. *FEBS Lett.* 329:171–77

Otten U, Ehrhard P, Peck R. 1989. Nerve growth factor induces growth and differentiation of human B lymphocytes. *Proc. Natl. Acad. Sci. USA* 86:10059–63

Palmer MR, Eriksdotter-Nilsson M, Henschen A, Ebendal T, Olson L. 1993. Nerve growth factor-induced excitation of selected neurons in the brain which is blocked by a low-affinity receptor antibody. *Exp. Brain. Res.* 93: 226–30

Pandiella-Alonso A, Malgaroli A, Vicentini LM, Meldolesi J. 1986. Early rise of cytosolic Ca^{2+} induced by NGF in PC12 and chromaffin cells. *FEBS Lett.* 208:48–51

Pulido D, Campuzano S, Koda T, Modolell J, Barbacid M. 1992. Dtrk, a *Drosophila* gene related to the trk family of neurotrophin receptors, encodes a novel class of neural cell adhesion molecule. *EMBO J.* 11:391–404

Puma P, Buxser SE, Watson L, Kelleher DJ, Johnson GL. 1983. Purification of the receptor for nerve growth factor from A875 melanoma cells by affinity chromatography. *J. Biol. Chem.* 258:3370–75

Rabiadeh S, Oh J, Zhong L, Yang J, Bitler CM, Butcher LL, Bredesen DE. 1993. Induction of apoptosis by the low-affinity NGF receptor. *Science* 261:345–48

Radeke MJ, Misko TP, Hsu C, Herzenberg LA, Shooter EM. 1987. Gene transfer and molecular cloning of the rat nerve growth factor receptor. *Nature* 325:593–97

Radziejewski C, Robinson RC, DiStefano PS, Taylor JW. 1992. Dimeric structure and conformational stability of brain-derived neurotrophic factor and neurotrophin-3. *Biochemistry* 31:4431–36

Ridgway RL, Syed NI, Lukowiak K, Bulloch AG. 1991. Nerve growth factor (NGF) induces sprouting of specific neurons of the snail, *Lymnaea stagnalis*. *J. Neurobiol.* 22: 377–90

Ritter AM, Mendell LM. 1992. Somal membrane properties of physiologically identified sensory neurons in the rat: effects of nerve growth factor. *J. Neurophysiol.* 68:2033–41

Rodriguez-Tébar A, Dechant G, Barde YA. 1990. Binding of brain-derived neurotrophic factor to the nerve growth factor receptor. *Neuron* 4:487–92

Rodriguez-Tébar A, Dechant G, Gotz R, Barde YA. 1992. Binding of neurotrophin-3 to its neuronal receptors and interactions with nerve growth factor and brain-derived neurotrophic factor. *EMBO J.* 11:917–22

Rosenthal A, Goeddel DV, Nguyen T, Lewis M, Shih A, et al. 1990. Primary structure and biological activity of a novel human neurotrophic factor. *Neuron* 4:767–73

Ross AH, Daou M-C, Kaplan DR, Lachyankar MB, McKinnon CA, et al. 1993. TrkA interacts with low-affinity NGF receptor and regulates neural differentiation. *Abstr. Soc. Neurosci.* 19:1476

Rozakis-Adcock M, Fernley R, Wade J, Pawson T, Bowtell D. 1993. The SH2 and SH3 domains of mammalian Grb2 couple the EGF receptor to the Ras activator mSos1. *Nature* 363:83–85

Schechter AL, Bothwell MA. 1981. Nerve growth factor receptors on PC12 cells: evidence for two receptor classes with differing cytoskeletal association. *Cell* 24:867–74

Schneider R, Schweiger M. 1991. A novel mosaic of cell aldhesion motifs in the extracellular domains of the neurogenic trk and trkB tyrosine kinase receptors. *Oncogene* 6:1807–11

Sehgal A, Wall DA, Chao MV. 1988. Efficient processing and expression of human nerve growth factor receptors in *Xenopus laevis* oocytes: effects on maturation. *Mol. Cell. Biol.* 8:2242–46

Sieber-Blum M. 1991. Role of the neurotrophic factors BDNF and NGF in the commitment of pluripotent neural crest cells. *Neuron* 6: 949–55

Snider WD. 1988. Nerve growth factor enhances dendritic arborization of sympathetic ganglion cells in developing mammals. *J. Neurosci.* 8:2628–34

Soppet D, Escandon E, Maragos J, Middlemas DS, Reid SW, et al. 1991. The neurotrophic

factors brain-derived neurotrophic factor and neurotrophin-3 are ligands for the trkB tyrosine kinase receptor. *Cell* 65:895–903

Spoerri PE, Romanello S, Petrelli L, Negro A, Guidolin D, Skaper SD. 1993. Neurotrophin-3 upregulates NGF receptors in a central nervous system glial cell line. *Neuroreport* 4: 33–36

Squinto SP, Stitt TN, Aldrich TH, Davis S, Bianco SM, et al. 1991. trkB encodes a functional receptor for brain-derived neurotrophic factor and neurotrophin-3 but not nerve growth factor. NT3 does not act on PC12. *Cell* 65:885–93

Steininger TL, Wainer BH, Klein R, Barbacid M, Palfrey HC. 1993. High-affinity nerve growth factor receptor (Trk) immunoreactivity is localized in cholinergic neurons of the basal forebrain and striatum in the adult rat brain. *Brain Res.* 612:330–35

Suda T, Takahashi T, Golstein P, Nagata S. 1993. Molecular cloning and expression of the Fas ligand, a novel member of the tumor necrosis factor family. *Cell* 75:1169–78

Suen KL, Bustelo XR, Pawson T, Barbacid M. 1993. Molecular cloning of the mouse grb2 gene: differential interaction of the Grb2 adaptor protein with epidermal growth factor and nerve growth factor receptors. *Mol. Cell. Biol.* 13:5500–12

Suter U, Angst C, Tien CL, Drinkwater CC, Lindsay RM, Shooter EM. 1992. NGF/BDNF chimeric proteins: analysis of neurotrophin specificity by homolog-scanning mutagenesis. *J. Neurosci.* 12:306–18

Sutter A, Riopelle RJ, Harris-Warrick RM, Shooter EM. 1979. Nerve growth factor receptors. Characterization of two distinct classes of binding sites on chick embryo sensory ganglia cells. *J. Biol. Chem.* 254:5972–82

Tanaka H, Agata A, Ohata K. 1989. A new membrane antigen revealed by monoclonal antibodies is associated with motoneuron axonal pathways. *Dev. Biol.* 132:419–35

Tancredi V, D'Arcangelo G, Mercanti D, Calissano P. 1993. Nerve growth factor inhibits the expression of long-term potentiation in hippocampal slices. *Neuroreport* 4: 147–50

Taniuchi M, Clark HB, Johnson EM, Jr. 1986a. Induction of nerve growth factor receptor in Schwann cells after axotomy. *Proc. Natl. Acad. Sci. USA* 83:4094–98

Taniuchi M, Johnson EM Jr, Roach PJ, Lawrence JC Jr. 1986b. Phosphorylation of nerve growth factor receptor proteins in sympathetic neurons and PC12 cells. In vitro phosphorylation by the cAMP-independent protein kinase FA/GSK-3. *J. Biol. Chem.* 261: 13342–49

Thorpe LW, Perez-Polo JR. 1987. The influence of nerve growth factor on the in vitro proliferative response of rat spleen lympho-cytes. *J. Neurosci. Res.* 18:134–39

Tsoulfas P, Soppet D, Escandon E, Tessarollo L, Mendoza-Ramirez JL, et al. 1993. The rat trkC locus encodes multiple neurogenic receptors that exhibit differential response to neurotrophin-3 in PC12 cells. *Neuron* 10: 975–90

Ueyama T, Hano T, Hamada M, Nishio I, Masuyama Y. 1991. New role of nerve growth factor—an inhibitory neuromodulator of adrenergic transmission. *Brain Res.* 559:293–96

Ullrich A, Schlessinger J. 1990. Signal transduction by receptors with tyrosine kinase activity. *Cell* 61:203–12

Vale RD, Shooter EM. 1983. Conversion of nerve growth factor-receptor complexes to a slowly dissociating, Triton X-100 insoluble state by anti nerve growth factor antibodies. *Biochemistry* 22:5022–28

Valenzuela DM, Maisonpierre PC, Glass DJ, Rojas E, Nunez L, et al. 1993. Alternative forms of rat TrkC with different functional capabilities. *Neuron* 10:963–74

van-Calker D, Takahata K, Heumann R. 1989. Nerve growth factor potentiates the hormone-stimulated intracellular accumulation of inositol phosphates and Ca^{2+} in rat PC12 pheochromocytoma cells: comparison with the effect of epidermal growth factor. *J. Neurochem.* 52:38–45

Venkatakrishnan G, McKinnon CA, Ross AH, Wolf DE. 1990. Lateral diffusion of nerve growth factor receptor: modulation by ligand-binding and cell-associated factors. *Cell. Regul.* 1:605–14

Verdi JM, Birren SJ, Ibáñez CF, Persson H, Kaplan DR, et al. 1994. p75LNGFR regulates Trk signal transduction and NGF-induced neuronal differentiation in MAH cells. *Neuron* 12:733–45

Verge VM, Merlio JP, Grondin J, Ernfors P, Persson H, et al. 1992. Colocalization of NGF binding sites, trk mRNA, and low-affinity NGF receptor mRNA in primary sensory neurons: responses to injury and infusion of NGF. *J. Neurosci.* 12:4011–22

Vetter ML, Martin-Zanca D, Parada LF, Bishop JM, Kaplan DR. 1991. Nerve growth factor rapidly stimulates tyrosine phosphorylation of phospholipase C-γl by a kinase activity associated with the product of the trk protooncogene. *Proc. Natl. Acad. Sci. USA* 88: 5650–54

Volonté, C, Angelastro JM, Greene LA. 1993a. Association of protein kinases ERK1 and ERK2 with p75 nerve growth factor receptors. *J. Biol. Chem.* 268:21240–45

Volonté C, Loeb DM, Greene LA. 1993b. A purine analog-sensitive protein kinase activity associates with Trk nerve growth factor receptors. *J. Neurochem.* 61:664–72

Volonté, C, Ross AH, Greene LA. 1993c. Association of a purine-analogue-sensitive protein kinase activity with p75 nerve growth factor receptors. *Mol. Biol. Cell* 4:71–78

von Bartheld CS, Kinoshita Y, Prevette D, Yin Q-W, Oppenheim RW, Bothwell M. 1994. Positive and negative effects of neurotrophins on the isthmo-optic nucleus in chicken embryos. *Neuron* 12:639–54

Welcher AA, Bitler CM, Radeke MJ, Shooter EM. 1991. Nerve growth factor binding domain of the nerve growth factor receptor. *Proc. Natl. Acad. Sci. USA* 88:159–63

Weskamp G, Reichardt LF. 1991. Evidence that biological activity of NGF is mediated through a novel subclass of high affinity receptors. *Neuron* 6:649–63

Widmer HR, Kaplan DR, Rabin SJ, Beck KD, Hefti F, Knüsel B. 1993. Rapid phosphorylation of phospholipase Cγl by brain-derived neurotrophic factor and neurotrophin-3 in cultures of embryonic rat cortical neurons. *J. Neurochem.* 60:2111–23

Wilson C, Goberdhan DC, Steller H. 1993. Dror, a potential neurotrophic receptor gene, encodes a *Drosophila* homolog of the vertebrate Ror family of Trk-related receptor tyrosine kinases. *Proc. Natl. Acad. Sci. USA* 90:7109–13

Wood JN, Lillycrop KA, Dent CL, Ninkina NN, Beech MM, et al. 1992. Regulation of expression of the neuronal POU protein Oct-2 by nerve growth factor. *J. Biol. Chem.* 267: 17787–91

Woost PG, Jumblatt MM, Eiferman RA, Schultz GS. 1992. Growth factors and corneal endothelial cells: I. Stimulation of bovine corneal endothelial cell DNA synthesis by defined growth factors. *Cornea* 11:1–10

Yamashita A, Kawana A. 1991. Nerve growth factor-induced intracellular calcium ion release in chick dorsal root ganglion neurons. *Neurosci. Lett.* 128:147–49

Yan H, Chao MV. 1991. Disruption of cysteine-rich repeats of the p75 nerve growth factor receptor leads to loss of ligand binding. *J. Biol. Chem.* 266:12099–104

Yan H, Schlessinger J, Chao MV. 1991. Chimeric NGF-EGF receptors define domains responsible for neuronal differentiation. *Science* 252:561–63

Annu. Rev. Neurosci. 1995. 18:255–81

NEUROMORPHIC ANALOGUE VLSI

Rodney Douglas[1,2]*, Misha Mahowald*[1]*, and Carver Mead*[2]

[1]MRC Anatomical Neuropharmacology Unit, Oxford OX1 3TH, England and
[2]Computation and Neural Systems Program, California Institute of Technology, Pasadena, California 91125

KEYWORDS: CMOS, computation, model, network, neuron

INTRODUCTION

Neuromorphic systems emulate the organization and function of nervous systems. They are usually composed of analogue electronic circuits that are fabricated in the complementary metal-oxide-semiconductor (CMOS) medium using very large-scale integration (VLSI) technology. However, these neuromorphic systems are not another kind of digital computer in which abstract neural networks are simulated symbolically in terms of their mathematical behavior. Instead, they directly embody, in the physics of their CMOS circuits, analogues of the physical processes that underlie the computations of neural systems. The significance of neuromorphic systems is that they offer a method of exploring neural computation in a medium whose physical behavior is analogous to that of biological nervous systems and that operates in real time irrespective of size. The implications of this approach are both scientific and practical. The study of neuromorphic systems provides a bridge between levels of understanding. For example, it provides a link between the physical processes of neurons and their computational significance. In addition, the synthesis of neuromorphic systems transposes our knowledge of neuroscience into practical devices that can interact directly with the real world in the same way that biological nervous systems do.

NEURAL COMPUTATION

The enormous success of general purpose digital computation has led to the application of digital design principles to the explanation of neural computa-

255

0147-006X/95/0301-0255$05.00

tion. Direct comparisons of operations at the hardware (Koch & Poggio 1987, McCulloch & Pitts 1943, von Neumann 1958) and algorithmic (Marr 1982) levels have encouraged the belief that the behavior of nervous systems will, finally, be adequately described within the paradigm of digital computation. That view is changing. There is growing recognition that the principles of neural computation are fundamentally different from those of general purpose digital computers (Hertz et al 1991, Hopfield 1982, Rumelhart & McClelland 1986). They resemble cooperative phemomena (Haken 1978, Julesz 1971, Marr & Poggio 1976) more closely than theorem proving (McCulloch & Pitts 1943, Newell et al 1958, Winston & Shellard 1990). The differences between the two kinds of computation are most apparent when their performance is compared in the real world, rather than in a purely symbolic one. General purpose digital approaches excel in a symbolic world, where the algorithms can be specified absolutely and the symbols of the computation can be assigned unambiguously. The real world is much less rigid. There the solution must emerge from a complex network of factors all operating simultaneously with various strengths, and the symbols must be extracted from input data that are unreliable in any particular but meaningful overall. Under these fluid conditions neural computations are vastly more effective than general purpose digital methods.

The reasons for the superior performance of nervous systems in the real world are not completely understood. One possibility rests in their different strategies for obtaining precision in the face of the noise inherent in their own components. Digital systems eliminate noise at the lowest level, by fully restoring each bit to a value of zero or one at every step of the computation. This representation is the foundation of Boolean logic. Consistent with this approach, digital architectures rely on each bit in the computation to be correct, and a fault in a single transistor can bring the entire computation to a halt. In contrast, neural systems use unrestored analogue quantities at the base level of their computations, which proceed according to elementary physical laws. Noise is eliminated from these analogue signals by feedback from the next level in the computation, where many signals have had an opportunity to combine and achieve precision at a collective rather than individual level. This collective signal is then used to optimize the performance of individual components by adaptation. The same architectures and adaptive processes that neural systems use to generate coherent action in the presence of imperfect components may also enable them to extract precise information from a noisy and ambiguous environment.

The fundamental difference in these computational primitives is reflected in the way that general purpose digital computers simulate neural models (Figure 1). The binary representation requires that logical functions form the foundation of the symbols and operations needed to model the biological

a.

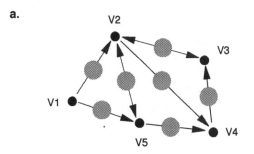

b.

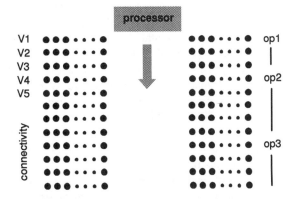

Figure 1 Comparison of the organization of a model on an analogue and on a general purpose digital simulator. (*a*) The analogue in this example represents five state variable nodes (*black filled circles*) of the modeled system. The variables at these nodes are continuous and can encode many bits of information. The variables interact along the paths indicated by arrows. The operations (*gray filled circles*) that occur along the paths are physical processes analogous to those of the modeled system (see Figure 5). The computations of all the state variables proceed in parallel, synchronized by their interactions and mutual dependence on time. (*b*) The digital system comprises many isolated nodes (*black filled circles*), each of which has a binary state, which represents only one bit. The multibit state variables of the modeled system are encoded on group nodes shown to the right of each variable (V1–V5). The connectivity of the modeled system is similarly encoded, as are numerical abstractions of the operations (op1–op*n*) performed along each path. This binary encoding scheme requires many more nodes than does the analogue implementation. In the digital system the interaction between nodes is not continuous but sampled. At each simulated time step, the digital processor evaluates the operations along each of the paths and updates the variables at each node. In the digital case, time is not a physical characteristic of the computation. Instead, it is an abstract variable similar to V1–V5 and must be updated by the processor after the completion of each time step.

system. It follows that the modeled system must be translated explicitly into a mathematical form. This representation is not efficient in terms of the time and number of devices required to perform the computation. First, the state variables and parameters of the model are encoded as abstract binary numbers, which occupy a large area on the surface of a chip relative to a single analogue variable. Even time must be abstracted so that the natural temporal relationships between processes are not guaranteed. Because the primitives are not natural physical processes, every relationship between the system variables must be explicitly defined. Even the most natural properties, such as statistical variation and noise, must be specified.

The efficiency of neuromorphic analogue VLSI (aVLSI) rests in the power of analogy, the isomorphism between physical processes occurring in different media (Figure 1). In general, neural computational primitives such as conservation of charge, amplification, exponentiation, thresholding, compression, and integration arise naturally out of the physical processes of aVLSI circuits. Calculation of these functions using digital methods would require many amplifiers for storage and many digital clock cycles to execute the algorithms that generate the functions (Hopfield 1990). Thus, the computational advantage of aVLSI and neural systems is due to the spatially dense, essentially parallel and real-time nature of analogue circuits (Hopfield 1990, Mead 1989). Of course, only a small number of functions can be generated directly by the physics of the analogue circuits. The analogue approach scores where those functions are the ones required, and where only modest precision ($1:10$–$1:10^3$) is necessary. Fortunately, the functions and precision of aVLSI circuits have proven effective for emulating neural phenomena.

At present, both digital and analogue technologies are orders of magnitude less powerful than is necessary to embody the complete nervous system of any animal, even if we knew how. The problem at this time is not one of complete understanding but of how best to embody the principles of simpler neural subsystems. The choice between digital simulation and analogue emulation depends on a number of factors, such as ease of implementation and the goal of the experiment. Digital simulation is precise, the results are easily stored and displayed, and the configuration of the simulation is as easy as recompiling a program. Digital simulation is effective in many applications where the size of the machine, its power consumption, and its time of response are not important. Analogue emulation is an attractive alternative to simulation if the performance of the system in simulation would be so slow that it is impractical, and as in the case of prosthetic devices, the size and power-consumption of the device are critical. Analogue VLSI emulations at present require specialized skills on the part of the designer. This requirement may be less of an obstacle in the future as the potential of aVLSI systems is developed.

EARLY SENSORY PROCESSING

The extraction of behaviorally relevant signals from sensory data is a natural starting point for neuromorphic aVLSI design. Sensory systems have many similar elements laid out in regular spatial arrays of one or two dimensions, with a high proportion of local interactions between elements. They must provide noise-resistant, alias-free operation over a wide range of stimulus intensities by using physical elements of finite dynamic range. The aVLSI medium is well suited to the synthesis of such systems. By creating neuromorphic sensory systems, we build a concrete ground that can be used to explore the interaction of medium and computation in both biological and silicon systems. We also create processors that can provide input to behaviorally functional neuromorphic systems and that could be developed as prosthetic input devices for humans.

Sensors are the critical link to the world. If the source data about the world are not properly captured and transmitted to later processing stages, no amount of computation can recover them. One approach to this problem is to collect and transmit as much raw data as possible. This strategy, which is adopted in traditional artificial sensors such as charge-coupled device (CCD) cameras, makes strong demands on the dynamic range of the sensors and the communication bandwidth of the transmission channel to the later processing stages. For example, many natural visual scenes have dynamic ranges of three orders of magnitude during steady illumination, and the dynamic range of the same scene viewed from sunny afternoon to dusk may vary by seven orders of magnitude. Not only do these ranges exceed the limits of current digital imaging technology, but the torrent of low-level data generated by simple gray-level imagers cannot be analyzed in real-time even by multi-processor general purpose computers (Scribner et al 1993).

Biological systems have evolved an alternative approach to sensing. Rather than transmitting all possible information to the brain, they extract only that which is salient to the later processing stages. One of the advantages of computing high-order invariants close to the sensors is that the bandwith of communication to subsequent processors is reduced to a minimum. Only salient data are transmitted to the next stage of processing. The degree of invariance achieved by early sensory processing is a compromise that is made differently by various modalities and species. For example, the space available for additional circuitry restricts the sophistication of signal processing that can be accomplished near to an array of sensors such as the retina. Also, invariance of response implies that sensory information has been discarded. The question of what to discard cannot be context sensitive if invariants are hard wired into early sensory processing. Understanding the trade-offs made by biological sensors requires a knowledge of a broad range of constraining factors, from the bandwidth of the individual computational elements to the behavior of the organism as a whole.

Vision

Early visual processing is a popular topic for research, and a number of algorithms have been described for detecting various features of the visual scene. The huge flux of information in algorithmic visual processing has, until recently, restricted these computations to high-performance digital computers. But mass-market applications are now driving the search for implementations of these algorithms on compact, low-power chips (Koch & Li 1994). In general, these VLSI circuits aim to capture the mathematics of vision, rather than emulate the biological methods of visual processing.

One of the earliest attempts to use aVLSI technology to emulate the principles of biological sensory processing outlined above was an electronic retina that emulates the outer plexiform layer of the mud puppy (Mahowald & Mead 1989, Mead & Mahowald 1988). This silicon retina includes components that represent the photoreceptor, horizontal cell, and bipolar cell layers of the retina. The photoreceptor transduces light into a voltage that is logarithmic in the intensity of the stimulus. The synaptic interactions between the cell types are implemented in analogue circuits. The photoreceptors drive the horizontal cells (a noninverting synapse) via a transconductance (Figure 2a). A resistive network emulates the gap junction connections between horizontal cells (Figure 2b). Thus, the voltage at each node of the horizontal cell network represents a spatially weighted average of the photoreceptor inputs to the network. As in the biological retina, the electrotonic properties give rise to an exponentially decreasing spatial receptive field in the horizontal cell network. The antagonistic center-surround receptive field of the bipolar cell is implemented with a differential amplifier that is driven positively by the photoreceptors but inhibited by the horizontal cell output.

The silicon retina performs like a video camera in that the image is transduced directly on the chip and the output is displayed on a monitor screen continuously in time. However, unlike the CCD camera, the silicon retina reports contrast rather than absolute brightness and so it is able to see comparable detail in shaded and bright areas of the same scene. This contrast encoding emulates an essential property of retinae and is due to the center-surround receptive field of the bipolar cell, which computes the difference of the logarithmic photoreceptor and horizontal cell outputs. By using the local average of the horizontal cells as a reference signal, the redundant features of the image are suppressed while novel features are enhanced. This occurs because large areas of uniform luminance produce only weak visual signals. In these regions the output from any single photoreceptor is canceled by the spatial average

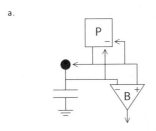

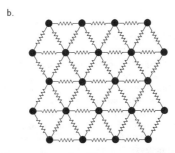

Figure 2   The silicon retina is a two-dimensional hexagonal array of pixels on a chip. The retina transduces a light image focussed directly on its surface, processing the stimulus by principles found in the outer plexiform layer of the vertebrate retina. Various versions of silicon retinae contain from 2304 (Mahowald & Mead 1989) to 8250 pixels (Boahen & Andreou 1992) on a 6.8 mm × 6.9 mm square of silicon using a 2-μm fabrication process. (*a*) A pixel from the silicon retina produced by Mahowald (1991) contains an analogue of a photoreceptor, P, a bipolar cell, B, and a node of the horizontal cell network (*filled circle*). (*b*) A hexagonal resistive network emulates the horizontal cells of the retina. The horizontal cells extend parallel to the surface of the retina and are resistively coupled to each other by gap junctions. The values of these resistors is controlled by applying an analogue voltage to the chip, so the degree of connectivity between horizontal cells can be varied after the chip is fabricated. In addition to the retinal array, the retina chips also have instrument circuits that monitor the outputs of the pixels. The pixels can be monitored individually or scanned out sequentially to a video monitor.

signal from the horizontal cell network. Novel luminance features, such as edges, evoke strong retinal signals because the receptors on either side of the edge receive significantly different luminance. A similar principle applies in the time domain, where the relatively slow temporal response of the horizontal cell network enhances the visual system's response to moving images. In fact, as in the physiology of biological retinal cells, the spatial and temporal characteristics of the retinal processing are simultaneously expressed, as can be seen in the response of the silicon retina to flashed lights of varying spatial extent illustrated in Figure 3.

The silicon retina encourages a functional interpretation of retinal processing. From an engineering point of view, the purpose of the outer-plexiform

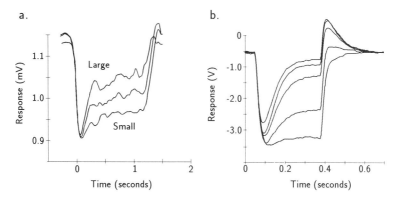

Figure 3   Comparison of the response of bipolar cells in a vertebrate and a silicon retina to a flashing light stimulus of varying spatial extent. (*a*) Response of a bipolar cell in the mud puppy, *Necturus maculosus* (data from Werblin 1974). (*b*) Output of a pixel in the silicon retina. Test flashes of the same intensity but of different diameters are centered on the receptive field of the unit. Larger flashes increase the excitation of the surround. The surround response takes time to build up to the capacitance of the resistive network. Because the surround level is subtracted from the center response, the output shows a decrease for long times. This decrease is larger for larger flashes. The overshoot at stimulus offset decays as the surround returns to its resting level. The peak response diminishes as the test flash becomes larger spatially. The larger test flash charges the network to a higher value. The charging of the network works against the finite response time of the receptor to produce the observed decrease in peak response at the bipolar cell level.

layer is to keep the output of retina from saturating over several orders of magnitude change in background illumination while allowing it to report the small contrast differences in a typical scene reliably. The encoded retinal output still contains sufficient information to fully support general visual processing in the brain, but it can be transmitted at a much lower bandwidth than can a gray-scale image. The efficient encoding of visual information is necessitated by a simple physical constraint, the finite range of retinal cell outputs. The hypothesis underlying the early silicon retina architecture was that lateral inhibition was sufficient to deal with the problem of dynamic range in retinal processing (Barlow 1981). Lateral inhibition removes redundant information caused by correlation of luminance across the image. However, lateral inhibition exacerbates another form of image redundancy, noise. Noise is a form of redundancy because it does not supply additional information about the world, so bandwidth is wasted by transmitting it (Atick & Redlich 1990, 1992). In the retina, noise arises from photon fluctuations at low light levels. But, it can also arise from the photoreceptors themselves, as dynamic thermal noise and as static miscalibration of photoreceptors' gains and operating points. The early silicon retinae were plagued by static mismatches between transistors, which caused each photoreceptor to give a different response to identical inputs. The solution to this problem was to incorporate more of the features of outer plexiform processing that had been observed in biological systems.

New generations of silicon retinae remove the intrinsic sources of noise by using adaptive elements (Mead 1989, Mahowald 1994) and resistive coupling between the receptors (Boahen & Andreou 1992). These retinae include feedback from the horizontal cells to the photoreceptors. This feedback allows the photoreceptor to respond over a large input range with higher gain than the photoreceptors in the earlier feedforward retinae. The feedback between the photoreceptors and the horizontal cells affects the time constants of the cells and the electrotonic spread of signals in the restive networks.

The inner-plexiform layer of the retina is more poorly understood than the outer one. It is believed to play a role in motion processing, and in lower vertebrates, such as the frog, it has complex synaptic interactions that give rise to invariant feature detectors. This complex processing is desirable when the output of the retina is used directly to mediate a specialized behavior (Horiuchi et al 1991, 1992). The cost of incorporating more complex processing in the retinal array is that the photodetectors are spread farther apart to make room for additional circuitry. An example of a complex silicon retina is the velocity-tuned retina of Delbrück (1993). This retina contains analogue delay elements that correlate the spatio-temporally distributed signals that arise from moving targets. At each pixel location, the retina generates three independent outputs that are tuned to a particular speed and direction of motion, irrespective of stimulus contrast. This retina contains 676 pixels on a 6.8 mm × 6.9 mm die fabricated in a 2 μm process. The computation is energy efficient. The analogue processing in the retina (not including video display) consumes 1.5 mW in the dark to 8 mW in the light, which is less than 5 μW per pixel, compared with 10 μW per pixel for a typical CCD pixel that does no computation.

Audition

Auditory processing is another interesting example of nonlinear sensory transduction. Here too, insights have been gained by the development of neuromorphic auditory processors. Early efforts focussed on emulating the cochlea (Lyon & Mead 1988, Mead 1989), the fluid-mechanical traveling wave structure that encodes sound into spatio-temporal patterns of discharge in the auditory neurons.

The sound pressure in the outer ear is coupled mechanically into vibrations of the cochlea's basilar membrane by the ossicles of the middle ear. The membrane behaves like a spatially distributed low-pass filter with a resonant peak. The resonant frequency of the filter decreases exponentially along the basilar membrane from the oval window toward the apex of the cochlea, so the optimal response of the membrane is spread in space according to frequency. This response can be modified by the outer hair cells, which are motile and able to apply forces to the basilar membrane at frequencies throughout the auditory spectrum. Hair cells provide mechanical feedback that modifies the damping of the basilar membrane, so changing its resonance. The cochlea can use this mechanism to preferentially amplify low-amplitude traveling waves. Thus, the cochlea acts as an automatic gain control mechanism for sound, analogous to the gain control for luminance in the outer-plexiform layer of the retina.

The silicon cochlea of Lyon & Mead (1988) uses a unidirectional cascade of second-order sections with exponentially scaled time constants to approximate the traveling wave behavior of the basilar membrane. Because the sections can be tuned to generate a gain greater than one over a range of frequencies, they can approximate the resonance behavior caused by the outer hair cells. The behavior of this silicon cochlea is remarkably similar to its biological counterpart. However, the degree of resonance of each stage of the filter cascade is quite sensitive to device mismatches. To investigate this problem, a more detailed aVLSI cochlea has been made that uses a resistive network to model the cochlea fluid, and a special purpose circuit to model regions of the basilar membrane (Watts 1993). The collective nature of resistive grid makes this cochlea less sensitive to device mismatch than the filter cascade design. The price of this robustness is an increase in circuit complexity and size.

This chip highlights an interesting methodological point—because neuromorphic models are constrained by what can be implemented in a physical medium, they can provide insight into biological design. The response of the basilar membrane scales—that is, the spatial pattern of the response is invariant with frequency except for a displacement along the membrane. There are two physical models that give rise to scaling. In the first—constant mass scaling—the mass of the membrane and the density of the fluid are constant, but their stiffness changes exponentially along its length. In the second model—increasing mass scaling—all three change exponentially along the length of the membrane. Although the behavior of these two models is indistinguishable, their implications for physical implementation are radically different. The constant mass model requires that membrane stiffness change by a factor of about one million. Such a range of variation can be simulated on a digital computer, but it cannot be implemented easily in a physical device. This suggests that the increasing mass model, in which the range of variation can be absorbed by three parameters rather than one, should be adopted. In this case no single parameter need change by a factor of more than 100–1000, which is feasible in real devices such as neurons and chips. Moreover, each of the three parameters could be controlled by just one independent parameter, the width of the basilar membrane, which increases linearly with displacement along its length (Watts 1993). Watts' (1993) resistive network silicon cochlea follows this idea. The exponentially varying membrane stiffness, membrane mass, and participatory fluid mass are all controlled by a single control line whose voltage increases linearly with membrane length, as does width. This chip emulates well the basilar membrane performance predicted by numerical and analytical methods.

The resistive network cochlea does not yet contain active elements that could emulate the outer hair cells. But, a simple circuit that models the sensory transduction and motor feedback required to do this has been designed and tested separately (Watts 1993). Basically, the outer hair cell circuit is a feedback system containing a saturating nonlinear element in its feedback loop (Watts 1993, Yates 1990). The circuit senses the local basilar membrane displacement and applies motor feedback there. These circuits preferentially amplify low-amplitude waves, because under these conditions the feedback gain is high. As the wave becomes amplified, the feedback element is driven into saturation, and the gain becomes smaller.

The transduction of cochlea signals into the early stages of neural processing has also been investigated by using neuromorphic methods. In the biological cochlea the spatial patterns of vibrations of the basilar membrane are sensed by the inner hair cells, which transduce the mechanical displacement into the release of neurotransmitter and thereby affect the rate of action potential generation by their associated auditory neurons. Consequently, each neuron has a bandpass tuning for frequency whose optimal response depends on the position of the inner hair cell on the basilar membrane. Thus, the combined pattern of mean rate responses in auditory neurons could represent the spectral shape of the input signal. But this explanation is not sufficient to explain auditory perception, because the bandpass of the neurons is not amplitude invariant even at the moderate amplitudes occurring during normal speech.

The temporal structure of action potential discharge offers an alternative representation of frequencies below about ~5 kHz. Auditory neurons operating in this range tend to discharge on one polarity of an input waveform, so their probability density function for spike generation is a half-wave rectified version of the analogue signal of their hair cell. This phase encoding persists even during fully saturated discharge (Evans 1982). Since individual auditory neurons rarely fire above 300 Hz, the phase-dependent spikes cannot occur on every cycle. Instead, the outputs of several neurons driven by the same hair cell must be combined to provide a reliable signal. For example, the outputs could be combined by a postsynaptic neuron that acts as a matched filter for spike repetition rate (Carr 1993, Lazzaro 1991, Suga 1988).

The matched filter is an autocorrelator that receives the unmodified auditory fiber action potentials as input and the same signal delayed by the match interval. In principle, the output of this correlator could have strong peaks not only at the match frequency but also at its harmonics. This problem is resolved by the basilar membrane filter, which attenuates the harmonics leaving a single peak at the match frequency. Thus, the time-delay matched-filter hypothesis is attractive but leaves open the questions of how the various time delays could arise in biological circuits and how precise these delays need be.

By building a silicon auditory processor that uses the time-delay correlation, Lazzaro & Mead (Lazzaro 1991, Lazzaro & Mead 1989, Lyon & Mead 1988) have shown that this approach can work robustly despite component variability (Figure 4). Their chip is composed of the filter cascade cochlea and circuitry that emulates the auditory neurons and their postsynaptic matched filters. The filter sections of the cochlea have exponentially scaled time constants and span a frequency range of 0.4 Hz to 4 kHz. Each section drives a circuit that models inner hair cell transduction. Each silicon hair cell, in turn, excites a group of simple neurons. The action potentials of all the groups are led both directly and via time delays to an array of correlators. The time delays of the correlators in the array increase exponentially and are roughly matched to the time constants of the silicon basilar membrane.

The auditory processor correctly identifies the frequency of a test signal in real time, which suggests that time-delay correlation is a feasible explanation for the operation of the early stages of auditory perception. The chip also offers interesting prospects for prostheses and auditory processing for computers (Lazzaro et al 1993).

RECONFIGURABLE NETWORKS

The sensory aVLSI systems described above have specialized neurons with hard-wired connections. Fortunately, sensory subsystems have a very regular structure that is dominated by local connectivity. In these cases it is possible to satisfy the connections of many thousands of neural nodes on a single chip. But beyond the sensory processing stage, the connectivity quickly becomes more complex, as do the characteristics of the neurons, which receive thousands of inputs along their branching dendritic trees. Such highly connected networks can only be constructed by distributing the network across multiple chips. There are many strategies for dividing the network among chips. These strategies typically recognize three different elements: the synapses onto the neurons, the input-output relation of the neurons, and the communication pathways of the axons.

Whatever the strategy employed, it is desirable to reuse the same hardware for exploring various neural architectures. This task requires the development of general silicon neurons, whose biophysical properties and connectivity can be reconfigured to match a range of neuronal types, and a communication

a.

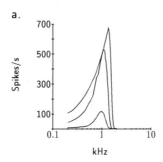

b.

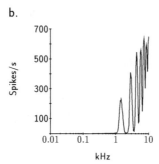

c.

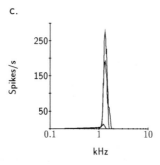

Figure 4 Response of a silicon auditory processor that emulates a cochlea, auditory neurons, and proposed postsynaptic neurons that act as correlators. (*a*) Average discharge rate of the auditory neurons in response to sinusoidal stimuli (2, 5, and 20 mV peak amplitude, bottom to top) applied to the cochlea. Notice that the bandwidth of the response is sensitive to input amplitude. (*b*) Output from a correlator postsynaptic to the auditory fibers shown in *a*, with cochlea filtering disabled. The correlator responds at its optimal frequency (about 1.5 kHz) with narrow bandwidth, but it also responds at higher harmonics. (*c*) Output of the correlator to the stimuli described in *a*, with cochlea filtering enabled. The response is logarithmic over the stimulus range 2–20 mV. There are no harmonics. The bandwidth is insensitive to stimulus amplitude. The auditory processor responds to frequencies over a range of approximately 0.3–3 kHz (data not shown). Modified from Lazzaro 1991.

strategy that can handle the huge convergence and divergence of general neuronal systems.

Neurons

There have been a number of reports of electronic neurons built of discrete components that emulate real neurons with various degrees of accuracy (Hoppensteadt 1986, Keener 1983) and that could be wired together to form small networks. In recent years the development of VLSI technology has made it possible to envisage very large networks of realistic CMOS neurons that could operate in real time and interact directly with the world via sensors.

VLSI designers have taken a wide range of approaches to the design of neurons for artificial neural networks. Examples of recent work in the field can be found in Murray & Tarassenko (1994), Ramacher & Ruckert (1991), Zaghloul et al (1994), *IEEE Transactions on Neural Networks*, and the *Advances in Neural Information Processing Systems* series. Many of these designs use pulse-based encoding, similar to the action potential output encoding of real neurons (Murray et al 1989, 1991; Murray & Tarrasenko 1994). However, in most of these cases, the connectivity between neurons is the focus of effort. The responses of the individual neurons are simplified to fixed input-output relations that conform to the models used in artificial neural network research. Less effort is devoted to the study of the complex biophysical mechanisms that underly neuronal discharge, and what their significance might be.[1]

Neurons act as elaborate spatio-temporal filters that map the state of their approximately 10^4 input synapses into an appropriate pattern of output action potentials. This mapping is dynamic and arises out of the interaction of the many species of active and passive conductances that populate the dendritic membrane of the neuron and that serve to sense, amplify, sink, and pipe analogue synaptic currents en route to the spike generation mechanism in the soma. The members of this society of conductances are each effective for a specific ion, or profile of ions, and they interact via their mutual dependence on either membrane potential or intracellular free calcium or both. The macroscopic conductance for a particular ion depends on the states of individual channels embedded in the neuronal membrane. These channels are charged, thus their performance is sensitive to voltage gradients across the membrane. The fraction of the channels that are conductive (open) at any membrane potential can be described by the Boltzmann distribution (Hille 1992). The fraction of channels that are open is given by

$$O (O + C) = 1/(1 + e^{-}zV/kT),$$

---

[1]However, DeYong et al (1992) have designed a neural processing element that is more clearly neuromorphic.

where O is the number of open channels; C, the number of closed channels; V, the transmembrane voltage; z, the channel charge; and kT, the thermal energy per charge carrier. This activation function has a sigmoidal form, saturating when all of the channels are open. Its dynamics are controlled by a time constant that may itself be voltage dependent. The activation is often opposed by an inactivation process that attenuates the conductance.

aVLSI neurons are synthesized by combining modular circuits that emulate the behavior of biological conductances. As in the example in Figure 5, the prototypical conductance module consists of an activation and an inactivation circuit that compete for control of an element that emulates the membrane conductance. When CMOS circuits are operating in their subthreshold regime (Mead 1989), the transistors' gate voltages control the heights of the energy barriers over which the charge carriers with Boltzmann-distributed energies must pass in order to flow across the channels. This similarity between the physics of membrane and transistor conductance offers a method of constructing compact circuits that represent biological channel populations.

The usual operation of these analogue circuits is similar to Hodgkin-Huxley descriptions (Hodgkin & Huxley 1952, Hille 1992) of membrane conductance. If the membrane potential is driven into the range of activation, the conductance for that ion will increase and charge the membrane towards its associated reversal potential. During activation, inactivation will grow with a slower time constant, finally quenching the conductance and allowing the membrane to relax back to its resting state.

The prototypical circuits are modified in various ways to emulate the particular properties of a desired ionic conductance. For example, some conductances are sensitive to calcium concentration rather than membrane voltage and require a separate voltage variable representing free calcium concentration. Synaptic conductances are sensitive to ligand concentrations, and these circuits require a voltage variable representing neurotransmitter concentration. The dynamics of the neurotransmitter in the cleft is governed by additional time constant circuits. These special circuits are simple modifications of the general theme of the prototypical circuit. Circuit modules that have been designed and tested so far include the sodium and potassium spike currents, the persistent sodium current, various calcium currents, the calcium-dependent potassium current, the potassium A–current, the nonspecific leak current, the exogenous (electrode) current source, the excitatory synapse, the potassium-mediated inhibitory synapse, and the chloride-mediated (shunting) inhibitory synapse.

Once a repertoire of circuits is available, the desired profile of currents can be inserted into the neuronal compartments that are capacitance nodes separated by axial resistances, just as one might specify a neuron for digital simulation. The behavior of the individual circuits that compose the neuron can be controlled by setting the voltages applied to the gates of various transistors in the module. These parameters determine, for example, the temporal dynamics of activation and inactivation, the voltage at which activation or inactivation occurs, and the maximum conductance that can be obtained when the ion conductance is fully activated. The parameters can be varied to

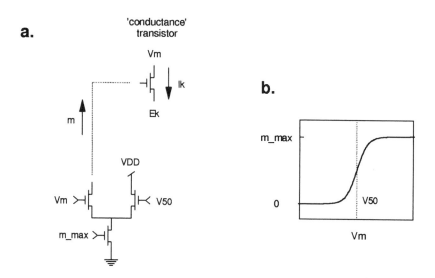

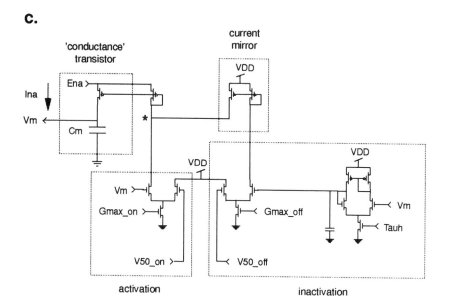

emulate neurons with a range of biophysical properties. The effect of changing these parameters is immediate, so the electrophysiological personality of the silicon neuron can be switched rapidly, for example, from a regular adapting to a bursting pyramidal cell (Figure 6). The fact that a range of neuronal subtypes with the same circuitry can be emulated suggests that the control parameters of the silicon neuron are analogous to those available in the development of real neurons. Of course, the range of reconfigurability is limited to the range of interaction of the circuits that are placed in the neurons at the time the chip is fabricated. Modules can be inactivated and so effectively removed from a compartment, but no new modules can be added. If additional properties are required, another silicon neuron with different morphology and different types of channels can be fabricated by using variations of the basic circuit modules.

To produce compact circuits it is often necessary to make approximations in the design of the circuit modules (Douglas & Mahowald 1994a). However, the striking results of the aVLSI emulations are the degree to which qualitative realistic neuronal behavior arises out of circuits that contain such approximations, and the fact that adjustments in the control parameters cause the neuron to change its behavior in a way that is consistent with our understanding of neuronal biophysics.

Communication

The degree of connectivity in neural systems and the real-time nature of neural processing demand different approaches to the problem of interchip communication than those used in traditional digital computers. VLSI designers have adopted several strategies for interchip communication in silicon neural net-

←——————————————————

Figure 5 Example of a neuromorphic CMOS aVLSI circuit. (*a, b*) Basic circuit that emulates transmembrane ion currents in the silicon neuron (Douglas & Mahowald 1994a,b; Mahowald & Douglas 1991). A differential pair of transistors have their sources linked to a single bias transistor (*bottom*). The voltage, m_{max}, applied to the gate of the bias transistor sets the bias current, which is the sum of the currents flowing through the two limbs of the differential pair. The relative values of the voltages, Vm and V_{50}, applied to the gates of the differential pair determine how the current will be shared between the two limbs. The relationship between Vm and the output current, m, in the left limb is shown in *b*. The current, m, is the activation variable that controls the potassium (in this example) current, Ik, that flows through the conductance transistor interposed between the ionic reversal potential, Ek, and the membrane potential. (*c*) The circuit that generates the sodium current of the action potential is composed of activation and inactivation subcircuits that are similar to those shown in *a*. The activation and inactivation circuits compete for control of the sodium conductance transistor by summing their output currents at the node marked by the asterisk. The current mirror is a technical requirement that permits a copy of the inactivation current to interact with the activation current. In this example, sodium current, Ina, flows from the sodium reversal potential, Ena, onto the membrane capacitance, Cm. The transconductance amplifier and capacitor on the right of the inactivation circuit act as a low pass filter, causing the inactivation circuit to respond to changes in membrane voltage with a time constant set by the voltage, τh.

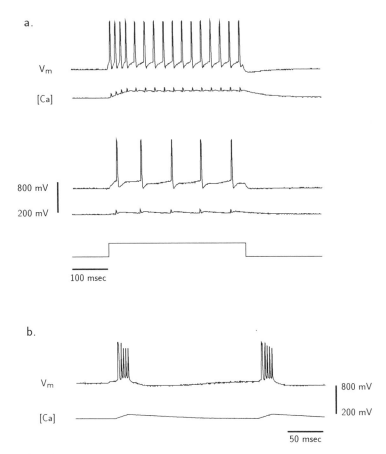

Figure 6   A silicon neuron that emulates the response of a cortical pyramidal neuron. (*a*) Response to two levels of intrasomatic current injection (time course shown below). The neuron is configured as a regular adapting pyramidal cell. Each pair of response traces consists of the somatic membrane potential, *Vm* (800 mV scale bar), and a voltage that is proportional to the intrasomatic free calcium concentration, [Ca] (200mV scale bar). Notice the adaptation of discharge and after-train hyperpolarization in the upper voltage trace. The voltage response of the silicon neuron is ten times larger than that of a real neuron (Douglas & Mahowald 1994a). (*b*) Response to sustained current injection of neuron configured as a bursting pyramidal cell.

works. Each strategy has advantages, and the choice of method depends on which factors are most crucial to the application.

One of the most literal approaches to interconnecting silicon neurons has been adopted by Mueller et al (1989a,b). They use a direct physical connection between nodes on different chips through a cross-bar switching array. A major advantage of this approach is that it allows continuous-time communication between nodes. In addition, the switching arrays provide flexible connectivity and can be programmed digitally by a host computer. The system is able to handle larger connectivities than are possible on a single chip because the dendrites of a single artificial neuron are extended over multiple chips. However, this approach requires many chips to model even a small number of neurons. The number of artificial neurons on each output chip is limited to less than 100 by the number of output pins available.

To increase the number of neurons that can be placed on a single chip, a strategy must be adopted that relies on the speed advantage of VLSI technology over biological neurons. Because neuromorphic systems are intended to interact with the real world, their dynamics should evolve concurrently with external events. To achieve speeds comparable to biological neurons, neuromorphic aVLSI circuits operate about 10^5 times more slowly than the maximal speed of CMOS. The excess bandwith is available for communication by multiplexing a single connection path in time. This communication strategy attempts to achieve in time what the axonal arborizations achieve in space.

A number of multiplexing techniques are currently in use in artificial neural systems (Brownlow et al 1990; Mahowald 1994; Murray et al 1989, 1991; Murray & Tarrasenko 1994; Sivilotti et al 1987). One communication strategy that is closely related to the action potential coding of biological neurons is an address-event representation (AER) (Mahowald 1992, 1994).

The address-event protocol works by placing the identity (a digital word) of the neuron that is generating an action potential on a common communications bus that is an effective universal axon. The bus broadcasts this address to all synapses, which decode the broadcast neuronal addresses. Those synapses that are connected to the source neuron detect that their source neuron has generated an action potential, and they initiate a synaptic input on the dendrite to which they are attached. Many silicon neurons or nodes can share a single bus because, like their biological counterparts, only a small fraction of the silicon neurons embedded in a network are generating action potentials at any time and because switching times in the bus are much faster than the switching times of neurons. At present, neuronal events can be broadcast and removed from the data bus at frequencies of 1 MHz. Therefore, about one thousand address-events can be transmitted in the time it takes 1 neuron to complete a single 1-ms action potential. And if, say, 10% of neurons discharge at 100 spikes/s, a single bus could support a population of about 10^5 neurons. Of course, many more neurons will be required to emulate even one cortical area. Fortunately, the limitation imposed by the bandwidth of the AER bus is not as discouraging as it may seem. The brain has a similar problem in space. If every axon from every neuron were as long as the dimensions of the brain, the brain would explode in size as the number of neurons increased. The brain avoids this fate by adopting a mostly local wiring strategy in which the average number of axons emanating from a small region decreases at least as the inverse square of their length (Mead 1990, Mitchison 1992, Stevens 1989). If the action potential traffic on the AER bus were similarly confined to a local population of neurons, the same bus could repeat in space and thus serve a huge number of neurons.

The AER communication protocol has a number of advantages. It preserves the temporal order of events taking place in the neural array as much as possible. Events are transmitted as they occur. AER has better temporal accuracy than standard sequential multiplexing because the bandwidth of the communication channel is devoted to the transmission of significant signals only. For example, an AER silicon retina generates address-events only at regions of the image where there is spatial or temporal change. In areas of uniform illumination, the neurons are generating only a few action potentials, so they do contribute much to the communication load. Finally, the AER scheme has the advantage that the mappings between source neurons and recipient synapses can be reprogrammed by inserting digital circuitry to map the output addresses to different synaptic codes.

Synapses

The synaptic couplings between neurons determine the function of the network. They are critical elements of artificial neural networks because it is believed that, in addition to establishing connections between neurons, they are the site of learning. Most VLSI synapse circuits have been developed for artificial neural networks and simply express a coupling weight, without concern for the temporal behavior of their biological counterparts.

Synaptic circuits fall into two broad classes: those that store their coupling weights locally and those that receive their weights together with each presynaptic input. Local weight storage facilitates local learning, which can occur in parallel at each synapse. The synaptic weight can be stored digitally[2] (Säckinger et al 1992) or in analogue form as the charge on a capacitor. Analogue storage may be volatile (Churcher et al 1993), in which case it must be refreshed periodically by some external source, or it may be a permanent charge stored on a well-insulated capacitor called a floating gate (Benson & Kerns 1993, Vittoz et al 1991). Permanent storage is usually preferable to

---

[2]Digital storage can be more precise than analogue storage, depending on the number of bits required. But for moderate precision (about six bits), digital storage requires a larger silicon surface area than analogue storage. Furthermore, digital arithmetic rules for synaptic update are cumbersome. Digital hardware is more appropriate for functions that are multiplexed, rather than those that occur massively in parallel.

volatile storage, but methods for updating the state of the floating gate are not yet flexible enough to implement many common learning rules.

Because the number of synapses per neuron is large ($5 - 10 \cdot 10^3$), they are the limiting spatial consideration in neuromorphic design. The synapses of individual neurons can be distributed across dendrites on multiple chips (Lansner & Lehmann 1993, Mueller et al 1989a), but this approach raises the problem of accurately transmitting the analogue voltages and currents at the compartment boundaries between chips. Alternatively, entire silicon neurons can be fabricated on the same chip so that only robust, action potential–like events need be transmitted between chips. In this case the synapses of the neuron are also limited to a single chip, and their number becomes a trade-off between space and biological versimilitude.

MULTICHIP SYSTEMS

Multichip neuromorphic systems are still in the early stages of development. The address-event protocol has been used to communicate between retinae and receiver chips, which contain an array of receiver cells that are scanned out onto a video monitor. Similar unidirectional connections have been developed for outputting address events from a digital computer to a receiver array (Mahowald 1992, 1994) and for receiving AER data from a silicon cochlea (Lazzaro et al 1993). These simple systems have been used to test the performance of the AER and explore the possibility of using neuromorphic preprocessing for sensory input to digital computers. The first steps toward more complex sensory-motor systems and multichip processing are described below.

Stereopsis

A functional multichip neuromorphic system has been described by Mahowald (1992, 1994). It is a stereopsis system consisting of two silicon retinae that transmit their image data to a third, stereocorrespondence, chip by using the address event protocol. This system computes the stereodisparity of objects continuously as the objects move around in the field of view of the retinae.

Stereodisparity arises from the projection of visual targets onto different positions in the images formed by the eyes. The stereodisparity of a visual target is proportional to the distance between the target and the viewer. To determine the distance of a target, the viewer must find the image features that correspond to each other and calculate their stereodisparity. Finding corresponding image features is difficult if many image features are similar to each other. The wallpaper illusion, in which a surface covered with identical, periodically arranged targets is perceived at the wrong depth, exemplifies the difficulty of determining stereocorrespondence.

The algorithms used by the cortex to compute stereodisparity are not known,

a.   b.

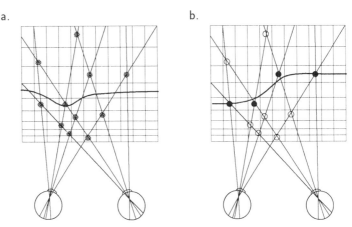

Figure 7   The cooperative multiresolution approach to the problem of stereocorrespondence. (*a*) The one-dimensional retinae receive disparate images from the projection of identical targets (*gray circles*) located in a two-dimensional space. The pattern of all possible targets consistent with the retinal images is shown. A two-dimensional array of correlators is indicated by the grid. A correlator is stimulated when a target lies within its grid position. The outputs of the activated correlators are averaged to provide an analogue estimate of the distance of the targets from the viewer (*heavy line*). (*b*) Positive feedback from the analogue distance estimate to the correlators most consistent with the analogue estimate, combined with competition between correlators, results in the enhancement of the correlators associated with some targets (*solid circles*) and the suppression of correlators associated with others (*open circles*). The analogue estimate of distance converges to a solution close to the enhanced correlators as the suppressed correlators are removed from the average. Targets associated with the enhanced correlators are sufficient to explain the images on the two retinae. They reflect the stereocorrespondence chip's estimate of the true state of the two-dimensional world.

but the response characteristics of certain elements that compute disparity have been recorded from cortex. For example, Poggio has described five distinct types of disparity-tuning reponses in cortical neurons of the primate (Poggio 1984, Poggio et al 1988). The principle innovation of the cooperative multi-resolution algorithm (Figure 7) is to introduce cooperative feedback interactions between disparity representations at different spatial scales. The high spatial frequency representation of disparity, analogous to Poggio's tuned-zero neurons, is spatially precise but susceptible to false correspondences. The low spatial frequency representation, analogous to Poggio's tuned-near and tuned-far neurons, makes an analogue estimate of disparity that averages out false correspondences at the expense of spatial blurring. Positive feedback between these two representations, coupled with nonlinear surround inhibition, allows the system to converge to a precise and correct determination of stereodisparity.

    The chip is able to fuse random dot stereograms and to solve other hard stereocorrespondence problems such as the fusion of surfaces tilted in depth.

The chip demonstrates that a network can use positive feedback to perform a significant computation, the suppression of false targets, while remaining continuously sensitive to retinal input. The biological relevance of the chip is that it expresses the stereofusion problem as just one instance of a general class of constraint-satisfaction problems in sensory perception and shows how this class of problems can computed with neuron-like elements.

Tectum

The real-time performance of neuromorphic systems suggests that the natural application of aVLSI is the investigation of neural computation in the context of behavioral functions. One of the behavioral functions that has been well studied in biological systems and that complements the aVLSI systems described above is the control of eye movement. Eye movement allows the visual system to actively seek new data based on previous data. It changes the fundamental character of the sensing process from a passive one to an active one (Ballard 1991).

The first steps towards synthesis of an aVLSI eye control system have been taken by DeWeerth, who constructed a primitive tecto-oculomotor system for tracking the brightness centroid of a visual image (DeWeerth 1992, Horiuchi et al 1994). Visual input to the system is provided by a two-dimensional retina with aggregation circuitry that computes the centroid of brightness in the horizontal and vertical dimensions. The magnitudes of the visual input signals are normalized before they are aggregated. This prevents changes in global illumination from affecting the localization of the centroid. The superior colliculus employs a similar computational principle to integrate visual information for oculomotor control (Sparks & Mays 1990). The silicon tectum resolves one example of a more general problem in neuroscience and artificial intelligence, that of reducing the dimensionality of complex data, such as audio and visual images, to low dimensional representations that can be used for motor control or decision making.

The aVLSI tecto-oculomotor system converts the output of the centroid computation into a motor signal and uses this signal to update the position of the retina. The aggregated output currents are converted into pulse trains that are directly applied to drive antagonistic motors in a bidirectional mechanical system that models the ocular muscles. Two motors are placed in an antagonistic configuration such that they produce torque on the eye in opposite directions. Each pulse output generated by the chip controls one of the motors. If the chip is stimulated by a bright spot that is not at the center of the photoreceptor array, the motors will produce a differential torque that rotates the eye so that the center of the photoreceptor array moves towards the stimulus. Because of the background firing rates of the neuron circuits, the motors will produce equal and opposite torques when the stimulus is centered.

Negative feedback is generated by the mechanical system to move the retina chip so that it faces the stimulus.

This simple system is a foundation for future development. The present tectum has photodetectors integrated in the two-dimensional array that computes the centroid. But, with the AER communication protocol, a two-dimensional image derived from a separate retina or an auditory localization chip could be transferred to the tectum to drive eye movements. The oculomotor system also presents an opportunity to explore hierarchical structures for motor control, such as the coordination of vergence and conjugate eye movements for pointing the eyes at targets in depth.

CONCLUSION

CMOS aVLSI has been used to construct a wide range of neural analogues, from single synapses to sensory arrays, and simple systems. The circuits in these examples are not completely general; rather, they are specialized to take advantage of the inherent physics of the analogue transistors to increase their computational efficiency.

These analogues emulate biological systems in real time, while using less power and silicon area than would an equivalent digital system. Conservation of silicon resource is not a luxury. Any serious attempt to replicate the computional power of brains must confront this problem. The brain performs about 10^{16} operations per second. Using the best digital technology, this performance would dissipate over 10 MW (Mead 1990) by comparison with the brain's consumption of only a few watts. Subthreshold analogue circuits are also no match for neuronal circuits, but they are a factor of 10^4 more power efficient than their digital counterparts. In the short term, small, low-power neuromorphic systems are ideal for prosthetic applications. Development of these applications is just beginning. For example, silicon retina projects have inspired work on a VLSI retinal implant (Wyatt et al 1993), and Fromherz et al (1991a,b) are studying silicon-neural interfaces. The development of other useful applications, such as the silicon cochlea for cochlear implants, will depend on the collaboration of neuromorphic silicon designers and biomedical engineers.

The specialized but efficient nature of neuromorphic systems causes analogue emulation to play a different role in the investigation of biological systems than does digital simulation. Analogue emulation is particularly useful for relating the physical properties of the system to its computational function because both levels of abstraction are combined in the same system. In many cases, these neuromorphic analogues make direct use of device physics to emulate the computational processes of neurons so that the base level of the analysis is inherent in the machine itself. Because the computation is cast as a physical process, it is relatively easy to move from emulation to physiological prediction.

Analogue VLSI systems emphasize the nature of computation as a physical process and focus attention on the need for computational strategies that take account of fundamental hardware characteristics. For example, the inherent variability between transistors operating in their subthreshold regime means that analogues cannot depend on finely tuned parameters. Instead, like biological systems, they must rely on processing strategies that minimize the effects of component variation. Biological systems operate with exquisite sensitivity over a wide range of physiological and environmental conditions by adjusting their operating points and combining multiple outputs to extract and amplify meaningful signals without amplifying component noise. Consequently, biological strategies such as signal aggregation, adaptation, and lateral inhibition also improve the performance of analogue silicon systems. By designing neuromorphic systems, we enlarge our vocabulary of computational primitives that provide a basis for understanding computation in nervous systems. This circuit vocabulary can be written into silicon systems to perform practical tasks and test our understanding. This is the quest of neuromorphic aVLSI research.

ACKNOWLEDGMENTS

The preparation of this manuscript, and much of the research discussed in it, were supported by grants to R Douglas, M Mahowald, and K Martin by the Office of Naval Research, and to C Mead by the Office of Naval Research and the Defense Advanced Research Projects Administration. Fabrication facilites were provided by MOSIS.

Literature Cited

Atick J, Redlich A. 1990. Towards a theory of early visual processing. *Neural Comput.* 2: 308–20

Atick J, Redlich A. 1992. What does the retina know about natural scenes? *Neural Comput.* 4:196–210

Ballard DH. 1991. Animate vision. *Artif. Intell.* 48:57–86

Barlow HB. 1981. Critical factors limiting the design of the eye and visual cortex. The Ferrier lecture. *Proc. R. Soc. London Ser. B* 212: 1–34

Benson RG, Kerns DA. 1993. UV-activated conductances allow for multiple time scale learning. *IEEE Trans. Neural Networks* 4: 434–40

Boahen K, Andreou A. 1992. A contrast sensitive silicon retina with reciprocal synapses. See Moody 1992, pp. 764–72

Brownlow M, Tarassenko L, Murray A, Hamilton A, Han I, Reekie HM. 1990. Pulse-firing neural chips for hundreds of neurons. See Touretzky 1990, pp. 785–92

Carr CE. 1993. Processing of temporal information in the brain. *Annu. Rev. Neurosci.* 16:223–43

Churcher S, Baxter DJ, Hamilton A, Murray AF, Reekie HM. 1993. Generic analog neural computation—the EPSILON chip. See Hanson 1993, pp. 773–80

Cowan J, Tesauro G, Alspector J, eds. 1994. *Advances in Neural Information Processing.* San Mateo, CA: Kaufmann

Delbrück T. 1993. Silicon retina with correlation-based, velocity-tuned pixels. *IEEE Trans. Neural Networks* 4:529–41

DeWeerth SP. 1992. Analog VLSI circuits for stimulus localization and centroid computation. *Int. J. Comp. Vision* 8:191–202

DeYong MR, Findley RL, Fields C. 1992. The design, fabrication, and test of a new VLSI hybrid analog-digital neural processing element. *IEEE Trans. Neural Networks* 3:363–74

Douglas R, Mahowald M. 1994a. A constructor set for silicon neurons. In *An Introduction to Neural and Electronic Networks,* ed. SF Zornetzer, JL Davis, C Lau, T McKenna. San Diego, CA: Academic. 2nd ed. In press

Douglas R, Mahowald M. 1994b. Silicon neurons. In *The Handbook of Brain Theory and Neural Networks,* ed. M Arbib. Cambridge, MA: Bradford. In press

Evans E. 1982. Functional anatomy of the auditory system. In *The Senses,* ed. H Barlow, J Mollon, pp. 251–63. Cambridge, UK: Cambridge Univ. Press

Fromherz P, Offenhäusser A, Vetter T, Weis J. 1991a. A neuron-silicon junction: a Retzius cell of the leech on an insulated-gate field-effect transistor. *Science* 252:1290–93

Fromherz P, Schaden H, Vetter T. 1991b. Guided outgrowth of leech neurons in culture. *Neurosci. Lett.* 129:77–80

Haken H. 1978. *Synergetics.* Berlin: Springer-Verlag. 2nd ed.

Hanson SJ, Cowan JD, Giles CL, eds. 1993. *Advances in Neural Information Processing Systems,* Vol. 5. San Mateo, CA: Kaufmann

Hertz J, Krogh A, Palmer RG. 1991. *Introduction to the Theory of Neural Computation* Redwood City, CA: Addison-Wesley

Hille B. 1992. *Ionic Channels of Excitable Membranes.* Sunderland, MA: Sinauer. 2nd ed.

Hodgkin AL, Huxley AF. 1952. A quantitative description of membrane current and its application to conduction and excitation in nerve. *J. Physiol.* 117:500–44

Hopfield JJ. 1982. Neural networks and physical systems with emergent collective computational abilities. *Proc. Natl. Acad. Sci. USA* 79:2554–58

Hopfield JJ. 1990. The effectiveness of analogue "neural network" hardware. *Network* 1:27–40

Hoppensteadt F. 1986. *An Introduction to the Mathematics of Neurons.* Cambridge, UK: Cambridge Univ. Press

Horiuchi T, Bair W, Bishofberger B, Moore A, Koch C, Lazzaro J. 1992. Computing motion using analog VLSI chips: an experimental comparison among different approaches. *Int. J. Comp. Vision* 8:203–16

Horiuchi T, Bishofberger B, Koch C. 1994. An analog VLSI saccadic eye movement system. See Cowan et al 1994, pp. 5821–89

Horiuchi T, Lazzaro JP, Moore A, Koch C. 1991. A delay-line based motion detection chip. See Lippman et al 1991, pp. 406–12

Julesz B. 1971. *Foundations of Cyclopean Perception.* Chicago, IL: Univ. Chicago Press

Keener J. 1983. Analogue circuitry for the FitzHugh-Nagumo equations. *IEEE Trans. Syst. Man Cybern.* 13:1010–14

Koch C, Li H, eds. 1994. *Vision Chips: Implementing Vision Algorithms with Analog VLSI Circuits.* Cambridge, MA: MIT. In press

Koch C, Poggio T. 1987. Biophysics of computation: Neurons, synapses, and membranes. In *Synaptic Function,* ed. GM Edelman, WE Gall, WM Cowan, pp. 637–97. New York: Wiley

Lansner JA, Lehmann T. 1993. An analog CMOS chip set for neural networks with arbitrary topologies. *IEEE Trans. Neural Networks* 4:441–44

Lazzaro J. 1991. Silicon model of an auditory neural representation of spectral shape. *IEEE J. Solid State Circuits* 26:772–77

Lazzaro J, Mead C. 1989. Silicon modeling of pitch perception. *Proc. Natl. Acad. Sci. USA* 86:9597–9601

Lazzaro J, Wawrzynek J, Mahowald M, Sivilotti M, Gillespie D. 1993. Silicon auditory processors as computer peripherals. *IEEE Trans. Neural Networks* 4:523–28

Lippman R, Moody J, Touretsky D, eds. 1991. *Advances in Neural Networks Information Processing Systems,* Vol. 3. San Mateo, CA: Kaufmann

Lyon RF, Mead CA. 1988. An analog electronic cochlea. *IEEE Trans. Acoustics Speech Signal Process.* 36:1119–34

Mahowald M. 1992. *VLSI analogs of neuronal visual processing: a synthesis of form and function.* PhD thesis. Calif. Inst. Technol., Pasadena, CA. 237 pp.

Mahowald M. 1994. *An Analog VLSI System for Stereoscopic Vision.* Boston, MA: Kluwer Academic

Mahowald M, Douglas R. 1991. A silicon neuron. *Nature* 354:515–18

Mahowald M, Mead C. 1989. Silicon retina. See Mead 1989, pp. 257–78

Marr D. 1982. *Vision.* San Francisco: Freeman

Marr D, Poggio T. 1976. Cooperative computation of stereo disparity. *Science* 194:283–87

McCullough W, Pitts W. 1943. A logical calculus of the ideas immanent in nervous activity. *Bull. Math. Biophys.* 5:115–33

Mead C. 1989. *Analog VLSI and Neural Systems.* Reading, MA: Addison-Wesley

Mead C. 1990. Neuromorphic electronic systems. *Proc. IEEE* 78:1629–36

Mead CA, Mahowald MA. 1988. A silicon model of early visual processing. *Neural Networks* 1:91–97

Mitchison G. 1992. Axonal trees and cortical architecture. *TINS* 15:122–26

Moody JE, Hanson SJ, Lippmann RP, eds. 1992. *Advances in Neural Information Processing Systems,* Vol. 4. San Mateo, CA: Kaufmann

Mueller P, Van der Spiegel J, Blackman D, Chiu T, Clare T, et al. 1989a. A programmable analog neural computer and simulator. See Touretzky 1989, pp. 712–19

Mueller P, Van der Spiegel J, Blackman D, Chiu T, Clare T, et al. 1989b. Design and fabrication of VLSI components for a general purpose analog neural computer. In *Analog VLSI Implementation of Neural Systems,* ed. C Mead, M Ismail, pp. 135–69. Boston, MA: Kluwer Academic

Murray A, Del Corso D, Tarassenko L. 1991. Pulse-stream VLSI neural networks mixing analog and digital techniques. *IEEE Trans. Neural Networks* 2:193–204

Murray A, Hamilton A, Tarassenko L. 1989. Programmable analog pulse-firing neural networks. See Touretzky 1989, pp. 712–19

Murray A, Tarasenko L. 1994. *Analogue Neural VLSI.* London: Chapman & Hall

Newell A, Shaw J, Simon H. 1958. Elements of a theory of human problem solving. *Psychol. Rev.* 65:151–66

Poggio G. 1984. Processing of stereoscopic information in primate visual cortex. In *Dynamic Aspects of Neocortical Function,* ed. GM Edelman, WE Gall, WM Cowan, pp. 613–35. New York: Wiley

Poggio G, Gonzales F, Krause F. 1988. Stereoscopic mechanisms in monkey visual cortex: binocular correlation and disparity selectivity. *J. Neurosci.* 8:4531–50

Ramacher U, Ruckert U, eds. 1991. *VLSI Design of Neural Networks.* Boston, MA: Kluwer Academic

Rumelhart D, McClelland J. 1986. *Parallel Distributed Processing.* Cambridge, MA: MIT Press

Säckinger E, Boser BE, Jackel LD. 1992. A neurocomputer board based on the ANNA neural network chip. See Moody 1992, pp. 773–80

Scribner D, Sarkady K, Kruer M, Caulfield J, Hunt J, Colbert M, Descour M. 1993. Adap-

tive retina-like preprocessing for imaging detector arrays. In *IEEE International Conference on Neural Networks,* 3:1955–60. San Francisco: IEEE

Sivilotti MA, Mahowald MA, Mead CA. 1987. Real-time visual computation using analog CMOS processing arrays. In *1987 Stanford Conference on Very Large Scale Integration,* ed. P Losleben, pp. 295–311. Cambridge, MA: MIT Press

Sparks D, Mays L. 1990. Signal transformations required for the generation of saccadic eye movements. *Annu. Rev. Neurosci.* 13: 309–36

Stevens C. 1989. How cortical connectedness varies with network size. *Neural Comput.* 1: 473–79

Suga N. 1988. Auditory neuroethology and speech processing: complex-sound processing by combination-sensitive neurons. In *Auditory Function,* ed. G Edelman, W Gall, W Cowan, pp. 679–720. New York: Wiley

Touretzky DS, ed. 1989. *Advances in Neural Information Processing Systems,* Vol. 1. San Mateo, CA: Kaufmann

Touretzky DS, ed. 1990. *Advances in Neural Information Processing Systems,* Vol. 2. San Mateo, CA: Kaufmann

Vittoz E, Oguey H, Maher M, Nuys O, Dijkstra E, Chevroulet M. 1991. Analog storage of adjustable synaptic weights. See Ramacher & Ruckert 1991, pp. 47–63

von Neumann J. 1958. *The Computer and the Brain.* New Haven, CT: Yale Univ. Press

Watts L. 1993. *Cochlea mechanics: analysis and analog VLSI.* PhD thesis. Calif. Inst. Technol., Pasadena, CA

Werblin FS. 1974. Control of retinal sensitivity. *J. Gen. Physiol.* 63:62–87

Winston P, Shellard S. 1990. *Artificial Intelligence at MIT. Expanding Frontiers,* Vol. 1. Cambridge, MA: MIT Press

Wyatt J, Edell D, Raffel J, Narayanan M, Grumet A, et al. 1993. Silicon retinal implant to aid patients suffering from certain forms of blindness. *Interim Progr. Rep.,* Mass. Inst. Technol., Boston, MA

Yates G. 1990. The basilar membrane non-linear input-output function. In *The Mechanics and Biophysics of Hearing,* ed. P Dalos, C Geisler, J Matthews, M Ruggero, C Steele. Berlin: Springer-Verlag

Zaghoul ME, Meador JL, Newcomb RW, eds. 1994. *Silicon Implementation of the Pulse Coded Neural Networks.* Boston, MA: Kluwer Academic

Annu. Rev. Neurosci. 1995. 18:283–317

SIGNAL TRANSDUCTION IN *DROSOPHILA* PHOTORECEPTORS

Rama Ranganathan[1], Denise M. Malicki[1], and Charles S. Zuker[1,2]

Howard Hughes Medical Institute and Departments of [1]Biology and [2]Neuroscience, University of California, San Diego, La Jolla, California 92093-0649

KEY WORDS: phototransduction, calcium, retinal degeneration, ion channels, genetics

INTRODUCTION

The ability to sense and to react to environmental stimuli is crucial for the survival of any organism. The primary event in sensory information processing is the transduction of external signals by specialized neurons into the electrical code that forms the input to processing centers in the brain. In the retina, photoreceptor neurons transduce the absorption rate of photons into a graded change in the ionic permeabilites of the cell membrane in a process known as phototransduction. Photoreceptor cells demonstrate beautiful properties of high sensitivity, rapid response kinetics, and broad dynamic response range that allow for efficient signal detection over a wide variety of light intensities. How are these properties encoded by the signal transduction machinery? The answer to this question is as yet incomplete, but recent advances in the study of phototransduction in the eye of the fruit fly *Drosophila melanogaster* have provided new insights into the molecular mechanisms. The goal of this review is to discuss these new advances in the context of existing models for phototransduction.

Phototransduction offers several advantages as a model system for the study of cellular signal transduction in general. First, photoreceptor cells are readily accessible for experimentation, and genetic, biochemical, and electrophysiological techniques for studying photoreceptor function have been well described (Zuker 1992). Second, functional constraints imposed by the dynamic nature of the visual input have forced the evolution of signaling mechanisms

283

0147-006X/95/0301-0283$05.00

that are optimized for speed and adaptability. Thus, phototransduction represents an excellent example of a highly regulated signaling process. Finally, phototransduction involves molecular mechanisms found in a wide variety of cellular signaling processes, suggesting that insight gained from the study of this system will be generally applicable. In both vertebrate and invertebrate photoreceptors, phototransduction is carried out by a G protein–mediated biochemical cascade characterized by the presence of a member of the seven–transmembrane domain receptor family, which regulates intracellular effector molecules through activation of a heterotrimeric G-protein. Such G protein-coupled transduction systems are found in signaling processes ranging from olfaction and neurotransmission to hormonal responses (reviewed in Stryer & Bourne 1986).

Although the biochemical and physiological study of phototransduction in vertebrate photoreceptors has contributed tremendously to our understanding of this process, a complete analysis of photoreceptor function in vivo requires a system also suitable for efficient genetic and molecular manipulation. The compound eye of *Drosophila* has proven to be an extremely useful model system in studying neuronal signal transduction processes; the powerful molecular genetic tools available in this system allow for the systematic dissection of even complex biochemical cascades in vivo. Since the eye is not required for viability (or fertility) in the laboratory setting, the mechanisms of visual transduction have been readily amenable to classical mutational analysis. This approach has resulted in the isolation of many *Drosophila* mutants defective in distinct aspects of photoreceptor function (reviewed in Ranganathan et al 1991b, Smith et al 1991b). In addition to the classical genetic approach, molecular screens designed to identify genes expressed solely in the eye have resulted in the isolation of DNA sequences from more than sixty unique chromosomal loci that encode eye-specific mRNA transcripts (Shieh et al 1989, Hyde et al 1990). As discussed below, molecular analyses of some of these mutants and eye-specific genes have helped to characterize some of the key molecules involved in *Drosophila* phototransduction and to formulate a working model for this process in invertebrates. Also, the ability to isolate null alleles of genes encoding products required in phototransduction provides the ideal genetic background for the rigorous structure-function analysis of important transduction proteins in vivo.

Significant advances in the study of *Drosophila* phototransduction include the development of preparations of dissociated photoreceptors suitable for electrophysiological analysis and the development of novel genetic suppressor and enhancer screens to identify additional components in the signaling cascade. Together with the traditional molecular genetic approach, these techniques are providing answers to previously intractable questions about photoreceptor function.

GENERAL PROPERTIES OF PHOTORECEPTOR FUNCTION

Excitation

Psychophysical studies carried out many years ago in humans showed that vertebrate photoreceptors have single-photon sensitivity (Hecht et al 1942). Subsequent electrophysiological studies by Baylor and colleagues (Baylor et al 1979) showed that in toad rods, absorption of a single photon by rhodopsin leads to the closure of about 500 ion channels at the peak of the response; this represents about 3% of the total current change possible. Thus vertebrate photoreceptors can generate a sizable output signal from activation of a single light-receptor molecule. Similarly, recent estimates of the single-channel conductance of the *Drosophila* light-activated channels based on stationary fluctuation analysis of channel noise (Hardie & Minke 1994a) suggest that single-photon responses represent the opening of roughly 650 channels at peak (~20 pA). This high sensitivity of transduction in both vertebrate and invertebrate photoreceptors is generated by a cascade of catalytic enzyme intermediates that amplify the input signal (Figure 1). In this scheme of catalytic amplification, the gain in visual excitation is determined by the kinetics of each enzymatic step and by the lifetime of each of the activated intermediates.

Deactivation

In addition to high sensitivity in signal detection, photoreceptors have evolved multiple regulatory mechanisms that allow them to maintain high temporal resolution and a broad dynamic range of response. High temporal resolution is ensured by rapid shut-off of each of the activated intermediates created during the excitation process so that the transduction machinery is reset quickly after generating a response. For example, the light-activated currents in *Drosophila* photoreceptors can shut off in less than 100 ms, following termination of the stimulus (Hardie 1991, Ranganathan et al 1991a). Although this deactivation could result from the inherent instability of the activated molecules involved in excitation, in many cases active mechanisms are clearly involved in the shut-off processes. For example, the termination of the enzymatic activity of the light receptor rhodopsin involves rapid phosphorylation followed by the stoichiometric interaction of an inhibitory protein called arrestin. As described below, recent work in *Drosophila* has begun to identify some of the molecules involved in inactivation at other levels in the transduction cascade and has suggested a major role for calcium-regulated protein phosphorylation in mediating this inactivation.

Adaptation

Given that the magnitude of the single-photon response is so large, the ability of photoreceptors to respond to over six logarithmic orders of light intensities

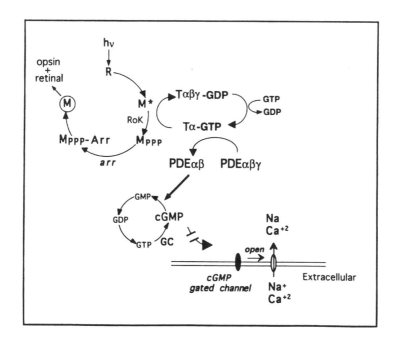

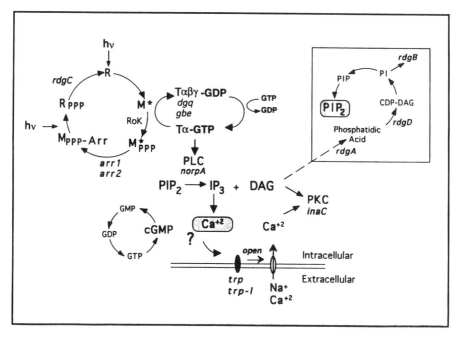

implies that the gain in the transduction system cannot be constant. If it were, even moderate intensity stimuli would saturate the transduction system, rendering the photoreceptor cell unable to respond to further stimuli. However, photoreceptor cells are capable of rapidly changing the gain in photoexcitation as a function of the background light intensity, and thus operate within the responsive range of the excitation machinery. This automatic gain control, generally referred to as photoreceptor adaptation, involves mechanisms that can sense the magnitude of photoreceptor activation and use this information as a feedback control signal to negatively regulate the excitation machinery. The study of the light-activated conductances in both vertebrates and invertebrates has led to the finding that the light-dependent change in intracellular calcium levels is the control signal for adaptation. Thus, both deactivation and adaptation of signal transduction involve biochemical mechanisms of stimulus-dependent negative feedback control.

A Comparison of Vertebrate and Invertebrate Phototransduction

An interesting result has emerged from the study of photoexcitation in vertebrates and invertebrates: Both of these systems share a great deal of similarity in overall strategy, but they differ in much of the underlying molecular machinery. In each case, photoexcitation begins with the light-dependent activation of the G protein–coupled receptor rhodopsin and the consequent activation of a G-protein, but then the pathways diverge and proceed via different intracellular signaling cascades that modulate the activity of cation-selective ion

←
___

Figure 1   Signal transduction in photoreceptor cells. The diagrams show, in expanded view, simplified models of phototransduction in both vertebrate (*top*)and invertebrate (*bottom*) cells. Absorption of a photon of light causes a conformational change in the rhodopsin molecule (R) and activates its catalytic properties. Active metarhodopsin (M*) catalyzes G protein–activation. The G-protein exchanges guanosine diphosphate (GDP) for guanosine triphosphate (GTP) and releases the inhibitory $\beta\gamma$ subunits. (*Top*) Active G-protein catalyzes the activation of a cGMP phosphodiesterase ($PDE_{\alpha\beta\gamma}$) by releasing the inhibitory constrains of the γ subunit. Activated $PDE_{\alpha\beta}$ hydrolyzes cGMP and causes a transient closure of the cGMP-gated channels, leading to the hyperpolarization of the photoreceptor cells. (*Bottom*) Active G-protein catalyzes the activation of the *norpA*-encoded PLC. PLC hydrolyzes PIP_2 into the intracellular messengers IP_3 and DAG. cGMP has also been implicated as a possible intracellular messenger mediating excitation. Extracellular sodium and calcium enter the cell through the light-activated conductance and cause the depolarization of the photoreceptor cells. The light-activated conductance appears to be composed of at least two types of channels. The *trp* gene is required for a class of channels with high calcium permeability. DAG is thought to modulate a photoreceptor cell–specific PKC (encoded by the *inaC* gene) that regulates deactivation and desensitization of the light response. Metarhodopsin is inactivated via phosphorylation by rhodopsin kinase (RoK) and arrestin binding (encoded by the *arr1* and *arr2* genes). Inactive metarhodopsin is photoconverted back to rhodopsin and then presumably dephosphorylated by the *rdgC*-encoded phosphatase. The box in the upper right hand (*bottom*) indicates a pathway likely to be required for the synthesis of PIP_2. *dgq* and *gbe* are the genes that encode the photoreceptor cell–specific isoforms of G_α and $G\beta$ subunits, respectively. See text for additional details.

channels (see below). This difference has important functional implications: Most invertebrate photoreceptors depolarize in response to light, but vertebrate photoreceptors hyperpolarize in response to light. A comparative study of these different systems of phototransduction offers the unique opportunity to understand how similar biological problems may be solved by different molecular mechanisms of signal transduction.

PHOTOTRANSDUCTION IN VERTEBRATES

Excitation

The sensory organelle of the vertebrate rod photoreceptor cell consists of a stack of membranous disks contained within a ciliary process known as the outer segment (Figure 1). This cellular compartment houses the phototransduction machinery (reviewed in Stryer 1986). The light receptor rhodopsin (R form) consists of a 40-kD apoprotein, opsin, covalently coupled to a chromophore, 11-*cis*-retinal, through a protonated Schiff base linkage at a lysine residue located in the seventh transmembrane domain. This lysine residue is conserved in all opsins studied thus far (Hargrave & McDowell 1992). Photoisomerization of the chromophore from the 11-*cis* form to the all-*trans* form causes a series of conformational changes in rhodopsin (Hargrave & McDowell 1992) and activates its catalytic properties. The photoisomerization of the chromophore occurs rapidly, in less than 1 ps (Schoenlein et al 1991), and efficiently, with a quantum yield of 0.6 (Wald 1968, Hayward et al 1981). Due to the very low frequency of spontaneous thermal isomerization of 11-*cis*-retinal, activation of rhodopsin is highly specific for light stimuli—a feature that allows for low inherent noise in the signaling cascade. This property is fundamental for the remarkably high signal-to-noise ratio necessary for single-photon sensitivity.

Activated rhodopsin [metarhodopsin II (MII) form] catalyzes the exchange of GTP for GDP on the α subunit of a specialized heterotrimeric G-protein known as transducin ($T_{\alpha\beta\gamma}$-GDP) and the dissociation of $T_{\beta\gamma}$. T_{α}-GTP activates a cGMP-phosphodiesterase (PDE) by relieving the inhibitory effect of the PDE-γ subunits (reviewed in Stryer 1991, Artemyev et al 1992). Active PDE hydrolyzes cGMP into GMP, and because the light-regulated channels are opened by cGMP, the transient reduction in cGMP levels causes these cation-selective channels to close. Thus, the absorption of a photon by the vertebrate photoreceptor causes a brief hyperpolarization of the cell.

The biophysical characterization of the cGMP-gated ion channels showed that they have evolved properties well suited for phototransduction. Experiments involving rapid perfusion of cGMP or flash photolysis of caged cGMP in inside-out patches from amphibian rods showed that the time course of current onset is only a few milliseconds and is in fact limited only by the kinetics of diffusion of cGMP (Karpen et al 1988). Also, these channels show no desensitization to prolonged cGMP exposure (Fesenko et al 1985, Karpen et al 1988). Thus, the cGMP-gated channels are able to rapidly and continuously report the activity of the underlying transduction cascade.

Another functionally important feature of these channels was revealed with the determination of the ionic selectivity of the cGMP-gated channels. Ion-permeation studies in excised patches (Fesenko et al 1985) and spectral analysis of noise fluctuations in intact cells (Bodoia and Detweiler 1985, Gray and Attwell 1985) indicated that the cation current carried by a single channel under physiological conditions is very small (~3 fA). Subsequent studies showed that this was the result of a voltage-dependent block by divalent cations of an otherwise large Na^+ current; recordings in excised patches in the absence of divalent cations revealed an unblocked single-channel current of about 1 pA (Haynes et al 1986, Zimmerman & Baylor 1986). Indeed, under physiological conditions the channel appears to be blocked >95% of the time by Ca^{2+} or Mg^{2+}. Interestingly, calcium ions are themselves also highly permeable through the channel (Haynes et al 1986, Zimmerman & Baylor 1986), with a 12.5-fold higher relative permeability in comparison to sodium (Nakatani & Yau 1988a). This simultaneous divalent permeation and block of the Na^+ current can be explained by models in which the divalent cations reside in the pore of the channel during transit much longer than monovalent ions, thus blocking the monovalent current. In functional terms, the small single-channel conductance during the physiological divalent block leads to low background noise from channel activity because the variances in membrane current depend on the square of the single-channel conductance (Hille 1992). This property greatly improves the signal-to-noise ratio in phototransduction and helps make the single-photon sensitivity possible.

Regulation

The calcium permeability of the cGMP-gated ion channel is fundamental for the regulation of vertebrate phototransduction, since several lines of evidence implicate changes in intracellular calcium in these processes. Calcium is an ideal candidate for a messenger of feedback regulation, since the Ca^{2+} permeability of the cGMP-gated channels makes $[Ca^{2+}]_i$ proportional to the number of recently opened channels and therefore a measure of the recent activity of the excitation machinery. Indeed, studies in salamander photoreceptors showed that the light-dependent reduction in the influx of calcium through the light-activated channels appears to mediate essentially all of the elegant feedback regulatory processes underlying photoreceptor adaptation and inactivation (Matthews et al 1988, Nakatani & Yau 1988b, Fain et al 1989; reviewed in Pugh & Lamb 1990).

Several molecular mechanisms appear to mediate calcium-dependent regulation in vertebrate photoreceptors. Hodgkin & Nunn (1988) showed that the change in $[Ca^{2+}]_i$ affects cGMP metabolism and that the enzyme guanylate cyclase, which synthesizes cGMP from GMP, is a site of calcium action. Recent studies have shown that the calcium sensitivity of guanylate cyclase activity is not direct; instead it is mediated through a calcium-binding protein that activates guanylate cyclase under low-calcium conditions (Dizhoor et al 1991, Lambrecht & Koch 1991, Polans et al 1991). Therefore, when a flash of light reduces the cGMP levels, the closure of the cGMP-gated channels causes a transient reduction in intracellular calcium concentration, and this reduction in $[Ca^{2+}]_i$ stimulates the resynthesis of cGMP by activating guanylate cyclase to restore the dark state. During a sustained stimulus, this mechanism also allows some closed channels to reopen because cGMP synthesis is up-regulated during the response. This extends the response range of the cell and contributes to light adaptation. In addition to this process, the lifetime of active rhodopsin is also controlled by calcium-dependent feedback regulation. A calcium-dependent modulation of rhodopsin phosphorylation in frog rods is mediated by a protein known as S-modulin (Kawamura & Murakami 1991, Kawamura 1993). Interestingly, S-modulin and a 26-kD calcium-binding bovine photoreceptor protein called recoverin (Dizhoor et al 1991) have similar structural and functional properties (Gray-Keller et al 1993, Kawamura et al 1993), suggesting that these proteins are identical. In addition to the modulation of the lifetime of active rhodopsin, recent studies show that the generation of activated rhodopsin is under the control of cytosolic calcium and may be mediated through a novel calcium-binding protein (Lagnado & Baylor 1994). Finally, the affinity of the cGMP-gated channels for cGMP may also be regulated by cytosolic calcium through the interaction of calmodulin with a channel-associated 240-kD protein (Hsu & Molday 1993). Thus, multiple steps in vertebrate phototransduction are controlled by calcium-dependent feedback regulation.

INVERTEBRATE PHOTOEXCITATION

As discussed above, most invertebrate photoreceptors depolarize rather than hyperpolarize in response to light through the opening of cation-selective ion channels. This difference implies some divergence in the underlying molecular mechanisms. Although several mechanisms could be proposed, one possibility is that light may control the generation, rather than the breakdown, of an excitatory messenger. As described below, current models of invertebrate phototransduction involve the light-dependent generation of inositol phosphates or cGMP. Inositol 1,4,5-trisphosphate (IP$_3$) is a well-characterized second messenger that controls the release of calcium from intracellular stores in many diverse systems (Berridge 1993). Intracellular calcium release in turn regulates the activity of many signaling molecules. IP_3 and Ca^{2+} have been implicated in the excitation of *Limulus* (Brown et al 1984; reviewed in Bacigalupo et al 1990), squid (Szuts et al 1986, Brown et al 1987), and fly (Devary et al 1987) photoreceptors.

Morphology of the Drosophila Compound Eye

The complete adult visual system of *Drosophila* is composed of two types of photoreceptive organs—the large compound eyes and three simple eyes located at the vertex of the head, called ocelli. The compound eyes are comprised of 700–800-unit eyes, called ommatidia, each of which contains 8 photoreceptor neurons and 12 accessory cells (reviewed in Tomlinson 1988). The photoreceptors lie at the core of the ommatidium and extend a photosensitive stack of microvilli into the extracellular space of a central canal. These stacks of microvilli, called rhabdomeres, are analogous to the outer segment of vertebrate photoreceptors and contain the phototransduction machinery (Smith et al 1991b). In contrast to the outer segment, which is essentially a ciliary body filled with isolated disks, rhabdomeres of invertebrate photoreceptors consist of true microvilli, which are contiguous with the plasma membrane (see Figure 1). The eight photoreceptors in each ommatidium can be divided both morphologically and functionally into three classes: R1–R6, R7, and R8 (reviewed by Hardie 1983, Franceschini 1985, Smith et al 1991b). R1–R6 cells represent the major class of photoreceptors in the compound eye and are maximally sensitive to blue light. These cells have rhabdomeres that are peripherally located in the central canal of each ommatidium and run the full length of the retina. R7 cells are maximally sensitive to ultraviolet light, and R8 cells are primarily blue-green sensitive. R7 and R8 cells have centrally located rhabdomeres; the R7 rhabdomere is located in the distal portion of the ommatidium, and the R8 rhabdomere is in the proximal portion. The axons of R1–R6 cells travel out of the retina and synapse onto neurons in the first optic ganglion, the lamina. Axons of R7 and R8 cells bypass the lamina and synapse onto neurons in the second optic ganglion, the medulla. Neurons in the lamina carry out the initial steps in the processing of motion detection, indicating that R1–R6 cells are the primary sensory cell type for this process. Indeed, most optomotor responses in *Drosophila* are mediated via input through the R1–R6 cells (reviewed by Heisenberg & Wolf 1984).

Photoexcitation in Drosophila

RHODOPSIN   As in vertebrates, the initial step in photoexcitation in invertebrates is the light-induced activation of rhodopsin. The major rhodopsin in the *Drosophila* eye, Rh1, is encoded by the *ninaE* gene (O'Tousa et al 1985, Zuker et al 1985) and is specifically expressed in the R1–R6 photoreceptor cells. This opsin shares not only primary sequence homology with bovine and human rhodopsins, but also structural features common to all G protein–coupled receptors. Rh1 is a blue-sensitive rhodopsin, with a λ_{max} of 480 nm for the rhodopsin (R) form, and a λ_{max} of 580 nm for the metarhodopsin (M) form. The finding that the expression of this rhodopsin is restricted to a subset of photoreceptors prompted the cloning of three other *Drosophila* opsins—Rh2 (Cowman et al 1986), Rh3 (Zuker et al 1987), and Rh4 (Montell et al 1987)—with the hope that these new opsins would represent the light receptors in the remaining classes of photoreceptor cells. Indeed, Rh2 was shown to be a violet-sensitive rhodopsin specifically expressed in the ocellar photoreceptors [$\lambda_{max}R$ = 420 nm; $\lambda_{max}M$ = 520 nm] (Feiler et al 1988, Mismer et al 1988, Pollock & Benzer 1988), while Rh3 and Rh4 were shown to be UV-sensitive rhodopsins [Rh3: $\lambda_{max}R$ = 345 nm, $\lambda_{max}M$ = 460 nm; Rh4: $\lambda_{max}R$ = 375 nm, $\lambda_{max}M$ = 460 nm] expressed in nonoverlapping subsets of R7 cells (Montell et al 1987, Feiler et al 1992). The gene encoding the R8-specific opsin has not yet been cloned.

Because the retinal chromophore has an inherent absorbance spectrum that is maximal in the UV, the interaction of the chromophore with the opsin must shift the absorbance peak into the visible spectrum. To study the structural features in opsin molecules that underlie spectral tuning in vivo, a recent study has examined the properties of chimeric Rh1-Rh2 proteins expressed in *ninaE* mutants (Britt et al 1993). Spectral-sensitivity recordings showed that multiple regions of the opsin protein are involved in spectral tuning of the R form but that the second putative transmembrane domain was sufficient for determining the spectral properties of the M form. Thus, the resting and photoactivated forms of rhodopsin have independent structural requirements for spectral tuning. This work sets the stage for a detailed molecular analysis of chromophore-opsin interactions during activation of rhodopsin.

G-PROTEINS   Several studies have provided good evidence for the existence of a light-regulated G-protein in fly photoreceptors (Blumenfeld et al 1985, Bentrop et al 1986, Devary et al 1987). For example, application of pharmacological agents that stimulate G-proteins enhance or mimic light-dependent activation of photoreceptor cells (Minke & Stephenson 1985). Two genes encoding visual system–specific G-proteins have been isolated in *Drosophila* (Lee et al 1990, Yarfitz et al 1991), and the relevance of the respective gene products for phototransduction is currently under investigation. These genes, *dgq* and *gbe,* encode a G_α and a G_β subunit, respectively. The eye-specific G_β ($G\beta e$) homologue shares only 50% amino acid identity with other G_β-proteins and may therefore represent a unique member of this highly conserved gene family. The recent analysis of mutants defective in $G\beta e$ showed that this protein is essential for coupling G_α with metarhodopsin (Dolph et al 1994). Interestingly, $G\beta e$ is also required for deactivation of the light response, suggesting an important role for the G_β subunit in terminating the active state of the signaling cascade. The eye-specific G_α gene product is most homologous to a vertebrate G-protein that activates phospholipase C (PLC); this homology is consistent with its role as activator of PLC in *Drosophila* photoexcitation (see below). Interestingly, the *dgq* gene produces two alternatively spliced transcripts that should generate two slightly different G_α products. Whether these two proteins are differentially regulated, have similar functions, or are expressed in the same or distinct subsets of cells is still unknown. However, recent mutational screens have identified hypomorphic alleles of *dgq* (R Hardy & CS Zuker, unpublished data), and preliminary data suggest a phenotype primarily characterized by a severe reduction in light sensitivity. These data are consistent with the concept that the *dgq* gene product is involved in photoexcitation in *Drosophila*.

PLC   Major insight into the mechanisms by which light activation triggers channel activity in invertebrates was gained by the isolation and characterization of the *no receptor potential A* (*norpA*) gene in *Drosophila* (Hotta & Benzer 1970, Pak et al 1970, Bloomquist et al 1988). Strong alleles of *norpA* completely abolish the light response (Hotta & Benzer 1970, Pak et al 1970). Moreover, electrophysiological studies of a temperature-sensitive allele of *norpA* (*norpA^{H52}*) showed that the electrical response to light can be rapidly abolished by shifting to the restrictive temperatures and that the response can be recovered by returning to the permissive temperature (Minke 1979). These experiments demonstrate that the *norpA* gene product is required for photoexcitation in *Drosophila*. The molecular cloning of the *norpA* locus revealed that it encoded the structural gene for the phospholipid-specific phosphodiesterase phospholipase C (PLC), which is abundantly expressed in the retina (Bloomquist et al 1988). These data implicate a light-dependent PLC activity as the basis for photoexcitation in *Drosophila* photoreceptors.

INTRACELLULAR MESSENGERS   Because PLC activity has been shown to mediate signaling in other systems through hydrolysis of the minor membrane phospholipid 4,5-phosphatidylinositol bisphosphate (PIP2) into the second messengers IP3 and diacylglycerol (DAG) (reviewed by Berridge 1987), a similar mechanism was suggested to mediate invertebrate phototransduction. Indeed, the light-dependent generation of IP3 seems to play a crucial role in the excitation of many invertebrate photoreceptors (Payne 1986, Devary et al 1987; and reviewed by Bacigalupo et al 1990), and there is some evidence that it does so through the mobilization of internal stores of Ca^{2+} from collections of smooth endoplasmic reticulum located at the base of the microvillar processes (Walz 1982a–c) (see Figure 1). Two studies provide specific evidence for the involvement of inositol phosphates in fly phototransduction. Spectral analysis of electrical noise induced by injection of IP$_3$ into fly photoreceptors showed that this noise has biophysical properties very similar to the response to dim illumination (Devary et al 1987). Also, biochemical experiments in preparations of fly photoreceptor membranes labeled with [^{3}H]-inositol showed that activated rhodopsin catalyzes the hydrolysis of inositol phosphates through the activation of a G-protein (Devary et al 1987). Thus, the light-dependent generation of IP$_3$ is likely to control photoexcitation in *Drosophila*.

What might be the mechanisms by which IP$_3$ opens the *Drosophila* light-activated channels? Until recently, the study of the terminal events in *Drosophila* photoexcitation and the biophysical characterization of the light-activated conductance had been hampered by the technical difficulty of recording electrical responses from the small photoreceptors present in these flies. The recent development of novel preparations of isolated *Drosophila* photoreceptors suitable for patch-clamp analysis is a significant advancement in the electrophysiological analysis of invertebrate phototransduction (Hardie 1991, Ranganathan et al 1991a). This preparation involves the mechanical dissociation of either late pupal or early adult retinas from *Drosophila*, and results in the isolation of clusters of R1–R8 photoreceptors, each of which represents one ommatidium stripped of all support and pigment cells (Figure 2). Although the clusters contain all three functionally distinct classes of photoreceptors, the R1–R6 cells can be

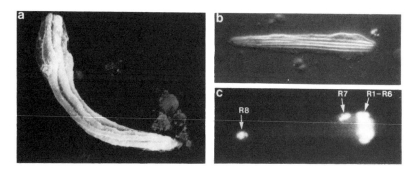

Figure 2 Isolated *Drosophila* photoreceptor clusters. Ommatidial clusters consisting of the eight photoreceptor neurons devoid of all of the support cells are the preparation used in the physiological characterization of the light response. These clusters can be isolated from wild-type and mutant flies. (*a*) A scanning electron micrograph (SEM) shows that each cluster is comprised of a group of tightly packed individual photoreceptors. (Figure kindly provided by G Harris.) (*b*) The distal end of this Nomarski interference contrast image of a photoreceptor cluster is to the right. (*c*) The same cluster as shown in *b* is stained with a fluorescent marker that is specific for nuclei. The arrows indicate the position of the R1–R6, R7, and R8 nuclei.

easily identified by the unique location of their nuclei at the distal margin of each cluster. Whole-cell patch-clamp recordings from these dissociated photoreceptors have provided a sensitive assay for the function of the underlying signal transduction cascade and have allowed the pharmacological manipulation of the photoreceptor cell environment.

An important and unresolved issue has been the identification of the intracellular messenger(s) that mediate the opening of plasma membrane conductances in *Drosophila* photoreceptors. Although the messengers that actually gate the plasma membrane ion channels remain elusive, patch-clamp techniques have now provided evidence implicating internal calcium stores in *Drosophila* photoexcitation. For example, photoreceptors recorded in a low-calcium environment show a stimulus-dependent loss of sensitivity that can be rescued by raising intracellular calcium levels (Hardie & Minke 1992). One possibility is that in the absence of external calcium, internal calcium pools required for excitation deplete rapidly in response to light stimuli. The reintroduction of an adequate calcium source would then restore sensitivity by refilling these intracellular calcium stores. In addition, studies investigating the maturation of the light-activated currents in *Drosophila* pupae showed that during a critical period prior to the natural development of the light response, light-activated currents can be induced by artificially raising intracellular calcium levels (Hardie et al 1993a). This result is unlikely to simply represent a requirement for cytosolic calcium, because buffering intracellular calcium levels with EGTA to as low as 25 nM in mature photoreceptors does not abolish the light response (Hardie et al 1993a). Thus the calcium-dependent maturation of the light response is most likely mediated by the filling of internal calcium stores required for excitation. Parenthetically, these results also suggest that the rate-limiting event in the final development of the phototransduction cascade may be the establishment of these internal calcium stores. Direct evidence for the presence of these internal stores comes from recent work using simultaneous high-speed, confocal–fluorescence imaging of calcium fluxes and whole-cell patch-clamp recordings in dissociated *Drosophila* photoreceptors (Ranganathan et al 1994). These experiments show that the application of thapsigargin, a drug known to deplete IP_3-sensitive calcium stores in many cell types (Inesi & Sagara 1992), causes a steady loss of calcium from an internal compartment. Depletion of these thapsigargin-sensitive stores only leads to a partial loss of light sensitivity, a result that raises the possibility that IP_3-sensitive stores are not required for photoexcitation. Alternatively, photoexcitation may depend on an IP_3-sensitive store that is only weakly sensitive to thapsigargin. Further studies using this calcium imaging technique should help resolve this issue.

Given the proposal that IP_3-mediated calcium release is the mechanism of photoexcitation, it is particularly interesting (and confusing) that the only messenger implicated in directly gating an invertebrate light-activated channel is cGMP. Studies involving direct application of cyclic nucleotides or calcium to excised inside-out patches of *Limulus* photoreceptors showed that ion channels with biophysical characteristics similar to the light-activated channels are opened by cGMP but not by Ca^{2+} (Bacigalupo et al 1991). In addition, pressure injection of either cGMP or hydrolysis-resistant analogues of cGMP excite *Limulus* photoreceptors (Johnson et al 1986, Feng et al 1991). The apparent involvement of cGMP in *Limulus* phototransduction and the absolute requirement for PLC demonstrated by the *Drosophila norpA* mutation could be reconciled by invoking the existence of calcium or IP_3-regulated forms of a guanylate cyclase or cGMP-phosphodiesterase. However, since no studies have yet suggested cGMP involvement in other invertebrate photoreceptors, including *Drosophila*, an alternative theory is that *Limulus* and *Drosophila* phototransduction systems have diverged and now utilize different intracellular signaling mechanisms.

Several possibilities may account for the mechanism of opening light-activated channels in *Drosophila*. First, calcium may directly gate plasma membrane channels, perhaps as a complex with calmodulin. In this regard, calmodulin-binding sites are found in putative light-activated channel proteins from *Drosophila* (Phillips et al 1992, Hardie & Minke 1993), although the function of these sites is still unclear. Second, the signal for the gating of the plasma membrane channels may be the depletion, rather than the release, of calcium from internal stores. Such "capacitative" mechanisms for channel activation have been proposed in several vertebrate cells (Putney 1990, Irvine 1992), and recent work has identified a novel intracellular messenger mediating this process (Randiramapita & Tsien 1993). Finally, in analogy with signal transduction in skeletal muscle, it has been proposed that the IP_3 receptor may physically interact with plasma membrane channels (Irvine 1991, Hardie & Minke 1993). This model suggests that store depletion may lead to direct gating of plasma membrane channels rather than the release of a diffusible messenger. In this model, intralumenal domains of the IP_3 receptor contain sensors for calcium levels that regulate the conformational state of the IP_3 receptor and thereby directly gate the plasma membrane channels through protein-protein interactions. Evidence that argues against these capacitative mechanisms comes from experiments in which depletion of intracellular stores by application of thapsigargin leads to a loss of the light response without any change in plasma membrane conductance (Ranganathan et al 1994). Further work should help distinguish these models for the terminal events in *Drosophila* phototransduction.

LIGHT-ACTIVATED CHANNELS   Ion-substitution experiments using whole-cell voltage-clamp techniques in wild-type photoreceptors demonstrated that the

Drosophila light-activated conductance is nonselective for cations and can support fluxes of even large monovalent ions such as $Tris^+$ and TEA^+ (Ranganathan et al 1991a). One particularly interesting finding was that this conductance is primarily permeable to Ca^{2+} ions, with an overall relative permeability of approximately 40:1 when compared to either Na^+ or Cs^+ (Hardie 1991, Ranganathan et al 1991a). As described below, this light-dependent calcium flux in *Drosophila* photoreceptors has an important role in mediating feedback regulation of the signaling cascade and is crucial for deactivation and adaptation. The light-dependent activation of calcium-permeable channels in *Drosophila* photoreceptors represents a model system for phosphatidyl inositol (PI)–dependent calcium influx, a poorly understood but important feature of all PI-signaling processes (Hardie & Minke 1993).

Recent work has shown that the *Drosophila* light-activated conductance is comprised of at least two biophysically distinct channels, one of which is responsible for the majority of the calcium permeability (Hardie & Minke 1992). The existence of more than one channel type was initially suggested by the finding that under certain conditions of voltage and calcium, photoreceptors fail to show a unique reversal potential and instead exhibit a biphasic reversal potential (Hardie & Minke 1992). Electrophysiological analysis of *transient receptor potential* (*trp*) mutants (Cosens & Manning 1969) showed that the biphasic reversal potential and the calcium-dependence of the reversal potential is greatly reduced in *trp* mutants, suggesting the absence of a class of calcium-permeable channels (Hardie & Minke 1992). Also, micromolar levels of extracellular La^{3+}, a known calcium-channel blocker, mimic the *trp* phenotype in wild-type photoreceptor cells, suggesting that a *trp*-dependent class of channels is specifically blocked by La^{3+} (Hochstrate 1989, Suss-Toby et al 1991, Hardie & Minke 1992). Molecular characterization of the *trp* gene showed a predicted protein product of 1275 amino acids with several putative transmembrane segments (Montell & Rubin 1989, Wong et al 1989). Examination of the Trp sequence has identified regions of weak similarity with neuronal voltage-gated Ca^{2+} channels in the transmembrane segments, an ankyrin-like repeat domain near the N-terminal, and a calmodulin-binding site (Phillips et al 1992). These data provide a working model that *trp* encodes a subunit of a plasma membrane channel with high calcium permeability.

Photoreceptor cells from *trp* mutants display normal responses to weak stimulation but are unable to maintain a receptor potential in response to maintained bright light stimuli. This results in a transient response that decays to baseline (thus the name *trp*). Also, *trp* photoreceptors are inactivated following an intense stimulus but recover from inactivation after a period of dark exposure (>1 min). The inactivation phenotype does not result from overdriven adaptation mechanisms; it is caused by a drastic decrease in the efficiency of the excitation process (Minke et al 1975b, Minke 1982). This conclusion is

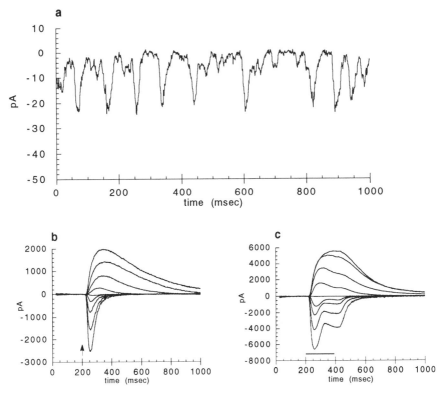

Figure 3 Light-activated electrical responses in voltage-clamped *Drosophila* photoreceptors. (*a*) These are single-photon responses (quantum bumps) in a dark-adapted wild-type photoreceptor. The recording was made at a holding potential of −80 mV during a continuous pulse of low-intensity red (>620 nm) light. (*b*) Currents were evoked by 10-ms flashes of white light at holding potentials ranging from −80 mV to +80 mV in steps of 20 mV. Leak subtractions were done off-line and all traces were normalized to zero at the pretrigger region just before onset of the light stimulus (*arrow*). Recordings were made with 1-mM Ca^{2+} in the bath (Ranganathan et al 1991a). Note the significant asymmetries in the current kinetics across the reversal potential. (*c*) Currents were evoked by 200-ms pulses of white light (*bar*) with the same protocol used in *b*.

based on the analysis of single-photon responses in *Drosophila* photoreceptors, which represent the quantal probabilistic electrical responses of the photoreceptor to single rhodopsin isomerizations (Figure 3*a*). These single-photon responses, called quantum bumps, statistically superimpose to generate the macroscopic light response. One important feature of adaptation mechanisms is a reduction in the amplitude of quantum bumps due to decreased gain in photoexcitation without a change in their frequency of occurrence (reviewed in Stieve 1986). Adaptation mechanisms also typically cause a decrease in the latency between photon absorption and current onset. Interestingly, the inactivation phenotype of *trp* mutants is associated with no change in the amplitude of quantum bumps and a profound decrease in the frequency of bump generation (Minke 1982). Furthermore, the latency of the light response is increased rather than decreased by the inactivation process. These data suggest a defect in the bump generation machinery in *trp* photoreceptors, resulting in exhaustion of the excitation process, and a complete lack of adaptation processes.

The demonstration that the *trp* gene product is required for plasma membrane calcium permeability has helped provide a likely mechanistic explanation for the *trp* phenotype. Given the apparent requirement of internal calcium pools for excitation, it has been suggested that the *trp* phenotype results from the depletion of intracellular calcium stores and low intracellular calcium levels during sustained intense stimulation (Hardie & Minke 1992). In this model, the *trp*-dependent influx of extracellular calcium would be required to rapidly refill internal stores necessary for maintaining excitation. In *trp* photoreceptors, slower processes would fill these stores, consistent with the slow recovery from inactivation. The low intracellular calcium concentrations are also consistent with the lack of adaptation. If this model is correct, then the *trp* gene product represents a critical component of the general mechanisms of PI-mediated calcium influx.

Although the molecular identity of the non-*trp*-dependent channels is unknown, a screen for calmodulin-binding proteins in the *Drosophila* head has identified a protein with significant sequence similarity to the Trp protein that is also expressed in photoreceptors (Phillips et al 1992). This gene, called *trp-like* (*trpl*), encodes a protein that displays 39% amino acid identity with Trp and weak similarity to vertebrate neuronal Ca^{2+} channels in the transmembrane segments. Although it is possible that *trpl* encodes a subunit of a light-activated channel, the role of this protein in phototransduction awaits mutational analysis of the *trpl* gene.

Detailed biophysical analyses of the light-activated channels have been complicated by the difficulty of single-channel recording, primarily because of the limited access to these channels with available recording configurations. Nevertheless, estimates of single-channel properties have been inferred from steady-state noise analysis in whole-cell voltage-clamp recordings (Hardie & Minke 1994a). Although the light-induced electrical noise in photoreceptors is typically dominated by fluctuations in the signaling cascade, these studies make use of a specific physiological state in which the spontaneous activation of light-activated channels reveals channel-induced noise in the absence of transduction-induced noise. Spectral analyses of this channel-induced noise suggest a relatively high unit conductance for the non-*trp* class of channels (12–30 pS) and a low unit conductance for the *trp*-dependent class (~0.7 pS) (Hardie & Minke 1994a). Because of their small single-channel conductance, the *trp* class of channels may represent a low-noise–conduction pathway, and like the vertebrate cGMP-gated channels, may help to improve the signal-to-noise ratio in phototransduction. Confirmation of these estimated single-channel properties will require direct recording of single-channel activity, perhaps through heterologous expression of the cloned channel genes.

REGULATION OF PHOTOTRANSDUCTION IN *DROSOPHILA*

Positive and Negative Feedback Regulation by Calcium Influx

The high calcium permeability of the *Drosophila* light-activated conductance is particularly interesting, since it suggests that extracellular calcium influx may play a role in regulation of phototransduction. Indeed, Ranganathan et al (1991a) and Hardie (1991) have demonstrated a role for extracellular calcium in mediating both positive- and negative-feedback regulation of phototransduction. Whole-cell voltage-clamp recordings in wild-type photoreceptors showed that the responses to light stimuli were strongly asymmetric across the reversal potential; the kinetics of activation, deactivation, and adaptation were all much faster during inward currents (Ranganathan et al 1991a) (see Figure 3*b* and 3*c*). These asymmetries do not result from a direct voltage dependence of the electrical repsonse; they are instead dependent on the direction of ion flow. These data suggest a model in which calcium entry through light-activated channels may regulate signal transduction and may be responsible for the rapid kinetics of the light response during inward currents (Hardie 1991, Ranganathan et al 1991a). Indeed, experiments measuring both activation and deactivation kinetics as a function of extracellular calcium ($[Ca^{2+}]_{out}$) showed that calcium influx was both sufficient and necessary to regulate these processes (Ranganathan et al 1991a) (Figure 4). In the absence of external calcium, photoreceptors display symmetric responses with slow activation and deactivation kinetics. In addition, rapid extracellular perfusion of high calcium solutions or hyperpolarizing voltage steps during a light response, both of which should increase calcium influx, cause a transient acceleration in activation kinetics followed by a rapid inactivation (Hardie 1991, Hardie & Minke 1994b). Thus the light-dependent calcium influx is required for mediating sequential positive and negative feedback-regulation of phototransduction.

Molecular Mechanisms of Feedback Regulation

POSITIVE FEEDBACK   What might be the molecular mechanisms for these calcium influx–dependent regulatory processes? Although little is known about the basis for the positive-feedback regulation, it is tempting to speculate that this may arise through a calcium-induced calcium release (CICR) mechanism,

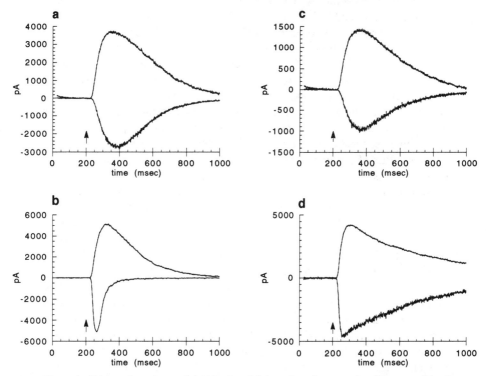

Figure 4   Calcium dependence of the kinetics of light-activated currents in wild-type and *inaC* mutant photoreceptors. Shown are responses of wild-type cells (*a,b*) and *inaC* mutant cells (*c,d*) to 10-ms flashes of white light at 0 and 1-mM Ca^{2+}$_{out}$, respectively. Recordings were taken at holding potentials of −80 mV and +80 mV, and the stimulus was applied at the time indicated by the arrow. Panels *a* and *b* show that extracellular calcium is both sufficient and necessary to regulate activation and deactivation of the light response primarily during inward currents. Thus extracellular calcium influx mediates sequential positive- and negative-feedback regulation of photoexcitation in *Drosophila* photoreceptors. Panels *c* and *d* show that the *inaC* mutation selectively removes the ability of extracellular calcium to cause normal negative-feedback regulation.

possibly analogous to that described in cardiac muscle (reviewed in Fabiato 1992). Another possibility is that because IP$_3$-receptor activity can be facilitated by calcium (Bezprozvanny et al 1991), influx of calcium may positively regulate release from internal stores. Further work should clarify whether such mechanisms operate in *Drosophila* photoreceptors.

PKC   The molecular basis for the calcium-dependent negative regulation in *Drosophila* photoreceptors has been the subject of much recent work, and these studies have provided a model for this process based on calcium-dependent protein phosphorylation. An electrophysiological screen for *Drosophila* phototransduction mutants with abnormal light-activated current kinetics demonstrated that photoreceptors from *inaC* mutants (Pak 1979) are specifically defective in the calcium-dependent negative regulatory mechanisms (Ranganathan et al 1991a). *InaC* photoreceptors show calcium dependence of activation kinetics similar to those of wild-type photoreceptors but have deactivation kinetics that are significantly slower and have lost the normal dependence on extracellular calcium (Figure 4). The expression of this phenotype is largely limited to inward currents and requires extracellular calcium, findings that strongly suggest that wild-type *inaC* activity is dependent on the light-dependent influx of calcium (Ranganathan et al 1991a). The molecular analysis of the *inaC* locus showed that it encodes a structural gene for an eye-specific isoform of protein kinase C (eye-PKC) (Schaeffer et al 1989, Smith et al 1991a), a class of serine-threonine kinases known to be activated by calcium and phospholipids (Nishizuka 1988). Eye-PKC is found only in photoreceptors, and within photoreceptors the protein localizes to the light-sensitive microvillar membranes (Smith et al 1991a). These results suggest a model in which the light-dependent generation of DAG and influx of external calcium activates eye-PKC, resulting in regulation of signal transduction through phosphorylation of one or more proteins involved in this signaling cascade (Smith et al 1991a). Recent biochemical experiments show that several retinal proteins in *Drosophila* are phosphorylated in a PKC-dependent manner (J Vinos & CS Zuker, unpublished data). Molecular characterization of these proteins may provide insight into the targets for eye-PKC in the phototransduction cascade.

The established concept that calcium-dependent negative regulatory mechanisms form the basis for adaptation mechanisms prompted the investigation of the role of eye-PKC in adaptation. Although initial data from whole-cell recordings in dissociated photoreceptors suggested a function for eye-PKC that was distinct from adaptation mechanisms (Smith et al 1991a), subsequent work using in vivo intracellular recordings from intact *Drosophila* heads has provided evidence for an important role for eye-PKC in adaptation (Hardie et al 1993b). Assays designed to measure the ability of photoreceptors to shift their operational range as a function of background illumination showed that *inaC* photoreceptors are essentially unable to carry out this process, and like *trp* photoreceptors, simply exhaust their excitatory mechanisms in response to increasing light intensities (Hardie et al 1993b). Thus, eye-PKC most likely mediates adaptation in *Drosophila* photoreceptors.

Interestingly, although bright stimuli cause loss of light sensitivity in *inaC* mutants because of exhausted excitatory mechanisms, responses to low-intensity stimuli are paradoxically enhanced in an extracellular calcium-dependent manner (Hardie et al 1993b). Also, as with the macroscopic light response, quantum bumps fail to terminate properly in *inaC* photoreceptors (Hardie et al 1993b), suggesting that eye-PKC functions at the level of the bump generation machinery. To explain these phenotypes, Hardie et al (1993b) have proposed a model in which eye-PKC acts at calcium release sites of the internal calcium pools to terminate the release of calcium, thereby terminating excitation. In *inaC* mutants, the inability to terminate calcium release would then lead to slower deactivation of quantum bumps and marcoscopic currents, enhancement of responses to weak stimuli due to failed deactivation, and inactivation to strong stimuli because of the depletion of calcium stores. Alternatively, eye-PKC may be required for resequestration of calcium into internal stores, since defects in these processes could also account for the *inaC* phenotype. The characterization of eye-PKC targets should distinguish between the above models and may help clarify the molecular basis for adaptation and quantum-bump generation.

CAM-KINASE   In addition to eye-PKC, there is evidence for other calcium-regulated kinases in the regulation of *Drosophila* phototransduction. In particular, Ca^{2+}-calmodulin-dependent protein kinase (CAM-kinase) has been implicated in the phosphorylation of a number of retinal specific proteins (LeVine et al 1990, Byk et al 1993). For example, proteins known as arrestins are phosphorylated in a Ca^{2+}- and calmodulin-dependent manner (Byk et al 1993). As described below, arrestins are proteins that are involved in the inactivation of metarhodopsin, suggesting that this process may be regulated by CAM-kinase. The recent isolation of calmodulin mutants (B Dixon, E Hafen & CS Zuker, unpublished data) in *Drosophila* should help to define the role of these potential feedback phosphorylation mechanisms in phototransduction.

Mechanisms of Metarhodopsin Inactivation

Several studies have provided evidence that metarhodopsin is an important target for regulatory mechanisms that terminate the active state of the light receptor long before its natural decay time. For example, metarhodopsin is functionally inactivated within milliseconds in vivo, although its natural lifetime in spectroscopic experiments ranges from minutes (Kuhn & Wilden 1987) to several hours (Schwemer 1984). This inherent stability of the activated receptor illustrates the requirement for mechanisms that inactivate metarhodopsin in order to avoid compromising the signaling process. Indeed, as described below, defects in these regulatory processes lead to abnormal termination of the light response, profound losses in light sensitivity, and photoreceptor degeneration (Dolph et al 1993). Interestingly, the molecules involved in these regulatory mechanisms are conserved in many diverse signaling cascades, suggesting that metarhodopsin inactivation represents a model system for the inactivation of all G protein–coupled receptors.

EXISTING MODELS IN VERTEBRATES   Current models for the molecular mechanisms mediating G protein–coupled receptor inactivation originated with studies investigating the biochemical basis of inactivation of vertebrate rhodopsin. Several groups suggested a role for kinases in this mediation process by demonstrating light-dependent phosphorylation of rhodopsin (Bownds et al 1972, Kuhn & Dreyer 1972, Frank et al 1973). This phosphorylation was demonstrated in vivo by direct injection of radiolabeled orthophosphate into frog eyes (Kuhn 1974) and shown to occur at a stretch of serines and threonines located at the carboxy terminus of the rhodopsin molecule, although other neighboring residues in the C terminus can be phosphorylated upon strong excitation (Thompson & Findlay 1984). The kinase that carries out this phosphorylation shows a high degree of specificity for the activated state of rhodopsin because both the resting state and the fully bleached state of rhodopsin are not substrates for this enzyme (McDowell & Kuhn 1977, Yamamoto & Shichi 1983). The enzyme, called rhodopsin kinase (RoK), is a 67-kD molecule (Palczewski 1988), which has been shown to bind to metarhodopsin-containing membranes (Kuhn et al 1984). Based on its ability to phosphorylate peptide substrates only after exposure to activated rhodopsin, RoK is thought to become activated through direct association with metarhodopsin (Palczewski et al 1991). Consistent with this model, studies have shown that RoK activity can be competed away with peptides that correspond to the putative association domains in rhodopsin (Kelleher & Johnson 1990).

The functional significance of the C-terminal phosphorylations of rhodopsin was demonstrated by in vitro assays of light-activated phosphodiesterase activity, which showed that the catalytic lifetime of metarhodopsin was greatly prolonged if phosphorylation was blocked by the removal of adenosine triphosphate (ATP) (Liebman & Pugh 1980) or RoK (Sitaramayya & Liebman 1983), or by proteolytic removal of the carboxy terminus of rhodopsin (Miller & Dratz 1984). Electrophysiological recordings of internally dialyzed vertebrate rod outer segments have supported this biochemical evidence for the role of phosphorylation of metarhodopsin in inactivation because the deactivation of the light-activated responses is significantly slowed if ATP is absent (Sather & Detweiler 1987, Nakatani & Yau 1988c) or if RoK activity is pharmacologically inhibited by sangivamycin (Palczewski et al 1992). These physiological studies must be interpreted with caution, however, since both the removal of ATP and the use of pharmacological agents such as sangivamycin are likely to have effects not limited only to the inhibition of RoK.

Although phosphorylation is required for the inactivation of metarhodopsin, this process is not sufficient for full inactivation (Wilden et al 1986, Bennet & Sitaramayya 1988). For instance, in vitro experiments measuring light-dependent PDE activity in isolated, washed preparations of rod outer-segment membranes showed that maximal phosphorylation of metarhodopsin can inhibit its activity several fold but that nearly complete inhibition could be achieved by the addition of an extremely abundant, soluble 48-kD protein (Wilden et al 1986). The ability to quench metarhodopsin-dependent PDE activity prompted the name arrestin for this 48-kD protein (Zuckerman & Cheasty 1986). The substrate specificity of arrestin is limited to phosphorylated metarhodopsin (all-*trans*-chromophore), since phosphorylated rhodopsin (11-*cis*-chromophore) did not exhibit high-affinity binding (Kuhn et al 1984). Other studies have shown that arrestin binding requires only a single phosphorylation of metarhodopsin (Bennet & Sitaramayya 1988). Kuhn's finding (1984) that arrestin could compete with the G-protein for binding to metarhodopsin led to a model in which partial phosphorylation of metarhodopsin followed by arrestin binding would terminate the interaction of G-protein and metarhodopsin, thereby quenching cascade activation (see Figure 1). Similar models have been proposed for the inactivation of other G protein–coupled receptors (Lefkowitz et al 1992).

METARHODOPSIN INACTIVATION IN *DROSOPHILA*   The mechanisms of metarhodopsin inactivation in *Drosophila* are thought to be fully conserved with vertebrates. Matsumoto & Pak (1984) have shown that the major *Drosophila* rhodopsin, Rh1, undergoes light-dependent phosphorylation of multiple serine and threonine residues at the C terminus. A study by Cassill and coworkers (1991) reports the isolation through use of polymerase chain reaction (PCR)– techniques of a group of genes expressed in the *Drosophila* retina that encode proteins with extensive homology to G protein–coupled receptor kinases. However, whether any of these proteins represent *Drosophila* RoK is still unknown. Biochemical experiments in retinal extracts from a closely related fly, *Musca domestica,* show an eye-specific kinase activity with substrate specificity for metarhodopsin that like vertebrate RoK, physically associates with metarhodopsin-containing membranes (Doza et al 1992).

Two arrestins have been isolated from *Drosophila*, arrestin 1 (Arr1) and arrestin 2 (Arr2) (Smith et al 1990, Hyde et al 1990, Levine et al 1990, Yamada et al 1990). *Arr1* was originally isolated by subtractive-hybridization techniques designed to identify genes preferentially expressed in the eye and was shown to encode a 364–amino acid protein that displays over 40% amino acid identity with human and bovine arrestins (Hyde et al 1990, Smith et al 1990). The Arr2 protein was purified on the basis that it is one of the most abundant light-dependent phosphoproteins in the eye, and using N-terminal sequence analysis, the gene encoding Arr2 was cloned (Levine et al 1990, Yamada et al 1990). Like Arr1, Arr2 also shows extensive amino acid similarity throughout its sequence with human and bovine arrestin (>40% identity) but contains the extended C-terminal region that is missing in Arr1. The functional significance of variation in this C-terminal domain is not understood. Quantitative

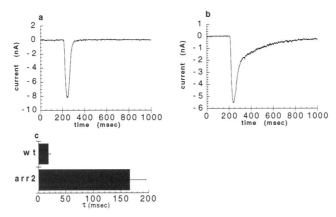

Figure 5 Arrestin is required for photoreceptor cell–deactivation. These are whole-cell voltage-clamp recordings of light-activated currents from wild-type (*top panel*) and *arr2* mutants (*right panel*). Responses are to 10-ms flashes of light at holding potentials of –40 mV. Preparations of isolated *Drosophila* photoreceptors and patch-clamp methods were as described (Dolph et al 1993). For quantitation of photoreceptor deactivation, tail currents of the light response were fitted to a single exponential function (*bottom panel*). The time constants (in milliseconds ± SD) are wt = 18.8 ± 2.9 and *arr2* = 166.9 ± 28.1.

Northern blot analysis (Levine et al 1990) and densitometric scanning of two-dimensional (2-D) protein gels (Matsumoto & Yamada 1991) have shown that the Arr2 protein is five- to sevenfold more abundant than Arr1. Interestingly, rhodopsin appears to be approximately fivefold more abundant than Arr2 (CS Zuker, unpublished data), suggesting that if the arrestin-metarhodopsin interaction is in fact stoichiometric, then this process could be saturated with stimuli that generate more metarhodopsin than available arrestin. If this interaction is required for metarhodopsin inactivation, such stimuli would be expected to cause serious defects in photoreceptor function.

The fact that two arrestin genes expressed in *Drosophila* photoreceptors have been isolated raises the issue of whether both arrestins are functionally redundant or whether each arrestin has a distinct function in regulating phototransduction. The possibility that the arrestin isoforms are required in different photoreceptor subtypes was eliminated by demonstrating that both Arr1 and Arr2 are expressed in all photoreceptors of the compound eye and the ocelli (Dolph et al 1993). Interestingly, in vivo phosphorylation studies have shown that although Arr1 and Arr2 become phosphorylated upon photoreceptor stimulation, the time course of Arr2 phosphorylation is much faster than that of Arr1 (Matsumoto & Yamada 1991). The phosphorylation of Arr1 and Arr2 is calcium dependent and is probably mediated through CAM-kinase (Byk et al 1993). These studies raise the interesting possibility that Arr1 and

Arr2 are differentially regulated by phosphorylation and that arrestin function may be subject to feedback regulation by calcium.

With regard to arrestin function in vivo, significant insight has been gained from the isolation and characterization of *Drosophila* mutants defective in arrestin expression (Dolph et al 1993). Because of the difficulty in predicting a reliable and easily scorable phenotype that defines the loss of arrestin function, Dolph et al (1993) used a genetic screen based on the loss of arrestin antigen on immunoblots to isolate several alleles of *arr1* and *arr2*. Electro-physiological analysis of strong *arr1* and *arr2* alleles revealed a set of phenotypes consistent with the stoichiometric requirement of arrestin binding for metarhodopsin inactivation in vivo. In whole-cell voltage-clamp experiments, a significant reduction in arrestin levels leads to abnormally slow deactivation of the light-activated currents (Dolph et al 1993) (Figure 5). Thus, arrestin is required for termination of the light response.

Direct evidence for defects in metarhodopsin inactivation in arrestin mutants came with the analysis of the prolonged depolarizing afterpotential (PDA). The PDA is a pathological state of the *Drosophila* photoreceptor in which a sustained photoresponse is triggered by substantial photoconversion (>20%) of rhodopsin to metarhodopsin (Minke et al 1975a, Hillman et al 1983) (Figure 6). During a PDA, photoreceptors are refractory to light stimuli and thus cannot respond to further visual input. Unlike vertebrate rhodopsins, most invertebrate photopigments are not bleached after light activation but can instead be photoconverted between the R and M forms. The wide separation of absorption maxima between these two forms permits the efficient experimental manipulation of these two states of the light receptor. Using this property, it has been shown that a PDA can be terminated by photoconverting metarhodopsin back to rhodopsin, suggesting that unattenuated metarhodopsin activity sustains the afterpotential. Thus a PDA is the consequence of the loss of normal metarhodopsin-inactivation mechanisms. If arrestin is involved in metarhodopsin inactivation, then defects in arrestin function may be expected to show defects in the PDA process. Interestingly, *arr2* photoreceptors undergo a PDA with 10 times less photoconversion of R to M (Dolph et al 1993) (Figure 6). This finding led to the following model: If arrestin interacts stoichiometrically to inactivate metarhodopsin, then saturation of arrestin function by excess metarhodopsin may represent the basis for the PDA (Dolph et al 1993). Thus, the ratio of free cytosolic arrestin to total rhodopsin would determine the threshold for PDA induction. Several findings are consistent with this model. First, the ratio of arrestin to rhodopsin in wild-type photoreceptors roughly predicts the requirement for photoconversion of at least 20% R to M for PDA induction. Second, *Drosophila* mutants with reduced rhodopsin levels are unable to generate a PDA, but a corresponding decrease in arrestin levels restores the PDA phenotype (Dolph et al 1993) (Figure 6). Finally, a theory that models

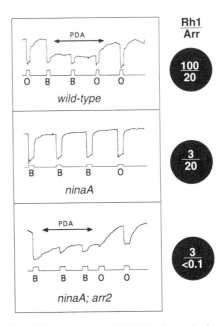

Figure 6 A prolonged depolarized afterpotential (PDA) is the result of excess metarhodopsin over available arrestin. Shown are electroretinogram recordings of wild-type flies (*top panel*), mutants with reduced rhodopsin levels (*ninaA*) (*middle panel*), and mutants with reduced levels of rhodopsin and arrestin (*ninaA* and *arr23* double mutants) (*bottom panel*). Responses are to maximum intensity 1-s flashes of 480-nm (R → M conversion) or 580-nm light (M → R conversion). PDA refers to the presence of a prolonged depolarizing afterpotential. The circles to the right indicate the approximate ratio of Rh1 rhodopsin to arrestin in the photoreceptor cells from the different genotypes.

the rate of metarhodopsin inactivation as a bimolecular interaction between free arrestin and metarhodopsin quantitatively predicts the PDA phenotype (R Ranganathan & CF Stevens, submitted). Also, this study showed that the kinetics of current deactivation in arrestin mutants can be fully described by the predicted rate of metarhodopsin inactivation based on remaining arrestin. Consistent with these data, in vitro biochemical studies in *Drosophila* show that light-dependent GTPase activity is not terminated in arrestin-depleted photoreceptor membranes but is restored upon addition of partially purified arrestin (Byk et al 1993). Thus the kinetics of metarhodopsin inactivation in vivo are most likely determined by the amount of free arrestin available to interact with metarhodopsin.

Rhodopsin Regeneration

An interesting issue is the mechanism of regeneration of rhodopsin from inactive arrestin-bound phosphorylated metarhodopsin. This process must involve the conformational change of metarhodopsin to rhodopsin, the dissociation of arrestin, and the dephosphorylation of rhodopsin. In this regard, recent studies have demonstrated a highly efficient calcium-dependent rhodopsin-phosphatase activity in the *Drosophila* eye that is absent in *retinal degeneration C* (*rdgC*) mutants (Byk et al 1993). The *rdgC* locus encodes a protein containing Ca^{2+}-binding sites and displaying sequence similarity to mammalian serine-threonine phosphatases (Steele et al 1992). A beautiful feature of *rdgC*-dependent rhodopsin dephosphorylation is that phosphatase activity is inhibited by arrestin binding (Byk et al 1993). This property is clearly crucial for photoreceptor function, since dephosphorylation of arrestin-bound metarhodopsin would presumably cause dissociation of arrestin and reactivation of metarhodopsin. Thus the current model for rhodopsin regeneration involves the photoisomerization of metarhodopsin to rhodopsin, which causes the dissociation of arrestin followed by dephosphorylation of rhodopsin by the RdgC protein (Figure 1). The calcium dependence of the phosphatase activity suggests that this step, like arrestin activity, may be subject to feedback regulation. The role of this putative feedback regulatory process awaits mutational analysis of functional domains of the *rdgC*-encoded phosphatase molecule.

RETINAL DEGENERATION

What may be the consequences of improper regulation of phototransduction? As previously discussed, defective regulatory processes can severely compromise photoreceptor function by reducing the overall speed of the light response, by removing adaptation, and by causing loss of light sensitivity. In addition, several studies have indicated that loss of regulation is often associated with photoreceptor degeneration in *Drosophila*, suggesting that photoreceptor viability also critically depends on these processes (see below). The mechanisms of stimulus-dependent degeneration in photoreceptors are of great interest because of possible similarities to mechanisms of excitotoxicity in central neurons and because of possible relevance to human diseases that are associated with retinal degeneration.

In *Drosophila*, a number of mutants have been isolated in which the morphogenesis of the eye is normal, but photoreceptors degenerate starting at the onset of adult life (Hotta & Benzer 1970, Harris et al 1976, Harris & Stark 1977; Steele & O'Tousa 1990). Although mutations in some genes required for normal rhodopsin levels (*ninaA, ninaE*) and for structural components (*ninaC, chp*) result in a light-independent rhabdomeric degeneration (reviewed in O'Tousa 1990), defects in some genes required for regulation of phototransduction result in a rapid, light-dependent photoreceptor degeneration. Three such mutations, *rdgB, rdgC,* and *arr2*, are discussed below.

rdgB   The phenotype of *rdgB* is characterized by a rapid, irreversible course of photoreceptor degeneration that can be triggered by as little as a single flash of light (Hotta & Benzer 1969, Harris & Stark 1977). *rdgB* photoreceptors do not degenerate if maintained in the dark or if transduction is blocked by the *norpA* mutation, suggesting a requirement for activation of the phototransduction cascade in mediating degeneration (Stark et al 1983). Interestingly, application of phorbol esters to eyes of *rdgB* mutants, but not wild-type, caused photoreceptor degeneration even in the dark, suggesting that the light-dependent activation of a PKC may mediate the degeneration process in *rdgB* cells (Minke et al 1990). This work raised the possibility that the eye-specific PKC encoded by the *inaC* locus may be involved in this process. Indeed, *rdgB*-dependent photoreceptor degeneration is suppressed in *inaC* mutants (Smith et al 1991a), implicating eye-PKC as a mediator of retinal degeneration in the absence of the *rdgB* gene product. Given that eye-PKC activity is controlled by the light-dependent influx of extracellular calcium (Ranganathan et al 1991a, Smith et al 1991a), agents that block the membrane permeability to calcium should also protect *rdgB* photoreceptors from degeneration. In fact, application of calcium channel blockers to eyes of *rdgB* mutants can delay, or even prevent, the light-dependent photoreceptor degeneration (Sahly et al 1992). These studies suggest that the *rdgB* gene product may somehow antagonize eye-PKC function, and the overactivity of eye-PKC in *rdgB* mutants may be responsible for photoreceptor degeneration.

A search for second-site suppressors of *rdgB* yielded the curious finding of *norpA* alleles that have no phenotype other than the ability to suppress *rdgB*-dependent photoreceptor degeneration (Harris & Stark 1977). Interestingly, this suppression was allele specific in that a given *norpA* allele would only suppress the *rdgB* allele that it was isolated with, suggesting a physical interaction of the two gene products. Further characterization of the targets of eye-PKC and the electrophysiological analysis of *rdgB* mutants may help to understand this complex interaction between *inaC, rdgB,* and *norpA*.

The cloning of the *rdgB* locus revealed a complex gene generating at least five separate transcripts expressed in the retina and other head structures (Vihtelic et al 1991). The *rdgB* gene product is an integral membrane protein with six predicted transmembrane domains, an ATP-binding site, and calcium-binding activity that may localize to the smooth endoplasmic reticulum found at the base of the rhabdomeric microvilli (Vihtelic et al 1993). In addition, the amino-terminal 281 residues of RdgB show significant (>40%) similarity to a rat brain phosphatidylinositol transfer protein (PI-TP) (Vihtelic et al 1993). Phospholipid transfer proteins are a class of soluble proteins that may be involved in shuttling phospholipids between membrane compartments (Kent et al 1991) and may also be involved in vesicular transport from the Golgi apparatus (Bankaitis et al 1989, Franzukoff & Schekman 1989). Biochemical studies have shown that this amino-terminal domain of RdgB has PI-TP activity in vitro (Vihtelic et al 1993). Together, these data suggest that RdgB may function to replenish rhabdomeric stores of PI metabolized during photoexcitation and may also function in transport of phototransduction proteins to the rhabdomeres. The function of the calcium-binding property of RdgB is unknown but may represent a mechanism for regulation of RdgB activity.

arr2   Like photoreceptor cells in *rdgB* mutants, photoreceptors with severe reduction in arrestin levels show a rapid light-dependent degeneration (Dolph et al 1993). The *norpA* mutation protects *arr2* photoreceptors from degeneration, indicating that PLC activation is required for this process. Because the functional defects in *arr2* mutants reflect the inability to inactivate metarhodopsin, these data argue that sustained cascade activation may irreversibly damage photoreceptors, possibly through calcium-dependent excitotoxicity. In support of this model, the *trp* mutation, which removes much of the light-dependent rise in intracellular calcium, also protects photoreceptor degeneration in *arr2* mutants (PJ Dolph & CS Zuker, unpublished data). Further characterization of this process may provide valuable insight into the mechanisms of neuronal excitotoxicity.

Photoreceptor degeneration in *arr2* mutants also provides an easily scorable phenotype for genetic screens for suppressors and enhancers of this process (PJ Dolph & CS Zuker, unpublished data). These studies should help to isolate interacting proteins that function in the arrestin-dependent regulatory pathway.

rdgC   The phenotype of *rdgC* photoreceptors is also characterized by light-dependent degeneration, but unlike in *rdgB* and *arr2,* this degeneration is not protected by the *norpA* mutation and requires significant light stimulation (Steele & O'Tousa 1990). This finding suggests that the pathophysiology of photoreceptor degeneration may be quite different in *rdgC* cells. As described above, the *rdgC* gene product is a calcium-dependent rhodopsin phosphatase, indicating that the inability to dephosphorylate rhodopsin may trigger degeneration through unknown mechanisms (Byk et al 1993). One possibility may be that the shorter half-lives of metarhodopsin (Schwemer 1984) and possibly phosphorylated rhodopsin may result in a breakdown of rhabdomeric membranes in excess of biogenesis in *rdgC* mutants, and may therefore cause a steady, light-dependent degeneration. Alternatively, it is also possible that the *rdgC* phosphatase has other substrates besides rhodopsin that mediate photoreceptor degeneration.

CONCLUSION

This review describes the combined application of molecular genetic, biochemical, and electrophysiological techniques in the analysis of photoreceptor function. Recent work utilizing these techniques in *Drosophila* has helped to further characterize several processes in phototransduction such as rhodopsin inactivation, photoreceptor deactivation and adaptation, and the terminal events in photoexcitation. Important issues for future work in this system include the identification of the mechanisms of channel activation, the genetic dissection of parallel feedback-regulatory processes, and the characterization of the precise molecular events that define the quantal response to light. Novel genetic screens designed to identify interacting proteins in the signaling cascade, and detailed biophysical analyses of wild-type and mutant photoreceptors will hopefully help us to understand these mechanisms that are fundamental for signal transduction.

ACKNOWLEDGMENTS

We would like to thank R Y Tsien, PJ Dolph, and MJ Kernan for critical reading of the manuscript, and RC Hardie and B Minke for communication of work in press. Work in the authors' lab was funded by grants from the National Eye Institute, the Pew Foundation, and the Howard Hughes Medical Institute. CSZ is an investigator of the Howard Hughes Medical Institute, and RR and DMM are trainees in the Medical Scientist Training Program.

Literature Cited

Artemyev NO, Rarick HM, Mills JS, Skiba NP, Hamm HE. 1992. Sites of interaction between rod G-protein α-subunit and cGMP-phosphodiesterase γ-subunit. Implications for the phosphodiesterase activation mechanism. *J. Biol. Chem.* 267:25067–72

Bacigalupo J, Johnson E, Robinson P, Lisman JE. 1990. Second messengers in invertebrate phototransduction. In *Transduction in Biological Systems,* ed. C Hidalgo, J Bacigalupo, E Jaimovich, J Vergara, pp. 27–45. New York: Plenum

Bacigalupo J, Johnson EC, Vergara C, Lisman J. 1991. Light-dependent channels from excised patches of *Limulus* ventral photoreceptors are opened by cGMP. *Proc. Natl. Acad. Sci. USA* 88:7938–42

Bankaitis VA, Malehorn DE, Emr SD, Greene R. 1989. The *Saccharomyces cerevisiae* SEC14 gene encodes a cytosolic factor that is required for transport of secretory proteins from the yeast Golgi complex. *J. Cell Biol.* 108:1271–81

Baylor DA, Lamb TD, Yau K-W. 1979. Re-sponses of retinal rods to single photons. *J. Physiol.* 288:613–34

Bennett N, Sitaramayya A. 1988. Inactivation of photoexcited rhodopsin in retinal rods: the roles of rhodopsin kinase and 48-kDa protein (arrestin). *Biochemistry* 27:1710–15

Bentrop J, Paulsen R. 1986. Light-modulated ADP-ribosylation, protein phosphorylation and protein binding in isolated fly photoreceptor membranes. *Eur. J. Biochem.* 161:61–67

Berridge MJ. 1987. Inositol trisphosphate and diacylglycerol: two interacting second messengers. *Annu. Rev. Biochem.* 56:159–93

Berridge MJ. 1993. Inositol trisphosphate and calcium signalling. *Nature* 361:315–25

Bezprozvanny I, Watras J, Ehrlich BE. 1991. Bell-shaped calcium-response curves of Ins(1,4,5)P3- and calcium-gated channels from endoplasmic reticulum of cerebellum. *Nature* 351:751–54

Bloomquist B, Shortridge R, Schneuwly S, Perdew M, Montell C, et al. 1988. Isolation of a putative phospholipase C gene, *norpA*, and its role in phototransduction. *Cell* 54: 723–33

Blumenfeld A, Erusalimsky J, Heichal O, Selinger Z, Meinke B. 1985. Light-activated guanosinetriphosphotase in *Musca* eye membranes resembles the prolonged depolarizing afterpotential in photoreceptor cells. *Proc. Natl. Acad. Sci. USA* 82:7116–20

Bodoia RD, Detwiler PB. 1985. Patch-clamp recordings of the light-sensitive dark noise in retinal rods from the lizard and frog. *J. Physiol.* 367:183–216

Bownds D, Dawes J, Miller J, Stahlman M. 1972. Phosphorylation of frog photoreceptor membranes induced by light. *Nature* 237: 125–27

Britt SG, Feiler R, Kirschfeld K, Zuker CS. 1993. Spectral tuning of rhodopsin and metarhodopsin in vivo. *Neuron* 11:29–39

Brown JE, Rubin LJ, Ghalayini AJ, Tarver AP, Irvine RF, et al. 1984. Evidence that myoinositol polyphosphate may be a messenger for visual excitation in *Limulus* photoreceptors. *Nature* 311:160–63

Brown JE, Watkins DC, Malbon CC. 1987. Light-induced changes in the content of inositol phosphates in squid *Loligo pealei* retina. *Biochem. J.* 247:293–97

Byk T, Bar-Yaacov M, Doza YN, Minke B, Selinger Z. 1993. Regulatory arrestin cycle secures the fidelity and maintenance of the photoreceptor cell. *Proc. Natl. Acad. Sci. USA* 90:1907–11

Cassill JA, Whitney M, Joazeiro CAP, Becker A, Zuker CS. 1991. Isolation of *Drosophila* genes encoding G protein-coupled receptor kinases. *Proc. Natl. Acad. Sci. USA* 88: 11067–70

Cosens D, Manning A. 1969. Abnormal electroretinogram from a *Drosophila* mutant. *Nature* 224:285–87

Cowman AF, Zuker CS, Rubin GR. 1986. An opsin gene expressed in only one photoreceptor cell type of the *Drosophila* eye. *Cell* 44: 705–10

Devary O, Heichal O, Blumenfeld A, Cassel D, Suss E, et al. 1987. Coupling of photoexcited rhodopsin to inositol phospholipid hydrolysis in fly photoreceptors. *Proc. Natl. Acad. Sci. USA* 84:6939–43

Dizhoor AM, Ray S, Kumar S, Neimi G, Spencer M, et al. 1991. Recoverin: a calcium sensitive activator of retinal rod guanylate cyclase. *Science* 251:915–18

Dolph PJ, Man-Son-Hing H, Yarfitz S, Colley NJ, Running Deer J, et al. 1994. An eye-specific Gβ subunit essential for termination of the phototranception cascade. *Nature* 370: 59–61

Dolph PJ, Ranganathan R, Colley NJ, Hardy RW, Socolich M, Zuker CS. 1993. Arrestin function in inactivation of G protein-coupled

receptor rhodopsin in vivo. *Science* 260: 1910–16

Doza YN, Minke B, Chorev M, Selinger Z. 1992. Characterization of fly rhodopsin kinase. *Eur. J. Biochem.* 209:1035–40

Fabiato A. 1992. Two kinds of calcium-induced release of calcium from the sarcoplasmic reticulum of skinned cardiac cells. *Adv. Exp. Med. Biol.* 311:245–62

Fain GL, Lamb TD, Matthews HR, Murphy RLW. 1989. Cytoplasmic calcium as the messenger for light adaptation in salamander rods. *J. Physiol.* 416:215–43

Feiler R, Bjornson R, Kirshfeld K, Mismer D, Rubin GM, et al. 1992. Ectopic expression of ultraviolet-rhodopsins in the blue photoreceptor cells of *Drosophila*: visual physiology and photochemistry of transgenic animals. *J. Neurosci.* 12:3862–68

Feiler R, Harris WA, Kirshfeld K, Wehrhan C, Zuker CS. 1988. Targeted misexpression of a *Drosophila* opsin gene leads to altered visual function. *Nature* 333:737–41

Feng JJ, Frank TM, Fein A. 1991. Excitation of *Limulus* photoreceptors by hydrolysis-resistant analogs of cGMP and cAMP. *Brain Res.* 552:291–94

Fesenko SS, Kolesnikou AL, Lyubarsky EE. 1985. Induction by cGMP of cationic conductance on the plasma membrane of the retinal rod outer segment. *Nature* 313:310–13

Franceschini N. 1985. Early processing of colour and motion in a mosaic visual system. *Neurosci. Res.* 2:S17–S49 (Suppl.)

Frank RN, Cavanagh HD, Kenyon KR. 1973. Light-stimulated phosphorylation of bovine visual pigments by adenosine triphosphate. *J. Biol. Chem.* 248:596–609

Franzusoff A, Schekmann R. 1989. Functional compartments of the yeast Golgi apparatus are defined by the *sec7* mutation. *Eur. Mol. Biol. Organ. J.* 8:2695–2702

Gray P, Atwell D. 1985. Kinetics of light-sensitive channels in vertebrate photoreceptors. *Proc. R. Soc. London Ser. B* 223:379–88

Gray-Keller MP, Polans AS, Palczewski K, Detwiler PB. 1993. The effect of recoverin-like calcium-binding proteins on the photoresponse of retinal rods. *Neuron* 10:523–31

Hardie RC. 1983. *Progress in Sensory Physiology*. New York: Springer

Hardie RC. 1991. Whole-cell recordings of the light induced current in dissociated *Drosophila* photoreceptors: evidence for feedback by calcium permeating the light-sensitive channels. *Proc. R. Soc. London Ser. B* 245:203–10

Hardie RC, Minke B. 1992. The *trp* gene is essential for a light-activated Ca^{2+} channel in *Drosophila* photoreceptors. *Neuron* 8: 643–51

Hardie RC, Minke B. 1993. Novel Ca^{2+} channels underlying transduction in *Drosophila* photoreceptors: implications for phosphoinositide-mediated Ca^{2+} mobilization. *Trends Neurosci.* 16:371–76

Hardie RC, Minke B. 1994a. Spontaneous activation of light-sensitive channels in *Drosophila* photoreceptors. *J. Gen. Physiol.* In press

Hardie RC, Minke B. 1994b. Calcium-dependent inactivation of light-sensitive channels in *Drosophila* photoreceptors. *J. Gen. Physiol.* In press

Hardie RC, Peretz A, Pollock JA, Minke B. 1993a. Ca^{2+} limits the development of the light response in *Drosophila* photoreceptors. *Proc. R. Soc. London Ser. B* 252:223–29

Hardie RC, Peretz A, Suss-Toby E, Rom-Glas A, Bishop SA, Selinger Z. 1993b. Protein kinase C is required for light adaptation in *Drosophila* photoreceptors. *Nature* 363:634–37

Hargrave PA, McDowell JH. 1992. Rhodopsin and phototransduction. *Int. Rev. Cytol.* 137B:49–97

Harris WA, Stark WS. 1977. Hereditary retinal degeneration in *Drosophila melanogaster*. A mutant defect associated with the phototransduction process. *J. Gen. Physiol.* 69:261–91

Harris WA, Stark WS, Walker JA. 1976. Genetic dissection of the photoreceptor system in the compound eye of *Drosophila melanogaster*. *J. Physiol.* 256:415–39

Haynes LW, Kay AR, Yau K-W. 1986. Single cyclic GMP-activated channel activity in excised patches of rod outer segment membrane. *Nature* 321:66–70

Hayward G, Carlsen W, Siegman A, Stryer L. 1981. Retinal chromophore of rhodopsin photoisomerizes within picoseconds. *Science* 211:942–44

Hecht S, Shlaer S, Pirenne MH. 1942. Energy, quanta, and vision. *J. Gen. Physiol.* 25:819–40

Heisenberg M, Wolf R. 1984. *Vision in Drosophila*. New York: Springer

Hille B. 1992. *Ionic Channels of Excitable Membranes*. Sunderland, MA: Sinauer

Hillman P, Hochstein S, Minke B. 1983. Transduction in invertebrate photoreceptors: role of pigment bistability. *Physiol. Rev.* 63:668–72

Hochstrate P. 1989. Lanthanum mimics the *trp* photoreceptor mutant of *Drosophila* in the blowfly *Calliphora*. *J. Comp. Physiol. A* 166:179–87

Hodgkin AL, Nunn BJ. 1988. Control of light-sensitive current in salamander rods. *J. Physiol.* 403:439–71

Hotta Y, Benzer S. 1969. Abnormal electroretinograms of visual mutants in *Drosophila*. *Nature* 222:354–56

Hotta Y, Benzer S. 1970. Genetic dissection of the *Drosophila* nervous system by means of

mosaics. *Proc. Natl. Acad. Sci. USA* 67:1156–63

Hsu Y-T, Molday RS. 1993. Modulation of the cGMP-gated channel of rod photoreceptor cells by calmodulin. *Nature* 361:76–79

Hyde DR, Mecklenburg KL, Pollock JA, Vihtelic TS, Benzer S. 1990. Twenty *Drosophila* visual system cDNA clones: one is a homolog of human arrestin. *Proc. Natl. Acad. Sci. USA* 87:1008–12

Inesi G, Sagara Y. 1992. Thapsigargin, a high affinity and global inhibitor of intracellular Ca^{2+} transport ATPases. *Arch. Biochem. Biophys.* 298:313–17

Irvine RF. 1991. Inositol tetrakisphosphate as a second messenger: confusions, contradictions, and a potential resolution. *Bioessays* 13:419–27

Irvine RF. 1992. Inositol phosphates and Ca^{2+} entry: toward a proliferation or a simplification? *FASEB J.* 6:3085–91

Johnson EC, Robinson PR, Lisman JE. 1986. Cyclic GMP is involved in the excitation of invertebrate photoreceptors. *Nature* 324:468–70

Karpen JW, Zimmerman AL, Stryer L, Baylor DA. 1988. Gating kinetics of the cyclic-CMP-activated channel of retinal rods: flash photolysis and voltage-jump studies. *Proc. Natl. Acad. Sci. USA* 85:1287–91

Kawamura S. 1993. Rhodopsin phosphorylation as a mechanism of cyclic GMP phosphodiesterase regulation by S-modulin. *Nature* 362:855–57

Kawamura S, Hisatomi O, Kayada S, Tokunaga F, Kuo C-H. 1993. Recoverin has S-modulin activity in frog rods. *J. Biol. Chem.* 268:14579–82

Kawamura S, Murakami M. 1991. Calcium-dependent regulation of cyclic GMP phosphodiesterase by a protein from frog retinal rods. *Nature* 349:420–23

Kelleher DJ, Johnson GL. 1990. Characterization of rhodopsin kinase purified from bovine rod outer segments. *J. Biol. Chem.* 265:2632–39

Kent C, Carman GM, Spence MW, Dowhan W. 1991. Regulation of eukaryotic phospholipid metabolism. *FASEB J.* 5:2258–66

Kuhn H. 1974. Light-dependent phosphorylation of rhodopsin in living frogs. *Nature* 250:588–90

Kuhn H. 1984. Interactions between photoexcited rhodopsin and light-activated enzymes in rods. In *Progress in Retinal Research*, ed. N Osborne, J Chader, 3:124–56. Oxford: Pergamon

Kuhn H, Dreyer WJ. 1972. Light dependent phosphorylation of rhodopsin by adenosine triphosphate. *FEBS Lett.* 20:1–6

Kuhn H, Hall SW, Wilden U. 1984. Light-induced binding of 48-kDa protein to photoreceptor membranes is highly enhanced by phosphorylation of rhodopsin. *FEBS Lett.* 176:473–78

Kuhn H, Wilden U. 1987. Deactivation of photoactivated rhodopsin by rhodopsin-kinase and arrestin. *J. Recept. Res.* 7:283–98

Lagnado L, Baylor DA. 1994. Calcium controls light-triggered formation of catalytically active rhodopsin. *Nature* 367:273–77

Lambrecht HG, Koch KW. 1991. Recoverin, a novel calcium-binding protein from vertebrate photoreceptors. *Eur. Mol. Biol. Organ. J.* 10:793–98

Lee Y-J, Dobbs MB, Verardi ML, Hyde DR. 1990. *dgq*: a *Drosophila* gene encoding a visual system-specific Ga molecule. *Neuron* 5:889–98

Lefkowitz RJ, Inglese J, Koch WJ, Pitcher J, Attramadal H, Caron MG. 1992. G-protein-coupled receptors: regulatory role of receptor kinases and arrestin proteins. *Cold Spring Harbor Symp. Quant. Biol.* 57:127–33

LeVine H, Smith DP, Whitney M, Malicki DM, Dolph PJ, et al. 1990. Isolation of a visual system specific arrestin: an in vivo substrate for light-dependent phosphorylation. *Mech. Dev.* 33:19–26

Liebman PA, Pugh EN. 1980. ATP mediates rapid reversal of cyclic GMP phosphodiesterase activation in visual receptor membranes. *Nature* 287:734–36

Matsumoto H, Pak WL. 1984. Light-induced phosphorylation of retina-specific polypeptides of *Drosophila* in vivo. *Science* 223: 184–86

Matsumoto H, Yamada T. 1991. Phosrestins I, II: arrestin homologs which undergo differential light-induced phosphorylation in the *Drosophila* photoreceptor in vivo. *Biochem. Biophys. Res. Commun.* 177:1306–12

Matthews HR, Murphy RLW, Fain GL, Lamb TD. 1988. Photoreceptor light adaptation is mediated by cytoplamic calcium concentration. *Nature* 334:67–69

McDowell JH, Kuhn H. 1977. Light-induced phosphorylation of rhodopsin in cattle photoreceptor membranes: substrate activation and inactivation. *Biochemistry* 16:4054–60

Miller JL, Dratz EA. 1984. Phosphorylation at sites near rhodopsin's carboxyl-terminus regulates light initiated cGMP hydrolysis. *Vis. Res.* 24:1509–21

Minke B. 1979. Transduction in photoreceptors with bistable pigments: intermediate processes. *Biophys. Struct. Mech.* 5:163–74

Minke B. 1982. Light-induced reduction in excitation efficiency in the *trp* mutant in *Drosophila*. *J. Gen. Physiol.* 79:361–85

Minke B, Rubinstein CT, Sahly I, Bar-Nachum S, Timberg R, Selinger Z. 1990. Phorbol ester induces photoreceptor-specific degeneration in a *Drosophila* mutant. *Proc. Natl. Acad. Sci. USA* 87:113–17

Minke B, Stephenson RS. 1985. The characteristics of chemically induced noise in *Musca* photoreceptors. *J. Comp. Physiol.* 156:339–56

Minke B, Wu C-F, Pak WL. 1975a. Isolation of light-induced response of central retinular cells from electroretinogram of *Drosophila*. *J. Comp. Physiol.* 98:345–55

Minke B, Wu C-F, Pak WL. 1975b. Induction of photoreceptor voltage noise in the dark in a *Drosophila* mutant. *Nature* 258:84–87

Mismer D, Michael WM, Laverty TR, Rubin GR. 1988. Analysis of the promoter of the Rh2 opsin gene in *Drosophila melanogaster*. *Genetics* 120:173–80

Montell C, Jones K, Zuker CS, Rubin GR. 1987. A second opsin gene expressed in the ultraviolet-sensitive R7 photoreceptor cells of *Drosophila melanogaster*. *J. Neurosci.* 7: 1558–66

Montell C, Rubin GM. 1989. Molecular characterization of the *Drosophila trp* locus: a putative integral membrane protein required for phototransduction. *Neuron* 2:1313–23

Nakatani K, Yau K-W. 1988a. Calcium and magnesium fluxes across the plasma membrane of the toad rod outer segment. *J. Physiol.* 395:695–729

Nakatani K, Yau K-W. 1988b. Calcium and light adaptation in retinal rods and cones. *Nature* 334:69–71

Nakatani K, Yau K-W. 1988c. Guanosine 3',5'-cyclic monophosphate-activated conductance studied in a truncated rod outer segment of the toad. *J. Physiol.* 395:731–53

Nishizuka Y. 1988. The molecular heterogeneity of protein kinase C and its implications for cellular regulation. *Nature* 334:661–65

O'Tousa JE. 1990. Genetic analysis of phototransduction in *Drosophila*. *Semin. Neurosci.* 2:207–15

O'Tousa JE, Baehr W, Martin RL, Hirsh J, Pak WL, Applebury ML. 1985. The *Drosophila ninaE* gene encodes an opsin. *Cell* 40:839–50

Pak WL. 1979. Study of photoreceptor function using *Drosophila* mutants. In *Neurogenetics, Genetic Approaches to the Nervous System*, ed. XO Breakfeld, pp. 67–99. New York: Elsevier

Pak WL, Grossfield J, Arnold K. 1970. Mutants of the visual pathway of *Drosophila melanogaster*. *Nature* 227:518–20

Palczewski K, Buczylko J, Kaplan MW, Polans AS, Crabb JW. 1991. Mechanism of rhodopsin kinase activation. *J. Biol. Chem.* 266: 12949–55

Palczewski K, McDowell JH, Hargrave PA. 1988. Purification and characterization of rhodopsin kinase. *J. Biol. Chem.* 263:14067–73

Palczewski K, Rispoli G, Detwiler PB. 1992. The influence of arrestin (48K protein) and rhodopsin kinase on visual transduction. *Neuron* 8:117–26

Payne R. 1986. Phototransduction by microvillar photoreceptors of invertebrates: mediation of a visual cascade by inositol trisphosphate. *Photobiochem. Photobiophys.* 13:373–97

Phillips AM, Bull A, Kelly LE. 1992. Identification of a *Drosophila* gene encoding a calmodulin-binding protein with homology to the *trp* phototransduction gene. *Neuron* 8:631–42

Polans AS, Buczylko J, Crabb J, Palczewski K. 1991. A photoreceptor calcium binding protein is recognized by autoantibodies obtained from patients with cancer-associated retinopathy. *J. Cell Biol.* 112:981–89

Pollock JA, Benzer S. 1988. Transcript localization of four opsin genes in the three visual organs of *Drosophila*; Rh2 is ocellus specific. *Nature* 333:779–82

Pugh EN, Lamb TD. 1990. Cyclic GMP and calcium: the internal messengers of excitation and adaptation in vertebrate photoreceptors. *Vis. Res.* 30:1923–48

Putney JW. 1990. Capacitative calcium entry revisited. *Cell Calcium* 11:611–24

Randriamampita C, Tsien RY. 1993. Emptying of intracellular Ca^{2+} stores releases a novel small messenger that stimulates Ca^{2+} influx. *Nature* 364:809–14

Ranganathan R, Bacskai BJ, Tsien RY, Zuker CS. 1994. Cytosolic calcium transients: spatial localization and role in *Drosophila* photoreceptor cell function. *Neuron.* In press

Ranganathan R, Harris GL, Stevens CF, Zuker CS. 1991a. A *Drosophila* mutant defective in extracellular calcium dependent photoreceptor inactivation and rapid desensitization. *Nature* 354:230–35

Ranganathan R, Harris WA, Zuker CS. 1991b. The molecular genetics of invertebrate phototransduction. *Trends Neurosci.* 14:486–93

Sahly I, Bar Nachum S, Suss-Toby E, Rom A, Peretz A, et al. 1992. Calcium channel blockers inhibit retinal degeneration in the *retinal-degeneration-B* mutant of *Drosophila. Proc. Natl. Acad. Sci. USA* 89:435–39

Sather WA, Detwiler PB. 1987. Intracellular biochemical manipulation of phototransduction in detached rod outer segments. *Proc. Natl. Acad. Sci. USA* 84:9290–94

Schaeffer E, Smith D, Mardon G, Quinn W, Zuker CS. 1989. Isolation and characterization of two new *Drosophila* protein kinase C genes, including one specifically expressed in photoreceptor cells. *Cell* 57:403–12

Schoenlein RW, Peteanu LA, Mathies RA, Shank CV. 1991. The first step in vision: femtosecond isomerization of rhodopsin. *Science* 254:412–15

Schwemer J. 1984. Renewal of visual pigment in photoreceptors of the blowfly. *J. Comp. Physiol. A* 154:535–47

Shieh B-H, Stamnes M, Seavello S, Harris G,

Zuker C. 1989. The *ninaA* gene required for visual transduction in *Drosophila* encodes a homologue of cyclosporin A-binding protein. *Nature* 338:67–70

Sitaramayya A, Liebman PA. 1983. Mechanism of ATP quench of phosphodiesterase activation in rod disc membranes. *J. Biol. Chem.* 258:1205–9

Smith DP, Ranganathan R, Hardy RW, Marx J, Tsuchida T, Zuker CS. 1991a. Photoreceptor deactivation and retinal degeneration mediated by a photoreceptor-specific protein kinase C. *Science* 254:1478–84

Smith DP, Shieh B-H, Zuker CS. 1990. Isolation and structure of an arrestin gene from *Drosophila. Proc. Natl. Acad. Sci. USA* 87: 1003–7

Smith DP, Stamnes MA, Zuker CS. 1991b. Signal transduction in the visual system of *Drosophila. Annu. Rev. Cell Biol.* 7:161–90

Stark WS, Chen D-M, Johnson MA, Frayer KL. 1983. The *rdgB* gene in *Drosophila*: retinal degeneration in several mutant alleles and inhibition by *norpA. J. Insect Physiol.* 29: 123–31

Steele F, O'Tousa JE. 1990. Rhodopsin activation causes retinal degeneration in *Drosophila rdgC* mutant. *Neuron* 4:883–90

Steele FR, Washburn T, Rieger R, O'Tousa JE. 1992. *Drosophila retinal degeneration C (rdgC)* encodes a novel serine/threonine protein phosphatase. *Cell* 69:669–76

Stieve H. 1986. *Bumps, The Elemental Excitory Responses of Invertebrates.* New York: Plenum

Stryer L. 1986. Cyclic GMP cascade of vision. *Annu. Rev. Neurosci.* 9:87–119

Stryer L. 1991. Visual excitation and recovery. *J. Biol. Chem.* 266:10711–14

Stryer L, Bourne HR. 1986. G-proteins: a family of signal transducers. *Annu. Rev. Cell Biol.* 2:391–419

Suss-Toby E, Selinger Z, Minke B. 1991. Lanthanum reduces the excitation efficiency in fly photoreceptors. *J. Gen. Physiol.* 98:849–68

Szuts EZ, Wood SF, Reid MS, Fein A. 1986. Light stimulates the rapid formation of inositol trisphosphate in squid retinas. *Biochem. J.* 240:929–32

Thompson P, Findlay JBC. 1984. Phosphorylation of ovine rhodopsin. Identification of the phosphorylation sites. *Biochem. J.* 220:773–80

Tomlinson A. 1988. Cellular interactions in the developing *Drosophila* eye. *Development* 104:183–93

Vihtelic TS, Goebl M, Milligan S, O'Tousa JE, Hyde DR. 1993. Localization of *Drosophila retinal degeneration B,* a membrane-associated phosphatidylinositol transfer protein. *J. Cell Biol.* 122:1013–22

Vihtelic TS, Hyde DR, O'Tousa 1991. Isolation

and characterization of the *Drosophila retinal degeneration B* (*rdgB*) gene. *Genetics* 127:761–68

Wald G. 1968. The molecular basis of visual excitation. *Nature* 219:800–7

Walz B. 1982a. Calcium sequestering smooth endoplasmic reticulum in the invertebrate photoreceptor. I. Intracellular topography as revealed by OsFeCN staining and in situ calcium accumulation. *J. Biol. Chem.* 93:839–48

Walz B. 1982b. Calcium sequestering smooth endoplasmic reticulum in the invertebrate photoreceptor. II. Its properties as revealed by microphotometric measurements. *J. Cell. Biol.* 93:849–59

Walz B. 1982c. Calcium sequestering smooth endoplasmic reticulum in retinula cells of the blowfly. *J. Ultrastruct. Res.* 81:240–48

Wilden U, Hall S, Kuhn H. 1986. Phosphodiesterase activation by photoexcited rhodopsin is quenched when rhodopsin is phosphorylated and binds 48-kDa protein. *Proc. Natl. Acad. Sci. USA* 83:1174–78

Wong F, Scharfer EL, Roop BC, LaMendola JN, Johnson-Seaton D, Shao D. 1989. Proper function of the *Drosophila trp* gene product during pupal development is important for normal visual transduction in the adult. *Neuron* 3:81–94

Yamada T, Takeuchi Y, Komori N, Kobayashi H, Sakai Y, et al. 1990. A 49-kilodalton phosphoprotein in the *Drosophila* photoreceptor is an arrestin homolog. *Science* 248:483–86

Yamamoto K, Shichi H. 1983. Rhodopsin phosphorylation occurs at metarhodopsin II level. *Biophys. Struct. Mech.* 9:259–67

Yarfitz S, Niemi GA, McConnell JL, Fitch CL, Hurley JB. 1991. A G beta protein in the *Drosophila* compound eye is different from that in the brain. *Neuron* 7:429–38

Zimmerman AL, Baylor DA. 1986. Cyclic GMP-sensitive conductance of retinal rods consists of aqueous pores. *Nature* 321:70–72

Zuckerman R, Cheasty JE. 1986. A 48 kDa protein arrests cGMP phosphodiesterase activation in retinal rod disk membranes. *FEBS Lett.* 207:35–41

Zuker CS. 1992. Phototransduction in *Drosophila*: a paradigm for the genetic dissection of sensory transduction cascades. *Curr. Opin. Neurobiol.* 2:622–27

Zuker CS, Cowman AF, Rubin GM. 1985. Isolation and structure of a rhodopsin gene from *Drosophila melanogaster*. *Cell* 40:851–58

Zuker CS, Montell C, Jones K, Laverty T, Rubin GR. 1987. A rhodopsin gene expressed in photoreceptor cell R7 of the *Drosophila* eye: homologies with other signal-transducing molecules. *J. Neurosci.* 7:1550–57

Annu. Rev. Neurosci. 1995. 18:319–57

LONG-TERM SYNAPTIC DEPRESSION

David J. Linden[1]

Department of Neuroscience, Johns Hopkins University School of Medicine, Baltimore, Maryland 21205

John A. Connor

Roche Institute of Molecular Biology, Roche Research Center, Nutley, New Jersey 07110

KEY WORDS: memory, synaptic plasticity, cerebral cortex, hippocampus, cerebellum

INTRODUCTION

It is widely assumed that long-term changes in synaptic strength underlie information storage in the brain and, ultimately, behavioral memory. Recent years have seen a major effort to identify and analyze electrophysiological model systems in which particular patterns of neural activity give rise to such enduring changes. Most of this attention has been focused upon hippocampal long-term synaptic potentiation (LTP), in which brief activation of a set of afferents gives rise to a persistent increase in synaptic strength of the activated synapses. LTP is a broad term. It has come to mean any persistent increase in synaptic strength induced by a variety of mechanisms in a large number of locations in the nervous system. The purpose of the present review is to examine the opposite phenomenon, use-dependent long-lasting decreases in synaptic strength, which have been collectively termed long-term synaptic depression (LTD). Also a blanket term, LTD denotes depression induced according to a variety of synaptic modification rules, mediated by various electrophysiological and biochemical events, and occurring extensively in the nervous system.

[1]Author to whom correspondence should be addressed.

0147-006X/95/0301-0319$05.00

The idea that particular patterns of synaptic activity might result in a selective strengthening of synapses was proposed by Cajal (1911). Hebb later formalized it, postulating that an increase in synaptic strength will occur when the presynaptic and the postsynaptic elements of a synapse are coactive (Hebb 1949). In the CA1 region of the hippocampus, LTP is typically induced by brief high-frequency activation of excitatory glutamatergic fibers; it is seen as an increase in synaptic strength that lasts from hours to days. It is attractive as a model system not only because of its duration but also because the induction of LTP requires a Hebbian coactivation. This coactivation results from the fact that LTP induction requires two separate conditions: The N-methyl-D-aspartate (NMDA) receptor must bind glutamate, and a threshold level of postsynaptic depolarization must be reached before ions can flow through the NMDA receptor-associated ion channel. Ca influx into the postsynaptic cell via NMDA receptor activation is thought to be necessary but not sufficient for induction of LTP (see Bliss & Collingridge 1993 for review). Certain forms of LTP are not dependent upon activation of NMDA receptors and demonstrate somewhat different properties as a result (see Johnston et al 1992 for review).

A variety of stimulation patterns are capable of inducing long-term changes in synaptic strength. Some of these are illustrated schematically in Figure 1. LTP is induced when presynaptic activity occurs together with strong postsynaptic activity. This persistent alteration is said to be homosynaptic because the increase in synaptic strength occurs in the same synapses where presynaptic activity was present during its induction. Associative LTP is a special case of homosynaptic LTP in which the same synaptic modification rule is operative, but postsynaptic activation is driven by neighboring afferents converging on the same postsynaptic cell.

Heterosynaptic LTD is a long-lasting decrease in synaptic strength induced when strong postsynaptic activity, driven by converging afferents, occurs in the absence of presynaptic activity. In contrast, homosynaptic LTD is a long-lasting decrease in synaptic strength induced when presynaptic activity occurs together with moderate postsynaptic activity. Like homosynaptic LTP, it is confined to synapses that were active during induction. Associative LTD is a persistent decrease in synaptic strength induced when presynaptic activity occurs explicitly out-of-phase with strong postsynaptic activity, the latter driven by converging afferents. While associative LTD may be considered a form of homosynaptic LTD, it should be considered a distinct form because the learning rules are different. While homosynaptic LTD depends upon presynaptic activity coupled with moderate postsynaptic activity, associative LTD presumably requires presynaptic activity occurring in the absence of postsynaptic activity, as the periods between strong activity are associated with a hyperpolarization of the postsynaptic cell. Cerebellar LTD is a form of asso-

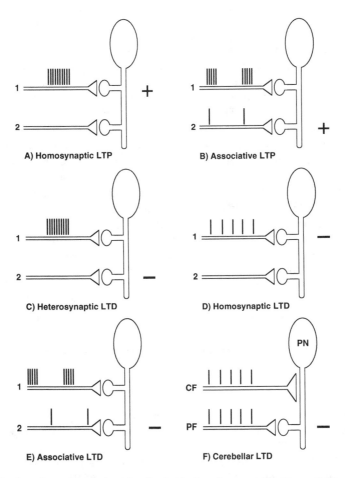

Figure 1 Several synaptic learning rules. Vertical hash marks represent pulses applied to afferent fibers at either low or high frequency. + and − signs indicate the sign and location of the relevant alteration in synaptic strength. (*A*) Homosynaptic LTP is induced by tetanic stimulation of one set of afferent fibers (*1*), typically 100Hz for 1 s, and results in a potentiation confined to stimulated synapses. (*B*) Associative LTP is induced by in-phase activation of a test pathway (*2*) at low frequency, typically 5 Hz, combined with tetanic stimulation of a conditioning pathway (*1*), and results in potentiation of the test pathway. The conditioning pathway is typically potentiated as well (not shown). (*C*) Heterosynaptic LTD is induced when tetanic stimulation to a conditioning pathway (*1*) is given in the absence of stimulation of a test pathway (*2*) and results in depression of the test pathway. Frequently, this treatment also results in homosynaptic LTP of the conditioning pathway (not shown). (*D*) Homosynaptic LTD is induced by prolonged low-frequency stimulation of one set of afferent fibers (*1*), typically 1 Hz for 10 min, and results in a depression confined to stimulated synapses. Other protocols have also been used to induce homosynaptic LTD (see text). (*E*) Associative LTD is induced by out-of-phase activation of a test pathway (*2*) at low frequency, typically 5 Hz, and tetanic burst stimulation of a conditioning pathway (*1*) resulting in a depression of the test pathway. (*F*) Cerebellar LTD is induced by in-phase activation of parallel-fiber (*PF*) and climbing-fiber (*CF*) inputs to a Purkinje neuron (*PN*) at low frequency, typically 2 Hz for 5 min, and results in depression confined to those parallel-fiber synapses stimulated during induction. (From Linden 1994b, reprinted by permission of the copyright holder, Cell Press.)

ciative LTD in which in-phase activation of two anatomically distinct inputs causes strong activation of the postsynaptic neuron and results in persistent depression of one particular input. While for consistency's sake this phenomenon should have a name like "in-phase associative LTD," it is commonly, as here, called cerebellar LTD.

In sum, depending on the modification rule(s) operative in a given synaptic array, correlated pre- and postsynaptic activity can give rise to either long-term synaptic potentiation or depression. However, uncorrelated pre- and postsynaptic activity may only give rise to synaptic depression. The term "anti-Hebbian" has been used to describe these different forms of synaptic depression. More precisely, the synaptic modification rules for associative LTD and heterosynaptic LTD, in which a lack of correlation between pre- and postsynaptic activity results in a decrease in synaptic strength, may be classified as "converse Hebbian" while the rules for homosynaptic LTD and cerebellar LTD, in which correlated pre- and postsynaptic activity results in a decrease, may be said to be "inverse Hebbian."

LTD OF EXCITATORY TRANSMISSION IN BRAIN

Some of the material in the following four sections has been adapted from Linden (1994b).

Heterosynaptic LTD

Heterosynaptic LTD was first seen as a correlate of homosynaptic LTP induced in the Schaffer collateral inputs to CA1 pyramidal cells in the hippocampus (Lynch et al 1977, Dunwiddie & Lynch 1978). In this preparation, heterosynaptic LTD was manifest as a decrease in the field response of commissural-CA1 synapses following LTP-inducing tetanization of Schaffer collateral inputs (or vice versa, as a decrease in Schaffer collateral-CA1 transmission following commissural fiber tetanization). However, over the years, heterosynaptic depression in area CA1 has proved difficult to study; some groups have failed to observe the phenomenon (Schwartzkroin & Wester 1975, Andersen et al 1977), while others have found only short-term depression (Alger et al 1978, Sastry et al 1984, Bashir & Collingridge 1992, Grover & Teyler 1992). Fortunately, heterosynaptic LTD has proved to be more reproducible using other preparations, including the hippocampal dentate gyrus (Levy & Steward 1979, Abraham & Goddard 1983), hippocampal area CA3 (Bradler & Barrionuevo 1989), and various cortical relays (Tsumoto & Suda 1979, Hirsch et al 1992).

From studies of these synaptic sites, a description of heterosynaptic LTD has emerged. As shown in chronic recordings from the perforant path–dentate gyrus synapses in vivo, heterosynaptic LTD persists for days to weeks (Krug et al 1985). This time course is similar to that of homosynaptic LTP in the same location. Additionally, heterosynaptic LTD lasting 15 hr has been demonstrated in an acute preparation (Colbert et al 1992). Heterosynaptic LTD is saturable upon repeated stimulation (Levy & Steward 1983, Christie & Abraham 1992a), is readily reversed by LTP induction (Abraham & Goddard 1983, Levy & Steward 1983, Christie & Abraham 1992a), and itself reverses previously established LTP (Levy & Steward 1979, 1983; Christie & Abraham 1992a, Kerr & Abraham 1993).

The induction of heterosynaptic LTD requires strong postsynaptic depolarization caused by synaptic activation and a consequent increase in postsynaptic Ca at neighboring inactive sites. This model is supported by several observations. Like that of homosynaptic LTP, induction of heterosynaptic LTD usually (Desmond et al 1991, Wickens & Abraham 1991, Christie & Abraham 1992a) but not always (Bradler & Barrionuevo 1990) requires activation of NMDA receptors; it is facilitated by agents that cause a reduction in GABAergic inhibition, such as picrotoxin and bicuculline (Bradler & Barrionuevo 1989, 1990; Abraham & Wickens 1991; Tomasulo et al 1993; Zhang & Levy 1993). These findings would seem to suggest that induction of LTP at active synapses is a prerequisite for the induction of heterosynaptic LTD; but examples of a dissociation between these two processes have been reported in which either heterosynaptic LTD is induced without homosynaptic LTP, or induction of homosynaptic LTP does not result in heterosynaptic LTD (Abraham & Goddard 1983, Bradler & Barrionuevo 1989). Further strengthening this dissociation is the observation that heterosynaptic LTD, but not homosynaptic LTP, may be prevented by blockade of voltage-gated, particularly L-type, Ca channels (Wickens & Abraham 1991, Christie & Abraham 1994). Perhaps the strongest evidence for the role of postsynaptic depolarization and consequent Ca increase in heterosynaptic LTD induction comes from a series of experiments in which tetanization of one set of synapses was replaced either by direct depolarization of the postsynaptic cell through somatic current injection, or by antidromic activation of the postsynaptic cell (Pockett & Lippold 1986, Pockett et al 1990, Christofi et al 1993). In these experiments, the induction of heterosynaptic LTD was facilitated by manipulations that blocked either spontaneous transmitter release, such as high external Mg, or ionotropic excitatory amino acid receptors, such as D-AP5 and CNQX, alone or in combination.

While the experiments described above indicate that evoked transmitter release from the depressed synapses is not necessary during LTD induction, it is still unclear whether some consequence of spontaneous release contributes to heterosynaptic LTD. Attempts to silence spontaneous firing of the unstimulated input have met with different results in different systems: Heterosynaptic LTD is not blocked by local application of tetrodotoxin in the perforant path–dentate gyrus synapses (Lopez et al 1990), but it is blocked at synapses between the lateral geniculate nucleus and primary visual cortex (Tamura et al 1992). If spontaneous release does contribute to induction of heterosynaptic LTD, it might function to activate postsynaptic metabotropic glutamate receptors, as suggested by the observation that LTD may be induced by postsynaptic activation during the simultaneous blockade of NMDA and AMPA (α-amino-3-hydroxy-5-methyl-4-isoxazolepropionate) receptors (Christofi et al 1993).

A requirement for postsynaptic Ca in LTD induction is indicated by the observations that LTD induction is blocked by removing external Ca, and that fura-2 imaging revealed increases in somatic and dendritic Ca during induction (Christofi et al 1993, Nowicky et al 1993). While by definition heterosynaptic LTD cannot be strictly input specific, it is only induced at synapses that are close enough to the stimulated input to be sufficiently depolarized, as determined by current-source density analysis (White et al 1988 1990). The current spread that determines the spatial parameters of this interaction is highly regulated by synaptic inhibition, as determined by application of a GABA$_A$ antagonist (Abraham & Wickens 1991, Tomasulo et al 1993, Zhang & Levy 1993).

How does the strong postsynaptic depolarization and consequent Ca influx caused by tetanization of one set of afferents result in depression at neighboring inactive sites? Ca influx via NMDA receptors at activated synapses might achieve a high Ca concentration there; this Ca load might diffuse passively to neighboring synapses and, by achieving a moderate concentration of Ca at those synapses, initiate a biochemical cascade resulting in heterosynaptic LTD. However, NMDA receptors are unlikely to be the sole source of Ca in LTD induction, because heterosynaptic LTD is not always blocked by NMDA receptor antagonists (Bradler & Barrionuevo 1990). An alternative explanation is that activation of NMDA receptors helps to provide depolarization that could spread either actively (Jaffe et al 1992) or passively to the region of neighboring inactive synapses, thereby triggering activation of voltage-gated Ca channels and consequent Ca influx. This model is consistent with the observations that induction of heterosynaptic LTD by afferent stimulation is blocked by L-type Ca channel antagonists (Wickens & Abraham 1991, Christie & Abraham 1994) and that heterosynaptic LTD may be induced by direct postsynaptic depolarization in the presence of an NMDA receptor antagonist (Christofi et al 1993). At present, little or no experimental evidence bears on either the molecular mechanisms of heterosynaptic LTD expression or the intermediate processes that might transduce a moderate rise in postsynaptic Ca at inactive synapses into this expression. It has recently been reported that bath application of a nitric oxide (NO) synthase inhibitor failed to alter heterosynaptic LTD in hippocampal area CA1 (Lum-Ragan & Gribkoff 1993).

Homosynaptic LTD

Homosynaptic LTD has been found in a number of brain structures using one of two induction protocols, either prolonged afferent stimulation at low frequency (typically 1–5 Hz for 5–15 min) or stimulation at high frequency (typically 50–100 Hz for 1–5 s). The latter is sometimes given together with some treatment that limits but does not eliminate postsynaptic activity. Low-frequency stimulation induces homosynaptic LTD in hippocampal synapses (Bramham & Srebro 1987, Dudek & Bear 1992, Mulkey & Malenka 1992), while high-frequency stimulation has been effective in striatal synapses (Calabresi et al 1992a, Lovinger et al 1993, Walsh 1993). Both low-frequency (Berry et al 1989, Kirkwood et al 1993) and high-frequency stimulation (Artola et al 1990, Hirsch & Crepel 1990, Kimura et al 1990, Kato 1993) have been used at neocortical synapses.

The first suggestion that low-frequency stimulation could induce a depression of hippocampal synapses was seen in a study by Dunwiddie & Lynch (1978) that used 100 pulses given at 1 Hz to the Schaffer collateral fibers. However, this depression was only monitored for a few minutes and did not appear to be input specific. More recently, Dudek & Bear (1992) and Mulkey & Malenka (1992) have shown that administration of 900 pulses at 1 Hz reliably induces persistent, input-specific LTD at this synapse. This form of LTD is saturable upon repeated stimulation, may be induced in previously potentiated pathways, and may be reversed by induction of LTP (Mulkey & Malenka 1992, Dudek & Bear 1993). It is blocked by NMDA receptor antagonists (Dudek & Bear 1992, Mulkey & Malenka 1992), strong hyper-polarization, and postsynaptic application of a Ca chelator (Mulkey & Malenka, 1992). These last findings indicate that, like induction of homosynaptic LTP, this form of LTD induction depends upon postsynaptic depolarization, activation of NMDA receptors, and consequent increases in internal Ca. However, it is likely that low-frequency stimulation that results in homosynaptic LTD causes a smaller postsynaptic Ca transient than the high-frequency stimulation that results in homosynaptic LTP. This is indicated by the observation that a stimulus barely able to induce LTP in normal external Ca (20 Hz, 30 s) will induce LTD when external Ca is lowered fivefold (Mulkey & Malenka 1992). The same type of low-frequency stimulation can also induce homosynaptic LTD of the responses recorded in layer IV of visual cortex to layer III stimulation. Like the hippocampal Schaffer collateral-CA1 synapse, homosynaptic LTD is this system is also input specific, reversible, and NMDA receptor dependent (Kirkwood et al 1993, Kirkwood & Bear 1994).

Three other protocols for induction of homosynaptic LTD using low-frequency stimulation have recently been reported. Low-frequency activation of

Schaffer collateral fibers in the presence of an AMPA receptor antagonist induces homosynaptic LTD when paired with mild depolarization of a CA1 neuron (Lin et al 1993). When strong postsynaptic depolarization is applied, homosynaptic LTP results. Another protocol relies on repeated paired-pulse stimulation of the commissural-CA1 synapse in the intact hippocampus (Thiels et al 1993). The pulse pairs are given at an interval that results in depression of the second pulse (25 ms). Application of 200 pairs at 0.5 Hz induces homosynaptic LTD that is reversible and NMDA receptor dependent. In a third protocol, when low-frequency test stimuli to Schaffer collateral fibers were applied 800–1600 ms after brief direct depolarizing pulses administered to a CA1 neuron, LTD of that same test stimulus developed after 50–100 out-of-phase pairings (Debanne et al 1994). This depression was not induced by depolarization alone or by pairing outside of a specific temporal window, was input specific, and was blocked by an NMDA receptor antagonist. It was noted that application of a shorter depolarizing pulse shifted the window of LTD induction to shorter pairing intervals, suggesting that a particular level of time-dependent postsynaptic signal (perhaps Ca entry via voltage-gated channels) was necessary at the time of arrival of the low-frequency test pulse. This finding suggests that LTD induced using this protocol may not be truly associative (as indicated by the authors); it does not appear to be the manifestation of a pre-not-post learning rule. We suggest that this phenomenon be provisionally grouped with other cases of homosynaptic LTD.

The involvement of NMDA receptors, postsynaptic depolarization, and postsynaptic Ca in the induction of homosynaptic LTD by low-frequency stimulation suggests that a number of LTP induction's features might also apply to induction of LTD. These include the phenomenon of cooperativity (a minimum number of fibers must be activated to induce LTP) and facilitation of induction by blockade of inhibition. At present, tests of these ideas have not been reported in the literature.

Some information is available about the second-messenger processes that might couple postsynaptic Ca influx to induction of homosynaptic LTD by low-frequency stimulation. Prolonged bath application of the protein phosphatase inhibitors calyculin A and okadaic acid blocked induction of LTD in both hippocampal (Mulkey et al 1993, O'Dell et al 1994) and neocortical (Kirkwood & Bear 1994) synapses. In addition, postsynaptic application of a membrane-impermeable phosphatase inhibitor (microcystin LR) also blocked LTD induction, indicating a postsynaptic site of action of these enzymes (Mulkey et al 1993). Furthermore, bath application of calyculin A reversed established LTD, suggesting that continued protein phosphatase activity might underlie LTD expression. Two lines of evidence suggest that activation of a calmodulin-dependent protein kinase may be necessary for induction of homosynaptic LTD in hippocampus. First, postsynaptic application of a mixture of calmodulin inhibitors blocked induction of LTD (Mulkey et al 1993). Second, LTD is impaired in αCaMKII mutant mice (CF Stevens, S Tonegawa & Y Wang, unpublished observations). In contrast, a γPKC mutant mouse shows unimpaired homosynaptic LTD (Abeliovich et al 1993).

It has been reported that bath application of compounds (such as Zn-protoporphyrin-IX) that block production of both the newly described gaseous messenger carbon monoxide and the enzyme soluble guanylyl cyclase do not block induction of homosynaptic LTD in hippocampus (Stevens & Wang 1993). On the other hand, a preliminary report notes that two compounds (hemoglobin and L-N-monomethyl-arginine) that interfere with signaling by another gaseous messenger, nitric oxide, do block such induction (Zorumski & Izumi 1993). As is par for the course with claims about nitric oxide, another group has reported no effect of nitric oxide synthase inhibitors on homosynaptic LTD induction (Lum-Ragan & Gribkoff 1993)

It should be noted that homosynaptic LTD induced by low-frequency stimulation in the hippocampus is more robust in young animals. One group claims that when orthodromic low-frequency stimulation to Schaffer collateral fibers is employed, the amplitude of depression in adult rats is only about 20%, while in younger animals (< 21 days old) it is up to 50% (Dudek & Bear 1993). Other investigators find that induction of homosynaptic LTD by low-frequency stimulation does not occur at all in slices of adult hippocampus (Staubli & Lynch 1990, Fujii et al 1991, Larson et al 1993, O'Dell & Kandel 1994). A developmental program probably underlies these differences, but at present it is unclear whether the program involves changes in intrinsic neuronal properties such as ion channel and receptor subtype expression, or whether the network activity of the younger animals is biased in a manner that favors induction of depression.

It is likely that homosynaptic LTD induced by low-frequency stimulation closely resembles depotentiation, a phenomenon (so far studied only in the hippocampus) in which a potentiated set of synapses are persistently reset to baseline values following stimulation at 1–5 Hz (Barrionuevo et al 1980, Staubli & Lynch 1990, Fujii et al 1991, Larson et al 1993, Wexler & Stanton 1993, O'Dell et al 1994). Depotentiation is blocked by NMDA receptor antagonists (Fujii et al 1991, O'Dell et al 1994) and by an antagonist of metabotropic receptors (Bashir et al 1993, Bortolotto et al 1994). Depotentiation most likely represents a sequence of events whereby induction of LTP facilitates the induction of subsequent homosynaptic LTD by low-frequency stimulation. Recently it has been shown that a similar facilitation is produced by repeated prior induction of short-term potentiation (STP, an increase in synaptic strength that typically returns to baseline within 30–60 min) as well (Wexler & Stanton 1993).

Homosynaptic LTD induced by high-frequency stimulation shares some but not all of the characteristics of low frequency–induced LTD. Like low frequency induced–LTD it is blocked by strong hyperpolarization, promoted by mild depolarization, and converted to LTP by strong depolarization of the postsynaptic neuron during the tetani (Artola et al 1990). High- or low-frequency stimulation of visual cortex via white matter produces no lasting changes of synaptic strength unless ancillary treatments are used. Stimuli of 50 Hz together with the $GABA_A$ blocker bicuculline (0.1–0.2 µM) generate approximately 20% depression, while in 0.3 µM bicuculline the same stimulus produced potentiation. A corresponding switch between depression and potentiation was also reported when varying postsynaptic depolarization accompanied the synaptic stimulus. That is, in 0.2 µM bicuculline, a 50-Hz stimulus produced LTP in neurons where strong depolarizing current was applied through an intracellular microelectrode, while with weaker postsynaptic depolarization the same tetanus produced LTD (Artola et al 1990).

However, NMDA receptor antagonists do not block induction of homosynaptic LTD in visual cortical (Artola et al 1990, Kato 1993) or corticostriatal synapses (Calabresi et al 1992a,b; Lovinger et al 1993, Walsh 1993) and even promote LTD induction at synapses of the prefrontal cortex (Hirsch & Crepel 1991). In visual cortex, LTD may be induced by transiently increasing external Ca to 4 µM in the absence of afferent stimulation, in a manner that is prevented by hyperpolarization of the postsynaptic neuron (Artola et al 1992). While postsynaptic injection of Ca chelators blocks high frequency–induced LTD in most cases (Brocher et al 1992, Hirsch & Crepel 1992), there are notable exceptions. In two studies involving visual cortex, EGTA and BAPTA injections blocked induction of LTP but not LTD (Kimura 1990, Yoshimura et al 1991). However, if one extrapolates from injection parameters, smaller amounts of the chelators were injected in these studies than in those where successful block was obtained. It is generally presumed that the levels of buffer injected abolish or at least greatly reduce the changes in dendritic Ca during induction. No one seems to have taken the trouble to show that this is the case, however, and it is risky to assume that because one Ca-dependent process (such as spike train adaptation) is blocked others will be also. Thus, the difference could arise from weak Ca buffering.

Several experiments have suggested that induction of homosynaptic LTD by high-frequency stimulation requires activation of metabotropic receptors. Metabotropic receptors are coupled to G-proteins and produce a variety of effects depending upon the subtype of receptor activated (see Schoepp 1993 for review). The metabotropic receptor subtypes mGLUR1 and mGLUR5 exert their effects via activation of the enzyme phospholipase C. Activation of phospholipase C cleaves the membrane phospholipid phosphatidylinositol-4,5-bisphosphate to yield 1,2-diacylglycerol and inositol-1,4,5-trisphosphate (IP_3). The former cleavage product is known to activate protein kinase C (PKC), and the latter causes mobilization of Ca from nonmitochondrial internal stores by binding specific intracellular receptors.

At corticostriatal synapses, LTD is blocked by either L-AP3, a putative metabotropic antagonist (Calabresi et al 1992a, 1993b), or chronic treatment with lithium salts, which interferes with phosphoinositide turnover (Calabresi et al 1993a). Neither treatment interferes with the ability of (1SR,3RS)-ACPD, a metabotropic agonist, to produce a presynaptically mediated short-term depression at this synapse (Calabresi et al 1993a,b; Lovinger et al 1993). In visual cortex, input-specific LTD could be induced following tetanization during simultaneous blockade of AMPA and NMDA receptors (Kato 1993). LTD induced using this protocol could be blocked by postsynaptic application of compounds that interfered with the G-protein linkage between the metabotropic receptor and its effectors (GDB-β-S), by binding of IP$_3$ to the IP$_3$ receptor (heparin), or by a Ca chelator. This finding suggests that activation of the postsynaptic metabotropic receptor is not only necessary for homosynaptic LTD induction but is also sufficient. A clear dichotomy is evident in homosynaptic LTD with respect to the glutamate receptor subtypes involved in induction. Some protocols require the activation of NMDA receptors, while others do not. Moreover, the protocols unaffected by NMDA receptor antagonists are blocked by inhibitors of metabotropic receptors (see also Yang et al 1994, discussed below).

Recently, several reports have shown that afferent stimulation in the presence of GABA receptor agonists induces homosynaptic LTD. LTD is induced in visual cortex neurons (layers II–IV) by tetanic stimulation of white matter in the presence of the GABA$_A$ agonist muscimol (Kato & Yoshimura 1993). This finding, together with that of Artola et al (1990), reflects the complexity of the signal pathways from the white matter to cortex and the difficulty of using pharmacology to analyze these circuits. In hippocampus, Yang et al (1994) have described a procedure in which local applications of GABA are made during low-frequency, subthreshold stimulation of Schaffer collateral inputs to CA1 neurons (0.1 Hz). The exogenous GABA is present before input volleys arrive and limits the postsynaptic depolarization to less than 10 mV measured intracellularly at the soma. Endogenous GABA release arrives several milliseconds after the excitatory volley. Five to ten minutes of this procedure (30–60 electrical stimuli) are sufficient to produce an average depression of 50%, measured intracellularly. In some cases, depression of up to 90% was noted. Either GABA$_A$ or GABA$_B$ activation could contribute to this form of LTD because depression was observed in the presence of either GABA$_A$ or GABA$_B$ blockers. GABA application without the accompanying low-frequency stimulation was without effect. LTD induction was not dependent upon NMDA receptor activation because the antagonist AP5 had no effect, but the metabotropic glutamate receptor antagonists L-AP3 (Schoepp et al 1990) and

MCPG (Eaton et al 1993) both blocked LTD induction. This protocol was effective in young rats (20–25 days) as reported, and also induced a similar degree of depression in older animals (S Otani & WB Levy, personal communication). Finally, a preliminary report has shown that in the medial vestibular nucleus, brief application of the GABA$_B$ receptor agonist baclofen induced LTD of EPSPs evoked by stimulation of nVIII (Kinney et al 1993).

Homosynaptic, input-specific LTD of a mixed electrotonic/glutamatergic synapse between the eighth nerve fibers and the Mauthner cell in goldfish brainstem can be produced by pairing brief, weak tetani of the eighth nerve with antidromic activation of the Mauthner cell that functions to recruit feedback interneurons (Yang & Faber 1991). LTD of both the glutamatergic EPSP and the electrotonic coupling efficacy were induced in 2 of 9 cells, while a transient depression was induced in 4 of 9 cells. The LTD induced in this manner can be reversed by tetani applied using a stronger stimulus intensity. When potentiating stimuli were given prior to the weak tetanization/inhibition protocol described above, the probability of inducing LTD increased. This increase is reminiscent of the phenomenon of depotentiation in hippocampal synapses (discussed above) in which sustained low-frequency stimulation (1–5 Hz for 5–15 min) appears to reduce the efficacy of previously potentiated synapses more effectively than it does that of "baseline" unpotentiated synapses.

Even though homosynaptic LTD induced by low-frequency stimulation is blocked by NMDA receptor antagonists, while that induced by high-frequency stimulation or the GABA-activating protocols detailed above is not, both forms probably represent the same phenomenon. On balance, the evidence indicates that postsynaptic Ca increases are necessary for homosynaptic LTD induction. However it is well established that Ca increases are necessary for LTP induction as well. In all cases, it is likely that a moderate increase in postsynaptic Ca, whether it arises from Ca influx via NMDA receptors, voltage-gated channels, internal Ca release, or some combination of these signals, is a common trigger for, or at least provides a necessary element of, homosynaptic LTD induction. The fine structure of this "moderate" Ca increase, however, requires further examination. First, one must consider compartmentalization of signals in dendritic spines and shafts. In pyramidal cells of hippocampal areas CA1 and CA3, tetanic stimuli of the type that trigger LTP result in large, widespread Ca increases in dendrites of the stratum radiatum and stratum oriens (Regehr et al 1989, Muller & Connor 1991, Regehr & Tank 1992). Grossly considered, most of these dendritic Ca changes result from activation of voltage-gated Ca channels (Miyakawa et al 1992b, Regehr & Tank 1992), but a more careful examination has demonstrated a strong, localized NMDA receptor–mediated component (Alford et al 1993, Perkel et al 1993). Weak tetanic stimulation that activates only a few presynaptic fibers has been shown tó cause significant AP5-blockable elevations of Ca only in spines, leaving dendritic shaft levels largely unchanged (Muller & Connor 1991). Second, one must consider time-course differences: LTP induction seems to require a large Ca elevation lasting 1–2 s (Malenka et al 1992), and the standard high-frequency induction protocol supplies this through the summation of pulses at each stimulus. During low-frequency stimulation without auxiliary depolarization, summation is greatly reduced, as small pulses of Ca are largely buffered/sequestered within a few hundred milliseconds in neuronal dendrites (Pozzo-Miller et al 1993). Thus, while large, local Ca changes might occur in the low-frequency LTD-induction protocols, their lifetimes should be short.

Associative LTD

The first description of associative LTD was in the Schaffer collateral-CA1 synapses of the hippocampus. In these experiments, a conditioning pathway received short high-frequency trains at 200 ms intervals while single pulses to a non-overlapping test pathway were paired in an explicitly out-of-phase fashion. This resulted in homosynaptic LTP of the conditioning pathway and associative LTD of the test pathway (Stanton & Sejnowski 1989, Stanton et al 1990). It was proposed that associative LTD resulted from pairing of the test input with hyperpolarization following the out-of-phase conditioning input, because pairing test stimuli with direct hyperpolarizing current induced LTD of the test input, while stimuli to the test input had no effect when applied alone (Stanton & Sejnowski 1989). NMDA receptor blockade did not alter the induction of associative LTD, although it did block induction of homosynaptic LTP in the conditioning pathway, showing that these two events are dissociable. Associative LTD induced by pairing test stimuli with either out-of-phase conditioning trains or direct postsynaptic hyperpolarization was blocked by bath application of L-AP3, a presumed antagonist of metabotropic receptors that could be acting either presynaptically or postsynaptically to cause this blockade (Stanton et al 1990).

Unfortunately, attempts to replicate associative LTD using the protocol of Stanton & Sejnowski (1989) in hippocampal area CA1 have failed (Paulsen et al 1993, Kerr & Abraham 1993). However, associative LTD has been demonstrated in the hippocampal perforant path–dentate gyrus synapse, under some special conditions. One line of evidence for associative LTD in this synapse comes from the study of NMDA receptor–mediated synaptic potentials, which may be recorded in isolation by the application of an AMPA receptor antagonist (CNQX) together with reduction of external Mg. Using this protocol, stimulation of perforant path fibers at 10 Hz coupled with direct postsynaptic hyperpolarization induced LTD of NMDA receptor–mediated EPSPs (Xie et al 1992). LTD was attenuated when the postsynaptic cell was loaded with a Ca chelator. While these experiments would appear to confirm those of Stanton & Sejnowski (1989) in area CA1, one must be aware that reduction of external Mg would allow greater postsynaptic Ca influx at more negative membrane potentials. It will be interesting to determine if this hyperpolarization protocol can be used to induce LTD under more physiological conditions.

While associative LTD cannot be induced in naive perforant path synapses by an out-of-phase conditioning protocol (Christie & Abraham 1992a), this same protocol will elicit associative LTD when the test pathway has been "primed" by 80 pulses delivered at the theta rhythm frequency, 5 Hz (Christie & Abraham 1992b). NMDA receptor blockade prior to priming, but not during the interval between priming and out-of-phase conditioning, will block induction of associative LTD. This suggests that priming, but not LTD induction per se, is NMDA receptor dependent. Priming stimuli will not facilitate associative LTD induction in Schaffer collateral-CA1 synapses (Kerr & Abraham 1993).

Our consideration of multi-synaptic forms of LTD such as heterosynaptic LTD and associative LTD has largely ignored the issue of which synapses were being stimulated in various protocols. In some cases, subsets of a homogeneous population of fibers are stimulated to serve as two separate inputs, while in other cases anatomically distinct inputs are used. For example, associative LTD in area CA1 has been reported when separate subsets of Schaffer collaterals are used for test and conditioning stimulation (Stanton & Sejnowski 1989), while associative LTD in area CA3 is seen when mossy fiber input was used for conditioning train stimulation and commissural input for test pulses (Chattarji et al 1989). The opposite protocol did not result in associative LTD of mossy fibers. Tetanization of commissural fibers causes heterosynaptic LTD of mossy fibers, but tetanization of mossy fibers does not induce the converse effect (Bradler & Barrionuevo 1989,1990). In the perforant path–dentate gyrus relay, heterosynaptic LTD can be induced bidirectionally: LTD of medial perforant path inputs to the dentate gyrus granule cells (which are pharmacologically different from lateral perforant path fibers) results from lateral perforant path tetanization or vice versa (White et al 1990). In sum, multi-synaptic forms of LTD may be induced using both heterogeneous and homogeneous inputs, but not all combinations will work. Constraints on the induction of LTD using heterogeneous inputs are likely related to the presence or absence of receptors and effector systems at the different synaptic relays. For example, in the case of the CA3 pyramidal cell, the constraints on heterosynaptic LTD could result from the absence of NMDA receptors at the mossy fiber but not at the commissural synapses.

Cerebellar LTD

Cerebellar Purkinje neurons, which are the sole output of the cerebellar cortex, receive excitatory inputs from two sources, parallel fibers originating from cerebellar granule neurons, and climbing fibers from the inferior olivary nucleus. The parallel fibers make highly divergent synapses on secondary and tertiary dendrites, each contacting many Purkinje neurons and each Purkinje neuron receiving inputs from many parallel fibers (~150,000 synapses). In contrast, each Purkinje neuron receives a powerful contact from a single climbing fiber on the soma and proximal dendritic arbor. The parallel-fiber synapse is thought to use glutamate as a transmitter and contains metabotropic (primarily mGLUR1) and AMPA type receptors. The transmitter of the climbing fibers is an excitatory amino acid but has yet to be definitively identified. Both types of synapse lack NMDA receptors in the adult (see Linden & Connor 1993 for a more extensive review).

Cerebellar LTD was first described in the intact cerebellum by Ito's laboratory (Ito et al 1982) and since that time has been analyzed in both acute slice preparations and primary cultures. In slice or in situ, the induction protocol consists of stimulating the parallel and climbing-fiber inputs together at low frequency (1–4 Hz) for a period of 4–6 min. This procedure results in a selective attenuation of the parallel-fiber–Purkinje neuron synapse (typically a 20–50% reduction of baseline synaptic strength) that reaches its full extent in 5–15 min and persists for the duration of the experiment, typically longer than 1 hr (Figure 2). No changes occur in the amplitude of the climbing-fiber response as a result of the stimulus pairing. In addition, LTD is input specific: It occurs only in those parallel-fiber synapses stimulated during climbing-fiber activation (Ito et al 1982, Ekerot & Kano 1985).

Cerebellar LTD may also be seen in primary culture. Hirano (1990a) used co-cultures of cerebellum and explants of inferior olivary nucleus and showed that Purkinje neurons in these cultures received characteristic excitatory inputs from both the olivary and granule neurons. By conjunctively stimulating a presynaptic granule neuron and the olivary explant he was able to elicit a 25–30% reduction in the EPSP from the granule neuron. In a second tissue-culture model, Linden et al (1991) found that glutamate pulses delivered from a micropipette, coupled with current injection to Purkinje neurons, produced a depressed response to further glutamate pulses that persisted for the experimental lifetime of the cells.

Like that of most forms of LTD discussed above, the induction of cerebellar LTD requires an increase in postsynaptic Ca. The climbing fiber contributes to LTD induction by causing sufficient postsynaptic depolarization to strongly activate voltage-gated Ca channels in the dendrites, thereby causing a massive action potential (Eccles et al 1966) and a large Ca influx. In fact, climbing-fiber activation may be replaced in the LTD induction protocol by direct depolarization of the postsynaptic neuron both in the cerebellar slice (Crepel & Krupa 1988, Glaum et al 1992) and in tissue culture models where there are no synaptic inputs of this type (Hirano 1990b, Linden et al 1991). Furthermore,

(A)

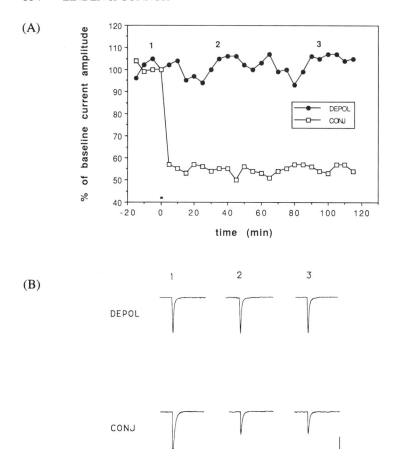

(B)

LTD induction is blocked by postsynaptic application of a Ca chelator (Sakurai 1990, Linden & Connor 1991, Konnerth et al 1992), electrical inhibition of Purkinje neurons during parallel-fiber/climbing-fiber conjunctive stimulation (Ekerot & Kano 1985, Hirano 1990b, Crepel & Jaillard 1991), or removal of external Ca (Linden & Connor 1991). Finally, studies using optical indicators have shown large Ca accumulations in Purkinje neuron dendrites following climbing-fiber stimulation (Ross & Werman 1987, Knopfel et al 1990, Miyakawa et al 1992a) and following the climbing-fiber/parallel-fiber conjunctive stimulation used to generate LTD (Konnerth et al 1992).

Parallel fiber activation appears to exert its effects on LTD induction through the activation of non-NMDA, excitatory amino acid receptors. In LTD induction protocols, parallel-fiber activation may be replaced by pulses of exogenous

(C)

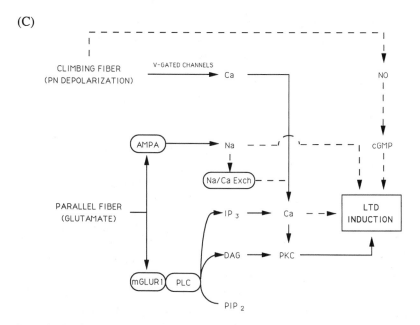

Figure 2 Cerebellar LTD. (*A*) LTD of glutamate currents in a cultured Purkinje neuron is induced by glutamate/depolarization conjunction. Six depolarizing pulses to −10 mV applied together with six glutamate pulses (heavy bar at t = 0) induce LTD of glutamate currents. Both test pulses and conjunctive pulses were applied at 0.05 Hz. (*B*) Corresponding glutamate current traces before and after LTD induction. Scale bars = 2 s, 100 pA. (*C*) A model of cerebellar LTD induction. *Solid lines* represent mechanisms believed to be well established, whereas *dotted lines* represent mechanisms that are still controversial. See text for explanation. mGLUR1 = metabotropic glutamate receptor, type 1; AMPA = AMPA receptor; DAG = 1,2 diacylglycerol; PKC = protein kinase C; IP3 = inositol-1,4,5-trisphosphate; PIP2 = phosphatidyl-inositol-4,5-bisphosphate; PLC = phospholipase C; NO = nitric oxide. (Parts *A* and *B* of this figure are from Linden & Connor 1992, reprinted by permission of Oxford University Press. Part *C* is from Linden 1994b, reprinted by permission of the copyright holder, Cell Press.)

glutamate, quisqualate (an agonist of both AMPA and metabotropic receptors) or AMPA together with ACPD (the latter being a selective agonist of metabotropic receptors). Aspartate, kainate, or AMPA alone is insufficient (Kano & Kato 1987, Crepel & Krupa 1988, Linden et al 1991). This agonist profile suggests that activation of metabotropic receptors is required. In a manner consistent with this notion, a putative metabotropic receptor antagonist, L-AP3, blocks LTD induced by quisqualate/depolarization conjunction, as does pretreatment with pertussis toxin (Linden et al 1991). This finding has been confirmed and extended by the report that inactivating antibodies against mGluR1 blocked the induction of LTD using a similar protocol (Shigemoto et al 1994).

Although voltage-gated Ca influx coupled with metabotropic receptor activation is necessary for LTD induction, present evidence suggests that it is not sufficient. Application of CNQX (an antagonist of AMPA, but not of meta-botropic receptors) blocks LTD induction when applied during quisqualate/de-polarization stimulation in cultured Purkinje neurons (Linden et al 1991) and when applied during parallel fiber/depolarization stimulation in cerebellar slice (Hémart et al 1994). This finding is consistent with an earlier observation in the intact cerebellum that application of kynurenate (also an antagonist of AMPA receptors but not of metabotropic receptors) blocks LTD produced by parallel-fiber/climbing-fiber stimulation (Kano & Kato 1988). Iontophoretic application of a metabotropic receptor agonist, (1SR,3RS)-ACPD, coupled with depolarization sufficient to cause Ca influx is not sufficient to induce LTD in cultured Purkinje neurons when voltage-gated Na channels are blocked with tetrodotoxin (Linden et al 1993). This same agonist, applied in the bath during parallel-fiber stimulation also failed to induce LTD (Crepel et al 1991). The AMPA receptor appears to exert its effect on LTD induction through a specific chemical consequence of Na influx, because replacement of external Na with other permeant cations such as Li or Cs during quisqualate/depolar-ization conjunction blocked the induction of LTD (Linden et al 1993). Fur-thermore, with no blockade of voltage-gated Na channels, LTD may be induced infrequently by (1SR,3RS)-ACPD/depolarization conjunction. Promoters of voltage-gated Na influx such as veratridine increased the probability of induc-tion. This last finding is somewhat consistent with another observation using the cerebellar slice preparation in which bath application of (1SR,3RS)-ACPD together with depolarization sufficient to activate voltage-gated Ca channels induced LTD in the majority of cells tested (Daniel et al 1992, but see Glaum et al 1992). Taken together, these results suggest that Na influx is necessary for LTD induction and that while influx via voltage-gated Na channels may provide a sufficient signal in some cases, the flux through AMPA receptors is much more effective. Although it is not clear what intracellular processes are engaged by an increase in postsynaptic Na to contribute to LTD induction, one possibility is that the high levels of intracellular Na slow the Na/Ca exchanger, thereby reducing the ability of the cell to extrude Ca and leading to increased internal Ca for a given influx (see also Kiedrowski et al 1994).

There is good evidence that the expression of LTD is mediated entirely postsynaptically by alterations in either the number or the sensitivity of AMPA receptors. While exogenous AMPA pulses are not sufficient to substitute for parallel-fiber activation during LTD induction (because they fail to activate the metabotropic receptor mGLUR1), as test stimuli they suffice to detect LTD once it has been induced. LTD may be detected with pulses of AMPA, gluta-mate, or quisqualate (Ito et al 1982, Crepel & Krupa 1988, Linden et al 1991, Linden & Connor 1991). In one type of experiment in tissue culture, LTD was induced by quisqualate/depolarization conjunction and detected with test pulses of AMPA in TTX-containing external saline (Linden et al 1991, Linden & Connor 1991). This protocol minimizes the potential contribution of pre-synaptic processes. More recently, in order to eliminate any presynaptic contribution to LTD induction this protocol has been amended to include physical isolation of the Purkinje neuron by the scraping away of adjacent cells, combined with inclusion of adenosine in the external medium to suppress spontaneous release of neurotransmitter in order to eliminate any presynaptic contribution to LTD induction (Linden 1994a). These findings are consistent with a report showing a decrease in quantal size following LTD induction in cultured Purkinje neurons (Hirano 1991).

The data suggest that three postsynaptic factors are necessary for cerebellar LTD induction: Ca influx through voltage-gated channels, activation of metabotropic receptors, and Na influx mediated by AMPA receptors or secondarily by voltage-gated Na channels. Of the many intermediates that might couple these factors to the alteration of AMPA receptors, the necessity of mGLUR1 activation suggests that phosphoinositide (PI) turnover and PKC activation compose one relevant step. At least two of the currently identified subtypes, mGLUR1 and mGLUR5, are linked directly to PI turnover; and cerebellar Purkinje neurons contain mGLUR1 in unusually high quantities (Masu et al 1991), particularly in the distal dendritic spines where the parallel-fiber synapses are received (Martin et al 1992). The involvement of PKC in LTD induction was suggested by experiments in which PKC inhibitors blocked induction when applied during glutamate/depolarization conjunction (Linden & Connor 1991). Application of these compounds after LTD had been induced had no effect, suggesting that continued PKC activation is not required for LTD to persist. This observation is complemented by the finding that bath application of PKC-activating phorbol esters induces a depression of Purkinje neuron responses to exogenous glutamate or AMPA (Crepel & Krupa 1988, Linden & Connor 1991).

Several studies using the slice preparation have suggested that release of nitric oxide (NO) and the consequent production of cGMP by soluble guanylyl cyclase are necessary for cerebellar LTD induction. Bath application of either NO synthase inhibitors or an extracellular NO scavenger (hemoglobin) was shown to block LTD induction by either parallel-fiber/climbing-fiber conjunctive stimulation (Shibuki & Okada 1991) or parallel-fiber/depolarization conjunctive stimulation (Crepel & Jaillard 1990, Daniel et al 1993). Likewise, an LTD-like phenomenon could be produced when climbing-fiber stimulation was replaced by bath application of NO donor molecules or membrane permeable analogs of cGMP (Shibuki & Okada 1991, Daniel et al 1993). Application of cGMP or NO donors directly to the Purkinje neuron (via a patch pipette) also resulted in depression of PF responses, while postsynaptic application of an NO synthase inhibitor did not block LTD induction (Daniel et al

1993). In contrast, Glaum et al (1992) found no effect of a bath-applied NO donor on parallel fiber-evoked EPSPs recorded in Purkinje neurons. In cultured Purkinje neurons, LTD of glutamate currents was unaffected by NO synthase inhibitors, an NO scavenger, or an NO donor (Linden & Connor 1992).

Assigning a role in LTD induction to NO signaling is further complicated by three biochemical observations. First, while neuronal NO synthase is present in parallel fibers, it is notably absent in Purkinje neurons and climbing fibers (Bredt et al 1990, Vincent & Kimura 1992). Second, following addition of NO donors, cGMP accumulations may be observed in Bergmann glia but not in Purkinje neurons (de Vente et al 1990, de Vente & Steinbusch 1992). Third, soluble guanylyl cyclase is inhibited by high concentrations of Ca (Olson et al 1976) such as are present in Purkinje neuron dendrites during LTD induction (Konnenth et al 1992). Together with the observation that LTD may be induced when climbing-fiber activation is replaced by direct depolarization of the Purkinje neuron, these findings make it difficult to design a mechanism by which climbing-fiber activation triggers NO release and NO once released acts on Purkinje neurons. Recently it has been demonstrated that NO synthase inhibitors and NO scavengers fail to block LTD induction in slices treated with a gliotoxin, leading to the proposal that Bergmann glia exert a tonic negative regulation of LTD induction that is relaxed by NO (Shibuki 1993).

Intracellular recording in slice has indicated that LTD is optimally induced when climbing-fiber stimulation precedes parallel-fiber stimulation by 125–250 ms (Ekerot & Kano 1989). Another study using slice preparation has shown that LTD is induced by climbing-fiber/parallel-fiber stimulation with an interval of 50 ms but that LTD induction will fail unless disynaptic inhibition is blocked by addition of a $GABA_A$ antagonist (Schreurs & Allcon 1993). Finally, a preliminary report using field-potential recording has indicated that the optimal interval is in the opposite direction, with parallel-fiber stimulation preceding climbing-fiber stimulation by 250 ms (Chen & Thompson 1992). The latter interval would require that some persistent signal from parallel fibers, such as a consequence of metabotropic receptor activation, linger for at least 250 ms to interact with the climbing-fiber signal.

Are there morphological or diffusional constraints that serve to compartmentalize some of the signals involved in cerebellar LTD induction and thereby restrict their interaction? For example, the Ca signal may derive from as many as three different sources. Are increases in internal Ca caused by IP_3 liberation (via the metabotropic receptor) or via the Na/Ca exchanger similar to those caused by the activation of voltage-gated Ca channels? If so, then how does direct depolarization of the Purkinje neuron (or climbing-fiber activation) provide a unique signal that is necessary for LTD induction? In other words, if dendritic Ca is increased during parallel-fiber stimulation alone (or glutamate application alone) then why isn't LTD induced by these treatments? One possibility is that both parallel fiber–activated and climbing fiber–activated Ca sources are present in the same compartment, and both sources must be activated simultaneously to produce a level of free Ca sufficient to induce LTD. Another possibility is that Ca from climbing-fiber activation and Ca from parallel-fiber activation are differently compartmentalized, and that separate targets of Ca action in different compartments must both be activated before LTD can be induced. The identity of these compartments is still a matter of speculation. They could represent different cells, as in the glial NO hypothesis mentioned above. More likely they represent subcellular compartments of the Purkinje neuron dendrite (such as dendritic spines vs shafts) or microdomains within the dendritic spine head.

LTD and Extrinsic Modulatory Transmitters

Many synaptic relays in the brain are influenced not only by intrinsic activity in afferent fibers but also by extrinsic modulatory fibers that are frequently monoaminergic or cholinergic. A substantial literature describes the effects of modulatory-input stimulation on homosynaptic LTP; likewise much has been written on how application of these neurotransmitters affects homosynaptic LTP; but very few reports have appeared on any form of LTD. One recent study in the intact hippocampus showed that stimulation of the medial septal area, which sends cholinergic fibers to the dentate gyrus, attenuated homosynaptic LTP of the lateral PP-DG synapse (induced by tetanic stimulation of lateral PP fibers) but enhanced heterosynaptic LTD of the medial PP-DG synapse induced by this same treatment (Pang et al 1993).

While septal activation alone produces no alteration of the strength of the PP-DG synapses, application of β-adrenergic agonists does. In hippocampal slices, bath application of isoproterenol (1 μM), which presumably mimics the activation of noradrenergic fibers that project from the dorsal locus coeruleus, produced a long-lasting depression of lateral PP synapses and a simultaneous potentiation of medial PP synapses (Dahl & Sarvey 1989). These synaptic alterations were produced when isoproterenol was applied in the absence of test pulses and was blocked by D-AP5, an NMDA receptor antagonist (Dahl & Sarvey 1990). It is interesting that prior application of isoproterenol and D-AP5 together blocked the subsequent production of lateral PP depression by isoproterenol alone but left the medial PP potentiation unaltered. This dissociation suggests that while the depression produced in the lateral PP may be independently modulated, its occurrence may still require potentiation of the medial PP. Because bath application of a modulatory neurotransmitter is not equivalent to activation of the fibers that release that transmitter, one cannot necessarily conclude from these studies that activation of extrinsic fibers alone is sufficient to induce LTD (see below for a similar issue in the *Aplysia* sensory-motor synapse).

LTD OF INHIBITORY TRANSMISSION IN BRAIN

To date, most studies examining the electrophysiological mechanisms of neuronal information storage have focused on potentiation or depression of excitatory, usually glutamatergic, synapses. Although inhibitory synapses are ubiquitous and crucial to nervous system function, long-term alterations of these synapses have received relatively little attention, in spite of their enormous computational importance. Inhibitory synapses often exert powerful control over synaptically driven cell firing. In part, this control derives from the spatial distribution of excitatory versus inhibitory synapses. Inhibitory synapses are often found on the soma, proximal dendrites, and axons of a target neuron whereas excitatory synapses are found on the distal dendrites. This interposed configuration allows a single inhibitory synapse to negate by shunting up to hundreds of integrated EPSPs. Long-term changes of inhibitory synaptic strength are therefore likely to have an important role in the storage of information by neural circuits. These changes include LTP of the synapses between inhibitory interneurons and principal cells of the cerebellar cortex (Kano et al 1992), visual cortex (Komatsu & Iwakiri 1993), and hippocampus (Morashita & Sastry 1991, Xie & Sastry 1991) as well as LTD of inhibitory synapses, described below.

The evidence for LTD of inhibitory synapses in hippocampus comes primarily from studies that utilized an in vitro kindling protocol. Repeated tetanic stimulation of stratum radiatum fibers resulted in an increase in the strength of excitatory synapses on CA1 pyramidal neurons (presumably induction of homosynaptic LTP) that ultimately led to the development of epileptiform discharges. This treatment resulted in a persistent depression of both the rate and amplitude of spontaneous GABAergic IPSPs; it also produced persistent depression of both the early and late phases of monosynaptic evoked IPSPs as well as responses to exogenous GABA pulses; all of these effects could be blocked by application of an NMDA receptor antagonist during tetanic stimulation (Stelzer et al 1987). This finding was largely confirmed by a study that showed LTD of monosynaptic evoked IPSPs recorded in CA3 pyramidal cells following tetanization of either mossy fibers or stratum radiatum; it was extended by the observation that this depression of IPSPs allowed for the recruitment of latent excitatory connections that led to synchronous firing (Miles & Wong 1987). More recently, using a slightly modified procedure, investigators have shown that a selective antagonist of AMPA receptors does not block induction of this form of LTD (Merlin & Wong 1993).

Induction of epileptiform discharges by bath application of a selective metabotropic receptor agonist (1S,3R-ACPD, 10 µM, 20 min) also resulted in a persistent depression of both the early and late components of evoked GABAergic IPSPs (Liu et al 1993). LTD could be blocked by intracellular application of GTP-γ-S but not by external application of an NMDA receptor antagonist, an inhibitor of protein kinase C (calphostin C), or a treatment that results in depletion of intracellular Ca stores (thapsigargin). This depression was not associated with any persistent change in the pharmacologically isolated AMPA receptor–mediated EPSP. Taken together, these observations suggest that ACPD acts postsynaptically, through a G-protein linkage, to depress GABA sensitivity. It should be noted that since activation of metabotropic receptors produces a reversible potentiation of NMDA receptor–mediated responses in CA1 neurons (Aniksztejn et al 1991), activation of postsynaptic metabotropic receptors could cause a bifurcating signal that results in depression of IPSPs through both NMDA receptor–dependent and –independent pathways.

It should be noted that some investigations have failed to find a decrease in inhibitory transmission following tetanization. Bekenstein & Lothman (1993) compared slices from control rats to those in which repeated tetanic stimulation of stratum radiatum (in the intact hippocampus) had resulted in a previous episode of self-sustaining limbic status epilepticus. They found that disynaptic IPSPs recorded in CA1 pyramidal neurons were reduced in epileptic slices but that monosynaptic IPSPs evoked by direct activation of basket cells in the presence of blockers of excitatory transmission were unaltered. This finding is in accord with a previous study that found CA1 responses to exogenous pulses of GABA (as assessed by the GABA-induced reduction of the extracellularly recorded population spike) to be unaltered following induction of homosynaptic LTP in the SC-CA1 pathway (Scharfman & Sarvey 1985). Another study failed to find an alteration in the slope conductance of the late inhibitory component of a mossy fiber–evoked PSC recorded in area CA3 following induction of mossy fiber–CA3 LTP by tetanization (Griffith et al 1986).

Recently a novel form of LTD has been reported in the cerebellum. Stimulation of the white matter underlying the cerebellar cortex results in monosynaptic $GABA_A$ergic IPSPs recorded in neurons of the cerebellar deep nuclei, presumably through activation of the axons of Purkinje neurons. These IPSPs were persistently depressed following a brief high-frequency tetanus (Morashita & Sastry 1993). This depression was not associated with alterations in postsynaptic resting membrane potential, input resistance, or IPSP reversal potential. It is not clear whether glutamatergic fibers were also being activated during the tetanic stimulus and might have contributed to LTD induction. It is interesting to consider that this mechanism might interact with other forms of synaptic plasticity in the cerebellum, including LTD of the parallel fiber-Purkinje neuron synapse, discussed above, and LTP of the synapses between inhibitory interneurons of the cerebellar cortex and Purkinje neurons (Kano et al 1992).

LTD IN SPINAL CORD AND NEUROMUSCULAR JUNCTION

A form of homosynaptic LTD induced by high-frequency stimulation has recently been reported in several synaptic relays of vertebrate spinal cord. In one study, field potentials recorded in the ventral horn were evoked by stimulation of the dorsal-horn/intermediate-nucleus region. Tetanic stimulation (100 Hz × 50 pulses × 6 bursts) induced a range of alterations of synaptic strength, including a significant LTD in 33% of slices that persisted for the 2.5 hr duration of the recording session (Pockett & Figurov 1993). A similar phenomenon has been reported in the synapses between dorsal root afferents and neurons in the superficial laminae of the dorsal horn. A brief tetanus delivered to the afferent fibers also produced a range of alterations in synaptic strength, including LTD of intracellularly recorded EPSP amplitude in 41% and LTP in 45% of cells tested (Randic et al 1993). The properties of LTD at this synapse are quite similar to those reported for LTD induced by high-frequency stimulation in cortex (Artola et al 1990). The probability of inducing LTP or LTD could be modified by manipulating the membrane potential of the postsynaptic neuron; cells held at −85 mV were more likely to show LTD while cells held at −70 mV were more likely to show LTP. Likewise, when tetanic stimulation given at the test pulse intensity evoked LTP, a second train given at a higher intensity also induced LTP; but when LTD was evoked by the first train, a second train at higher intensity induced either LTP or LTD. Furthermore, LTD could be induced in the presence of NMDA and GABA$_A$ receptor antagonists. Together these observations suggest that, like induction of homosynaptic LTD by high-frequency stimulation in brain, induction of LTD at these spinal synapses depends upon some intermediate level of postsynaptic activation that does not require activation of NMDA receptors.

A form of heterosynaptic LTD has been shown using a preparation in which a single embryonic *Xenopus* muscle cell is innervated by two spinal neurons in tissue culture (Lo & Poo 1991). Stimulation of one synapse at moderate frequency (2 Hz × 80 pulses) resulted in a persistent heterosynaptic depression of the other. Like heterosynaptic depression in the hippocampus (Abraham & Goddard 1983), depression of the silent synapse was not dependent upon the establishment of potentiation at the stimulated synapse, and heterosynaptic depression could only be induced if the two synapses were within a certain distance of each other on the postsynaptic cell (White et al 1988 1990, Tomasulo et al 1993). Synchronous stimulation of both synapses resulted in no change of synaptic strength.

In addition, LTD has been observed using a similar preparation in which a single spinal neuron innervates a muscle cell and the activation of a second converging neuron is mimicked by application of exogenous acetylcholine (ACh) pulses (Dan & Poo 1992). Repeated application of ACh pulses alone, or of Ach pulses explicitly out-of-phase with presynaptic activity but not in-phase with presynaptic activity, resulted in a persistent depression of synaptic transmission that could be blocked by postsynaptic loading with a Ca chelator. However, presynaptic activity does not appear to be necessary for the induction of LTD, because repetitive depolarization of the muscle cell alone was sufficient to cause LTD (Lo et al 1994; see Christofi et al 1993 for a similar protocol in hippocampus). Several lines of evidence suggested that this form of LTD was mediated by a reduction in ACh release: There was no reduction in postsynaptic sensitivity to ACh pulses either at the site of conditioning pulse application or at an adjacent site; there was no alteration in the amplitude distribution of mEPSCs; and quantal analysis was consistent with a presynaptic reduction in quantal content. Taken together these findings suggest a model similar in some respects to one proposed for hippocampal homosynaptic LTP (reviewed in Williams et al 1993): Ca influx into the postsynaptic cell causes production of a retrograde messenger that acts upon the presynaptic terminal to alter (in this case, reduce) transmitter release. However, because in-phase presynaptic activation and ACh pulses result in no change in synaptic strength, one must propose the additional condition that presynaptic activation recruits yet another process that blocks the effect of ACh pulses. This process could involve a transient elevation in presynaptic Ca concentration; alternatively it might involve an additional postsynaptic signal, such as a further increase in Ca that drives the concentration from an LTD-inducing range to a null range.

What retrograde messenger mediates this form of heterosynaptic inhibition? Activation of one potential retrograde messenger system, a postsynaptic G protein–linked arachidonate cascade, resulted in an increase in the frequency of spontaneous synaptic currents (Harish & Poo 1992) and is therefore unlikely to mediate synaptic depression. However, several lines of evidence from a recent study (Wang & Lu 1994) implicate nitric oxide (NO) in this process. First, external application of an NO synthase inhibitor or an NO scavenger (hemoglobin) blocked the induction of LTD by repetitive postsynaptic depolarization. Second, application of NO donors produced a reversible depression of the frequency of spontaneous synaptic currents. Third, a similar depression was produced by treatments that increase intracellular cGMP (cGMP analogs, activators of soluble guanylate cyclase), an increase that is one of the consequences of NO production. cGMP analogs also produced a depression of the amplitude of spontaneous synaptic currents, and NO donors produced a similar decrement in the amplitude of evoked synaptic currents—observations suggesting that the NO/cGMP cascade may produce some postsynaptic alterations as well.

LTD IN SYNAPSES OF INVERTEBRATES

Examination of long-term changes in the synaptic strength of invertebrates has the advantage that in many cases the synaptic alteration may be directly correlated with learning as measured behaviorally. One case where this claim has been made is the habituation of the gill withdrawal reflex evoked by siphon skin stimulation in the marine mollusk *Aplysia*, which can demonstrate both a short-term and a long-term component and in both cases is thought to result from a depression in synaptic strength between a sensory neuron and an identified motoneuron, L7 (Castellucci et al 1978). In the intact abdominal ganglion, repeated stimulation of the sensory neuron gives rise to a short-term synaptic depression that is mediated by a decrease in the number of transmitter vesicles per impulse, as revealed by quantal analysis (Castellucci & Kandel 1974). A similar short-term depression may be produced by application of the molluscan neuropeptide FMRFamide together with activation of the presynaptic neuron (Small et al 1989); this depression is mediated, at least in part, by lipoxygenase metabolites of arachidonic acid (Piomelli et al 1987), which depress neurotransmitter release by means of a variety of alterations in both ion channels and the secretory apparatus (see Byrne et al 1993, for review). FMRFamide has been reported to cause a decrease in the input resistance of the postsynaptic motoneuron, and such a decrease may also contribute to synaptic depression (Pieroni & Byrne 1992; but see Small et al 1989).

When this sensory-motor synapse is formed in a dissociated cell culture preparation, 5 applications of FMRFamide given over a 2-hr period produce synaptic depression that persists for at least 24 hr (Montarolo et al 1988). Because the application of exogenous FMRFamide is thought to mimic the effect of stimulating one or more FMRFamide-containing interneurons in the intact ganglion (Buonomano et al 1992, Small et al 1992), this phenomenon has been called "long-term heterosynaptic inhibition." Two different homosynaptic stimulation protocols designed to mimic the firing pattern of the sensory neuron during repeated siphon stimulation induced short-term synaptic depression (60–90 min duration) but failed to produce a long-lasting effect. Thus, induction of a long-term effect may require activation of FMRFamide-containing interneurons or some other signal present in the intact ganglion. A role for protein synthesis in this form of synaptic depression is suggested by the observation that the protein synthesis inhibitor anisomycin blocks the long-term, but not the short-term depression produced by repeated FMRFamide exposure. Further investigations have supported the idea that persistent synaptic depression of this synapse requires macromolecular synthesis, because induction of depression by repeated FMRFamide application in culture results in structural changes observable at the light-microscopic level. These changes include a significant loss of sensory cell varicosities and neurites that correlates with the degree of depression induced (Schacher & Montarolo 1991). These structural changes are also blocked by inhibitors of protein or RNA synthesis (Bailey et al 1992).

In keeping with the proposed biochemical route of action of FMRFamide, prolonged application of arachidonic acid can also produce long-lasting hetero-synaptic inhibition and some of the associated alterations in sensory cell morphology (Schacher et al 1993). Furthermore, an inhibitor of the lipoxy-genase branch of arachidonic acid metabolism (but not an inhibitor of the cyclooxygenase branch) blocked both the expression of long-term hetero-synaptic inhibition and some of the associated changes in presynaptic mor-phology (Wu & Schacher 1994). FMRFamide may produce its effect on presynaptic function through the dephosphorylation of the substrates of cAMP-dependent protein kinase (Sweatt et al 1989, Ichinose & Byrne 1991), the activation of which produces a potentiation of this same synapse (see Byrne et al 1993 for review).

Another interesting invertebrate model system for the study of long-term synaptic modification is the neuromuscular junction of the crayfish, as investi-gated by Atwood and coworkers. Motoneurons of the crayfish may be classified into two types, which correspond to different functional requirements. Phasic motoneurons produce large EPSPs that are rapidly and transiently depressed during repetitive activation, while tonic motoneurons produce smaller EPSPs more resistant to depression. The fast closer excitor of the crayfish claw is a phasic neuron that normally fires few impulses. When, in young crayfish, a nerve containing the axon of this neuron was stimulated for 2 hr/day at 5 Hz for 7–14 days through chronic implanted electrodes, the motoneuron was transformed to display more tonic properties: the initial EPSP was 44% smaller, and the degree of synaptic fatigue produced by a sustained stimulus (5 Hz $\times$ 30 min) was 91% lower (Lnenicka & Atwood 1985). These alterations, which were mediated by a decrease in neurotransmitter release, were seen to persist for 10 days following the termination of tonic stimulation and were consequently given the name "long-term adaptation." This phenomenon was much smaller in adult crayfish. Like the FMRFamide-induced LTD in the *Aplysia* sensory-motor synapse described above, long-term adaptation in the crayfish was accompanied by distinct morphological alterations (Lnenicka et al 1986). These alterations caused the phasic presynaptic terminal to more closely resemble a tonic terminal (larger cross-sectional area, larger mitochondria, greater synaptic contact area, greater varicosity). Given these observations, it is not surprising that application of a protein synthesis inhibitor blocked the induction of long-term adaptation (Nguyen & Atwood 1990).

A good deal of work has been done to determine the requirements for the induction of long-term adaptation. First, selective stimulation of the fast closer excitor motoneuron without activation of other fibers running in the same nerve is sufficient to induce long-term adaptation (Lnenicka & Atwood 1988). Second, stimulation of sensory inputs to the fast closer excitor motoneuron that are subthreshold for spike generation is also sufficient to induce this phenomenon. Third, experiments utilizing localized blockade of spike conduction by application of tetrodotoxin to the motoneuron axon showed that neither muscle activity nor transmitter release from neuromuscular synapses was necessary for induction. However, stimulation at a point peripheral to the block was also capable of inducing long-term adaptation (Lnenicka & Atwood 1989). Taken together, these findings suggest that tonic depolarization of any portion of this motoneuron can induce long-term adaptation.

Long-term adaptation differs from LTD in that it represents two separate changes in synaptic transmission: depression and fatigue resistance. Recently it has been found that these two components of long-term adaptation may be dissociated. When either stimulation frequency or number of impulses per day is reduced, fatigue resistance may be induced without a depression of initial transmitter release (Mercier et al 1992). The converse dissociation has yet to be demonstrated.

The two examples of persistent synaptic depression in invertebrates discussed above are fundamentally different from any form of LTD reported in vertebrates. Unlike that of heterosynaptic LTD in the vertebrate brain or developing neuromuscular junction, the induction of long-term heterosynaptic inhibition of the *Aplysia* sensory-motor synapse does not represent the implementation of a post-not-presynaptic (converse-Hebbian) modification rule. Rather, it is likely to rely on the recruitment of a specific type of interneuron by repetitive stimulation and the consequent inhibition of transmitter release. Similarly, long-term adaptation of the crayfish neuromuscular junction has no obvious counterpart in vertebrate brain. Because it may be induced by repeated activation of the presynaptic motoneuron regardless of whether the postsynaptic muscle cell is activated, it does not seem to follow either a converse-Hebbian pre-not-postsynaptic modification rule (as claimed for hippocampal associative LTD) or an inverse-Hebbian rule (as in hippocampal homosynaptic LTD). The finding that subthreshold depolarization of any portion of the motoneuron can induce long-term adaptation of the crayfish neuromuscular junction makes it an entirely unique form of synaptic modification.

While these two invertebrate synaptic depression paradigms are phenomenologically different from those so far reported in vertebrate brain, they may nonetheless share certain molecular mechanisms, such as recruitment of an arachidonic acid cascade.

CONCLUSION

If LTP and STP were the only forms of synaptic alteration, it would still be possible to store information as varying synaptic weights. LTP undergoes passive decay over a period of days (Castro et al 1989), and therefore synaptic weights are not driven irreversibly to their maximal values. However, neural circuits containing synapses that can actively decrease their strength possess a distinct computational advantage. First, active resetting of potentiated synapses prevents saturation of LTP and makes the synapse both more responsive and more temporally flexible than does passive decay. Second, LTD of previously potentiated synapses could serve as a "forgetting" mechanism (Tsumoto 1993) for information stored by increases in synaptic strength. In addition, Christie et al (1994) have noted that LTP could just as easily serve as a forgetting mechanism for information stored by decreases in synaptic strength. Third, in a network of synaptic contacts, LTD could serve to attenuate the signal from neighboring potentiated synapses in a manner analogous to the way lateral inhibition promotes edge detection in the visual system.

A comparison of several aspects of the different forms of LTD induction discussed in this review is presented in Table 1. From examination of this table, together with Figure 1, it is clear that the four forms of LTD examined in vertebrate synapses represent four separate synaptic modification rules. However, the patterns of stimulation associated with different rules may engage common mechanisms of LTD induction and possibly expression. This may occur even if the initial events in induction (such as activation of postsynaptic receptors) are different.

Associative LTD, if it exists in its "pure" form as the implementation of a pre-not-postsynaptic modification rule, likely engages mechanisms distinct from those in other forms of LTD because it does not appear to require postsynaptic activity and consequent increases in postsynaptic Ca at the altered synapses. Homosynaptic LTD and heterosynaptic LTD are likely to engage similar induction mechanisms; there is good evidence that they both result from moderate increases in postsynaptic Ca at the altered synapses, and that treatments that further increase postsynaptic Ca cause LTP to be induced instead. This notion has been suggested in a number of recent reviews (Artola & Singer 1993, Christie et al 1994, Linden 1994b, Malenka & Nicoll 1993). Cerebellar LTD is also likely to engage different mechanisms: It requires large increases in postsynaptic Ca and explicitly requires the coactivation of two morphologically distinct inputs.

While the notion that four different learning rules impinge upon three separate induction mechanisms is conceptually attractive, relevant supporting evidence is currently scant. Experiments that would enable the necessary comparison are primarily of three types. First is a comparison of the locus of LTD expression, particularly a presynaptic/postsynaptic distinction, through such measures as paired-pulse facilitation (to detect one type of presynaptic alteration), replacement of test pulses with exogenous transmitter, and quantal analysis. Second, in those locations where multiple forms of LTD may be

Table 1   Induction of various forms of long-term synaptic alteration

Induction protocol	Activity at the altered synapse		Presumed Postsynaptic [Ca]	Block by postsynaptic Ca chelator?
	Presynaptic	Postsynaptic		
Homosynaptic LTP	+ +	+ +	High	Yes
Associative LTP	+	+ +	High	Yes
Heterosynaptic LTD	0	+ +	Moderate	Yes
Homosynaptic LTD	+	+	Moderate	Yes
Associative LTD	+	0	Low	?
Cerebellar LTD	+	+ +	High	Yes
LTA (crayfish)[a]	+	+	Moderate	?
LTHI *(Aplysia)*[b]	+	+	Moderate	?

[a] Long-term adaptation, does not require postsynaptic activation.
[b] Long-term heterosynaptic inhibition is likely to require activation of a modulatory interneuron.

induced, occlusion/saturation experiments may be used to address common modes of expression. Third, different forms of LTD may be compared for their ability to reverse LTP.

The idea that moderate postsynaptic activation, as occurs in homosynaptic LTD induction, results in persistent synaptic depression while stronger postsynaptic activation results in persistent potentiation was formalized by Bienenstock, Cooper, and Munro (1982) and has since come to be known as the BCM rule. Another aspect of this rule is that the two thresholds for synaptic modification, the lower one for inducing LTD and the higher one for inducing LTP, are not fixed and can vary depending upon the recent patterns of synaptic use (Bienenstock et al 1982, Bear et al 1987). The first part of this rule has now received experimental confirmation in both hippocampal (Dudek & Bear 1992, Mulkey & Malenka 1992) and neocortical (Kirkwood et al 1993) synapses: Low-frequency stimulation of afferents gives rise to LTD, high-frequency stimulation causes LTP, and stimulation at some intermediate frequency does not alter synaptic strength (it will be interesting to determine the molecular events that underlie this null-point phenomenon).

The second part of the BCM rule has also been experimentally confirmed. First, in slices of adult hippocampus, treatments that failed to induce any change in synaptic strength, such as weak intermediate-frequency stimulation (30 Hz, 0.15 s) of afferents (Huang et al 1992), or stimulation at low frequency (O'Dell & Kandel 1994), could produce a refractory period of about 60 min during which LTP could not be induced by strong high-frequency stimulation. Second, while some investigators find that induction of homosynaptic LTD by low-frequency stimulation does not occur in slices of adult hippocampus (Staubli & Lynch 1990, Fujii et al 1991, Larson et al 1993, O'Dell & Kandel

1994), others find that it is merely reduced in amplitude when compared to juvenile tissue (Dudek & Bear 1993). In either case, robust LTD (or depotentiation) may be induced following either previous potentiation of adult hippocampal synapses (Barrionuevo et al 1980, Staubli & Lynch 1990, Fujii et al 1991, Larson et al 1993, O'Dell & Kandel 1994) or priming with weak 30-Hz stimulation that failed to induce LTP (Wexler & Stanton 1993). It is important to note that the alteration of thresholds for LTP and LTD by prior activity does not explicitly require a change in synaptic strength.

A molecular mechanism that could underlie the BCM rule has also been proposed: Moderate increases in postsynaptic Ca, as occur in homosynaptic LTD induction, might alter the relative activities of Ca-calmodulin-dependent protein kinase II and protein phosphatase I and thereby mediate a persistent synaptic depression through an action on postsynaptic receptors (Lisman 1989)—perhaps a direct dephosphorylation of the receptors themselves. While some of the experimental evidence discussed above tends to support such a model (Mulkey et al 1993, Kirkwood et al 1994, O'Dell & Kandel 1994), there is reason to believe that the model requires elaboration. To this point, the focus of the discussion of homosynaptic LTD induction has been on postsynaptic Ca concentration at the altered synapse. In addition, postsynaptic Ca at altered synapses has been discussed as if it were a single compartment to which many signals could contribute. However, the various sources of Ca could contribute to separate pools having limited interactions with each other and with various second messengers (see the similar case of cerebellar LTD, above), and the timing of signals in homosynaptic LTD induction might constrain the range of these interactions.

NOTE ADDED IN PROOF

Since the initial submission of this manuscript, several important findings have appeared in the literature. These include a report indicating that hippocampal homosynaptic LTD induction requires Ca influx via voltage-gated channels and metabotropic glutamate receptor activation and that this results in a depression of transmitter release as determined by quantal analysis (Bolshakov & Siegelbaum 1994). Another report has indicated that in a single cell the same change in postsynaptic Ca can have opposite effects on different components of the synaptic response. High-frequency stimulation applied to the glutamatergic synapse between prelimbic cortical afferents and core neurons in the nucleus accumbens induces both homosynaptic LTP of non-NMDA receptor–mediated currents and homosynaptic LTD of NMDA receptor–mediated currents (Kombian & Malenka 1994). Both alterations are blocked by NMDA receptor blockade during the tetanus or postsynaptic loading with a Ca chelator. This result suggests that mechanisms downstream from postsynaptic Ca increases may be of critical importance in determining whether synaptic potentiation or depression occurs.

ACKNOWLEDGMENTS

Thanks to K Takahashi, who helped to write a portion of the section on inhibitory LTD, and to W Singer, A Artola, S Nakanishi, G Barrionuevo, F Crépel, M Bear, R Malenka, R Nicoll, L Bindman, N Kato, W Abraham, W Levy, NT Slater, T Tsumoto, T Teyler, D Lovinger, B Gähwiler, B Lu, T O'Dell, and P Calabresi for sharing manuscripts prior to publication. NT Slater and S Schacher read portions of an earlier version of this manuscript and offered helpful comments. This work was supported in part by PHS grant MH51106. DJL is a Klingenstein Fellow, an Alfred P Sloan Research Fellow, and a McKnight Scholar

---

Literature Cited

Abeliovich A, Chen C, Goda Y, Silva AJ, Stevens CF, Tonegawa S. 1993. Modified hippocampal long-term potentiation in PKCγ mutant mice. *Cell* 75:1253–62

Abraham WC, Goddard GV. 1983. Asymmetric relationships between homosynaptic long-term potentiation and heterosynaptic long-term depression. *Nature* 305:717–19

Abraham WC, Wickens JR. 1991. Heterosynaptic long-term depression is facilitated by blockade of inhibition in area CA1 of the hippocampus. *Brain Res.* 546:336–40

Alford S, Frenguelli BG, Schofield JG, Collingridge GL. 1993. Characterization of Ca²⁺ signals induced in hippocampal CA1 neurons by the synaptic activation of NMDA receptors. *J. Physiol. London* 469:693–716

Alger BE, Megela AL, Teyler TJ. 1978. Transient heterosynaptic depression in the hippocampal slice. *Brain Res. Bull.* 3:181–84

Andersen P, Sundberg SH, Sveen O, Wigstrom H. 1977. Specific long-lasting potentiation of synaptic transmission in hippocampal slices. *Nature* 266:736–37

Anikstejn L, Bregestovski, P, Ben-Ari Y. 1991. Selective activation of quisqualate metabotropic receptor potentiates NMDA but not AMPA responses. *Eur. J. Pharmacol.* 205: 327–28

Artola A, Singer W. 1993. Long-term depression of excitatory synaptic transmission and its relationship to long-term potentiation. *Trends Neurosci.* 16:480–87

Artola A, Brocher S, Singer W. 1990. Different voltage-dependent thresholds for inducing long-term depression and long-term potentiation in slices of rat visual cortex. *Nature* 347:69–72

Artola A, Hensch TK, Singer W. 1992. A rise of [Ca²⁺]ᵢ in the postsynaptic cell is necessary and sufficient for the induction of long-term depression (LTD) in neocortex. *Soc. Neurosci.* 18:1351

Bailey CH, Montarolo P, Chen M, Kandel ER, Schacher S. 1992. Inhibitors of protein and RNA synthesis block structural changes that accompany long-term synaptic plasticity in *Aplysia. Neuron* 9:749–58

Barrionuevo G, Shottler F, Lynch G. 1980. The effects of repetitive low-frequency stimulation on control and "potentiated" synaptic responses in the hippocampus. *Life Sci.* 27: 2385–91

Bashir ZI, Collingridge GL. 1992. NMDA receptor-dependent transient homo- and heterosynaptic depression in picrotoxin-treated hippocampal slices. *Eur. J. Neurosci.* 4:485–90

Bashir ZI, Jane DE, Sunter DC, Watkins JE, Collingridge GL. 1993. Metabotropic glutamate receptors contribute to the induction of long-term depression in the CA1 region of the hippocampus. *Eur. J. Pharmacol.* 239: 265–66

Bear MF, Cooper LN, Ebner FF. 1987. A physiological basis for a theory of synaptic modification. *Science* 237:42–48

Bekenstein JW, Lothman EW. 1993. Dormancy of inhibitory interneurons in a model of temporal lobe epilepsy. *Science* 259:97–100

Berry RL, Teyler TJ, Taizhen H. 1989. Induction of LTP in rat primary visual cortex: tetanus parameters. *Brain Res.* 481:221–27

Bienenstock EL, Cooper LN, Munro PW. 1982. A theory for the development of neuron selectivity: orientation specificity and binocular interaction in visual cortex. *J. Neurosci.* 2: 32–48

Bliss TVP, Collingridge GL. 1993. A synaptic model of memory: long-term potentiation in the hippocampus. *Nature* 361:31–39

Bolshakov VY, Siegelbaum SA. 1994. Postsynaptic induction and presynaptic expression of hippocampal long-term depression. *Science* 264:1148–52

Bortolotto ZA, Bashir ZI, Davies CH, Collingridge GL. 1994. A molecular switch activated by metabotropic glutamate receptors regulates induction of long-term potentiation. *Nature* 368:740–43

Bradler JE, Barrionuevo G. 1989. Long-term potentiation in hippocampal CA3 neurons: tetanized input regulates heterosynaptic efficacy. *Synapse* 4:132–42

Bradler JE, Barrionuevo G. 1990. Heterosynaptic correlates of long-term potentiation induction in hippocampal CA3 neurons. *Neuroscience* 35:265–71

Bramham CR, Srebro B. 1987. Induction of long-term depression and potentiation by low- and high-frequency stimulation in the dentate area of the anesthetized rat: magnitude, time course and EEG. *Brain Res.* 405:100–7

Bredt DS, Hwang PM, Snyder SH. 1990. Localization of nitric oxide synthase indicating a neural role for nitric oxide. *Nature* 347: 768–70

Brocher S, Artola A, Singer W. 1992. Intracellular injection of Ca^{2+} chelators blocks induction of long-term depression in rat visual cortex. *Proc. Natl. Acad. Sci. USA* 89:123–27

Buonomano DV, Cleary LJ, Byrne JH. 1992. Inhibitory neuron produces heterosynaptic inhibition of the sensory-to-motor neuron synapse in *Aplysia. Brain Res.* 577:147–50

Byrne JH, Zwartjes R, Homayouni R, Critz SD, Eskin A. 1993. Roles of second messenger pathways in neuronal plasticity and in learning and memory. Insights gained from *Aplysia. Adv. Second Messenger Phosphoprotein Res.* 2:47–108

Cajal SR. 1911. *Histologie du Système Nerveux de l'Homme et des Vertebrés.* Paris: Malone. Vol. 2

Calabresi P, Maj R, Pisani A, Mercuri NB, Bernardi G. 1992a. Long-term synaptic depression in the striatum: physiological and pharmacological characterization. *J. Neurosci.* 12:4224–33

Calabresi P, Pisani A, Mercuri NB, Bernardi, G. 1992b. Long-term potentiation in the striatum is unmasked by removing the voltage-dependent magnesium block of NMDA receptor channels. *Eur. J. Neurosci.* 4:929–35

Calabresi P, Pisani A, Mercuri NB, Bernardi G. 1993a. Lithium treatment blocks long-term synaptic depression in the striatum. *Neuron* 10:955–62

Calabresi P, Pisani A, Mercuri NB, Bernardi G. 1993b. Heterogeneity of metabotropic glutamate receptors in the striatum: electrophysiological evidence. *Eur. J. Neurosci.* 5: 1370–77

Castellucci VF, Kandel ER. 1974. A quantal analysis of the synaptic depression underlying habituation of the gill-withdrawal reflex in *Aplysia. Proc. Natl. Acad. Sci. USA* 71: 5004–8

Castellucci VF, Carew TJ, Kandel ER. 1978. Cellular analysis of long-term habituation of the gill-withdrawal reflex in *Aplysia. Science* 202:1306–8

Castro CA, Silbert LH, McNaughton BL, Barnes CA. 1989. Recovery of spatial learning deficits after decay of electrically induced synaptic enhancement in the hippocampus. *Nature* 342:545–48

Chattarji S, Stanton PK, Sejnowski TJ. 1989. Commissural synapses, but not mossy fiber synapses, in hippocampal field CA3 exhibit associative long-term potentiation and depression. *Brain Res.* 495:145–50

Chen C, Thompson RF. 1992. Associative long-term depression revealed by field potential recording in cerebellar slice. *Soc. Neurosci.* 18:1215

Christie BR, Abraham WC. 1992a. NMDA-dependent heterosynaptic long-term depression in the dentate gyrus of anesthetized rats. *Synapse* 10:1–6

Christie BR, Abraham, WC. 1992b. Priming of associative long-term depression in the dentate gyrus by theta frequency activity. *Neuron* 9:79–84

Christie BR, Abraham WC. 1994. L-type voltage-sensitive calcium channel anagonists block heterosynaptic long-term depression in the dentate gyrus of anesthetized rats. *Neurosci. Lett.* 167:41–45

Christie BR, Kerr DS, Abraham WC. 1994. The flip side of synaptic plasticity: long-term depression mechanisms in the hippocampus. *Hippocampus* 4:127–35

Christofi G, Nowicky AV, Bolsover SR, Bindman LJ. 1993. The postsynaptic induction of nonassociative long-term depression of excitatory synaptic transmission in rat hippocampal slices. *J. Neurophysiol.* 69:219–29

Colbert CM, Burger BS, Levy WB. 1992. Longevity of synaptic depression in the hippocampal dentate gyrus. *Brain Res.* 571:159–61

Crepel F, Daniel H, Hemart N, Jaillard D. 1991. Effects of ACPD and AP3 on parallel fibre-mediated EPSPs of Purkinje cells in cerebellar slices in vitro. *Exp. Brain Res.* 86:402–6

Crepel F, Jaillard D. 1990. Protein kinases, nitric oxide and long-term depression of synapses in the cerebellum. *NeuroReport* 1: 133–36

Crepel F, Jaillard D. 1991. Pairing of pre- and postsynaptic activities in cerebellar Purkinje cells induces long-term changes in synaptic efficacy in vitro. *J. Physiol. London* 432: 123–41

Crepel F, Krupa M. 1988. Activation of protein kinase C induces a long-term depression of glutamate sensitivity of cerebellar Purkinje cells. An in vitro study. *Brain Res.* 458:397–401

Dahl D, Sarvey JM. 1989. Norepinephrine induces pathway-specific long-lasting potentiation and depression in the hippocampal dentate gyrus. *Proc. Natl. Acad. Sci. USA* 86:4776–80

Dahl D, Sarvey JM. 1990. β-adrenergic agonist-induced long-lasting synaptic modifications in hippocampal dentate gyrus requires activation of NMDA receptors, but not electrical activation of afferents. *Brain Res.* 526:347–50

Dan Y, Poo M-M. 1992. Hebbian depression of isolated neuromuscular synapses in vitro. *Science* 256:1570–73

Daniel H, Hemart N, Jaillard D, Crepel F. 1992. Coactivation of metabotropic glutamate receptors and of voltage-gated calcium channels induces long-term depression in cerebellar Purkinje cells in vitro. *Exp. Brain Res.* 90:327–31

Daniel H, Hemart N, Jaillard D, Crepel F. 1993. Long-term depression requires nitric oxide and guanosine-3,-5-cyclic monophosphate production in cerebellar Purkinje cells. *Eur. J. Neurosci* 5:1079–82

Debanne D, Gahwiler BH, Thompson SM. 1994. Asynchronous pre- and postsynaptic activity induces associative long-term depression in area CA1 of the rat hippocampus in vitro. *Proc. Natl. Acad. Sci. USA* 91:1148–52

Desmond NL, Colbert CM, Zhang DX, Levy WB. 1991. NMDA receptor antagonists block the induction of long-term depression in the hippocampal dentate gyrus of the anesthetized rat. *Brain Res.* 552:93–98

de Vente J, Bol JGJM, Berkelmans HS, Schipper J, Steinbusch HWM. 1990. Immunocytochemistry of cGMP in the cerebellum of the immature, adult, and aged rat: the involvement of nitric oxide. A micropharmacological study. *Eur. J. Neurosci.* 2:845–62

de Vente J, Steinbusch HWM. 1992. On the stimulation of soluble and particulate guanylate cyclase in the rat brain and the involvement of nitric oxide as studied by cGMP immunocytochemistry. *Acta Histochem.* 92: 13–38

Dudek SM, Bear MF. 1992. Homosynaptic long-term depression in area CA1 of hippocampus and effects of N-methyl-D-aspartate receptor blockade. *Proc. Natl. Acad. Sci. USA* 89:4363–67

Dudek SM, Bear MF. 1993. Bidirectional long-term modification of synaptic effectiveness in the adult and immature hippocampus. *J. Neurosci.* 13:2910–18

Dunwiddie T, Lynch G. 1978. Long-term potentiation and depression of synaptic responses in the rat hippocampus: localization and frequency dependency. *J. Physiol. London* 276:353–67

Eaton SA, Jane DE, Jones PLSJ, Porter RHP, Pook PC-K, et al. 1993. Competitive antagonism at metabotropic glutamate receptors by (S)-4-carboxyphenylglycine and (RS)-alpha-methyl-4-carboxy-phenylglycine. *Eur. J. Pharmacol.* 244:195–97

Eccles JC, Llinas R, Sasaki K. 1966. The excitatory synaptic action of climbing fibres on the Purkinje cells of the cerebellum. *J. Physiol. London* 182:268–96

Ekerot C-F, Kano M. 1985. Long-term depression of parallel fibre synapses following stimulation of climbing fibres. *Brain Res.* 342:357–60

Ekerot C-F, Kano M. 1989. Stimulation parameters influencing climbing fibre induced long-term depression of parallel fibre synapses. *Neurosci. Res.* 6:264–68

Fujii S, Saito K, Miyakawa H, Ito K, Kato H. 1991. Reversal of long-term potentiation (depotentiation) induced by tetanus stimulation of the input to CA1 neurons of guinea pig hippocampal slices. *Brain Res.* 555:112–22

Glaum SR, Slater NT, Rossi DJ, Miller RJ. 1992. The role of metabotropic glutamate receptors at the parallel fiber-Purkinje cell synapse. *J. Neurophysiol.* 68:1453–62

Griffith WH, Brown TH, Johnston D. 1986. Voltage-clamp analysis of synaptic inhibition during long-term potentiation in hippocampus. *J. Neurophysiol.* 55:767–75

Grover LM, Teyler TJ. 1992. N-methyl-D-aspartate receptor-independent long-term potentiation in area CA1 of rat hippocampus: input-specific induction and preclusion in a non-tetanized pathway. *Neuroscience* 49:7–11

Harish OE, Poo M-M. 1992. Retrograde modulation at developing neuromuscular synapses: involvement of G protein and arachidonic acid cascade. *Neuron* 9:1201–9

Hebb DO. 1949. *The Organization of Behavior.* New York: Wiley

Hémart N, Daniel H, Jaillard D, Crépel F. 1994.

Receptors and second messengers involved in long-term depression in rat cerebellar slices in vitro: a reappraisal. *Eur. J. Neurosci.* In press

Hirano T. 1990a. Effects of postsynaptic depolarization in the induction of synaptic depression between a granule cell and a Purkinje cell in rat cerebellar culture. *Neurosci. Lett.* 119:145–47

Hirano T. 1990b. Effects of postsynaptic depolarization in the induction of synaptic depression between a granule cell and a Purkinje cell in rat cerebellar culture. *Neurosci. Lett.* 119:145–47

Hirano T. 1991. Differential pre- and postsynaptic mechanisms for synaptic potentiation and depression between a granule cell and a Purkinje cell in rat cerebellar culture. *Synapse* 7:321–23

Hirsch JC, Crepel F. 1990. Use-dependent changes in synaptic efficacy in rat prefrontal neurons in vitro. *J. Physiol. London* 427:31–49

Hirsch JC, Crepel F. 1991. Blockade of NMDA receptors unmasks a long-term depression in synaptic efficacy in rat prefrontal neurons in vitro. *Exp. Brain Res.* 85:621–24

Hirsch JC, Crepel F. 1992. Postsynaptic Ca is necessary for the induction of LTP and LTD of monosynaptic EPSPs in prefrontal neurons: an in vitro study in the rat. *Synapse* 10:173–75

Hirsch JC, Barrionuevo G, Crepel F. 1992. Homo- and heterosynaptic changes in efficacy are expressed in prefrontal neurons: an in vitro study in the rat. *Synapse* 12:82–85

Huang Y-Y, Colino A, Selig DK, Malenka RC. 1992. The influence of prior synaptic activity on the induction of long-term potentiation. *Science* 255:730–33

Ichinose M, Byrne JH. 1991. Role of protein phosphatases in the modulation of neuronal membrane currents. *Brain Res.* 549:146–50

Ito M, Sakurai M, Tongroach P. 1982. Climbing fibre induced depression of both mossy fiber responsiveness and glutamate sensitivity of cerebellar Purkinje cells. *J. Physiol. London* 324:113–34

Jaffe DB, Johnston D, Lasser-Ross N, Lisman JE, Miyakawa H, Ross WN. 1992. The spread of Na spikes determines the pattern of dendritic Ca entry into hippocampal neurons. *Nature* 357:244–46

Johnston D, Williams S, Jaffe D, Gray R. 1992. NMDA-receptor-independent long-term potentiation. *Annu. Rev. Physiol.* 54:489–85

Kano M, Kato M. 1987. Quisqualate receptors are specifically involved in cerebellar synaptic plasticity. *Nature* 325:276–79

Kano M, Kato M. 1988. Mode of induction of long-term depression at parallel fibre-Purkinje cell synapses in rabbit cerebellar cortex. *Neurosci. Res.* 5:544–56

Kano M, Rexhausen U, Dreessen J, Konnerth A. 1992. Synaptic excitation produces a long-lasting rebound potentiation of inhibitory synaptic signals in cerebellar Purkinje cells. *Nature* 356:601–4

Kato N. 1993. Dependence of long-term depression on postsynaptic metabotropic glutamate receptors in visual cortex. *Proc. Natl. Acad. Sci. USA* 90:3650–54

Kato N, Yoshimura H. 1993. Tetanization during $GABA_A$ receptor activation induces long-term depression in visual cortex slices. *Neuropharmacology* 32:511–13

Kerr DS, Abraham WC. 1993. A comparison of associative and non-associative conditioning procedures in induction of LTD in CA1 of the hippocampus. *Synapse* 14:305–13

Kiedrowski L, Brooker G, Costa E, Wroblewski JT. 1994. Glutamate impairs neuronal calcium extrusion while reducing sodium gradient. *Neuron* 12:295–300

Kimura F, Tsumoto T, Nishigori A, Yoshimura Y. 1990. Long-term depression but not potentiation is induced in Ca^{2+}-chelated visual cortex neurons. *NeuroReport* 1:65–68

Kinney GA, Peterson BW, Slater NT. 1993. Long-term synaptic plasticity in the rat medial vestibular nucleus studied using patch-clamp recording in an in vitro brain slice preparation. *Soc. Neurosci.* 19:1993

Kirkwood A, Bear MF. 1994. Homosynaptic long-term depression in the visual cortex. *J. Neurosci.* 14:3404–12

Kirkwood A, Dudek SM, Gold JT, Aizenman CD, Bear MF. 1993. Common forms of synaptic plasticity in the hippocampus and neocortex in vitro. *Science* 260:1518–21

Knopfel T, Vranesic I, Staub C, Gahwiler BH. 1990. Climbing fibre responses in olive-cerebellar slice cultures. II. Dynamics of cytosolic calcium in Purkinje cells. *Eur. J. Neurosci.* 3:343–48

Komatsu Y, Iwakiri M. 1993. Long-term modification of inhibitory synaptic transmission in developing visual cortex. *NeuroReport* 4:907–10

Kombian SB, Malenka RC. 1994. Simultaneous LTP of non-NMDA and LTD of NMDA-receptor-mediated responses in the nucleus accumbens. *Nature* 368:242–46

Konnerth A, Dreessen J, Augustine GJ. 1992. Brief dendritic calcium signals initiates long-lasting synaptic depression in cerebellar Purkinje cells. *Proc. Natl. Acad. Sci. USA* 89:7051–55

Krug M, Muller-Welde P, Wagner M, Ott T, Matthies H. 1985. Functional plasticity in two afferent systems of the granule cells in the rat dentate area: frequency-related changes, long-term potentiation and heterosynaptic depression. *Brain Res.* 360:264–72

Larson J, Xiao P, Lynch G. 1993. Reversal of

LTP by theta frequency stimulation. *Brain Res.* 600:97–102

Levy WB, Steward O. 1979. Synapses as associative memory elements in the hippocampal formation. *Brain Res.* 175:233–45

Levy WB, Steward O. 1983. Temporal contiguity requirements for long-term associative potentiation/depression in the hippocampus. *Neuroscience* 8:791–97

Lin J-H, Way L-J, Gean P-W. 1993. Pairing of pre- and postsynaptic activities in hippocampal CA1 neurons induces long-term modifications of NMDA receptor-mediated synaptic potential. *Brain Res.* 603:117–20

Linden DJ. 1994a. Input-specific induction of cerebellar long-term depression does not require presynaptic alteration. *Learn. Mem.* 1:121–28

Linden DJ. 1994b. Long-term synaptic depression in the mammalian brain. *Neuron* 12:457–72

Linden DJ, Connor JA. 1991. Participation of postsynaptic PKC in cerebellar long-term depression in culture. *Science* 254:1656–59

Linden DJ, Connor JA. 1992. Long-term depression of glutamate currents in cultured cerebellar Purkinje neurons does not require nitric oxide signalling. *Eur. J. Neurosci.* 4:10–15

Linden DJ, Connor JA. 1993. Cellular mechanisms of long-term depression in the cerebellum. *Curr. Opin. Neurobiol.* 3:401–6

Linden DJ, Smeyne M, Connor JA. 1993. Induction of cerebellar long-term depression in culture requires postsynaptic action of sodium ions. *Neuron* 10:1093–1100

Linden DJ, Dickinson MH, Smeyne M, Connor JA. 1991. A long-term depression of AMPA currents in cultured cerebellar Purkinje neurons. *Neuron* 7:81–89

Lisman J. 1989. A mechanism for the Hebb and the anti-Hebb processes underlying learning memory. *Proc. Natl. Acad. Sci. USA* 86:9574–78

Liu Y-B, Disterhoft JF, Slater NT. 1993. Activation of metabotropic glutamate receptors induces long-term depression of GABAergic inhibition in the hippocampus. *J. Neurophysiol.* 69:1000–4

Lnenicka GA, Atwood HL. 1985. Age-dependent long-term adaptation of crayfish phasic motor axon synapses to altered activity. *J. Neurosci.* 5:459–67

Lnenicka GA, Atwood HL. 1988. Long-term changes in neuromuscular synapses with altered sensory input to a crayfish motoneuron. *Exp. Neurol.* 100:437–47

Lnenicka GA, Atwood HL. 1989. Impulse activity of a crayfish motoneuron regulates its neuromuscular synaptic properties. *J. Neurophys.* 61:91–96

Lnenicka GA, Atwood HL, Marin L. 1986. Morphological transformation of synaptic terminals of a phasic motoneuron by long-term tonic stimulation. *J. Neurosci.* 6:2252–58

Lo Y-J, Poo M-M. 1991. Activity-dependent synaptic competition in vitro: heterosynaptic suppression of developing synapses. *Science* 254:1019–22

Lo Y-J, Lin Y, Sanes DH, Poo M-M. 1994. Depression of developing neuromuscular synapses induced by repetitive postsynaptic depolarizations. *J. Neurosci.* 14:4694–704

Lopez H, Burger B, Dickstein R, Desmond NL, Levy WB. 1990. Long-term potentiation and long-term depression in the dentate gyrus: quantification of dissociable synaptic modifications in the hippocampal dentate gyrus favors a particular class of synaptic modification equations. *Synapse* 5:33–47

Lovinger DM, Tyler C, Merritt A. 1993. Short- and long-term synaptic depression in rat neostriatum. *J. Neurophysiol.* 70:1937–49

Lum-Ragan JT, Gribkoff VK. 1993. The sensitivity of hippocampal long-term potentiation to nitric oxide synthase inhibitors is dependent upon the pattern of conditioning stimulation. *Neuroscience* 57:973–83

Lynch GS, Dunwiddie T, Gribkoff V. 1977. Heterosynaptic depression: a postsynaptic correlate of long-term potentiation. *Nature* 266:737–39

Malenka RC, Nicoll RA. 1993. NMDA receptor-dependent synaptic plasticity: multiple forms and mechanisms. *Trends Neurosci.* 16:521–27

Malenka RC, Lancaster B, Zucker RS. 1992. Temporal limits on the rise in postsynaptic calcium required for the induction of long-term potentiation. *Neuron* 9:121–28

Martin LJ, Blackstone CD, Huganir RL, Price DL. 1992. Cellular localization of a metabotropic glutamate receptor in rat brain. *Neuron* 9:259–70

Masu M, Tanabe Y, Tsuchida K, Shigemoto R, Nakanishi S. 1991. Sequence and expression of a metabotropic glutamate receptor. *Nature* 349:760–65

Mercier AJ, Bradacs H, Atwood HL. 1992. Long-term adaptation of crayfish neurons depends on the frequency and number of impulses. *Brain Res.* 598:221–24

Miles R, Wong RKS. 1987. Latent synaptic pathways revealed after tetanic stimulation in the hippocampus. *Nature* 329:724–26

Miyakawa H, Lev-Ram V, Lasser-Ross N, Ross WN. 1992a. Calcium transients evoked by parallel fiber and climbing fiber synaptic input in guinea pig cerebellar Purkinje neurons. *J. Neurophysiol.* 68:1178–89

Miyakawa H, Ross WN, Jaffe D, Callaway JC, Lasser-Ross N, Lisman JE, Johnston D. 1992b. Synaptically activated increases in Ca^{2+} concentration in hippocampal CA1 pyramidal cells are primarily due to voltage-gated Ca^{2+} channels. *Neuron* 9:1163–73

Montarolo PG, Kandel ER, Schacher S. 1988. Long-term heterosynaptic inhibition in *Aplysia*. *Nature* 333:171–74

Morishita W, Sastry BR. 1991. Chelation of postsynaptic Ca^{2+} facilitates long-term potentiation of hippocampal IPSPs. *NeuroReport* 2:389–92

Morishita W, Sastry BR. 1993. Long-term depression of IPSPs in rat deep cerebellar nuclei. *NeuroReport* 4:719–22

Mulkey RM, Herron CE, Malenka RC. 1993. An essential role for protein phosphatases in hippocampal long-term depression. *Science* 261:1051–55

Mulkey RM, Malenka RC. 1992. Mechanisms underlying induction of homosynaptic long-term depression in area CA1 of the hippocampus. *Neuron* 9:967–75

Muller W, Connor JA. 1991. Synaptic Ca^{2+} responses in dendritic spines: the spine as an individual neuronal compartment. *Nature* 354:73–76

Nguyen PV, Atwood HL. 1990. Expression of long-term adaptation of synaptic transmission requires a critical period of protein synthesis. *J. Neurosci.* 10:1099–1109

Nowicky AV, Christofi G, Barry MF, Bolsover SR, Bindman LJ. 1993. Activity-dependent calcium changes imaged during postsynaptic induction of long-term depression of synaptic transmission in rat hippocampal slices. *Proc. 23rd Congr. Int. Union Physiol. Sci.*, p. 174. (Abstr.)

O'Dell TJ, Kandel ER. 1994. Low-frequency stimulation erases LTP through an NMDA receptor-mediated activation of protein phosphatases. *Learn. Mem.* 1:129–39

Olson DR, Kon C, Breckenridge B. 1976. Calcium ion effects on guanylate cyclase of brain. *Life Sci.* 18:935–40

Pang K, Williams MJ, Olton DS. 1993. Activation of the medial septal area attenuates LTP of the lateral perforant path and enhances heterosynaptic LTD of the medial perforant path in aged rats. *Brain Res.* 632:150–60

Paulsen O, Li Y-G, Hvalby O, Andersen P, Bliss TVP. 1993. Failure to induce long-term depression by an anti-correlation procedure in area CA1 of the rat hippocampal slice. *Eur. J. Neurosci.* 5:1241–46

Perkel DJ, Petrozzino JJ, Nicoll RA, Connor JA. 1993. The role of Ca^{2+} entry via synaptically activated NMDA receptors in the induction of long-term potentiation. *Neuron* 11:817–23

Pieroni JP, Byrne JH. 1992. Differential effects of serotonin, FMRFamide, and small cardioactive peptide on multiple, distributed processes modulating sensorimotor synaptic transmission in *Aplysia*. *J. Neurosci.* 12:2633–47

Piomelli D, Volterra A, Dale N, Siegelbaum SA, Kandel ER, et al. 1987. Lipoxygenase metabolites of arachidonic acid as second messengers for presynaptic inhibition of *Aplysia* sensory cells. *Nature* 328:38–43

Pockett S, Figurov, A. 1993. Long-term potentiation and depression in the ventral horn of rat spinal cord in vitro. *NeuroReport* 4:97–99

Pockett S, Lippold OJC. 1986. Long-term potentiation and depression in hippocampal slices. *Exp. Neurol.* 91:481–87

Pockett S, Brookes NH, Bindman LJ. 1990. Long-term depression at synapses in slices of rat hippocampus can be induced by bursts of postsynaptic activity. *Exp. Brain Res.* 80: 196–200

Pozzo-Miller LD, Petrozzino JJ, Mahanty NK, Connor JA. 1993. Optical imaging of cytosolic calcium, electrophysiology, and ultrastructure in pyramidal neurons of organotypic slice cultures from rat hippocampus. *Neuroimage* 1:109–20

Randic M, Jiang MC, Cerne R. 1993. Long-term potentiation and long-term depression of primary afferent neurotransmission in the rat spinal cord. *J. Neurosci.* 13:5228–41

Regehr WG, Tank DW. 1992. Calcium concentration dynamics produced by synaptic activation of CA1 hippocampal pyramidal cells. *J. Neurosci.* 12:4202–23

Regehr WG, Connor JA, Tank DW. 1989. Optical imaging of calcium accumulation in hippocampal pyramidal cells during synaptic activation. *Nature* 341:533–36

Ross WN, Werman R. 1987. Mapping calcium transients in the dendrites of Purkinje cells from the guinea-pig cerebellum in vitro. *J. Physiol. London* 389:319–36

Sakurai M. 1990. Calcium is an intracellular mediator of the climbing fiber in induction of cerebellar long-term depression. *Proc. Natl. Acad. Sci. USA* 87:3383–85

Sastry BR, Chirwa SS, Goh JW, Maretic H, Pandanaboina MM. 1984. Verapamil counteracts depression but not long-lasting potentiation of the hippocampal population spike. *Life Sci.* 34:1075–86

Schacher S, Montorolo PG. 1991. Target-dependent structural changes in sensory neurons of *Aplysia* accompany synaptic inhibition. *Neuron* 6:679–90

Schacher S, Kandel ER, Montorolo PG. 1993. cAMP and arachidonic acid simulate long-term structural and functional changes produced by neurotransmitters in *Aplysia* sensory neurons. *Neuron* 10:1079–88

Scharfman HE, Sarvey JM. 1985. γ-Aminobutyrate sensitivity does not change during long-term potentiation in rat hippocampal slices. *Neuroscience* 15:695–702

Schoepp DD. 1993. The biochemical pharmacology of metabotropic glutamate receptors. *Biochem. Soc. Trans.* 21:97–102

Schoepp D, Johnson BG, Smith ECR, McQuaid LA. 1990. Stereoselectivity and mode of inhibition of phosphoinositide-coupled excitatory amino acid receptors by 2-amino-3-phosphonopropionic acid. *Mol. Pharmacol.* 38:222–28

Schreurs BG, Alkon DL. 1993. Rabbit cerebellar slice analysis of long-term depression and its role in classical conditioning. *Brain. Res.* 631:235–40

Schwartzkroin PA, Wester K. 1975. Long-lasting facilitation of a synaptic potential following tetanization in the in vitro hippocampal slice. *Brain Res.* 89:107–19

Shibuki K. 1993. Nitric oxide: a multi-functional messenger substance in cerebellar synaptic plasticity. *Seminars Neurosci.* 5:217–23

Shibuki K, Okada D. 1991. Endogenous nitric oxide release required for long-term synaptic depression in the cerebellum. *Nature* 349:326–28

Shigemoto R, Abe T, Nomura S, Nakanishi S, Hirano T. 1994. Antibodies inactivating mGluR1 metabotropic glutamate receptor block long-term depression in cultured Purkinje cells. *Neuron* 12:1245–55

Small SA, Kandel ER, Hawkins RD. 1989. Activity-dependent enhancement of presynaptic inhibition in *Aplysia* sensory neurons. *Science* 243:1603–6

Small SA, Cohen TE, Kandel ER, Hawkins RD. 1992. Identified FMRFamide-immunoreactive neuron LPL16 in the left pleural ganglion of *Aplysia* produces presynaptic inhibition of siphon sensory neurons. *J. Neurosci.* 12:1616–27

Stanton PK, Sejnowski TJ. 1989. Associative long-term depression in the hippocampus induced by Hebbian covariance. *Nature* 339:215–18

Stanton PK, Chattarji S, Sejnowski TJ. 1990. 2-amino-3-phosphonopropionic acid, an inhibitor of glutamate-stimulated phosphoinositide turnover, blocks induction of homosynaptic long-term depression, but not potentiation in rat hippocampus. *Neurosci. Lett.* 127:61–66

Staubli U, Lynch G. 1990. Stable depression of potentiated synaptic responses in the hippocampus with 1–5 Hz stimulation. *Brain Res.* 513:113–18

Stelzer A, Slater T, ten Bruggencate G. 1987. Activation of NMDA receptors blocks GABAergic inhibition in an in vitro model of epilepsy. *Nature* 326:698–701

Stevens CF, Wang Y. 1993. Reversal of long-term potentiation by inhibitors of haem oxygenase. *Nature* 364:147–49

Sweatt JD, Volterra A, Edmonds B, Karl KA, Siegelbaum SA, Kandel ER. 1989. FMRFamide reverses protein phosphorylation produced by 5-HT and cyclic AMP in *Aplysia* sensory neurones. *Nature* 342:275–78

Tamura H, Tsumoto T, Hata Y. 1992. Activity-dependent potentiation and depression of visual cortical responses to optic nerve stimulation in kittens. *J. Neurophysiol.* 68:1603–12

Thiels E, Barrionuevo G, Berger TW. 1993. LTD in intact hippocampus: use and NMDA receptor dependence. *Soc. Neurosci.* 19:1324

Tomasulo RA, Ramirez JJ, Steward O. 1993. Synaptic inhibition regulates associative interactions between afferents during the induction of long-term potentiation and depression. *Proc. Natl. Acad. Sci. USA* 90:11578–82

Tsumoto T. 1993. Long-term depression in cerebral cortex: a possible substrate of "forgetting" that should not be forgotten. *Neurosci. Res.* 16:263–70

Tsumoto T, Suda K. 1979. Cross-depression: an electrophysiological manifestation of binocular competition in the developing neocortex. *Brain Res.* 168:190–94

Vincent SR, Kimura H. 1992. Histochemical mapping of nitric oxide synthase in the rat brain. *Neuroscience* 46:755–84

Wang T, Lu B. 1994. Nitric oxide mediates activity-dependent synaptic suppression at developing neuromuscular synapse. *Soc. Neurosci.* 20:633

Walsh JP. 1993. Depression of excitatory synaptic input in rat striatal neurons. *Brain Res.* 608:123–28

Wexler EM, Stanton PK. 1993. Priming of homosynaptic long-term depression in hippocampus by previous synaptic activity. *NeuroReport* 4:591–94

White G, Levy WB, Steward O. 1988. Evidence that associative interactions between synapses during the induction of long-term potentiation occur within local dendritic domains. *Proc. Natl. Acad. Sci. USA* 85:2368–72

White G, Levy WB, Steward O. 1990. Spatial overlap between populations of synapses determines the extent of their associative interaction during the induction of long-term potentiation and depression. *J. Neurophysiol.* 64:1186–98

Wickens JR, Abraham WC. 1991. The involvement of L-type calcium channels in heterosynaptic long-term depression in the hippocampus. *Neurosci. Lett.* 130:128–32

Williams JH, Errington ML, Li Y-G, Lynch MA, Bliss TVP. 1993. The search for retrograde messengers in long-term potentiation. *Seminars Neurosci.* 5:149–58

Wu F, Schacher S. 1994. Pre- and postsynaptic changes mediated by two second messengers contribute to expression of *Aplysia* long-term heterosynaptic inhibition. *Neuron* 12:407–21

Xie X, Berger TW, Barrionuevo G. 1992. Isolated NMDA receptor-mediated synaptic responses express both LTP and LTD. *J. Neurophysiol.* 67:1009–13

Xie Z, Sastry BR. 1991. Inhibition of protein kinase activity enhances long-term potentiation of hippocampal IPSPs. *NeuroReport* 2:389–92

Yang X-D, Faber DS. 1991. Initial synaptic efficacy influences induction and expression of long-term changes in transmission. *Proc. Natl. Acad. Sci. USA* 88:4299–303

Yang X-D, Connor JA, Faber DS. 1994. Weak excitation and simultaneous inhibition induce long-term depression in hip-

pocampal CA1 neurons. *J. Neurophysiol.* 71:1586–90

Yoshimura Y, Tsumoto T, Nishigori A. 1991. Input-specific induction of long-term depression in Ca^{2+}-chelated visual cortex neurons. *NeuroReport* 2:393–96

Zhang DX, Levy WB. 1993. Bicuculline permits the induction of long-term depression by heterosynaptic, translaminar conditioning in the hippocampal dentate gyrus. *Brain Res.* 613:309–12

Zorumski CF, Izumi Y. 1993. Nitric oxide and hippocampal synaptic plasticity. *Biochem. Pharmacol.* 46:777–85

Annu. Rev. Neurosci. 1995. 18:359–83

LOCALIZATION OF BRAIN FUNCTION: The Legacy of Franz Joseph Gall (1758–1828)[1]

S. Zola-Morgan

Veterans Affairs Medical Center, San Diego, and University of California, San Diego, La Jolla, California 92093-0603

KEY WORDS: craniology, cognition, organology, phrenology, soul

A BRIEF HISTORY OF IDEAS ABOUT LOCALIZATION OF BRAIN FUNCTION

The First Era: Antiquity to the Second Century AD

The history of ideas about localization of brain function can be divided roughly into three eras. During the first era, which spans from antiquity to about the second century AD, debate focused on the location of the soul, i.e. what part of the body housed the essence of being and the source of all mental life (for reviews, see Finger 1994, Gross 1987b, Star 1989). In an early and particularly prophetic Greek version of localization of function, the soul was thought to be housed in several body parts, including the head, heart, and liver, but the portion of the soul associated with intellect was located in the head (McHenry 1969).

The individual who has been viewed by many historians as having the greatest influence during this era was Galen (130–200 AD), an anatomist of Greek origin. Using animals, he performed experiments that provided evidence that the brain was the center of the nervous system and responsible for sensation, motion, and thinking (Finger 1994, Gross 1987b, Hall 1968).

The Second Era: The Second Century to the Eighteenth Century

In this era the debate focused on whether cognitive functions were localized in the ventricular system of the brain or in the brain matter itself. It appears that Galen associated higher cognitive functions with the brain matter (McHenry 1969, Gross 1993, Finger 1994), although some of his later writings continued to emphasize the importance of the ventricles (Duckworth 1962). The influence of the Church during this era cannot be overstated; for example, ethereal spirits and ideas were believed to flow through the empty spaces of the brain's ventricles. Nevertheless, by the fifteenth and sixteenth centuries individuals such as da Vinci (1452–1519) and Vesalius (1514–1564) were questioning the validity of ventricular localization (McHenry 1969). Finally, during the seventeenth century, partly as a result of the strongly held views and prolific writings of Thomas Willis (1621–1675), and during the eighteenth century, with the publication of clinical descriptions of cognitively impaired patients accompanied by crude descriptions of brain damage (Baader 1762, Clendening 1942), the view that intellectual function was localized in brain matter and not in the ventricles became solidified.

The Third Era: The Nineteenth Century to the Present

During this era, the debate has focused on how mental activities (or cognitive processes) are organized in the brain. An early idea, which became known as the localizationist view, proposed that specific mental functions were carried out by specific parts of the brain. An alternative idea, which became known as the equipotential view, held that large parts of the brain were equally involved in all mental activity and that there was no specificity of function within a particular brain area (Squire 1987, Clark & Jacyna 1987).

The present review focuses on a portion of the current era, i.e. the period from the 1790s to the 1860s. Specifically, it considers the impact on ideas about localization of brain function made by the early nineteenth century localizationist Franz Joseph Gall. The movement associated with Gall came historically and generically to be known as phrenology (although as I describe below, Gall did not favor this term). This review first describes the European zeitgeist within which Gall's localizationist views emerged. It then considers the points of view held by the two major protagonists of the localizationist movement during the first half of the nineteenth century, Franz Joseph Gall and Johann Spurzheim. I suggest that Gall has not been given appropriate credit and recognition for his views, which were more fundamental to the development of contemporary ideas about localization of function than is usually recognized. Although Gall was a good scientist and a superb neuroanatomist, his reputation became mixed with and was transcended by that of

Spurzheim. A new hypothesis is presented to explain the major schism that occurred between them. Next, I discuss the reaction against the localization movement during the nineteenth century. Finally, I identify several specific contributions to the brain sciences made by Gall and the legacy of his doctrine to contemporary neuroscience.

FRANZ JOSEPH GALL AND EIGHTEENTH CENTURY LOCALIZATIONIST VIEWS

Gall's Early Years

It is somewhat ironic that Gall became one of the most influential scientists of the nineteenth century based not on his reputation as an outstanding scholar, physician, and anatomist, but instead on his identity by the scientific community and the general population as the founder of "phrenology." As I describe below, Gall neither invented nor approved of the term phrenology, nor was he pleased to be associated with the phrenological movement. This section briefly describes Gall's early life and then his activities as a young physician and scientist in Vienna.

Franz Joseph Gall (Figure 1) was born in 1758 in Tiefenbronn, Badenia (Germany), and died in 1828 in Paris. By the early part of his life, there had

Figure 1   Franz Joseph Gall (1758–1828). (From Haymaker & Schiller 1953.)

already developed a popular view in Europe that temperament and psycholog-ical characteristics could be explained by certain physical characteristics of the body and face (Finger 1994). By the age of nine, Gall had already formu-lated a hypothesis, based upon anecdotal evidence from the performance of his schoolmates, that good verbal memory was associated with bulging eyes. (The explanation for this relationship that Gall later developed was that the specialized faculty of verbal memory was located in the front part of the brain, and when verbal memory was especially developed, the orbits were pushed forward.)

Gall moved to Vienna and graduated from medical school in 1785. During the 20 years he spent in Vienna, Gall developed the ideas that led to his doctrine of localization. Though he labeled his contributions "physiology," they were not physiology in the sense that we now use the term; by current terminology, it would be more appropriate to describe his doctrine of localization as a combination of psychology and functional anatomy (Acker-knecht 1973). Gall apparently did not publish any formal versions of his doctrine while in Vienna, although he did give public lectures in which he elaborated its principal ideas.

After moving to Paris at the turn of the century, Gall became well known as a descriptive anatomist. A distinction must be made, however, between two kinds of anatomy associated with Gall. On the one hand, as a result of his anatomical dissections, Gall made several important anatomical discoveries that stand today. On the other hand, Gall also attempted to develop a functional anatomy that served the purpose of his doctrine of localization. Gall's func-tional anatomy was not grounded in empirical analyses and bore no relationship to his very careful and thorough descriptive anatomy. For Gall, descriptive anatomy and functional anatomy were entirely separate from each other. He argued that it was rare for (descriptive) anatomy to lead to the elucidation of function: "...it is thus without the help of any anatomical dissection that we have ourselves made most of our physiological discoveries" (Gall & Spurzheim 1809, Clarke & Jacyna 1987; the term physiological in this context refers to relationships between structure and function in the brain).

Gall's ill-fated association with the doctrine of phrenology unfortunately overshadowed his important and lasting contributions to anatomy. Even Flour-ens, who severely criticized Gall and his doctrine, wrote, "I shall never forget the feeling I experienced the first time I saw Gall dissect a brain. It seemed to me that I had never seen this organ before" (Flourens 1845).

Period of Transition: The Late Eighteenth Century

Gall's ideas about the localization of cognitive functions began to tear at the religious and social fabric of the late eighteenth century. In particular, his

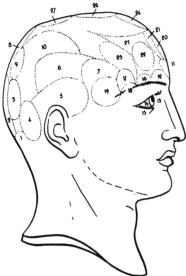 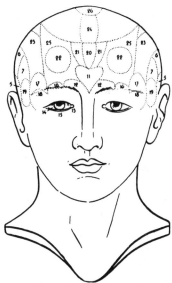

Fig. 1.–The phrenological system of Gall seen from profile. For numbers see text.

Fig. 2.–The phrenological system of Gall, frontal view.

Figure 2   Gall's system of organology seen from right profile and frontal views. The organs were, for the most part, bilateral. Gall's 27 organs representing specific functions of the mind were divided into two groups. *A* There were 19 organs common to humans and animals: 1. instinct of reproduction; 2. love of offspring; 3. affection, friendship; 4. instinct of self-defense, courage; 5. carnivorous instinct, tendency to murder; 6. guile, acuteness, cleverness; 7. the feeling of property, the instinct of stocking up on food (in animals), covetousness, the tendency to steal; 8. pride, arrogance, haughtiness, love of authority, loftiness; 9. vanity, ambition, love of glory; 10. circumspection, forethought; 11. memory of things, memory of facts, educability, perfectibility; 12. sense of places, space proportions; 13. memory and sense of people; 14. memory of words; 15. sense of language and speech; 16. sense of colors; 17. sense of sounds, gift of music; 18. sense of connections between numbers; 19. sense of mechanics of construction, talent for architecture. *B* There were eight organs that differentiated the human species from other animal species: 20. comparative sagacity (wisdom); 21. sense of metaphysics; 22. sense of satire, witticism; 23. poetic talent; 24. kindness, benevolence, gentleness, compassion, sensitivity, moral sense; 25. faculty to imitate, mimic; 26. religion, knowing God; 27. firmness of purpose, constancy, perseverance, obstinacy. (From Ackernecht & Vallois 1956)

notion that various mental faculties were represented in different places in the brain was seen by the Austrian government as in conflict with moral and religious views of the unity of the soul and mind.

Due in part to his open letter in 1798 to Baron Joseph Francois de Retzer, in which Gall revealed his intention to construct a physiology of the brain and describe the functions of the brain in humans and animals, and due also to the growing popularity of his public lectures, Gall lost his permission to publish and to give lectures. "This doctrine concerning the head, which is talked about with enthusiasm will perhaps cause a few to loose their heads and it leads to materialism, therefore is opposed to the first principles of morals and religion...", the Emperor Francis wrote on Christmas Eve, 1801 (Ackerknecht & Vallois 1956) (Figure 2).

Before examining the doctrine of "organology" (the term that was originated and used by Gall; the term phrenology was developed later by Spurzheim), I review briefly the context of the late eighteenth century within which Gall generated the early versions of his doctrine. Gall began developing the formal ideas about organology primarily during the 1790s. During this time it was believed that a science of human behavior was on the horizon, based on a constant, knowable correlation between the external surfaces of the body and the internal spirit (Pogliano 1991). For example, many people embraced the doctrine of physiognomy, the idea that the face properly understood would reveal the character and abilities of a person.

Perhaps the most influential idea of the time about localization of brain function, an idea that had a significant impact on Gall's thinking, was that of Albrecht von Haller (Clarke & Jacyna 1987). In the mid-eighteenth century, Haller had developed his doctrine of brain equipotentiality or a type of *action commune*. Haller attempted to describe the physiology of the major components of the brain on the basis of his experimental results. In Haller's view, white matter and gray matter, for example, were distinguishable components; his experiments showed the gray matter of the cerebral cortex to be inexcitable, and the white matter highly irritable. From these flawed experiments he concluded that the white matter represented the seat of sensations, the *sensorium commune*.

By placing higher mental functions in a particular region of the brain, Haller had made a major contribution to the development of ideas about localization of function. He did not attempt to subdivide the brain into regions of specific mental or cognitive functions. Instead he believed that the parts of a distinguishable anatomical component, the white matter for instance, performed as a whole, each part having equivalent functional significance (Clarke & Jacyna 1987). This theory of the equipotentiality of brain regions prevailed into the early nineteenth century, and it was on this background that Gall developed his ideas. Indeed, one might characterize Gall's organology as a reaction against the equipotential view of Haller. Gall's insight was that, despite its similarity in appearance, brain tissue was not equipotential but instead was actually made up of many discrete areas that had different and separate functions.

LOCALIZATION OF FUNCTION IN THE NINETEENTH CENTURY: GALL'S ORGANOLOGY AND SPURZHEIM'S PHRENOLOGY

Collaboration of Gall and Spurzheim

Gall began work on his doctrine on the brain in 1792 or earlier, and he gave public courses on the subject beginning in 1796 or earlier (Ackerknecht & Vallois 1956). In 1800 he was joined in his work by Johann Gaspard Spurzheim (1776–1832) (Figure 3), who had attended a course of Gall's lectures and had become drawn to his ideas about the brain. They left Vienna together in 1805, motivated by the Austrian government's growing displeasure with Gall's doctrine, and they traveled for two years, lecturing in the major cities of Europe. In 1807 Gall and Spurzheim settled in Paris, where they worked and published together, mainly on topics of descriptive neuroanatomy, until their rift in 1813.

In 1809, Gall and Spurzheim published a book on descriptive anatomy of the brain, *Recherches sur le Système Nerveux*. Gall was a gifted anatomist and developed innovative and refined dissection procedures. There is some debate about whether Gall carried out his descriptive anatomy of the brain while still in Vienna during the last part of the 1700s, either before or during the time he was developing the ideas for his doctrine, or whether the descriptive anat-

Figure 3   Johann Gaspard Spurzheim (1776–1832). (From Finger 1994.)

omy actually followed the establishment of the doctrine of organology (Clark & Jacyna 1987, Temkin 1947, Pogliano 1991). Careful reading of Gall's and Spurzheim's papers, however, appears to resolve this question. Gall believed that physiological research, in the sense of psychology and functional anatomy, should precede anatomical research. "I owe nearly all my anatomical discoveries to my physiological or pathological concepts" (Gall & Spurzheim 1809). Moreover, Spurzheim's own observations suggest that the work on descriptive anatomy came later. "After finishing my studies in 1800, I joined M. Gall in order to pursue in a special manner the anatomical part of the researches" (Spurzheim 1818). Nevertheless, when presenting his results to the public, Gall would begin with his anatomical discoveries, and they were the exclusive subject of *Recherches sur le Système Nerveux* with Spurzheim in 1809 (Ackerknecht & Vallois 1956).

Gall published the first two volumes of *Anatomie et Physiologie du Système Nerveux* in 1810–1812. Volume 1 was devoted to additional anatomical discoveries, and Volume 2 began to describe the doctrine of organology. Spurzheim assisted with these two volumes, although as a result of his growing disagreement with Gall, Volume 2 was completed mainly by Gall. Volumes 3 and 4 were published in 1818 and 1819 after Gall's rift with Spurzheim. Volumes 2–4 contained the first comprehensive and detailed presentation of the organologic doctrine in both words and images. They contained hundreds of plates of detailed illustrations of brains upon which were mapped Gall's physiology, i.e. his ideas about the relationship between anatomy and psychology. Subsequently, Gall published a six-volume series, *Sur les Fonctions du Cerveau,* from 1822 to 1825. These volumes contained all of Gall's anatomic and physiological discoveries. In addition, they contained his fully established theory of the relationship between humans and nature.

The Essence of Organology

Gall's conclusions about the relationship between brain and behavior rested on four assumptions (Gall 1822). The first two assumptions were somewhat metaphysical. Although not the focus of the present review, I touch on them briefly because they represent an aspect of Gall's thinking that has been viewed by historians as contributing substantially to his schism with Spurzheim. (In a later section I suggest that a close reading of the evidence supports a different view.)

Gall was substantially influenced by nativist thinking and by the writings in the 1780s of Herder, who believed that all phenomena of animated nature were dependent on biological organs (Gall & Spurzheim 1809, Pogliano 1991). Accordingly, the first two assumptions of Gall's doctrine were 1. Moral and intellectual faculties are innate, and 2. Their exercise and manifestation depend on organic, specifically cerebral, structures.

The second two assumptions are more germane to the issue of localization of function. As noted earlier, Gall had become convinced that the brain was the organ of the mind, the seat of all faculties, tendencies, and feelings. These ideas were based on his observations in normal individuals as well as cases of pathology. He presented many examples in favor of his thesis: studies of twins that showed the same brain organs to result in the same personalities; observed differences in the mentalities of men and women, who also differed physically; observations of the subnormal intellectual capacities and behavior of idiots, who had deformed brains; and clinical cases in which a blow on the head had changed the personality (Ackerknecht & Vallois 1956). Accordingly, Gall's second two assumptions were 3. The brain is the organ of all faculties, of all tendencies, of all feelings ("the organ of the soul"; Gall 1822), and 4. The brain is composed of as many particular organs as there are faculties, tendencies, and feelings.

It was Gall's intent to provide empirical evidence to show that the brain was the undisputed organ of the mind (Clarke & Jacyna 1987) and to identify the fundamental faculties and organs of the brain. As an example of his approach, he recommended examining people who were gifted with a talent, and as a counterproof, examining people who lacked this talent. A comparison of the cranial structures of the two individuals might reveal prominent differences. Gall was never convinced that comparisons between individuals were completely valid, however, and he cautioned against the indiscriminate use of his doctrine.

Gall relied mainly on comparative anatomy to develop his doctrine of organology. He collected hundreds of skulls and plaster casts of the heads of living and deceased humans and animals. Because he never belonged to any official institution that owned such collections, he created his own collection at rather substantial personal expense (Ackerknecht & Vallois 1956). His collection, which at times contained several hundred pieces, included skulls and casts from many famous people (e.g. Voltaire, Descartes) as well as from lunatics, criminals, and several species of animals.

Gall attempted to correlate physical aspects of the skulls and casts with prominent characteristics of human and animal behavior or human personality. He believed that—in the same individual, at least—larger organs showed greater potential than did smaller organs for expressing behavior (Spurzheim 1832). He eventually determined that there were 27 separate faculties (Figure 2). It is important to underscore Gall's view that the relevant measurements were those that compared different structures within the same brain, not those that compared the brains of different species or even those that compared different members of the same species (Robinson 1978). In this sense, as described below, he was far more cautious than Spurzheim.

Nevertheless, despite his cautiousness in some areas, Gall made a leap of faith from the idea that specific regions of the brain were the centers for specific faculties to the idea that one could reliably determine which faculties were highly developed in a particular individual by palpating the skull. At this point in time, there were, of course, no means by which living brains could be directly examined. Thus, the development of the 27 faculties within an individual had to be inferred by supposing that "…the skull is molded on the brain…" (Gall 1822). Accordingly, the surface of the skull would mirror the development of the underlying organ of the brain. In other words, he deduced the intellectual and moral character of an individual by palpating the contours of the cranium, a method he came to call "cranioscopy" (Clarke & Jacyna 1987). This aspect of his doctrine came to be the most heavily criticized on scientific grounds.

In several of the six volumes of *Sur les Fonctions du Cerveau,* Gall specified the 27 fundamental faculties and corresponding organs. Gall understood that his list of faculties and organs might not be complete, and he suggested that other organs might be added in the future. Indeed, there was considerable brain surface that was uncommitted (Figure 2). Spurzheim eventually revised the system of organs and increased their number to 35 (Figure 4). [Some historians have indicated that Spurzheim listed as few as 32 faculties (for example, see Pogliano 1991) or as many as 37 faculties (for example, see Clarke & Jacyna 1987, p. 224). This confusion might stem in part from the fact that Spurzheim was a prolific writer and lecturer and modified or reorganized his views from one presentation to the next. In addition, much of the information about Spurzheim's ideas and views comes from an indirect source, i.e. the writings of George Combe (1788–1858), a Scottish lawyer who had abandoned his profession to dedicate himself entirely to Spurzheim and phrenology. It is clear, however, from Spurzheim's principal book in 1832, *Outlines of Phrenology,* that his framework consisted of 35 "special faculties of the mind" (Spurzheim 1832, pp. 23–70).] Other phrenologists suggested an increasing number of faculties. Vimont in 1835 brought the number to 42; a few years later Barthel, a Belgian phrenologist, described 46 faculties (Ackerknecht & Vallois 1956); and in some hands the number of faculties grew to 50 or more (Peacock 1982, Finger 1994).

As the number of faculties increased, so did the dissimilarities between Gall's original doctrine of organology and the newer localizationist schemes. Accordingly, in evaluating Gall's contributions to ideas about localization of brain function, it is necessary to consider Gall separately from individuals who developed other schemes of localization. In particular, it is important to distinguish Gall's doctrine from a view that eventually became far more prominent and influential than organology, i.e. the phrenology of Spurzheim.

The Rift Between Gall and Spurzheim in 1813

Spurzheim's name was inseparably connected to Gall's with the popularization of the doctrine of organology and subsequently the even more popular phrenology, i.e. Spurzheim's modification of Gall's system. In 1813 Spurzheim separated from Gall as a result of growing disagreement between the two. Several explanations have been suggested by historians to account for the rift. One important difference between the two collaborators involved their respective views about the fundamental nature of humans (for example, see Clarke & Jacyna 1987, Pogliano 1991). Gall was said to have accepted the existence of evil propensities in humans (e.g. organ 5, the tendency to murder). Spurzheim has been described by some historians as believing that humans were created potentially good, and in his revision of Gall's doctrine was described as omitting evil propensities (Clarke & Jacyna 1987). However, in Spurzheim's *Outline of Phrenology* (1832), he did include an organ of destructiveness (organ 1 in his 1832 publication; see Figure 4, organ 6, of this review). Moreover, Spurzheim's description of the organ seems consistent with Gall's view of evil propensities: "Some highwaymen are satisfied with stealing, others show the most sanguinary inclination to kill without necessity...The primary nature of this propensity is a simple impulse to destroy; it does not consider the object of its application, nor the manner of destroying" (Spurzheim 1832).

Spurzheim eventually used phrenology to argue for social reform, and it is possible that the two collaborators disagreed on this point. However, it turns out that in many ways Gall had anticipated Spurzheim's views with respect to phrenology as the basis for a movement for social reform. Although Gall believed that the faculties he was describing were innate, he was as fervent a social reformer as he was a nativist. Specifically, he believed that through the development of the organ of educability (organ 11, Figure 2), the "good" organs could be reinforced and counteract the effect of the "bad" ones (Ackerknecht & Vallois 1956). "Educability perfects, curbs, directs the innate faculties,..." (Gall 1822). This theme appears substantially in the first volume of *Sur les Fonctions du Cerveau* (1822) and is considered in the later volumes of *Anatomie et Physiologie du Système Nerveux*. Thus, the two authors appeared to share, at least to some extent, similar views on this issue.

Spurzheim did disagree with Gall on the emphasis that should be placed on certain faculties. For example, as described by Flourens (1846), while Gall saw "hope" as nothing more than an attribute (associated with organ 26, the organ of religion; see Figure 2), Spurzheim made it a primary organ (organ 17, Figure 4) and insisted on three organs of religion (organs 14, 16, and 18, Figure 4).

Although these differences in views about human nature were significant, they seem in themselves insufficient to cause a major rift between these two long-term and productive collaborators. Perhaps a more plausible explanation for the rift involved the fundamental difference in the way the two viewed the relationship of scientific investigation to the development of the doctrine of localization. Gall and Spurzheim completed publication of *Recherches sur le*

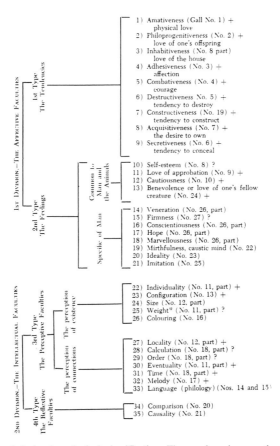

Figure 4 Spurzheim's phrenological classification. The numbers in parentheses indicate the corresponding faculties in Gall's classification. "+" indicates organs common to humans and animals. Spurzheim believed the other organs to be specific to humans except for a few that were doubtful. (From Ackerknecht & Vallois 1956.)

Système Nerveux in 1809 and Volume 1 of *Anatomie et Physiologie du Système Nerveux* in 1810. Both volumes were devoted exclusively to descriptive anatomy, including the methods and the results of their many brain dissections. Their next enterprise, Volume 2 of *Anatomie et Physiologie du Système Nerveux,* was to be fundamentally different from the two previous publications. This volume was to contain the first formal description of the doctrine of organology, and it was during the preparation of this volume that Gall and Spurzheim had their falling out. They disagreed over several important scientific aspects of the doctrine.

First, Gall had tried to determine the existence of faculties through observation and experimentation, while Spurzheim leaned more toward speculation and introspection. Second, it is rarely appreciated that Gall remained skeptical about the general applicability of correlations between mental faculties and cranial shape. He was comfortable only with the idea that in some cases of particularly well-developed organs, a relationship appeared to hold. By contrast, Spurzheim apparently had no such reservations (Clarke & Jacyna 1987). Third, Gall used a retrospective approach to determine the correlations between an individual's faculties and cranial shape. That is, he measured skulls and casts (mostly postmortem material) of individuals for whom he had a priori knowledge about their behavioral characteristics. Spurzheim, by contrast, leaned toward a prospective approach, viewing phrenology as a way of predicting behavior. Finally, unlike Gall, Spurzheim became more concerned with the popularization of phrenology and with the development of a social movement based on phrenology than with further scientific investigation and evaluation of the doctrine.

In the view of Flourens (1846), a contemporary of Gall and Spurzheim, "Spurzheim never would have imagined the doctrine: he found it already concocted;... He did not imagine it; and perhaps never could have had the facilities enjoyed by Gall for carrying it successfully into the world." Thus, it seems reasonable to suppose that, more than any other reason, the crux of the rift was due to Gall's imposing a level of scientific rigor on the continued development and refinement of the doctrine. Spurzheim resisted continued scientific investigation. He favored immediate popularization of the doctrine without further validation, and he set about using the doctrine to justify social reform.

Emergence of Phrenology

After the rift, Spurzheim departed for England, where he published his version of localization and successfully created the phrenological movement. Spurzheim modified Gall's original organology in terms of both the number and character of the faculties (Figure 4). In addition, some of the locations of the organs were changed. In many ways, however, Spurzheim's overall scheme was not very different from Gall's original organology. Referring to two published works by Spurzheim in which the doctrine of phrenology is described, Flourens (1846) states, "...and these two works are merely a reproduction of the doctrine of Gall. Spurzheim makes Gall's book over again—the same book that they commenced together—and abridges it." Flourens's statement lends further support to the hypothesis that the rift arose from disagreements about the process of science rather than from disagreements about the doctrine itself.

Nevertheless, despite his lack of scientific rigor with respect to the doctrine of localization, Spurzheim can be given substantial credit for popularizing the localizationist point of view. Even his choice of terminology overshadowed that of his mentor. After experimenting with several terms, Gall had adopted the term organology to emphasize the idea that the mental faculties were housed in cortical organs (Gall 1825). Spurzheim preferred the term phrenology: "The name *Phrenology* is derived from two Greek words, [phrene] mind, and [logos] discourse. I have chosen it to designate the doctrine of the special manifestations of the mind, and of the bodily conditions under which they have place" (Spurzheim 1832). Spurzheim's term became popularly accepted to represent both Gall's and his own versions of the doctrine of localization, and it became a generic term applied to other versions of localization as well. In this way, Gall historically became linked to theories and ideas that were different from his.

Spurzheim was a champion of social reform, and one cannot overstate the impact that phrenology had on social reform movements in Europe and the United States during the nineteenth century. Mental illness was one of the areas to which phrenology could be readily applied, i.e. by identifying and categorizing mental deficiency as well as providing an explanation for it. The French and US versions of Spurzheim's volume, *Observations sur la Phrénologie, ou la Naissance de l'Homme* (1818), had a wide circulation in the early nineteenth century, and became a standard for the treatment of madness (Pogliano 1991) (there have been many reviews of the application of phrenology to social reform; see for example, Combe 1827, Pogliano 1991, Star 1989).

Twenty years after his split from Gall, Spurzheim died during a visit to the United States, where the phrenological movement had become popular and was being applied in medical and clinical settings (see Freemon 1992 for a review of how US academic physicians in the 1820s and 1830s used phrenological theory to understand neurological symptoms).

NINETEENTH CENTURY REACTION AGAINST GALL AND THE LOCALIZATIONIST VIEWPOINT

Eventually, the phrenological movement, particularly as popularized by Spurzheim, met with considerable criticism and resistance around the world. In Europe, criticism often centered first on issues of metaphysics, philosophy, and religion and later on issues of science. In the United States, criticisms centered mainly on issues of science, and particularly questioned the validity of phrenological theory's relation between brain anatomy and skull shape.

As noted, Gall's name remained linked to most of the subsequent phrenological systems that were developed, even those that were developed after his death. This association between Gall and the phrenological movement occurred despite the fact that in many cases the development of individual phrenological schemes bore no relation to the scientific observations that shaped and constrained Gall's thinking and his doctrine of organology.

Yet there is no question that Gall's own conceptualization of how the brain was organized with respect to behavior provided sufficient reason on its own to draw criticism from his scientific colleagues. This criticism took several forms, but can be divided roughly into two categories, i.e. criticisms motivated by issues of philosophy and politics, and criticisms based on issues of science. (For two recent reviews of the criticisms directed at the phrenological movement, see Clarke & Jacyna 1987, and Star 1989.)

Philosophical and Political Criticisms

Gall's organology, and later versions of phrenology, faced similar critiques from philosophy and politics. Clerics and metaphysicians were concerned with the larger theological implications of the phrenological system. For example, in Flourens's critique of phrenology in 1846 (dedicated to Descartes), Gall and his followers were declared guilty of undermining the unity of the soul, human immortality, free will, and the very existence of God (Harrington 1991). As did the Emperor of Austria, Napoleon saw in Gall's system a tendency toward materialism. In Napoleon's *Mémorial de Sainte-Helene,* one can read: "I contributed to ruin Gall. Corvisart was his ardent follower; he and his likes have a strong tendency toward materialism" (Ackerknecht & Vallois 1956). Napoleon as well as prominent personages from several disciplines, including Cuvier, Pinel, Portal, Sabatier, and Tenon, felt that the anatomical location of the soul was to be found in a definite spot, an idea directly challenged by Gall's doctrine. These attitudes might have accounted for the fact that in 1821 Gall obtained only the vote of his friend Saint-Hilaire for his candidacy to the Académie des Sciences (Ackerknecht & Vallois 1956).

In Great Britain, where Gall's ideas had taken hold in their revised form of Spurzheim's phrenology, there was a sense that the ideas promoted a libertarianism like that responsible for the political context that had led to the French Revolution. In particular, the doctrine in some ways appeared to eliminate freedom of choice and personal responsibility because of its innate and deterministic aspects (Knight 1986). Gall and the phrenologists countered this criticism with the idea that they were describing only tendencies toward particular faculties or behaviors. The doctrine was not deterministic, they explained, in the sense that any one faculty could be affected by other faculties and dispositions.

Scientific Criticisms

Criticism of Gall did not come only from philosophic and political arguments by powerful and authoritative individuals. For example, Rolando, the famous Italian neuroanatomist (1809), recognized the elegance of Gall's dissection techniques and his tracing of fiber tracts from the spinal cord to the cerebrum. However, he found no logical connection between the tracings of the fibers and the distinct organs in the convolutions of the brain proposed to house particular mental faculties. In his *Saggio* of 1809, Rolando wondered why such an "excessive uproar" had been raised in Europe about Gall's clearly illogical ideas (Pogliano 1991).

Another scientific criticism had to do with the questionable way in which Gall had determined the locus and extent of each of the 27 organs. For example, Gall had localized the carnivorous instinct and the tendency to murder (organ 5, Figure 2) above the ear for three reasons: (*a*) This was the widest part of the skull in carnivores; (*b*) A prominence was found there in a student who was fond of torturing animals; and (*c*) This region was well developed in an apothecary who later became an executioner (Barker 1897). As another example, Gall had localized the memory of words (organ 14, Figure 2) and the sense of language (organ 15, Figure 2) in the orbital portion of the inferior surface of the frontal lobe based upon his observation that people with protruding eyes were particularly good at memorization of text and other aspects of language. Gall supposed that the portion of the brain concerned with language must be especially well developed in these cases. Because the eyes were being pushed out, he hypothesized that the region just behind the eyes was the area of the brain concerned with language.

It should be noted that Gall did not rely entirely on skull features or bulging eyes as the proof of the importance of the anterior lobes for speech (Gross 1987a, Finger 1994). He cited several clinical case studies. For example, he described an inability to remember the names of friends and family members in two men who had sustained wounds about the eye (Finger 1994). As it turned out, a region of the frontal lobe of the brain was later identified by Broca as a speech area.

Another scientific issue raised by critics during the nineteenth century was the fact that Gall never specified the precise extent or the anatomical borders of any of the organs. This lack of rigor, it was argued, made it impossible to correlate a specific faculty with the size of an organ or cranial capacity (Harrison 1825, Sewall 1939). Related criticisms involved Gall's seeming failure to acknowledge that there were variations in the thickness of the skull, i.e. variations from one individual specimen to another and from one locus to another within the same skull (Sewall 1839). Although Gall had attempted to show through studies of embryonic development that cranial protuberances might reflect underlying brain volume (Pogliano 1991), there was already good evidence that the size of the brain could not be predicted from the volume of intracranial space, suggesting that the size of bumps on the skull did not predict the extent of underlying brain tissue (Freemon 1992).

Gall's 27 faculties were also criticized as internally inconsistent and without logic. For example, Gall ascribed to organ 7 (Figure 2) the desire to own property as well as the tendency to steal. The concatenation of these two characteristics scandalized some of Gall's contemporaries (Ackerknecht & Vallois 1956), and there appeared to be no logic behind the overlap of these two characteristics. As another example, organ 4 was involved with instincts that some critics thought to be contradictory, i.e. the instinct of self-defense and also the tendency to get into fights. Moreover, if bumps on the cranium indicated only tendencies toward behavior, then the doctrine of phrenology appeared to be scientifically questionable, because it no longer generated simple predictions that could be tested (Knight 1986).

Criticism by Pierre Flourens

Often, philosophical criticisms were inseparable from scientific ones. For example, the most influential of all the critics of Gall's organology, and of phrenology in general, was the respected French experimental physiologist Pierre Flourens (1794–1867). Flourens was a member of the Académie des Sciences and had published many papers on comparative anatomy, anesthesiology, embryology, and physiology. Beginning in 1822 and continuing for the next 40 years, Flourens wrote a long series of reviews critical of Gall's doctrine and of phrenology. He proposed to undermine the phrenologists on both rational and empirical grounds (Harrington 1991).

Initially, Flourens was an enthusiastic supporter of Gall and, as noted earlier, a great admirer of Gall's descriptive anatomy and his techniques for brain dissection. However, between 1819 and 1822, although he continued to admire Gall's anatomical work, he renounced Gall's organology. It is said that Flourens had come under the influence and patronage of Cuvier, a comparative anatomist and physiognomist. Cuvier viewed the whole brain as the material instrument of an indivisible soul (Clarke & Jacyna 1987). Though purporting to attack Gall's science and experimental methods (see below), Flourens was, in reality, more interested in supporting the view of Cartesian philosophy that there was no equation between mind and brain, and that the soul was unitary and indivisible. His determination to revive Cartesian philosophy was maintained for more than 40 years, and one of his last major critical reviews of phrenology, *Phrenology Examined,* was dedicated to Descartes, "I frequently quote Descartes: I even go further; for I dedicate my work in his memory. I am writing in opposition to a bad philosophy, while I am endeavoring to recall a sound one" (Flourens 1846).

It is now understood that Flourens's experimental work contained many errors, as did the conclusions he derived from his work on brain stimulation. In his physiological experiments, Flourens, like many before him, failed to excite the cerebral cortex because he used inadequate and inappropriate forms of stimulation. In his lesion experiments, he chose to make cuts through the brain that were indiscriminate and that ignored anatomical boundaries that might have reflected underlying functional specificity, leading him to such conclusions as, "One can remove a rather large part of an animal's brain, either in front, or in the back or on top, without his losing any of his faculties" (Flourens 1851). Accordingly, Flourens asserted that the cerebral hemispheres, like other subdivisions of the brain, carried out specific functions, their *action propre,* but that these functions were distributed throughout their individual structures, with no precise localization as Gall had insisted (Clarke & Jacyna 1987). Flourens was influential and had a significant negative impact on the credibility of Gall's organology. At the same time, he held back for more than 40 years the correct physiological concept of discrete localization of brain function (Clarke & Jacyna 1987). Nevertheless, Gall's ideas survived and experienced a renaissance in the 1860s, mainly as a result of the efforts of a French neurologist, Bouillaud. Moreover, with the finding in 1861 of the first example of the modern era of localization of function, i.e. a "speech center" in the brain, Gall's reputation was somewhat restored. More importantly, the idea of localization of function regained credibility.

THE IMPACT OF GALL ON BRAIN SCIENCE

As described in previous sections, there was a lack of acceptance of Gall's ideas by most of Gall's scientific colleagues during the first half of the nineteenth century. Moreover, there has been a long-standing contemporary view that Gall's ideas were unorthodox and had little empirical credibility. As I have indicated, the current view of Gall has been based in part on the historical neglect of his work as an anatomist and the historically erroneous superimposition of Spurzheim's phrenological movement onto Gall's doctrine of organology. Of course, Gall himself contributed to the unfavorable contemporary view of his work by moving from organology to the advocacy of cranioscopy. Nevertheless, in this section, I indicate that Gall's contributions to the evolution of ideas about localization of brain function ought to be understood as substantial in today's light. As described next, Gall had profound effects on the course of brain sciences. These fall into two categories: (*a*) specific contributions made by Gall to particular areas of neuroscientific investigation, including anatomy, experimental research, and clinical research; and (*b*) general effects that Gall had on the course of scientific investigation and the evolution of neuroscience.

Specific Contributions to Neuroscience

Even today, the association of Gall with the phrenological movement has served to overshadow his many other scientific activities. In this section I identify some of the specific contributions Gall made to several fields of study.

The examples presented here are meant to be representative and not exhaustive (for more extensive discussions of this topic, see for example Flourens 1846, Ackerknecht & Vallois 1956).

Without question, Gall's most recognized and lasting contributions were in the realm of neuroanatomy. As noted, Gall's descriptive anatomy, based upon his radically new and informative dissection techniques, was superior to that of both his predecessors and his contemporaries. He saw in the brain a meaningful structure, not just forms and mechanical connections (Gross 1987b, Ackerknecht 1973). Rather than sectioning the brain in arbitrary and unsystematic planes, Gall, working together with Spurzheim at the start of the nineteenth century, dissected along a hypothesized ascending path from the peripheral nervous system to the spinal cord and the medulla oblongata and then to the cortex (Pogliano 1991). Using this approach, Gall eliminated the commonly held view that the spinal cord was a "tail" of the brain, and established instead that the brain was a later development of the spinal cord (Meyer 1971).

Gall and Spurzheim also established the basic division between gray and white matter. All fibers, they wrote, originated and took their "nourishment" from gray matter (Gall & Spurzheim 1809). Gall, in particular, recognized the white matter as made up of fibers having a conductive property. He carefully studied the origin of cranial nerves I–VIII and traced their course from the medulla. He also provided the first clear description of the commissures, demonstrated the existence of the crossing of the pyramidal tracts, and differentiated between divergent fibers ("*fibres sortants ou divergents*") that ascended through the cord and brain stem and radiated into the cortex, and convergent fibers ("*fibres rentrants ou convergents*") sent back from the cortex to form the commissures (Ackerknecht & Vallois 1956, Meyer 1971).

As a result of his careful anatomy and dissection techniques, Gall established not only the importance of the brain with respect to intellectual function, but also—more importantly—the role of the cortex (Ackernecht & Vallois 1956, Young 1970, Gross 1987a). Until Gall, the cortex was generally regarded as merely vascular tissue or as a glandular structure (Ackerknecht 1973, Ackerknecht & Vallois 1956). As described earlier, Haller and even Flourens had neglected the importance of the cortex because of their flawed experiments that failed to stimulate the cortex.

Gall also had a significant impact on anthropology, which developed as a discipline in the latter part of the nineteenth century. Gall and the phrenologists who followed him were the first scientists to take numerous measurements of the skull and to develop the measuring instruments that anthropology would come to use. In addition, Gall popularized the idea of a framework for understanding behavior that depended significantly on instinct and inheritance but that could be modified by experiential influences.

Gall, Bouillaud, and Broca

An oft-cited example of a specific contribution made by Gall to our understanding of brain function is the idea that he anticipated the discovery by Broca
in 1861 of a specific speech area of the brain (Bouillaud 1848, Ackerknecht
& Vallois 1956). However, I believe that a careful reading of the facts surrounding this discovery tells a somewhat different story. In fact, Broca never
mentioned Gall's name in his 1861 report. Moreover, he referred to Gall's
doctrine in a rather negative way. Nevertheless, Broca's work stands as a clear
example of a modern idea of localization of function built upon the foundation
and fundamental idea, i.e. that specific parts of the brain mediate specific
behaviors, established by Gall a half century earlier.

As noted above, two of Gall's 27 faculties were the memory for words
(organ 14, Figure 2) and the adjacent center of language (organ 15, Figure 2).
These two language-related faculties were located, according to Gall, in the
posterior region of the orbital part of the inferior surface of the frontal lobe.
The reasons Gall placed these two faculties in the frontal lobe have been
described above.

The French neurologist Jean-Baptiste Bouillaud (1796–1881), an enthusiastic supporter of Gall's work, had accumulated a large number of clinical cases
in which he established what he believed to be a link between frontal lobe
damage and loss of speech (Bouillaud 1825). Thus, Bouillaud believed he had
substantial evidence that Gall had correctly localized the faculty of verbal
memory in the frontal lobes. For a variety of reasons, however, including the
fact that by this time the influential Flourens seemed to have demonstrated
that localization of cerebral functions did not exist, Bouillaud's views were
not welcomed. Bouillaud, on the other hand, was so convinced that he and
Gall were correct that in 1848 he offered 500 francs to anyone who could
show him a brain from an individual who had suffered from speech disturbance
and did not have damage to the left frontal lobe (Ackerknecht & Vallois 1956,
Finger 1994).

Both Gall and Bouillaud seemed to be vindicated in 1861 with the publication of the proceedings from a meeting of the Societé d'Anthropologie de
Paris. Paul Broca (1824–1880), assisted by Alexandre Ernest Aubertin (1825–
1893), Bouillaud's son-in-law and a strong believer in localization and in
Bouillaud's hypothesis, presented the neuropathological findings from the
brain of Broca's patient, Monsieur Leborgne. [This patient subsequently was
referred to by the name "Tan," the only utterance Broca ever heard Monsieur
Leborgne make (Broca 1861).] Prior to his death from a diffuse gangrene
infection, this patient had been hospitalized for 21 years and suffered from
loss of speech, among other neurologic conditions.

Examination of Tan's brain revealed many infarcts, including a large area of damage in the left frontal lobe. However, the damage observed in Tan's brain did not correspond to Gall's cranial prominences labeled 14 and 15 (Figure 2) on the orbital surface of the frontal lobes. Instead, the lesion was in the second and third convolution of the frontal lobe, "...the third convolution has suffered the greatest loss of substance, having been transected near the anterior limit of the Sylvian fissure; its entire posterior extent has been destroyed..." (Broca 1861; see Benton 1991, Figures 1 and 2).

Broca cautiously took up the localizationist view and generously acknowledged the validity of what he referred to as Bouillaud's doctrine, "This case, therefore, confirms the opinion of M. Bouillaud, who localizes the faculty of articulate speech in these lobes" (Broca 1861). At the same time, Broca appeared to distance himself from any association with Gall's doctrine, "At all events, one need only compare the present case with those previously reported to dismiss the notion that the faculty of articulate speech resides at a fixed point or is located under some bump on the skull." In clarifying his own position with respect to the idea of localization, Broca (1861) stated, "Whether the faculty of articulate speech resides within the frontal lobe taken as a whole or within one of the frontal convolutions...is a much more difficult question to answer. Additional cases must be collected to resolve this issue. The exact name and location of the affected convolutions must be specified..."

Broca's finding from his patient Tan has been regarded by some historians as the most important clinical discovery in the history of cortical localization (Sondhaus & Finger 1988). Moreover, within the decade, what some historians regard as the most important laboratory discovery pertaining to cortical localization was reported, i.e. Gustav Fritsch (1838–1927) and Eduard Hitzig (1838–1907) discovered the cortical motor area in the dog and proved that cortical localization was not restricted to a single function (Finger 1994). The discoveries of the speech area by Broca and the motor area by Fritsch and Hitzig were seen as vindications for Gall's ideas and reestablished him as the father of localization.

We should not lose sight of the fact that early in the development of organology Gall had pointed to several sword-wound cases of the anterior region of the head that were accompanied by language problems as supportive of his doctrine. However, the association between Gall's early findings and the discovery by Broca nearly 50 years later was more a matter of coincidence than an example of slow incremental scientific work eventually coming to verify the accuracy of an earlier point of view. The reality is that none of the brain areas identified by Gall, including areas 14 and 15, ever came to be associated with the functions that Gall attributed to them. Yet, Gall's fundamental idea that certain regions of the brain were specialized for mediating specific functions was eventually embraced by Broca, Fritsch, and Hitzig, and by most brain scientists by the latter part of the nineteenth century. As I indicate in the next section, in this sense Gall's fundamental idea remains a substantial legacy to contemporary neuroscience.

Overall Impact of Gall on the Course of Science

In one sense, it was the controversy that surrounded Gall and his ideas about organology and craniology rather than the actual content of his ideas that had a major effect on the course of science. That is, the controversy served to focus considerable attention on the role of the brain in cognitive function. Edwin G. Boring (1942) recognized this idea when he wrote, "The theory of Gall…is an instance of a theory that is essentially wrong but just right enough to further scientific thought…"

Following Gall's death in 1828, brain scientists rarely considered again the possibility that had been extant for centuries before, i.e. that cognitive function could be localized in a body organ other than the brain. During the early part of the nineteenth century, some scientists did continue to place some cognitive functions outside the brain. For example, Pinel & Bichat localized mental disease mostly in the abdominal ganglia (Clarke & Jacyna 1987). Gall, however, unswervingly insisted that all aspects of mind be located in the brain, a view that has remained intact and has continued to guide the course of contemporary neuroscience. Flourens, Gall's great opponent, was able to characterize Gall's contribution correctly, "the idea that the brain is the exclusive location of the soul was in science before Gall; one can say that since Gall, this idea reigns in science" (Flourens 1846).

An additional effect that Gall had on the course of science stemmed from the intrinsically exciting and intellectually interesting theory of organology itself. The "…era of localization begins with Gall…" (Soury 1897) in the sense that his idea about the multiplicity of organs that mediate specific functions proved far more fertile than the age-old sterile search for the seat of the *sensorium commune* in the brain (Ackerknecht 1973). Gall's new framework for describing and understanding human behavior was as experimentally challenging as it was controversial. Thus, the introduction of Gall's theory stimulated considerable research on the brain, including experimental work with animals and clinical research with patients. This work, which was motivated in part by attempts to either support or refute the localizationist view, spanned much of the nineteenth century, and it resulted in the development of fundamental facts and ideas about the relationship between the brain and cognitive function that have stood the test of time.

By the view developed here, the true legacy of Gall was not necessarily his tangible findings, although these were significant. Rather, Gall's real contribution was that he helped establish a new level of argument, i.e. the issue moved from one that asked whether the brain supports cognitive function to the question of what specific parts of the brain support specific cognitive abilities. A substantial portion of contemporary neuroscientific research continues to be guided by this fundamental question.

Lastly, it is important to acknowledge an additional overall contribution made by Gall and his work. Gall helped popularize brain science and made it accessible and interesting to the general public. Despite political, religious, and philosophical concerns, Gall's ideas and framework, with the help of Spurzheim, eventually penetrated the culture, including both the scientific and popular literature as well as the news media, in both Europe and the United States.

It seems fitting to comment on Gall's character and on how it may have motivated him to carry out his descriptive anatomical work as well as his work that led to the development of his doctrine of organology. Prior to his death from a cerebral hemorrhage at age 70, Gall had arranged to have his skull and brain preserved and analyzed by his colleagues. Examination of his skull, which is housed in the collection of the Musée de l'Homme in Paris, revealed that "...colors, music, mathematics, and mechanics and chiefly poetry were very weak; the last one actually so weak that he had a kind of dislike of poetry" (Ottin 1834). He was also found to have well-developed perseverance. The accuracy of the description of Gall based upon cranial analysis has to be tempered by the fact that the examiners had known Gall very well (Ackerknecht & Vallois 1956).

As described earlier, Gall was a careful observer and scientist. Accordingly, perhaps the clearest insight into his motivation comes from Gall himself, "How did I achieve what I did? I never would make plans, I never knew what I would come to. I have been guided by the purest, the most innocent instinct. I was not lead by interest, nor by honors, nor by money. The sole blind impulse to force the secrets of nature, animals, and men, did it" (Neuburger 1897).

ACKNOWLEDGMENTS

This work was supported by the Medical Research Service of the Department of Veterans Affairs and National Institutes of Health grant NS19063. I thank my colleagues Larry Squire and Charles Gross for their many helpful comments.

Literature Cited

Ackerknecht EH. 1973. Contributions of Gall and the phrenologists to knowledge of brain function. In *The Brain and Its Functions: An Anglo-American Symposium, London, 1957,* pp. 149–53. Amsterdam: Israel

Ackerknecht EH, Vallois HV. 1956. *Franz Joseph Gall, Inventor of Phrenology and His Collection.* Transl. C St. Leon, 1956. Madison, WI: Dept. Hist. Med., Univ. Wis. Med. Sch. (From French)

Baader J. 1762 (1778). Observationes medicae, incisionibus cadaverum anatomicis illustrae. In *Thesaurus Dissertationium,* ed. E Sandifort, 3:1–62 (In Latin)

Barker LF. 1897. The phrenology of Gall and Flechsig's doctrine of association centres in the cerebrum. *Bull. Johns Hopkins Hosp.* 8: 7–14

Benton A. 1991. The prefrontal region: its early history. In *Frontal Lobe Function and Dysfunction*, ed. HS Levin, HM Eisenberg, AL Benton, pp. 3–32. New York: Oxford Univ. Press

Boring EG. 1942. *Sensation and Perception in the History of Experimental Psychology*, p. 55. New York: Appleton-Century-Crofts

Bouillaud JB. 1825. Recherches cliniques propres à demontrer que la perte de la parole correspond à la lesion des lobules antérieurs du cerveau et à confirmer l'opinion de M. Gall sur le siège de l'organe du langage articule. *Arch. Gen. Med.* 8:25–45 (In French)

Bouillaud J. 1848. *Recherches Cliniques Propres à Demontrer que le Sens du Langage Articule et le Principe Coordinateur des Mouvements de la Parole Resident dans les Lobules Antérieurs du Cerveau.* Paris: Balliere (In French)

Broca P. 1861. Remarks on the seat of the faculty of articulate speech, followed by the report of a case of aphemia (loss of speech). Transl. C Wasterlain, DA Rottenberg, in *Bull. Soc. Anat. Paris* 6:332–33, 343–57 (From French)

Clarke E, Jacyna LS. 1987. In *Nineteenth-Century Origins of Neuroscientific Concepts.* Berkeley/Los Angeles/London: Univ. Calif. Press

Clendening L. 1942. *Source Book of Medical History.* New York: Dover

Combe G. 1827. *The Constitution of Man, Considered in Relation to External Objects.* Edinburgh: Neill. 5th Am. ed.

Corsi P, ed. 1991. *The Enchanted Loom.* New York/Oxford: Oxford Univ. Press. 383 pp.

Duckworth WLH. 1962. *Galen on Anatomical Procedures. The Later Books.* Cambridge: Cambridge Univ. Press. 279 pp.

Finger S. 1994. *Origins of Neuroscience: A History of Explorations into Brain Function*, pp. 32–62. New York/Oxford: Oxford Univ. Press

Flourens P. 1846. *Phrenology Examined.* Transl. C de L Meigs, 1846. Philadelphia: Hogan & Thompson (From French)

Flourens P. 1851. *Examen de la Phrénologie.* Paris (In French)

Freemon FR. 1992. Phrenology as a clinical neuroscience: how American academic physicians in the 1820s and 1830s used phrenological theory to understand neurological symptoms. *J. Hist. Neurosci.* 1:131–43

Gall F. 1822–1825. *Sur les Fonctions du Cerveau.* Paris: Shoell (In French)

Gall F, Spurzheim J. 1809. *Recherches sur le Système Nerveux.* Paris: Shoell (In French)

Gall F, Spurzheim J. 1810–1819. *Anatomie et Physiologie du Systeme Nerveux.* Paris: Shoell (In French)

Gross CG. 1987a. Phrenology. In *Encyclopedia of Neuroscience*, ed. G Adelman, pp. 948–50. Boston: Birkhauser

Gross CG. 1987b. Early history of neuroscience. See Gross 1987a, pp. 843–46

Harrington A. 1991. Beyond phrenology: localization theory in the modern era. See Corsi 1991, pp. 207–15

Harrison JP. 1825. Observations on Gall & Spurzheim's theory. *Philadelphia J. Med. Phys. Sci.* 11:233–49

Haymaker W, Schiller F, eds. 1953. *The Founders of Neurology.* Springfield, IL: Thomas. 2nd ed.

Knight D. 1986. *The Age of Science.* New York/Oxford: Basil Blackwell

May MT. 1968. *Galen on the Usefulness of the Parts of the Body.* Ithaca: Cornell Univ. Press. 461 pp.

McHenry LC, Jr. 1969. *Garrison's History of Neurology.* Springfield, IL: Thomas

Meyer A. 1971. *Historical Aspects of Cerebral Anatomy.* London/New York: Oxford Univ. Press

Neuburger M. 1987. *Die historische Entwicklung der experimentellen Gehirn und Rückenmarksphysiologie vor Flourens.* Stuttgart: Enke (In German)

Ottin NJ. 1834. *Précis Analytique et Raisonné du Système du Docteur Gall*, pp. 234–35. Paris: Crochard. 5th ed. (In French)

Peacock A. 1982. The relationship between the soul and the brain. In *Historical Aspects of the Neurosciences*, ed. FC Rose, WF Bynum, pp. 83–98. New York: Raven

Pogliano C. 1991. Between form and function: a new science of man. See Corsi 1991, pp. 144–203

Robinson DN, ed. 1978. *Significant Contributions to the History of Psychology, Ser. E, Physiological Psychology.* Vol II: *X. Bichat, J.G. Spurzheim, P. Flourens.* Washington DC: Univ. Publ. Am.

Rolando L. 1809. Saggio sopra la vera struttura del cervello dell'uomo e degl'animali e sopra le funzioni del sistema nervoso. Sassari (In Italian)

Sewall T. 1839. *Examination of Phrenology.* Boston: King

Sondhaus E, Finger S. 1988. Aphasia and the C.N.S. from Imhotep to Broca. *Neuropsychology* 2:87–110

Soury J. 1897. In *Dictionnaire de Physiologie*, ed. C Richet, p. 611. (In French)

Spurzheim J. 1818. *Observations sur la Phrénologie, ou la Naissance de l'Homme.* Paris: Treuttel et Weurtz (In French)

Spurzheim J. 1832. *Outlines of Phrenology.* Boston: March, Capen & Lyon

Star SL. 1989. *Regions of the Mind: Brain Research and the Quest for Scientific Certainty.* Stanford, CA: Stanford Univ. Press. 278 pp.

Temkin O. 1947. Gall and the phrenological movement. *Bull. Hist. Med.* 21:275–321

Young RM. 1970. *Mind, Brain and Adaptation in the Nineteenth Century.* Oxford: Oxford Univ. Press

Annu. Rev. Neurosci. 1995. 18:385–408

MECHANISMS OF NEURAL PATTERNING AND SPECIFICATION IN THE DEVELOPING CEREBELLUM

Mary E. Hatten[1] and Nathaniel Heintz[1,2]

[1]The Rockefeller University and [2]Howard Hughes Medical Institute, 1230 York Avenue, New York, New York 10021-6399

KEY WORDS: development, granule cell, Purkinje cell, gene regulation

INTRODUCTION

The cerebellar cortex is one of the best-studied regions of the CNS. For nearly a century, all of the cerebellar cell types and the patterns of their synaptic connections have been known. Much of this wealth of information comes from Ramon y Cajal's work (1889, 1911, 1960) with Golgi studies. Further information on the development (Altman & Bayer 1985a–c, Rakic 1971, Miale & Sidman 1961), anatomy (Palay & Chan-Palay 1974), fiber tracts (Brodal 1981), and circuitry (Llinas & Hillman 1969) of the cerebellar cortex has emerged over the past several decades. The cerebellum provides a unique system for studying CNS development, combining the three classic patterns of CNS development—morphogenetic movements, the formation of ganglionic structures, and the establishment of neuronal layers—within one brain region. Remarkably simple in its basic plan, the adult cerebellar cortex contains only three layers and two principal classes of neurons. The abundance of one of these principal neurons, the granule cell, has enabled detailed analyses of the molecular mechanisms that underlie the basic steps in neuronal differentiation and has pointed out the role of local community factors in CNS neuronal development. Moreover, studies of naturally occurring mutations (Heintz et al 1993, Sidman 1968) and of targeted gene disruptions that block discrete steps in the development of this region (McMahon 1993, Joyner & Hanks 1991,

385

0147-006X/95/0301-0385$05.00

McMahon & Bradley 1990) provide a method for understanding the genetic control of cerebellar development.

Experiments discussed in this review indicate that a complex series of morphogenetic movements establishes the domain of cells normally fated to form the cerebellum. Studies of molecular mechanisms underlying this initial patterning of the cerebellar plate implicate a role for murine homologues of the Drosophila *engrailed* genes in cerebellar development; *Engrailed*[+] (*En*[+]) cells mark the position of cells fated to produce cerebellar cells. The establishment of the cerebellar territory is followed by further specification of particular classes of cerebellar cells; changing patterns of gene expression restrict the fate of immature precursor cells to form Purkinje cells, granule cells, or interneurons. During this latter phase, individual cells commence programs of neural differentiation that are controlled by epigenetic cues.

ESTABLISHMENT OF CEREBELLAR CELL–FATE MAPS

Early development of the anterior portion of the neural tube involves the formation of the three brain vesicles: the prosencephalon, mesencephalon, and rhombencephalon. Further patterning subdivides the rhombencephalon into the metencephalic and myelincephalic vesicles. While this region is forming [embryonic day 9 (E9)], a failure of neural-tube closure creates a gap along the dorsal aspect of the neural tube, which bows into a mouth-like structure as the tube bends to establish the pontine flexure. Further deepening of the flexure brings the mesencephalon (midbrain) closer to the primordium of the cerebellum (metencephalon); anterior aspects of the myelencephalon (brainstem) fold underneath the developing cerebellar plate (Figure 1). To define the ontogeny of cerebellar cells, Alvarado-Mallart (Otero et al 1993; Alvarado-Mallart & Sotelo 1992, 1982; Martinez et al 1991; Alvarado-Mallart et al 1990; Martinez & Alvarado-Mallart 1989) and Le Douarin (Hallonet & Le Douarin 1993, Hallonet et al 1990) grafted portions of the mes- and metencephalic vesicles and used the chick-quail marking system (Balaban et al 1988, Le Douarin 1969) to localize the donor cells. In this type of grafting experiment, the rostro-caudal extent of the cerebellar territory was established by exchanging the rostral half of the metencephalon with the caudal half of the mesencephalon (or the caudal half of the metencephalon with the rostral half of the mesencephalon). Surprisingly, these experiments provided evidence that cells fated to form the cerebellar anlage are derived from both the mes- and metencephalic vesicles. The neuroepithelium of the metencephalon generated the majority of cells in the cerebellar cortex; a V-like area of the mediodorsal aspect of the anlage arose from a caudal movement of cells from the mesencephalon (Hallonet & Le Douarin 1993). The mesencephalic contribution to the cerebellum therefore refines the classical view that each of the primitive brain

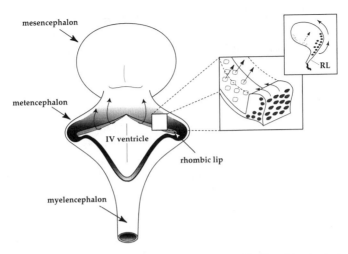

Figure 1 Formation of the murine cerebellar anlage. At E14, rapid cell proliferation in the roof of the IVth ventricle thickens the cerebellar primordium (*darkly shaded area*). In the frontal view the EGL cells (*filled*) move up and over the rhombic lip (*lightly shaded area*), spreading across the surface of the metencephalon in a rostro-medial path (*arrows*). A cutaway view of this region (*boxed area*) shows EGL cells leaving the fringe-like rhombic lip (*RL*) in a rostral direction; postmitotic cells in the ventricular zone (VZ) (*open circles*) migrate through the wall of the anlage to occupy a position above their site of origin. In a sagittal view (*second box*), cells in the medial aspect of the VZ (*open circles*) move out radially, with cells in the lateral aspect (*filled cells*), i.e. the rhombic lip (*RL*) spilling over the edge of the neuroepithelium to form the EGL. The V-shaped mesencephalic component of the cerebellar territory is shown as a striped triangle in the mediodorsal aspect of the underlying neuroepithelium. (For details see Hallonet & Le Douarin 1993, Alvarado-Mallart et al 1990, Altman & Bayer 1985a–c.)

vesicles gives rise to one of the main regions of the brain, because the cerebellar territory apparently arises from both the mes- and metencephalic vesicles.

Chick-quail chimera experiments have revealed that interneurons of the molecular layer of the cerebellar cortex do not originate from the external germinal layer (EGL), as was previously believed. Instead fate-mapping experiments show that cells fated to generate the interneurons of the molecular layer, the Purkinje cells, and the glial cells of the cortex arise via radial migrations out from the neuroepithelium. As discussed below, the EGL arises solely from the rostral portion of the metencephalon, via tangential movements across the surface of the anlage (Hallonet & Le Douarin 1993).

Fate-mapping experiments have located the cerebellar territory within the dorso-ventral axis of the neural tube. To define the lateral extent of the cerebellar territory, Hallonet & Le Douarin (1993) grafted neural territories that extended from 0 to 180° in the sagittal plane (i.e. from the roof plate of the neural tube to the floor-plate region) and analyzed the source of cerebellar cells. This type of grafting experiment demonstrated that the cerebellar primordium is located in the alar plate of the lateral wall of the neuroepithelium (Hallonet & Le Douarin 1993, Hallonet et al 1990); it extends from 25 to 120° in the sagittal plane of the neural tube.

GENETIC CONTROL OF CEREBELLAR PATTERNING

One general method of identifying genes important for cerebellar development has been the analysis of neurological mutant mice with defects in the patterning and histogenesis of the cerebellar cortex (Heintz et al 1993, Ross et al 1989). Two of these mutants, *meander tail* (Fletcher et al 1991, Ross et al 1989) and *leaner* (Herrup & Wilcyzynski 1982), show alterations in the anterior-posterior patterning of the cerebellum. Histological studies of *meander tail* indicate a severe disruption of the organization of cells within the anterior lobe of the cerebellar cortex, resulting in a loss of laminar structure. The discrete patterning defect seen in *meander tail* suggests that the anterior aspect of the cerebellar anlage is a separate developmental unit (Ross et al 1989). This view is supported by studies of the expression of both *Pcp2/lacZ* (Vandaele et al 1991) and *En-2/lacZ* transgenes (Logan et al 1993), for which expression is restricted to the anterior lobe of the adult cerebellum.

Because isolating and analyzing developmental mutants in mice is difficult compared with the ease of isolating developmental mutants in invertebrates, a number of laboratories have isolated vertebrate homologues of Drosophila segment-polarity genes and examined their expression in developing murine brain. Among Drosophila genes thought to provide positional information along the anterior-posterior axis, three classes of genes are expressed in overlapping regions of the mid-hindbrain region of the mouse neural tube. These include vertebrate homologues of *Engrailed (En)*, a gene that encodes a homeobox-containing transcription factor (Desplan et al 1985), and of two types of genes thought to control the spatial regulation of *En* expression. The latter include *paired box*–containing genes (including *paired* and *gooseberry*) and *wingless* genes (Heemskerk et al 1991, Morrissey et al 1991, DiNardo et al 1988). Two murine *En* genes, *En-1* and *En-2*, are expressed in the presumptive cerebellar territory. Expression of *En-1* commences early in the murine embryo, at E8 (1 somite), when transcripts are seen in the mes- and metencephalon but not the anterior prosencephalon or the posterior hindbrain (McMahon et al 1992, Davis & Joyner 1988). By E9, *En-1* expression extends over the cerebellar territory. Although *En-2* is also expressed in the cerebellar primordium, it is expressed later than *En-1*; it has been detected at E9.5 (5 somites) in the murine embryo (McMahon et al 1992, Davis & Joyner 1988). The expression of *En* genes across the cerebellar primordium therefore suggests that they function in the regional specification of the CNS. The timing of the expression of the two genes, with *En-1* being expressed at earlier stages and across a larger area than *En-2*, suggests that *En-1* is required for the generation of the cerebellar territory and that *En-2* is required for later stages of cerebellar development.

To analyze the function of the mouse *En* genes and to determine the differential roles of *En-1* and *En-2*, Joyner and colleagues used homologous recombination in embryonic stem (ES) cells to generate mice that were homozygous for targeted deletions of *En*. As expected, *En-1* mutant mice have severe developmental defects; most of the colliculi of the midbrain and the cerebellum are missing (Wurst et al 1994). The loss of mid-hindbrain in *En-1* mice is seen as early as E9.5, suggesting a role for *En-1* in the generation and/or survival of cells normally fated to form the cerebellum. By contrast, mice homozygous for null alleles of *En-2* have a less severe phenotype; defects are primarily restricted to the foliation of the early postnatal cerebellum (Millen et al 1994). Mutational analyses of *En* in regions of the embryonic brain that is fated to form the cerebellar cortex therefore suggest that *En-1* transcription defines an entry point into the cerebellar lineage; *En-2* functions in later phases of cerebellar development.

Analyses of sequences regulating expression of the *En-2* gene in transgenic mice reveal an enhancer region in the *En-2* locus that directs expression of the gene to the mes- and metencephalic domains, which give rise to the cerebellar anlage (Logan et al 1993). The presence of regulatory sequences that direct expression of particular genes in the anterior cerebellar cortex was also noted by Vandaele et al (1991), who showed that a fragment of the 5′ flanking sequences of the Purkinje cell protein-2 gene restricts expression only to cerebellar anterior Purkinje cells. These findings are consistent with studies of *meander tail* (Ross et al 1989) and *leaner* mutant mice, which have disrupted patterning of the anterior lobe. Together these studies suggest that the anterior lobe of the cerebellar domain is a developmental unit. Early steps in brain development thus generate a domain of the CNS normally fated to give rise to the cerebellum; subdomains along the A-P axis are revealed by *En-2* and *PCP-2* transgene expression and by *meander tail* and *leaner* mutant phenotypes.

By analogy to the control of *En* expression during Drosophila development (Heemskerk et al 1991, Morrissey et al 1991, DiNardo et al 1988), it was anticipated that murine *paired box*–containing genes, the *Pax* genes and homologues of the wingless gene *Wnt-1*, would initiate and maintain the spatial pattern of *En* expression seen in the presumptive cerebellar territory. Studies of the expression of *Pax* genes indicate that among the eight *Pax* genes isolated, *Pax-2, -5,* and *-8* transcripts are localized to the mid-hindbrain region (Fritsch & Gruss 1993, Gruss & Walther 1992). Thus *Pax-2, -5,* and *-8* may play a role in initiating *En* expression. *Wnt-1* has a role in cerebellar development;

animals homozygous for null alleles of the *Wnt-1* gene showed a complete loss of the cerebellum and other tissues (McMahon & Bradley 1990, Thomas & Capecchi 1990). In *Wnt-1⁻* null mutant mice, a loss of *Wnt-1* expression leads to a loss of *En* expression in cells of the presumptive cerebellar territory within 48 h of neural-plate induction (McMahon et al 1992). A role for *Wnt-1* in *En* expression is also supported by transplantation of *Wnt-1⁺* regions into the prosencephalon (Bally-Cuif et al 1992), which results in maintained *En* expression in the donor tissue and induction of *En* in the host neuroepithelium. The latter shift is associated with a change in the tissue's presumptive thalamic fate to a cerebellar fate (Gardner & Barald 1991, Itasaki et al 1991, Martinez et al 1991). All of these experiments and the direct mutational analyses discussed above consistently demonstrate a role for the *En* transcription factor in the specification and development of the cerebellar territory.

DEVELOPMENT OF THE CEREBELLAR ANLAGE

While early patterning events establish the cerebellar territory—including several distinct developmental units, or compartments, within the territory—later stages of cerebellar histogenesis involve the generation and differentiation of specific neural cell types. Within the cerebellar territory, this second phase of development commences as precursor proliferation begins to thicken the zone between E10 and E14 (Miale & Sidman 1961). The first pool of precursors cells to become postmitotic in the cerebellar plate are fated to give rise to precursors of the cerebellar deep nuclei (Pierce 1975). As these cells begin to exit the cell cycle in the medial aspect of the metencephalic neuroepithelium (E10 and E11), they form a mantle above the ventricular zone (VZ) (Altman & Bayer 1985a). From E13 onwards, a second population of precursor cells, which includes Purkinje cell precursors (CA Mason & R Blazeski, submitted) and other cortical precursor cells, exits the cell cycle in the VZ (Hallonet & Le Douarin 1993, Alvarado-Mallart & Sotelo 1992, Altman & Bayer 1985a) and migrates along the radial-glial-fiber system (Misson et al 1988) to establish a dense zone of immature neuronal cells in the middle of the anlage (Yuasa et al 1991, Hatten & Mason 1990) (CA Mason & R Blazeski, submitted). While immature cortical cells continue to settle in this layer, the mantle of cells fated to become neurons of the deep nuclei move to the rostro-medial portion of the anlage (Hallonet & Le Douarin 1993, Altman & Bayer 1985c). Therefore, the directed migration of postmitotic precursors of Purkinje cells and interneurons along the glial-fiber system layers a cortical component above the precursors of cerebellar deep nuclei.

While these events are occurring, a third pool of precursor cells undergoes a highly unusual developmental scheme. Proliferating at the edge of the neuroepithelium (rhombic lip), where there is no overlying cellular structure, this pool of dividing precursor cells streams rostrally up the lip and onto the surface of the anlage, where they establish a displaced germinal zone, the EGL (Ramon y Cajal 1889). Continued proliferation of EGL cells generates an expansive pool of precursors that spread rostro-medially across the roof of the anlage (Figure 1). After birth, rapid proliferation in the EGL expands the zone from a single cell layer to a layer eight cells in thickness (Fugita 1967, Fugita et al 1966). Precursor cell proliferation continues within the EGL until about post-natal day 15 (P15), when the zone disappears owing to the inward migration of postmitotic cells to form the internal granule cell layer (IGL). The prolonged period of clonal expansion of EGL precursor cells, from E13 until P15, generates a huge population of cells, which outnumbers the Purkinje neurons by ~250:1 in the adult cerebellar cortex.

EPIGENETIC REGULATION OF CEREBELLAR NEURAL DIFFERENTIATION

Since the pioneering studies by Sidman (1968), histogenesis in the developing cerebellum has been recognzed as responsive to environmental cues and local cell-cell interactions that are most easily interpreted as epigenetic mechanisms for neural differentiation (reviewed in Heintz et al 1993, Sidman 1983, Mullen 1982). The general principle that derives from these genetic and experimental neuroanatomical studies is that histogenesis in the developing CNS is controlled by spatially and temporally dynamic epigenetic cues that program specific stages in neuronal or glial differentiation. One might consider such a cue, its transducing mechanisms, and the resultant pleiotrophic functions elicited owing to its action as molecular subroutines of neural differentiation. As an example, the molecular subroutine for glial-guided migration (Sidman & Rakic 1973) might consist of a signal generated by interactions between a differentiating neuron and radial glia, transduction of that signal to the nucleus, production of proteins important for the mechanics of neuronal migration or the maintenance of glial function, reception of a "stop" signal upon arrival in the appropriate laminar position, and cessation of the synthesis of proteins specific to glial-guided migration.

Three important points concerning the nature of epigenetic regulation and the functions of subroutines deserve further consideration. First, differentiation of a particular cell type likely results from cascades of several epigenetic subroutines. Second, although these subprograms are shared across cell types in the CNS, local signals in different environments may induce their expression in different orders. In turn, the order of subroutines can organize the program for differentiation of a particular class of neurons. The timing of cell migration and axon outgrowth among cerebellar precursor cells provides an example. Whereas precursors of the Purkinje neuron, like neural precursors in the developing neocortex, migrate just after exit from the cell cycle and form axonal connections later in their program of differentiation, granule cell precursors extend axons just after exit from the cell cycle and migrate along Bergmann glia at later times in development. Thus the temporo-spatial ordering of local cues for specific molecular subroutines can generate multiple programs for neuronal differentiation.

A third feature of the epigenetic regulation of neuronal differentiation is the fact that the choice of a particular subroutine can limit the potential of a cell to respond to other epigenetic signals in the environment, a process termed "progressive restriction" (reviewed in Anderson 1989). The best-studied example of progressive restriction in neural precursors has come from studies of a bipotential precursor cell in the neural crest population of the sympathoadrenal lineage. In vitro studies of these cells by Anderson & Axel (1986) demonstrated that the choice of cell fate is influenced by the concentration of diffusible signals in the local environment. In this system, high concentrations of glucocorticoid favor development of the cells along a chromaffin cell pathway and lower concentrations favor development along a neuronal pathway. Growth factor responsiveness of partially committed precursors then involves a cascade of changes in growth factor responsiveness that progressively restricts the potential of the cells to enter other differentiation pathways (Birren & Anderson 1990, Stemple et al 1988).

Although the general hypothesis that dynamic environmental cues can be critical in eliciting appropriate subroutines of cellular differentiation that may progressively restrict the potential of differentiating cells, many of these mechanisms may also serve to regulate the timing of critical events in an intrinsic cellular differentiation program. To gain an understanding of these critical developmental mechanisms and their particular roles in histogenesis, it is crucial to obtain a complete picture of the molecular subroutines for differentiation of single cell types in the developing CNS. In the remaining sections of this essay, we emphasize studies aimed at understanding the differentiation of specific cerebellar cell types and the definition of specific molecules and mechanisms that play a role in these developmental events.

DEVELOPMENT OF THE PURKINJE CELL

The Purkinje cell is the primary neuron of the cerebellar cortex, providing its sole efferent pathway. Since the early studies of Ramon y Cajal (1911), it has been recognized that Purkinje cell differentiation occurs in several recognizable stages: neurogenesis, migration, and axon-target interactions. Studies of early events in the neurogenesis and migration of Purkinje cells that used antibodies to Purkinje cell–specific proteins spot-35 (Yuasa et al 1991) and Calbindin (Chedotal & Sotelo 1992, Mason et al 1990; CA Mason & R

Blazeski, submitted) indicate that immature Purkinje cells begin expressing cell markers at E14, just after their final mitosis (Altman & Bayer 1985c, Miale & Sidman 1961). Between E14 and E17, Calbindin-positive, immature Purkinje cells migrate along radial-glial processes (Misson et al 1988) through the "mantle" of deep nuclei precursor cells (Chedotal & Sotelo 1992, Altman & Bayer 1985a). Local signals elicited by the cell-cell interactions of immature Purkinje cells contacting radial glia and ingrowing axons at early phases of their development, as well as traversing a zone of differentiating deep nuclear cells, could influence early steps in the Purkinje cell developmental program.

Classical experiments on Purkinje cell development have focused on the postnatal period, when granule cell migration is ongoing and axon-target interactions between parallel fibers and Purkinje cells are forming. Although analysis of a number of mouse mutants has emphasized the interdependence of Purkinje cells and granule cells during this later phase of development, studies of the agranular cerebellar cortex of the *weaver* (*wv*) mutant mouse (Sotelo 1980) or of the x-irradiated cerebellum of wild-type mice (Sotelo 1977, Altman 1976, Llinas et al 1973) demonstrate that early steps in Purkinje development can occur in the absence of local signals from granule cells. This conclusion is supported by primary-cell culture experiments on Purkinje cells purified in the late embryonic period, prior to interactions with granule cells; these experiments demonstrate that the expression of Calbindin and axon formation occurs in the absence of other cell types (Baptista et al 1994). Final stages of Purkinje cell development appear to involve signals from granule cells; parallel fiber–Purkinje cell interactions promote dendritic spine formation (Figure 2).

TERMINAL DIFFERENTIATION OF PURKINJE CELLS: AXON-TARGET INTERACTIONS

The climbing fiber–Purkinje cell relationship has held a special place in developmental neurobiology. Indeed it was from studies of the cerebellum that Ramon y Cajal first proposed that neurotrophic influences appear late in neuronal development and that initial phases of neurite extension occur in the absence of such a "force" (Ramon y Cajal 1911). To examine the coordination of afferent ingrowth with Purkinje cell development, Sotelo (Chedotal & Sotelo 1992, Wassef et al 1992, Armengol & Sotelo 1991) and Mason (Hatten & Mason 1990) have analyzed the time course of Purkinje cell development and climbing-fiber maturation in embryonic and postnatal mouse cerebellum. These studies demonstrate that Purkinje cells migrate along radial glia between E14 and E17 (Yuasa et al 1991, Hatten & Mason 1990) and settle into a zone just beneath the zone of proliferating granule cells. Remarkably, although climbing fibers appear quite primitive while they are growing into the cere-

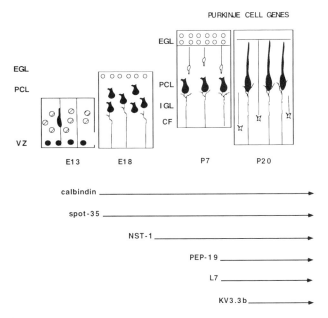

Figure 2 Changing patterns of gene expression during Purkinje cell development in the murine cerebellum. Precursors of the Purkinje neuron (*filled circles*) arise in the ventricular zone (*VZ*) of the neuroepithelium along the roof of the anterior aspect of the metencephalon. At E13, postmitotic precursor cells express Calbindin and spot-35. Calbindin$^+$ cells exit the VZ, traversing the mantle of postmitotic precursors of the deep nuclei (*striped circles*), by migrating along the radial glial-fiber system. In this same period (E14–E15), afferent axons invade the anlage, extending into the collapsing VZ, thereby forming *en passant* contacts with migrating cells. By E16–18, postmigratory, immature Purkinje cells have settled into a dense zone. After birth, as the anlage expands, the zone of immature Purkinje cells spreads into a single cell layer (*P7*). Terminal differentiation of the Purkinje cell relies on the establishment of axon-target interactions, with the parallel fibers of the granule cell (*open circles*), arising in the external germinal layer (*EGL*) and undergoing extension during granule cell development in the early postnatal period, and with afferent axons, the climbing fibers (*CF*). In this later phase of development, when synaptic activity commences, several classes of genes are expressed, including L7 and a specific K$^+$ channel subunit, Kv3.3b. (For details, see Goldman-Wohl et al 1994, Sangameswaran et al 1989, Nordquist et al 1988, Oberdick et al 1988.)

bellar anlage, they segregate in a manner reflecting the topography of the brainstem (Wassef et al 1992). By E18 or 19, most olivary axons project into the wide band of Purkinje cells, now several cells thick, as simple unbranched fibers with small growing cones (Hatten & Mason 1990; CA Mason & R Blazeski, submitted).

The cell-cell contacts formed between ingrowing axons and immature Purkinje cells presumably act to induce specific steps in Purkinje cell maturation. Within a day or two after the afferents have contacted the Purkinje cells, the growing cones of olivary afferents make simple *punctum adherens* contacts, some of which include small accretions of vesicles and postsynaptic densities, characteristic of immature synapses. These contacts generally occur along the Purkinje cell somata, a relationship that continues from E18 to P3, when olivary axons begin to form extensive climbing arborizations (Mason & Gregory 1984, Palay & Chan-Palay 1974). The recent cloning and characterization of a novel heat shock family member (*NST-1*) that is expressed only in Purkinje cells in the cerebellar cortex provides the first example of a gene whose expression correlates with the establishment of climbing fiber–Purkinje cell contact during development (C Yang & D Colman, personal communication). Although the function of this protein in the cerebellum is not known, one might anticipate that its induction at this time is due to climbing-fiber contact, revealing an epigenetic mechanism that responds to climbing-fiber afferents. It is possible that the transsynaptic regulation of cyclic nucleotide phosphodiesterase expression revealed in Purkinje cells in adult rats following climbing-fiber ablation (Balaban et al 1988) is also indicative of this program.

The final stage in Purkinje cell differentiation occurs during the second and third postnatal weeks, when Purkinje cells undergo a period of rapid growth that results in elaboration of their characteristic dendritic arbor and formation of synaptic contacts at specializations along these arbors, the dendritic spines. Although the molecular basis for the interaction between granule cells and Purkinje cells is not understood, several molecular markers are expressed during the period of dendritic spine formation (Goldman-Wohl et al 1994, Sangameswaran et al 1989, Oberdick et al 1988, Nordquist et al 1988). The coordinate timing of axon-target interactions with expression of genes that mark terminal differentiation of the Purkinje cells, including expression of specific ion channels, suggests that these interactions regulate patterns of gene expression associated with synaptic activity.

Although the appearance of Purkinje cell dendritic spines requires granule cell interactions in vivo and in vitro (see Baptista et al 1994), analysis of the expression of both the L7 (Oberdick et al 1993) and Pep 19 (Sangameswaran et al 1989) genes in deafferented cerebellum or in the agranular cerebellum of *wv* mutant mice has demonstrated that these genes are expressed in Purkinje cells that do not receive normal afferent input, suggesting that their expression may be controlled by an intrinsic program. Further support for this conclusion comes from the expression of the Kv3.3b Purkinje cell–specific potassium channel, whose expression in vivo correlates with elaboration of synaptic spines (Goldman-Wohl et al 1994). In this case, it has been demonstrated that purified Purkinje cells can express the Kv3.3b gene in isolation. However, the timing of expression of the Kv3.3b gene can be regulated in culture by interactions with both differentiated granule cells and cerebellar glia.

The terminal differentiation of Purkinje cells can occur according to a cell intrinsic program, but the precise timing of expression of genes involved in synaptogenesis in single cells may be regulated by both positive and negative influences from other cerebellar cell types. The existence of epigenetic programs that fine tune the production of proteins involved in establishment of the functional circuitry of the cerebellum may be important for the proper integration of individual cells into these circuits. Further analysis of mechanisms regulating the expression of these genes in vivo and in vitro should result in the identification of specific mechanisms that govern synaptogenesis in response to both intrinsic and extrinsic signals. One might expect that these mechanisms will also play a role in the functional plasticity of the developing and adult nervous system.

PURKINJE CELL COMPARTMENTATION: THE ORGANIZATION OF THE CEREBELLUM INTO SAGGITAL COMPARTMENTS

Studies of the molecular and biochemical heterogeneity of Purkinje cells have revealed that mechanisms for generation of at least three different classes of Purkinje cells are superimposed upon a general program of Purkinje cell differentiation. The discovery of parasagittal compartments of biochemically distinct Purkinje cells and of transient biochemical heterogeneity in the Purkinje cell population during differentiation has led to the exciting possibility that functional maps of the cerebellum may be identifiable through analysis of these molecular markers and their programs of expression. Because these studies have recently been reviewed in detail by both Hawkes (1993) and Sotelo & Wassef (1991), we refer the reader to those authors for further discussion. This work provides the first comprehensive description of the parcellation of a morphologically monotonous CNS neuronal population into molecularly identifiable subclasses that may be functionally distinct. It seems probable that this type of heterogeneity will be essential for the establishment of distinct functional compartments, perhaps subserving topographical patterns of connnections, in many morphologically uniform neuronal populations.

DEVELOPMENT OF THE GRANULE NEURON

The cerebellar granule cell represents one of the most amenable classes of CNS neurons for developmental studies. Arising in a displaced germinal zone, the EGL, granule neurons undergo the major steps of neural development in a strict spatio-temporal pattern; dividing cells form a compact superficial zone, and cells undergo subsequent steps in development striated in three underlying layers (Figure 3). A key feature of granule cell development, revealed by fate mapping of the cerebellar cell population (Hallonet et al 1990), is that the

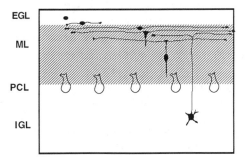

Figure 3   Spatio-temporal pattern of granule cell differentiation and migration in vivo. Granule cell precursors undergo proliferation in the superficial aspect of the external germinal layer (*EGL*), after which postmitotic cells descend into the deeper aspect of the EGL and extend two long axons parallel to the cerebellar laminae. Thereafter, a descending migratory process emerges, which directs the the cell body through the molecular layer (*ML*), along the radially aligned Bergmann glial fibers. Postmigratory cells settle in a dense neuronal layer, the internal granule cell layer (*IGL*), beneath the Purkinje cells (*PCL*) and extend short dendrites that interact with ingrowing afferent fibers (not shown). (For details, see Gao & Hatten 1993, 1994.)

granule neuron arises from a germinal zone that gives rise to only one class of neuron. The homogeneity of the precursor cells in the germinal zone and of cells undergoing specific steps of development in underlying zones provides the opportunity to harvest large numbers of cells in any given stage of development for cell biological or molecular biological analyses. Indeed, the small size and extraordinary number of cerebellar granule neurons, which approximate the number of neurons in the cortex, has allowed the purification of enough precursor cells to carry out molecular biological analyses on an identified CNS cell.

A major issue in CNS development is whether neural identity is controlled by a progressive restriction in cell potential. To compare the potential of EGL precursors generated in the early postnatal period with that of VZ precursors generated in the early embryonic period, we recently implanted EGL precursors and cells from the E13 cerebellar anlage into the EGL of early postnatal mice and compared their fates after short survival times. Whereas EGL cells gave rise exclusively to granule neurons, precursor cells taken from the cerebellar primordium on E13 generated Purkinje neurons, interneurons, granule neurons, and astroglia, after implantation into the early postnatal EGL. These results indicate that early postnatal EGL cells have a restricted potential; they produce only granule neurons (Gao & Hatten 1994).

In addition to revealing the differences in potential of two pools of cerebellar precursor cells, an implantation assay provides a test of whether the EGL contains signals that inhibit the differentiation of other types of cerebellar neurons, including the Purkinje neurons. The ability of E13 cells to generate all of the major types of cerebellar cells, when transplanted into the EGL, argues against inhibitory signals as a control of cell potential in the EGL. Instead, local positive signals for neuronal differentiation appear to induce at least partially committed precursors to undergo terminal steps in their development. Thus early patterning events in the neuroepithelium generate partially fated precursors, with signals for further development, including those needed to induce expression of genes required for neurogenesis and migration along the radial glial-fiber system that are localized to the germinal zones. Migration of the cell through the anlage thereafter provides a strict spatio-temporal program of epigenetic regulation (reviewed in Hatten 1993).

CHANGING PATTERNS OF GENE EXPRESSION DEFINE FOUR STAGES OF GRANULE NEURON DEVELOPMENT

To determine whether there are nonoverlapping molecular markers for particular stages of granule cell development, Kuhar et al (1993) constructed a cDNA expression library from immature granule cells purified from mouse cerebellum on P3–P5 and used a novel antibody-screening method to identify cDNA clones expressed during transit of the granule cell population through these phases. This approach has yielded several dozen cDNA clones that are differentially regulated at the transcriptional level during cerebellar development. In situ localization of select clones reveals that granule neuron differentiation can be divided into at least four discrete stages: neurogenesis, initiation of neuronal differentiation, axon outgrowth and migration, and formation of synaptic connections (Figure 4).

LOCAL SIGNALS CONTROL EGL PRECURSOR CELL PROLIFERATION

In vitro experiments purifying EGL progenitor cells and assaying the contribution of specific cell-cell interactions to DNA synthesis indicate that local, homotypic interactions among EGL precursor cells stimulate precursor cell proliferation (Gao et al 1991). The finding that local environmental cues control cerebellar EGL cell neurogenesis is supported by in vitro studies of neurogenesis of cortical precursor cells by Temple & Davis (1994), showing that local interactions influence entry into neuronal lineages, and by in vivo transplantation experiments by McConnell & Kaznowski (1991), showing that the penultimate cell cycle influences the commitment of cortical precursor cells to a particular neuronal layer of the developing cortex. Thus studies of EGL precursor cells reveal a general role for local signals in CNS neurogenesis. All of these findings are consistent with the location of neuronal precursor cell

GRANULE CELL GENES

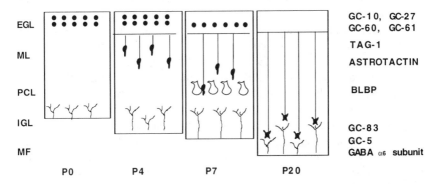

Figure 4 Changing patterns of gene expression define four stages of granule cell development. In situ hybridization analysis of the expression of genes in immature granule cells undergoing development indicates a discrete pattern of localization for four classes of genes. At P0 several genes, including GC10, GC27, GC60, and GC61, are localized to proliferating granule cells in the superficial layer of the EGL (*solid circles in shaded zone*). By P4, postmitotic granule cells in the deeper aspect of the EGL express the axonal glycoprotein TAG-1, as they undergo extension of the parallel fibers, and astrotactin, a gene required for migration along the Bergmann glial system. As the immature granule cells reach the end of the glial fibers, in the Purkinje cell layer, a brain-specific lipid-binding protein (*BLBP*) is expressed in both granule cells and Bergmann glia. The upper aspect of the internal granule cell layer (*IGL*) therefore represents a previously undescribed transient layer. After traversing the Purkinje cell zone, as well as the upper aspect of the IGL, and settling in deeper regions of the IGL, mature granule cells express the GABAα6 subunit, GC-83. A subpopulation of cells expresses the marker GC-5. Ingrowing afferent axons, the mossy fibers, are shown at the bottom of each panel. Thus changing patterns of gene expression define at least four stages of granule cell development. (For details, see Feng et al 1994, Kuhar et al 1993.)

proliferation within compact ventricular zones in the developing mammalian brain, zones where close cell appositions are maintained (Hinds & Ruffett 1971, Boulder Committee 1970, His & Ruffett 1889, Schaper, 1897).

Because local signals participate in the regulation of both cell proliferation and differentiation in the EGL, one might expect that genes expressed within these proliferative zones would include those important for proliferation in all cell types, as well as those required for the generation and interpretation of local cues that are specific to neurogenesis. Of the cDNAs that have been localized to the superficial EGL in the developing cerebellum, two are of particular interest because they demonstrate that the proliferation of neuronal precursors may involve neuron-specific variants of known cell cycle regulatory molecules. Thus, analysis of the GC34 cDNA isolated by Kuhar et al (1993) established that it encodes a brain-specific regulatory subunit (PP2ABβ) for protein phosphatase 2A (PP2A). This confirmed earlier studies of multiple isoforms of this PP2A regulatory subunit in human cell lines, where it was shown that the human homologue of PP2ABβ is expressed only in human neuronal cell lines (Mayer et al 1991). PP2AB subunits form trimeric complexes with the catalytic subunits of PP2A, thereby inhibiting enzyme activity. The role of PP2A in regulating cdc2 kinase (Félix et al 1990), the displacement of PP2AB from the catalytic complex of PP2A by DNA tumor virus oncoproteins (Scheidtmann et al 1991, Yang et al 1991), and the perturbations of neuroblast proliferation in PP2AB subunit mutants in Drosophila (Gomes et al 1993, Mayer-Jaekel et al 1993) all suggest that these B subunits of PP2A can be critically important in cellular growth control. The existence of a neuron-specific isoform of the PP2AB subunit that is coexpressed with the ubiquitous isoform is intriguing, since it raises the possibility that PP2A catalytic activity is regulated in a cell-specific manner during neurogenesis and differentiation. The recent cloning of a neuron-specific form of mouse cyclin D2 mRNA and its localization to proliferating granule cell precursors in the EGL (Ross & Risken 1994) is also interesting in this context. In this case, the neuron-specific transcript is altered only in the 3' untranslated region of the mRNA, indicating that the function of its encoded cyclin D2 is probably not altered but suggesting the possibility that its expression may be regulated differently in neurons. The functional consequences of neuron-specific isoforms of common signal transduction molecules or of neuron-specific regulatory schemes for expression of common enzymes are not yet understood but are of obvious interest in deciphering cell-specific mechanisms for proliferation and differentiation.

CONTROL OF GRANULE CELL DIFFERENTIATION: THE *WEAVER* GENE

To provide a genetic analysis of the molecular signals that induce the early steps of granule neuron differentiation, including cell migration, we utilized cells from the neurological mutant mouse *weaver.* Although the proliferation of EGL precursor cells is normal in *weaver* cerebellum (Rezai & Yoon 1972), the induction of granule neuron differentiation, including expression of late neuronal markers, axon extension, and glia-guided neuronal migration fails (Hatten et al 1984, Sotelo & Changeux 1974, Rakic & Sidman 1973) (Figure 5). The rescue of phenotypic defects in granule cell differentiation demonstrated by in vitro analyses, including migration along glial fibers, reveals that early steps in granule cell differentiation require local, membrane-bound signals (Gao et al 1992). The nonautonomous action of the *weaver* gene emphasizes the importance of community effects, involving local interactions among like precursor cells, in early steps of CNS neuronal differentiation. The requirement for interactions between neuronal precursors contrasts with the

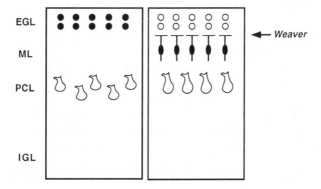

Figure 5 The *weaver* gene disrupts the differentiation of granule cells in the murine cerebellar cortex. Whereas wild-type cells undergo each of the major steps in development, neurogenesis, axon outgrowth, migration, and synaptogenesis in discrete layers of the cerebellar cortex, *weaver* EGL development is arrested prior to axon extension (*arrow*). (*Left*) EGL precursor cells undergoing mitosis (*filled cells*) are located at the pial surface. (*Right*) Postmitotic cells undergoing axon extension and glia-guided migration (*filled cells*) through the molecular layer (*ML*) are located below the proliferating precursor cell population. (For details, see Gao and Hatten 1993, Gao et al 1992.)

earlier hypothesis that the failure of granule cell migration is secondary to an abnormality of the glial-fiber system (Rakic & Sidman 1973). Although in vitro assays provide evidence as to whether specific signaling pathways can function in neuronal development, it is critical to carry out parallel in vivo analyses. Toward that end, we labeled *weaver* EGL precursor cells, reimplanted the cells into the EGL, and examined their differentiation in situ (Gao & Hatten 1993). After implantation into the wild-type EGL, *weaver* cells progressed through all of the steps of granule neuron differentiation seen in vivo, including axon guidance, migration, laminar positioning, and dendritic elaboration (Gao & Hatten 1993).

GRANULE CELL MIGRATION ALONG GLIAL FIBERS

The third stage of granule cell development involves the control of cell positioning, via migration along the glial-fiber system (Sidman & Rakic 1973). The close apposition of migrating cells to glial fibers (Gregory et al 1988, Edmondson & Hatten 1987, Rakic 1971) has suggested that membrane-adhesion receptor systems provide the molecular basis for CNS migrations along glial fibers (Fishman & Hatten 1993). Antibody-perturbation studies of granule neuron migration in vitro (Edmondson et al 1988) and of glass fibers coated with glial membranes (Fishman & Hatten 1993) demonstrate that the neural glycoprotein astrotactin provides a neural receptor system for migration along glial fibers. Expression of the major component of the astrotactin activity, a cell-surface glycoprotein with an apparent molecular weight of approximately 100 kDa, is maximal during the period of granule cell migration and assembly into the internal granule cell layer of the cerebellar cortex in vivo (ME Hatten & T Stitt, unpublished observation).

Molecular cloning of the major component of the astrotactin activity by screening a granule cell cDNA library with affinity-purified antisera indicates that the gene encodes a protein that contains EGF repeats and fibronectin type III domains (C Zheng, N Heintz & ME Hatten, submitted). Affinity-purified antibodies against expressed fusion protein inhibit neuron-glial interactions in an in vitro assay system (Edmondson et al 1988). By using Northern blot analysis, cDNAs for astrotactin encode a brain-specific transcript of approximately 7.0 kb, which is developmentally regulated, showing high levels of expression in developing brain and low levels in adult brain. By using in situ hybridization, astrotactin is expressed by migrating granule cells and post-migratory cells during the period when the internal granule cell layer forms. The structure of the astrotactin protein, having EGF and FN III repeats, suggests that it represents a novel family of adhesion receptor systems (C Zheng, N Heintz & ME Hatten, in preparation).

Although little new information is available concerning mechanisms regulating differentiation of radial glial substrates for migration, recent studies implicate a new, brain-specific member of the lipid-binding protein family (BLBP) (Feng et al 1994) in this process. Members of this protein family, such as cellular retinoic acid–binding proteins (CRABP), have been shown to carry small, hydrophobic signaling molecules between cellular compartments. The expression of BLBP is spatially and temporally correlated with glial differentiation in many parts of the mouse CNS, including the postnatal cerebellum, the embryonic spinal cord, and the cerebral cortex. In situ hybridization and immunocytochemistry show that BLBP is transiently expressed in radial glia in both the embryonic ventricular zone and the postnatal cerebellum. Subcellular localization studies by immunoelectron microscopy demonstrate that BLBP is present in the nucleus as well as the cytoplasm. Affinity-purified anti-BLBP antibodies block glial and neuronal differentiation in primary cell cultures but have no effect on cell proliferation or adhesion. These results demonstrate that BLBP participates in a novel signaling pathway that plays an important role in the establishment of the radial-glial-fiber system that is required for the migration of immature neurons to establish the neural layers throughout the CNS.

TERMINAL DIFFERENTIATION OF GRANULE NEURONS

As the granule cells enter the IGL, two classes of axon-target interactions affect terminal differentiation of the granule cell. First, as described above, the parallel fibers of the granule cell, extended as a T-shaped axon in the molecular layer, begin to form connections with the extending Purkinje cell dendrites. Second, granule cells begin to form connections with ingrowing afferent axons, the mossy fibers. Studies by Mason (reviewed in Hatten & Mason 1990) provide evidence that mossy fibers enter the murine cerebellum at the onset of granule cell migration, often projecting up toward the EGL (Mason & Gregory 1984). By P3–P7, mossy fibers form globular *en passant* boutons in the internal granule layer and send fine branches upward into the Purkinje cell layer. The growth cones of these fine branches contact both Purkinje cells and granule cells transiting the Purkinje cell layer en route to the internal granule layer (Mason & Gregory 1984). By P15, mossy fibers have formed large terminals among small sets of granule neurons, the synaptic glomeruli. Vellate astroglial cells are thought to organize these glomeruli, enwrapping the granule neurons and mossy fiber terminals after synaptogenesis has been completed. In vitro studies on purified granule cells cocultured with pontine explants, a source of mossy fibers, provide evidence that granule cells provide a signal that arrests the ingrowth of the mossy fibers (Baird et al 1992), a signal that is mediated by synaptic activity (DA Baird, personal communication) that appears to be specific to mossy fiber–granule neuron connections. Thus axon-target interactions are likely to provide epigenetic signals for the terminal differentiation of the granule neuron.

Terminal differentiation of granule cells results in expression of a great many molecules involved in the function of granule cells in the cerebellum. As anticipated for any differentiated neuronal cell type, these genes participate in different aspects of mature granule cell function and display differences in the mechanisms of regulation. For example, genes such as those encoding the $\alpha6$ subunit of the GABA receptor (Kato 1990, Schofield et al 1994) and glutamic acid dehydrogenase (GAD) (Wuenschell & Tobin 1988) are specifically involved in granule cell function as presynaptic partners for Purkinje cells. Expression of the αG subunit of the GABA receptor is restricted to cells in the IGL, and it is the only gene yet characterized that is expressed exclusively in cerebellar granule neurons. Detailed analysis of in situ hybridization patterns demonstrate that the $\alpha6$ GABA receptor gene is expressed in all granule cells in the IGL, providing a definitive marker for this cell type. In contrast, the expression of a second class of markers has been shown to be restricted to a subpopulation of differentiated granule cells. Thus, the GC5 gene (Kuhar et al 1993) encodes a novel transcript that is expressed in only approximately 50% of the mature granule cell population (J Miwa, ME Hatten & N Heintz, unpublished data). The expression of the GC5 gene demonstrates either molecular heterogeneity within the adult granule cell population or temporally dynamic control of this gene in granule cells. For example, GC5 gene expression could result from patterned firing of a subset of granule cells. Although direct evidence for activity-dependent gene expression in cerebellar granule cells is lacking, recent studies have identified a cerebellum-like gene, whose expression in granule cells is dependent on direct interactions with Purkinje neurons (AM Smith & RJ Mullen, personal communication). These examples suggest that the patterns of gene expression in mature granule cells are likely to reflect diverse epigenetic pathways for regulation of granule cell function.

CONCLUSIONS

Early patterning of the cerebellar primordium follows mechanisms that set forth a general axial regionalization of the brain. Current evidence implicates a redundant role for two murine *En* genes in the formation of the cerebellar territory. Although these early patterning events define the cerebellar region within the anterior-posterior axis of the developing brain, later events involving the development of individual cells within the region underlie the histogenesis of the cerebellar cortex. Cell-fate decisions in cerebellum appear to require a series of steps that progressively restrict the potential of precursor cells. At embryonic periods, three general pools of partially committed precursors are generated: the deep nuclei, the cortical cells, and the EGL cells. Studies of EGL cells indicate that local signals among neuronal precursor cells induce the early steps in neuronal differentiation, including axon outgrowth and migration, with a program of changing gene expression leading to a granule cell identity. Further characterization of genes that mark specific stages in cerebellar histogenesis and an understanding of how these genes are expressed will be essential to a molecular understanding of the development of this critical region of the brain.

ACKNOWLEDGMENTS

This work was supported by NIH grants NS15429 (MEH), Program Project P01 30532 (MEH, NH), the Howard Hughes Medical Institute (NH), and the McKnight Foundation. We thank our colleagues who participated in experiments reported here and Drs. Sally Temple, Kathy Zimmerman, Alex Joyner, Carol Mason, and Robert Darnell for reading the manuscript.

Literature Cited

Altman J. 1976. Experimental reorganization of the cerebellar cortex. VII. Effects of late X-irradiation schedule to interfere with cell acquisition after stellate cells are formed. *J. Comp. Neurol.* 165:65–75

Altman J, Bayer SA. 1985a. Embryonic development of the rat cerebellum I. Delineation of the cerebellar primordium and early cell movement. *J. Comp. Neurol.* 231:1–26
Altman J, Bayer S. 1985b. Embryonic develop-

ment of the rat cerebellum II. Translocation and regional distribution of the deep neurons. *J. Comp. Neurol.* 231:27–41

Altman J, Bayer S. 1985c. Embryonic development of the rat cerebellum III. Regional differences in the time of origin, migration and settling of Purkinje neurons. *J. Comp. Neurol.* 231:42–65

Alvarado-Mallart R-M, Martinez S, Jones CL. 1990. Pluripotentiality of the 2-day-old avian germinative neuroepithelium. *Dev. Biol.* 139: 75–88

Alvarado-Mallart R-M, Sotelo C. 1982. Differentiation of cerebellar anlage heterotypically transplanted to adult rat brain. *J. Comp. Neurol.* 212:247–67

Alvarado-Mallart R-M, Sotelo C. 1992. New insights on the factors orienting the axonal outgrowth of grafted Purkinje cells in the *pcd* cerebellum. *Dev. Neurosci.* 14: 153–65

Anderson DJ. 1989. The neural crest lineage problem: neuropoiesis? *Neuron* 3:1–12

Anderson DJ, Axel R. 1986. A bipotential neuroendocrine precursor whose choice of cell fate is determined by NGF and glucocorticoids. *Cell* 47:1079–90

Armengol JA, Sotelo C. 1991. Early dendritic development of Purkinje cells in the rat cerebellum: a light and electron microscopic study using axonal tracing in 'in vitro' slices. *Dev. Brain. Res.* 64:95–114

Baird DH, Hatten ME, Mason CA. 1992. Cerebellar target neurons provide a stop signal for afferent neurite extension in vitro. *J. Neurosci.* 12:619–34

Balaban E, Teillet M-A, Le Douarin N. 1988. Application of the quail-chick chimera system to the study of brain development and behavior. *Science* 214:1330–42

Bally-Cuif L, Alvarado-Mallart R-M, Darnell DK, Wassef M. 1992. Relationship between *Wnt-1* and *En-2* expression domains during early development of normal and ectopic met- and mesencephalon. *Development* 115: 999–1009

Baptista CA, Hatten ME, Blazeski R, Mason CA. 1994. Cell-cell interactions influence survival and differentiation of purified Purkinje cells in vitro. *Neuron* 12:1–20

Birren SJ, Anderson DJ. 1990. A v-myc-Immortalized sympathoadrenal progenitor cell line which in neuronal differentiation is initiated by FGF but not by NGF. *Neuron* 4: 189–201

Boulder Committee. 1970. Embryonic vertebrate central nervous system: revised terminology. *Anat. Rec.* 166:257–62

Brodal A. 1981. *Neurological Anatomy in Relation to Clinical Medicine.* New York: Oxford Univ. Press. 3rd ed.

Chedotal A, Sotelo C. 1992. Early development of olivocerebellar projections in the fetal rat using CGRP immunocytochemistry. *Eur. J. Neurosci.* 4:1159–79

Davis CA, Joyner AL. 1988. Expression patterns of the homeobox containing genes *En-1* and *En-2* and the proto-oncogene diverge during mouse development. *Genes Dev.* 2:1736–44

Desplan C, Theis J, O'Farrell PH. 1985. The *Drosophila* developmental gene *engrailed* encodes a sequence-specific DNA binding activity. *Nature* 318:630–35

DiNardo S, Sher E, Heemskerk-Jongens J, Kassis JA, O'Farrell PH. 1988. Two-tiered regulation of spatially patterned engrailed gene expression during *Drosophila* embryogenesis. *Nature* 332:604–9

Edmondson JC, Hatten ME. 1987. Glial-guided granule neuron migration in vitro: a high-resolution time-lapse video microscopic study. *J. Neurosci.* 7:1928–34

Edmondson JC, Liem RKH, Kuster JE, Hatten ME. 1988. Astrotactin, a novel cell surface antigen that mediates neuron-glia interactions in cerebellar microcultures. *J. Cell Biol.* 106:505–17

Félix MA, Cohen P, Karsenti E. 1990. Cdc2 H1 kinase is negatively regulated by a type 2A phosphatase in the *Xenopus* early embryonic cell type: evidence from the effects of okadaic acid. *EMBO J.* 9:675–83

Feng L, Hatten ME, Heintz N. 1994. Brain-specific lipid binding protein (BLBP). *Neuron* 12:895–908

Fishman R, Hatten ME. 1993. Multiple receptor systems promote CNS neuronal migration. *J. Neurosci.* 13:3485–95

Fletcher C, Norman DJ, Heintz N. 1991. Genetic mapping of meander tail, a mouse mutation affecting cerebellar development. *Genomics* 9:647–55

Fritsch R, Gruss P. 1993. Murine paired box containing genes. In *Cell-Cell Signaling in Vertebrate Development,* ed. EJ Robertson, FR Maxfield, HJ Vogel, pp. 229–45. San Diego: Academic

Fugita S. 1967. Quantitative analysis of cell proliferation and differentiation in the cortex of the postnatal mouse cerebellum. *J. Cell Biol.* 32:277–87

Fugita S, Shimada M, Nakanuna T. 1966. ^{3}H-thymidine autoradiographic studies on the cell proliferation and differentiation in the external and internal granular layers of the mouse cerebellum. *J. Comp. Neurol.* 128: 191–209

Gao WQ, Hatten ME. 1993. Neuronal differentiation rescued by implantation of *weaver* granule cell precursors into wild-type cerebellar cortex. *Science* 260:367–69

Gao WQ, Hatten ME. 1994. Immortalizing oncogenes subvert the control of granule cell lineage in developing cerebellum. *Development* 120:1059–70

Gao WQ, Heintz N, Hatten ME. 1991. Cerebellar granule cell neurogenesis is regulated by cell-cell interactions in vitro. *Neuron* 6:705–15

Gao WQ, Liu XL, Hatten ME. 1992. The *weaver* gene encodes a non-autonomous signal for CNS neuronal differentiation. *Cell* 68:841–54

Gardner CA, Barald KF. 1991. The cellular environment controls the expression of engrailed-like protein in the cranial neuroepithelium of quail-chick chimeric embryos. *Development* 113:1037–48

Goldman-Wohl DS, Chan E, Baird D, Heintz N. 1994. Kv3.3b: a novel *shaw* type potassium channel expressed in terminally differentiated cerebellar Purkinje cells and deep cerebellar nuclei. *J. Neurosci.* 14:511–22

Gomes R, Karess RE, Ohkura H, Glover DM, Sunkel CE. 1993. Abnormal anaphase resolution (*aar*): a locus required for progression through mitosis in *Drosophila. J. Cell Sci.* 104:583–93

Gregory WA, Edmondson JC, Hatten ME, Mason CA. 1988. Cytology and neuron-glial apposition of migrating cerebellar granule cells in vitro. *J. Neurosci.* 8:1728–38

Gruss P, Walther C. 1992. *Pax* in development. *Cell* 69:719–22

Hallonet MER, Le Douarin NM. 1993. Tracing neuroepthelial cells of the mesencephalic and metencephalic alar plates during cerebellar ontogeny in quail-chick chimaeras. *Eur. J. Neurosci.* 5:1145–55

Hallonet MER, Teillet M-A, Le Douarin NM. 1990. A new approach to the development of the cerebellum provided by the quail-chick marker system. *Development* 108:19–31

Hatten ME. 1993. The role of migration in central nervous system neuronal development. *Curr. Opin. Neurobiol.* 3:38–44

Hatten ME, Liem RKH, Mason CA. 1984. Two forms of cerebellar glial cells interact differently with neurons in vitro. *J. Cell Biol.* 98:193–204

Hatten ME, Mason CA. 1990. Mechanisms of glial-guided neuronal migration in vitro and in vivo. *Experientia* 46:907–16

Hawkes R, Blyth S, Chockkan V, Tano D. 1988. Structural and molecular compartmentalization in the cerebellum. *Can. J. Neurol. Sci. Suppl.* 3:S29–S35

Heemskerk J, DiNardo S, Kostikan R, O'Farrell PH. 1991. Multiple modes of *engrailed* regulation progression cell fate determination. *Nature* 552:404–10

Heintz N, Norman DJ, Gao, W-Q, Hatten ME. 1993. Neurogenetic approaches to mammalian brain development. In *Genome Analysis,* 6:19–44. Cold Spring Harbor: Cold Spring Harbor Lab.

Herrup K, Wilcyzynski SL. 1982. Cerebellar degeneration in *leaner* mutant mice. *Neuroscience* 7:2185–96

Hinds JW, Ruffett TL. 1971. Cell proliferation in the neural tube: an electron microscopic and Golgi analysis in the mouse cerebral vesicle. *Z. Zellforsch. Mikrosk. Anat.* 115:226–64

His JW, Ruffett TL. 1889. Die Neuroblasten und deren Entstehung im embryonalen Mark. *Abh. Math. Phys. Kl. Konigl. Saechs. Akad. Wiss.* 26:313–72

Itasaki N, Ichijo H, Hama C, Matsuno T, Nakamura H. 1991. Establishment of rostrocaudal polarity in tectal primordium: *engrailed* expression and subsequent tectal polarity. *Development* 113:1133–44

Joyner AL, Hanks M. 1991. The *engrailed* genes, evolution of function. *Semin. Dev. Biol.* 2:435–45

Kato K. 1990. Novel GABAA receptor α subunit is expressed only in cerebellar granule cells. *J. Mol. Biol.* 214:619–24

Kuhar S, Feng L, Vidan S, Ross ER, Hatten ME, Heintz N. 1993. Developmentally regulated cDNAs define four stages of cerebellar granule neuron differentiation. *Development* 112:97–104

Le Douarin N. 1969. Particularites du noyau interphasique chez la caille japonaise Coturnix coturnix japonica. Utilisation de ces particularites come 'marquage biologique' dans les recherches sur les interactions tissulaires et les migrations cellulaires au cours de l'ontogenese. *Bull. Biol. Fr. Belg.* 103:435–52

Llinas R, Hillman DE. 1969. Physiological and morphological organization of the cerebellar circuits in various vertebrates. In *Neurobiology of Cerebellar Evolution and Development,* ed. R Llinas, pp. 43–73. Chicago: AMA-ERF Inst. Biomed. Res.

Llinas R, Hillman DE, Precht W. 1973. Neuronal circuit reorganization of mammalian agranular cerebellar cortex. *J. Neurobiol.* 4:69–94

Logan C, Khoo WK, Cado D, Joyner AL. 1993. Two enhancer regions in the mouse *en-2* locus direct expression to the mid-hindbrain region and mandibular myoblasts. *Development* 117:905–16

Martinez S, Alvarado-Mallart R-M. 1989. Rostral cerebellum originates from the caudal portion of the so-called mesencephalic vesicle: a study using chick/quail chimeras. *Eur. J. Neurosci.* 1:549–60

Martinez S, Wassef M, Alvarado-Mallart R-M. 1991. Induction of a mesencephalic phenotype in the 2-day-old chick prosencephalon is preceded by the early expression of the homeobox gene *en. Neuron* 6:971–81

Mason CA, Christakos S, Catalano SM. 1990. Early climbing fibers interactions with Purkinje cells in postnatal mouse cerebellum. *J. Comp. Neurol.* 297:77–90

Mason CA, Gregory E. 1984. Postnatal maturation of cerebellar mossy and climbing fibers: transient expression of dual features on single axons. *J. Neurosci.* 4:1715–35

Mayer RE, Hendrix P, Cron P, Matthies R, Stone SR, et al. 1991. Structure of the 55-kDa regulatory subunit of protein phosphatase 2A: evidence for a neuronal-specific isoform. *Biochemistry* 30:3589–97

Mayer-Jaekel RE, Ohkura H, Gomes R, Sunkel CE, Baumgartner S, et al. 1993. The 55-kd regulatory subunit of *Drosophila* protein phosphatase 2A is required for anaphase. *Cell* 72:621–33

McConnell SK, Kaznowski CE. 1991. Cell cycle dependence of laminar determination in developing neocortex. *Science* 254:282–85

McMahon AP. 1993. Cell signaling in induction and anterior-posterior patterning of the vertebrate central nervous system. *Curr. Opin. Neurobiol.* 3:4–7

McMahon AP, Bradley A. 1990. The *Wnt-1* (*int-1*) proto-oncogene is required for development of a large region of the mouse brain. *Cell* 62:1073–85

McMahon AP, Joyner AL, Bradley A, McMahon JA. 1992. The midbrain-hindbrain phenotype of *Wnt-1/Wnt-1-*mice results from stepwise deletion of engrailed-expressing cells by 9.5 days postcoitum. *Cell* 69:581–95

Miale I, Sidman RL. 1961. An autoradiographic analysis of histogenesis in the mouse cerebellum. *Exp. Neurol.* 4:277–96

Millen KJ, Wurst W, Herrup K, Joyner AL. 1994. Abnormal embryonic cerebellar development and patterning of postnatal foliation in two mouse *Engrailed-2* mutants. *Development* 120:695–706

Misson J-P, Edwards MA, Yamamoto M, Caviness VS Jr. 1988. Identification of radial glial cells within the developing murine central nervous system: studies based upon a new immunohistochemical marker. *Dev. Brain. Res.* 44:95–108

Morrissey D, Askew D, Raj L, Weir M. 1991. Functional dissection of the paired segmentation gene in *Drosophila* embryos. *Genes Dev.* 5:1684–96

Mullen RJ. 1982. Analysis of CNS development with mutant mice and chimeras. In *Genetic Approach to Developmental Neurobiology,* ed. Y Tsukada, pp. 183–93. Berlin/Heidelberg/New York: Springer-Verlag

Nordquist DT, Kozak CA, Orr HT. 1988. cDNA cloning and characterization of three genes uniquely expressed in cerebellum by Purkinje neurons. *J. Neurosci.* 8:4780–89

Oberdick J, Levinthal F, Levinthal C. 1988. A Purkinje cell differentiation marker shows a partial DNA sequence homology to the cellular *sis/PDGF2* gene. *Neuron* 1:367–76

Oberdick J, Schilling K, Smeyne RJ, Corbin JG, Bocchiaro C, Morgan JI. 1993. Control of segment-like patterns of gene expression in the mouse cerebellum. *Neuron* 10:1007–18

Otero RM, Sotelo CM, Alvarado-Mallart R-M. 1993. Chick/quail chimeras with partial cerebellar grafts: an analysis of the origin and migration of cerebellar cells. *J. Comp. Neurol.* 333:597–615

Palay SF, Chan-Palay V. 1974. *Cerebellar Cortex, Cytology and Organization.* Berlin: Springer-Verlag

Pierce ET. 1975. Histogenesis of the deep cerebellar nuclei in the mouse, an autoradiographic study. *Brain Res.* 95:503–18

Rakic P. 1971. Neuron-glia relationship during granule cell migration in developing cerebellar cortex. A Golgi and electron microscopic study in *Macacus rhesus. J. Comp. Neurol.* 141:283–312

Rakic P, Sidman RL. 1973. Sequence of developmental abnormalities leading to granule cell deficit in cerebellar cortex of *weaver* mutant mice. *J. Comp. Neurol.* 152:103–32

Ramon y Cajal S. 1889. Sobre las fibras nerviosas de la capa granulosa del cerebelo. *Rev. Trim. de Histol. Norm. y Pathol.* 3,4:5

Ramon y Cajal S. 1911. *Histologie du Susteme Nerveux de l'Homme et des Vertebres.* Paris: Maloine (Reprinted by Consejo Superior de Investigaciones Cientificas, Madrid, 1955)

Ramon y Cajal S. 1960. *Studies on Vertebrate Neurogenesis.* Transl. L Guth. Springfield, IL: Thomas, p. 423 (From Spanish)

Rezai Z, Yoon CH. 1972. Abnormal rate of granule cell migration in cerebellum of "weaver" mutant mice. *Dev. Biol.* 29:17–26

Ross ME, Fletcher C, Mason CA, Hatten ME, Heintz N. 1989. Meander tail reveals a developmental unit in mouse cerebellum. *Proc. Natl. Acad. Sci. USA* 87:4189–92

Ross ME, Risken M. 1994. MN20, a D2 cyclin found in brain, is implicated in neural differentiation. *J. Neurosci.* In press

Sangameswaran L, Hempstead J, Morgan JI. 1989. Molecular cloning of a neuron-specific transcript and its regulation during normal and aberrant cerebellar development. *Proc. Natl. Acad. Sci. USA* 86:5651–55

Schaper A. 1897. The earliest differentiation in the central nervous system of vertebrates. *Science* 5:430–31

Scheidtmann KH, Mumby MC, Rundell K, Walter G. 1991. Dephosphorylation of simian virus 40 large-T antigen and p53 protein by protein phosphatase 2A: inhibition by small-t antigen. *Mol. Cell. Biol.* 11:1996–2003

Schofield PR, Darlison MG, Fujita N, Burt DR, Stephenson FA, et al. 1994. Sequence and functional expression of the GABA$_A$ receptor shows a ligand-gated receptor super-family. *Nature* 328:221–27

Sidman RL. 1968. Development of interneuronal connections in brains of mutant mice. In *Physiological and Biochemical Aspects of Nervous Integration,* ed. FD Carlson, pp. 163–93. Englewood Cliffs, NJ: Prentice-Hall

Sidman RL. 1983. Experimental neurogenetics. In *Genetics of Neurological and Psychiatric Disorders,* ed. SS Kety, LP Rowland, RL Sidman, SW Matthysse, pp. 19–46. New York: Raven

Sidman RL, Rakic P. 1973. Neuronal migration with special reference to developing human brain: a review. *Brain Res.* 62:1–35

Sotelo C. 1977. Formation of presynaptic dendrites in the rat cerebellum following neonatal x-irradiation. *Neuroscience* 2:275–83

Sotelo C. 1980. Mutant mice and the formation of cerebellar circuitry. *Trends Neurosci.* 20: 33–35

Sotelo C, Changeux P. 1974. Bergmann fibers and granule cell migration in the cerebellum of homozygous weaver mutant mouse. *Brain Res.* 77:484–91

Sotelo C, Wassef M. 1991. Cerebellar development: afferent organization and Purkinje cell heterogeneity. *Phil. Trans. R. Soc. London Ser. B* 331:307–13

Stemple DL, Mahanthappa NK, Anderson DJ. 1988. Basic FGF induces neuronal differentiation, cell division, and NGF dependence in chromaffin cells: a sequence of events in sympathetic development. *Neuron* 1:517–25

Temple S, Davis AA. 1994. Isolated rat cortical progenitor cells are maintained in division in vitro by membrane-associated factors. *Development* 120:999–1008

Thomas KR, Capecchi MR. 1990. Targeted disruption of the murine *int-1* proto-oncogene resulting in severe abnormalities in midbrain and cerebellar development. *Nature* 346: 847–50

Vandaele S, Nordquist DT, Fedderson RM, Tretjakoff I, Peterson AC, Orr HT. 1991. Purkinje cell protein-2 regulatory regions and transgene expression in cerebellar compartments. *Genes Dev.* 5:1136–48

Wassef M, Cholley B, Heizmann CW, Sotelo C. 1992. Development of the olivocerebellar projection in the rat. II. Matching of the developmental compartmentations of the cerebellum and inferior olive through the projection map. *J. Comp. Neurol.* 323: 519–36

Wuenschell CW, Tobin AJ. 1988. The abnormal cerebellar organization of *weaver* and *reeler* mice does not affect the cellular distribution of three neuronal mRNAs. *Neuron* 1:805–15

Wurst W, Auerbach AB, Joyner AL. 1994. Multiple developmental defects in *Engrailed-1* mutant mice: an early mid-hindbrain deletion and patterning defects in forelimbs and sternum. *Development* 120:2065–75

Yang SI, Lickteig RL, Estes R, Rundell K, Walter W, Mumby MC. 1991. Control of protein phosphatase 2A by simian virus 40 small-t antigen. *Mol. Cell Biol.* 11:1988–95

Yuasa S, Kawamura K, Ono K, Yamakuni T, Takahashi Y. 1991. Development and migration of Purkinje cells in the mouse cerebellar primordium. *Anat. Embryol.* 184: 195–212

Annu. Rev. Neurosci. 1995. 18:409–41

LEARNING AND MEMORY IN THE VESTIBULO-OCULAR REFLEX

Sascha du Lac[1]*, Jennifer L. Raymond*[1]*, Terrence J. Sejnowski*[2],
and Stephen G. Lisberger[1]

[1]Department of Physiology, W.M. Keck Foundation Center for Integrative Neuroscience, and Neuroscience Graduate Program, University of California, San Francisco, San Francisco, California 94143 and [2]Howard Hughes Medical Institute, Department of Biology, and Institute for Neural Computation, University of California, San Diego, La Jolla, California 92186

KEY WORDS: plasticity, vestibular system, eye movements, cerebellum, temporal dynamics

---

Studies of the neural basis of learning and memory in intact animals must, by their nature, start "from the top" by choosing a behavior that can be modified through learning, revealing how neuronal activity gives rise to that behavior, and then investigating, in the awake, behaving animal, changes in neural signaling that are associated with learning. Such studies also must recognize that the learning and memory expressed in the behavior of an animal will reflect both the properties of the neural network that mediates the behavior and the nature of the underlying changes in the operation of cells or synapses. In the past 10 years, there has been an explosion of information about learning and memory in the vestibulo-ocular reflex (VOR) of the awake, behaving monkey. At the same time, there have been unprecedented advances in understanding mechanisms of cellular plasticity such as long-term potentiation (LTP) in the hippocampus and long-term depression (LTD) in the cerebellum. A prerequisite for understanding learning and memory is to elevate specific mechanisms of cellular plasticity into cellular mechanisms of learning by establishing their function in the context of a neural system that mediates learning and memory in a particular behavior. Our review synthesizes the

*The first two authors contributed equally to the ideas and writing of this paper.

combined behavioral, physiological, anatomical, cellular, and computational analyses needed to understand learning and memory in the VOR.

INTRODUCTION TO THE VOR AND RELATED BEHAVIORS

Under normal behavioral conditions, the VOR prevents images of the stationary world from slipping across the retina. Inertial sensors in the vestibular apparatus detect head motion and send signals into the brain to generate compensatory eye movements that are opposite in direction to head motion. In the laboratory, the VOR is evoked by passive head rotation in darkness. The behavior is quantified by measuring the evoked eye motion and computing the *gain* of the VOR, defined as eye speed divided by angular head speed in darkness. We define the gain of the *normal* VOR as that recorded during passive head rotation in a naive subject that has not yet been subjected to conditions that cause learning.

The VOR is a fast reflex that operates without visual feedback, at least on the time scale of individual head turns. In monkeys, this is ideal because the normal gain of the VOR is near 1.0, and the VOR alone is nearly sufficient to stabilize retinal images. In other species, including humans, the normal gain of the VOR is less than one, so visual-tracking systems must cooperate with the VOR to prevent retinal-image motion during head turns. One such system, the optokinetic response (OKR), operates in all species. It responds to smooth motion of images that cover a large portion of the visual field and generates compensatory eye movements in the same direction as the visual stimulus. A second tracking mechanism, smooth-pursuit eye movements, operates effectively only in primates. It responds to the motion of small targets by keeping the eyes moving at approximately the same speed as that of the target. The OKR and pursuit must be included in any discussion of neural expressions of learning and memory in the VOR because these two visual-tracking systems share circuitry with the VOR and place constraints on the sites of the neural changes, and possibly on the mechanisms that underlie learning in the VOR.

For the purposes of this review, we define *learning* as the acquisition of behavioral changes and *memory* as the changes themselves. Learning occurs in the VOR under most conditions that provide persistent image motion during head turns. If monkeys wear spectacles that magnify or miniaturize vision, then the gain of the VOR that is measured in darkness increases or decreases over a time course of hours or days (Miles & Fuller 1974). Learning occurs during the vestibular stimulation provided either by the subject's active head turns or by passive oscillation. The visual conditions provided by spectacles can be mimicked by arranging for a large visual stimulus to move either exactly with or opposite to the rotation of the subject's head (e.g. Collewijn & Groot-endorst 1979). The learned changes in the gain of the VOR are remembered for days if the subject is deprived of either visual (Robinson 1976) or vestibular (Miles & Eighmy 1980) stimuli after the VOR has been modified.

PROPOSED SITES OF MEMORY IN THE VOR

A decade ago, the location of memory in the VOR was highly controversial, and two reviews by different groups of investigators espoused very different views (Miles & Lisberger 1981, Ito 1982). The past ten years, however, have seen a large increase in the amount of data and models relevant to the sites of memory in the VOR. In this section of our review, we outline the available data on the sites of memory and present a testable hypothesis that accounts for available data. We also discuss the remaining areas of disagreement and outline the kinds of experiments that would resolve these outstanding issues.

Behavioral Analysis of Memory in the VOR

We can learn a good deal about sites of memory from a careful analysis of behavioral changes, even before examining the organization of the essential neural network or the responses of its constituent neurons. For example, be-havioral studies have shown that learning in the VOR is more complex (and interesting) than a simple scaling of reflex commands.

LATENCY OF THE MODIFIED COMPONENT OF THE VOR   The use of brief pulses of head motion as a vestibular stimulus revealed that the first 5 ms of the VOR do not change even after large changes in the steady-state gain of the VOR (Lisberger 1984). These data divided the VOR into separate modified and unmodified components and demonstrated that the earliest part of the VOR is driven entirely by the unmodified component. For the stimulus used by Lisber-ger (1984), the latencies of the unmodified and modified components are 14 and 19 ms, respectively. For other stimuli, however, the unmodified and modified components of the VOR cannot be distinguished based on their latencies. Broussard et al (1992) found that changes in the gain of the VOR caused small but consistent changes in the earliest portion of the eye move-ments evoked by electrical stimulation of the vestibular apparatus with a single pulse. Khater et al (1993) found that learning-related changes could be detected as soon as the eyes started to move when a natural vestibular stimulus provided extremely rapid head accelerations.

Because changes in the gain of the VOR appear in the earliest eye move-ments evoked by some stimuli (Broussard et al 1992, Khater et al 1993), one site of memory for the VOR is likely to be found in the shortest-latency VOR pathways, which reside entirely in the brainstem and include just two synapses (e.g. Precht & Baker 1972). If the disynaptic VOR pathways contain a site of memory, why did Lisberger (1984) not find any changes in the first 5 ms of the responses to a pulse of head velocity? We suggest that the answer lies in the latencies of the responses of vestibular primary afferents for the vestibular stimuli used in different studies. Primary afferents respond with latencies that range from 5 to 18 ms (Lisberger & Pavelko 1986) for the relatively low head accelerations in the stimulus used by Lisberger (1984). If the afferents with shorter latencies (5 ms) project into unmodified pathways and those with longer latencies (>10 ms) project into modified pathways, then the first 5 ms of the VOR should not be modified even if the site of memory is in disynaptic VOR pathways. In contrast, we know that all afferents respond synchronously within 1 ms for stimulation with single electrical pulses (Brönte-Stewart & Lisberger 1994), and we presume that all afferents also respond synchronously for the very rapid head acceleration used by Khater et al (1993). For these stimuli, a site of memory in disynaptic modified pathways should cause learning to be expressed in the earliest part of the evoked eye movement. Even though this logic suggests that one site of memory is in the disynaptic brainstem VOR pathways, the experiments of Brönte-Stewart & Lisberger (1994) suggest that there may be additional sites of memory in pathways that have additional intervening synapses.

DYNAMICS OF THE MODIFIED COMPONENT OF THE VOR   The time course, or "temporal dynamics," of the VOR changes as a function of the gain of the VOR. For the natural stimulus provided by a ramp of head velocity, the evoked eye velocity only slightly overshoots a final steady eye velocity when the gain of the VOR is normal (Figure 1). The overshoot is much larger after the gain of the VOR has been lowered, and it is barely evident when the gain of the VOR is high (Lisberger & Pavelko 1986). For electrical stimulation with single pulses, the effects of changing the gain of the VOR were larger in the later portion of the evoked eye movements than at their onset or peak (Broussard et al 1992). For electrical stimulation with trains of pulses, there was a complex temporal structure in the relationship between the gain of the VOR and the evoked eye movements. The magnitude and time course of the effects depended critically on which afferents were activated by the stimulus, and the learning-related changes were larger in the later portions of the evoked eye movements, growing over a time course of about 40 ms (Brönte-Stewart & Lisberger 1994). These data imply that the memory of a modified VOR cannot be implemented simply as a scale factor in one or more VOR pathways. If it were, learning in the VOR would cause simple increases or decreases in the speed of compensatory eye movement at all times, both a few milliseconds and tens of milliseconds after the onset of the modified component of the response. If, for example, the mechanism responsible for memory were a change in strength of transmission (e.g. LTP or LTD) at the synapse from

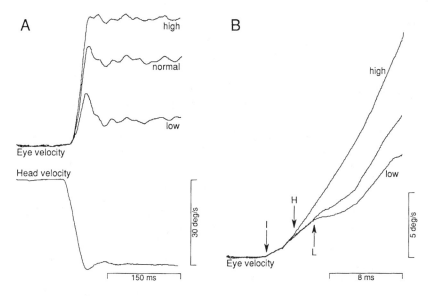

Figure 1 Effect of learning in the VOR on the eye movements evoked by ramps of head velocity. Each trace shows the average of 10 traces of eye or head velocity. (*A*) Slow-sweep records showing the VOR before learning (*normal*) and after learning induced by magnifying (*high*) or miniaturizing (*low*) spectacles. (*B*) Fast-sweep records showing the events at the initation of the VOR for the data in *A*. The arrow labeled "I" indicates the initation of the VOR, and those labeled "H" and "L" point out the times when the high-gain and low-gain records of eye velocity diverge from the control record. The gain of the VOR was 0.32, 1.05, and 1.57 for the records labeled low, normal, and high, respectively. Upward deflections are rightward motion. Reprinted with permission from Lisberger et al (1990).

primary afferents onto secondary vestibular neurons, then the eye movements evoked by electrical stimuli should scale uniformly as a function of the gain of the VOR. Neither of these predictions is supported by data.

Both the architecture of the neural network that mediates the VOR and the properties of the cellular mechanisms of learning may contribute to the complex relationship between the gain of the VOR and the temporal dynamics of the evoked eye movements. 1. Changes in the VOR could result from changes in the relative strengths of pathways with different dynamics (Lisberger et al 1983, Minor & Goldberg 1991, Quinn et al 1992a) or changes in the dynamics of the VOR pathways themselves (Lisberger & Sejnowski 1992). 2. Memory may reside in a small amount of synaptic potentiation or depression that is amplified over a time course of tens of milliseconds by neural feedback loops. 3. Memory could result from changes in cellular properties that have a long time course, such as the ionic conductances that determine the repetitive firing properties of the relevant neurons or a long-duration synaptic potential, rather than in the strength of fast synaptic transmission.

Essential Circuit for the VOR

The most direct VOR pathway comprises a three-neuron reflex arc that includes afferents from the vestibular nerve, interneurons in the vestibular nucleus, and extraocular motoneurons (e.g. Precht & Baker 1972, Highstein 1973). In addition to this basic pathway, there are a number of less direct pathways, including projections across the midline that relay vestibular afferent information between the vestibular nuclei on the two sides of the brainstem (Shimazu & Precht 1966) and projections from the vestibular nucleus to the motoneurons through the nucleus prepositus (Baker & Berthoz 1975). Important side loops include projections from the nucleus prepositus and the vestibular nucleus to a portion of the cerebellum called the flocculus/ventral paraflocculus (F/VPF) (Langer et al 1985b) and back from the F/VPF to the vestibular nucleus (Langer et al 1985a). We know these side loops are important because complete bilateral ablations of the F/VPF or the whole cerebellum abolish learning in the VOR while having little or no effect on the normal VOR (Robinson 1976, Nagao 1983, Flandrin et al 1983, Lisberger et al 1984).

Much is known about the connections, signal processing, and firing properties of three types of neurons whose firing patterns change consequent to learning in the VOR. Based on their responses during a variety of behavioral paradigms and some direct evaluations of their connections, the neurons appear to be interconnected in the pattern illustrated by Figure 2. Position-Vestibular-Pause cells (PVPs), so-named because they fire in relation to eye position and vestibular rotation and they pause during saccades, are in the vestibular nuclei and are some of the principal interneurons in the disynaptic VOR pathways. PVPs receive monosynaptic inputs from the vestibular nerve (Scudder & Fuchs 1992) and project monosynaptically to extraocular motoneurons (Scudder & Fuchs 1992, McCrea et al 1987). Flocculus Target Neurons (FTNs), so-named because they are the targets of monosynaptic inhibition from the F/VPF (Lisberger & Pavelko 1988, Lisberger et al 1994b), also are in the vestibular nuclei and also receive monosynaptic inputs from the vestibular nerve (Broussard & Lisberger 1992). Available data imply that at least some FTNs project directly to ocular motoneurons (Scudder & Fuchs 1992, Lisberger et al 1994b). Horizontal-Gaze Velocity Purkinje cells (HGVPs) are so-named because they discharge in relation to horizontal-gaze velocity during interactions of visual and vestibular stimuli (Miles et al 1980b, Lisberger & Fuchs 1978a) and are Purkinje cells in the flocculus and ventral paraflocculus. HGVPs project directly to the vestibular nucleus (Langer et al 1985a), where they monosynaptically inhibit interneurons in the brainstem VOR pathways (e.g. Baker et al 1972, Highstein 1973), almost certainly including FTNs.

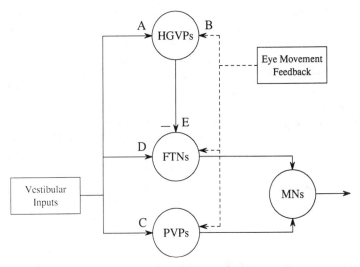

Figure 2 Schematic diagram showing the key neurons that participate in learning and memory in the VOR and the flow of signals within the neural network. The large circles represent horizontal-gaze velocity Purkinje cells (*HGVPs*) in the flocculus/ventral paraflocculus, the flocculus target neurons (*FTNs*) and position-vestibular-pause cells (*PVPs*) in the vestibular nucleus, and extraocular motoneurons (*MNs*). The letters (*A, B, C, D, E*) provide a vocabulary for discussing possible sites of learning and memory. Dashed lines refer to feedback connections, and solid lines show feed-forward connections. Note that HGVPs have an inhibitory influence on FTNs, as shown by the minus sign.

Because most of the electrophysiological recordings in awake, behaving animals have been made in rhesus monkeys, we discuss data primarily from this species. However, because the cell types and connections in the VOR pathways appear to be highly conserved across vertebrate animals (e.g. Dieringer 1986, du Lac & Lisberger 1992, Pastor et al 1994), we expect that much of what we describe applies to other species as well.

A major advance in localizing sites of memory in the VOR has come from the recognition that primary changes in neuronal activity due to local changes in synaptic transmission or cellular sensitivity must be distinguished from secondary changes that are simply transmitted from the primary site via changes in the activity of inputs to the secondary site. It is difficult to distinguish primary from secondary effects of learning because the extensive feedback in the VOR pathways makes it impossible to interpret the responses of neurons in the context of serial connections from the vestibular apparatus to the motoneurons. In the oculomotor system, feedback provides signals related to eye movement. These eye-movement signals are thought to be feedback of the motor command and to arise from neurons that drive eye movement, rather than from proprioceptors in the eye muscles or the orbital tissues (Keller & Robinson 1971, Lisberger & Fuchs 1978b). For example, PVPs, FTNs, and HGVPs all receive inputs related to eye movement, as evidenced by their responses during pursuit eye movements with the head stationary. Accordingly, during the VOR, the responses of neurons in the VOR pathways result from a combination of head- and eye-movement inputs, even for PVPs and FTNs, which receive monosynaptic inputs from vestibular primary afferents.

With the feedback organization of the VOR pathways in mind, consider the problem of comparing a neuron's responses during the VOR before and after changes in the gain of the VOR. Suppose that none of the synaptic connections or intrinsic properties of the neuron become modified consequent to learning. Learning, by definition, causes a change in the eye movement evoked by a given head movement. The amplitude of the response will be changed in the afferents that provide eye-movement inputs to a neuron, and that change will be reflected in the neuron's response during the learned VOR, even though there have not been any cellular changes in the neuron or its afferent synapses. In PVPs, FTNs, and HGVPs, the component of neuronal firing rate that is due to eye-movement feedback can be dissociated from that component due to head-movement inputs by taking advantage of the fact that different oculomotor behaviors produce different combinations of head-movement signals and eye-velocity feedback signals. During smooth pursuit, the eyes move, but the head does not. During the VOR, both the eyes and the head move. During a paradigm called "cancellation of the VOR," the head moves, but the eyes are stationary in the orbit because the subject tracks a target that moves exactly with the head. Cancellation of the VOR provides a means to behaviorally eliminate the eye-movement feedback to a neuron. Therefore, the components of neuronal firing caused by eye-movement feedback and vestibular inputs can be distinguished by comparing neuronal responses during pursuit (eye movement only) and VOR cancellation (head movement only) to those during the VOR (eye and head movement). This method of analysis is based on the assumption that neuronal firing rates during the VOR can be predicted by the sum of firing during pursuit and during cancellation of the VOR, an assumption that has been verified for HGVPs (Lisberger & Fuchs 1978a, Lisberger et al 1994a) and FTNs (Lisberger et al 1994c).

Neural Correlates of Memory in the VOR Pathways

The responses of FTNs, PVPs, and HGVPs during the VOR are each modified after changes in the gain of the VOR. In describing these changes in neuronal responses, we address the following issues: Are the changes in the responses measured during the VOR in the correct direction to account for the changes in the learned behavior? Is the latency from head movement to neuronal response short enough to account for the short-latency component of the learned response? Do the changes associated with learning simply correlate with the altered eye movement or do they reflect alterations in the sensitivity to head-movement inputs? Because of the technical difficulties associated with recording from individual neurons during the several hours required to get a large change in the gain of the VOR, most studies of learning-related changes in neuronal responses have compared populations of neurons recorded while the gain of the VOR is low, normal, and high.

We define the "correct" direction of changes in neuronal responses according to the connections of each class of neurons. FTNs and PVPs are in the direct pathways that drive the VOR and during the normal VOR show responses that will drive the associated eye movement. Therefore, increases in the amplitude of the responses of FTNs or PVPs would be in the correct direction to drive increases in the gain of the VOR. Decreases in the amplitude of the responses of FTNs or PVPs would be in the correct direction to reduce the gain of the VOR. A change in the sign of the response would cause FTNs or PVPs to counteract rather than drive the VOR and would be in the correct direction to reduce the gain of the VOR. Because HGVPs inhibit their target neurons in the brainstem (presumably FTNs), different logic must be used to define the correct direction of changes in the responses of HGVPs. Consider a head turn that causes an increase in the firing of the vestibular inputs to FTNs. If the same stimulus normally causes little or no response in HGVPs, then FTNs will increase their firing and the vestibular signal will be forwarded to ocular motoneurons to drive the VOR. If, however, the firing of HGVPs increases at the same time as the firing in the vestibular inputs to FTNs increases, then the responses of FTNs will be reduced by inhibition from the HGVPs, and the gain of the VOR will be lower than normal. If the firing of HGVPs decreases at the same time as the firing in the vestibular inputs to FTNs increases, then the responses of FTNs will be amplified, and the gain of the VOR will be higher than normal. Thus, increases in the size of the vestibular responses of HGVPs would be in the correct direction to cause decreases in the gain of the VOR, and vice versa.

FLOCCULUS TARGET NEURONS   Changes in the gain of the VOR cause dramatic changes in the responses of FTNs during the VOR (Lisberger & Pavelko 1988, Lisberger et al 1994c). When the gain of the VOR is normal, FTNs exhibit large responses that consist of increases in their firing rate during the VOR evoked by head turns toward the side of recording (henceforth called *ipsiversive* head turns). When the gain of the VOR is high, the responses are larger but maintain the same direction selectivity so that firing rate still increases during ipsiversive head turns. When the gain of the VOR is low, the sign of the responses of FTNs reverses so that firing rate decreases during the same ipsiversive head rotation. As outlined above, the changes in FTN responses recorded by Lisberger et al (1994c) are in the correct direction to mediate the associated changes in the gain of the VOR. Recordings from individual FTNs during brief periods of learning (Partsalis et al 1993) have suggested that individual FTNs undergo changes similar to those documented by comparing populations of FTNs recorded when the gain of the VOR was low, normal, and high (Lisberger et al 1994c).

The latencies of FTN responses during the VOR make them good candidates to mediate the earliest modified component of the VOR. Table 1 provides the logic on which we base this suggestion. During ramps of head velocity (cf Figure 1), FTNs responded with a median latency of 11 ms (Lisberger et al 1994c). Since at least some FTNs project monosynaptically to ocular motoneurons, we assume that a response in FTNs will influence motoneuronal firing within 1 ms. In turn, motoneurons respond an average of 7 ms before the onset of the eye movements evoked by ramps of head velocity (Lisberger et al 1994c). Thus, signals transmitted from the vestibular apparatus through FTNs to motoneurons will introduce a latency of about 19 ms ($11 + 1 + 7$ ms) between the onset of head motion and the onset of eye motion. Because the latency of the VOR is 14 ms for ramps of head velocity, the pathway through FTNs has a latency that is too long to drive the initial unmodified eye velocity of the VOR. However, the median latency of the pathway through FTNs agrees well with the latency of 19 ms measured for the modified component of the VOR induced by ramps of head velocity.

Comparison of the firing rates of FTNs when VOR gain was high and when it was low indicates that the effects of motor learning on the responses of FTNs are too large to be explained by eye-movement feedback signals alone (Lisberger et al 1994c). Moreover, recordings from FTNs during cancellation of the

Table 1 Latencies from head turn to eye movement during ramps of head velocity[a]

	PVPs	FTNs	HGVPs
Head turn to neuron[b]	7	11	23
Neuron to motoneuron	1	1	2
Motoneuron to eye movement	7	7	7
Head turn to eye movement[c]	15 ms	19 ms	32 ms

[a] Responses to ramps of head velocity ($600°/s^2$). Columns show the latencies for pathways through each of three different interneurons. The numbers in the first three rows add up to the total latency for each pathway, given in the bottom row.
[b] This row gives the median latencies for the full sample of each class of neuron.
[c] These latencies should be compared with the latency of 14 ms for the unmodified, earliest component of the VOR and 19 ms for the modified component of the VOR, using the same ramp of head velocity.

VOR reveal that motor learning causes a change in the sensitivity to head-velocity inputs that is in the correct direction to account for motor learning (Lisberger et al 1994c). From these data we conclude that changes in the responses of FTNs are caused partly by a primary change in the transmission of vestibular inputs to FTNs.

POSITION-VESTIBULAR-PAUSE CELLS   Changes in the gain of the VOR cause small changes in the responses of PVPs under some behavioral conditions. These changes in the responses of PVPs are in the correct direction to support the change in the gain of the VOR. During contraversive head turns (away from the side of recording), PVPs have larger responses when the gain of the VOR is high than when it is low (Lisberger et al 1994c). During ipsiversive head turns, however, the amplitude of the responses of PVPs does not depend on the gain of the VOR.

The latency of their responses make PVPs good candidates to subserve the earliest unmodified component of the VOR. During the VOR induced by the ramps of head velocity used in our experiments, PVPs respond with a median latency of 7 ms. Since they project directly to extraocular motoneurons, PVPs, like FTNs, will influence eye movements after a latency of about 8 ms (1 ms to influence motoneuronal firing, plus 7 ms from motoneuronal firing to the onset of eye movement). Thus, signals transmitted from the vestibular apparatus through PVPs to extraocular motoneurons will introduce a latency of about 15 ms from head movement to eye movement (Table 1); this latency is in good agreement with the 14 ms latency of the VOR evoked by ramps of head velocity.

Two lines of evidence suggest that the learning-associated changes in the responses of PVPs are secondary to feedback of eye-velocity signals, rather than a reflection of primary changes in the strength of the head-velocity inputs to PVPs (Lisberger et al 1994c). 1. The responses of PVPs during cancellation of the VOR are not modified in association with changes in the gain of the VOR. Because the responses during cancellation of the VOR provide a direct estimate of the strength of the vestibular input to these cells, we would have expected a change in the strength of vestibular transmission to PVPs to cause a measurable change in responses under this condition. 2. The changes in the responses of the PVPs can be explained simply by taking into account their innate sensitivity to eye velocity, measured during pursuit with the head stationary, and the learning-associated changes in the eye velocity during the VOR. We conclude that PVPs contribute to the expression of memory in the VOR but only by virtue of their inputs from eye-velocity feedback. The data are not consistent with the conclusion that there is a primary site of memory in the vestibular inputs to PVPs.

HORIZONTAL-GAZE VELOCITY PURKINJE CELLS HGVPs express marked changes in their responses after learning has induced increases or decreases in the gain of the VOR. When the gain of the VOR is normal, HGVPs show little or no response during the VOR (Lisberger & Fuchs 1974). After the gain of the VOR has become low or high, HGVPs exhibit pronounced responses during the VOR. During a low-gain VOR, HGVPs and the vestibular inputs to FTNs show increased firing at the same time. During a high-gain VOR, the firing of HGVPs decreases at the same time as the firing in the vestibular inputs increases. According to the logic outlined at the start of this section, the changes in the responses of HGVPs, if measured during the VOR, are in the correct direction to cause changes in the gain of the VOR. These data are not controversial: Similar results have been obtained by multiple investigators in monkeys (Miles et al 1980a, Watanabe 1984, Lisberger & Pavelko 1988, Lisberger et al 1994a) and rabbits (Dufosse et al 1978). In addition, recordings that followed individual Purkinje cells during brief periods of learning (Watanabe 1984) have revealed the same neural expression of memory found by comparing large populations of HGVPs recorded at different gains of the VOR (Lisberger et al 1994a).

The modified responses of most HGVPs occur with a latency that is too long to contribute to the earliest modified component of the VOR. During the VOR induced by ramps of head velocity, HGVPs respond with a median latency of 23 ms. Stimulation of the F/VPF produces eye movements after latencies of at least 9 ms, consistent with a latency of 2 ms for the firing of HGVPs to affect the firing of motoneurons. The total latency from head to eye movement for signals transmitted from the vestibular apparatus through HGVPs to extraocular motoneurons is 32 ms (Table 1). We conclude that most HGVPs respond too late to contribute to the earliest expression of memory in the VOR[1] but that changes in the responses of HGVPs during the VOR do contribute to later components of the modified VOR.

Analysis of the firing of HGVPs during cancellation of the VOR has demonstrated changes in the amplitude of responses to vestibular inputs that are in the wrong direction to account for learning-induced changes in VOR gain or in the firing of HGVPs during the VOR (Miles et al 1980a, Lisberger et al 1994a). If the sole site of the cellular changes associated with VOR memory

---

[1]In the case of HGVPs, the distribution of latencies is quite broad. The median latency is not a good estimate of the responses of the full population and it may be important to consider the possibility that individual cells with different latencies subserve different functions. A small fraction (5%) of the HGVPs recorded during the VOR induced by ramps of head velocity (Lisberger et al 1994a) responded with latencies of 10 ms or less and therefore had the potential to contribute to the earliest part of the modified VOR. In the case of FTNs and PVPs, however, the distribution of latencies was much narrower and the median latency provided a good estimate of the latency of the full sample (Lisberger et al 1994c).

were in the vestibular inputs to HGVPs, then we would have expected the responses of HGVPs during cancellation of the VOR to become larger after the gain of the VOR had been lowered and smaller after the gain of the VOR had been raised. The opposite occurs. The same "wrong-way" results were obtained when the sensitivity to vestibular inputs was estimated by subtracting the eye-velocity component of the firing rate from the firing of HGVPs during the VOR (Lisberger et al 1994a).

Controversy has surrounded interpretation of the recordings from HGVPs because of the paradox that HGVPs show changes in the correct direction to support the modified VOR if measured during the VOR (Miles et al 1980a, Watanabe 1984, Lisberger et al 1994a) and in the wrong direction to cause the modified VOR if measured during cancellation of the VOR (Miles et al 1980a, Lisberger et al 1994a). Computer modeling (Lisberger & Sejnowski 1992, Lisberger 1994) has now demonstrated that that the paradoxical data are predicted if (a) there are sites of memory in the vestibular inputs to both FTNs and HGVPs, and (b) the memory in the inputs to HGVPs consists of changes in the time course and strength of vestibular transmission. Explanation of the paradox hinges on the assumption that changes in the responses of neurons that transmit eye-velocity feedback to HGVPs are responsible for some of the changes in the firing of HGVPs measured during the VOR. New data have also excluded a number of previous explanations for this paradox. The possibility of species differences (Lisberger 1982, Kawato & Gomi 1992) has been ruled out by the similarity of the data from monkeys (Watanabe 1984) and rabbits (Dufosse et al 1978). The possibility that eye-velocity feedback is important in monkeys but not in rabbits is negated by the finding of pronounced eye-movement responses in Purkinje cells in the flocculus of rabbits (Leonard & Simpson 1985, Nagao 1991). The possibility that different studies were recorded from different populations of Purkinje cells is negated by the finding that the paradox exists in almost all individual HGVPs (Lisberger et al 1994a).

A current controversy stems from the question of whether it is correct to judge the function of the F/VPF from the responses of HGVPs. Some HGVPs have been recorded in the flocculus (Lisberger et al 1994a), but most of the HGVPs in the literature have been recorded in the ventral paraflocculus. There is now evidence of some anatomical differences in the source of the visual inputs to the flocculus vs the ventral paraflocculus (Gerrits & Voogd 1989). New experiments are required to determine whether the flocculus performs a function that is different from that of the HGVPs. This issue could be addressed by extensively recording from the flocculus and/or by determining the effects on learning of lesions that remove the flocculus but spare the ventral paraflocculus. Nagao (1992) has claimed to find differences in the responses of Purkinje cells recorded in the flocculus and the ventral paraflocculus in monkeys. However, his conclusions are compromised by a number of technical problems (for a more extensive discussion, see Lisberger et al 1994a). Even if future recordings reveal that non-HGVPs in the F/VPF also contribute to memory in the VOR, the demonstration of learning-related changes in the responses of HGVPs during the VOR in the dark (Miles et al 1980a, Lisberger et al 1994a) requires that HGVPs be included in any theory of learning and memory in the VOR.

Further Constraints on the Sites of Memory in the VOR

The evidence that we have presented so far implicates the vestibular inputs to FTNs as one candidate locus of memory in the VOR. FTNs receive information about head movement from multiple sources: monosynaptic input from afferents traveling in the ipsilateral vestibular nerve, polysynaptic inputs from the contralateral vestibular nerve (Broussard & Lisberger 1992), and inhibition from Purkinje cells in the F/VPF (Lisberger et al 1994b). In addition, FTNs probably receive disynaptic inputs from the ipsilateral vestibular nerve (Broussard & Lisberger 1992). In theory, changes in the amplitude of any of these input signals or in the strength of synapses from any or all of these inputs, or postsynaptic changes in the intrinsic properties of FTNs could produce the changes observed in FTN response properties. However, the fact that the VOR circuitry also mediates visual-tracking eye movements of pursuit and the OKR places significant constraints on which of these potential candidates could mediate learning-related changes in the responses of FTNs during the VOR.

Because pursuit is driven, at least in part, by inputs to FTNs from the F/VPF, we would expect that changes in the strength of transmission from HGVPs onto FTNs would result in parallel changes in the pursuit eye movements. Such changes do not occur (Lisberger 1994). From this finding, we conclude that the synapse between HGVPs and FTNs is not a site of memory in the VOR. This conclusion is supported by the finding that changes in the gain of the VOR affect neither the magnitude nor the time course the eye movements evoked by electrical stimulation of the F/VPF (Lisberger 1994).

In monkeys, changes in the gain of the VOR cause parallel changes in a part of the OKR that has a very slow time course, building up over 5 to 15 s and lasting as long as 1 min in the dark after the moving stimulus is turned off. This "long time constant" component of the OKR becomes smaller than normal when the gain of the VOR is low and larger than normal when the gain of the VOR is high (Lisberger et al 1981). Because vestibular primary afferents do not carry signals related to the OKR (Buttner & Waespe 1981), changes in the synapses between vestibular afferents and FTNs would not affect the OKR. Therefore, these synapses cannot be the sole site of memory in the brainstem. However, secondary vestibular neurons do carry signals related to the long time constant component of OKR (Henn et al 1974), and changes in transmission from secondary neurons to FTNs would affect the

OKR. For the remainder of the paper, we use the term *vestibular inputs to FTNs* to refer to the full set of possible brainstem vestibular inputs and not just to the inputs from primary afferents.

Use of Computer Models to Reveal Possible Sites of Memory

In neural circuits that contain feedback (as do most circuits in the brain), the expression of memory in the responses of cells is determined partly by the site and nature of the cellular mechanism of memory and partly by the architecture of the neural network in which the memory mechanism is embedded. Positive feedback, such as exists in the oculomotor system, can act as an amplifier to convert small cellular changes into large neural and behavioral expressions of memory (Miles et al 1980a,b) or as an integrator to convert transient changes in input signals into sustained changes in behavioral output (Lisberger & Sejnowski 1992). The possible effects of positive feedback on the operation of a neural network invalidate reasoning that is based on purely feedforward neural connections and demand the quantitative analysis that is provided by dynamic, recurrent models to form hypotheses about the site of memory. In addition, the relationship between the dynamics and gain of the VOR is an important constraint on the sites and mechanisms of memory, but it is too complex to be evaluated without quantitative modeling.

Lisberger & Sejnowski (1992) and Lisberger (1994) used computer simulations to search for a combination of sites of memory that could account for the gain and dynamics of the VOR in the dark; the gain and dynamics of pursuit with the head stationary; and the responses of FTNs, PVPs, and HGVPs during the VOR, cancellation of the VOR, and pursuit when the gain of the VOR was high, normal, or low. Figure 2 summarizes the possible sites of memory that were tested by the modeling and provides a vocabulary for expressing the results of the computer simulations. The models were able to reproduce available behavioral and neural data only if there were sites of memory in the brainstem vestibular inputs to FTNs (site D) and in the vestibular inputs to HGVPs (site A). Part of the change in the gain of the VOR was accomplished by varying the strength of transmission in parallel at these sites. The amount of change at these sites in the model was selected to reproduce the measured effect of changes in the gain of the VOR on the responses of HGVPs during cancellation of the VOR. Changes in the strength of vestibular inputs to FTNs were in the correct direction to cause changes in the gain of the VOR. Changes in the strength of vestibular inputs to HGVPs were in the wrong direction to cause changes in the gain of the VOR but were in the right direction to maintain stability in the VOR.

Further changes in the gain of the VOR and the required change in dynamics were attained by altering the time course but not the strength of the vestibular inputs to HGVPs. When the input was made more transient, either by adding some overshoot to the vestibular input (Lisberger 1994) or by shortening the time constant of filtering at that site (Lisberger & Sejnowski 1992), the gain of the VOR was reduced and the response to ramps of head velocity became more transient. In principle, a change in the time course of a neural input could result from changes in cellular or circuit properties. A faster time course could be obtained by (a) differentially changing the weights of inputs with different time courses (e.g. Lisberger 1994), (b) using changes in strength of transmission to alter the temporal filtering properties of a local neural network (e.g. Fujita 1982, Lisberger & Sejnowski 1992), or (c) altering a cellular mechanism in a way that changes the temporal filtering properties of a synapse or the spike generator in the postsynaptic cell.

Within the context of the connections shown in Figure 2, models were created that could reproduce available data only if they implemented sites of memory at both A and D. Changes in the strength of transmission at sites A or D alone caused the model to exhibit unstable runaway behavior, because of the eye-velocity feedback pathway through HGVPs. Parallel changes in the strength of transmission at sites A and D maintained stability but did not produce either changes in the dynamics of the VOR or large enough changes in the gain of the VOR without requiring changes in the strength of transmission at site D much larger than demonstrated in recordings from HGVPs during cancellation of the VOR. Changes in the strength of transmission within the eye-velocity feedback pathway (site B) caused changes in the gain and dynamics of pursuit eye movements, contradicting the lack of effect of changes in the gain of the VOR on pursuit (Lisberger 1994). Altering the time course of the vestibular inputs to FTNs (site D) caused the dynamics of the VOR to vary in the wrong direction: Decreases in the gain of the VOR were associated with decreased rather than increased overshoot during ramps of head velocity. Altering the strength or time course of the vestibular inputs to PVPs (site C) incorrectly predicted that changes in the gain of the VOR should be associated with changes in the responses of PVPs during cancellation of the VOR. As discussed above, learning-related changes in the inhibitory inputs to FTNs from HGVPs (site E) can be ruled out by the failure of changes in the gain of the VOR to cause parallel changes in pursuit eye movements or in the eye movements evoked by electrical stimulation of the F/VPF.

Other attempts have been made to model the site or sites of memory in the VOR, but none have been successful at reproducing the internal signals recorded from neurons at different gains of the VOR, and some have ignored important constraints imposed by well-documented parts of the essential neural network. One class of models (Lisberger et al 1983, Quinn et al 1992a, Minor & Goldberg 1991) was based on a model proposed by Skavenski & Robinson (1973). Each of these models consists of two or more parallel VOR pathways with different temporal-filtering properties, and each accomplishes changes in the gain and temporal dynamics of the VOR by adjusting the weights differentially in the pathways. Lisberger et al (1983) used this approach to reproduce data showing that learning in the VOR is frequency-selective when the vestibular stimulus for adaptation is a sine wave at a single frequency. Minor & Goldberg (1991) demonstrated that it was possible to reproduce the general temporal trajectory of eye velocity evoked by ramps of head velocity when the gain of the VOR was low, normal, and high. Quinn et al (1992a) reproduced changes in the gain and phase of the VOR during sinusoidal rotation of the visual scene and the animal at single frequencies. Each of these models demonstrates the feasibility of accomplishing some aspects of adaptive changes in the VOR by distributing sites of memory across parallel neural pathways that have different temporal dynamics. Quinn et al (1992b) demonstrated that it is feasible to implement this class of model as a neural network that has an architecture derived from the basic organization of the brainstem VOR pathways. However, none of these models were helpful in localizing the sites of memory in the brain because they did not attempt to emulate the flow of neural signals in the biological VOR pathways, and they did not contain nodes that represented PVPs, FTNs, and HGVPs.

A second class of models was based on the original proposal by Ito (1972) that the sole site of memory in the VOR is in a pathway from the vestibular labyrinth through the flocculus to the vestibular nucleus. Fujita (1982) demonstrated that this model can produce realistic learning and memory in the eye movements of the VOR, if eye-velocity positive feedback to the F/VPF plays little or no role in the operation of the system. Gomi & Kawato (1992) used a model that included the possibility of eye-velocity positive feedback through the F/VPF and employed an automatic-optimization algorithm to adjust the weights of transmission of vestibular and eye-movement signals through the node that represented Purkinje cells in the flocculus. The model selected a large weight for the eye-velocity input to the flocculus. Without systematically exploring why eye-velocity inputs to the flocculus were strong in their model, Gomi & Kawato (1992) concluded that eye-velocity positive feedback was not important for the operation of the biological system. Because they were based on the preconception that the sole site of memory for the VOR is in the flocculus, these models have not elucidated the sites of memory in the VOR pathways in the brain.

The model of Lisberger (1994) is one of a third class of models that attempts to represent the known architecture of the biological neural network for the VOR. One of the deficiencies of Lisberger's model is that it lumps the two sides of the brain together and fails to represent the commissural connections between the two vestibular nuclei. Galiana (1986) used a model that included both sides of the brain and explicitly represented many of the neurons that are known to participate in the VOR. Her model demonstrated that the commissural connections between the vestibular nuclei would be feasible sites of memory in the VOR. Available data are compatible with Galiana's suggestion. Lisberger et al (1994c) have provided evidence that one site of memory is in the vestibular inputs to FTNs, and Broussard & Lisberger (1992) have shown that one potential vestibular input to FTNs is an excitatory connection that is transmitted from the contralateral vestibular nucleus. However, there are two problems with Galiana's (1986) model: 1. The model fails to replicate the equal amplitude eye-velocity and head-velocity inputs to HGVPs in monkeys (Lisberger & Fuchs 1978a, Miles et al 1980b), and 2. Galiana's (1986) contention that the gain-of-pursuit eye movement should not change even if the site of memory in the VOR is in feedback loops overlooks the temporal dynamics of pursuit evoked by target motion at constant speed. Lisberger (1994) has shown that the temporal dynamics of eye velocity during pursuit are affected by changing the gain of positive feedback in the model, not by changing the gain of the VOR in monkeys.

A Unifying Hypothesis for the Sites of Memory in the VOR

We suggest that there are multiple sites of memory in the VOR and that each site performs a different function. A site of memory in the brainstem appears to drive the earliest modified component of the VOR. Because Lisberger & Miles (1980) looked for and did not find changes in the vestibular responses of non-FTN vestibular neurons in the vestibular nucleus, the primary site of memory in the brainstem is likely to be in synapses onto FTNs or in the intrinsic properties of the FTNs themselves.[2] A second site of memory, in the vestibular inputs to HGVPs, may be in the F/VPF. The memory in the vestibular inputs to HGVPs has been modeled so that separate changes in the strength and time course of vestibular inputs have different functions. Although the change in the strength of vestibular inputs is in the wrong direction to cause the associated changes in the gain of the VOR, it is in the correct direction to maintain stability in the system in the face of changes at the brainstem site of memory. The change in the time course, once amplified and integrated by the rest of the VOR pathways, is converted into a signal that would cause part of the associated changes in the gain of the VOR.

POSSIBLE MECHANISMS OF LEARNING IN THE VOR

Conclusions about the mechanisms of learning require an integrated understanding not only of the details of mechanisms of cellular plasticity operation,

---

[2]If the site of memory is in the intrinsic properties of FTNs, then it must be localized so that changes in the gain of the VOR do not affect the responses of FTNs to inhibitory inputs from F/VPF. Otherwise, changing the gain of the VOR would affect the eye movements evoked by electrical stimulation in the F/VPF as well as the gain of pursuit, which it does not (Lisberger 1994).

but also of the operation of those mechanisms in the context of the presynaptic and postsynaptic activity at a putative site of memory. In particular, the mere existence of a mechanism for cellular plasticity does not constitute evidence that the mechanism participates in a specific form of behavioral learning. In this section of our review, we evaluate how known mechanisms of cellular plasticity might work in the context of the presynaptic and postsynaptic activity at the putative sites of memory in the VOR. For learning in the vestibular inputs to HGVPs, we evaluate the available presynaptic signals in the context of cerebellar LTD, which has been proposed as a specific cellular mechanism for learning in the VOR (Ito 1989). For learning in the vestibular inputs to FTNs, where mechanisms of cellular plasticity have not yet been described, we evaluate the available presynaptic and postsynaptic activity in the context of current ideas about mechanisms of cellular plasticity in the hippocampus and cerebral cortex. This part of our paper is, by its nature, speculative, and we conclude that available data are far too fragmentary to allow any firm conclusions about mechanisms of learning in the VOR. We present this section as a framework for designing experiments that will examine the mechanisms of learning in the realistic conditions imposed by the operation of the neural network for the VOR in awake, behaving animals.

Anatomical Structures Involved in Learning in the VOR

Ablation studies have identified anatomical structures that may be involved in learning in the VOR. Removal of the whole cerebellum or bilateral ablation of the entire flocculus and ventral paraflocculus prevented learning in the VOR but had relatively little effect on the normal VOR (e.g. Robinson 1976, Lisberger et al 1984, Barmack & Pettorossi 1985, Nagao 1983). In contrast, ablation of the uvula and nodulus in the midline vestibulo-cerebellum had no effect on learning or memory in the VOR (Cohen et al 1992). One possibility is that ablation of the F/VPF abolishes learning in the VOR because it removes the site(s) of memory. However, this is not consistent with evidence that at least one site of memory for the VOR is in the brainstem. As a resolution to this problem, Miles & Lisberger (1981) suggested that output signals from the F/VPF were essential as "teachers" to guide learning, independent of the role of the F/VPF as a site of memory for the VOR.

A recent experiment by Luebke & Robinson (1994) provides strong support for the idea that the output from the F/VPF might guide learning. They first demonstrated that stimulation of the inferior olive at 7 Hz causes Purkinje cells in the cat's flocculus to cease firing simple spikes, thereby eliminating all functional output from the flocculus. They then used this paradigm to reversibly inactivate the flocculus in cats that had been preadapted to have increased or decreased VOR gains. Luebke & Robinson (1992) found that the memory of the adapted VOR was retained during inactivation of the F/VPF and that rotation with normal viewing did not cause the VOR to relearn a gain of 1.0. Thus, inactivation of the flocculus had no effect on memory, but it prevented learning in the VOR. Surgical lesions of the inferior olive also prevent learning (Barmack & Simpson 1980, Tempia et al 1991). This result could reflect a direct contribution of climbing-fiber inputs to learning in the cerebellar cortex, but it might also reflect disruption of a direct contribution of climbing fibers to learning in the brainstem or alteration of the normal operation of Purkinje cells.

Behavioral Rules for Learning in the VOR

At the behavioral level, an adequate condition for learning in the VOR is the association of visual and vestibular inputs. The gain of the VOR can be modified if visual experience is altered so that the directions of image motion and head motion are correlated consistently during head turns. If a subject wears magnifying spectacles, for example, then the normal VOR will be too small, images will move in the opposite direction from each head turn, and the gain of the VOR will increase. Likewise, if a subject wears miniaturizing glasses, then the normal VOR is too large, images will move in the same direction as the head turn, and the gain of the VOR will decrease. These behavioral considerations suggest a head-plus-image-motion learning rule for the VOR:

> If image motion is in the same direction as head turns, then the gain of the VOR should decrease.
>
> If image motion is in the opposite direction from head motion, then the gain of the VOR should increase.

Possible Rules for Learning in the Brainstem

At the neural level, the visual and vestibular sensory inputs that guide learning must be represented in the discharge of neurons that converge on the sites of memory. As we have already mentioned, FTNs receive vestibular inputs from multiple sources. Because of the clear necessity of the F/VPF for learning, we focus on the potential visual error signals from the HGVPs, even though visual inputs reach the vestibular nucleus from a variety of sources. Two previous papers (Miles & Lisberger 1981, Lisberger 1988) have proposed that correlated changes in the activity of HGVPs and vestibular inputs provide error signals that guide cellular mechanisms of learning at FTNs in the vestibular nuclei. To analyze this hypothesis, we consider the neural error signals available as inputs to FTNs from HGVPs and from the vestibular system when the gain of the VOR is 1.0 and the monkey is subjected to head turns under altered visual conditions that, if prolonged, would cause learning. The left-most column in Table 2 describes the visual conditions used to cause learning as $\times N$, where N is the ideal gain of the VOR for that visual condition. From left to right, the next four columns represent the direction of the change in the gain of the VOR required to eliminate image motion during head turns, and the direction of the change of firing in vestibular inputs, HGVPs, and FTNs for an ipsiversive head turn during each of the conditions used to cause learning.[3] The right-most column represents the absolute firing rate of FTNs during ipsiversive head turns for each adapting condition. During ipsiversive head turns, the vestibular inputs to FTNs will show increases in firing rate. Under conditions that call for a decrease in the gain of the VOR (×0, ×0.4, ×0.7), ipsiversive head turns are associated with increases in the firing rates of HGVPs. Under conditions that call for an increase in the gain of the VOR (×2), ipsiversive head turns are associated with decreases in the firing rate of HGVPs. Under normal (×1) viewing conditions, an ipsiversive head turn does not cause any change in the simple-spike firing rate of HGVPs, and no learning occurs. Thus, the neural signals available as inputs to FTNs suggest a neural learning rule for the VOR:

If changes in the firing rate of vestibular and HGVP inputs to FTNs are in the same direction, then the gain of the VOR should decrease.
If changes in the firing rate of vestibular and HGVP inputs to FTNs are in opposite directions, then the gain of the VOR should increase.

Table 2 Evaluation of possible rules for learning in the vestibular inputs to FTNs[a]

Behavioral condition[b]	Required change in VOR gain	Modulation of vestibular inputs[c]	Modulation of HGVP firing	Modulation of FTN firing	Absolute firing rate of FTNs[d]
×0	−	+ +	+ + +	−	***
×0.4	−	+ +	+ +	0	****
×0.7	−	+ +	+	+	*****
×1	0	+ +	0	+ +	******
×2	+	+ +	− − −	+ + + + +	*********
Pursuit	0	0	+ + +	− − −	*
Baseline	0	0	0	0	****

[a] The table represents responses of vestibular afferents, HGVPs, and FTNs to ipsiversive head turns under various adapting conditions when the gain of the VOR is 1.0. The entries were inferred from average responses measured during sinusoidal vestibular rotation under behavioral conditions that simulated ×0, ×1, and ×2 adapting conditions or during pursuit with the head stationary (Lisberger & Fuchs 1978a, Lisberger et al 1994b).

[b] Behavioral conditions are indicated as ×N, where N is the gain of the VOR required to eliminate image motion during head turns. Also shown are the responses during ipsiversive pursuit eye movements and the baseline firing rate in the resting condition with the head stationary and eyes stationary at straight-ahead gaze.

[c] Modulation of firing is represented by "−" signs for a decrease, "+" signs for an increase, and zero for no change in firing rate from baseline.

[d] The number of asterisks indicates the absolute firing rate of the FTNs.

[3] Because HGVPs, FTNs, and PVPs are all spontaneously active, signals related to the sensory stimuli are encoded in changes in the firing rate from the resting rate.

CONSTRAINTS ON CELLULAR MECHANISMS OF LEARNING AT FTNs   At least two classes of mechanisms of cellular plasticity would allow correlated changes in firing rate of the vestibular and cerebellar inputs to FTNs to guide changes in the strength of vestibular inputs to FTNs. 1. Purkinje cell terminals could release modulatory substances (e.g. Chan-Palay et al 1982) that would interact directly in an activity-dependent way with the terminals of neurons that provide vestibular inputs to FTNs to cause synaptic potentiation or depression. 2. Because HGVPs directly inhibit the FTNs, activity in HGVPs could guide learning through its effects on the level of activity in the postsynaptic neurons, the FTNs. Following the examples of activity-dependent plasticity now known at numerous sites in the brain, modification of the synapses between vestibular inputs and FTNs might depend on the relationship between activity in the presynaptic vestibular axons and some aspect of activity in FTNs.

A mechanism based on comparison of the absolute firing rate of FTNs with that of vestibular inputs would provide consistent guidance for learning in the VOR. Consideration of Table 2 reveals that the gain of the VOR increases when the firing rates are high both in FTNs and in their vestibular inputs. The gain of the VOR decreases when firing rates are low in FTNs and high in their vestibular inputs. In contrast, any variable correlated with the direction of modulation of the firing of FTNs cannot control learning because the direction of modulation of FTN firing is not consistently related to the direction of change in the gain of the VOR. For example, ipsiversive head turns under $\times 2$ and $\times 0.7$ viewing conditions would be associated with an increase in both the postsynaptic activity of FTNs and the presynaptic activity of vestibular inputs, but the changes in the gain of the VOR associated with these conditions are in opposite directions. Furthermore, a comparison of the changes in the firing rate of the FTNs during $\times 0$, $\times 0.4$, and $\times 0.7$ viewing conditions shows that a decrease in the gain of the VOR can occur under conditions associated with either a decrease, an increase, or no change in the firing rate of the FTNs during an ipsiversive head turn. Because of the high level of spontaneous activity in FTNs and their vestibular inputs, a simple correlation of presynaptic and postsynaptic activity cannot provide a complete learning rule because the resting activity would cause transmission to become potentiated maximally, even in the absence of the requisite association of vestibular and cerebellar input signals. Instead, a neural mechanism that depends on activity in the FTNs and their vestibular inputs must involve set points or thresholds as reference points to control whether transmission is potentiated or depressed. In the case of FTNs, the set point for postsynaptic activity must be at a firing rate above the resting rate, corresponding to an absolute firing rate of +6 in Table 2. If the firing rate of the vestibular inputs to FTNs is above resting rate and the firing of FTNs is above the set point of +6, then the gain of the VOR increases. If the firing rate of the vestibular inputs to FTNs is above resting rate and the firing of FTNs is below the set point of +6, then the gain of the VOR decreases. If the firing rate of FTNs is at the set point of +6, then the gain of the VOR does not change. Several authors have proposed mechanisms that could be used to implement a set point to regulate the levels of synaptic depression and potentiation in neurons that are spontaneously active (reviewed by Artola & Singer 1993).

OUTPUT OF HGVPs AS AN ERROR SIGNAL TO GUIDE LEARNING?   The firing of HGVPs has a feature that may prove to be necessary for a neural error signal that guides learning in the VOR. The simple-spike firing of HGVPs contains useful information about the need to change the gain of the VOR whether or not the subject is using the OKR and/or pursuit eye movements to eliminate the image motion caused by a VOR that is too large or too small. The activity of HGVPs has separate components driven by visual motion and eye motion (Stone & Lisberger 1990a). Either ipsiversive image motion or ipsiversive eye motion alone is sufficient to increase the simple-spike firing rate of HGVPs. When a subject executes a head turn under conditions that cause learning in the VOR, image motion will accompany at least the first 100 ms of the head turn. If the direction of the image motion is ipsiversive, then the firing rate of HGVPs will increase. If the subject fails to initiate visual tracking, then the image motion will persist throughout the head turn, and the firing of HGVPs will remain high. If the subject does initiate visual tracking, then the image motion may disappear, but the ipsiversive smooth eye velocity initiated by visual tracking will cause the firing of HGVPs to remain high for the duration of the head turn. Even during sinusoidal oscillation at low frequencies, when image motion is eliminated almost completely by the visual-tracking system, the output of HGVPs continues to provide useful information about the direction of errors in the VOR. The consistent modulation of HGVPs would be one way to explain the finding that the VOR undergoes learning during sinusoidal oscillation at low frequencies, even if the visual stimulus is a small spot that is tracked almost perfectly (Lisberger et al 1984).

The firing of HGVPs also has a feature that may not be appropriate for an error signal that guides learning. This problem does not arise when the gain of the VOR is 1.0, because the simple-spike firing rate of HGVPs is unmodulated during head rotation in the dark or in normal visual conditions (×1). In some conditions, however, HGVPs provide an error signal even though the gain of the VOR is appropriate and visual inputs do not provide an adequate condition for learning. After the gain of the VOR has been adapted to be high or low, for example, the simple-spike firing rate of HGVPs is modulated consistently during the VOR in the dark (Miles et al 1980a, Watanabe 1984, Lisberger et al 1994a). The direction of the response of HGVPs is such that if simple-spike firing guides learning in the vestibular inputs to FTNs, then the combination of vestibular inputs and simple-spike firing would cause a VOR with a low gain to get still lower and a VOR with a high gain to get still higher. In practice, this situation could prevent forgetting by causing automatic reinforcement of short-term potentiation or depression in the vestibular inputs to FTNs. In principle, however, it contradicts our assumption that a useful error signal for guiding learning in the VOR should be present only when there is visual feedback and not during the VOR in the dark.

A POSSIBLE ROLE FOR CLIMBING FIBERS IN LEARNING IN THE BRAINSTEM? *Climbing fibers* and *mossy fibers* provide two different kinds of inputs to the cerebellum. Mossy fibers synapse on granule cells. The axons of granule cells ascend in the cerebellar cortex and form *parallel fibers,* which make excitatory connections onto Purkinje cells. Mossy fiber inputs to the cerebellum cause Purkinje cells to emit simple spikes, which fire at rates as high as 200 or 300 spikes/s (until now our discussion of the firing of HGVPs has concerned only the simple spikes). Climbing fibers make extensive synaptic contacts directly on Purkinje cells and cause them to emit *complex spikes,* which fire at low rates that seldom exceed 1 or 2 spikes/s. In many parts of the cerebellum, climbing fibers send collaterals to the regions of the deep cerebellar nucleus that are related to the Purkinje cells that are the primary targets of the climbing fibers.

Climbing-fiber collaterals to the FTNs provide a potential solution to the problem outlined in the previous section, i.e. that a cellular learning rule based on the simple-spike firing of HGVPs might guide inappropriate learning during head turns in the dark when the gain of the VOR is high or low. The complex spike activity of Purkinje cells in the F/VPF is driven by visual inputs related to the motion of small targets or large textures in primates (Stone & Lisberger 1990b) and to the motion of large textures in rabbits (Alley et al 1975, Graf et al 1988). Available anatomical evidence is consistent with the possibility that the climbing-fiber inputs to the F/VPF send collaterals to FTNs (Balaban et al 1981). Therefore, climbing fibers may transmit information about the presence and direction of image motion to the FTNs. Since the visual inputs provided by climbing fibers would not be modulated consistently during the VOR in the dark or in the absence of visual image motion, they could serve as an absolute indicator of the need for learning in the VOR. They could operate either as a primary error signal to guide learning or as a permissive influence that would enable a learning mechanism based on activity in FTNs, HGVPs, and the vestibular inputs to FTNs.

LTD as a Cellular Mechanism of Learning in the Cerebellar Cortex

Ito (1972, 1982) proposed the "flocculus hypothesis" of motor learning in the VOR, based on a model of associative learning in the cerebellum suggested by Brindley (1964) and developed more thoroughly by Marr (1969) and Albus (1971). In this class of models, the climbing fiber acts as a teacher that instructs the synapses from parallel fibers to Purkinje cells by a mechanism that depends on paired activity between these two inputs. According to the flocculus hypothesis, climbing-fiber activity related to contraversive image motion causes long-term depression (LTD) at the synapses from vestibular parallel fibers onto Purkinje cells. Under stimulation conditions that require an increase in the gain of the VOR, vestibular inputs to Purkinje cells originating from the ipsilateral vestibular labyrinth would be more active during complex spikes and would undergo LTD (Ito 1982). Under conditions that require decreases in the gain of the VOR, vestibular inputs originating from the contralateral labyrinth would be more active during complex spikes and would undergo LTD (Ito 1993).

LTD IN THE CEREBELLUM   Experiments in a variety of preparations have provided evidence that a cellular mechanism for LTD exists in the cerebellar cortex. Conjunctive stimulation of parallel- and climbing-fiber inputs to a Purkinje cell causes depression of transmission in the synapses from the parallel fibers to the Purkinje cell (see Ito 1989, Linden & Connor 1993).

We must assume that cerebellar LTD has a companion LTP and that the absence of conjunction between climbing- and parallel-fiber inputs potentiates a synapse, while conjunction depresses the synapse. If LTD existed without either LTP or, at least, decay of LTD, then the spontaneous activity of parallel fibers and climbing fibers would cause parallel-fiber synapses onto Purkinje cells to become terminally depressed. In cerebellar slices, LTP can be induced in parallel fiber–Purkinje cell synapses if the Purkinje cell is hyperpolarized and/or loaded with the Ca^{2+} chelator EGTA and the parallel fibers are activated (Sakurai 1987, Crepel & Jaillard 1991, Shibuki & Okada, 1992). We also adopt the suggestion by Ito (1993) that cerebellar LTD operates reciprocally on mossy-fiber inputs that respond to a given input but have opposite direction preferences. Thus, decreases in the amplitude of the responses of HGVPs to ipsiversive head motion could be mediated either by LTD of synapses from parallel fibers that show increased firing for ipsiversive head motion or by LTP of synapses from parallel fibers that show decreased firing for ipsiversive head motion. Likewise, increases in the amplitude of the responses of HGVPs to ipsiversive head motion could be mediated by LTD of synapses from parallel fibers that show decreased firing during ipsiversive head motion or by LTP of synapses from parallel fibers that show increased firing during ipsiversive head motion.

PREDICTIONS OF THE FLOCCULUS HYPOTHESIS FOR LEARNING AT PARALLEL-FIBER INPUTS TO HGVPs   Evaluated in the context of the known mossy-fiber inputs to HGVPs, the flocculus hypothesis lacks the specificity needed to cause changes in the gain of the VOR without affecting other functions of the HGVPs. In monkeys, HGVPs receive three separate mossy-fiber inputs related to head velocity, eye velocity, and image motion (Miles & Fuller 1975, Lisberger & Fuchs 1978b, Miles et al 1980b, Noda 1986, Stone & Lisberger 1990a), and the simple-spike firing of HGVPs is approximately equal to the sum of these three inputs (Lisberger & Fuchs 1978a, Miles et al 1980b). During conditions that cause learning, not only the vestibular parallel-fiber inputs, but also the eye-movement and visual parallel-fiber inputs to the HGVPs fire in conjunction with climbing-fiber activity. Thus, the flocculus hypothesis predicts that learning in the VOR would be associated with changes in the strength of eye-movement and visual parallel-fiber inputs as well as vestibular parallel-fiber inputs to HGVPs.

The predictions of the flocculus hypothesis for the effect of different adapting conditions on the strength of vestibular inputs to HGVPs are inconsistent with the observations by Miles et al (1980b) and Lisberger et al (1994). After monkeys had been exposed to conditions that caused learning in the VOR (Table 3), increases in the gain of the VOR were associated with increases in the vestibular sensitivity of HGVPs, and decreases in the gain of the VOR were associated with decreases in the vestibular sensitivity of the HGVPs. According to the flocculus hypothesis, the conjunction of increased activity in the vestibular parallel-fiber inputs and visual climbing-fiber inputs under conditions that require the gain of the VOR to be increased ($\times$2) should cause decreases in the amplitude of the vestibular responses of HGVPs. The absence of conjunction of climbing-fiber inputs and vestibular parallel-fiber inputs during conditions that require the gain of the VOR to be decreased ($\times$0) should cause increases in the amplitude of the vestibular responses of HGVPs. Although available data on the amplitude of the sustained vestibular inputs to HGVPs contradict the flocculus hypothesis, modeling studies by Lisberger (1994) and Lisberger & Sejnowski (1992) have raised the possibility that changes in the amplitude of transient vestibular responses of HGVPs, in the direction predicted by the flocculus hypothesis, could participate in changing the gain of the VOR. New experiments will be needed to test for the postulated changes in the transient vestibular responses of HGVPs and to understand whether oppositely directed changes in the transient and sustained vestibular sensitivity of HGVPs are compatible with learning mediated by LTD in the cerebellar cortex of the F/VPF.

Analysis of the flocculus hypothesis for eye-movement mossy-fiber inputs yields different predictions depending on whether or not visual-tracking mechanisms such as pursuit or the OKR are used during learning to generate smooth eye movements that reduce the image motion. When the subject does not track the moving images seen under the adapting conditions (Table 3), eye movements are always in the opposite direction from head movements: The direction

Table 3 Evaluation of cerebellar LTD as a rule for learning in the parallel fiber inputs to HGVPs[a]

| | Vestibular inputs | | | | Eye-movement inputs | | | | | | | |
| | | | | | Not tracking | | | | Tracking | | | |
Behavioral condition[b]	Vestibular mossy fiber[c]	Climbing fiber[d]	Predicted change[e]	Recorded change[e]	Eye-movement mossy fiber[c]	Climbing fiber	Predicted change	Recorded change	Eye-movement mossy fiber	Climbing fiber	Predicted change	Recorded change
LR reversal	+	−	+	−	−	−	−	+	+	−	+	+
×0	+	−	+	NA[f]	−	−	−	NA	0	−	0	NA
×0.25	+	−	+	−	−	−	−	NA	−	+	−	NA
×2	+	+	−	+	−	+	+	+	+	+	+	+
Pursuit	0	−	0	NA	0	−	0	NA	+	−	+	NA
×1 (Rapid)	+	+	−	NA	−	+	+	NA	−	+	+	NA

[a]The table compares predicted and observed changes in the sensitivity of HGVPs to vestibular and eye-movement mossy-fiber inputs produced by various adapting conditions. Entries in the table are extrapolated from averaged responses of HGVPs (Lisberger & Fuchs 1978a).

[b]LR reversal indicates left-right reversal of vision with prisms. Other adapting conditions are indicated as ×N, where N is the gain of the VOR required to eliminate image motion during head turns. Also shown are the responses during ipsiversive pursuit eye movements and during rapid head turns with normal vision [×1 (Rapid)].

[c]"+" and "−" signs indicate increases and decreases in the firing rates of the mossy fibers during ipsiversive head turns during the adapting conditions.

[d]"+" and "−" signs indicate increases and decreases in the firing rates of the climbing fibers during ipsiversive head turns during the adapting conditions.

[e]"+" and "−" signs indicate increases and decreases in the sensitivity of HGVPs to retinal-image motion produced by head turns under the adapting conditions. "+" and "−" signs indicate increases and decreases in the sensitivity of HGVPs to the mossy-fiber inputs after exposure to the adapting conditions.

[f]Data not available.

of changes in the amplitude of the responses of HGVPs to eye velocity should be opposite that predicted for the responses to head velocity. The flocculus hypothesis predicts decreases in the strength of the eye-movement inputs for left-right reversal (LR reversal), ×0, and ×0.25 viewing conditions and increases for ×2 viewing conditions. When the subject does track the moving images seen under the adapting conditions (Table 3), the flocculus hypothesis predicts that the responses of HGVPs to eye velocity should get larger after learning in the LR reversal and ×2 viewing conditions, that they should get smaller after adaptation with ×0.25 viewing, and that they should not change after adaptation with ×0 viewing. In a partial test of these predictions, Miles et al (1980) found that both LR reversal of vision and ×2 viewing conditions caused small increases in the amplitude of the eye-movement responses of HGVPs during pursuit with the head stationary. These data would be consistent with the predictions of the flocculus hypothesis if the monkey were actually tracking the visual world during LR reversal of vision. However, the data are not conclusive, because it is unclear whether or not the monkeys were tracking the visual scene during adaptation.

Visual simple-spike responses of HGVPs are always modulated in the opposite direction from visual climbing fibers, and therefore the flocculus hypothesis predicts that the visual simple-spike responses of HGVPs will be maximally facilitated by any condition that causes image motion. No one has looked for increases in the amplitude of the image-motion response of HGVPs after learning. If such a change occurs, it should cause increases that were looked for but not seen in eye acceleration at the initiation of pursuit after the gain of the VOR had become high or low (Lisberger 1994). Either Lisberger's data contradict the predictions of the flocculus hypothesis or the visual parallel-fiber inputs to HGVPs are maximally potentiated even at the normal gain of the VOR.

Analysis of the parallel- and climbing-fiber inputs to HGVPs under tracking conditions that do not cause changes in the gain of the VOR reveals additional conditions under which the flocculus hypothesis lacks the specificity needed to control the gain of the VOR. During pursuit with the head stationary (Table 3), climbing fibers are silent when simple-spike activity increases (Lisberger & Fuchs 1978a, Stone & Lisberger 1990b), so the flocculus hypothesis predicts that the responses of HGVPs to eye velocity should get stronger. During rapid head turns under normal visual conditions with the lights on (Table 1), the 14-ms latency from the onset of head motion to the onset of the VOR causes a brief, transient image motion that affects climbing-fiber activity even if the gain of the VOR is 1.0 (Stone & Lisberger 1990b). According to the flocculus hypothesis, the conjuction of climbing- and parallel-fiber activity under this condition should cause depression of the vestibular responses of HGVPs and enhancement of their eye-movement responses. There is no evidence that these changes occur.

The problems of specificity in the flocculus hypothesis could be circumvented by recent findings that the cellular mechanisms of LTD are more complex than assumed by the flocculus hypothesis. Simple conjunctive activation of parallel fibers and climbing fibers is not adequate for the induction of LTD. For example, Ekerot & Kano (1985) showed in an in vivo preparation that the induction of LTD by conjunctive stimulation of parallel fibers and climbing fibers was blocked if cerebellar inhibitory neurons were simultaneously activated. In vitro studies have confirmed their observation that inhibitory inputs to the Purkinje cells can prevent the induction of LTD (Crepel & Jaillard 1991, Shibuki & Okada 1992).

In the in vitro preparations used to study LTD, many cellular conditions do not pertain to those in the intact animal. Inhibitory inputs were blocked, many of the normals inputs to Purkinje cells were physically missing or damaged, and Purkinje cells were below threshold for firing. These factors must be considered in deciding whether the LTD studies in vitro can contribute to learning in the behaving animal, when Purkinje cells and their mossy-fiber inputs fire spontaneously at rates of about 100 spikes/s. In the vestibular nucleus, for example, the temporal dynamics of neurons depends critically on whether the membrane is above or below the threshold for repetitive firings of action potential (du Lac & Lisberger 1993).

We conclude that LTD remains a candidate mechanism for learning in the cerebellar cortex. However, the existence of multiple mossy-fiber inputs to the F/VPF requires a precisely regulated form of LTD and not a form that is invoked whenever there is conjunctive activation of parallel- and climbing-fiber inputs to a given Purkinje cell. Several questions must be answered before cerebellar LTD can be elevated from the status of a mechanism of cellular plasticity with unknown function: 1. Does learning in the VOR cause the changes predicted by Table 3 in the responses of HGVPs to eye movement and visual inputs? 2. What precisely are the cellular requirements for LTD and under what behavioral conditions does the activity in the relevant neurons meet those cellular requirements? 3. Can LTD provide the requisite synaptic specificity, and what are the factors that contribute to specificity in this form of cellular plasticity? and 4. How do the results obtained in brain slices and tissue culture relate to the conditions in vivo?

CONCLUSIONS

We have presented a new hypothesis concerning the sites of memory in the VOR and the contribution of network dynamics to the expression of memory. Based on extensive behavioral data and single-unit recordings, we suggest that one site of memory is in the vestibular inputs onto FTNs in the brainstem. We propose that a second site of memory is in the vestibular inputs to HGVPs in the flocculus and ventral paraflocculus of the cerebellum. We have considered the details of possible mechanisms of cellular plasticity in the brainstem and in the cerebellum in relation to the neural signals produced by behavioral conditions that cause learning. Details of the cellular mechanisms of synaptic plasticity are of critical importance in the induction and specificity of learned behavioral changes in the VOR. Therefore, we suggest that future research on possible cellular mechanisms of learning in the VOR be performed in conditions that mimic as nearly as possible the neural activity that is present under behavioral circumstances that cause learning. We also suggest that future behavioral and neural analyses of learning in the VOR explicitly test the predictions made by possible cellular mechanisms of learning. An understanding of learning and memory in the VOR will result only from integration of the constraints and complexities provided by the real-world environment of the functioning brain with a detailed understanding of the in vivo operation of possible cellular mechanisms of learning.

ACKNOWLEDGMENTS

We thank Dean Buonomono, Gal Cohen, Vince Ferrera, Maninder Kahlon, and Yookyung Selig for helpful comments. Research from SG Lisberger's laboratory was supported by NIH grants EY03878 and EY10198, by a contract (N-00014-89-J-3094) from DARPA and the Office of Naval Research, and by Scholars and Development Awards from the McKnight Neuroscience Endowment Fund.

Literature Cited

Albus JS. 1971. A theory of cerebellar function. *Math. Biosci.* 10:25–61

Alley K, Baker R, Simpson JI. 1975. Afferents to the vestibulo-cerebellum and the origin of the visual climbing fibers in the rabbit. *Brain Res.* 98:582–89

Artola A, Singer W. 1993. Long-term depression of excitatory synaptic transmission and its relationship to long-term potentiation. *Trends Neurosci.* 16:480–87

Baker R, Berthoz A. 1975. Is the prepositus hypoglossi nucleus the source of another vestibulo-ocular pathway? *Brain Res.* 86:121–27

Baker R, Precht W, Llinas R. 1972. Cerebellar modulatory action on the vestibulo-trochlear pathway in the cat. *Exp. Brain Res.* 15:364–85

Balaban CD, Kawaguchi Y, Watanabe E. 1981. Evidence of a collateralized climbing fiber projection from the inferior olive to the flocculus and vestibular nuclei in rabbits. *Neurosci. Lett.* 22:23–29

Barmack NH, Pettorossi VE. 1985. Effects of unilateral lesions of the flocculus on optokinetic and vestibuloocular reflexes of the rabbit. *J. Neurophysiol.* 53:481–96

Barmack NH, Simpson JI. 1980. Effects of microlesions of dorsal cap of inferior olive of rabbits on optokinetic and vestibuloocular reflexes. *J. Neurophysiol.* 43:182–206

Brindley GS. 1964. The use made by the cerebellum of the information that it receives from sense organs. *Int. Brain Res. Org. Bulletin* 3:80

Brönte-Stewart HM, Lisberger SG. 1994. Physiological properties of vestibular primary afferents that mediate motor learning and normal performance of the vesibulo-ocular reflex in monkeys. *J. Neurosci.* 14:1290–1308

Broussard DM, Bronte-Stewart HM, Lisberger SG. 1992. Expression of motor learning in the response of the primate vestibulo-ocular reflex pathway to electrical stimulation. *J. Neurophysiol.* 67:1493–1508

Broussard DM, Lisberger SG. 1992. Vestibular inputs to brain stem neurons that participate in motor learning in the primate vestibulo-ocular reflex. *J. Neurophysiol.* 68:906–9

Buttner U, Waespe W. 1981. Vestibular nerve activity in the alert monkey during vestibular and optokinetic nystagmus. *Exp. Brain Res.* 41:310–15

Chan-Palay V, Ito M, Tongroach PM, Palay S. 1982. Inhibitory effects of motilin, somatostatin, [Leu]enkephalin, [Met]enkephalin, and taurine on neurons of the lateral vestibular nucleus: interactions with gamma-aminobutyric acid. *Proc. Nat. Acad. Sci.* 79:3355–59

Cohen H, Cohen B, Raphan T, Waespe W. 1992. Habituation and adaptation of the vestibuloocular reflex: a model of differential control by the vestibulocerebellum. *Exp. Brain Res.* 90:526–38

Collewijn H, Grootendorst AF. 1979. Adaptation of optokinetic and vestibulo-ocular reflexes to modified visual input in the rabbit. In *Reflex Control of Posture and Movement. Progress in Brain Research*, ed. R Granit, O Pompeiano, 50:772–81. Amsterdam: Elsevier

Crepel F, Jaillard D. 1991. Pairing of pre- and postsynaptic activities in cerebellar Purkinje cells induces long-term changes in synaptic efficacy in vitro. *J. Physiol.* 432:123–41

Dieringer N. 1986. Comparative neurobiology of the organization of gaze-stabilizing reflex systems in vertebrates. *Naturwissenschaften* 73:299–304

Dufosse M, Ito M, Jastreboff PJ, Miyashita Y. 1978. A neuronal correlate in rabbit's cerebellum to adaptive modification of the vestibulo-ocular reflex. *Brain Res.* 150:611–16

du Lac S, Lisberger SG. 1992. Eye movements and brainstem neuronal responses evoked by cerebellar and vestibular stimulation in chicks. *J. Comp. Physiol.* 171:629–38

du Lac S, Lisberger SG. 1993. Input-output transformations in medial vestibular nucleus neurons in vitro. *Soc. Neurosci. Abstr.* 19:1491

Ekerot CF, Kano M. 1985. Long-term depression of parallel fibre synapses following stimulation of climbing fibres. *Brain Res.* 342:357–60

Fujita M. 1982. Simulation of adaptive modification of the vestibulo-ocular reflex with an adaptive filter model of the cerebellum. *Biol. Cybern.* 45:207–14

Galiana HL. 1986. A new approach to understanding adaptive visual-vestibular interactions in the central nervous system. *J. Neurophysiol.* 55:349–74

Gerrits NM, Voogd J. 1989. The topographical organization of climbing and mossy fiber afferents in the flocculus and ventral paraflocculus in rabbit, cat, and monkey. *Exp. Brain Res. Suppl.* 17:26–29

Gomi H, Kawato M. 1992. Adaptive feedback control models of the vestibulocerebellum and spinocerebellum. *Biol. Cybern.* 68:105–14

Graf W, Simpson JI, Leonard CS. 1988. Spatial organization of visual messages of the rabbit's cerebellar flocculus. II. Complex and simple spike responses of Purkinje cells. *J. Neurophysiol.* 60:2091–2121

Henn V, Young LR, Finley C. 1974. Vestibular nucleus neurons in alert monkeys are also influenced by moving visual fields. *Brain Res.* 71:144–49

Highstein SM. 1973. Synaptic linkage in the vestibulo-ocular and cerebello-vestibular pathways to the VIth nucleus in the rabbit. *Exp. Brain Res.* 17:301–14

Ito M. 1972. Neural design of the cerebellar motor control system. *Brain Res.* 40:81–84

Ito M. 1982. Cerebellar control of the vestibulo-ocular reflex—around the flocculus hypothesis. *Annu. Rev. Neurosci.* 5:275–98

Ito M. 1989. Long-term depression. *Annu. Rev. Neurosci.* 12:85–102

Ito M. 1993. Cerebellar flocculus hypothesis. *Nature* 363:24–25

Kawato M, Gomi H. 1992. The cerebellum and VOR/OKR learning models. *Trends Neurosci.* 15:445–53

Keller EL, Robinson DA. 1971. Absence of a stretch reflex in extraocular muscles of the monkey. *J. Neurophysiol.* 34:908–19

Khater TT, Quinn KJ, Pena J, Baker JF, Peterson BW. 1993. The latency of the cat vestibulo-ocular reflex before and after short- and long-term adaptation. *Exp. Brain Res.* 94:16–32

Langer T, Fuchs AF, Chubb MC, Scudder CA, Lisberger SG. 1985a. Floccular efferents in the rhesus macaque as revealed by autoradiography and horseradish peroxidase. *J. Comp. Neurol.* 235:26–37

Langer T, Fuchs AF, Scudder CA, Chubb MC. 1985b. Afferents to the flocculus of the cerebellum in the rhesus macaque as revealed by retrograde transport of horseradish peroxidase. *J. Comp. Neurol.* 235:1–25

Leonard CS, Simpson JI. 1985. Purkinje cell activity in the flocculus of the alert rabbit during natural visual and vestibular stimulation. *Soc. Neurosci. Abstr.* 4:162

Linden DJ, Connor JA. 1993. Cellular mechanisms of long-term depression in the cerebellum. *Curr. Op. Neurobiol.* 3:401–6

Lisberger SG. 1982. Role of the cerebellum during motor learning in the vestibulo-ocular reflex: different mechanisms in different species? *Trends Neurosci.* 12:437–41

Lisberger SG. 1984. The latency of pathways containing the site of motor learning in the vestibulo-ocular reflex. *Science* 225:74–76

Lisberger SG. 1988. The neural basis for learning of simple motor skills. *Science* 242:728–35

Lisberger SG. 1994. Neural basis for motor learning in the vestibulo-ocular reflex of primates: III. Computational and behavioral analysis of the sites of learning. *J. Neurophysiol.* 72:974–99

Lisberger SG, Broussard DM, Brönte-Stewart HM. 1990. Properties of pathways that mediate motor learning in the vestibulo-ocular reflex of monkeys. *Cold Spring Harbor Symp. Quant. Biol.* 50:813–22

Lisberger SG, Fuchs AF. 1974. Response of flocculus Purkinje cells to adequate vestibular stimulation in the alert monkey: fixation vs. compensatory eye movements. *Brain Res.* 69:347–53

Lisberger SG, Fuchs AF. 1978a. Role of primate flocculus during rapid behavioral modification of vestibulo-ocular reflex. I. Purkinje cell activity during visually guided horizontal smooth-pursuit eye movements and passive head rotation. *J. Neurophysiol.* 41:733–63

Lisberger SG, Fuchs AF. 1978b. Role of primate flocculus during rapid behavioral modification of vestibulo-ocular reflex. II. Mossy fiber firing patterns during horizontal head rotation and eye movement. *J. Neurophysiol.* 41:764–77

Lisberger SG, Miles FA. 1980. Role of primate medial vestibular nucleus in adaptive plasticity of vestibulo-ocular reflex. *J. Neurophysiol.* 43:1725–45

Lisberger SG, Miles FA, Optican LM. 1983. Frequency-selective adaptation: evidence for channels in the vestibulo-ocular reflex? *J. Neurosci.* 3:1234–44

Lisberger SG, Miles FA, Optican LM, Eighmy BB. 1981. The optokinetic response in monkey: underlying mechanisms and their sensitivity to long-term adaptive changes in the vestibulo-ocular reflex. *J. Neurophysiol.* 45:869–90

Lisberger SG, Miles FA, Zee DS. 1984. Signals used to compute errors in monkey vestibulo-ocular reflex: possible role of flocculus. *J. Neurophysiol.* 52:1140–53

Lisberger SG, Pavelko TA. 1986. Vestibular signals carried by pathways subserving plasticity of the vestibulo-ocular reflex. *J. Neurosci.* 6:346–54

Lisberger SG, Pavelko TA. 1988. Brainstem neurons in modified pathways for motor learning in the primate vestibulo-ocular reflex. *Science* 242:771–73

Lisberger SG, Pavelko TA, Bronte-Stewart HM, Stone LS. 1994a. Neural basis for motor learning in the vestibulo-ocular reflex of primates: II. Changes in the responses of Horizontal Gaze Velocity Purkinje cells in the cerebellar flocculus and ventral paraflocculus. *J. Neurophysiol.* 72:954–73

Lisberger SG, Pavelko TA, Broussard DM. 1994b. Responses during eye movements of brainstem neurons that receive monosynaptic inhibition from the flocculus and ventral paraflocculus in monkeys. *J. Neurophysiol.* 72:909–27

Lisberger SG, Pavelko TA, Broussard DM. 1994c. Neural basis for motor learning in the vestibulo-ocular reflex of primates: I. Changes in the responses of brainstem neurons. *J. Neurophysiol.* 72:928–53

Lisberger SG, Sejnowski TJ. 1992. Motor learning in a recurrent network model based on the vestibulo-ocular reflex. *Nature* 360:159–61

Luebke AE, Robinson, DA. 1994. Climbing fiber intervention blocks plasticity of the vestibuloocular reflex. *Ann. NY Acad. Sci.* 656:428–30

Marr D. 1969. A theory of cerebellar cortex. *J. Physiol.* 202:437–70

McCrea RA, Strassman A, May E, Highstein SM. 1987. Anatomical and physiological characteristics of vestibular neurons mediating the horizontal vestibulo-ocular reflex of the squirrel monkey. *J. Comp. Neurol.* 264:547–70

Miles FA, Braitman DJ, Dow BM. 1980a. Long-term adaptive changes in primate vestibulo-ocular reflex. IV. Electrophysiological observations in flocculus of adapted monkeys. *J. Neurophysiol.* 43:1477–93

Miles FA, Eighmy BB. 1980. Long-term adaptive changes in primate vestibulo-ocular reflex. I. Behavioral observations. *J. Neurophysiol.* 43:1406–25

Miles FA, Fuller JH. 1974. Adaptive plasticity in the vestibulo-ocular responses of the rhesus monkey. *Brain Res.* 80:512–16

Miles FA, Fuller JH. 1975. Visual tracking and the primate flocculus. *Science* 189:1000–2

Miles FA, Fuller JH, Braitman DJ, Dow BM. 1980b. Long-term adaptive changes in primate vestibulo-ocular reflex. III. Electrophysiological observations in flocculus of normal monkeys. *J. Neurophysiol.* 43:1437–76

Miles FA, Lisberger SG. 1981. Plasticity in the vestibulo-ocular reflex: a new hypothesis. *Annu. Rev. Neurosci.* 4:273–99

Minor LB, Goldberg JM. 1991. Vestibular-nerve inputs to the vestibulo-ocular reflex: a functional ablation study in the squirrel monkey. *J. Neurosci.* 11:1636–48

Nagao S. 1983. Effects of vestibulocerebellar lesions upon dynamic characteristics and adaptation of vestibulo-ocular and optokinetic responses in pigmented rabbits. *Exp. Brain Res.* 53:152–68

Nagao S. 1991. Contribution of oculomotor signals to the behavior of rabbit floccular Purkinje cells during reflex eye movements. *Neurosci. Res.* 12:169–84

Nagao S. 1992. Different roles of flocculus and ventral paraflocculus for oculomotor control in the primate. *NeuroReport* 3:13–16

Noda H. 1986. Mossy fibres sending retinal-slip, eye, and head velocity signals to the flocculus of the monkey. *J. Physiol.* 379:39–60

Partsalis AM, Zhang Y, Highstein SM. 1993. The Y group in vertical visual-vestibular interactions and VOR adaptation in the squirrel monkey. *Soc. Neurosci. Abstr.* 19:138

Pastor AM, De La Cruz RR, Baker R. 1992. Characterization and adaptive modification of the goldfish vestibuloocular reflex by sinusoidal and velocity step vestibular stimulation. *J. Neurophysiol.* 68:2003–15

Pastor AM, De La Cruz RR, Baker R. 1994. Cerebellar role in adaptation of the goldfish vestibulo-ocular reflex. *J. Neurophysiol.* In press

Precht W, Baker R. 1972. Synaptic organization of the vestibulo-trochlear pathway. *Exp. Brain Res.* 15:158–84

Quinn KJ, Schmajuk N, Baker JF, Peterson BW. 1992a. Simulation of adaptive mechanisms in the vestibulo-ocular reflex. *Biol. Cybern.* 67:103–12

Quinn KJ, Schmajuk N, Jain A, Baker JF, Peterson BW. 1992b. Vestibulo-ocular reflex arc analysis using an experimentally constrained neural network. *Biol. Cybern.* 67:113–22

Robinson DA. 1976. Adaptive gain control of the vestibulo-ocular reflex by the cerebellum. *J. Neurophysiol.* 39:954–69

Sakurai M. 1987. Synaptic modification of parallel fibre-Purkinje cell transmission in in vitro guinea-pig cerebellar slices. *J. Physiol.* 394:463–80

Scudder CA, Fuchs AF. 1992. Physiological and behavioral identification of vestibular nucleus neurons mediating the horizontal vestibuloocular reflex in trained rhesus monkeys. *J. Neurophysiol.* 68:244–64

Shibuki K, Okada D. 1992. Cerebellar long-term potentiation under suppressed postsynaptic Ca^{2+} activity. *NeuroReport* 3:231–34

Shimazu H, Precht W. 1966. Inhibition of central vestibular neurons from the contralateral labyrinth and its mediating pathway. *J. Neurophysiol.* 29:467–92

Skavenski AA, Robinson DA. 1973. Role of abducens neurons in vestibuloocular reflex. *J. Neurophysiol.* 36:724–38

Snyder LH, King WM. 1992. Effect of viewing distance and location on the axis of head rotation on the monkey's vestibuloocular reflex. I. Eye movement responses. *J. Neurophysiol.* 67:861–74

Stone LS, Lisberger SG. 1990a. Visual responses of Purkinje cells in the cerebellar flocculus during smooth pursuit eye movements in monkeys. I. Simple spikes. *J. Neurophysiol.* 63:1241–61

Stone LS, Lisberger SG. 1990b. Visual responses of Purkinje cells in the cerebellar flocculus during smooth pursuit eye movements in monkeys. II. Complex spikes. *J. Neurophysiol.* 63:1262–75

Tempia F, Dierenger N, Strata P. 1991. Adaptation and habituation of the vestibulo-ocular reflex in intact and inferior olive-lesioned rats. *Exp. Brain Res.* 86:568–78

Watanabe E. 1984. Neuronal events correlated with long-term adaptation of the horizontal vestibulo-ocular reflex in the primate flocculus. *Brain Res.* 297:169–74

Annu. Rev. Neurosci. 1995. 18:443–62

THE ROLE OF AGRIN IN SYNAPSE FORMATION

Mark A. Bowe and Justin R. Fallon

Neurobiology Group, Worcester Foundation for Experimental Biology, 222 Maple Avenue, Shrewsbury, Massachusetts 01545

KEYWORDS: synaptogenesis, acetylcholine receptor, neuromuscular junction, agrin receptor, neural development

INTRODUCTION

The precise, rapid, and ordered communication between neurons and their targets is the distinguishing characteristic of the nervous system. Chemical synapses are the primary locus of this information exchange, and the structural and molecular specializations essential for their proper functioning are understood in considerable detail (reviewed in Jessell & Kandel 1993, Stevens 1993). In contrast, much less is known about how synaptic structure is regulated during development and regeneration. Such structural changes may also mediate important aspects of synaptic plasticity in learning and memory (Bliss & Collingridge 1993, Lisman & Harris 1993).

Most of our knowledge of synaptogenesis has come from studying the neuromuscular junction. A search for the mechanisms underlying differentiation at this synapse has focussed on three molecular systems: the cytoskeleton and its associated membrane proteins, trophic factors, and a specific extracellular matrix component, agrin. Recent reviews have dealt with the overall process of synaptogenesis and with the contribution of the cytoskeleton and trophic factors (Salpeter 1987, Falls et al 1993, Froehner 1993, Hall & Sanes 1993). Here we focus on agrin, which is likely to play a central role in signaling and regulating the differentiation of the postsynaptic apparatus at the neuromuscular junction (Nastuk & Fallon 1993). In addition, we also discuss the mounting evidence suggesting an analogous role for agrin in the CNS.

443

0147-006X/95/0301-0443$05.00

THE MATURE SYNAPSE

The mature nerve-muscle synapse is comprised of pre- and postsynaptic specializations separated by a synaptic cleft. The presynaptic apparatus is characterized by ordered arrays of synaptic vesicles poised to fuse with the plasma membrane at active zones. The postsynaptic membrane is encrusted with acetylcholine receptors (AChRs), which can reach densities of >10,000 molecules/μm^2. The synaptic cleft is occupied by a discrete synaptic basal lamina, a highly specialized extracellular matrix domain that is discussed in more detail below. The localized and tightly regulated secretion of acetylcholine into the confined space of the synaptic cleft, coupled with the high AChR concentration in the postsynaptic membrane, ensures rapid and reliable synaptic transmission between cells. The signal is efficiently terminated by acetylcholinesterase (AChE), which is also localized to the synapse. Perturbation of these specializations, such as the decrease in the number of functional AChRs seen in myasthenia gravis, can lead to debilitating and often fatal clinical outcomes.

One of the most important and provocative features of synapses in both the periphery and the CNS is their restricted subcellular distribution: A given synapse typically occupies <0.1% of the cell surface. This topological restriction lies at the heart of synaptic differentiation—how to make one domain of a cell distinct from others. This arrangement offers obvious advantages for economical, efficient, and rapid signaling: All the necessary machinery is highly concentrated at the appropriate site. In neurons, this restricted distribution also facilitates the cell's ability to receive multiple (up to 20,000), diverse inputs in an ordered and segregated manner. Finally, the segregation of these components to distinct domains serves to maximize the independent regulation of synapses on the same cell, which is likely to be crucial for processes such as long-term potentiation on neurons (Bliss & Collingridge 1993).

The molecular machinery of synaptic transmission at the neuromuscular junction is understood in considerable detail. The major proteins of synaptic vesicles have been characterized, and a picture of how the docking, exocytosis, and retrieval of these vesicles is achieved and regulated is emerging (Sollner et al 1993, Sudhof et al 1993). There has been a parallel explosion in our molecular and functional understanding of the AChR (Hille 1992). In addition to the constituents directly involved in synaptic transmission, there are a host of proteins likely to be important for the structure and regulation of the synapse. The cytoskeletal proteins at the synapse can be broadly grouped into two sets: (*a*) dystrophin, utrophin (dystrophin-related protein), and their associated elements such as the 58-kD and 87-kD proteins; and (*b*) several molecules that are associated with focal adhesions, including α-actinin, vinculin, talin, paxillin, and filamin. The former group may be linked to laminin in the synaptic basal lamina via the dystrophin-associated glycoprotein complex (Bloch et al

1991, Ervasti & Campbell 1993). The latter proteins may communicate with the extracellular matrix (ECM) through integrins (Bozyczko et al 1987). Specialized forms of intermediate filament protein and β-spectrins are also localized at synapses (Burden 1982, Bloch & Morrow 1989). The postsynaptic 43-kD protein probably anchors AChRs to as yet unidentified components of the underlying cytoskeleton (Maimone & Merlie 1993, Scotland et al 1993; reviewed in Froehner 1993).

The synaptic basal lamina contains not only agrin but also high levels of laminin A chain, a heparan sulphate proteoglycan, and collagen IV (Sanes et al 1990). The asymmetric form of AChE is also localized in this structure (Anglister & McMahan 1985). These and other proteins in the synaptic ECM are likely to play diverse roles, including mediating cell-cell adhesion, directing the differentiation of pre- and postsynaptic specializations, and serving as a repository of factors regulating gene expression (Brenner et al 1992, Jo & Burden 1992).

GENERAL CONSIDERATIONS OF POSTSYNAPTIC DIFFERENTIATION

How do synapses form? Synaptic differentiation can be considered on several distinct, yet interrelated, levels. For example, the pre- and postsynaptic cells must express the specific gene products that constitute the synapse, such as synaptic vesicle proteins and neurotransmitter receptors. Recent studies have begun to reveal the mechanisms underlying the differentiation of muscle cells and motor neurons (Tapscott et al 1990, Smith 1994). Moreover, the cell needs to express these proteins in a quantitatively and temporally regulated manner. Finally, these components must be marshaled to the site on the cell where they will be used, and then they must be precisely organized into a mature synapse.

The regulation of the expression and the spatial organization of synaptic components at neuromuscular junctions are intimately related, yet are likely to be subserved by distinct mechanisms. For example, two classes of stimuli are known to regulate AChR expression: electrical activity and neuronally derived trophic factors, which apparently work in complementary fashions (Witzemann et al 1991, Simon et al 1992). In mature muscle, electrical activity causes a global decrease in AChR expression, while trophic factors provide both quantitative and qualitative regulation of AChR-subunit gene expression at the synapse. One such factor is likely to be ARIA (AChR-inducing activity), a member of the heregulin family (Falls et al 1993). [Unidentified trophic factors are likely to play an analogous role in neurons (Role 1988, Schwarz-Levey et al 1994).] However, neither electrical activity nor ARIA induce AChR clustering in the initial stages of synaptogenesis. Conversely, there is no evidence that agrin regulates AChR-subunit gene expression. However, it must be stressed that these results stem from work on model systems that probably reflect early events in synapse formation. These factors could have different effects later in development; for example, global and local differences in electrical activity have a dramatic influence on synaptic stabilization (Lichtman & Balice-Gordon 1990).

AGRIN AND POSTSYNAPTIC DIFFERENTIATION AT THE NEUROMUSCULAR JUNCTION

Several important rules governing the early events in postsynaptic differentiation were established in pioneering nerve-muscle coculture experiments (Anderson & Cohen 1977, Frank & Fischbach 1979; reviewed in Hall & Sanes 1993). Foremost among these is that motor neurons induce the formation of the postsynaptic apparatus. Second, this induction is limited to the appropriate synaptic partners and is not a function of simple adhesion: Neurites from sensory, cerebral cortical, and other neurons form extensive myotube contacts but effect little AChR clustering (see also Lieth & Fallon 1993). Third, although motor neurons promote AChR expression at developing synapses (Role et al 1985), initial clustering occurs by lateral migration of preexisting (diffusely distributed) AChRs; new synthesis is not required. Finally, electrical activity is not necessary for nerve-induced AChR clustering. Together, these findings indicate that there are specific factors supplied by motor neurons that induce AChR aggregation at developing neuromuscular junctions.

The coculture results suggested that neurally derived factors could be distinguished on the basis of their ability to induce AChR clusters on myotubes. Indeed, neural extracts and conditioned medium were shown to contain high molecular-weight factors with such activity (Christian et al 1978, Podleski et al 1978). Although these factors have not been identified, their biochemical properties and activity profiles (Olek et al 1986, Podleski & Salpeter 1988) indicate that they are likely to be agrin. Many of these extracts also contained activities that upregulate AChR expression; two of these molecules, calcitonin gene–related peptide (CGRP) and ARIA, are synthesized by motor neurons and are good candidates for mediating neural effects on AChR gene expression (Jessell et al 1979, New & Mudge 1986, Peng et al 1989, Changeux et al 1992, Falls et al 1993). Two other factors were identified as ascorbic acid and transferrin; their role in synaptogenesis is currently unclear (Markelonis et al 1982, Knaack et al 1986, Horovitz et al 1989).

A series of elegant in vivo experiments by McMahan and colleagues showed that another, quite unexpected structure could also induce postsynaptic differentiation on regenerating myofibers: the synaptic basal lamina (Burden et al 1979, McMahan & Slater 1984; reviewed in McMahan & Wallace 1989).

When myofibers regenerate into preexisting basal lamina sheaths, postsynaptic specializations form at the site of the original neuromuscular junction. These specializations form in the absence of reinnervation and display many features of mature endplates, including high concentrations of AChRs and AChE, and postsynaptic folds (Anglister & McMahan 1985). These experiments demonstrated that molecules stably associated with the synaptic basal lamina are sufficient to induce postsynaptic differentiation. They also suggested that this specialized ECM segment offered a relatively simple source from which to purify this molecule (Rubin & McMahan 1982).

ECM extracts from the synapse-rich *Torpedo californica* electric organ were used as a source to purify this factor. These extracts induced AChR aggregation on cultured myotubes (Godfrey et al 1984). Monoclonal antibodies that were raised against this clustering activity recognized an antigen stably associated with the synaptic basal lamina (Fallon et al 1985) and were used in conjunction with several biochemical methods to identify the factor, which was named agrin (from the Greek *ageirein,* meaning to assemble) (Nitkin et al 1987). Biochemically purified agrin is comprised of four polypeptides of 150, 135, 95, and 70 kD, which are stable fragments of a native 200-kD molecule (Smith et al 1992; reviewed in McMahan 1990).

Several lines of evidence support a central role for agrin in directing postsynaptic differentiation. Agrin is stably associated with the mature synaptic basal lamina (Reist et al 1987) and is present at developing synapses in vivo (Godfrey et al 1988b, Fallon & Gelfman 1989, Hoch et al 1993). Agrin is synthesized by motor neurons and released by growing neurites, where it is associated with nerve-induced AChR aggregates (Magill-Solc & McMahan 1988, Cohen & Godfrey 1992). Antibodies against agrin block the formation of motor neuron–induced AChR clusters (Reist et al 1992). Recombinant agrin induces AChR clusters on muscle cells (Campanelli et al 1991, Ruegg et al 1992). Interestingly, agrin is also synthesized by muscle cells, and this muscle agrin is concentrated at postsynaptic specializations (Fallon & Gelfman 1989, Lieth et al 1992; and see below). Finally, an agrin receptor is present on developing myotubes and has been purified from *Torpedo* electric organ postsynaptic membranes (Nastuk et al 1991, Ma et al 1993, Bowe et al 1994). Together, these data indicate that agrin directs the aggregation of AChRs at both developing and regenerating nerve-muscle synapses.

Although AChR clustering is agrin's benchmark activity, it is important to appreciate that agrin also induces the redistribution of a dozen or more proteins that are also concentrated at synapses (Table 1). Moreover, these molecules are functionally diverse, ranging from integral membrane proteins such as the agrin receptor to peripheral membrane proteins, cytoskeletal elements, and basal lamina constituents. Agrin is therefore likely to play an overarching regulatory role in postsynaptic differentiation.

Table 1 Agrin induces the redistribution of diverse components of the neuromuscular junction

Synaptic component	Phase of distribution[a]	References
Plasma membrane		
Acetylcholine receptor	Early	Godfrey et al 1984
Acetylcholinesterase (globular)	Early	Wallace 1989
Butyrylcholinesterase (globular)	Early	Wallace 1989
Agrin receptor	Early	Nastuk et al 1991
43-kD protein	Early	B Wallace, personal communication
Extracellular matrix		
Heparan sulfate proteoglycan	Late	Nitkin & Rothschild 1990, Wallace 1989
Laminin	Late	Nitkin & Rothschild 1990
Muscle agrin	Late	Lieth et al 1992
Acetylcholinesterase (A_{12} asymmetric)	?	Wallace 1989
Cytoskeleton		
α-actinin	Late	Shadiack & Nitkin 1991
Filamin	Late	Shadiack & Nitkin 1991
Vinculin	Late	Shadiack & Nitkin 1991

[a] Early: ~1–4 h; Late: >4 h

AGRIN EXPRESSION

Agrin has been detected at neuromuscular junctions in every species for which probes are available (electric ray, chick, frog, rat) (Fallon et al 1985, Reist et al 1987, Hoch et al 1993); whether agrin is expressed in invertebrates is unknown. Agrin is present at virtually all developing nerve-muscle synapses from the time they first form in embryonic chicks (Godfrey et al 1988b, Fallon & Gelfman 1989). In contrast, agrin is only detected at a subset of nerve-muscle synapses early in development in rats (Hoch et al 1993). It is unknown if this finding reflects a species difference or is the result of insufficiently sensitive immunological reagents. Agrin mRNA is expressed in motor neurons and muscle of both chick and rat from the time of initial synapse formation (Tsim et al 1992, Hoch et al 1993).

Agrin is expressed by both pre- and postsynaptic cells at neuromuscular junctions. Aneural muscle cells in vivo and in vitro express agrin (Fallon & Gelfman 1989, Lieth et al 1992). Moreover, muscle agrin is expressed at AChR clusters induced either by coculture with neurons or by incubation with purified *Torpedo* agrin (Lieth & Fallon al 1993). Muscle agrin is expressed in the late phase of agrin-induced AChR clustering (Table 1) and thus does not appear to be involved in initial AChR aggregation. Rather, muscle agrin may play a role in the maturation or stabilization of AChR clusters.

Agrin is also detected in the basal laminae of CNS capillaries and smooth muscle cells, kidney, and cardiac cells (Godfrey et al 1988a). Expression at other sites is species dependent. For example, agrin immunoreactivity surrounds Schwann cells in myelinated nerve in *Torpedo,* chick, and rat but is restricted to nodes of Ranvier in frog (Reist et al 1987). Agrin's function outside the neuromuscular junction is unknown.

AGRIN STRUCTURE

Agrin is a single chain, multidomain extracellular matrix protein of ~200 kD (Figure 1). Analysis of the full-length rat and chick and partial electric ray–deduced amino acid sequences shows 60% identity and an additional 20% conservation among species (Rupp et al 1991, Smith et al 1992, Tsim et al 1992). Several domains, particularly in the carboxy terminus, show a greater than 80% identity. From both functional and structural viewpoints it is useful to consider the amino and carboxyl halves of agrin separately.

The carboxyl-terminal half of agrin (which roughly corresponds to the biochemically purified 95-kD polypeptide) (Nitkin et al 1987) binds to the agrin receptor (Ma et al 1993, Bowe et al 1994) and embodies the full range of agrin's clustering activity (Wallace 1989; and see below). Prominent in this half of the molecule are four EGF-like (epidermal growth factor–like) cysteine repeats that are similar to one another but not to those found in the amino-ter-

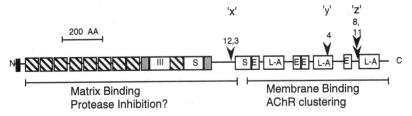

Figure 1   The structural organization of rat agrin. A 38–amino acid hydrophobic region (*filled*) at the extreme N terminus is thought to be a signal sequence. The remaining N-terminal portion of the molecule has nine Kazel-type protease inhibitor/EGF repeats (*cross hatch*), a region homologous to domain III of laminin B chains (*III*), a serine- and threonine-rich domain (*S*), and two internal cysteine-rich repeats (*stippling*). The C-terminal half, which is sufficient by itself for binding to the plasma membrane and inducing AChR clustering, has four EGF-like cysteine repeats (*E*), a serine-and-threonine-rich domains, and three regions homologous to the G-domain of the laminin A chain (*L-A*). Arrowheads indicate some sites of alternative exon splicing. At the *x* site alternative splice-acceptor-site usage results in 12- or 3–amino acid inserts; at the *y* site a 4–amino acid insert can be present; and at the *z* site there can be the 8–, 11–, 19– (both), or no amino acid insert(s). The 8– and 11–amino acid inserts are uniquely expressed in the nervous system. The organization of chick agrin is very similar (Ruegg et al 1992).

minal half. Analysis of exon structure (Rupp et al 1992) and protein sequence (Patthy & Nikolics 1993) also revealed three ~100–amino acid repeats that share significant homology to the G-domain of the laminin A chain, and to similar regions in perlecan (Noonan et al 1991, Murdoch et al 1992, Rothberg & Artavanis 1992).

The amino-terminal half of agrin is sufficient for its association with the extracellular matrix, and in some cases it modulates agrin's biological activity (Ferns et al 1993). One region of particular interest is ~40% identical to domain III of the laminin B1, B2, A, and S chains. In laminin this domain is involved in binding to the basal lamina component nidogen (entactin) (Timpl 1989). The homologous domain in agrin could play a similar role, since agrin externalized by growing neurites binds to a nidogen-rich ECM extract but not to laminin (Cohen et al 1994). The amino-terminal half of agrin also has nine tandem repeats; the carboxyl and amino portions of each are homologous to Kazel-type protease-inhibitor domains and EGF, respectively (Rupp et al 1992). Biochemical studies have confirmed that agrin has serine protease–inhibitor activity with specificity for trypsin, chymotrypsin, and plasmin (Biroc et al 1993). Interestingly, proteolysis has been implicated in synapse formation and in elimination of polyneuronal innervation (Connold et al 1986, Champaneria et al 1992). On the other hand, alternative sequence analyses have suggested that these nine repeats are homologous to follistatin, which is known to bind growth factors. This similarity raises the intriguing possibility that agrin presents trophic factors to motor neurons and/or muscle cells (Patthy & Nikolics 1993).

Agrin is encoded by a single gene located on human chromosome 1 region pter-p32 and on mouse chromosome 4 (Rupp et al 1992). There is no indication that the gene is associated with known mutations resulting in neuromuscular disorders or neurological diseases. The agrin gene consists of 37 exons. The exons and introns that have been sequenced are rather small, ranging in size from 12 to 333 and 65 to 712 base pairs, respectively.

ALTERNATIVE SPLICING OF AGRIN

Agrin mRNA is alternatively spliced at three sites, yielding several forms. Ruegg et al (1992) first recognized the importance of these forms when they observed that the AChR-clustering activity of a recombinant chick agrin depended upon the presence of an alternatively spliced 11–amino acid insert. Further analysis of rat and chick agrin showed that an eight–amino acid insert could also be included at this site, denoted z (Figure 1) (Rupp et al 1992, Hoch et al 1993, Thomas et al 1993). All combinations at this z site are expressed, yielding forms with no or 8–, 11–, or 19–amino acid inserts. Splicing at the z site is always accompanied by a four–amino acid insert at the y site, but the latter insert can be expressed independently. A third site, x, can have a 12– or a 3–amino acid insert, depending upon splice-acceptor-site usage. The sequences of the y and z inserts do not show strong homology to any entries in the protein database. However, the size, location, and sequence of these inserts is conserved among species: The rat and chick 4–, 8–, and 11–amino acid inserts are 100, 75, and 45% identical, respectively.

The biological activity of recombinant rat and chick agrin alternatively spliced at the y and z sites has been assayed (Table 2) (Ferns et al 1992, 1993; Ruegg et al 1992). Both full-length, cell-attached agrin and truncated, soluble agrin (which roughly corresponds to the carboxyl half of the molecule) have

Table 2 Functional consequences of alternative agrin mRNA splicing

		Myotube type			
	Chick	Mouse and rat			
Agrin	primary	primary	C2	S26	S27
Full-length (cell-attached)					
Rat[a]					
12, 0, 0[b]	ND[c]	ND	+	−	−
12, 4, 0	−	+	+	−	−
12, 4, 8	+	+	+	+	+
12, 4, 11	(+)	+	+	−	−
12, 4, 19	+	+	+	+	+
Truncated (C-terminal)[d]					
Rat[a]					
12, 0, 0	ND	ND	−	−	−
12, 4, 0	ND	ND	−	−	−
12, 4, 8	+	+	+	+	(+)
12, 0, 8	ND	ND	+	+	−
Chick[e]					
12, 0, 0	−	ND	ND	ND	ND
12, 4, 0	−	ND	ND	ND	ND
12, 4, 11	+	ND	ND	ND	ND
Torpedo[f]					
(?)[g]	+	+	+	ND	−
Truncated (N-terminal)[h]					
Rat	ND	ND	−	−	−

[a] Campanelli et al 1992; Ferns et al 1992, 1993; M Ferns & Z Hall, unpublished observations.

[b] Numbers refer to alternatively expressed amino acid inserts at the x, y, and z sites, respectively (see Figure 1 and text).

[c] Symbols for AChR clustering activity: −, none; +, high activity; (+), low activity; and ND, not determined.

[d] Corresponding to carboxyl-terminal half (see Figure 1).

[e] Ruegg et al 1992.

[f] Godfrey et al 1984, Nastuk et al 1990, Gordon et al 1993.

[g] Splice form unknown.

[h] Corresponding to amino-terminal half (see Figure 1).

been used in these studies (Figure 1). All of the tested full-length molecular splice forms induce AChR clustering with equal potency when assayed on either mouse primary or C2 myotubes. However, truncated, soluble agrin was only active if at least one insert at the y or z site was present. Moreover, there were major differences in effectiveness: Truncated forms containing the eight–amino acid insert were >1000-fold more potent than those with only the four–amino acid insert. Deletion of the four–amino acid insert did not diminish the activity of the forms containing the eight–amino acid insert. In contrast, when chick agrin was assayed on chick myotubes, there was an absolute requirement for an 8– and/or 11–amino acid insert at the z site (Ruegg et al 1992). Soluble forms of rat agrin containing the eight–amino acid insert are also active on these cells (M Ferns & Z Hall, unpublished observations).

Further insight into the role of alternative splicing and the function of the N- and C-terminal agrin domains has come from studies using muscle cells deficient in proteoglycan expression (Table 2). These cell lines, S26 and S27, have moderate and severe proteoglycan defects, respectively. Both show nerve-induced but not spontaneous AChR clustering (Gordon & Hall 1989, Gordon et al 1993). Truncated, soluble rat agrin induces AChR clustering on S26 cells but only if the eight–amino acid insert is present. Full-length, cell-attached agrin with the eight–amino acid insert induces clustering on S27 cells, but little activity is observed with even the most potent soluble forms. These results suggest that proteoglycans may play a role in agrin's AChR-clustering activity (Ferns et al 1993).

Although agrin is present in many tissues, expression of the 8– and 11–amino acid inserts is restricted to the nervous system (Ruegg et al 1992, Smith et al 1992, Biroc et al 1993, Hoch et al 1993). The pattern of alternative splicing at the z site changes during development (Hoch et al 1993, Thomas et al 1993). Splicing of the four–amino acid insert occurs in the CNS, PNS, skeletal and cardiac muscle, and to a lesser extent in kidney. A recent elegant study using electrophysiological recording followed by RT-PCR (reverse translation–polymerase chain reaction) analysis of individual cells has shown that in ciliary ganglia, over 70% of the neurons express agrin with the 8– and 11–amino acid inserts (Smith & O'Dowd 1994). In contrast, agrin expressed by glial cells was devoid of inserts at the z site. Further, individual neurons varied in their patterns of alternative agrin mRNA splicing and often expressed more than one splice type. Interestingly, over one third of these neurons expressed forms with no splicing at the z site. These results indicate that although alternative splicing at the z site is a property of neurons, it may not be necessary for all the functions of neuronally expressed agrin.

How these inserts regulate agrin's activity is unknown. They apparently do not form part of agrin's binding site for its receptor, since excess synthetic peptides corresponding to these inserts neither inhibit nor stimulate AChR clustering. It seems that these inserts cause a conformational change that augments agrin's activity or renders it independent of accessory molecules such as proteoglycans (Yayon et al 1991). The differential activity of these forms could indicate the existence of multiple agrin receptors, but only a single receptor has been detected to date (see below). These studies also reveal that the biological activity of the agrin forms is target dependent. Myotubes may therefore vary in response to different forms depending, for example, on their developmental stage or physiological state (e.g. during synapse elimination or after denervation). Finally, these observations suggest that there may be more than one intracellular signaling pathway for agrin-induced AChR clustering. The existence of such multiple pathways is consistent with some of the findings concerning agrin-induced tyrosine phosphorylation of the AChR (discussed below).

AGRIN'S MECHANISM OF ACTION

There are at least two distinct phases of agrin-induced molecular redistribution (Table 1). The early stage begins 1–2 h after agrin treatment of cultured myotubes. At that time there is clustering of several membrane proteins, including AChRs, globular cholinesterases (Wallace 1989), agrin receptors (Nastuk et al 1991), and the 43-kD protein (BG Wallace, personal communication). The aggregation of these molecules is contemporaneous and does not require new protein synthesis.

A second set of molecules becomes concentrated at these induced AChR aggregates after several hours of agrin treatment. These late-phase elements include cytoskeletal molecules and basal lamina components such as laminin and muscle-derived agrin (Wallace 1989, Nitkin & Rothschild 1990, Shadiack & Nitkin 1991, Lieth et al 1992). The appearance of the basal lamina molecules requires protein synthesis. The expression of the late-phase molecules correlates well with increased AChR cluster stability, suggesting that these molecules may be important for synaptic maturation. This biphasic cellular response also suggests that agrin initiates a cascade of signaling events, with both short- and long-term consequences.

Agrin's mechanism of action has been studied most with regard to AChR redistribution. Agrin induces AChR clustering at concentrations in the 0.1–1 pM range (Nitkin et al 1987, Ferns et al 1993). Signaling is initiated by its calcium-dependent binding to a high affinity plasma membrane receptor (see below). Formation of AChR aggregates requires metabolic energy and is inhibited by activators of protein kinase C (Wallace 1988). On the other hand, altering cyclic nucleotide levels or inhibiting protein synthesis, glycosylation, or calmodulin function does not affect agrin's activity (Wallace 1988). Further, the stoichiometry of agrin receptors to AChRs is in the range of 1:50–1:100

(Bowe et al 1994). These results indicate that intracellular signal-transduction mechanisms mediate agrin's effects.

Several lines of evidence indicate that tyrosine phosphorylation is involved in agrin-induced AChR clustering. Junctional AChRs are phosphorylated on tyrosine (Qu et al 1990). Agrin elicits an increase in tyrosine phosphorylation on AChR β-subunits in cultured myotubes (Wallace et al 1991). In addition, agrin-induced AChR clustering is temporally correlated with tyrosine phosphorylation of AChRs and is blocked by tyrosine kinase inhibitors (Wallace et al 1991, Wallace 1992, Qu & Huganir 1993). Further, AChR clustering induced by overexpression of the 43-kD protein is accompanied by tyrosine phosphorylation of unidentified non-AChR proteins (Dai et al 1993). These results raise the possibility of a link between agrin's signaling pathway and the 43-kD protein. The tyrosine kinase(s) involved in any of these cases have not been identified, but one transmembrane and two cytoplasmic tyrosine kinases present in *Torpedo* electric organ are candidates (Jennings et al 1993, Swope & Huganir 1993).

Other pathways may also mediate agrin's activity. For example, unlike in chick, tyrosine phosphorylation of AChRs in rat is not detected until after birth, several days after synapses form and agrin accumulates at them (Qu et al 1990). This observation suggests alternative or cooperative mechanisms of agrin-induced AChR clustering or a biphasic nature of agrin's action. The presence of more than one such mechanism could also account for the target-dependent bioactivity of agrin isoforms.

THE AGRIN RECEPTOR

A primary function of agrin is to convey information from the neuron to the muscle cell, information that instructs the muscle cell to form a postsynaptic apparatus in a specific location. The muscle cell in turn secretes agrin at the site of nerve contact, perhaps forming a feedback loop to amplify or stabilize this signal (Lieth & Fallon 1993). A crucial component in these interactions is the agrin cell surface receptor. Using ligand-binding techniques, our laboratory characterized agrin receptors expressed on cultured myotubes and on postsynaptic membranes from *Torpedo* electric organ. Agrin binding to these receptors occurs in physiological buffers, is saturable, and is of high affinity; half maximal binding occurs at $\sim 10^{-10}$ M (Nastuk et al 1991, Ma et al 1993). Significantly, like nerve- and agrin-induced AChR clustering, ligand binding requires extracellular calcium (Henderson et al 1984, Wallace 1988). Additionally, both agrin binding to postsynaptic membranes and agrin-induced AChR clustering on cultured myotubes exhibit a similar pH dependence; they are inhibited at pH < 7 (Wallace 1988, Ma et al 1993). Ligand binding is unaffected by pretreatment of membranes with chondroitinase, heparatinase,

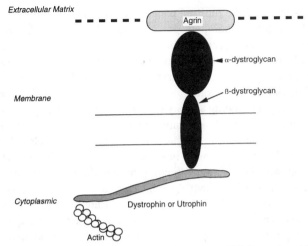

Figure 2 A schematic representation of the agrin receptor (α- and β-dystroglycan) from *Torpedo* postsynaptic membranes. The purified receptor is a complex of α-dystroglycan, which is sufficient for ligand binding, and β-dystroglycan (Bowe et al 1994). The cytoplasmic portion of β-dystroglycan can bind to dystrophin (Suzuki et al 1994). In muscle the dystroglycans are part of a larger complex consisting of adhalin, the syntrophins, and 25- and 35-kDa dystrophin-associated glycoproteins (Ervasti & Campbell 1994, Peters et al 1994; see Fallon & Hall 1994 for review).

keratinase, neuraminidase, O-glycanase, and N-glycanase (Nastuk et al 1991, Ma et al 1993, Bowe et al 1994). Neither peptides containing the RGD (arginine-glycine-aspartic acid) sequence, which inhibit many integrin-based interactions, nor the LRE (leucine-arginine-glutamic acid) tripeptide, which is involved in S-laminin interactions, inhibit agrin binding (Hunter et al 1991, Hynes 1992, Ma et al 1993, Bowe et al 1994).

Ligand-binding studies also provided the first evidence that the agrin receptor is distinct from the AChR. As described above, agrin receptors coaggregate with AChRs on agrin-stimulated myotubes. However, anti-AChR antibody-induced clustering of AChRs does not cause aggregation of agrin receptors, and AChR internalization driven by these antibodies does not deplete surface agrin receptors (Nastuk et al 1991). Further, AChRs can be separated from agrin receptors in solubilized postsynaptic membranes (Ma et al 1993).

These ligand-binding studies laid the groundwork for the identification and purification of an agrin receptor (Bowe et al 1994) (Figure 2). Monocolonal antibodies were raised against *Torpedo* electric organ postsynaptic membrane proteins and screened for their ability to immunoprecipitate solubilized agrin-binding proteins. Biochemical and immunoaffinity purification methods, together with agrin-affinity chromatography, demonstrated that the agrin receptor is a complex of two membrane glycoproteins of 190 kD and 50 kD.

Microsequencing of the purified material yielded the surprising result that the 190-kD and 50-kD polypeptides correspond to α- and β-dystoglyan, respectively. α-Dystroglycan is a highly glycosylated extrinsic peripheral membrane glycoprotein. Both proteins were originally characterized as dystrophin-associated glycoproteins (Ervasti & Campbell 1993; reviewed in Fallon & Hall 1994). When solubilized membranes were size-fractionated by gel filtration, all agrin-binding activity coeluted with the dystroglycan complex in a single symmetrical peak, indicating that this complex is the major agrin receptor in these membranes. A ligand-overlay method revealed that α-dystroglycam is sufficient to bind agrin in a calcium-dependent fashion. Moreover, α-dystroglycan was the only agrin-binding protein detected with this method. Together, these date indicate that agrin binding to α-dystroglycan is a fundamental event in the induction of postsynaptic differentiation. The associated β-dystroglycan, which has several phosphotyrosine consensus sequences and proline-rich regions in its cytoplasmic tail, could mediate the signal-transduction cascade initiated by ligand binding to α-dystroglycan.

Independently, three other laboratories used a monoclonal antibody against α-dystroglycan to show that it is the major agrin-binding protein on muscle cells and *Torpedo* membrane (Campanelli et al 1994, Gee et al 1994, Sugiyama et al 1994). They all reported that agrin binding to an α-dystroglycan, like nerve- and agrin-induced AChR clustering, is calcium dependent and is blocked by heparan. Moreover, the monoclonal antibody blocked agrin binding to α-dystroglycan on Western blots. The next step in these studies is a direct demonstration of the dystroglycan complex as the receptor that signals agrin-induced AChR clustering.

Dystroglycan as an agrin receptor has intriguing implications for our understanding of how agrin works to cluster AChRs. In the muscle cell membrane, α- and β-dystroglycan are likely to form a critical transmembrane link between dystrophin in the cytoskeleton and agrin and/or laminin in the basal lamina (Ervasti & Campbell 1993). Since dystrophin in turn binds actin, the dystroglycans could be part of the long-sought cytoskeletal mechanism underlying AChR clustering. An earlier finding by Burden and colleagues that dystrophin can be chemically cross-linked to the AChR-associated 43-kD protein (Burden et al 1983) supports an attractive model, whereby agrin binding to dystroglycan is coupled to the immobilization of AChRs.

Finally, these results provide a new way of thinking about the pathogenesis of Duchenne Muscular Dystrophy (DMD). It has been postulated that the dystrophin-associated protein complex serves a purely structural role in linking the basal lamina to the cytosekelton. Disruption of this complex following dystrophin loss may thus lead to a weakening of the cell membrane and subsequent rupture (Matsumura & Campbell 1994). The placement of the dystroglycans in a signal-transduction pathway raises the possibility that the loss of dystrophin or dystrophin-associated proteins could also lead to defective signaling in DMD or other neuromuscular disorders.

AGRIN IN THE CNS

Although neuronal synapse formation and regulation is far more complex than that observed at neuromuscular junctions, these events share important similarities. For example, postsynaptic densities, high concentrations of neurotransmitter receptors, extracellular material in the synaptic cleft, and submembranous cytoskeletal networks are characteristic of synapses in both the PNS and the CNS (McMahan & Kuffler 1971, Fertuck & Salpeter 1976, Peters et al 1976, Triller et al 1985, Jacob et al 1986, Bekkers & Stevens 1989, Killisch et al 1991, Craig et al 1993, Petralia et al 1994). Synaptic conservation at the molecular level is manifested in the observation that all the ligand-gated ion channels characterized to date, whether in the CNS or PNS, are members of the same gene superfamily (Olsen & Tobin 1990, Sargent 1993, Seeburg 1993). The key structural and functional attributes shared by these receptors suggest that widely conserved mechanisms underlie their organization in the postsynaptic membrane.

Consistent with the functional conservation of synapses across the nervous system, evidence suggests that agrin's role in directing synaptic differentiation at neuromuscular junctions may be recapitulated in the CNS. Agrin mRNA and protein are present in the brain (Rupp et al 1991, Ruegg et al 1992, Smith et al 1992, Hoch et al 1993). Agrin mRNA expression is ontogenetically regulated such that peak levels coincide with periods of maximal synaptogenesis among young neurons (Hoch et al 1993, Thomas et al 1993). In situ hybridization shows that agrin mRNA is present in subpopulations of cortical and subcortical neurons (O'Connor et al 1994). It is notable that the hippocampus and olfactory bulb, two regions distinguished by lifelong synaptic plasticity, contain neurons expressing some of the highest levels of agrin mRNA in the adult brain. Moreover, agrin message levels in the hippocampus and several linked brain regions are altered following induced epileptiform seizures (O'Connor et al 1992). Taken together, these findings suggest the involvement of agrin in synaptic differentiation and regulation in the brain.

Recent findings indicate that neurons express agrin receptors. For example, agrin binds to dendrites and cell bodies of cultured hippocampal neurons and is concentrated at some of the synapses on these cells (M Nastuk, G Banker & J Fallon, unpublished observations). Agrin also binds to isolated rat brain synaptosomes. Moreover, as observed on myotubes, agrin also induces the redistribution of its receptor on cultured neurons (Nastuk & Fallon 1991). The study of neuronal agrin receptors promises to yield important insights into the molecular mechanisms underlying synapse formation and plasticity in the brain.

ACKNOWLEDGMENTS

We thank Drs. K Deyst, M Nastuk, and M Jacob for helpful comments on this manuscript, and M Ferns and Z Hall for useful insights and for sharing unpublished observations. We also thank many colleagues who provided manuscripts to us prior to publication. The work from this laboratory was supported by NIH grant HD23924 and a postdoctoral fellowship from the Muscular Dystrophy Association to MAB.

> Any *Annual Review* chapter, as well as any article cited in an *Annual Review* chapter, may be purchased from the Annual Reviews Preprints and Reprints service. 1-800-347-8007; 415-259-5017; email: arpr@class.org

---

Literature Cited

Anderson MJ, Cohen MW. 1977. Nerve-induced and spontaneous redistribution of acetylcholine receptors on cultured muscle cells. *J. Physiol. London* 268:757–73

Anglister L, McMahan UJ. 1985. Basal lamina directs acetylcholinesterase accumulation at synaptic sites in regenerating muscle. *J. Cell Biol.* 101:735–43

Bekkers JM, Stevens CF. 1989. NMDA and non-NMDA receptors are co-localized at individual excitatory synapses in cultured rat hippocampus. *Nature* 341:230–33

Biroc SL, Payan DG, Fisher JM. 1993. Isoforms of agrin are widely expressed in the developing rat and may function as protease inhibitors. *Dev. Brain Res.* 75:119–29

Bliss TV, Collingridge GL. 1993. A synaptic model of memory: long-term potentiation in the hippocampus. *Nature* 361:31–39

Bloch RJ, Morrow JS. 1989. An unusual β-spectrin associated with clustered acetylcholine receptors. *J. Cell Biol.* 108:481–93

Bloch RJ, Resneck WG, O'Neill A, Strong J, Pumplin DW. 1991. Cytoplasmic components of acetylcholine receptor clusters of cultured rat myotubes: the 58-kD protein. *J. Cell Biol.* 115:435–46

Bowe MA, Deyst KA, Fallon JR. 1994. Identification and purification of an agrin receptor from *Torpedo* postsynaptic membranes: a heteromeric complex related to the dystroglycans. *Neuron* 12:1173–80

Bozyczko D, Decker C, Muschler J, Horwitz AF. 1987. Integrin on developing and adult skeletal muscle. *Exp. Cell Res.* 183:72–91

Brenner HR, Herczeg A, Slater CR. 1992. Synapse-specific expression of acetylcholine receptor genes and their products at original synaptic sites in rat soleus muscle fibres regenerating in the absence of innervation. *Development* 116:41–53

Burden S. 1982. Identification of an intracellular postsynaptic antigen at the frog neuromuscular junction. *J. Cell Biol.* 94: 521–30

Burden SJ, DePalma RL, Gottesman GS. 1983. Crosslinking of proteins in acetylcholine receptor-rich membranes: association between the beta-subunit and the 43 kd subsynaptic protein. *Cell* 35:687–92

Burden SJ, Sargent PB, McMahan UJ. 1979. Acetylcholine receptors in regenerating muscle accumulate at original synaptic sites in the absence of the nerve. *J. Cell Biol.* 82: 412–25

Campanelli JT, Ferns M, Hoch W, Rupp F, von Zastrow M, et al. 1992. Agrin: a synaptic basal lamina protein that regulates development of the neuromuscular junction. *Cold Spring Harbor Symp. Quant. Biol.* 57:461–72

Campanelli JT, Hoch W, Rupp F, Kreiner T, Scheller RH. 1991. Agrin mediates cell contact–induced acetylcholine receptor clustering. *Cell* 67:909–16

Campanelli JT, Roberds SL, Campbell KP, Scheller RH. 1994. A role for dystrophin-associated glycoproteins and utrophin in agrin-induced AChR clusatering. *Cell.* 77: 663–74

Champaneria S, Swenarchuk LE, Anderson MJ. 1992. Increases in pericellular proteolysis at developing neuromuscular junctions in culture. *Dev. Biol.* 149:261–77

Changeux JP, Duclert A, Sekine S. 1992. Calcitonin gene–related peptides and neuromuscular interactions. *Ann. NY Acad. Sci.* 657: 361–78

Christian CN, Daniels MP, Sugiyama H, Vogel Z, Jacques L, Nelson PG. 1978. A factor from neurons increases the number of acetylcholine receptor aggregates on cultured muscle cells. *Proc. Natl. Acad. Sci. USA* 75: 4011–15

Cohen MW, Godfrey EW. 1992. Early appearance of and neuronal contribution to agrin-like molecules at embryonic frog nerve-muscle synapses formed in culture. *J. Neurosci.* 12:2982–92

Cohen MW, Moody-Corbett F, Godfrey EW. 1994. Neuritic deposition of agrin on culture substrate: implications for nerve-muscle synaptogenesis. *J. Neurosci.* 14:3293–3303

Connold AL, Evers JV, Vrbova G. 1986. Effect of low calcium and protease inhibitors on synapse elimination during postnatal development in the rat soleus muscle. *Brain Res.* 393:99–107

Craig AM, Blackstone CD, Huganir RL, Banker G. 1993. The distribution of glutamate receptors in cultured rat hippocampal neurons: postsynaptic clustering of AMPA-selective subunits. *Neuron* 10:1055–68

Dai Z, Scotland PB, Froehner SC, Peng HB. 1993. Expression of the postsynaptic 43-kD protein in *Xenopus* embryonic cells and its association with phosphotyrosine. *Soc. Neurosci. Abstr.* 19(2):1273

Ervasti JM, Campbell KP. 1993. A role for the dystrophin-glycoprotein complex as a transmembrane linker between laminin and actin. *J. Cell Biol.* 122:809–23

Fallon JR, Gelfman CE. 1989. Agrin-related molecules are concentrated at acetylcholine receptor clusters in normal and aneural developing muscle. *J. Cell Biol.* 108:1527–35

Fallon JR, Hall ZW. 1994. Building synapses: Agrin and dystroglycan stick together, *Trends Neurosci.* 17:469–73

Fallon JR, Nitkin RM, Reist NE, Wallace BG, McMahan UJ. 1985. Acetylcholine receptor-aggregating factor is similar to molecules concentrated at neuromuscular junctions. *Nature* 315:571–74

Falls DL, Rosen KM, Corfas G, Lane WS, Fischbach GD. 1993. ARIA, a protein that stimulates acetylcholine receptor synthesis, is a member of the Neu ligand family. *Cell* 72:801–15

Ferns MJ, Campanelli JT, Hoch W, Scheller RH, Hall Z. 1993. The ability of agrin to cluster AChRs depends on alternative splicing and on cell surface proteoglycans. *Neuron* 11:491–502

Ferns MJ, Hoch W, Rupp F, Kreiner T, Scheller RH. 1992. RNA splicing regulates agrin-mediated acetylcholine receptor clustering activity on cultured myotubes. *Neuron* 8: 1079–86

Fertuck HC, Salpeter MM. 1976. Quantitation of junctional and extrajunctional acetylcholine receptors by electron microscope autoradiography after 125I-α bungarotoxin binding at mouse neuromuscular junctions. *J. Cell Biol.* 69:144–58

Frank E, Fischbach GD. 1979. Early events in neuromuscular junction formation in vitro: induction of acetylcholine receptor clusters in the postsynaptic membrane and morphology of newly formed synapses. *J. Cell Biol.* 83:143–58

Froehner SC. 1993. Regulation of ion channel distribution at synapses. *Annu. Rev. Neurosci.* 16:347–68

Gee SH, Montanaro F, Lindenbaum MH, Carbonetto S. 1994. Dystroglycan-alpha, a dystrophin-associated glycoprotein, is a functional agrin receptor. *Cell* 77:675–86

Godfrey EW, Dietz ME, Morstad AL, Wallskog PA, Yorde DE. 1988a. Acetylcholine receptor–aggregating proteins are associated with the extracellular matrix of many tissues in *Torpedo. J. Cell Biol.* 106:1263–72

Godfrey EW, Nitkin RM, Wallace BG, Rubin LL, McMahan UJ. 1984. Components of *Torpedo* electric organ and muscle that cause aggregation of acetylcholine receptors on cultured muscle cells. *J. Cell Biol.* 99:615–27

Godfrey EW, Siebenlist RE, Wallskog PA, Walters LM, Bolender DL, Yorde DE. 1988b. Basal lamina components are concentrated in premuscle masses and at early acetylcholine receptor clusters in chick embryo hindlimb muscles. *Dev. Biol.* 130:471–86

Gordon H, Hall ZW. 1989. Glycosaminoglycan variants in the C2 muscle cell line. *Dev. Biol.* 135:1–11

Gordon H, Lupa M, Bowen D, Hall Z. 1993. A muscle cell variant defective in glycosaminoglycan biosynthesis forms nerve-induced, but not spontaneous clusters of the acetylcholine receptor and the 43-kD protein. *J. Neurosci.* 13:586–95

Hall ZW, Sanes JR. 1993. Synaptic structure and development: the neuromuscular junction. *Neuron* 10:99–121 (Suppl.)

Henderson LP, Smith MA, Spitzer NC. 1984. The absense of calcium blocks impulse-evoked release of acetylcholine but not de novo formation of functional neuromuscular synaptic contacts in culture. *J. Neurosci.* 4: 3140–50

Hille B. 1992. *Ionic Channels of Excitable Membranes.* Sunderland, MA: Sinauer Assoc.

Hoch W, Ferns M, Campanelli JT, Hall ZW, Scheller RH. 1993. Developmental regulation of highly active alternatively spliced forms of agrin. *Neuron* 11:479–90

Horovitz O, Knaack D, Podleski TR, Salpeter MM. 1989. Acetylcholine receptor α-subunit mRNA is increased by ascorbic acid in cloned L5 muscle cells: Northern blot analysis and in situ hybridization. *J. Cell Biol.* 108:1823–32

Hunter DD, Cashman N, Morris VR, Bulock JW, Adams SP, Sanes JR. 1991. An LRE-(leucine-arginine-glutamate-) dependent mechanism for adhesion of neurons to S-laminin. *J. Neurosci.* 11:3960–71

Hynes RO. 1992. Integrins: versatility, modulation, and signaling in cell adhesion. *Cell* 69:11–25

Jacob MH, Lindstrom JM, Berg DK. 1986. Surface and intracellular distribution of a putative neuronal nicotinic acetylcholine receptor. *J. Cell Biol.* 103:205–14

Jennings CB, Dyer SM, Burden SJ. 1993. Muscle-specific *trk*-related receptor with a kringle domain defines a distinct class of receptor tyrosine kinases. *Proc. Natl. Acad. Sci. USA* 90:2895–99

Jessell TM, Kandel ER. 1993. Synaptic transmission: a bidirectional and self-modifiable form of cell-cell communication. *Neuron* 10(Suppl.):1–30

Jessell TM, Siegel RE, Fischbach GD. 1979. Induction of acetylcholine receptors on cultured skeletal muscle by a factor extracted from brain and spinal cord. *Proc. Natl. Acad. Sci. USA* 76:5397–5401

Jo SA, Burden SJ. 1992. Synaptic basal lamina contains a signal for synapse-specific transcription. *Development* 115:673–80

Killisch I, Dotti CG, Laurie DJ, Luddens H, Seeburg PH. 1991. Expression patterns of GABAA receptor subtypes in developing hippocampal neurons. *Neuron* 7:927–36

Knaack D, Shen I, Salpeter MM, Podleski TR. 1986. Selective effects of ascorbic acid on acetylcholine receptor number and distribution. *J. Cell Biol.* 102:795–802

Lichtman JW, Balice-Gordon RJ. 1990. Understanding synaptic competition in theory and in practice. *J. Neurobiol.* 21:99–106

Lieth E, Cardasis CA, Fallon JR. 1992. Muscle-derived agrin in cultured myotubes: expression in the basal lamina and at induced acetylcholine receptor clusters. *Dev. Biol.* 149:41–54

Lieth E, Fallon JR. 1993. Muscle agrin: neural regulation and localization at nerve-induced acetylcholine receptor clusters. *J. Neurosci.* 13:2509–14

Lisman JE, Harris KM. 1993. Quantal analysis and synaptic anatomy—integrating two views of hippocampal plasticity. *Trends Neurosci.* 16:141–47

Ma J, Nastuk MA, McKechnie BM, Fallon JR. 1993. The Agrin receptor: localization in the

postsynaptic membrane, interaction with Agrin, and relationship to the acetylcholine receptor. *J. Biol. Chem.* 268:25,108–17

Magill-Solc C, McMahan UJ. 1988. Motor neurons contain agrin-like molecules. *J. Cell Biol.* 107:1825–33

Maimone MM, Merlie JP. 1993. Interaction of the 43-kD postsynaptic protein with all subunits of the muscle nicotinic acetylcholine receptor. *Neuron* 11:53–66

Markelonis GJ, Oh TH, Eldefrawi ME, Guth L. 1982. Sciatin: a myotrophic protein increases the number of receptors and receptor clusters in cultured skeletal. *Dev. Biol.* 89: 353–61

Matsumura K, Campbell KP. 1994. Dystrophin-glycoprotein complex: its role in the molecular pathogenesis of muscular dystrophies. *Muscle Nerve* 17:2–15

McMahan UJ. 1990. The Agrin hypothesis. *Cold Spring Harbor Symp. Quant. Biol.* 50: 407–18

McMahan UJ, Kuffler SW. 1971. Visual identification of synaptic boutons on living ganglion cells and of varicosities in postganglionic axons in the heart of the frog. *Proc. R. Soc. London Ser. B* 177:485–508

McMahan UJ, Slater CR. 1984. The influence of basal lamina on the accumulation of acetylcholine receptors at synaptic sites in regenerating muscle. *J. Cell Biol.* 98:1453–73

McMahan UJ, Wallace BG. 1989. Molecules in basal lamina that direct the formation of synaptic specializations at neuromuscular junctions. *Dev. Neurosci.* 11:227–47

Murdoch AD, Dodge GR, Cohen I, Tuan RS, Iozzo RV. 1992. Primary structure of the human heparan sulfate proteoglycan from basement membrane (HSPG2/perlecan). A chimeric molecule with multiple domains homologous to the low density lipoprotein receptor, laminin, neural cell adhesion molecules, and epidermal growth factor. *J. Biol. Chem.* 267:8544–57

Nastuk MA, Fallon JR. 1991. Agrin binding to neurons of the embryonic chick spinal cord. *Soc. Neurosci. Abstr.* 17(1):219

Nastuk MA, Fallon JR. 1993. Agrin and the molecular choreography of synapse formation. *Trends Neurosci.* 16:72–76

Nastuk MA, Gordon H, Fallon JR. 1990. C2 Muscle cells express agrin binding molecules. *Soc. Neurosci. Abstr.* 16:1003

Nastuk MA, Lieth E, Ma J, Cardasis CA, Moynihan EB, et al. 1991. The putative agrin receptor binds ligand in a calcium-dependent manner and aggregates during agrin-induced acetylcholine receptor clustering. *Neuron* 7: 807–18

New HV, Mudge AW. 1986. Calcitonin gene-related peptide regulates AChR synthesis. *Nature* 323:809–11

Nitkin RM, Rothschild TC. 1990. Agrin-induced reorganization of extracellular matrix components on cultured myotubes: relationship to AChR-aggregation. *J. Cell Biol.* 111: 1161–70

Nitkin RM, Smith MA, Magill C, Fallon JR, Yao YM, et al. 1987. Identification of agrin, a synaptic organizing protein from *Torpedo* electric organ. *J. Cell Biol.* 105:2471–78

Noonan DM, Fulle A, Valente P, Cai S, Horigan E, et al. 1991. The complete sequence of perlecan, a basement membrane heparan sulfate proteoglycan, reveals extensive similarity with laminin A chain, low density lipoprotein-receptor, and the neural cell adhesion molecule. *J. Biol. Chem.* 266: 22,939–47

O'Connor LT, Lauerborn JC, Gall CM, Smith MA. 1992. Agrin gene expression in adult rat CNS and its regulation by seizure activity. *Soc. Neurosci. Abstr.* 18(1):414

O'Connor LT, Lauterborn JC, Gall CM, Smith MA. 1994. Localization and alternative splicing of agrin mRNA in adult rat brain: transcripts encoding active isoforms are not restricted to cholinergic regions. *J. Neurosci.* 14:1141–51

Olek AJ, Ling A, Daniels MP. 1986. Development of ultrastructural specializations during the formation of acetylcholine receptor aggregates on cultured myotubes. *J. Neurosci.* 6:487–97

Olsen RW, Tobin AJ. 1990. Molecular biology of GABAₐ receptors. *FASEB. J.* 4:1469–80

Oppenheim RW. 1991. Cell death during development of the nervous system. *Annu. Rev. Neurosci.* 14:453–501

Patthy L, Nikolics K. 1993. Functions of agrin and agrin-related proteins. *Trends Neurosci.* 16:76–81

Peng HB, Chen QM, de Biasi S, Zhu DL. 1989. Development of calcitonin gene–related peptide (CGRP) immunoreactivity in relationship to the formation of neuromuscular junctions in *Xenopus* myotomal muscle. *J. Comp. Neurol.* 290:533–43

Peters A, Palay SL, Webster H de F. 1976. *The Fine Structure of the Nervous System.* Philadelphia: Saunders

Peters MF, Kramarcy NR, Sealock R, Froehner SC. 1994. β-2-Syntrophin: localization at the neuromuscular junction in skeletal muscle. *NeuroReport* In press

Petralia RS, Yokotani N, Wenthold RJ. 1994. Light and electron microscope distribution of the NMDA receptor subunit NMDAR1 in the rat nervous system using a selective anti-peptide antibody. *J. Neurosci.* 14:667–96

Podleski TR, Axelrod D, Ravdin P, GreenbergI, Johnson MM, Salpeter MM. 1978. Nerve extract induces increase and redistribution of acetylcholine receptors on cloned muscle cells. *Proc. Natl. Acad. Sci. USA* 75:2035–39

Podleski TR, Salpeter MM. 1988. Acetylcholine receptor clustering and triton solubility: neural effect. *J. Neurobiol.* 19:167–85

Qu Z, Huganir RL. 1993. Tyrosine kinase inhibitors block agrin-induced acetylcholine receptor clustering on cultured chick myotubes. *Soc. Neurosci. Abstr.* 19(2):1340

Qu ZC, Moritz E, Huganir RL. 1990. Regulation of tyrosine phosphorylation of the nicotinic acetylcholine receptor at the rat neuromuscular junction. *Neuron* 4:367–78

Reist NE, Magill C, McMahan UJ. 1987. Agrin-like molecules at synaptic sites in normal, denervated, and damaged skeletal muscles. *J. Cell Biol.* 105:2457–69

Reist NE, Werle MJ, McMahan UJ. 1992. Agrin released by motor neurons induces the aggregation of acetylcholine receptors at neuromuscular junctions. *Neuron* 8:865–68

Role LW. 1988. Neural regulation of acetylcholine sensitivity in embryonic sympathetic neurons. *Proc. Natl. Acad. Sci. USA* 85: 2825–29

Role LW, Matossian VR, OBrien RJ, Fischbach GD. 1985. On the mechanism of acetylcholine receptor accumulation at newly formed synapses on chick myotubes. *J. Neurosci.* 5: 2197–204

Rothberg JM, Artavanis TS. 1992. Modularity of the slit protein. Characterization of a conserved carboxyl-terminal sequence in secreted proteins and a motif implicated in extracellular protein interactions. *J. Mol. Biol.* 227:367–70

Rubin LL, McMahan UJ. 1982. Regeneration of the neuromuscular junction: steps toward defining the molecular basis of the interaction between nerve and muscle. In *Disorders of the Motor Unit*, ed. DL Schotland, pp. 187–97. New York: Wiley

Ruegg MA, Tsim KW, Horton SE, Kroeger S, Escher G, et al. 1992. The agrin gene codes for a family of basal lamina proteins that differ in function and distribution. *Neuron* 8:691–99

Rupp F, Ozçelik TH, Linial M, Peterson K, Francke U, Scheller RH. 1992. Structure and chromosomal localization of the mammalian agrin gene. *J. Neurosci.* 12:3535–44

Rupp F, Payan DG, Magill-Solc C, Cowan DM, Scheller RH. 1991. Structure and expression of a rat Agrin. *Neuron* 6:811–23

Salpeter MM. 1987. Development and neural control of the neuromuscular junction and of the junctional acetylcholine receptor. In *The Vertebrate Neuromuscular Junction,* ed. MM Salpeter, pp. 55–115. New York: Liss

Sanes JR, Engvall E, Butkowski R, Hunter DD. 1990. Molecular heterogeneity of basal laminae: isoforms of laminin and collagen IV at the neuromuscular junction and elsewhere. *J. Cell Biol.* 111:1685–99

Sargent PB. 1993. The diversity of neuronal

nicotinic acetylcholine receptors. *Annu. Rev. Neurosci.* 16:403–43

Schwarz-Levey M, Brumwell CL, Dreyer SE, Jacob MJ. 1994. Innervation and target tissue interactions differentially regulate actylcholine receptor subunit transcript levels in developing neurons in situ. *Neuron* In press

Scotland PB, Colledge M, Melnikova I, Dai Z, Froehner SC. 1993. Clustering of the acetylcholine receptor by the 43-kD protein: involvement of the Zinc finger domain. *J. Cell Biol.* 123:719–28

Seeburg PH. 1993. The molecular biology of mammalian glutamate receptor channels. *Trends Neurosci.* 16:359–65

Shadiack AM, Nitkin RM. 1991. Agrin induces α-actinin, filamin, and vinculin to co-localize with AChR clusters on cultured myotubes. *J. Neurobiol.* 22:617–28

Simon AM, Hoppe P, Burden SJ. 1992. Spatial restriction of AChR gene expression to subsynaptic nuclei. *Development* 114:545–53

Smith JC. 1994. Hedgehog, the floor plate, and the zone of polarizing activity. *Cell* 76:193–96

Smith MA, Magill-Solc C, Rupp F, Yao Y-M, Schilling JW, et al. 1992. Isolation and characterization of a cDNA that encodes an agrin homolog in the marine ray. *Mol. Cell. Neurosci.* 3:406–17

Smith MA, O'Dowd DK. 1994. Cell-specific regulation of agrin RNA splicing in the chick ciliary ganglion. *Neuron* 12:795–804

Sollner T, Bennett JK, Whitehaeart SW, Scheller RH, Rothman JE. 1993. A protein assembly-dissembly pathway in vitro that may correspond to sequential steps of synaptic vesicle docking, activation, and fusion. *Cell* 75:409–18

Stevens CF. 1993. Quantal release of neurotransmitter and long-term potentiation. *Neuron* 10(Suppl. 1):55–64

Sudhof TC, De Camilli P, Niemann H, Jahn R. 1993. Membrane fusion machinery: insights from synaptic proteins. *Cell* 75:1–4

Sugiyama J, Bowen DC, Hall ZW. 1994. Dystroglycan binds nerve and muscle agrin. *Neuron* 13:103–15

Suzuki A, Yoshida M, Hayashi K, Mizuno Y, Hagiwara Y, Ozawa E. 1994. Molecular organization at the glycoprotein-complex-binding site of dystrophin. Three dystrophin-associated proteins bind directly to carboxy-terminal portion of dystrophin. *Eur. J. Biochem.* 220:283–92

Swope SL, Huganir RL. 1993. Molecular cloning of two abundant protein tyrosine kinases in *Torpedo* electric organ that associate with the acetylcholine receptor. *J. Biol. Chem.* 268:25,152–61

Tapscott SJ, Davis RL, Lassar AB, Weintraub H. 1990. MyoD: a regulatory gene of skeletal myogenesis. *Adv. Exp. Med. Biol.* 280:3–5

Thomas WS, O'Dowd DK, Smith MA. 1993. Developmental expression and alternative splicing of chick agrin RNA. *Dev. Biol.* 158: 525–35

Timpl R. 1989. Structure and biological activity of basement membrane proteins. *Eur. J. Biochem.* 180:487–502

Triller A, Cluzeaud F, Pfeiffer F, Betz H, Korn H. 1985. Distribution of glycine receptors at central synapses: an immunoelectron microscopy study. *J. Cell Biol.* 101:683–88

Tsim KW, Ruegg MA, Escher G, Kroeger S, McMahan UJ. 1992. cDNA that encodes active agrin. *Neuron* 8:677–89

Wallace BG. 1988. Regulation of agrin-induced acetylcholine receptor aggregation by Ca^{++} and phorbol ester. *J. Cell Biol.* 107:267–78

Wallace BG. 1989. Agrin-induced specializations contain cytoplasmic, membrane, and extracellular matrix-associated components of the postsynaptic apparatus. *J. Neurosci.* 9:1294–1302

Wallace BG. 1992. Mechanism of agrin-induced acetylcholine receptor aggregation. *J. Neurobiol.* 23:592–604

Wallace BG, Qu Z, Huganir RL. 1991. Agrin induces phosphorylation of the nicotinic acetylcholine receptor. *Neuron* 6:869–78

Witzemann V, Brenner HR, Sakmann B. 1991. Neural factors regulate AChR subunit mRNAs at rat neuromuscular synapses. *J. Cell Biol.* 114:125–41

Yayon A, Klagsbrun M, Esko JD, Leder P, Ornitz DM. 1991. Cell-surface, heparin-like molecules are required for binding of basic fibroblast growth factor to its high-affinity receptor. *Cell* 64:841–48

Annu. Rev. Neurosci. 1995. 18:463–95
Copyright © 1995 by Annual Reviews Inc. All rights reserved

MOLECULAR MECHANISMS OF DRUG REINFORCEMENT AND ADDICTION

David W. Self and Eric J. Nestler

Laboratory of Molecular Psychiatry, Departments of Psychiatry and Pharmacology, Yale University School of Medicine and Connecticut Mental Health Center, 34 Park Street, New Haven, Connecticut 06508

KEY WORDS: reward, self-administration, drug craving, nucleus accumbens, cAMP-dependent protein phosphorylation, G proteins

INTRODUCTION

Drug *reinforcement* is a form of behavioral plasticity in which behavioral changes occur in response to acute exposure to a reinforcing drug. Drugs are classified as reinforcers if the probability of a drug-seeking response is increased when the response is temporally paired with drug exposure. Such rapid and powerful associations between a drug reinforcer and a drug-seeking response probably reflect the drug's ability to directly modulate preexisting brain-reinforcement systems. These preexisting systems normally mediate the reinforcement produced by natural reinforcers such as food, sex, and social interaction. Upon acute exposure most abused drugs function as positive reinforcers, presumably because they produce a positive affective state (e.g. euphoria).

Chronic exposure to reinforcing drugs can lead to drug *addiction,* which is also characterized by an increase in drug-seeking behavior. Clinically, this sustained increase in drug-seeking behavior (i.e. *craving*) is a core feature of drug addiction. However, unlike drug reinforcement in nonaddicted subjects, addicted subjects exhibit a sustained increase in drug-seeking behavior even when the drug is absent or withdrawn. In some of these situations drugs can function as negative reinforcers in that they offer relief from a negative affective state (e.g. dysphoria). Since acute drug reinforcement and the addicted state are both characterized by an increase in drug-seeking behavior, common

463

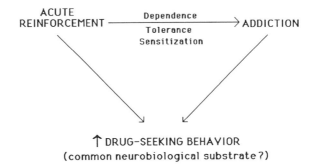

Figure 1   Diagrammatic representation of behavioral plasticity associated with drug reinforcement and addiction. Acute drug reinforcement produces an increase in drug-seeking behavior during drug exposure. Chronic drug reinforcement, through processes including dependence, tolerance, and sensitization, produces an addicted state characterized by an increase in drug-seeking behavior during drug withdrawal or even after prolonged abstinence. Similar behavioral changes during drug reinforcement and drug addiction suggest the involvement of common neurobiological substrates.

neurobiological substrates may mediate the motivational changes associated with both states (Figure 1). Therefore, persistent drug-seeking behavior in addicted subjects suggests that drug addiction could involve relatively long-lasting adaptations in brain-reinforcement systems.

Attempts to understand the mechanisms underlying drug addiction have focused on the phenomena of drug dependence, tolerance, and sensitization. Drug *dependence* is defined pharmacologically as the need for continued drug exposure to avoid a *withdrawal* syndrome, which is characterized by physical or motivational disturbances when the drug is withdrawn. We now know a great deal about the molecular and cellular adaptations in specific brain regions that underlie many of the physical symptoms of opiate dependence and withdrawal (Nestler 1992, Nestler et al 1993). In an attempt to regain homeostasis, these adaptations restore normal brain function during opiate exposure but produce overt electrophysiological and behavioral disturbances during opiate withdrawal. The motivational symptoms of drug withdrawal are also presumably mediated by molecular and cellular adaptations to chronic drug exposure; however, they may occur in brain regions associated with reinforcement. In addition to dependence, chronic drug exposure can result in *tolerance* or *sensitization,* in which specific effects of the drug decrease or increase, respectively, in response to a constant drug dose. The same drug can produce tolerance and sensitization to its many, varied effects; these changes are presumably also mediated via molecular and cellular adaptations in specific brain regions.

The extent to which these different types of long-term adaptations contribute to the overall motivational changes associated with repeated drug use remains unknown. Persistent drug craving may involve adaptations that underlie the aversive motivational symptoms associated with drug dependence and withdrawal. It may also involve adaptations that underlie tolerance to drug reinforcement, which could contribute to the pattern of escalating drug intake seen in clinical situations. These long-term adaptations can be thought of as reinforcement-opposing processes, since they oppose the acute effects of drug reinforcers. In contrast, craving could involve adaptations that underlie sensitization to drug reinforcement and to the priming effects of drugs or conditioned cues. These long-term adaptations can be thought of as reinforcement-enhancing processes, since they enhance or mimic the acute effects of drug reinforcers.

In this chapter we review the current status of molecular mechanisms of drug reinforcement and addiction. These studies have only recently begun to provide evidence that certain intracellular proteins associated with the adenosine-3′,5′-cyclic-monophosphate (cAMP) second-messenger and protein-phosphorylation system may mediate acute drug reinforcement, as well as show adaptative up-regulation following chronic drug exposure. Previous studies had demonstrated similar molecular adaptations in the cAMP second-messenger system of locus ceruleus neurons following chronic opiate exposure (Nestler 1992, Nestler et al 1993). These molecular adaptations have been shown to contribute to the increase in activity of locus ceruleus neurons during opiate withdrawal. Since increased activity of locus ceruleus neurons produces many of the physical withdrawal behaviors characteristic of opiate dependence, a direct causal link between the molecular, cellular, and behavioral changes associated with opiate physical dependence has been established. Therefore, the locus ceruleus can serve as a model system to guide investigations into the molecular mechanisms underlying drug reinforcement and other motivational aspects of addiction, such as drug craving. However, unlike the locus ceruleus the relationship between neuronal activity in reinforcement-related brain regions and the motivational symptoms of drug addiction has not yet been established. In these studies, behavioral measures of drug reinforcement and drug craving must be utilized as endpoints to directly establish the involvement of molecular mechanisms in drug addiction. Despite the inherent difficulties of attempting to modulate intracellular molecular systems in vivo, current studies are providing the first direct evidence that the same molecular systems that show adaptations to chronic drug exposure may play a role in mediating certain measures of drug reinforcement and drug craving. Furthermore, these studies suggest that both physical and motivational aspects of drug addiction are caused by similar molecular adaptations that occur in specific brain regions associated with either of the two phenomena.

THE ROLE OF THE MESOLIMBIC DOPAMINE SYSTEM IN DRUG REINFORCEMENT

Most abused drugs are considered positive reinforcers in that acute exposure leads to further drug use. Virtually all drugs that are reinforcing in people are also reinforcing in laboratory animals. A substantial behavioral and pharmacological literature has established the mesolimbic dopamine system as a major neural substrate of the reinforcement produced by opiates, psychostimulants, ethanol, nicotine, and cannabinoids (Wise 1990, Koob 1992). This system consists of dopaminergic neurons in the ventral tegmental area (VTA) and their target neurons in regions such as the nucleus accumbens (NAc). In the case of psychostimulants, rats will self-administer amphetamine (Hoebel et al 1983) and dopamine (Dworkin et al 1985) directly into the NAc. Although cocaine is not self-administered directly into the NAc (Goeders & Smith 1983), bilateral injections of dopamine-receptor antagonists into the NAc (Phillips et al 1983, Maldonado et al 1993) or lesions of the VTA-NAc dopamine neurons (Roberts et al 1977, Roberts & Koob 1982, Koob & Goeders 1989) attenuate the reinforcing effects of intravenously self-administered cocaine. These studies provide strong evidence that dopamine receptors in the NAc mediate the reinforcing effects of psychostimulants. In contrast, opiates are self-administered directly into the VTA (Bozarth & Wise, 1981), where they activate dopamine neurons via inhibition of inhibitory interneurons (Johnson & North 1992), and subsequently increase dopaminergic transmission to the NAc (Leone et al 1991). Other drugs of abuse, such as ethanol (Imperato & Di Chiara 1986) and cannabinoids (Chen et al 1990), also cause increased dopamine efflux in the NAc, leading some investigators to suggest that dopamine may be a final common neurotransmitter in opiate, psychostimulant, and other drug reinforcement (Bozarth & Wise 1986, Di Chiara & Imperato 1988).

However, opiates may also act independently of dopamine neurons. This idea is supported by the findings that opiates are self-administered directly into the NAc (Olds 1982, Goeders et al 1984) and that NAc injections of opiate antagonists (Vaccarino et al 1985) or lesions of neurons within the NAc (Zito et al 1985, Dworkin et al 1988) reduce the reinforcing efficacy of intravenously self-administered opiates. In contrast to the effects on psychostimulant self-administration, lesions of the mesolimbic dopamine projection to the NAc (Pettit et al 1984) and a blockade of dopamine receptors with dopamine antagonists (Ettenberg et al 1982) generally fail to affect intravenous heroin self-administration. These latter findings have led to the hypothesis that opiates and dopamine can act directly on separate postsynaptic receptors in the NAc and thereby independently activate a common reinforcement substrate within this brain region (Ettenberg et al 1982, Koob 1988, Cunningham & Kelley 1992). Other dopamine-independent reinforcement mechanisms in the NAc may exist, since drugs that act directly on glutamate receptors, such as phen-cyclidine (PCP) and MK-801, are self-administered into the NAc (Carlezon & Wise 1993).

Other brain regions may also be targets for drugs of abuse. Cocaine is self-administered directly into the prefrontal cortex (Goeders & Smith 1983), although dopaminergic innervation of this structure is not essential for main-taining intravenous cocaine self-adminstration (Martin-Iverson et al 1986). Opiates are self-administered into the CA3 region of the hippocampus (Stevens et al 1991, Self & Stein 1993) and the lateral hypothalamus (Olds & Williams 1980, Cazala 1990). As in the NAc, opiate reinforcement in the hippocampus is apparently dopamine independent. Although other afferent and efferent connections to these brain regions are important for the processing of reinforc-ing signals, the VTA, NAc, prefrontal cortex, hippocampus, and lateral hypo-thalamus contain the target receptors that are directly activated by most drugs of abuse. Moreover, these brain regions are all connected in a functional circuit (Trujillo et al 1993) that allows drug reinforcement to occur through simulta-neous activation of multiple reward sites. Interestingly, the NAc is a central hub in this circuit, receiving direct monosynaptic inputs from each of the other reinforcing nuclei. The NAc, therefore, may be a final common anatomical target for drugs that act on receptors in other reinforcing nuclei.

RECEPTOR SUBTYPES IN OPIATE AND PSYCHOSTIMULANT REINFORCEMENT

The pharmacological substrates of opiate and psychostimulant reinforcement have been studied with receptor subtype–selective agonists and antagonists (see Self & Stein 1992a). These studies have utilized three principal behavioral methods that measure drug reinforcement either directly or indirectly. The self-administration paradigm provides a direct test of a drug's reinforcing properties, or the ability of an antagonist to reduce drug reinforcement. In contrast, the brain stimulation reward paradigm provides an indirect test of drug reinforcement by measuring a drug's ability to enhance or reduce the reinforcing effects of electrical brain stimulation. Another indirect test of drug reinforcement is the ability of drugs to produce a conditioned place preference to a specific environment that was previously associated with drug exposure. Generally the results from studies that use these methods have provided a consistent body of evidence regarding the receptor subtypes that mediate opiate and psychostimulant reinforcement (see Table 1).

Opiates activate three major types of opioid receptors: the mu, delta, and kappa opioid receptors. Studies generally indicate that central mu and delta, but not kappa, opioid receptors mediate opiate reinforcement (Koob 1992, Self & Stein 1992a). Psychostimulants mediate reinforcement through direct or

Table 1 Receptor subtype involvement in opiate and psychostimulant reinforcement[a]

		Self-administration		Self-stimulation		Conditioned place preference	
		Agonist	Antagonist	Agonist	Antagonist	Agonist	Antagonist
Opioid	Mu	+	−	+	NT	+	−
	Delta	+	−	+	NT	+	−
	Kappa	0	0	0	NT	0	NT
Dopamine	D_1-like	+	−	0	−	0/+	−
	D_2-like	+	−	+	−	+	−

[a] (+), agonist-induced reinforcement; (−), antagonist-induced blockade of reinforcement; 0, no effect; and NT, not tested.

indirect activation of dopamine receptors. Originally, studies suggested that only the D_2-like dopamine receptors, which include the D_2, D_3, and D_4 receptors subtypes (Sibley & Monsma 1992), could directly mediate psychostimulant reinforcement (Self & Stein 1992a, Woolverton & Johnson 1992). Given the fact that D_2-like dopamine receptors and mu and delta opioid receptors all inhibit cAMP formation via G_i and G_o proteins, a common mechanism of drug reinforcement has been hypothesized at the postreceptor, signal-transduction level (Nestler 1992, Self & Stein 1992a, Self et al 1994b). Further support for a common signal-transduction mechanism in drug reinforcement comes from the finding that cannabinoid receptors also inhibit cAMP formation via G_i and G_o proteins (Howlett et al 1986, Matsuda et al 1990). In addition to inhibition of cAMP formation, these G proteins can inhibit neuronal activity through direct G_i- or G_o-activation of K^+ channels (Brown 1990) or inhibition of voltage-sensitive Ca^{2+} channels (Dolphin 1990). Therefore, the G_i and G_o proteins, and possibly the cAMP second-messenger and protein-phosphorylation system, may represent a final common intracellular pathway utilized by reinforcing drugs as diverse as opiates, psychostimulants, and cannabinoids.

However, D_1-like dopamine receptors, which include the D_1 and D_5 subtypes (Sibley & Monsma 1992), may also mediate reinforcement. Although earlier studies reported the prototypical D_1 dopamine agonist (SKF 38393) to be without reinforcing effects, two highly lipophilic D_1-selective dopamine agonists (SKF 82958 and SKF 77434) are avidly self-administered by rats (Self & Stein 1992b), and the reinforcement produced by one of these D_1 dopamine agonists was selectively antagonized by a D_1 but not a D_2 dopamine-selective antagonist (Self et al 1993b). These data provide strong evidence that D_1-like dopamine receptors can mediate direct reinforcing effects that are independent of D_2-reinforcement mechanisms. Given that D_1 dopamine receptors are coupled to the stimulatory G protein G_s, which stimulates (rather than inhibits) cAMP formation, it may be that G_s and G_i proteins can mediate reinforcement through opposing signal-transduction pathways.

This apparent contradiction can be reconciled if the two pathways operate in separate populations of neurons. This situation occurs in the caudate-putamen (dorsal striatum), where D_1 and D_2 dopamine-receptor mRNAs appear to be largely expressed by separate neuronal populations with distinct peptide content and projection regions (Gerfen 1992), and may also occur in the NAc (ventral striatum). However, functional studies have demonstrated a biochemical interaction between D_1 and D_2 dopamine receptors on cAMP accumulation in the striatum (Onali et al 1985), suggesting that these receptors may also be colocalized on the same neurons (Surmeier et al 1993). If they are, then D_1 and D_2 dopamine receptor–mediated reinforcement, if produced via opposing effects on the cAMP path- way, could depend on different temporal properties of these responses, with a sustained increase in D_1-mediated cAMP activity augmenting the consequences of a subsequent and transient decrease in cAMP activity mediated by D_2 receptors. D_1 and D_2 dopamine receptor–mediated reinforcement mechanisms may also utilize separate second-messenger systems in the same neurons. For example, in one report the ability of D_1 dopamine agonists to modulate electrophysiological responses in the NAc does not correlate with their efficacy in stimulating adenylyl cyclase (Johansen et al 1991), and Andersen et al (1990) claim that some central D_1 dopamine receptors may be coupled to other second-messenger systems. In any event, all of the opioid- and dopamine-receptor subtypes implicated in drug reinforcement are capable of modulating cAMP levels, which indirectly suggests that the G protein–cAMP system may play a role in drug reinforcement.

DIRECT EVIDENCE FOR INVOLVEMENT OF INHIBITORY G PROTEINS IN THE NUCLEUS ACCUMBENS IN DRUG REINFORCEMENT

Pertussis toxin (PTX) blocks receptor-mediated responses by selectively and irreversibly inactivating G_i and G_o proteins through catalyzing the ADP-ribosylation of a specific cysteine residue on the α subunit of the proteins (Gilman 1987). By using PTX, numerous biochemical, electrophysiological, and behavioral studies have established that mu and delta opioid and D_2 dopamine receptor–mediated responses utilize G_i and G_o proteins (see Self et al 1994b).

Recently, PTX has been used to study the role of G_i and G_o proteins in drug reinforcement. Intracerebroventricular administration of PTX prevented the development of a conditioned place preference to mu- and delta-selective opioid-receptor agonists (Suzuki et al 1991), suggesting that inhibitory G proteins mediate the positive motivational effects of opiates. Direct evidence for involvement of these G proteins in opiate reinforcement was found when PTX pretreatment in the hippocampus prevented the acquisition of intra-hippocampal morphine self-administration, and PTX injections into the VTA reduced intra-VTA morphine self-administration to levels maintained by saline (Self & Stein 1993). Thus, as in other measures of opiate action, PTX also attenuates opiate reinforcement.

A general role for G_i and G_o proteins in drug reinforcement is supported by a recent study in which intravenous self-administration of either cocaine or heroin was antagonized by bilateral intra-NAc injections of PTX (Self et al 1994b). In this study, PTX injections into the NAc increased the rate of both cocaine and heroin self-administration in a manner identical to reducing the unit dose per injection of the drug or to pretreating animals with a dopamine- or opioid-receptor antagonist, respectively. These increases in drug self-administration are thought to reflect the animal's attempt to compensate for reduced or shortened reinforcement produced by the intravenous drug injections (Koob & Goeders 1989, Self & Stein 1992a). Moreover, the NAc-PTX injections produced rightward shifts in the self-administration dose-response curves (Figure 2), a further indication that PTX antagonizes cocaine and heroin reinforcement. The onset of PTX-induced increases in cocaine and heroin self-administration coincided with the onset of decreases in G protein–ADP ribosylation in the NAc (Figure 3), suggesting that initially, the increased drug self-administration results directly from a reduction in G-protein function. The partial recovery of self-administration baselines at later time points, when ADP-ribosylation levels remained altered, may represent secondary adaptations to a primary reduction in G_i and G_o function. Further biochemical analyses found that PTX did not affect levels of $G_{s\alpha}$ and G_β subunits; thus, the PTX-induced increases in drug self-administration were likely caused by a selective effect on specific G-protein subunits and not by a generalized toxic effect of PTX. These results provide direct evidence of a common role of G_i and G_o proteins in the NAc in mediating the acute reinforcing properties of psychostimulant and opiate drugs.

In contrast, stimulation of G_s proteins in the NAc with cholera toxin (CTX), which irreversibly activates G_s proteins through the ADP-ribosylation of specific arginine residues (Gilman 1987), produces prolonged increases in cocaine self-administration (DW Self & L Stein, unpublished observations). Thus, either activation of G_s proteins with CTX or inactivation of G_i and G_o proteins with PTX produces an antagonist-like effect on cocaine self-administration. Since both manipulations ultimately lead to increased intracellular cAMP levels, these findings indirectly suggest that cAMP may antagonize drug reinforcement.

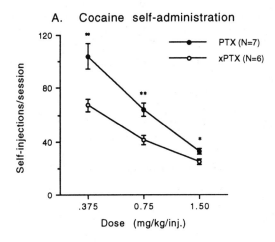

A.   Cocaine self-administration

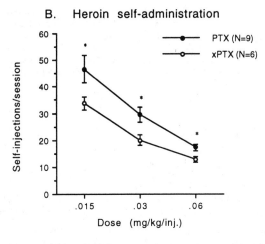

B.   Heroin self-administration

Figure 2 Effects of bilateral PTX injections (0.1 µg/1 µl/side) into the NAc on the dose-response relationship of intravenous (*A*) cocaine and (*B*) heroin self-administration 14–16 days later. The data are expressed as the mean ± the standard error of the mean (SEM) of self-administration totals as a function of injection dose for daily three-hour test sessions. Control rats received similar injections of inactive PTX (xPTX). Asterisks indicate that values differ from xPTX-treated controls: *p ≤ .05, **p ≤ .01, by Students t-test. (From Self et al 1994b.)

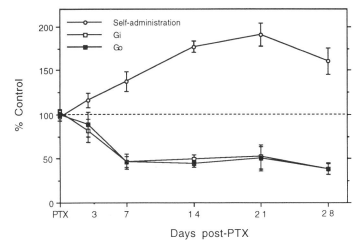

Figure 3 Time course of PTX-induced increases in cocaine self-administration and inactivation of G_i and G_o proteins in the NAc as assessed by ADP-ribosylation. A decrease in ADP-ribosylation reflects a decrease in the levels of functional G_i and G_o proteins. The data for cocaine self-administration (0.75 mg/kg/injection) are expressed as the mean ± SEM percentage of individual pre-PTX baselines ($N = 6$). The data for ADP-ribosylation experiments are expressed as the mean ± SEM percentage of the contralateral NAc injected with inactive PTX ($N = 6$ for each time point). (From Self et al 1994b.)

DIRECT EVIDENCE FOR INVOLVEMENT OF THE cAMP SYSTEM IN THE NUCLEUS ACCUMBENS IN DRUG REINFORCEMENT

The cAMP system modulates neuronal activity through a complex cascade of intracellular messengers involving protein kinases, protein phosphatases, and phosphoproteins (Nestler & Greengard 1984, 1994), and opiates and dopamine can have diverse effects on neuronal excitability and plasticity through their regulation of these intracellular messengers. We have directly tested the role of cAMP-dependent protein kinase in mediating the reinforcing effects of opiates and psychostimulants by infusing activators or inhibitors of the protein kinase into the NAc of rats prior to self-administration tests (Self et al 1993a). Rp- and Sp-cAMPS are membrane-permeable, phosphodiesterase-resistant cAMP analogues. Rp-cAMPS acts as an inhibitor and prevents endogenous cAMP from activating the kinase; Sp-cAMPS acts as an activator and mimics the effects of cAMP in activating the kinase.

Figure 4 shows the effects of infusion of either of these compounds into the NAc of rats 30 minutes prior to cocaine self-administration tests. The cAMP-dependent protein kinase inhibitor, Rp-cAMPS, dose-dependently reduced cocaine self-administration in a manner similar to increasing the unit dose per injection of cocaine or to pretreating the animal with a dopamine agonist, suggesting that this compound facilitates or prolongs the reinforcing effects of cocaine. This finding agrees with the idea that D_2 dopamine receptors produce reinforcement through inhibition of cAMP formation and, thus, through reduced cAMP-dependent protein kinase activity (see Figure 8, below). In contrast, the cAMP-dependent protein kinase activator, Sp-cAMPS, dose-dependently increased cocaine self-administration in a manner similar to reducing the self-administered dose of cocaine or to pretreating with a dopamine antagonist, suggesting that this compound reduces or shortens the reinforcing effects of cocaine. Some animals displayed bursts of cocaine self-administration; these burst-like response patterns resemble extinction-like responding, suggesting that the Sp-cAMPS produced an insurmountable antagonism of cocaine reinforcement in these animals. Other cAMP analogues that activate cAMP-dependent protein kinase also produced increases in cocaine self-administration (DW Self & EJ Nestler, unpublished observations). Taken together, the self-administration rate increases produced by the protein kinase activators, by CTX (which stimulates cAMP formation), and by PTX (which can increase cAMP levels by removing the inhibitory influence of G_i and G_o proteins on adenylyl cyclase) suggest that acute activation of cAMP-dependent protein kinase in the NAc reduces the acute reinforcing effects of cocaine. Parallel studies suggest that cAMP analogues may have similar effects

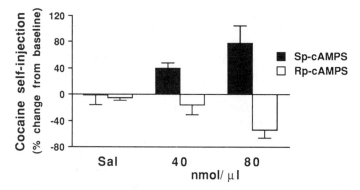

Figure 4 Effects of bilateral NAc injections of the cAMP analogues Sp-cAMPS (cAMP-dependent protein kinase activator), Rp-cAMPS (cAMP-dependent protein kinase inhibitor), or saline (Sal) on cocaine self-administration (0.5 mg/kg per injection). Each dose of the cAMP analogues was injected in 1 μl/side 30 min prior to the start of the test session. Data are taken from the first hour of the test session for Rp-cAMPS and the second hour for Sp-cAMPS, when each compound produced its maximal effects. The data are expressed at the mean ± SEM percentage change in cocaine self-administration from baseline values obtained at the preceding test session. (Based on data in Self et al 1993a.)

on heroin self-administration (DW Self & EJ Nestler, unpublished observations), thereby supporting the idea that cAMP has a common role in both psychostimulant- and opiate-reinforcement mechanisms.

The mechanism through which acute elevations in cAMP activity reduce drug reinforcement remains unknown, but current efforts are aimed at identifying specific phosphoprotein substrates that may be involved. The ability of cAMP-dependent protein kinase activators to attenuate cocaine reinforcement is consistent with the idea that D_2 dopamine receptors mediate reinforcement through inhibition of the cAMP pathway. However, as stated above in our discussion of receptor mechanisms of reinforcement, our findings are paradoxical given the role of the D_1 dopamine receptors, which stimulate cAMP formation, in promoting drug reinforcement. Possible explanations include different sites of action of the exogenously applied cAMP analogues and D_1-receptor ligands, different temporal properties of D_1- and D_2-mediated regulation of the cAMP pathway, or the involvement of other second messengers. One additional possibility, which requires further analysis, is that the protein kinase activator Sp-cAMPS could antagonize D_1 receptor–mediated responses by promoting phosphorylation and desensitization of the D_1 receptor, which has been demonstrated in vitro (Landau et al 1993). This desensitization could occur gradually within the time frame of a self-administration test session (Balmforth et al 1990, Barton & Sibley 1990) and explain why the Sp-cAMPS is most effective during the second hour of testing (see Figure 4 legend). Such D_1-receptor desensitization could also explain the progressive, within-session rate increases reported for D_1 dopamine-agonist self-administration (Self & Stein 1992b). Thus, experimental increases in cAMP activity could antagonize drug reinforcement by opposing both D_1 and D_2 dopamine-receptor mechanisms.

The cAMP analogues could also be acting presynaptically in the NAc to influence neurotransmitter release. For instance, cAMP analogues increase dopamine release in the striatum (Santiago & Westerink 1990) and may increase glutamate release from cortical inputs, but these effects would tend to augment rather than antagonize drug reinforcement. The antagonist-like effect of Sp-cAMPS on cocaine self-administration cannot be explained by possible actions on extracellular adenosine receptors because the effect is not produced immediately following injection into the NAc and because CTX, which lacks adenosine-receptor activity, produces similar results. On the other hand, the agonist-like effects of Rp-cAMPS on cocaine self-administration, which occur more rapidly after injection into the NAc (see Figure 4 legend), cannot conclusively be attributed to inhibition of cAMP-dependent protein kinase. Further studies that modulate cAMP-dependent protein kinase activity through different means (e.g. antisense oligonucleotides) are needed to strengthen the results of these initial experiments.

ADAPTATIONS IN THE MESOLIMBIC DOPAMINE SYSTEM FOLLOWING CHRONIC DRUG EXPOSURE

Although physical disturbances produced by cessation of chronic drug exposure may contribute to the increased level of drug craving seen during withdrawal, drug addicts report continued drug craving long after physical symptoms have subsided. This suggests that different brain regions mediate physical and motivational symptoms of drug dependence, a view supported by direct experimental evidence (e.g. Koob 1992). The motivational symptoms associated with drug addiction could result from drug-induced adaptations in the normal functioning of reinforcement-related brain regions, such as the VTA and NAc. Behaviorally, adaptations in the VTA-NAc pathway could produce reinforcement-enhancing processes such as sensitization or reinforcement-opposing processes such as tolerance and withdrawal-induced aversion as outlined above. At present, it is unclear whether reinforcement-enhancing processes or reinforcement-opposing processes actually trigger drug craving. Nonetheless, recent studies have found that chronic drug exposure produces adaptations at the molecular and cellular levels in VTA dopamine neurons, and in their target neurons in the NAc, that may underlie aspects of behavioral sensitization, tolerance, and aversion associated with drug addiction (Koob & Bloom 1988, Nestler 1992, Nestler et al 1993). The results from these studies have provided the basis for specific hypotheses that will now guide future investigations to more directly test the role of specific adaptations in mediating drug craving in addicted subjects.

Adaptations in VTA Dopamine Neurons

Chronic treatment with psychostimulants can produce long-term increases in drug-induced dopamine release in the NAc (e.g. Kalivas & Duffy 1993a, Robinson 1993, Wolf et al 1993). Although long-term changes in basal dopamine release were not found in these studies, the long-term increases in drug-induced dopamine release may represent reinforcement-enhancing processes that lead to compulsive drug-taking during relapse. In contrast, other studies have found short-term decreases in basal and drug-induced dopamine release following chronic treatments with psychostimulants, opiates, or ethanol (e.g. Parsons et al 1991, Pothos et al 1991, Imperato et al 1992, Rossetti et al 1992, Segal & Kuczenski 1992). These short-term decreases in dopamine release are hypothesized to underlie the aversive motivational symptoms associated with drug withdrawal and thus may represent reinforcement-opposing processes that could contribute to drug craving in the early stages of withdrawal.

Several studies have demonstrated that repeated administration of opiate and psychostimulant drugs produces a progressive sensitization to the locomotor-activating (e.g. Kalivas & Stewart 1991, Robinson 1993) and -reinforcing

(Piazza et al 1990; Horger et al 1990, 1992) effects of these drugs. Presynaptic mechanisms may play a role in the induction of behavioral sensitization, since repeated injections of opiates or psychostimulants into the VTA, but not into the NAc, mimic the sensitization seen with systemic drugs (Kalivas 1985, Kalivas & Weber 1988, Vezina & Stewart 1990, Hooks et al 1992). However, a recent report found that repeated opiate injections into the NAc produced behavioral sensitization to systemic amphetamine (Cunningham & Kelley 1992), indicating that opiate drugs can induce cross-sensitization to psychostimulants, possibly through postsynaptic mechanisms in the NAc (see below).

Repeated cocaine treatments produce transient changes in VTA function that may be involved in the development of behavioral sensitization. Chronic cocaine and amphetamine administration produce a transient increase in the spontaneous firing rate of VTA dopamine neurons in the early stages of withdrawal; the increased neuronal activity is apparently caused by an autoreceptor subsensitivity that may be due to chronic overstimulation of D_2 dopamine receptors on the perikarya (Henry et al 1989, Ackerman & White 1990, Wolf et al 1993). This subsensitivity does not appear to involve a down-regulation of D_2 dopamine autoreceptors because autoradiographic studies have failed to find changes in D_2-receptor binding in the VTA following cocaine treatments (Peris et al 1990). However, chronic cocaine administration produces transient decreases in the level of the inhibitory G proteins $G_{i\alpha}$ and $G_{o\alpha}$, which couple to D_2 autoreceptors in the VTA (Nestler et al 1990, Striplin & Kalivas 1992). The level of inhibitory G proteins in the VTA is negatively correlated with the initial level of behavioral responsiveness to cocaine (Striplin & Kalivas 1992). In addition, PTX injections into the VTA to functionally inactivate the G proteins produce behavioral sensitization to cocaine (Steketee et al 1991, 1992). Although the PTX-induced reduction in G-protein levels returns to baseline after 30 days, a sensitized behavioral response is still observed. These studies suggest that transient adaptations at the G-protein level that lead to autoreceptor subsensitivity may be involved in the induction but not the expression of behavioral sensitization to psychostimulants.

Adaptations in G-protein levels in VTA neurons may result from chronic overstimulation of VTA autoreceptors during the cocaine treatments. Studies from our and other laboratories have found that chronic cocaine (or morphine) administration increases levels of tyrosine hydroxylase (TH) in the VTA (Beitner-Johnson & Nestler 1991, Sorg et al 1993, Vrana et al 1993). Interestingly, these cocaine treatments are reported to either increase (Vrana et al 1993) or have no effect (Sorg et al 1993) on TH mRNA levels in the VTA, depending on the treatment schedule employed, raising at least the possibility that TH levels may be regulated independently of TH expression. Furthermore, the increased TH levels are only observed at short withdrawal times, similar to the transient decrease in G-protein levels. The increase in TH levels resulting in increased dopamine synthesis and the decrease in inhibitory G-protein levels resulting in autoreceptor subsensitivity could both lead to the transient increase in dopamine release in the VTA reported by Kalivas & Duffy (1993b). Chronic morphine also induces TH in the VTA (Beitner-Johnson & Nestler 1991), although the electrophysiological state of VTA neurons under these conditions has not yet been determined.

More recently, longer withdrawal from chronic cocaine treatment (10–14 days) was found to decrease, rather than increase, the spontaneous activity of VTA neurons (Ackerman & White 1992). These changes could underlie the initial decrease in basal dopamine release in the NAc seen following the chronic drug treatment as mentioned above, although longer withdrawal times have been associated with increased drug-induced dopamine release in the NAc (Kalivas & Duffy 1993a). Persistent reductions in dopamine transporters in the NAc following cessation of chronic cocaine treatments may contribute to the increased effect of cocaine on dopamine levels (Cerruti et al 1994). Thus, VTA neurons exhibit an increase followed by a decrease in spontaneous activity, whereas dopamine release in the NAc first decreases, then increases. Since these various opposing changes in VTA dopaminergic function occur while behavioral sensitization persists, chronic adaptations in VTA neurons (and in dopamine release in the NAc) apparently cannot fully account for behavioral sensitization to cocaine. Nonetheless, these adaptations may play a role in other aspects of cocaine addiction.

Adaptations in NAc Neurons

In contrast to transient adaptations in the VTA, a long-term supersensitivity (lasting up to one month) has been reported for postsynaptic D_1 dopamine receptor–mediated responses in the NAc following repeated chronic cocaine or morphine treatments (Henry & White 1991, Tjon et al 1994). It is important to distinguish these long-term adaptations in postsynaptic responsiveness in the NAc from the short-term adaptations in the VTA, where psychostimulants are believed to induce behavioral sensitization. Moreover, long-term adaptations in the NAc may be secondary to the more transient adaptations that occur in the VTA. Long-term supersensitivity of D_1-mediated responses is one possible mechanism to explain persistent drug craving during withdrawal. While chronic cocaine or amphetamine administration are reported to increase D_2 receptor–binding density in the NAc (Goeders & Kuhar 1987, Kleven et al 1990, Ziegler et al 1991), decreases in D_1 receptor–binding density have been reported for the dorsal striatum but not the NAc (Kleven et al 1990, Peris et al 1990, Farfel et al 1992, Mayfield et al 1992, Alburges et al 1993). However, one recent report found that chronic cocaine administration given by continuous infusion, rather than intermittent injections, reduced levels of D_1-receptor binding in the NAc (Neisewander et al 1994). Thus, changes at the receptor level cannot explain the functional D_1-receptor supersensitivity, suggesting that there may be an up-regulation at the postreceptor signal-transduction level. Indeed, studies from our laboratory have found that chronic cocaine treatment increases levels of adenylyl cyclase and cAMP-dependent protein kinase in the NAc (Terwilliger et al 1991), which could account for the observed D_1-receptor supersensitivity. Other reports failed to find changes in basal or dopamine-stimulated adenylyl cyclase activity following different treatment regimens and withdrawal periods from chronic cocaine administration (Mayfield et al 1992, Unterwald et al 1993). Chronic cocaine treatment also decreases the levels of G_i and G_o proteins in the NAc (Nestler et al 1990, Terwilliger et al 1991, Striplin & Kalivas 1993), which could further contribute to up-regulated cAMP activity and to D_1-receptor supersensitivity in this brain region. Studies are needed to test whether increased levels of cAMP-dependent protein kinase activity are responsible for the supersensitive electrophysiological responses to D_1 receptors in the NAc of cocaine-treated animals.

Interestingly, there is behavioral evidence for involvement of G_s proteins and cAMP-dependent protein kinase in the sensitization to psychomotor stimulants. Repeated stimulation of D_1 dopamine receptors produces a transient increase in D_1-mediated behaviors (Hu et al 1992). Cunningham & Kelley (1993) reported increased locomotor responses to psychostimulants following CTX infusions into the NAc, which persistently activates G_s proteins and stimulates cAMP formation. Recent experiments from our laboratory have found that repeated injections of the cAMP analogue 8-bromo cAMP into the NAc enhances the development of locomotor sensitization to low doses of cocaine, further suggesting that chronic stimulation of cAMP-dependent protein kinase is involved in this behavioral sensitization (Miserendino & Nestler 1995). These latter results implicate the cAMP system in NAc neurons in the expression of sensitization to psychostimulants.

Chronic morphine treatment has also been shown to up-regulate the NAc-cAMP system in a manner similar to chronic administration of cocaine (Terwilliger et al 1991). These adaptations could be thought of as homeostatic compensatory adaptations to chronic opioid receptor–induced inhibition of cAMP activity in the NAc that lead to overstimulation of the NAc-cAMP system during withdrawal. By analogy to the locus ceruleus (see Introduction), these adaptations could represent a biochemical mechanism of drug dependence that underlies the aversive motivational symptoms produced by antagonist-precipitated withdrawal in the NAc of opiate-dependent rats (Koob et al 1989, Stinus et al 1990).

Related studies have found that chronic ethanol administration produces many intracellular adaptations in the VTA and NAc similar to chronic morphine and cocaine adminstration (Nestler et al 1994). None of the other brain regions tested showed a similar response to morphine, cocaine, and ethanol,

NORMAL STATE

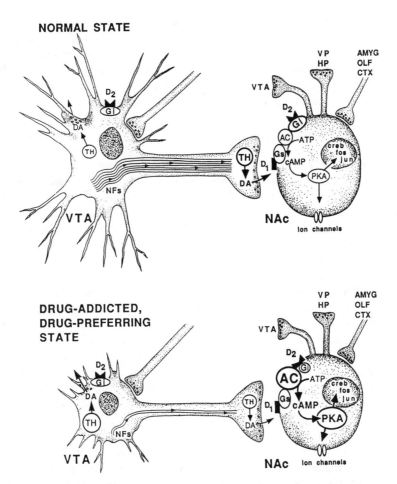

Figure 5 Schematic summary of similar biochemical manifestations hypothesized to be associated with the drug-addicted and genetically drug-preferring state. The top panel depicts a normal VTA neuron projecting to an NAc neuron. Shown in the VTA neuron are TH, dopamine (DA), presynaptic dopamine receptors (D_2) coupled to G proteins (G_i), and neurofilaments (NFs). Shown in the NAc neuron are dopamine receptors (D_1 and D_2), G proteins (G_i and G_s), components of the intracellular cAMP system [adenylyl cyclase (AC), cAMP-dependent protein kinase (PKA), and possible substrates for the kinase-ion channels and the nuclear transcription factors CREB, fos and jun], and major inputs and outputs of this region [ventral pallidum (VP), hippocampus (HP), amygdala (AMYG), olfactory cortex (OLF), and other cortical regions (CTX)]. The bottom panel depicts a VTA neuron projecting to the NAc after chronic morphine or cocaine treatment, or from a Lewis (drug-preferring) rat as compared with a relatively non-drug-preferring Fischer 344 rat. (From Beitner-Johnson & Nestler 1992.)

indicating that only the mesolimbic dopamine system responds to several abused drugs in this manner. Moreover, chronic treatment with several non-abused drugs fails to produce these effects (Terwilliger et al 1991). The fact that three main drugs of abuse produce such selective and similar effects in brain-reinforcement nuclei suggests that these adaptations play a general role in the motivational symptoms related to drug abuse (Figure 5).

Evidence for Structural Changes in the VTA-NAc Pathway

Chronic regimens of morphine or cocaine have also been shown to decrease the levels of the three major neurofilament proteins in the VTA (Beitner-Johnson et al 1992). Lower levels of neurofilament proteins, a major element of the neuronal cytoskeleton, have been associated with decreased axonal caliber and transport (see Hammerschlag & Brady 1989). Indeed, direct evidence for impairment of axonal transport following chronic morphine treatment has been demonstrated (Beitner-Johnson & Nestler 1993b). Decreased axonal transport rates could decrease the amount of TH transported from dopamine cell bodies in the VTA to nerve terminals in the NAc. At a constant rate of TH synthesis, this would tend to lead to the buildup of TH in the VTA as mentioned above. In contrast, decreased rates of TH transport to the NAc may be expected to reduce levels of TH in the NAc. Although lower levels of TH in the NAc were not found in drug-treated rats, the degree of phosphorylation of the enzyme was reduced (Beitner-Johnson & Nestler 1991). Because dephosphorylation of TH decreases its catalytic activity, the drug-induced reduction in TH phosphorylation in the NAc is probably associated with decreased enzyme activity in this brain region. Such changes in NAc-TH activity could explain the reduced levels of in vivo dopamine synthesis (Brock et al 1990) and reports of short-term reductions in in vivo levels of basal and stimulated dopamine release, during drug withdrawal. Moreover, drug-induced changes in the level of these VTA proteins and the consequent reduction in dopaminergic transmission in the NAc may lead to the drug-induced up-regulation observed in the NAc-cAMP system (Figure 5).

Other phosphoproteins in the VTA are regulated by both chronic cocaine and morphine exposure, including glial fibrillary acidic protein (GFAP; a glia-specific intermediate filament protein) and α-internexin [a neurofilament (NF)-like protein also referred to as NF-66], while other major cytoskeletal proteins in this brain region are unaffected (Beitner-Johnson et al 1992, 1993b; Nestler et al 1993). Although chronic administration of morphine and cocaine exert different effects on GFAP and on α-internexin, common drug-induced alterations in neurofilament proteins have led to the hypothesis that morphological changes in the VTA may be associated with the drug-addicted state (see Figure 5).

Neurofilament and GFAP synthesis, as well as neurite outgrowth and astrocytosis, can be influenced by a variety of neurotrophins and other growth factors. This knowledge raises the possibilities that growth factors might pharmacologically modify the actions of drugs of abuse in the mesolimbic dopamine system and that adaptations in the growth factors themselves might even contribute to some of the other adaptations observed. Preliminary evidence for both possibilities has been obtained recently. Chronic morphine administration was found to selectively decrease levels of insulin-like growth factor I (IGF I) in the VTA but not in the adjacent substantia nigra (Beitner-Johnson & Nestler 1993a). Moreover, chronic infusion into the VTA of brain-derived neurotrophic factor (BDNF) or neurotrophin-4 (NT-4), but not several related neurotrophins, can prevent the ability of morphine to both increase levels of TH and GFAP in this brain region and increase activity of the cAMP system in the NAc (Berhow et al 1994). The finding that local infusion of neurotrophins into the VTA can prevent some of the effects of systemic morphine in the NAc supports the view, stated above, that adaptations that occur in the NAc may be secondary to adaptations that occur in the VTA. Together these studies on growth factors promise to yield fundamentally new information concerning the mechanisms underlying drug addiction and its possible treatment.

Molecular Mechanisms Underlying Adaptations in the VTA-NAc System

The precise molecular mechanisms by which chronic drug treatment alters levels of specific proteins in the VTA-NAc pathway is still unknown, but studies are beginning to show that gene expression can be regulated by drug exposure (see Nestler 1992, Nestler et al 1993). Studies of the regulation of gene expression by drugs of abuse have focused on two families of transcription factors: cAMP response element–binding protein (CREB) and CREB-like proteins and the products of certain immediate early genes (IEGs), such as c-fos and c-jun. The CREB protein is phosphorylated by cAMP- or calcium-dependent protein kinases and binds to specific DNA sequences [cAMP response elements (CREs)] to regulate transcription. CREB phosphorylation does not appear to dramatically alter its DNA-binding activity but instead increases the ability of CREB bound to DNA to regulate transcription. c-Fos, c-Jun, and related IEG products can form a homodimeric or heterodimeric complex that binds to specific DNA sequences referred to as AP-1 (Activating Protein-1)-binding sites to regulate transcription (Figure 6). Most genes probably contain numerous response elements for these and many other transcription factors, suggesting that complex interactions among multiple mechanisms control the expression of a given gene.

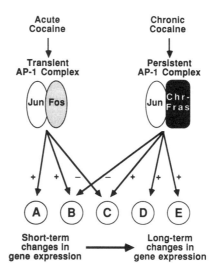

Figure 6   Scheme illustrating the regulation of AP-1-binding proteins by acute and chronic cocaine. Acute cocaine administration transiently induces several AP-1-binding proteins in the NAc, including c-Fos, c-Jun, FosB, and JunB. Induction of these transcription factors could mediate some of the rapid stimulatory (+) and inhibitory (−) effects of cocaine on the expression of neural genes (A–E) that contain AP-1-binding sites. In contrast, chronic cocaine administration results in the persistent induction of AP-1-binding activity, which is accounted for by different Fos- and/or Jun-like proteins, including recently identified proteins designated chronic Fras (Chr-Fras; Fos-related antigens). Such chronic, persistent AP-1 complexes of altered composition might exert different transcriptional effects compared with the acute AP-1 complexes. (From Nestler et al 1993.)

Acute administration of cocaine or amphetamine is known to increase the expression of c-fos, c-jun, and related IEGs and to increase AP-1 binding activity in the NAc and caudate-putamen (Graybiel et al 1990, Young et al 1991, Cole et al 1992, Hope et al 1992, Nguyen et al 1992). One possible mechanism of cocaine action is that the drug induces expression of c-fos via dopamine activation of D_1 receptors and the subsequent activation of the cAMP pathway (Young et al 1991). Cocaine and amphetamine also induce zif268 in these brain regions. Zif268 is a transcription factor that binds to a distinct response element but is regulated as an IEG product in a fashion similar to Fos- and Jun-like proteins. There is a preliminary report that acute morphine also induces IEGs in the NAc (Liu et al 1993).

The ability to induce c-fos and the other IEGs in the NAc is attenuated upon repeated cocaine treatments, while the increased AP-1-binding activity persists for at least one week after the cocaine treatments cease (Hope et al 1992). Continued AP-1 binding would, therefore, appear to reflect the induction of

IEGs—other than c-fos, c-jun, and related genes—that code for proteins with AP-1-binding activity. Indeed, recent studies from our laboratory have demonstrated that chronic cocaine administration leads to the induction and accumulation of several novel Fos-like proteins in the NAc that are candidates for the persistent AP-1-binding activity (Hope et al 1994). These novel Fos-like proteins could form a conformationally distinct AP-1 complex with unique transcriptional activity and/or specificity and thereby lead to a profile of gene expression after chronic cocaine that is qualitatively as well as quantitatively different from that induced by an acute cocaine injection (Figure 6). Such differences could conceivably account for the dramatic adaptations in G proteins and the cAMP system in the NAc following chronic cocaine exposure.

Since chronic cocaine and morphine treatments up-regulate the cAMP second-messenger system in the NAc, the CREB family of transcription factors should also be influenced by these drugs. Indeed, preliminary studies have found that these chronic drug treatments regulate CREB phosphorylation, immunoreactivity, and CRE-binding in this brain region (see Nestler et al 1993). The consequences of drug regulation of CREB phosphorylation and expression require further investigation. Similarly, studies of drug regulation of CREB-like and Fos- and Jun-like transcription factors in the VTA are now needed to identify the molecular mechanisms by which chronic drug exposure may lead to altered levels of TH, neurofilament proteins, and GFAP in this brain region.

INDIRECT EVIDENCE FOR INVOLVEMENT OF ADAPTATIONS IN THE MESOLIMBIC DOPAMINE SYSTEM IN THE MOTIVATIONAL SYMPTOMS OF DRUG ADDICTION

Certain genetically inbred rat strains exhibit differences in their tendency to self-administer opiates, cocaine, and alcohol. For example, Lewis rats self-administer each of these drugs at much higher rates than Fischer rats (Suzuki et al 1988, George & Goldberg 1989), and they also develop greater degrees of conditioned place preference to morphine and cocaine (Guitart et al 1992, Kosten et al 1994). Studies of intracellular messengers in drug-naive Lewis and Fischer rats have shown that the NAc of the Lewis rats contains lower levels of $G_{i\alpha}$, higher levels of adenylyl cyclase and cAMP-dependent protein kinase, and lower levels of TH compared to Fischer rats. In the VTA, Lewis rats have higher levels of TH and GFAP and lower levels of neurofilament proteins (Beitner-Johnson et al 1991, 1993; Guitart et al 1992, 1993). In each case, levels of these specific intracellular signaling proteins in the VTA-NAc pathway of the relatively drug-preferring Lewis rats, as compared to Fischer rats, resemble the morphine- and cocaine-induced changes in outbred Spra-gue-Dawley rats (Figure 5). Thus, while drug-naive Lewis rats are clearly not drug dependent, differences in the biochemical profile of the mesolimbic dopamine system could possibly underlie their predilection to self-administer drugs. This speculation suggests that adaptations in the mesolimbic dopamine system following chronic drug exposure could play a role in compulsive drug taking and increased levels of drug craving in addicted subjects. However, it is important to caution that Lewis and Fischer rats also differ in other behavioral responses to drugs of abuse, e.g. in stimulant-induced locomotor activation and sensitization (Kosten et al 1994). Thus, the biochemical differences between the strains could reflect a strain difference in inherent responsiveness to drugs of abuse, rather than a specific difference in drug reinforcement mechanisms per se.

Similar inherent differences in levels of these intracellular messenger proteins have been demonstrated among groups of outbred Sprague-Dawley rats. These groups of rats are distinguished by low versus high locomotor responses to a novel environment, which has been related to their reactivity to mild stress (see Piazza et al 1989). As in Lewis rats and chronically treated Sprague-Dawley rats, low-responding rats exhibited higher levels of cAMP-dependent protein kinase in the NAc and higher levels of TH and GFAP and lower levels of neurofilament proteins in the VTA, when compared to high-responding rats (Miserendino et al 1993). Based on these biochemical data, we would predict that low-responding rats resemble Lewis rats and chronically treated Sprague-Dawley rats in some drug-related behaviors regulated by the VTA and NAc. For example, the low-responding rats might be inherently drug preferring compared to the high-responding animals. However, Piazza et al (1989) have reported that high-responding animals more readily acquire amphetamine self-administration at low doses. A difficult and important question regarding these studies is whether or not an animal's inherent sensitivity to a drug's reinforcing effects is involved in the animal's vulnerability to develop drug dependence.

Another important question regarding the molecular adaptations observed in response to chronic drug treatments is whether or not similar adaptations would occur in self-administering animals. In a preliminary study, we have found that most of the same molecular adaptations occur, albeit with less magnitude, in the mesolimbic dopamine system in rats self-administering heroin, as observed previously with "forced" morphine treatments. It is important to note that while these animals are exposed to daily, self-regulated doses of opiate in six-hour self-administration test sessions, based on measures of antagonist-precipitated withdrawal, they do not display marked signs of physical dependence as do animals chronically treated with morphine pellets. As in the morphine-pelleted rats, however, levels of cAMP-dependent protein kinase activity were increased in the NAc, whereas TH levels were increased and neurofilament levels were decreased in the VTA, when compared to animals receiving similar yoked injections of saline (McClenahan et al 1994). These results indicate that lower, self-regulated doses of opiate may produce molecular adaptations in the mesolimbic dopamine system without the simultaneous development of physical dependence, suggesting that motivational symptoms of opiate addiction can develop independently of, and perhaps even more readily than, physical symptoms of opiate dependence.

POSSIBLE ROLE OF THE cAMP SYSTEM IN THE NUCLEUS ACCUMBENS IN THE MOTIVATIONAL SYMPTOMS OF DRUG ADDICTION

As stated above, chronic exposure to drugs of abuse produces motivational symptoms of drug addiction, characterized by compulsive drug taking or, in the absence of the drug, intense drug craving. Animal experiments have shown that chronic drug exposure can produce both reinforcement-enhancing and reinforcement-opposing processes that involve VTA dopamine neurons and their target neurons in the NAc, although the relative contribution of the two types of processes to drug craving remains controversial (see Introduction). Several studies suggest that reinforcement-enhancing processes actually trigger relapse to drug self-administration in addicted subjects. For example, acute morphine injections into the VTA (Stewart 1984) or amphetamine injections into the NAc (Stewart & Vezina 1988) can reinstate drug-seeking behavior in rats with previous drug self-administration experience, and amphetamine injections into the NAc can enhance the conditioned secondary reinforcement produced by cues associated with primary reinforcers (Kelley & Delfs 1991). These findings suggest that activation of the VTA-NAc pathway and the subsequent increase in dopamine levels in the NAc triggers drug craving and relapse to drug self-administration.

Conversely, opiate antagonists suppress drug-seeking behavior in rats with previous experience in self-administering heroin (Stewart & Wise 1992), and they produce an aversive stimulus that disrupts reinforced responding when injected into the NAc in opiate-dependent rats (Koob et al 1989, Stinus et al 1990). Furthermore, the threshold level of electrical brain stimulation required to reinforce responding is elevated for up to three days following a two-day cocaine self-administration binge (Markou & Koob 1991). These latter results suggest that acute drug withdrawal may be associated with a reinforcement-opposing state, in which addicted subjects are less interested in seeking drug reinforcement. Thus, while reinforcement-opposing processes may be involved in the aversive consequences of drug withdrawal, such processes may not be involved in relapse to drug self-administration, at least during early withdrawal periods. Research is now needed to test the role of the various molecular adaptations that occur in the mesolimbic dopamine system following chronic drug exposure (e.g. Figure 5) in mediating drug craving. These studies could also yield important information regarding the involvement of reinforcement-enhancing processes, such as sensitization, or reinforcement-opposing processes, such as tolerance and aversion, to the mechanisms underlying drug craving.

As with mechanisms of acute drug reinforcement, the known signal-transduction mechanisms for dopamine and opioid receptors implicate certain G proteins and the cAMP second-messenger system in drug craving and relapse to drug self-administration. Unfortunately, however, very few studies using receptor-selective ligands in animal models of drug craving or relapse have been published to date. Wise et al (1990) demonstrated that the D_2-like dopamine receptor agonist bromocriptine powerfully reinstates drug-seeking behavior in animals with experience in self-administering either heroin or cocaine. We have begun to use a similar reinstatement paradigm to test the ability of the D_2 and D_3 dopamine-selective agonists quinpirole and 7-OH-DPAT and the full D_1-like dopamine agonist SKF 82958 to reinstate drug-seeking behavior in rats with experience self-administering cocaine. Initial results from these studies suggest that the D_2 and D_3 agonists markedly and dose-dependently reinstate drug-seeking behavior, while the D_1 dopamine agonist has only minimal effects at higher, and possibly less selective, doses (DW Self, WJ Barnhart & EJ Nestler, unpublished observations). These results suggest that drug-seeking behavior can be induced by activation of D_2 and D_3 but not D_1 dopamine receptors. Given that these D_1, D_2, and D_3 dopamine-receptor agonists are each self-adminstered by rats (Wise et al 1990, Self & Stein 1992b, Caine & Koob 1993), it is interesting that they would have different abilities to reinstate drug-seeking behavior. Ironically, the D_2 dopamine-receptor agonist bromocriptine has been used clinically to treat psychostimulant addiction, despite its reinstatement properties in animals. Since D_2-like dopamine receptors inhibit cAMP formation, it could be hypothesized that inactivation of cAMP-dependent protein kinase activity in the NAc generates a priming stimulus that reinstates cocaine-seeking behavior and possibly drug craving (see Figure 8, below). We have recently provided evidence to support this hypothesis by injecting the cAMP-dependent protein kinase inhibitor Rp-cAMPS into the NAc of rats during the reinstatement paradigm (Self et al 1994a). Rp-cAMPS dose-dependently reinstates drug-seeking behavior in rats with experience self-administering cocaine, while injections of vehicle are without effect (Figure 7).

In contrast, pretreatment with D_1 but not D_2 dopamine agonists delays the initiation of cocaine self-administration (Self 1992) and blocks the ability of

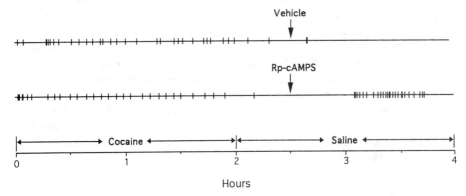

Figure 7 Effects of intra-NAc injections of vehicle or the cAMP-dependent protein kinase inhibitor Rp-cAMPS (80nmol/1 μl/side), on the reinstatement of nonreinforced responding in a representative rat. In this paradigm, animals are trained to self-administer intravenous cocaine (0.5 mg/kg/injection) for 2 h, followed by a 2-h period when only saline injections are available. Thirty minutes into the saline period, the rats receive the intra-NAc injections. The hatch marks denote the times of each self-injection response. (Based on data in Self et al 1994a.)

cocaine to reinstate drug-seeking behavior in animals maintained on cocaine (DW Self & EJ Nestler, unpublished observations). These studies suggest that D_1 dopamine agonists may actually suppress cocaine craving in addicted subjects. D_1 dopamine agonists could suppress cocaine craving through D_1 receptor–mediated activation of cAMP-dependent protein kinase in the NAc. If so, then acute activation of the NAc-cAMP system could oppose cocaine craving, as well as cocaine reinforcement (Figure 8). Furthermore, these studies suggest that D_1 dopamine agonists may have clinical value as replacement therapy in the long-term treatment of cocaine addiction, similar to methadone's use in the treatment of heroin addiction, although the effects of long-term exposure to D_1 agonists on drug craving have not yet been studied. Drugs that more directly elevate cAMP levels, such as phosphodiesterase inhibitors, could also be of clinical benefit.

A remaining central question is how the chronic drug-induced adaptations in the NAc-cAMP system relate to drug craving. Chronic drug exposure produces an up-regulation in the NAc-cAMP system as indicated by reduced amounts of G_i proteins and increased amounts of adenylyl cyclase and cAMP-dependent protein kinase. Whether drug-induced adaptations in these proteins actually reflect a functional increase in the activity of the cAMP system is unknown, since these changes could represent an attempt by NAc neurons to compensate for a persistent decrease in cAMP formation. An up-regulated NAc-cAMP system could account for tolerance to the acute reinforcement produced by drugs that inhibit cAMP activity and could account for the aversive consequences of drug withdrawal. Alternatively, chronic up-regulation of the NAc-cAMP system may produce different, even opposite effects than acute activation of the same system (as accomplished by intra-NAc injections of Sp-cAMPS). These questions can now be addressed directly by measuring the phosphorylation state of substrate proteins following acute and chronic drug exposure, and during extended withdrawal periods. Such phosphoproteins (i.e. ion channels, enzymes, receptors, cytoskeletal proteins, transcription factors) ultimately regulate the changes in neuronal responsiveness and gene expression that underlie the behavioral plasticity of drug reinforcement and addiction.

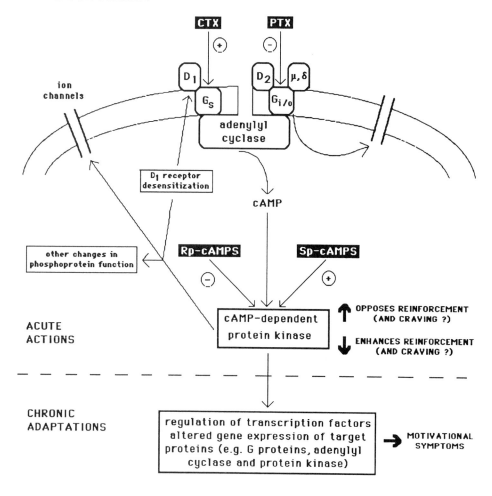

CONCLUSIONS

The studies outlined here have provided the first direct evidence for the involvement of the G protein–cAMP system in the NAc in acute drug reinforcement. Increased cAMP activity acutely opposes drug reinforcement mechanisms in the NAc while decreased cAMP activity acutely enhances it. Chronic drug exposure produces adaptations in the G protein–cAMP system in this brain region. These adaptations could represent part of the pathophysiological mechanisms underlying changes in drug-reinforcement mechanisms associated with drug addiction. The extent to which these adaptations reflect dependence, tolerance, or sensitization remains unknown. Further behavioral work is needed in fact to determine the relative contributions of each type of change in causing compulsive drug taking and drug craving. As these behavioral phenomena are better defined, it will be possible to establish the precise role of molecular adaptations in G proteins and the cAMP system in drug addiction. Ultimately, the increased understanding of the molecular basis of drug addiction will lead to improved pharmacotherapies for addictive disorders.

ACKNOWLEDGMENTS

This work was supported by USPHS grants DA 07359, DA 08220, DA 00223, DA 04060, and DA 05603, by the VA-Yale Alcoholism Research Center, and by the Abraham Ribicoff Research Facilities of the Connecticut Mental

←———————————————————————————————

Figure 8 Hypothetical model of the molecular mechanisms underlying acute drug reinforcement and the motivational symptoms of drug addiction in NAc neurons. Drug reinforcement is mediated by D_2 dopamine and μ and δ opioid receptors that activate G_i and G_o proteins, resulting in decreased adenylyl cyclase activity, cAMP formation, and cAMP-dependent protein kinase activity. G_i and G_o proteins can also mediate direct regulation of K^+ and Ca^{2+} channels (*arrow*). Decreased cAMP-dependent protein kinase activity results in a decreased phosphorylation state of target proteins that modulate neuronal activity (e.g. ion channels, enzymes, other receptors). Rp-cAMPS enhances reinforcement signals by further decreasing the phosphorylation state of these proteins. PTX blocks these reinforcement signals by inactivating G_i and G_o proteins and preventing signal transduction. CTX (through G_s stimulation of adenylyl cyclase and cAMP formation) and Sp-cAMPS (through direct stimulation of the protein kinase) also block these signals by increasing cAMP-dependent protein kinase activity and the phosphorylation state of target proteins. In this model, D_1 and D_2 receptors may be located on separate populations of NAc neurons, such that D_1 receptors produce drug-reinforcement signals independent of the D_2 receptor-cAMP cascade. However, exogenous stimulation of cAMP-dependent protein kinase activity in D_1 receptor-containing neurons could block D_1 reinforcement signals through a phosphorylation-mediated desensitization of the D_1 receptor.

The model also speculates that acute stimulation of D_1 receptors and possibly cAMP-dependent protein kinase activity blocks drug craving as well as drug reinforcement, while acute stimulation of D_2 receptors increases drug craving by decreasing cAMP-dependent protein kinase activity. Chronic drug exposure is known to up-regulate the cAMP system in the NAc (Figure 5), through mechanisms that might involve changes in gene expression (Figure 6). These and as of yet unidentified molecular adaptations may mediate the motivational symptoms (e.g. drug craving) associated with drug addiction.

Health Center, State of Connecticut Department of Mental Health. We would like to thank WJ Barnhart for his help in the preparation of this manuscript.

Literature Cited

Ackerman JM, White FJ. 1990. A10 somatodendritic dopamine autoreceptor sensitivity following withdrawal from repeated cocaine treatment. *Neurosci. Lett.* 117:181–87

Ackerman JM, White FJ. 1992. Decreased activity of rat A10 dopamine neurons following withdrawal from repeated cocaine. *Eur. J. Pharmacol.* 218:171–73

Alburges ME, Narang N, Wamsley JK. 1993. Alterations in the dopaminergic receptor system after chronic administration of cocaine. *Synapse* 14:314–23

Anderson PH, Gingrich JA, Bates MD, Dearry A, Falardeau P, et al. 1990. Dopamine receptor subtypes: beyond the D_1/D_2 classification. *Trends Pharmacol. Sci.* 11:231–36

Balmfourth AJ, Warburton P, Ball SG. 1990. Homologous desensitization of the dopamine D_1 receptor. *J. Neurochem.* 55:2111–16

Barton AC, Sibley DR. 1990. Agonist-induced sensitization of D_1-dopamine receptors linked to adenylyl cyclase activity in cultured NS20Y neuroblastoma cells. *J. Pharmacol. Exp. Ther.* 38:531–41

Beitner-Johnson D, Guitart X, Nestler EJ. 1991. Dopaminergic brain reward regions of Lewis and Fischer rats display different levels of tyrosine hydroxylase and other morphine- and cocaine-regulated phosphoproteins. *Brain Res.* 561:146–49

Beitner-Johnson D, Guitart X, Nestler EJ. 1992. Neurofilament proteins and the mesolimbic dopamine system: common regulation by chronic morphine and chronic cocaine in the rat ventral tegmental area. *J. Neurosci.* 12:2165–76

Beitner-Johnson D, Guitart X, Nestler EJ. 1993. Glial fibrillary acidic protein and the mesolimbic dopamine system: regulation by chronic morphine and Lewis-Fischer strain differences in the rat ventral tegmental area. *J. Neurochem.* 61:1766–73

Beitner-Johnson D, Nestler EJ. 1991. Morphine and cocaine exert common chronic actions on tyrosine hydroxylase in dopaminergic brain reward regions. *J. Neurochem.* 57:344–47

Beitner-Johnson D, Nestler EJ. 1992. Common intracellular actions of chronic morphine and chronic cocaine in dopaminergic brain reward regions. *Ann. NY Acad. Sci.* 654:70–87

Beitner-Johnson D, Nestler EJ. 1993a. Chronic morphine decreases insulin-like growth factor-I levels in the ventral tegmental area of the rat brain. *Ann. NY Acad. Sci.* 692:246–48

Beitner-Johnson D, Nestler EJ. 1993b. Chronic morphine impairs mesolimbic transport in the rat mesolimbic dopamine system. *NeuroReport* 5:57–60

Berhow MT, Russell DS, Self DW, Lindsay RM, Nestler EJ. 1994. Influence of neurotrophic factors on biochemical changes in the mesolimbic dopamine system associated with drugs of abuse. *Soc. Neurosci. Abstr.* 20:1102

Bozarth MA, Wise RA. 1981. Intracranial self-administration of morphine into the ventral tegmental area. *Life Sci.* 28:551–55

Bozarth MA, Wise RA. 1986. Involvement of the ventral tegmental dopamine system in opioid and psychomotor stimulant reinforcement. In *Problems of Drug Dependence: NIDA Research Monograph,* ed. LS Harris, 67:190–96. Washington, DC: NIDA

Brock JW, Ng JP, Justice JB Jr. 1990. Effect of chronic cocaine on dopamine synthesis in the nucleus accumbens as determined by microdialysis perfusion with NSD-1015. *Neurosci. Lett.* 117:234–39

Brown DA. 1990. G-proteins and potassium currents in neurons. *Annu. Rev. Physiol.* 52:215–42

Caine SB, Koob GF. 1993. Modulation of cocaine self-administration in the rat through D-3 dopamine receptors. *Science* 260:1814–16

Carlezon WA Jr., Wise RA. 1993. Rats self-administer the non-competitive NMDA receptor antagonists phencyclidine (PCP) and MK-801 directly into the nucleus accumbens. *Soc. Neurosci. Abstr.* 19:830

Cazala P. 1990. Dose-dependent effects of morphine differentiate self-administration elicited from lateral hypothalamus and mesencephalic central gray area in mice. *Brain Res.* 527:280–85

Cerruti C, Pilotte NS, Uhl G, Kuhar MJ. 1994. Reduction in dopamine transporter mRNA after cessation of repeated cocaine administration. *Mol. Brain Res.* 22:132–38

Chen J, Paredes W, Li J, Smith D, Lowinson J, Gardner EL. 1990. Δ⁹-tetrahydrocannabinol produces naloxone blockable enhancement of presynaptic dopamine eflux in nucleus accumbens of conscious, freely-moving rats as measured by intracerebral microdialysis. *Psychopharmacology* 102:156–62

Cole AJ, Bhat RV, Patt C, Worley PF, Baraban JM. 1992. D1 dopamine receptor activation of multiple transcription factor genes in rat striatum. *J. Neurochem.* 58:1420–26

Cunningham ST, Kelley AE. 1992. Evidence for opiate-dopamine cross-sensitization in nucleus accumbens: studies of conditioned reward. *Brain Res. Bull.* 29:675–80

Cunningham ST, Kelley AE. 1993. Hyperactivity and sensitization to psychostimulants following cholera toxin infusion into the nucleus accumbens. *J. Neurosci.* 13:2342–50

Di Chiara G, Imperato A. 1988. Drugs abused by humans preferentially increase synaptic dopamine concentrations in the mesolimbic system of freely moving rats. *Proc. Natl. Acad. Sci. USA* 85:5274–78

Dolphin AC. 1990. G protein modulation of calcium currents in neurons. *Annu. Rev. Physiol.* 52:243–55

Dworkin SI, Geoders NE, Smith JE. 1986. The reinforcing and rate effects of intracranial dopamine administration. In *Problems of Drug Dependence: NIDA Research Monograph*, ed. LS Harris, 67:242–48. Washington, DC: NIDA

Dworkin SI, Guerin GF, Goeders NE, Smith JE. 1988. Kainic acid lesions of the nucleus accumbens selectively attenuate morphine self-administration. *Pharmacol. Biochem. Behav.* 29:175–81

Ettenberg A, Pettit HO, Bloom FE, Koob GF. 1982. Heroin and cocaine intravenous self-administration in rats: mediation by separate neural systems. *Psychopharmacology* 78:204–9

Farfel GM, Klevin MS, Woolverton WL, Seiden LS, Perry BD. 1992. Effects of repeated injections of cocaine on catecholamine receptor binding sites, dopamine transporter binding sites and behavior in rhesus monkeys. *Brain Res.* 578:235–43

George FR, Goldberg SR. 1989. Genetic approaches to the analysis of addiction processes. *Trends Pharmacol. Sci.* 10:78–83

Gerfen CR. 1992. The neostriatal mosaic: multiple levels of compartmental organization in the basal ganglia. *Annu. Rev. Neurosci.* 15:285–320

Gilman AG. 1987. G proteins: transducers of receptor-generated signals. *Annu. Rev. Biochem.* 56:615–49

Goeders NE, Kuhar MJ. 1987. Chronic cocaine administration induces opposite changes in dopamine receptors in the striatum and nucleus accumbens. *Alcohol Drug Res.* 7:207–16

Goeders NE, Lane JD, Smith JE. 1984. Self-administration of methionine enkephalin into the nucleus accumbens. *Pharmacol. Biochem. Behav.* 20:451–55

Goeders NE, Smith JE. 1983. Cortical dopaminergic involvement in cocaine reinforcement. *Science* 221:773–75

Graybiel AM, Moratalla R, Robertson HA. 1990. Amphetamine and cocaine induce drug-specific activation of the c-*fos* gene in striosome-matrix compartments and limbic subdivisions of the striatum. *Proc. Natl. Acad. Sci. USA* 87:6912–16

Guitart X, Beitner-Johnson D, Nestler EJ. 1992. Fischer and Lewis rat strains differ in basal levels of neurofilament proteins and in their regulation by chronic morphine. *Synapse* 12:242–43

Guitart X, Kogan JH, Berhow M, Terwilliger RZ, Aghajanian GK, Nestler EJ. 1993. Lewis and Fischer rat strains show differences in biochemical, electrophysiological, and behavioral parameters: studies in the nucleus accumbens and locus coeruleus of drug naive and morphine-treated animals. *Brain Res.* 611:7–17

Hammerschlag R, Brady ST. 1989. Axonal transport and the neuronal cytoskeleton. In *Basic Neurochemistry*, ed. G Siegel, B Agranoff, RW Albers, P Molinoff, pp. 457–78. New York: Raven. 4th ed.

Henry DJ, Greene MA, White FJ. 1990. Electrophysiological effects of cocaine in the mesoaccumbens dopamine system: repeated administration. *J. Pharmacol. Exp. Ther.* 251:833–39

Henry DJ, White FJ. 1991. Repeated cocaine administration causes persistent enhancement of D1 dopamine receptor sensitivity within the rat nucleus accumbens. *J. Pharmacol. Exp. Ther.* 258:882–90

Hoebel BG, Monaco AP, Hernandez L, Aulisi EF, Stanley BG, Lenard L. 1983. Self-injection of amphetamine directly into the brain. *Psychopharmacology* 81:158–63

Hooks MS, Jones GH, Liem BJ, Justice JB Jr. 1992. Sensitization and individual differences to IP amphetamine, cocaine, or caffeine following repeated intracranial amphetamine infusions. *Pharmacol. Biochem. Behav.* 43:815–23

Hope BT, Kosofsky B, Hyman SE, Nestler EJ. 1992. Regulation of IEG expression and AP-1 binding by chronic cocaine in the rat nucleus accumbens. *Proc. Natl. Acad. Sci. USA* 89:5764–68

Hope BT, Nye HE, Kelz MB, Self DW, Iadarola MJ, et al. 1994. Induction of long-lasting AP-1 complex composed of altered

Fos-like proteins in brain by chronic cocaine and other chronic treatments. *Neuron* In press

Horger BA, Giles MK, Schenk S. 1992. Preexposure to amphetamine and nicotine predisposes rats to self-administer a low dose of cocaine. *Psychopharmacology* 107:271–76

Horger BA, Shelton K, Schenk S. 1990. Preexposure sensitizes rats to the rewarding effects of cocaine. *Pharmacol. Biochem. Behav.* 37: 707–11

Howlett AC, Qualy JM, Khachatrian LL. 1986. Involvement of G_i in the inhibition of adenylate cyclase by cannabimimetic drugs. *Mol. Pharmacol.* 29:307–13

Hu X-T, Brooderson RJ, White FJ. 1992. Repeated stimulation of D_1 dopamine receptors causes time dependent alterations in the sensitivity of both D_1 and D_2 dopamine receptors within the rat striatum. *Neuroscience* 50: 137–47

Imperato A, Di Chiara G. 1986. Preferential stimulation of dopamine release in the nucleus accumbens of freely moving rats by ethanol. *J. Pharmacol. Exp. Ther.* 239:219–28

Imperato A, Mele A, Scrocco MG, Puglici-Allegra S. 1992. Chronic cocaine alters limbic extracellular dopamine. Neurochemical basis for addiction. *Eur. J. Pharmacol.* 212:299–300

Johansen PA, Hu X-T, White FJ. 1992. Relationship between D1 dopamine receptors, adenylate cyclase, and the electrophysiological responses of rat nucleus accumbens neurons. *J. Neural Transm.* 86:97–113

Johnson SW, North RA. 1992. Opioids excite dopamine neurons by hyperpolarization of local interneurons. *J. Neurosci.* 12:483–88

Kalivas PW. 1985. Interactions between neuropeptides and dopamine neurons in the ventral medial mesencephalon. *Neurosci. Biobehav. Rev.* 9:573–87

Kalivas PW, Duffy P. 1993a. Time course of extracellular dopamine and behavioral sensitization to cocaine. I. Dopamine axon terminals. *J. Neurosci.* 13:266–75

Kalivas PW, Duffy P. 1993b. Time course of extracellular dopamine and behavioral sensitization to cocaine. II. Dopamine perikarya. *J. Neurosci.* 13:276–84

Kalivas PW, Stewart J. 1991. Dopamine transmission in the initiation and expression of drug- and stress-induced sensitization of motor activity. *Brain Res. Rev.* 16:223–44

Kalivas PW, Weber B. 1988. Amphetamine injection into the ventral mesencephalon sensitizes rats to peripheral amphetamine and cocaine. *J. Pharmacol. Exp. Ther.* 245:1095–1102

Kelley AE, Delfs JM. 1991. Dopamine and conditioned reinforcement. I. Differential effects of amphetamine microinjections into striatal

subregions. *Psychopharmacology* 103:187–96

Kleven MS, Pery BD, Woolverton WL, Seiden LS. 1990. Effects of repeated injections of cocaine on D_1 and D_2 dopamine receptors in rat brain. *Brain Res.* 532:265–70

Koob GF. 1988. Separate neurochemical substrates for cocaine and heroin reinforcement. In *Quantitative Analysis of Behavior: Biological Determinants of Reinforcement and Memory*, ed. ML Commons, RM Church, JR Stellar, AR Wagner, pp. 139–56. Hillsdale, NJ: Erlbaum

Koob GF. 1992. Drugs of abuse: anatomy, pharmacology and function of reward pathways. *Trends Pharmacol. Sci.* 13:177–84

Koob GF, Bloom FE. 1988. Cellular and molecular mechanisms of drug dependence. *Science* 242:715–23

Koob GF, Goeders NE. 1989. Neuroanatomical substrates of drug self-administration. In *The Neuropharmacological Basis of Reward*, ed. JM Liebman, SJ Cooper, pp. 214–63. Oxford: Clarendon. 433 pp.

Koob GF, Wall TL, Bloom FE. 1989. Nucleus accumbens as a substrate for the aversive stimulus effects of opiate withdrawal. *Psychopharmacology* 98:530–34

Kosten TA, Miserendino MJD, Chi S, Nestler EJ. 1994. Fischer and Lewis rats strains show differential cocaine effects in conditioned place preference and behavioral sensitization but not in locomotor activity or conditioned taste aversion. *J. Pharmacol. Exp. Ther.* 269:137–44

Landau EM, Ma CL, Healy EC, Blitzer RD, Schmauss C. 1993. The role of cAMP-dependent protein kinase in the desensitization of the human D_1 dopamine receptor. *Soc. Neurosci. Abstr.* 19:1371

Leone P, Pocock D, Wise RA. 991. Morphinedopamine interaction: ventral tegmental morphine increases nucleus accumbens dopamine release. *Pharmacol. Biochem. Behav.* 39:469–72

Liu J, Nickolenko J, Sharp FR. 1993. Morphine induction of Fos in striatum is mediated by NMDA and D1 dopamine receptors. *Soc. Neurosci. Abstr.* 19:1022

Maldonado R, Robledo P, Chover AJ, Caine SB, Koob JF. 1993. D_1 dopamine receptors in the nucleus accumbens modulate cocaine self-administration in the rat. *Pharmacol. Biochem. Behav.* 45:239–42

Markou A, Koob GF. 1991. Postcocaine anhedonia. An animal model of cocaine withdrawal. *Neuropsychopharmacology* 4:17–26

Martin-Iverson MT, Szostak C, Fibiger HC. 1986. 6-Hydroxydopamine lesions of the medial prefrontal cortex fail to influence intravenous self-administration of cocaine. *Psychopharmacology* 88:310–14

Matsuda LA, Lolait SJ, Brownstein MJ, Young

AC, Bonner TI. 1990. Structure of a cannabinoid receptor and functional expression of the cloned cDNA. *Nature* 346:561–64

Mayfield RD, Larson G, Zahniser NR. 1992. Cocaine-induced behavioral sensitization in D1 dopamine receptor function in rat nucleus accumbens and striatum. *Brain Res.* 573: 331–35

McClenahan AW, Self DW, Beitner-Johnson D, Terwilliger RZ, Nestler EJ. 1994. Heroin self-administration produces biochemical adaptations similar to chronic morphine treatment in brain reward regions. *Soc. Neurosci. Abstr.* 20:1229

Miserendino MJD, Guitart X, Terwilliger R, Chi S, Nestler EJ. 1993. Individual differences in locomotor activity are associated with levels of tyrosine hydroxylase and neurofilament proteins in the ventral tegmental area of Sprague Dawley rats. *Mol. Cell. Neurosci.* 4:440–48

Miserendino MJD, Nestler EJ. 1995. Behavioral sensitization to cocaine: modulation by the cyclic AMP system in the nucleus accumbens. *Brain Res.* In press

Neisewander JL, Lucki I, McGonigle P. 1994. Time-dependent changes in sensitivity to apomorphine and monoamine receptors following withdrawal from continuous cocaine administration in rats. *Synapse* 16:1–10

Nestler EJ. 1992. Molecular mechanisms of drug addiction. *J. Neurosci.* 12:2439–50

Nestler EJ, Greengard P. 1984. *Protein Phosphorylation in the Nervous System.* New York: Wiley

Nestler EJ, Greengard P. 1994. Protein phosphorylation and the regulation of neuronal function. In *Basic Neurochemistry: Molecular, Cellular and Medical Aspects,* ed. GJ Siegel, BW Agranoff, RW Albers, PB Molinoff, pp. 449–74. New York: Raven. 5th ed.

Nestler EJ, Guitart X, Ortiz J, Trevisan L. 1994. Second messenger and protein phosphorylation mechanisms underlying possible genetic vulnerability to alcoholism. *Ann. NY Acad. Sci.* 708:108–18

Nestler EJ, Hope BT, Widnell KL. 1993. Drug addiction: a model for the molecular basis of neural plasticity. *Neuron* 11:995–1006

Nestler EJ, Terwilliger RZ, Walker JR, Sevarino KA, Duman RS. 1990. Chronic cocaine treatment decreases levels of the G-protein subunits $G_{i\alpha}$ and $G_{o\alpha}$ in discrete regions of the rat brain. *J. Neurochem.* 55: 1079–82

Nguyen TV, Kosofsky BE, Birnbaum R, Cohen BM, Hyman SE. 1992. Differential expression of c-Fos and Zif268 in rat striatum after haloperidol, clozapine, and amphetamine. *Proc. Natl. Acad. Sci. USA* 89:4270–74

Olds ME. 1982. Reinforcing effects of morphine in the nucleus accumbens. *Brain Res.* 237:429–40

Olds ME, Williams KN. 1980. Self-administration of D-Ala²-Met-enkephalinamide at hypothalamic self-stimulation sites. *Brain Res.* 194:155–70

Onali P, Olianas MC, Gessa GL. 1985. Characterization of dopamine receptors mediating the inhibition of adenylate cyclase in rat striatum. *Mol. Pharmacol.* 28:138–45

Parsons LH, Smith AD, Justice JB Jr. 1991. Basal extracellular dopamine is decreased in the rat nucleus accumbens during abstinence from chronic cocaine. *Synapse* 9:60–65

Peris J, Boyson SJ, Cass WA, Curella P, Dwoskin LP, et al. 1990. Persistence of neurochemical changes in dopamine systems after repeated cocaine administration. *J. Pharmacol. Exp. Ther.* 253:38–44

Pettit HO, Ettenberg A, Bloom FE, Koob GF. 1984. Destruction of dopamine in the nucleus accumbens selectively attenuates cocaine but not heroin self-administration in rats. *Psychopharmacology* 84:167–73

Phillips AG, Broekkamp CL, Fibiger HC. 1983. Strategies for studying the neurochemical substrates of drug reinforcement in rodents. *Prog. Neuropsychopharmacol. Biol. Psychiatr.* 7:585–90

Piazza PV, Deminiere JM, Le Moal M, Simon H. 1989. Factors that predict individual vulnerability to amphetamine self-administration. *Science* 245:1511–13

Piazza PV, Deminiere JM, Le Moal M, Simon H. 1990. Stress- and pharmacologically-induced behavioural sensitization increases vulnerability to acquisition of amphetamine self-administration. *Brain Res.* 514:22–26

Pothos E, Rada P, Mark GP, Hoebel BG. 1991. Dopamine microdialysis in the nucleus accumbens during acute and chronic morphine, naloxone-precipitated withdrawal and clonidine treatment. *Brain Res.* 566:348–50

Roberts DCS, Corcoran ME, Fibiger HC. 1977. On the role of ascending catecholaminergic systems in intravenous self-administration of cocaine. *Pharmacol. Biochem. Behav.* 6: 615–20

Roberts DCS, Koob GF. 1982. Disruption of cocaine self-administration following 6-hydroxydopamine lesions of the ventral tegmental area in rats. *Pharmacol. Biochem. Behav.* 17:901–4

Robinson TE. 1993. Persistent sensitizing effects of drugs on brain dopamine systems and behavior: implications for addiction and relapse. See Dworkin & Smith 1993, pp. 373–402

Rosetti ZL, Hmaidan Y, Gessa GL. 1992. Marked inhibition of mesolimbic dopamine release: a common feature of ethanol, morphine, cocaine and amphetamine abstinence in rats. *Eur. J. Pharmacol.* 221:227–34

Santiago M, Westerink BHC. 1990. Role of adenylate cyclase in the modulation of the release of dopamine: a microdialysis study in the striatum of the rat. *J. Neurochem.* 55: 169–74

Segal DS, Kuczensky R. 1992. Repeated cocaine administration induces behavioral sensitization and corresponding decrease in extracellular dopamine responses in caudate and accumbens. *Brain Res.* 577:351–55

Self DW. 1992. *The role of dopamine receptors and inhibitory G proteins in stimulant and opioid reinforcement.* PhD thesis. Univ. Calif., Irvine. 231 pp.

Self DW, Barnhart WJ, Nestler EJ 1994a. Reinstatement of cocaine self-administration by a protein kinase inhibitor in the nucleus accumbens. *Soc. Neurosci. Abstr.* In press

Self DW, Chi S, Nestler EJ. 1993a. Modulation of cocaine self-administration by cyclic AMP analogues in the nucleus accumbens. *Soc. Neurosci. Abstr.* 19:1858

Self DW, Lam DM, Kossuth SR, Stein L. 1993b. Effects of D_1- and D_2-selective antagonists on SKF 82958 self-administration. In *College on the Problems of Drug Dependence: NIDA Research Monograph,* ed. L. Harris, 132:230. Washington, DC: NIDA

Self DW, Stein L. 1992a. Receptor subtypes in opioid and stimulant reward. *Pharmacol. Toxicol.* 70:87–94

Self DW, Stein L. 1992b. The D_1 agonists SKF 82958 and SKF 77434 are self-administered by rats. *Brain Res.* 582:349–52

Self DW, Stein L. 1993. Pertussis toxin attenuates intracranial morphine self-administration. *Pharmacol. Biochem. Behav.* 46:689–95

Self DW, Terwilliger RZ, Nestler EJ, Stein L. 1994b. Inactivation of G_i and G_o proteins in nucleus accumbens reduces both cocaine and heroin reinforcement. *J. Neurosci.* 14:6239–47

Sibley DR, Monsma FJ Jr. 1992. Molecular biology of dopamine receptors. *Trends Pharmacol. Sci.* 13:61–69

Sorg BA, Chen S-Y, Kalivas PW. 1993. Time course of tyrosine hydroxylase expression after behavioral sensitization to cocaine. *J. Pharmacol. Exp. Ther.* 266:424–30

Steketee JD, Striplin CD, Murray TF, Kalivas PW. 1991. Possible role for G-proteins in behavioral sensitization to cocaine. *Brain Res.* 545:287–91

Steketee JD, Striplin CD, Murray TF, Kalivas PW. 1992. Pertussis toxin in the A10 region increases dopamine synthesis and metabolism. *J. Neurochem.* 58:811–16

Stevens KE, Shiotsu G, Stein L. 1991. Hippocampal μ-receptors mediate opioid reinforcement in the CA3 region. *Brain Res.* 545:8–16

Stewart J. 1984. Reinstatement of heroin and cocaine self-administration behavior in the rat by intracerebral application of morphine in the ventral tegmental area. *Pharmacol. Biochem. Behav.* 20:917–23

Stewart J, Vezina P. 1988. A comparison of the effects of intra-accumbens injections of amphetamine and morphine on reinstatement of heroin intravenous self-administration behavior. *Brain Res.* 457:287–94

Stewart J, Wise RA. 1992. Reinstatement of heroin self-administration habits: morphine prompts and naltrexone discourages renewed responding after extinction. *Psychopharmacology* 108:79–84

Stinus L, Le Moal M, Koob GF. 1990. Nucleus accumbens and amygdala are possible substrates for the aversive stimulus effects of opiate withdrawal. *Neuroscience* 37:767–73

Striplin CD, Kalivas PW. 1992. Correlation between behavioral sensitization to cocaine and G protein ADP-ribosylation in the ventral tegmental area. *Brain Res.* 579:181–86

Striplin CD, Kalivas PW. 1993. Robustness of G protein changes in cocaine sensitization shown with immunoblotting. *Synapse* 14:10–15

Surmeier DJ, Reiner A, Levine MS, Ariano MA. 1993. Are neostriatal dopamine receptors co-localized? *Trends Neurosci.* 16:299–305

Suzuki T, Funada M, Narita M, Misawa M, Nagasi H. 1991. Pertussis toxin abolishes μ- and δ-opioid agonist-induced place preference. *Eur. J. Pharmacol.* 205:85–88

Suzuki T, George FR, Meisch RA. 1988. Differential establishment and maintainence of oral ethanol reinforced behavior in Lewis and Fischer 344 inbred rat strains. *J. Pharmacol. Exp. Ther.* 245:164–70

Terwilliger RZ, Beitner-Johnson D, Sevarino KA, Crain SM, Nestler EJ. 1991. A general role for adaptations in G-proteins and the cyclic AMP system in mediating the chronic actions of morphine and cocaine on neuronal function. *Brain Res.* 548:100–10

Trujillo KA, Herman JP, Schafer MK-H, Mansour A, Meador-Woodruff JH, et al. 1993. Drug reward and brain circuitry: recent advances and future directions. In *Biological Basis of Substance Abuse,* ed. SG Korenman, JD Barchas, pp. 119–42. New York/Oxford: Oxford Univ. Press. 516 pp.

Tjon GHK, De Vries TJ, Ronken E, Hogenboom F, Wardeh G, et al. 1994. Repeated and chronic morphine administration causes differential long-lasting changes in dopaminergic neurotransmission in rat striatum without changing its δ- and κ-opioid regulation. *Eur. J. Pharmacol.* 252: 205–12

Unterwald EM, Cox BM, Creek MJ, Cote TE, Izenwasser S. 1993. Chronic repeated cocaine administration alters basal and opioid-regulated adenylyl cyclase activity. *Synapse* 15:33–38

Vaccarino FJ, Bloom FE, Koob GF. 1985. Blockade of nucleus accumbens opiate receptors attenuates intravenous heroin reward in the rat. *Psychopharmacology* 86:37–42

Vezina P, Stewart J. 1990. Amphetamine administered to the ventral tegmental area but not to the nucleus accumbens sensitizes rats to systemic morphine: lack of conditioned effects. *Brain Res.* 516:99–106

Vrana SL, Vrana KE, Koves TR, Smith JE, Dworkin SI. 1993. Chronic cocaine administration increases CNS tyrosine hydroxylase enzyme activity and mRNA levels and tryptophan hydroxylase enzyme activity levels. *J. Neurochem.* 61:2262–68

Wise RA. 1990. The role of reward pathways in the development of drug dependence. In *Psychotropic Drugs of Abuse,* ed. DJK Balfour, pp. 23–57. Oxford: Pergamon

Wise RA, Murray A, Bozarth MA. 1990. Bromocriptine self-administration and bromocriptine-reinstatement of cocaine-trained and heroin-trained lever pressing in rats. *Psychopharmacology* 100:355–60

Wolf ME, White FJ, Nassar R, Brooderson RJ, Khansa MR. 1993. Differential development of autoreceptor subsensitivity and enhanced dopamine release during amphetamine sensitization. *J. Pharmacol. Exp. Ther.* 264: 249–55

Woolverton WL, Johnson KM. 1992. Neurobiology of cocaine abuse. *Trends Pharmacol. Sci.* 13:193–200

Young ST, Porrino LJ, Iadarola MJ. 1991. Cocaine induces striatal c-Fos-immunoreactive proteins via dopaminergic D1 receptors. *Proc. Natl. Acad. Sci. USA* 88:1291–95

Ziegler S, Lipton J, Toga A, Ellison G. 1991. Continuous cocaine administration produces persisting changes in brain neurochemistry and behavior. *Brain Res.* 552:27–35

Zito KA, Vickers G, Roberts DCS. 1985. Disruption of cocaine and heroin self-administration following kainic acid lesions of the nucleus accumbens. *Pharmacol. Biochem. Behav.* 23:1029–36

Annu. Rev. Neurosci. 1995. 18:497–529
Copyright © 1995 by Annual Reviews Inc. All rights reserved

THE ROLE OF THE FLOOR PLATE IN AXON GUIDANCE

Sophia A. Colamarino and Marc Tessier-Lavigne
Howard Hughes Medical Institute, Programs in Cell Biology, Developmental
Biology, and Neuroscience, Department of Anatomy, University of California, San
Francisco, California 94143-0452

KEY WORDS: commissural axon, circumferential migration, longitudinal migration, netrin,
             chemotropism

INTRODUCTION

Integration of information from the two sides of the body is mediated by axons
that project across the midline of the central nervous system through the
numerous commissures present at all axial levels. Establishment of these
so-called commissural projections is one of the earliest events in the wiring
of the CNS, starting soon after neural tube closure. Cells at the midline of the
embryonic neural tube are strategically positioned to influence the direction
of growth of axons near the midline and their decision whether or not to cross
to the other side. In this review we focus on the guidance role played by the
floor plate—the structure that comprises the cells occupying the ventral midline
of the developing spinal cord, hindbrain, midbrain, and caudal diencephalon.

In early development, many axonal projections are organized in an orthog-
onal grid; axons grow either longitudinally (i.e. rostrally or caudally, parallel
to the floor plate) or circumferentially (i.e. in the transverse plane along the
circumference of the neural tube, towards or away from the floor plate). Recent
studies have provided evidence that the floor plate plays important roles in
organizing both longitudinal projections near the midline and some circumferential projections, including the commissural projections. The reason for
much of the current interest in the function of the floor plate is that it appears
to provide a variety of attractive and inhibitory guidance cues, both diffusible
and short-range, that selectively influence the growth of different populations
of axons. This property makes the floor plate a potentially rich source for the
molecular identification of guidance cues.

497

Here we consider a series of choices an axon must make when navigating the midline of the embryo: whether to migrate longitudinally or circumferentially towards or away from the midline, whether to turn longitudinally before or after crossing the midline, and whether to grow rostrally or caudally along the midline. We emphasize these issues at the level of the spinal cord, highlighting the role of the floor plate in guiding spinal commissural axons, but we also review what is known about the role of the floor plate in directing axon projections at other axial levels, and its potential role in guiding cell migrations. We focus on the cellular aspects of the guidance events because few specific molecules have been implicated, though we also summarize relevant molecular data. This includes the recent identification of a family of molecules, the netrins, which are good candidates for directing circumferential migrations in the spinal cord. Remarkably, these molecules are vertebrate homologues of the UNC-6 gene product in the nematode *Caenorhabditis elegans,* which is required for circumferential migrations of axons and cells in that species (Ishii et al 1992).

THE FLOOR PLATE

The Floor Plate in Warm-Blooded Vertebrates

Morphologically the floor plate is made up of columnar ependymal cells, recognizable by their characteristic wedge-shaped appearance, that span the width of the neural tube at its ventral midline (His 1888). Floor plate cells are among the first to differentiate in the neural tube, apparently under the influence of inductive signals from the notochord (reviewed in Jessell & Dodd 1992). Morphological criteria alone have not permitted a completely unambiguous delimitation of the dimensions of the floor plate, either in the transverse plane or along the rostrocaudal axis, but recent molecular studies have suggested more precise criteria for defining these dimensions.

In the transverse plane, many molecular markers appear to be expressed by a group of cells that closely corresponds to what has previously been defined as the floor plate on morphological grounds (Chuang & Lagenaur 1990, Echelard et al 1993, Keshet et al 1991, Monaghan et al 1993, Placzek et al 1993, Riddle et al 1993, Roelink et al 1994, Ruiz i Altaba et al 1993, Sasaki & Hogan 1993). In the rat at embryonic day (E) 11-12 (i.e. when commissural axons begin to cross the midline), this group is about 15–20 cells wide (Placzek et al 1993). There may be heterogeneity within this group, however, as some other markers have either a more restricted or a somewhat greater domain of expression than the floor plate proper (McKanna 1992, McKanna & Cohen 1989, Ruiz i Altaba et al 1993).

The anterior boundary of the floor plate has historically been placed at the hindbrain-midbrain junction (i.e. the fovea isthmii) (Kingsbury 1920, 1930), but there are several reasons for believing that the floor plate actually extends through the midbrain into the caudal diencephalon and ends near the mammillary region. A variety of markers that define the floor plate at spinal cord levels appear to have a rostral limit of expression approximately at the mammillary area (Klar et al 1992, Matsui et al 1990, Placzek et al 1993, Puelles et al 1987, Ruiz i Altaba et al 1993, Sasaki & Hogan 1993). Likewise, floor plate chemo-attractant activity (see below) is found at all of these axial levels (Placzek et al 1993), as is netrin-1, a floor plate-derived chemoattractant (Kennedy et al 1994, Serafini et al 1994). In addition, prior to the bending of the neural tube to form the cephalic flexure, the mammillary area also marks the rostral extent of the notochord (Puelles et al 1987), which induces the floor plate.

The Floor Plate in Zebrafish

The floor plate in zebrafish extends from the spinal cord through the hindbrain and midbrain but not apparently into the caudal diencephalon (Hatta et al 1991). In the transverse plane of the spinal cord, the floor plate comprises a single very large cell at the ventral midline that is distinct morphologically and antigenically from other spinal cord cells (Hatta et al 1991). It had been proposed that the floor plate also includes the cells immediately flanking this midline floor plate cell, which were termed lateral floor plate cells, so that in cross section the floor plate would comprise three cells (Kuwada et al 1990). However, more recent molecular data (Krauss et al 1993; A Klar, K Hatta & T Jessell, personal communication) and the fact that neuronal cells can apparently be intercalated between the midline and these lateral cells (Bernhardt et al 1992b) suggest that the midline cell alone is the closer analogue of the floor plate in warm-blooded vertebrates.

TRAJECTORIES OF AXONS WHOSE PROJECTIONS MAY BE INFLUENCED BY THE FLOOR PLATE

The floor plate is a key landmark for many developing axons as they project to their ultimate targets during embryonic development. Given its midline location and its early differentiation, the floor plate is in a position to act as an intermediate cellular target for extending axons. In this manner it can be thought of as a vertebrate equivalent to the guidepost cells that help direct axon targeting in the invertebrate nervous system (Bovolenta & Dodd 1990, Palka et al 1992).

Any discussion of axonal trajectories in relation to the floor plate is complicated by the fact that slightly different classes of neurons interacting with the floor plate have apparently been identified in different species. For clarity, each species is treated separately. Furthermore, we restrict our discussion to neurons that have been shown to be (or are likely to be) influenced by the floor plate, with a particular focus on the earliest developing projections. Readers should refer to the Figure 1 legend for a detailed description of these embryonic neuronal populations and axonal trajectories. In brief, two types of growth patterns—circumferential and longitudinal—are characteristic of early neuronal populations of the vertebrate neural tube. Circumferential neurons are found in the more dorsal aspects of the spinal cord and comprise two classes of cells, commissural and association, whose axons initially extend in the same direction ventrally along the lateral edge of the spinal cord but that diverge upon reaching the lateral or ventrolateral marginal zone (depending on the species, see Figure 1). Axons from the association neurons turn at right angles to join the ipsilateral longitudinal pathway whereas axons from the commissural neurons cross the basal portion of the floor plate and turn longitudinally. Upon executing the turn, the axons then fasciculate amongst themselves and with other longitudinally oriented axons running in the ventral marginal zone.

Commissural neurons apparently may take two different trajectories to reach the floor plate. In chick the earliest commissural axons follow the lateral edge of the neural tube until they arrive at the ventral midline (Holley 1982; Holley & Silver 1987; Yaginuma et al 1990, 1991). Similarly, axons of the commissural neurons in zebrafish (the CoPA and CoSA neurons) (Kuwada et al 1990) and those of the commissural neurons in *Xenopus* (the dorsolateral commis-

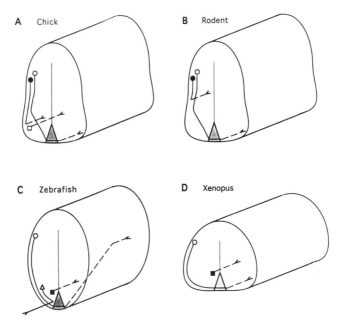

sural interneurons and commissural interneurons) (Jacobson & Huang 1985, Roberts & Clarke 1982, Roberts et al 1988) grow along the circumference of the neural tube to reach the floor plate. However, the later-born commissural neurons in chick and most of those in rodent can follow a slightly different route (Altman & Bayer 1984, Holley 1982, Wentworth 1984, Yaginuma et al 1991) (see Figure 1). Although these neurons also extend ventrally close to the lateral edge of the neural tube in the dorsal spinal cord, they break away from the edge of the spinal cord upon reaching the nascent motor column (which by this stage has become a distinct cellular mass) and course ventro-medially to reach the floor plate (see figures 9E–G in Oppenheim et al 1988). Most of these later-born axons eventually appear to fasciculate to some extent with other axons, some of which may be the early-born commissural axons

←————————————————————————————————————

Figure 1   Schematic diagram summarizing early axonal populations in the developing spinal cord whose growth may be affected by the floor plate. Note the two patterns of axonal projections—circumferential and longitudinal—characteristic of the earliest neurons. The shaded area represents the floor plate. Dorsal is up; caudal is to the left. Dashed lines indicate rostrally directed projections. The developmental stage listed for each species is that at which the first axons begin to extend. (*A*) Chick (stage 15) (Holley 1982; Holley & Silver 1987; Oppenheim et al 1988; Yaginuma et al 1990, 1991): Axons of the circumferential neurons grow ventrally along the lateral margin of the spinal cord (but not in direct contact with the external limiting membrane) in the transverse plane. Upon reaching the ventrolateral cord, axons of the ipsilaterally projecting association neurons (●) turn at right angles and project longitudinally. Axons of the contralaterally projecting commissural neurons (○) grow to the floor plate and cross the ventral midline before turning to project longitudinally. The earliest-born commissural neurons have axons that reach the floor plate by growing along the edge of the spinal cord (like the zebrafish commissural neurons in *C*). The later-born commissural axons (illustrated here) break away from the edge in the ventral spinal cord and grow ventro-medially to the floor plate. Primitive longitudinal (PL) cells (*open squares*) have longitudinally directed axons that extend either rostrally or caudally in the ventrolateral spinal cord. It is unknown whether they contact the floor plate or instead grow parallel to them at some distance. (*B*) Rodent (rat E11; mouse E9) (Altman & Bayer 1984, Holley 1982, Wentworth 1984): Commissural neurons (○) in rodent are similar to the later-born commissural neurons described in *A* for the chick, except that those in rodent have only been documented to project rostrally, while those in chick have been found to turn both rostrally and caudally. Association neurons (●) are also similar except that the earliest ones in rodent project longitudinally in the lateral, rather than ventrolateral, marginal zone. (*C*) Zebrafish (16h) (Bernhardt et al 1992, Kuwada et al 1990): Commissural (CoPA and CoSA) neurons (○) project circumferentially along the edge of the spinal cord to the ventral midline. After crossing the floor plate they turn rostrally and ascend obliquely (dorso-rostrally) for approximately one segment to join the dorsal longitudinal pathway. Association-like neurons (not shown), whose axons turn ipsilaterally, have been described (CiA and CiD neurons) but are not well characterized. Axons of the VeLD neurons (*open triangles*) initially extend circumferentially to contact the floor plate but do not cross the midline and instead turn to project caudally. Kolmer-Agduhr (KA) neurons (■) have longitudinally directed axons that extend rostrally in the ventral spinal cord. Whether they too contact the floor plate before turning is unknown. (*D*) Xenopus (stage 25) (Dale et al 1987, Jacobson & Huang 1985, Roberts & Clarke 1982, Roberts et al 1988): Commissural neurons (dorsolateral commissural interneurons and commissural interneurons) (○) project circumferentially along the edge of the spinal cord to the ventral midline. After crossing the floor plate they project longitudinally either rostrally, caudally, or with branches in both directions. See *C* for description of KA neurons (■).

which have become passively displaced away from the pial surface in the ventral spinal cord due to the birth and lateral migration of the motoneurons and end up in a more medial position (Holley 1982, Oppenheim et al 1988).

Longitudinal neurons differentiating in the ventral neural tube have been described in most species. Their axons extend in the ventral or ventrolateral marginal zone, parallel to the floor plate. The different types of longitudinal neurons include the primitive longitudinal (PL) neurons in chick (Yaginuma et al 1990), the Kolmer-Agduhr (KA) neurons in *Xenopus* (Dale et al 1987), and the KA and VeLD neurons in zebrafish (Bernhardt et al 1992b, Kuwada et al 1990). Longitudinal neurons might also be present in rodents, but they have not yet been documented.

Axons that grow to the ventral midline, cross the midline, and then turn to project longitudinally are also found at higher axial levels and have been particularly well characterized in zebrafish. Identified vestibulospinal and reticulospinal neurons in the brainstem project to the spinal cord by sending axons in the bilateral pair of medial longitudinal fascicli (MLF), which run adjacent to the ventral midline (Kimmel et al 1982, Metcalfe et al 1986). The majority contribute descending axons to the ipsilateral MLF, but several cell types project in the contralateral MLF. Among those that project contralaterally are the axons of the Mauthner neurons, which are the first to cross the midline and to pioneer the contralateral MLF in their region of the hindbrain.

What Do These Neurons Become in the Adult?

Each of these classes of neurons described above is likely to comprise a heterogenous population of cells. Detailed morphological and immunocyto-chemical analyses have been used to define many different subtypes of commissural neurons in the cold-blooded vertebrates (Kuwada et al 1990, Roberts et al 1988) and of commissural and association neurons in the warm-blooded vertebrates (Oppenheim et al 1988, Silos-Santiago & Snider 1992, Silos-Santiago & Snider 1994). This heterogeneity within populations no doubt reflects an underlying heterogeneity in function. However, although the axonal trajectories of the early neurons detailed above have been studied in the embryo, their eventual identity in the adult has not been fully established. Oppenheim et al (1988) suggested that some of the commissural axons may serve as intersegmental spinal cord interneurons. In addition, many of these neurons may serve as supraspinal projection neurons. Long-survival [^{3}H] thymidine radiograms in rat identified the early commissural neurons as belonging to the spinothalamic, spinoreticular, and spinocerebellar projection neurons (Altman & Bayer 1984). Modern axon tracing techniques should allow for a more precise determination of the final targets of these neuronal populations.

CUES

Despite superficial differences between species, there are two basic patterns of anatomical projections in the early spinal cord—circumferential and longitudinal. These growth patterns exist from the earliest stages of development, and the decision to send axons into either one or the other of these two modes of growth appears to be a general feature of axon extension in the developing vertebrate embryo. Cues responsible for generating this stereotyped pattern of axon trajectories not only need to be present very early on, but also must be capable of directing multiple classes of cells. Moreover although many of the late-projecting axons appear to fasciculate with earlier ones, the earliest axonal projections occur in terrain devoid of other axons, so the cues that direct circumferential and longitudinal projections and the switching between these modes of extension must derive from the early neuroepithelium or adjacent tissues. These cues must instruct the axon (a) to initiate growth in the proper direction (either circumferentially or longitudinally) and (b) once at the midline to make the proper choice concerning whether to cross the midline or to remain ipsilateral and then to turn longitudinally in the proper direction and remain in the appropriate fascicle. We now review the experimental evidence indicating that the floor plate is one source of cues responsible for directing several of these decisions.

GROWING TO THE MIDLINE ALONG A CIRCUMFERENTIAL TRAJECTORY

Almost all of the different populations of neurons discussed in detail above grow circumferentially at least over a portion of their trajectory, the majority growing all the way to the ventral midline. Only the PL cells in the chick (and possibly the KA cells in *Xenopus* and zebrafish) are known to initiate longitudinal growth directly (Yaginuma et al 1990). In this section we therefore focus on the cues that guide circumferential growth in the direction of the ventral midline. There is evidence that the mechanisms involved include both short-range local guidance cues derived from non–floor plate cells and long-range chemoattractants emanating from floor plate cells. Short-range cues may serve primarily to direct growth along the edge of the spinal cord, while at least one role for the chemoattractant may be to direct axons that have to grow to the ventral midline through the cellular environment of the motor column.

Attraction of Commissural Axons to the Midline by a Floor Plate–Derived Chemoattractant

In warm-blooded vertebrates there is clear evidence that the floor plate has a potent chemotropic effect on commissural axons, a mechanism suggested by

Ramon y Cajal (Ramon y Cajal 1909). In 1938 Weber obtained evidence for this type of mechanism through experimental manipulations in chick embryos (Weber 1938). Rough transection of the early developing spinal cord was followed by healing of the spinal cord into separate vesicles, such that in some cases the dorsal aspect of the spinal cord was separated from the ventral regions by a blood clot. Nonetheless, commissural axons originating from the dorsal cord appeared to home in on the floor plate, growing through the clot as necessary (Weber 1938). These early studies were subsequently forgotten but were revived in recent years when in vitro experiments showed that explants of rat floor plate tissue can promote the outgrowth of commissural axons from rat dorsal spinal cord explants into three-dimensional collagen matrices (Tessier-Lavigne et al 1988). Furthermore, in vitro experiments with both rodent dorsal spinal cord (Placzek et al 1990a, Tessier-Lavigne et al 1988) and chick dorsal spinal cord (SA Colamarino & M Tessier-Lavigne, unpublished observations) have shown that floor plate cells not only possess an outgrowth-promoting activity but can also attract commissural axons, causing them to reorient growth within dorsal explants over about 100–250µm towards an ectopically located floor plate. There is a high degree of specificity in the interaction between the floor plate and commissural neurons: The floor plate's ability to induce turning of commissural axons is not mimicked by explants of any other portion of the neural tube (although a small amount of activity can be found in the ventral spinal cord at later stages) (Placzek et al 1990a, Tessier-Lavigne et al 1988). Moreover, within the spinal cord the chemoattractant selectively affects commissural axons. The axons of association neurons and motoneurons, which do not project to the floor plate in vivo, do not reorient in response to the floor plate in culture (Placzek et al 1990a, Tessier-Lavigne et al 1988).

Studies in vivo have confirmed the existence of a chemotropic activity capable of attracting commissural neurons at a distance. In chick embryos, rotation of a segment of the spinal cord such that the floor plate becomes apposed to the dorsal half of the remainder of the neural tube elicits redirection of commissural axons towards the ectopic floor plate (Yaginuma & Oppenheim 1991). Similarly, in Danforth short-tail, a mouse mutant lacking the floor plate at the most caudal levels, commissural axons at levels 120µm caudal to the end of the floor plate turn rostrally to grow to the floor plate (Bovolenta & Dodd 1991). More dramatically, a floor plate grafted alongside the spinal cord of a developing chick embryo in ovo causes commissural axons to pierce the external limiting membrane and project abnormally out of the spinal cord towards the graft (Placzek et al 1990b).

Thus, both in vitro and in vivo experiments have demonstrated the ability of the floor plate to attract commissural neurons over hundreds of microns, a mechanism that is likely to contribute to the guidance of commissural axons in the circumferential plane to the ventral midline. It has been suggested that use of a long-range diffusible factor to induce growth of axons to the midline may not be specific to vertebrates. Klämbt et al (1991) have proposed that in *Drosophila,* where midline cells of the CNS are important in the formation of the commissures, attraction at a distance occurs for at least some sets of axons. In grasshopper, time-lapse recordings have revealed the apparently directed behavior of Q1 axons approaching the midline, leading to the hypothesis that a diffusible cue emanating from the midline region may be responsible for inducing the medial growth of these axons (see Myers & Bastiani 1993b).

Growth Parallel to the Edge of the Spinal Cord Does Not Require the Floor Plate

Although the floor plate clearly has a chemotropic activity, there is also evidence for the existence of a default pathway for axon growth independent of the floor plate along the edge of the spinal cord. In vitro experiments have demonstrated that cues specifying a ventral migration must be intrinsic to the dorsal neuroepithelium. Commissural axons in rat dorsal spinal cord explants cultured in the absence of floor plate still grow straight along their correct dorsoventral trajectory, which is near the edge of the spinal cord, until eventually reaching the cut edge of the tissue (Placzek et al 1990a, Tessier-Lavigne et al 1988; see also Nornes et al 1990). Evidence for the presence of non–floor plate–derived cues has also come from studies in vivo in embryos lacking a floor plate. At levels that lack a floor plate in the mouse Danforth short-tail mutant, commissural axons still succeed in reaching the ventral midline by growing along the circumference of the neural tube (Bovolenta & Dodd 1991). In analogous experiments in chick, where floor plate differentiation is prevented by notochord removal, commissural axons still reach the midline, again by growing along the edge of the spinal cord (Yamada et al 1991). However, it is important to note that, in both the mouse mutant and the chick notochord-removal studies, the commissural axons were navigating in a spinal cord devoid of motoneurons (i.e. a dorsalized spinal cord). This occurs because the notochord and/or floor plate are also necessary for motoneuron induction (Bovolenta & Dodd 1991, Yamada et al 1991). It is not possible to infer from these studies what the behavior of the commissural neurons would have been had they been challenged to reach the ventral midline in the presence of a differentiated motor column (but see below).

Similar observations were made in the zebrafish mutant *cyc-1,* which lacks the midline floor plate cell at all axial levels (Hatta et al 1991). Although navigation does go wrong at the midline, almost all circumferential projections to the midline appear the same as in wildtype. In the brainstem, many of the reticulospinal neurons apparently grow to the midline in the same way as in wildtype (Hatta et al 1991). In the spinal cord, the initial short circumferential extension of the VeLD neurons is also normal. Similarly, virtually all CoPA and CoSA commissural neurons grow normally around the outer edge of the marginal zone to arrive at the midline (Bernhardt et al 1992a,b). In fact, all CoPA and all but 17% of the CoSA axons are directed circumferentially from the point at which the axons exited the cell bodies (Bernhardt et al 1992b). In other words, with the exception of 17% of CoSA axons, the trajectories do not display any signs consistent with a wandering behavior, giving the best indication that these commissural axons are actively guided circumferentially even in the absence of the floor plate. The same result was obtained in UV-irradiated *Xenopus* embryos lacking both notochord and floor plate, where commissural neurons likewise maintained the ability to circumnavigate the outer edge of the spinal cord as usual to arrive at the ventral midline (Clarke et al 1991). Other dorsally situated cells that undergo axonogenesis at the same time as commissural neurons have axons that grow longitudinally, indicating that circumferential growth of the commissural axons does not reflect the presence of absolute barriers to longitudinal migration (Kuwada et al 1990). Instead, cues specifying a circumferential migration must be present.

What is the source of the information that directs continued ventral growth along the edge of the spinal cord? It is unlikely that other ventrally situated cell populations serve to attract the commissural axons, since growth along the edge can occur in isolated dorsal spinal cord explants (Placzek et al 1990a, Tessier-Lavigne et al 1988) and in embryos that lack motoneurons owing to genetic defects or experimental ablation (Bernhardt et al 1992a, Bovolenta & Dodd 1991, Yamada et al 1991). Electron microscopy (EM) studies of the early neural tube have found no evidence of any guidepost cells along the edge of the spinal cord (Holley 1982, Holley & Silver 1987). These same studies have also ruled out the existence of preformed channels of extracellular space as a possible mechanism of circumferential guidance to the midline. Although axons extend close to the pial surface of the spinal cord, their growth cones rarely, if ever, contact the basal lamina, indicating that it too is not responsible for steering the axons ventrally (Holley & Silver 1987, Yaginuma et al 1991). Finally, the possibility that other axons are providing the cues is equally unlikely as all studies of axon extension in the circumferential plane at early time points have found these axons to be unfasciculated and independently migrating to the midline (Holley 1982; Holley & Silver 1987; Jacobson & Huang 1985; Kuwada et al 1990; Yaginuma et al 1990, 1991).

The local cues directing ventral migration must therefore apparently derive from the neuroepithelial cells themselves and be present on their surface. Examination of the circumferentially growing axons en route to the midline reveals that they make extensive contacts with the neuroepithelial cells, while also growing through randomly aligned extracellular spaces (Holley & Silver 1987, Yaginuma et al 1991). Growth cones of the earliest extending axons exhibit complex morphologies (often taken as a sign that they are actively searching their environment for appropriate signals) during the dorsoventral migration (Yaginuma et al 1991), which supports the notion that circumferenti-ally directed axons may be reading cues distributed in neuroepithelial cells. By contrast, growth cones of axons extending late in development are simpler in form (Yaginuma et al 1991), perhaps indicating that they are relying more heavily on fasciculation with other axons as a means of pathfinding along the dorsoventral axis. Although the cues that direct migration along the edge are unknown, the simplest model is that positive cues are immobilized in a narrow corridor that parallels the edge of the spinal cord and/or that negative cues lie in the region from which they are excluded (see below). Any mechanism that is postulated, however, must also explain why the axons only grow ventral, not dorsal, and why their growth is confined to the transverse plane.

Early Versus Late Extension of Commissural Axons—A Hypothesis

From the evidence above it would seem that some circumferential projections do not require long-range cues from the floor plate to grow along the edge to the midline but that the floor plate is also a source of a diffusible chemotropic factor that can attract some or all axons at a distance. Is there a reason for this apparent redundancy in guidance mechanisms? One hypothesis consistent with the data is that the floor plate–derived chemoattractant is not required for growth along the edge of the spinal cord but is essential for directing the commissural axons that grow to the midline through the developing motor column. In this manner it could provide these axons with a mechanism of pathfinding at a time when the environment of the ventral spinal cord has become sufficiently complex that simple growth along the edge is no longer adequate.

There has been no direct test of this hypothesis, which would require ex-amining the ability of later-born commissural axons (that grow when the motor column is already of substantial size) to navigate through motoneurons in the absence of the floor plate. In mouse and chick, the successful arrival of commissural axons at the ventral midline in the absence of the floor plate occurred in a spinal cord lacking motoneurons (Bovolenta & Dodd 1991, Yamada et al 1991). The axons in this case grew along the edge of the spinal cord, which is analogous to pathfinding at the earliest stages of neural tube development and commissural axon extension. In the cold-blooded vertebrates the only trajectories of commissural neurons described so far are early projec-tions involving growth to the midline along the edge of the spinal cord (Jacob-son & Huang 1985, Kuwada et al 1990, Roberts & Clarke 1982), which by the model described above, would not be expected to require the presence of the floor plate. This raises the interesting question of whether any later-ex-

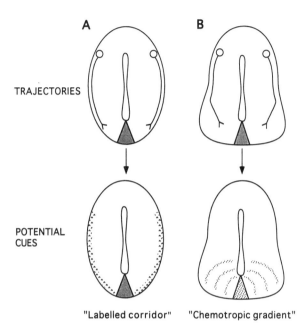

TRAJECTORIES

POTENTIAL
CUES

"Labelled corridor"   "Chemotropic gradient"

Figure 2   Different types of cues may collaborate to guide commissural axons to the ventral midline. (*A*) Commissural axons that grow along the edge (such as the earliest born in chick and those in zebrafish and *Xenopus,* as well as all circumferential axons in the dorsal spinal cord) may be primarily directed by an immobilized corridor of molecules distributed along the lateral edge of the spinal cord. The ventricular zone may also contain inhibitory cues restricting axon growth to the edge (not shown). (*B*) Commissural axons that break away from the edge in the ventral spinal cord (such as those in rat and the later-born ones in chick) may be directed by a gradient of a diffusible chemoattractant released by the floor plate. The operation of these cues is not mutually exclusive.

tending commmissural axons in these species will be found to require the presence of the floor plate to arrive at the ventral midline.

These considerations suggest a model in which growth along the edge occurs in response to floor plate–independent signals, but breaking away from the edge to grow through the motor column requires floor plate-derived signals (Figure 2). One limitation of the model is that there is at least one known case, that of the CoSA axons, in which initial projections along the edge are disrupted in the absence of the floor plate. Thus, axons growing along the edge may in some cases be simultaneously responding to the chemoattractant as well.

Molecules Involved in Directing Circumferential Projections to the Midline

NETRIN-1 IS A FLOOR PLATE–DERIVED CHEMOATTRACTANT   Recent studies have led to the identification of a candidate for the floor plate chemoattractant, netrin-1. The netrins were identified during the purification of an activity in an embryonic chick brain extract that, like the floor plate, can promote outgrowth of commissural axons from explants of rat dorsal spinal cord into collagen gels (Serafini et al 1994). The outgrowth-promoting activity was due to two proteins, netrin-1 and netrin-2, which each individually possess the activity observed for floor plate in vitro. In situ hybridization studies showed that netrin-1 is expressed by floor plate cells, while netrin-2 is expressed at lower levels in roughly the ventral two thirds of the spinal cord, excluding the floor plate (Kennedy et al 1994).

Netrin-1 has both the outgrowth-promoting and orienting activities of the floor plate (Serafini et al 1994, Kennedy et al 1994). Since it is expressed by floor plate cells, it is likely to contribute to the chemoattractant activity of the floor plate. Biochemically, the netrins partition between the soluble fraction and the membrane fraction of cells. Thus, although netrin-1 can function as a diffusible chemoattractant, its diffusion is likely to be retarded by interactions with the environment. This could conceivably contribute to stabilizing or sharpening the gradient of netrin-1 that is presumed to emanate from the floor plate.

The netrins define a family of vertebrate homologues of the UNC-6 protein of *C. elegans* (Ishii et al 1992), which is required for circumferential migrations of cells and axons in both dorsal and ventral directions along the body wall in that species. The homology between the netrins and UNC-6 is illustrated in Figure 3, which also shows their homology to the amino-terminal end of the B2 chain of laminin. In *unc-6* mutants, both dorsal and ventral circumferential migrations fail at a certain frequency, although axon growth (as opposed to guidance) is not impaired (Hedgecock et al 1990). Preliminary evidence suggests that *unc-6* is expressed in a ventral domain of the worm (W Wadsworth & E Hedgecock, personal communication). If so, UNC-6 could be present in a gradient with its high point at the ventral midline and could act as an attractant for cells and axons that migrate ventrally and as a repellent for cells and axons that migrate dorsally. This parallel between UNC-6 and netrin-1, not just in sequence but also in potential function, indirectly supports the hypothesis that netrin-1 is required for accurate guidance of commissural axons to the ventral midline in vertebrates.

The homology of netrin-1 to UNC-6 has also raised the intriguing possibility that netrin-1 may influence dorsal circumferential migrations by acting as a chemorepellent. One population of axons that extend dorsally in vivo away from the ventral midline are motoneurons of the trochlear nucleus (lying near the hindbrain-midbrain junction). Preliminary data have indicated that the floor plate can inhibit trochlear motoneuron outgrowth in vitro and that netrin-1 mimics this effect (SA Colamarino & M Tessier-Lavigne, unpublished data). Thus, netrin-1 appears to be a bifunctional molecule—functioning as an attractant for some axons and as an inhibitor (and possibly a repellent) for others depending, presumably, on the netrin receptor expressed by the cell.

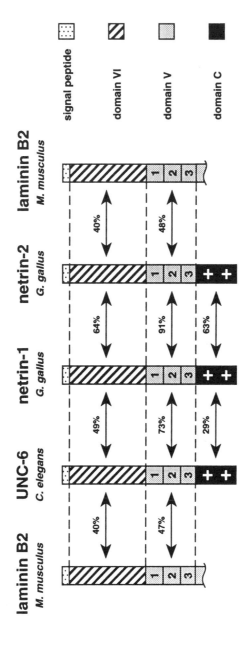

Figure 3   Structures of the netrins [from chicken (*Gallus gallus*)] compared to UNC-6 and the B2 chain of mouse (*Mus musculus*) laminin. All proteins are homologous in domains VI and the three EGF-like repeats of domain V. The C-terminal domain of the netrins and UNC-6 are homologous but diverge completely from the laminin sequence. Percent identity between sequences is indicated. (See Serafini et al 1994 for details.)

THE IDENTITY OF THE CUES THAT DIRECT GROWTH ALONG THE EDGE IS UNKNOWN
As mentioned, cues specifying growth along the edge could be permissive molecules with a restricted distribution parallel to the edge, or inhibitory molecules distributed in the region from which the axons are excluded (i.e. the ventricular zone). One candidate for a permissive cue is a complex of laminin and heparan sulfate proteoglycan (Ln-HSPG), detected by the INO antibody (Matthew & Patterson 1983). In the rat, INO labels a corridor along the edge of the spinal cord in vivo. Moreover, a substrate containing Ln-HSPG supports commissural axon outgrowth in vitro, and INO is capable of inhibiting this outgrowth, while antibodies against laminin do not (D Karagogoes & J Dodd, personal communication). These observations are consistent with the possibility that a Ln-HSPG complex (or a molecule that cross-reacts immunologically) is a component of a mechanism for guidance along the edge of the spinal cord. More detailed studies of this point are required, however, as laminin itself is not detected in the rat spinal cord (Hunter et al 1992), nor is INO immunoreactivity detected in the chick spinal cord (Shiga & Oppenheim 1991). A substantial number of other cell-surface and extracellular matrix molecules have been documented in the early spinal cord (see Table 1). Although the presence of these molecules has been described, few have been shown to directly affect the growth of dorsal spinal cord neurons. F-spondin, which is expressed by floor plate cells in vivo, can mediate attachment of commissural cell bodies to a substrate in vitro but does not promote commissural axon outgrowth (Klar et al 1992). While netrin-1 is likely to be a chemoattractant, netrin-2 could conceivably contribute to guidance along the edge. Netrin-2 can promote commissural axon outgrowth into collagen gels (Serafini et al 1994, Kennedy et al 1994), and its mRNA is expressed widely by the neuroepithelial cells of the spinal cord (Kennedy et al 1994). Although its protein distribution is unknown, it may have a laterally restricted distribution and contribute to guidance of the axons. Many of the other molecules listed in Table 1 may well have effects on commissural neurons, but they have yet to be tested.

Table 1 Expression of selected cell-surface and extracellular matrix molecules in the early developing spinal cord

Molecule	Neuroepithelial cells[a]	FP	Commissural cells	References
TAG-1	−	−	++[b,c]	Dodd et al 1988, Shiga & Oppenheim 1991
L1	−	−	++[d]	Dodd et al 1988
NgCAM/G4	+[e]	+[e]	++	Shiga & Oppenheim 1991, Shiga et al 1993
Nr-CAM	−	++	++[f]	Krushel et al 1993

Table 1   (*continued*)

Molecule	Neuroepithelial cells[a]	FP	Commissural cells	References
Neurofascin	+	−	+ +[g]	Shiga & Oppenheim 1991
F11	−	−	+ +[d]	Shiga & Oppenheim 1991
SC-1	−	+ +	−	Tanaka & Obata 1984, Yamada et al 1991
F84.1	−	+ +	−	Prince et al 1992
p84	−	+ +	−	Chuang & Lagenaur 1990
IGFBP-2	+[h]	+ +	−	Wood et al 1992
C-kit	−	−	+ +	Keshet et al 1991
Steel factor	−	+ +	−	Keshet et al 1991, Matsui et al 1990
BMP-1	−	+ +	−	Sasaki & Hogan 1994
BMP-6/DVR-6	−	−	+ +	Wall et al 1993
Sonic hedgehog/ vhh1	−	+ +	−	Echelard et al 1993, Krauss et al 1993, Riddle et al 1993, Roelink et al 1994
T-Cadherin	+ +	+ +	+ +[g]	Kanekar & Ranscht 1992; B Ranscht, personal communication
N-Cadherin	+ +	+[i]	+ +[g]	Hatta et al 1987, Shiga & Oppenheim 1991
Netrin-1	−	+ +	−	Kennedy et al 1994, Serafini et al 1994
Netrin-2	+	−	−	Kennedy et al 1994, Serafini et al 1994
N-CAM (PSA)	−	+ +	+ +	Boisseau et al 1991, Dodd et al 1988
F-spondin	+[j]	+ +	−	Klar et al 1992
Thrombospondin	+ +	ND[k]	ND	O'Shea & Dixit 1988
S-laminin	−	+ +[l]	−	Hunter et al 1992
Fibronectin	−	+ +[l,m]	−	ffrench-Constant 1989, Shiga & Oppenheim 1991
Collagen IV	−	+ +[l,m]	−	Shiga & Oppenheim 1991
β-1 integrin	+[n]	+ +[m]	+ +[f]	Shiga & Oppenheim 1991
Keratan sulfate proteoglycan	ND	+ +	ND	Hatta 1992
tPA	−	+ +	−	Sumi et al 1992

[a] Excludes differentiated neurons
[b] "+" indicates relative amount of expression for each molecule
[c] Concentrated on precommissural and commissural segments (i.e. circumferential portion) of axons
[d] Concentrated on postcommissural segment (i.e. longitudinal portion) of axons
[e] Only when contacted by NgCAM-positive axon or growth cone
[f] Concentrated on commissural segment of axons (i.e. under the floor plate)
[g] Concentrated on commissural and postcommissural segments of axons
[h] At very early stages of development, persists slightly longer in caudal neural tube
[i] Weakly present on basal surface of floor plate cells
[j] In ventral ventricular zone
[k] ND, not determined
[l] On basal lamina underlying floor plate as well as on floor plate cells
[m] Concentrated on apical surfaces of floor plate cells
[n] Concentrated on ventricular and pial endfeet

GUIDANCE AT THE MIDLINE

Guidance events at the midline are more complex than simple growth in the transverse plane and, as a result, are less well understood. The major experimental approach to studying guidance at the midline has been to examine the trajectory of commissural axons in embryos lacking a floor plate. As discussed in the next several sections, the axon behavior at the midline is abnormal under these conditions. However, the degree to which selective perturbation of the floor plate has been achieved has varied considerably. The most selective perturbation has been obtained in the case of the zebrafish *cyc-1* mutant, in which the major (and perhaps only) defect is the absence of floor plate cells (Hatta et al 1991). In these embryos, all aspects of guidance at the midline are affected, but the severity is relatively modest. At the other extreme, in the mouse Danforth short-tail mutant and experimentally manipulated chick embryos, where both the floor plate and notochord are lacking, axon behavior at the ventral midline is seriously disrupted. The generalized perturbation of the entire ventral spinal cord in these embryos may be the cause of the more severe guidance defects. Alternatively, the difference in severity of guidance defects could indicate that the floor plate is simply not as essential for midline guidance events in cold-blooded vertebrates as it is in warm-blooded vertebrates. In support of this possibility, UV-irradiated *Xenopus* embryos have a severely compromised ventral midline but may have only relatively minor defects in axon trajectories at the midline (Clarke et al 1991).

The decision to cross the midline is discussed here separately from the decision to initiate longitudinal growth. This is not to imply that the two decisions are necessarily independent. Crossing and turning may be separable processes that involve distinct cues, but they may also be linked. At the end of these sections we discuss two models for guidance at the midline in which single cues can direct both the decisions to cross and to initiate longitudinal growth.

The Initial Decision to Cross or Not to Cross the Midline

In contrast to the ability of most early circumferential axons to successfully reach the midline in the absence of the floor plate, the decision whether to cross the ventral midline is in fact perturbed in embryos lacking a floor plate. In the Danforth short-tail mouse mutant and in the chick notochord removal studies a general confusion in axon behavior is seen at the ventral midline. While some axons do stay in the spinal cord and cross to the contralateral side, many instead project out at the midline, forming an abnormal ventral root (Bovolenta & Dodd 1991, Yamada et al 1991).

In the zebrafish *cyc-1* mutant, in which the floor plate is more selectively removed, consistent though less severe navigation errors at the midline are observed. These abnormal trajectories can be phenocopied by laser ablation of the midline floor plate cell in normal embryos (Bernhardt et al 1992a, Hatta 1992). Of these errors, most are mistakes in midline crossing behavior. In the spinal cord, approximately 25% of the commissural axons fail to cross the midline, and instead project longitudinally on the ipsilateral side of the cord (Bernhardt et al 1992a,b). In the case of noncrossing axons, 15% of the VeLD axons incorrectly cross the midline in the mutant, suggesting a role for the floor plate in normally inhibiting the crossing of these axons (Bernhardt et al 1992a) This inhibition would likely be through a contact-mediated mechanism, given that as part of their normal course the VeLD axons appear to contact the floor plate before turning away from the midline (Kuwada et al 1990).

In the brainstem of *cyc-1* animals errors in initial crossing behavior likewise appear to be made by some identified reticulospinal neurons. However, when the Mauthner cells in the hindbrain were analyzed in detail for defects in axonal trajectories it appeared that although there is a gross disorganization of longitudinal pathways (the normally separate and symmetric fascicles of the medial longitudinal fasiculi (MLF) are often fused into one diffuse bundle at the midline) these axons actually made only a very few errors in initial crossing decisions (Hatta 1992). This disorganization of the MLF arises principally from the variable behavior of the axons after crossing the midline upon initiation of longitudinal decent (see the next section).

It is unclear why the neurons in the brainstem of the *cyc-1* mutant are, for the most part, able to make the correct midline crossing decision in the absence of the floor plate, while those in the spinal cord seem to be more prone to error. Furthermore, although errors are made in the mutant spinal cord, the fact that a greater percentage of axons makes the correct decision suggests the presence of a second source of signals directing midline crossings. One logical candidate for this source is the underlying notochord. Consistent with this possibility is the fact that laser ablation of the notochord in *cyc-1* mutants substantially increases the percentage of defects at the midline (Greenspan et al 1993). Brainstem pathfinding has yet to be analyzed under these conditions.

Although UV-irradiated *Xenopus* embryos lacking both the floor plate and the notochord have severe anatomical disruptions, they display apparently minor axon guidance defects (Clarke et al 1991). Despite the fusion of the entire ventral midline area, those commissural neurons that could be examined in detail (the glycine-containing commissural interneurons) all managed to cross the midline. This crossing did show many abnormal features, however. Instead of crossing the midline at right angles directly under the pial surface, the axons crossed at variable angles, in some cases running almost parallel with the midline for a distance in the newly created ventral marginal zone before crossing. The authors suggest that in *Xenopus* the floor plate is not needed so much to dictate midline crossing as to facilitiate it by directing the angle of crossing. However, since the trajectories of noncrossing axons, such as the KA axons, have not been examined in detail in these embryos, it is unknown whether they made crossing errors. Therefore, the floor plate or other missing structures may play a more extensive role in directing crossing behavior at the midline of UV-irradiated *Xenopus* embryos than is apparent from the analysis of one population of commissural neurons.

Cellular and Molecular Biology of Midline Crossing

The floor plate clearly provides some information necessary to direct the proper routing of axons at the midline. The cell biology of this interaction is only beginning to be understood. The first growth cones to reach the floor plate may traverse it by inserting themselves between the basal or most ventral surface of the floor plate and the surrounding basal lamina (Kuwada et al 1990, Yaginuma et al 1991, Yoshioka & Tanaka 1989). In the grasshopper, contact and fasciculation with the contralateral homologue is involved in successful growth-cone crossing of the midline (Myers & Bastiani 1993b), but this does not seem to be the case in vertebrates: Although clustering of axons in the floor plate along the rostral-caudal axis has been observed during the earliest periods of floor plate crossing, and growth cones frequently contact each other when in the floor plate, actual fasciculation between them and/or other axons (crossing in either direction) has not been observed (Yaginuma et al 1990). Furthermore, crossing of commissural axons in zebrafish is achieved in segments in which the contralateral commissural neuron is missing (Kuwada et al 1990).

Systematic observations of growth cones in fixed tissue have demonstrated significant changes in their shape and complexity as they contact the floor plate (Bovolenta & Dodd 1990, Yaginuma et al 1991), suggesting that midline crossing involves more than just passive channeling of axons. These changes could reflect an interaction with signals on the floor plate cells or on the basal lamina underlying the floor plate, which the growth cones contact during crossing (Yaginuma et al 1991). Processes from the basal surface of the floor plate do tightly enwrap the crossing axons (Glees & Le Vay 1964, Shiga & Oppenheim 1991, Yoshioka & Tanaka 1989). Recently it has been suggested that some signals received by the growth cones are transmitted via a novel mechanism of macromolecular transfer from the floor plate cells. The apparent operation of this mechanism was observed in a transgenic mouse line in which β-galactosidase expression is driven in a subset of floor plate cells; remarkably, the enzyme was also found within adjacent commissural axons (Campbell & Peterson 1993). Previously, dye coupling between axons and cells had been documented during midline crossing in the grasshopper (Myers & Bastiani 1993a), but the transfer of large cytoplasmic proteins such as β-galactosidase would presumably require a more complex mechanism than simple gap-junc-

tional coupling. A specialized secretory activity for the floor plate was previously suggested (Tanaka et al 1988, Yoshioka & Tanaka 1989) based on EM
observation of numerous vesicles and dense bodies within the cytoplasm of
floor plate cells (Glees & Le Vay 1964, Shiga & Oppenheim 1991, Tanaka et
al 1988, Yoshioka & Tanaka 1989).

The molecular nature of the signals that direct crossing behavior remains to
be determined. Keratan sulfate proteoglycans in the floor plate have been
suggested as possible repellents that prevent passage of noncrossing fibers
through the floor plate (Hatta 1992; see Snow et al 1990 for a discussion of
inhibition by proteoglycans). Many growth-promoting molecules are known
to be enriched in the floor plate (see Table 1) and could possibly aid midline
crossing by serving as permissive signals. N-Cadherin, neurofascin, N-CAM,
and Nr-CAM are examples of adhesion molecules particularly concentrated
on the commissural segment of the crossing axons that are also expressed by
the floor plate cells (Boisseau et al 1991, Dodd et al 1988, Krushel et al 1993,
Shiga & Oppenheim 1991). Molecules in the basal lamina—such as s-laminin,
laminin, heparan sulfate proteoglycan, fibronectin, and collagen type IV, as
well as other molecules secreted by the floor plate cells that may become
trapped in the basal lamina (e.g. F-spondin and netrin-1)—are also all potential
signals, either permissive or inhibitory (Hunter et al 1992, Kennedy et al 1994,
Klar et al 1992, Shiga & Oppenheim 1991). In the absence of the floor plate,
a disruption of the basal lamina or loss of the permissive signals could be
responsible for the tendency of some commissural axons to project out of the
spinal cord at the midline.

Role of the Floor Plate in Initiating and Maintaining Longitudinal Axonal Growth

For many of the early axons, regardless of whether they cross the midline or
not, the ventral midline marks the point of termination of a circumferential
growth pattern and the site of initiation of a longitudinal one (Bovolenta &
Dodd 1990, Kuwada et al 1990, Roberts & Clarke 1982, Yaginuma et al 1991,
Yaginuma et al 1990). Where studied in detail, the turn has been found to
occur precisely at the edge of the floor plate (Bovolenta & Dodd 1990). The
close correlation of the floor plate with the change in growth pattern suggests
a role for the floor plate in influencing longitudinal projections.

In embryos with a severely perturbed ventral midline, axons often fail to
turn longitudinally at all. As discussed, many commissural axons in the Danforth short-tail mouse mutant project out of the spinal cord at the midline rather
than turning 90° to project longitudinally (Bovolenta & Dodd 1991). In addition, those that stay in the cord can also fail to turn. In the Danforth short-tail
mutant and in UV-irradiated *Xenopus* embryos some axons can apparently be
traced that cross the midline and continue to project circumferentially back up

the edge of the contralateral spinal cord (Bovolenta & Dodd 1991, Clarke et al 1991).

Perhaps surprisingly, in cold-blooded vertebrates the vast majority of axons do in fact turn to project longitudinally (even if on the wrong side of the midline) in the absence of the floor plate. However, errors in the direction of the turns are frequently recorded. In the spinal cord of the zebrafish *cyc-1* mutant, many CoPA and CoSA axons turned caudally instead of rostrally (9% and 24%, respectively) and many VeLD axons turned rostrally instead of caudally (22%) (Bernhardt et al 1992a,b). Even the KA neurons, which have not yet been documented to contact the floor plate, can project in the wrong direction (17%) (Bernhardt et al 1992b). A small number of axons display more complicated errors, such as an abnormal bifurcation after crossing the midline and then longitudinal projections in both directions (Bernhardt et al 1992a). In the brainstem a few Mauthner axons were found to turn incorrectly (Hatta 1992). In UV-irradiated *Xenopus* embryos many more commissural axons are directed caudally than normal (34% vs 13%), and some KA neurons are misdirected as well (Clarke et al 1991). These data imply that the floor plate may be involved, at least to some extent, in determining not only when to begin longitudinal growth (before or after crossing) but also the polarity of this growth. Because the commissural axons of the Danforth short-tail mutants were not analyzed in detail for turning behavior, if any axons did turn in these embryos, it is unknown whether polarity errors were made.

Even after a correct turn has been made, the floor plate also appears to be involved in helping axons seek out and remain in the appropriate longitudinal fascicle. The most common error in axonal pathfinding in the spinal cord of *cyc-1* embryos is the inability of the CoPA axons to directly join the dorsal longitudinal fasciculus (Bernhardt et al 1992a). The axons instead ascend in the contralateral ventral cord for longer distances and linger in the ventral longitudinal fasciculus, which they normally ignore, before gradually shifting dorsally. Similarly, the principal error of the reticulospinal axons in the brainstem of the *cyc-1* embryos is a disruption of the ability to maintain fasciculation with the correct longitudinal tracts, resulting in disorganized formation of the MLF (Hatta 1992). Axons of the Mauthner cells begin descending as normal, but about half of them leave the fascicle at random locations and wander back across the midline. Other reticulospinal axons may join the incorrect longitudinal tract, failing to recognize the MLF as their appropriate target.

Transplantation of wild-type floor plate precursor cells into *cyc-1* embryos only rescues the disorganized phenotype of the MLF at the level of the injected cells (Hatta 1992). Contact with the floor plate is not sufficient to completely rectify an axon's projection, as axons can aberrantly recross the midline after passing through an axial level containing the transplanted floor plate. Continued presence of the floor plate therefore seems to be required to achieve normal longitudinal projections, suggesting that the floor plate must serve as a barrier inhibiting midline crossing at all times subsequent to the initial crossing. Further indirect evidence that the floor plate is a continuous source of signals for longitudinal axons can be deduced from chick spinal cord–rotation experiments where axons descending in the ventral funiculus also aberrantly crossed the midline upon reaching a discontinuity in the floor plate (see Yaginuma & Oppenheim 1991). These experiments may be interpreted to support a role for the floor plate in providing a short-range inhibitory signal necessary for maintaining laterality of longitudinal projections.

Cellular and Molecular Biology of Longitudinal Growth

Evidence from embryos lacking the floor plate thus implies that the floor plate has at least some effect on all decisions regarding longitudinal growth: its initiation, direction, and maintenance. The question remains how these influences are achieved.

The floor plate may signal crossing growth cones to change their substrate affinities. If this change were one that resulted in an increased propensity of the growth cone to fasciculate with longitudinal axons, then passage through the floor plate would be the key to converting from a circumferential growth mode to a longitudinal one. The discovery that in rat a switch in expression of adhesion molecules, from TAG-1 to L1, occurs as the commissural axons cross the floor plate has helped provide a framework for understanding how the commissural axons might successfully disregard all longitudinal pathways prior to crossing the midline (Dodd et al 1988). Although the appearance of these molecules on the neurons may, in part, be regulated autonomously (Karagogeos et al 1991), the sharp transition in their expression suggests that the floor plate may be involved in refining the timing or spatial extent of their expression.

Elucidation of signals that instruct the longitudinal axons to turn rostral or caudal, or both, has been more difficult. In rats, growth cones just exiting the floor plate are highly complex; many filopodia extend both rostrally and caudally in contact with the floor plate, as if they were examining its border for directional cues (Bovolenta & Dodd 1990). Upon executing their rostral turn, the axons extend adjacent to the floor plate edge for the first several hundred microns and then shift laterally in the ventral longitudinal fasciculus. The close association of the longitudinally extending axons with the floor plate border suggests that it may contain a graded distribution of some signal indicating the rostrocaudal polarity of the neural tube. One attempt to examine this possibility involved in vivo rotations of the neural tube. Yaginuma & Oppenheim (1991) found that commissural axons extended normally through a segment of spinal cord that had been inverted along the rostrocaudal axis. These experiments did not, however, distinguish whether directional cues derive from outside the spinal cord or whether the polarity of the neural tube became respecified after the rotation.

Finally, the signal to switch to a longitudinal growth mode cannot be as simple as a change that results in a fasciculation with just any longitudinal axons encountered. This is made obvious in the zebrafish spinal cord, where after crossing the midline, the commissural axons ignore any longitudinally directed axons in the ventral longitudinal fasciculus and instead extend dorsally to fasciculate in the dorsal longitudinal fasciculus (Kuwada et al 1990). A specificity must somehow be imparted to each axon regarding which longitudinal tract to join. In the absence of the floor plate the CoPA commissural axons appeared to have difficulty distinguishing between the ventral longitudinal fasciculus and the dorsal longitudinal fasciculus (Bernhardt et al 1992a). On the other hand, one cannot exclude that this defect (and all defects in longitudinal growth) arises not directly from the lack of a floor plate but secondarily from the loss or displacement of the longitudinal axons onto which the axons normally fasciculate (Clarke et al 1991) We know of many instances where turns to join longitudinal tracts occur independently of the floor plate (the decision of the association axons to join an ipsilateral longitudinal tract is one such case), which highlights the possible role of early longitudinally oriented axons, such as those of the PL neurons, in influencing the longitudinal growth of other axons.

Models to Interpret Midline Guidance Defects

The decisions made at the midline (whether to cross and whether to initiate longitudinal growth) may be directed by distinct cues. However, two simple models can be envisioned in which only one type of signal, either a crossing signal or a turning signal, is sufficient to direct both decisions.

Axon behavior at the midline could be entirely controlled by signals from the floor plate specifying whether the axons should enter (Figure 4A). In this model, the signal would specify the relative attractiveness of the floor plate over the surrounding neuroepithelium and would be positive (more attractive than the neuroepithelium) for the crossing axons and inhibitory (less attractive than the rest of the neuroepithelium) for the noncrossing axons. Axons barred from entering the floor plate would be forced to turn longitudinally at the ipsilateral floor plate border in order to remain in contact with the more favorable neuroepithelium. Crossing axons enticed to enter the floor plate would be forced to turn longitudinally at the contralateral border of the floor plate to remain in contact with the more favorable floor plate substrate (Bovolenta & Dodd 1990). In this manner, the decision to cross the floor plate would not be separated from the decision to turn, since it is the floor plate–derived crossing signal that would directly result in the choice of when to turn longitudinally. This model requires additional signals to instruct the axons to turn either rostral or caudal and cannot easily explain why virtually all axons do turn in the *cyc-1* mutant (which presumably lacks these crossing signals).

At the other extreme, a model can be proposed in which signals specifying turning behavior (and not crossing behavior) govern the proper routing of axons at the midline (Figure 4*B*). These cues do not necessarily need to be floor plate–derived, but in this model the reception of the signal to turn will directly influence whether the axon crosses the floor plate: If the turn signal

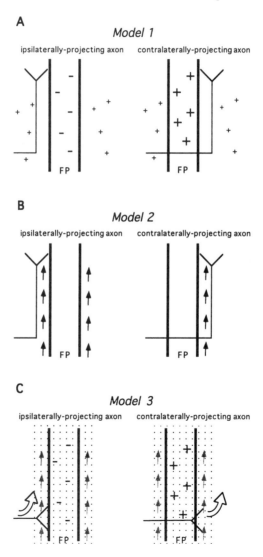

is received before crossing the floor plate, an axon will remain ipsilateral; if it is received at the contralateral border of the floor plate, the axon will cross the floor plate before turning. For this model to work, axons that turn only after crossing would somehow have to ignore the turn signal on the ipsilateral side. This would most easily be achieved if the floor plate modulates the ability of the crossing axons to recognize the turn signal (as discussed in the context of the TAG-1–L1 switch). The alternative, that the cues on the left and right sides of the embryo are different, seems quite implausible.

Irrespective of the form of cues the floor plate provides, none of these models require that the floor plate be their only source. In fact, from the data we have reviewed, some redundancy is expected, such that cells other than the floor plate (e.g. notochord cells or other axons) are also sources of cues. A strong prediction of both models, however, is that in the absence of all the sources of cues, axons should fail to turn. This is what is observed in the Danforth short-tail mutant and chick embryos lacking both floor plate and notochord. On the other hand, in the selective absence of the floor plate other sources would be expected to compensate to some extent, resulting in the array of less severe defects found in the zebrafish *cyc-1* embryos.

Regardless of whether all or some of the sources of signals are eliminated, it is more difficult to understand how either of these two models in their simplest form could possibly account for the fact that in the *cyc-1* mutant some commissural axons apparently turn longitudinally before crossing. In Model 1 (see Figure 4), loss of positive signals from the floor plate for crossing axons should not result in an ipsilateral turn. Likewise, in Model 2 the crossing axons should not be able to recognize an ipsilateral turning cue without first contacting the floor plate. One way around this problem would be if the apparently

---

Figure 4 Possible models for axon guidance at the ventral midline (see text for details). In *A* and *B*, crossing and turning events result from the operation of a single type of cue. In each diagram, an axon projecting circumferentially approaches the floor plate and either turns longitudinally on the ipsilateral side (*left*) or crosses the floor plate before turning (*right*). (*A*) Model 1: The floor plate (*thick vertical lines*) serves as a more (+) or less (−) favorable substrate than the surrounding neuroepithelium (+) for contralaterally and ipsilaterally projecting axons, respectively. Longitudinal turns occur because the axons grow to remain in contact with their most favored substrate. (*B*) Model 2: Signals that direct turning (*arrows*) on or near the floor plate borders indicate when the axons should turn longitudinally. Contralaterally projecting axons ignore turning signals on the ipsilateral side (which are therefore not represented), until passage through the floor plate somehow up-regulates their ability to respond to the signals. (*C*) Model 3: This model is designed to account for defects observed in the *cyc-1* mutant. Crossing, turning, and overriding signals together guide axon routing at the midline. Axons are sensitive to the turning signals (*arrows*) but initially ignore them owing to an overriding signal (*dots*) released by the floor plate that masks the turning signals and drives the axons to the midline. Contact with the midline then modifies the axons such that they are no longer sensitive to this overriding signal and can respond to the turning signals. The floor plate must also provide contact-dependent crossing signals that prevent the ipsilaterally projecting axons from entering.

ipsilaterally projecting commissural axons in the mutant had in fact initially crossed the midline and turned longitudinally before abberantly wandering back across the midline owing to the absence of a barrier to recrossing. In support of this possibility, aberrant wandering back and forth across the midline has been seen for the VeLD axons (J Kuwada, personal communication). If this interpretation is correct, then cues from remaining sources like the notochord must be sufficient to direct the correct crossing and turning decisions but not enough to provide the barriers to recrossing the midline. Although this situation may seem unlikely, it occurs in the brainstem of the zebrafish mutant, where the only serious defects in axon trajectories are errors of recrossing. Thus, if it were true that errors of recrossing are also the prime defect in the spinal cord, this would provide a unifying hypothesis about what occurs at all axial levels when the floor plate is selectively removed.

If, however, in the absence of the floor plate commissural axons do sometimes turn ipsilaterally without ever crossing the midline, then neither Model 1 nor Model 2 can be correct. A more accurate but complicated model would have to be postulated in which crossing axons are not impervious to the turn signal prior to crossing the floor plate: The floor plate under normal circumstances would provide another signal (perhaps the chemoattractant) that overrides sensitivity to the turning cues (Figure 4*C*). In this new model, passage through the midline would presumably need to modify the axons (perhaps through the TAG-1–L1 switch) such that they could now ignore the overriding signal and respond to the contralateral turning cues. This would explain why in *cyc-1* embryos, which lack the floor plate and its overriding signal, commissural axons can turn ipsilaterally. To account for all the data from *cyc-1*, however, Model 3 must incorporate several additional assumptions. To explain why only a fraction of the axons turn ipsilaterally in *cyc-1*, it would be necessary to postulate either that there is another source of the overriding signal (e.g. the notochord) or that the axons only have a low-level sensitivity to the turning cue prior to passage through the midline. To further explain why virtually all axons that cross in *cyc-1* do in fact turn, it is also necessary to postulate that the modification of the axons as they pass through the midline (which allows them to ignore the overriding signal and respond to turning signals) does not require the presence of floor plate cells. This model could also account for the routing of noncrossing axons, except that contact-dependent signals would be needed to prevent them from entering the floor plate so that they would respond to the ipsilateral turning cues. In this way, proper guidance at the midline would result from a combination of crossing, turning, and overriding signals.

These considerations show that if some commissural axons do indeed turn ipsilaterally in *cyc-1* embryos, then their normal routing probably requires the operation of multiple signals. The other interpretation we have suggested—that apparent ipsilateral turns actually result from an initial crossing followed by a recrossing—is more easily explained as resulting from simple absence of a barrier to recrossing in *cyc-1*. Time-lapse analysis of vitally labeled commissural axons could determine whether their aberrant ipsilateral projections do indeed result exclusively from initial crossing followed by recrossing of the axons, or whether some turn ipsilaterally without crossing.

CELL MIGRATIONS THAT MAY BE AFFECTED BY THE FLOOR PLATE

Pathfinding is usually discussed in the context of axonal elongation and growth cone guidance, but it also occurs at the level of cell-body migrations. There are suggestions that the floor plate may contribute to directing cell body as well as axonal migrations across the midline.

The fact that many targets receive bilateral innervation has long been appreciated. That the contralateral component can arise not by a crossing of the axon to the contralateral side, but instead by a midline crossing of the cell body would not, however, have been obvious a priori. Previously this had been inferred to occur for a subpopulation of oculomotor neurons (see for example Naujoks-Manteuffel et al 1991, Puelles & Privat 1977). A more direct observation of cell-body crossing comes from recent studies in chick on the sensory receptor fields of the inner ear, which receive bilateral innervation from the hindbrain. The efferent axons to the receptor fields arise from early-born neurons located in rhombomere 4, adjacent to the motor nuclei of the VII and VIII nerves, and they exit the brainstem through the same combined nerve root. A recent study found that the commissure of the contralateral component is generated by cell-body migration across the floor plate, which is initiated at some point after the axons have already reached the nerve-exit point (Simon & Lumsden 1993). Retrograde labeling with DiI application to the nerve root allowed the progression of the cell bodies to be followed from their original location in the ventral motor column near the ipsilateral floor plate border, through their presence in the floor plate, to their ultimate settling near the contralateral border of the floor plate.

Why only this select population initiates this migration and what cues instruct them to back across the floor plate and stop in the contralateral ventral cord are unresolved. The role the floor plate plays, if any, in providing these cues has yet to be examined, but given its central position in the path of this migration it is hard to imagine that the floor plate plays only a passive role in the event. Interestingly, the axons of these neurons are found to be fasciculated from the nerve-exit point to the ipsilateral ventral motor column but unfasciculated along the length through the floor plate to the cell body on the contralateral side. It would thus appear that when the cells are propelling themselves backward and adding new axon cylinder at the proximal end, they are probably not using other axons of the same class as a cue to their migration.

A more complex example of cell body migration is represented by the fascinating case of rat precerebellar neurons. Generated in the primary precerebellar neuroepithelium (the rhombic lip) located in the dorsal wall of the IVth ventricle, they actually migrate away from their cerebellar target and circumnavigate the entire ventrolateral aspect of the medulla before ultimately settling as the neurons of the inferior olivary nucleus (ION), the lateral reticular nucleus (LRN), and the external cuneate nucleus (ECN), among others (Bourrat & Sotelo 1990a,b). What is extraordinary about this extensive migration is that the cells that become olivary neurons migrate and coalesce to become the ION near the ipsilateral ventral midline, but those destined to become neurons of the LRN and ECN migrate across the midline and continue dorsally to populate the territories of the future LRN and ECN on the side contralateral to which they were born. By the time the migrations are completed, the LRN neurons will have migrated three quarters of the way around the circumference of the medulla to form the contralateral LRN (in the lateroventral medulla). Those of the ECN will have migrated around the entire margin of the brainstem to form the contralateral ECN, which actually lies dorsally alongside the rhombic lip, the site at which these cells are generated.

The purpose of this impressive migration is not clear at all, but it provides an accessible system to study some of the cellular and molecular mechanisms of pathfinding by cell bodies. In particular, given that the neurons of the ION, LRN, and ECN are generated at the same time and dorsal location, what are the cues that direct all the cells ventrally? Moreover, what cues are used to halt the ION neurons before they cross the midline, in the ipsilateral ION territory, yet ensure that the LRN and ECN neurons will bypass their domains ipsilaterally and only settle in them after they cross the midline? At this point there are no answers to these questions, but Bourrat & Sotello (1990a,b) have naturally proposed the floor plate's involvement in all of the above processes.

SUMMARY

The data reviewed above have implicated the floor plate in directing axonal growth towards the midline, in directing the behavior of axons at the midline, and finally, in directing the longitudinal growth of axons alongside the midline.

In the case of growth to the midline, there is clear evidence for the existence of two types of cues that collaborate to direct growth—short-range cues that can direct axons along the edge of the spinal cord and a long-range chemoattractant secreted by the floor plate cells whose main function may be to direct later-extending commissural axons that must migrate through the complex environment of the developing motor column. Determining the precise contribution of these cues will require identifying them and perturbing them in vivo. The cues that direct growth along the edge are unknown; their identification in the spinal cord would be of quite general significance, since the growth of axons parallel to (but not in contact with) the pial surface is a widespread feature of early axon growth at all axial levels of the neural tube. A strong candidate for the long-range chemoattractant is netrin-1, a homologue of the UNC-6 protein of *C. elegans* and a distant relative of laminin, which is expressed by floor plate cells and which can both promote and orient commissural axon outgrowth. Netrin-1 may also influence growth of other populations of neurons that exhibit stereotyped behaviors near the ventral midline.

Much less is known about the exact role of the floor plate in directing axon growth at the midline, though it is clearly required for accurate guidance. In the absence of the floor plate, a range of errors has been found, the most prominent of which are aberrant midline crossings and errors in longitudinal growth near the ventral midline. The severity of these errors varies with species, which could result from either the variable importance of the floor plate in the different species or the fact that, so far, quite different manipulations of the ventral midline region have been performed in different species. The most specific perturbation of the ventral midline occurs in the zebrafish *cyc-1* mutant, where the selective loss of the floor plate leads to stereotyped misrouting events. Perhaps surprisingly, virtually all axons that grow to the midline turn longitudinally (although sometimes in the wrong direction). The observed defects in the hindbrain can be interpreted to result primarily from the loss of a barrier to recrossing provided by the floor plate so that axons do not become confined to the appropriate side of the embryo after turning and wander across the midline at a certain frequency. We have suggested that perhaps this can account for most of the errors observed in the spinal cord too. However, other cues from the floor plate directing navigation at the midline may certainly be present, but the exact form of these cues (e.g. crossing, turning, or overriding signals, as we have discussed) remains to be resolved. If these other types of signals do exist, it seems unlikely that the floor plate is their sole source (at least in zebrafish), and there is in fact evidence that the notochord contributes at least some of these signals. There is at present no information on the molecular nature of any of the cues that direct guidance events at the midline.

Studies of the role of the floor plate in axon guidance have highlighted the difficulty of analyzing such intricate guidance events and elucidating their molecular basis. Sharp turns made by axons at particular landmarks like the floor plate are made in many regions of the nervous system that are less amenable to experimental analysis. Lessons that are learned from studying axon guidance at the floor plate are therefore likely to provide insights into similar types of decisons made in other regions of the nervous system.

ACKNOWLEDGMENTS

We thank Tom Jessell and Jane Dodd for many helpful discussions and comments on the manuscript and Erin Peckol, Timothy Kennedy, Kimberly Tanner, Jen Zallen, and Lindsay Hinck for critical reading of the manuscript. Many thanks to Caroline Murphy for assistance in preparing the manuscript. This work was supported by grants from the Lucille P. Markey Charitable Trust, the Searle Scholars' Program/Chicago Community Trust, the McKnight Endowment Fund for Neuroscience, and the Esther A. and Joseph Klingenstein Fund. SAC is a Howard Hughes Medical Institute Predoctoral Fellow; MT-L was a Lucille P. Markey Scholar in Biomedial Sciences and is currently an Assistant Investigator of the Howard Hughes Medical Institute.

Literature Cited

Altman J, Bayer SA. 1984. The development of the rat spinal cord. *Adv. Anat. Embryol. Cell Biol.* 85:1–164

Bernhardt RB, Nguyen N, Kuwada JY. 1992a. Growth cone guidance by floor plate cells in the spinal cord of zebrafish embryos. *Neuron* 8:869–82

Bernhardt RB, Patel CK, Wilson SW, Kuwada JY. 1992b. Axonal trajectories and distribution of GABAergic spinal neurons in wild-type and mutant zebrafish lacking floor plate cells. *J. Comp. Neurol.* 326:263–72

Boisseau S, Nedelec J, Poirier V, Rougon G, Simonneau M. 1991. Analysis of high PSA N-CAM expression during mammalian spinal cord and peripheral nervous system development. *Development* 112:69–82

Bourrat F, Sotelo C. 1990a. Early development of the rat precerebellar system: migratory routes, selective aggregation and neuritic differentiation of the inferior olive and lateral reticular nucleus neurons. An overview. *Arch. Ital. Biol.* 128:151–70

Bourrat F, Sotelo C. 1990b. Migratory pathways and selective aggregation of the lateral reticular neurons in the rat embryo: a horseradish peroxidase in vitro study, with special reference to migration patterns of the precerebellar nuclei. *J. Comp. Neurol.* 249:1–13

Bovolenta P, Dodd J. 1990. Guidance of commissural growth cones at the floor plate in embryonic rat spinal cord. *Development* 109: 435–47

Bovolenta P, Dodd J. 1991. Perturbation of neuronal differentiation and axon guidance in the

spinal cord of mouse embryos lacking a floor plate: analysis of Danforth's short-tail mutation. *Development* 113:625–39

Campbell RM, Peterson AC. 1993. Expression of a *lacZ* transgene reveals floor plate cell morphology and macromolecular transfer to commissural axons. *Development* 119:1217–28

Chuang W, Lagenaur CF. 1990. Central nervous system antigen P84 can serve as a substrate for neurite outgrowth. *Dev. Biol.* 137: 219–32

Clarke JDW, Holder N, Soffe SR, Storm-Mathisen J. 1991. Neuroanatomical and functional analysis of neural tube formation in notochordless *Xenopus* embryos: laterality of the ventral spinal cord is lost. *Development* 112:499–516

Dale N, Roberts A, Ottersen OP, Storm-Mathisen J. 1987. The morphology and distribution of "Kolmer-Agduhr cells," a class of cerebrospinal-fluid-contacting neurons revealed in the frog embryo spinal cord by GABA immunocytochemistry. *Proc. R. Soc. London Ser. B* 232:193–203

Dodd J, Morton SB, Karagogeos D, Yamamoto M, Jessell TM. 1988. Spatial regulation of axonal glycoprotein expression on subsets of embryonic spinal neurons. *Neuron* 1:105–16

Echelard Y, Epstein DJ, St-Jacques B, Shen L, Mohler J, et al. 1993. Sonic hedgehog, a member of a family of putative signaling molecules, is implicated in the regulation of CNS polarity. *Cell* 75:1417–30

ffrench-Constant C. 1989. Alternative splicing

of fibronectin is temporally and spatially regulated in the chicken embryo. *Development* 106:375–88

Glees P, Le Vay S. 1964. Ependymal cells of the chick embryo spinal cord. *J. Hirnfors.* 6:355–60

Greenspan S, Patel C, Hashmi S, Kuwada JY. 1993. The notochord guides specific growth cones in the zebrafish spinal cord. *Soc. Neurosci. Abstr.* 19:237

Hatta K. 1992. Role of the floor plate in axonal patterning in the zebrafish CNS. *Neuron* 9:629–42

Hatta K, Kimmel CB, Ho RK, Walker C. 1991. The cyclops mutation blocks specification of the floor plate of the zebrafish central nervous system. *Nature* 350:339–41

Hatta K, Takagi S, Fujisawa H, Takeichi M. 1987. Spatial and temporal expression pattern of N-cadherin cell adhesion molecules correlated with morphogenetic processes of chicken embryos. *Dev. Biol.* 120:215–27

Hedgecock EM, Culotti JG, Hall DH. 1990. The *unc-5, unc-6,* and *unc-40* genes guide circumferential migrations of pioneer axons and mesodermal cells on the epidermis in *C. elegans. Neuron* 2:61–85

His W. 1888. Zur Geschichte des Gehrins, sowie der centralen und peripherischen Nervenbahnen beim menschlichen Embryo. *Abh. d. math. phys. Kl. d. Königl. Sächsischen Gesellschaft d. Wiss.* Bd. 14, No. 8, pp. 341–92

Holley JA. 1982. Early development of the circumferential axonal pathway in mouse and chick spinal cord. *J. Comp. Neurol.* 205:371–82

Holley JA, Silver J. 1987. Growth pattern of pioneering chick spinal cord axons. *Dev. Biol.* 123:375–88

Hunter DD, Llinas R, Ard M, Merlie JP, Sanes JR. 1992. Expression of S-laminin and laminin in the developing rat central nervous system. *J. Comp. Neurol.* 323:238–51

Ishii N, Wadsworth WG, Stern BD, Culotti JG, Hedgecock EM. 1992. UNC-6, a laminin-related protein, guides cell and pioneer axon migrations in *C. elegans. Neuron* 9:873–81

Jacobson M, Huang S. 1985. Neurite outgrowth traced by means of horseradish peroxidase inherited from neuronal ancestral cells in frog embryos. *Dev. Biol.* 110:102–13

Jessell TM, Dodd J. 1992. Floor plate-derived signals and the control of neural cell pattern in vertebrates. *Harvey Lect.* 86:87–128

Kanekar SS, Ranscht B. 1992. Expression of T-cadherin in the floor plate correlates with commissural axon outgrowth across the ventral midline of the spinal cord. *Soc. Neurosci. Abstr.* 18:37

Karagogeos D, Morton SB, Casano F, Dodd J, Jessell TM. 1991. Developmental expression of the axonal glycoprotein TAG-1: differential regulation by central and peripheral neurons in vitro. *Development* 112:51–67

Kennedy TE, Serafini T, de la Torre JR, Tessier-Lavigne M. 1994. Netrins are diffusable chemotropic factors for commissural axons in the embryonic spinal cord. *Cell* 78:425–35

Keshet E, Lyman SD, Williams DE, Anderson DM, Jenkins NA, et al. 1991. Embryonic RNA expression patterns of the c-*kit* receptor and its cognate ligand suggest multiple functional roles in mouse development. *EMBO J.* 10:2425–35

Kimmel CB, Powell SL, Metcalfe WK. 1982. Brain neurons which project to the spinal cord in young larvae of the zebrafish. *J. Comp. Neurol.* 205:112–27

Kingsbury BF. 1920. The extent of the floor plate of His and its significance. *J. Comp. Neurol.* 32:113–35

Kingsbury BF. 1930. The developmental significance of the floor-plate of the brain and spinal cord. *J. Comp. Neurol.* 50:177–207

Klämbt C, Jacobs JR, Goodman CS. 1991. The midline of the *Drosophila* central nervous system: a model of the genetic analysis of cell fate, cell migration, and growth cone guidance. *Cell* 64:801–15

Klar A, Baldassare M, Jessell TM. 1992. F-spondin: a gene expressed at high levels in the floor plate encodes a secreted protein that promotes neural cell adhesion and neurite expression. *Cell* 69:95–110

Krauss S, Concordet J-P, Ingham PW. 1993. A functionally conserved homolog of the drosophila segment polarity gene *hh* is expressed in tissues with polarizing activity in zebrafish embryos. *Cell* 75:1431–44

Krushel LA, Prieto AL, Cunningham BA, Edelman GM. 1993. Expression patterns of the cell adhesion molecule Nr-Cam during histogenesis of the chick nervous system. *Neuroscience* 53:797–812

Kuwada JY, Bernhardt RR, Chitnis AB. 1990. Pathfinding by identified growth cones in the spinal cord of zebrafish embryos. *J. Neurosci.* 10:1299–1308

Matsui Y, Zsebo KM, Hogan BLM. 1990. Embryonic expression of a haematopoietic growth factor encoded by the *Sl* locus and the ligand for c-kit. *Nature* 347:667–69

Matthew WD, Patterson PH. 1983. The production of a monoclonal antibody that blocks the action of a neurite outgrowth-promoting factor. *Cold Spring Harbor Symp. Quant. Biol.* 2:625–31

McKanna JA. 1992. Optic chiasm and infundibular decussation sites in the developing rat diencephalon are defined by glial raphes expressing p35 (lipocortin 1, annexin I). *Dev. Dyn.* 195:75–86

McKanna JA, Cohen S. 1989. The EGF receptor kinase substrate p35 in the floor plate of the embryonic rat CNS. *Science* 243:1477–79

Metcalfe WK, Mendelson B, Kimmel CB. 1986. Segmental homologies among reticulospinal neurons in the hindbrain of the zebrafish larva. *J. Comp. Neurol.* 251:147–59

Monaghan AP, Kaestner KH, Grau E, Schütz G. 1993. Postimplantation expression patterns indicate a role for the mouse *forkhead*/HNF-3 α, β, γ genes in determination of the definitive endoderm, chordamesoderm, and neuroectoderm. *Development* 119: 567–78

Myers PZ, Bastiani MJ. 1993a. Cell-cell interactions during the migration of an identified commissural growth cone in the embryonic grasshopper. *J. Neurosci.* 13:115–26

Myers PZ, Bastiani MJ. 1993b. Growth cone dynamics during the migration of an identified commissural growth cone. *J. Neurosci.* 13:127–43

Naujoks-Manteuffel C, Sonntag R, Fritzsch B. 1991. Development of the amphibian oculomotor complex: evidences for migration of oculomotor motoneurons across the midline. *Anat. Embryol.* 183:545–52

Nornes HO, Knapik E, Kroger S. 1990. Polarized outgrowths of circumferential fibers from explants of avian embryonic spinal cord. *Soc. Neurosci. Abstr.* 16:316

O'Shea KS, Dixit VM. 1988. Unique distribution of the extracellular matrix component thrombospondin in the developing mouse embryo. *J. Cell Biol.* 107:2737–48

Oppenheim RW, Shneiderman A, Shimizu I, Yaginuma H. 1988. Onset and development of intersegmental projections in the chick embryo spinal cord. *J. Comp. Neurol.* 275: 159–80

Palka J, Whitlock KE, Murray MA. 1992. Guidepost cells. *Curr. Opin. Neurobiol.* 2: 48–54

Placzek M, Jessell TM, Dodd J. 1993. Induction of floor plate differentiation by contact-dependent, homeogenetic signals. *Development* 117:205–18

Placzek M, Tessier-Lavigne M, Jessell T, Dodd J. 1990a. Orientation of commissural axons in vitro in response to a floor plate-derived chemoattractant. *Development* 110:19–30

Placzek M, Tessier-Lavigne M, Yamada T, Dodd J, Jessell TM. 1990b. Guidance of developing axons by diffusible chemoattractants. *Cold Spring Harbor Symp. Quant. Biol.* 55:279–89

Prince JT, Nishiyama A, Healy PA, Beasley L, Stallcup WB. 1992. Expression of the F84.1 glycoprotein in the spinal cord and cranial nerves of the developing rat. *Dev. Brain Res.* 68:193–201

Puelles L, Amat JA, Martinez M. 1987. Segment-related, mosaic neurogenetic pattern in the forebrain and mesencephalon of early chick embryos: I. Topography of AChE-positive neuroblasts up to stage HH18. *J. Comp. Neurol.* 266:247–68

Puelles L, Privat A. 1977. Do oculomotor neuroblasts migrate across the midline in the fetal rat brain? *Anat. Embryol.* 150:187–206

Ramon y Cajal S. 1909. *Histologie du Systeme Nerveux de l'Homme et des Vertebres,* pp. 657–64. Madrid: Consejo superior de investigaciones cientificas

Riddle RD, Johnson RL, Laufer E, Tabin C. 1993. Sonic hedgehog mediates the polarizing activity of the ZPA. *Cell* 75:1401–16

Roberts A, Clarke JDW. 1982. The neuroanatomy of an amphibian embryo spinal cord. *Trans. R. Soc. Lond. Ser. B.* 296:195–212

Roberts A, Dale N, Ottersen OP, Storm-Mathisen J. 1988. Development and characterization of commissural interneurons in the spinal cord of *Xenopus laevis* embryos revealed by antibodies to glycine. *Development* 103:447–61

Roelink H, Augsburger A, Heemskerk J, Korzh V, Norlin S, et al. 1994. Floor plate and motor neuron induction by *vhh-1,* a vertebrate homolog of *hedgehog* expressed by the notochord. *Cell* 76:761–75

Ruiz i Altaba A, Prezioso VR, Darnell JE, Jessell TM. 1993. Sequential expression of HNF-3β and HNF-3α by embryonic organizing centers: the dorsal lip/node, notochord and floor plate. *Mech. Dev.* 44:91–108

Sasaki H, Hogan BLM. 1993. Differential expression of multiple fork head related genes during gastrulation and axial pattern formation in the mouse embryo. *Development* 118: 47–59

Sasaki H, Hogan BLM. 1994. HNF-3β as a regulator of floor plate development. *Cell* 76: 103–15

Serafini T, Kennedy T, Galko M, Mirzayan C, Jessell T, Tessier-Lavigne M. 1994. The netrins define a family of axon outgrowth-promoting proteins homologous to *C. elegans* UNC-6. *Cell* 78:409–24

Shiga T, Oppenheim RW. 1991. Immunolocalization studies of putative guidance molecules used by axons and growth cones of intersegmental interneurons in the chick embryo spinal cord. *J. Comp. Neurol.* 310:234–52

Shiga T, Shirai T, Grumet M, Edelman GM, Oppenheim RW. 1993. Differential expression of neuron-glia cell adhesion molecule (Ng-CAM) on developing axons and growth cones of interneurons in the chick embryo spinal cord: an immunoelectron microscopic study. *J. Comp. Neurol.* 329:512–18

Silos-Santiago I, Snider WD. 1992. Development of commissural neurons in the embryonic rat spinal cord. *J. Comp. Neurol.* 325: 514–26

Silos-Santiago I, Snider WD. 1994. Development of interneurons with ipsilateral projections in embryonic rat spinal cord. *J. Comp. Neurol.* 340:221–331

Simon H, Lumsden A. 1993. Rhombomere-specific origin of the contralateral vestibulo-acoustic efferent neurons and their migration across the embryonic midline. *Neuron* 11:209–20

Snow DM, Lemmon V, Carrino DA, Caplan A, Silver J. 1990. Sulfated proteoglycans in astroglial barriers inhibit neurite outgrowth in vitro. *Exp. Neurol.* 109:111–30

Sumi Y, Dent MAR, Owen DE, Seeley PJ, Morris RJ. 1992. The expression of tissue and urokinase-type plasminogen activators in neural development suggests different modes of proteolytic involvement in neuronal growth. *Development* 116:625–37

Tanaka H, Obata K. 1984. Developmental changes in unique cell surface antigens of chick embryo spinal motoneurons and ganglion cells. *Dev. Biol.* 106:26–37

Tanaka O, Yoshioka T, Shinohara H. 1988. Secretory activity in the floor plate neuroepithelium of the developing human spinal cord: morphological evidence. *Anat. Rec.* 222:185–90

Tessier-Lavigne M, Placzek M, Lumsden AGS, Dodd J, Jessell TM. 1988. Chemotropic guidance of developing axons in the mammalian central nervous system. *Nature* 336:775–78

Wall NA, Blessing M, Wright CVE, Hogan BLM. 1993. Biosynthesis and in vivo localization of the decapentaplegic-Vg-related protein, DVR-6 (bone morphogenetic protein-6). *J. Cell Biol.* 120:493–502

Weber A. 1938. Croissance des fibres nerveuses commissurales lors de lésions de la moelle

épinière chez de jeunes embryons de Poulet. *Biomorphosis* 1:30–35

Wentworth LW. 1984. The development of the cervical spinal cord of the mouse embryo. II. A Golgi analysis of sensory, commissural, and association cell differentiation. *J. Comp. Neurol.* 222:96–115

Wood TL, Streck RD, Pintar JE. 1992. Expression of the IGFBP-2 gene in post-implantation rat embryos. *Development* 114:59–66

Yaginuma H, Homma S, Künzi R, Oppenheim RW. 1991. Pathfinding by growth cones of commissural interneurons in the chick embryo spinal cord: a light and electron microscopic study. *J. Comp. Neurol.* 304:78–102

Yaginuma H, Oppenheim RW. 1991. An experimental analysis of in vivo guidance cues used by axons of spinal interneurons in the chick embryo: evidence for chemotropism and related guidance mechanisms. *J. Neurosci.* 11:2598–613

Yaginuma H, Shiga T, Homma S, Ishihara R, Oppenheim RW. 1990. Identification of early developing axon projections from spinal interneurons in the chick embryo with a neuron specific β-tubulin antibody: evidence for a new "pioneer" pathway in the spinal cord. *Development* 108:705–16

Yamada T, Placzek M, Tanaka H, Dodd J, Jessell TM. 1991. Control of cell pattern in the developing nervous system: polarizing activity of the floor plate and notochord. *Cell* 64:635–47

Yoshioka T, Tanaka O. 1989. Ultrastructural and cytochemical characterisation of the floor plate ependyma of the developing rat spinal cord. *J. Anat.* 165:87–100

Annu. Rev. Neurosci. 1995. 18:531–53

MOLECULAR NEUROBIOLOGY AND GENETICS OF CIRCADIAN RHYTHMS IN MAMMALS

Joseph S. Takahashi

NSF Center for Biological Timing, Department of Neurobiology and Physiology, Northwestern University, Evanston, Illinois 60208

KEY WORDS: circadian clock genes, neurogenetics, gene expression, suprachiasmatic nucleus, immediate-early genes, mouse genetics

INTRODUCTION

One of the goals of neuroscience is to understand behavior. How is it that genes ultimately interact with each other and with the environment to produce an organism's behavior? The mammalian circadian system provides an ideal model in which to study the physiological mechanisms underlying behavior, and in turn, genetics provides a means to unravel these mechanisms at the molecular level. As Seymour Benzer (1973) stated over twenty years ago, "Our objectives are to discern the genetic component of a behavior, to identify it with a particular gene and then to determine the actual site at which the gene influences behavior and learn how it does so." While Benzer was referring specifically to *Drosophila,* we reiterate his statement today for the mouse. In many ways, mouse genetics in the 1990s stands at the threshold of discovery of behavioral mechanisms just as *Drosophila* genetics did in the 1970s. Substantial classical genetic and molecular genetic resources are now available in the mouse to complement our existing neurobiological understanding of mammalian circadian rhythms (Takahashi et al 1994).

Temporal organization is now recognized as a fundamental feature of all living systems. Just as orderly developmental timing is essential for normal growth and differentiation, organization within the time domain is essential for the normal maintenance and behavior of the mature organism (Pittendrigh

531

1993). One of the best understood temporal domains is the daily environmental cycle of night and day. Over the last four decades, the field of circadian biology has made striking progress in defining the properties of rhythms, in understanding their regulation at both the formal and physiological levels, and more recently, in elucidating their underlying cellular and molecular mechanisms.

A central question in the field of circadian rhythms concerns the molecular mechanism of circadian clocks. How are circadian oscillations generated and at what level of cellular organization do they originate? On the basis of a wide variety of evidence, the mechanism of the circadian clock appears to be cell autonomous and to involve periodic gene expression. Two major approaches have been used to study the cellular and molecular mechanisms regulating circadian rhythms: (*a*) physiological analysis using perturbations with pharmacological and biochemical methods, and (*b*) genetic and molecular genetic analysis using circadian clock mutants. With the exception of work with *Drosophila, Neurospora,* and perhaps a few other organisms that are amenable to genetics, most work in the field has favored physiological approaches. The predominant strategy for studying circadian clocks has been to localize circadian pacemakers at the organismal level and then to isolate the pacemaker-containing tissues in vitro for more mechanistic analysis (Takahashi 1991). The success of this approach is exemplified by the analysis of circadian oscillators in mollusks (Jacklet 1984, 1989; Koumenis & Eskin 1992; Block et al 1993) and vertebrates (Meijer & Rietveld 1989, Takahashi et al 1989, Cahill et al 1991, Klein et al 1991, Miller 1993, Takahashi 1993, van den Pol & Dudek 1993).

In this review I summarize progress in our understanding of the cellular and molecular mechanisms underlying circadian behavior in mammals using both physiological and genetic approaches. Although past progress in the field has been primarily at the physiological level, we are entering a new era in which the genetic approach can be successfully applied to mammals. The goals of this review are threefold: first, to describe the general properties of circadian rhythms in animals; second, to emphasize the involvement of macromolecular synthesis at all levels of the circadian system; and third, to discuss genetic approaches to circadian rhythms in mammals.

A number of reviews have addressed various aspects of circadian rhythms (Meijer & Rietveld 1989; Rosbash & Hall 1989; Rusak 1989; Takahashi et al 1989, 1993; Dunlap 1990, 1993; Hall 1990; Hall & Kyriacou 1990; Morse et al 1990; Rusak & Bina 1990; Underwood 1990; Cahill et al 1991; Hastings et al 1991; Klein et al 1991; Ralph & Lehman 1991; Moore 1992; Rietveld 1992; Young 1993; Block et al 1993; Miller 1993; Pittendrigh 1993; Schwartz 1993; Takahashi 1993; van den Pol & Dudek 1993; Turek 1994; Turek & Van Cauter 1994), and I refer the reader to these reviews for more extensive coverage of specific topics.

GENERALIZATIONS ABOUT CIRCADIAN RHYTHMS

Circadian rhythms are defined by three general properties: (*a*) a *self-sustained oscillation* in an organism's behavior, physiology, or biochemistry with a period length close to 24 hours in constant environmental conditions; (*b*) *entrainment,* or synchronization, of the circadian rhythm (both period and phase control) by environmental cycles of light or temperature; and (*c*) *temperature compensation* of the period length of the rhythm with temperature coefficients (Q_{10}) within the range 0.8 to 1.3.

At a formal level, all circadian systems contain at least three elements: (*a*) an input pathway or set of input pathways that convey environmental information to the circadian pacemaker for entrainment, (*b*) a circadian pacemaker that generates the oscillation, and (*c*) an output pathway or set of output pathways through which the pacemaker regulates its various output rhythms. In all systems, a photic entrainment pathway is present as an input. Photic entrainment pathways differ markedly in various organisms, and these differences exist at both the receptor and signal-transduction levels. At the output level, diversity is even more extreme, and clock-regulated outputs span the molecular to behavioral levels. As discussed above, the input and output pathways of the circadian system are specific to each organism (Takahashi 1991). Despite these differences in the coupling pathways (inputs and outputs) of the circadian clock, the core mechanism of the pacemaker appears to be fundamentally similar in all organisms. Whether these similarities in pacemaker mechanisms are functionally analogous or phylogenetically homologous remains to be seen.

Circadian Pacemakers Are Discretely Localized

In animals, circadian rhythms expressed at the behavioral level are regulated by discrete central nervous system structures that function as circadian pacemakers. Among vertebrates the localization of circadian function within the nervous system has been remarkable. Circadian oscillators have been identified in three diencephalic structures: the hypothalamic suprachiasmatic nucleus (SCN) of mammals (Meijer & Rietveld 1989, Klein et al 1991); the pineal gland of birds, reptiles, and fish (Falcón et al 1989, Takahashi et al 1989, Underwood 1990); and the retinas of amphibians and birds (Underwood et al 1990, Cahill et al 1991, Pierce et al 1993). At the physiological level, both *oscillator* and *pacemaker* function have been defined. A *circadian oscillator* is a structure that expresses a self-sustained oscillation under constant conditions (the absolute minimum number of cycles is two). Circadian oscillators have been localized by isolating the tissue in question in vitro and demonstrating the persistence of circadian oscillations in the isolated tissue under constant conditions. Circadian oscillator function has been convincingly demonstrated for all three diencephalic structures listed above (Deguchi 1979, Takahashi et al 1980, Green & Gillette 1982, Besharse & Iuvone 1983, Earnest & Sladek 1986) as well as for ocular oscillators in molluscan eyes (Jacklet 1984, Block et al 1993). In contrast, a *circadian pacemaker* is a circadian oscillator that drives and therefore controls some overt rhythmic process, such as locomotor behavior. The distinction between pacemaker and oscillator function has become important because most circadian systems contain multiple oscillators. Pacemaker function has been experimentally demonstrated by transplanting the structure in question and then showing that a pacemaker property such as the phase or period of the recipient rhythm is regulated by the transplant. This demonstration has been accomplished in vertebrates for the pineal gland of sparrows (Zimmerman & Menaker 1979) and the SCN of hamsters (Ralph et al 1990), and in invertebrates for the brain of silkmoths (Truman & Riddiford 1970) and *Drosophila* (Handler & Konopka 1979) and for the optic lobes of cockroaches (Page 1982). The discrete localization of function found in the circadian system appears to be rare among neural systems regulating comparable physiological or behavioral processes.

Circadian Clocks Are Cell Autonomous

Although it has been known for over three decades that unicellular organisms can express circadian rhythms (reviewed in Edmunds 1988), the cellular nature of circadian oscillators in animals has only recently been established. Among vertebrates, circadian rhythms can be expressed by primary cell cultures derived from the pineal gland of birds (Deguchi 1979; Zatz et al 1988; Robertson & Takahashi 1988a,b), the SCN of rodents (Murakami et al 1991, Watanabe et al 1993), and the retina of birds (Pierce et al 1993). The strongest case is the chick pineal in which a circadian oscillator system with a photic entrainment pathway can be studied in cell culture (reviewed in Takahashi et al 1989). Photoreception, circadian oscillator function, and melatonin biosynthesis (the hormonal output of the pineal) all appear to be cellular properties of a single cell type, the pinealocyte.

Although primary cell cultures from the pineal, SCN, or retina can generate circadian oscillations, these experiments do not directly address whether the generation of a rhythm is an intercellular or intracellular phenomenon. In fact, several papers continue to resurrect the idea that circadian oscillations might be generated by coupling high-frequency oscillations (e.g. Miller 1993, van den Pol & Dudek 1993). In spite of the interest in such a mechanism, there remains no definitive evidence that circadian oscillations are based on a fundamental high-frequency oscillator. A direct test of whether the circadian pacemaker is cell autonomous in a multicellular organism was achieved using the marine mollusc *Bulla* (Michel et al 1993). A circadian rhythm in membrane conductance is expressed in *Bulla* basal retinal neurons maintained in single-cell microcultures. These experiments establish that isolated *Bulla* neurons are sufficient for the generation of circadian rhythms and that intercellular communication is not necessary for this phenomenon. Definitive experiments of this type have yet to be accomplished in vertebrates.

GENE EXPRESSION IS INVOLVED AT ALL LEVELS OF THE CIRCADIAN SYSTEM

A striking feature of the circadian system is the widespread involvement of macromolecular synthesis (Takahashi 1993). At the level of the circadian oscillator, macromolecular synthesis is necessary for progression of the oscillation through its cycle. At the input level, light induces a set of immediate-early genes (IEGs) in the mammalian SCN. At the output level, a number of target genes are transcriptionally regulated by the circadian clock system.

Circadian Oscillator: A Critical Period for Macromolecular Synthesis

In a wide variety of organisms, the mechanism of the circadian clock appears to involve transcriptional and translational processes (Edmunds 1988, Rosbash & Hall 1989, Hall 1990, Morse et al 1990, Young 1992, Dunlap 1993, Takahashi 1993, Takahashi et al 1993). This conclusion is based upon two types of results: the isolation of single-gene mutations that affect circadian period (discussed below) and the application of protein-synthesis inhibitors that cause phase shifts or period changes in circadian rhythms. In all organisms that have been examined, protein-synthesis inhibitors cause phase-dependent phase shifts. For example, in the golden hamster (Takahashi & Turek 1987, Inouye et al 1988, Wollnik et al 1989), chick pineal (Takahashi et al 1989), and *Aplysia* eye (Jacklet 1984, Koumenis & Eskin 1992), about half of the circadian cycle is sensitive to inhibitors of translation with maximal sensitivity occurring from about circadian time (CT) 18 to CT6. In addition, pulses of the reversible RNA synthesis inhibitor 5,5-dichloro-1-β-D-ribofuranosylbenzimida-zole (DRB) cause phase-dependent phase shifts in the *Aplysia* eye (Raju et al 1991). The DRB experiments suggest that there is also a critical period for transcription during the circadian cycle from CT20 to CT10 in *Aplysia.*

In addition to inhibitor experiments using short pulses (<6 h), longer treatment with inhibitors can arrest or stop the motion of the circadian pacemaker (Khalsa et al 1992). In *Bulla* eyes, long-duration application of protein-synthesis inhibitors delays the phase of the subsequent rhythm precisely by the duration that the pulse extends past CT0 of the cycle. These experiments suggest that the motion of the circadian pacemaker is arrested during these long inhibitor pulses and that the critical period for protein synthesis begins in the late subjective night (near CT0). By analogy to work in the cell cycle, there appears to be a phase in the circadian cycle that is similar to the restriction point of the animal cell cycle, in which partial inhibition of protein synthesis causes fibroblasts to arrest in G1 (Pardee 1989). This view of the circadian cycle could be useful in attempting to define a beginning point of the cycle, such as "START," which has been useful in unraveling cell-cycle mechanisms in budding and fission yeast (Hartwell et al 1974, Hartwell & Weinert 1989, Norbury & Nurse 1992, Reed 1992). Thus, both the inhibitor-induced phase shifts and the arrest of the circadian cycle suggest that the circadian clock requires macromolecular synthesis during a critical period for progression through the cycle.

Input: Light Regulates Immediate–Early Gene Expression

In *Neurospora* and *Aplysia,* the phase-shifting effects of light are blocked by inhibitors of protein synthesis, which suggests that entrainment of the circadian clock may also involve gene expression (Johnson & Nakashima 1990, Raju et al 1990). Consistent with this idea, photic stimuli strongly induce the expression of a set of IEGs in the mammalian SCN (reviewed in Kornhauser et al 1994). These include c-*fos, jun-B, NGFI-A* (also known as *zif268, egr-1,* and *krox-24*) and *NGFI-B* (also known as *nur77*) (Rea 1989; Aronin et al 1990; Earnest et al 1990; Kornhauser et al 1990, 1992; Rusak et al 1990, 1992; Sutin & Kilduff 1992; Takeuchi et al 1993; Schwartz et al 1994). All of these IEGs are transcription factors and are thought to act as signaling molecules to regulate downstream target genes (Sheng & Greenberg 1990, Morgan & Curran 1991). Fos is the most extensively studied IEG, and several features of the regulation of Fos in the SCN are noteworthy. First, the photic induction of Fos is anatomically specific and restricted to the retinorecipient regions of the SCN and the intergeniculate leaflet, which are both components of the mammalian circadian system (Rea 1989, Aronin et al 1990, Rusak et al 1990). Second, the photic induction of Fos is gated by the circadian clock and occurs only at phases of the cycle when light is capable of inducing behavioral phase shifts (Kornhauser et al 1990, Colwell & Foster 1992, Rea 1992, Sutin & Kilduff 1992, Chambille et al 1993, Schwartz et al 1994). Furthermore, the circadian gating of c-*fos* in the SCN appears to be regulated, at least in part, by the phosphorylation of the *trans*-acting factor CREB (cyclic-AMP response element–binding protein) (Ginty et al 1993). Third, the photic threshold for induction of c-*fos* is quantitatively correlated with the phase-shifting effects of light (Kornhauser et al 1990). Fourth, the Fos response in the SCN is modality specific and is associated with photic stimuli, but not with nonphotic stimuli (Janik & Mrosovsky 1992, Mead et al 1992, Cutera et al 1993, Zhang et al 1993a). This indicates that Fos induction is related to the photic input rather than to a phase-shifting mechanism per se. Fifth, Fos induction is attenuated by *N*-methyl-D-aspartate (NMDA) antagonists, non-NMDA antagonists and mecamylamine which all block behavioral phase shifting to light (Colwell et al 1990, 1991; Abe et al 1991, 1992; Ebling et al 1991; Vindlacheruvu et al 1992; Rea et al 1993; Zhang et al 1993b). Sixth, light induces AP-1 DNA-binding activity in the SCN, and the light-induced AP-1 complex appears to be primarily composed of Fos and Jun-B (Kornhauser et al 1992, Takeuchi et al 1993). Finally, a requirement for c-*fos* and *jun-B* gene expression in photically induced phase shifts has been shown by antisense oligonucleotide experiments (Wollnik et al 1994). Treatment of rats in vivo with antisense oligonucleotides against both IEGs blocked Fos and Jun-B induction in the SCN as well as light-induced behavioral phase shifts. Taken together these experiments strongly suggest that Fos and Jun-B gene expression and AP-1 activity are components of the photic-entrainment signaling pathway in rodents.

Output: Clock-Controlled Target Genes

There are a number of genes whose expression patterns have been shown to be regulated by the circadian clock. The abundance of specific mRNAs varies with a circadian rhythm in cyanobacteria (Kondo et al 1993), microorganisms (Loros et al 1989, Bell-Pedersen et al 1992), plants (Kay 1993), and animals (Wuarin et al 1992, Pierce et al 1993, Stehle et al 1993), and in some cases the RNA rhythms have been shown to be transcriptionally regulated (Loros & Dunlap 1991, Millar & Kay 1991, Pierce et al 1993). In mammals, clock control of transcription has been clearly demonstrated for the transcriptional activator DBP (D element binding protein in the albumin promotor) and the enzyme cholesterol 7-α hydroxylase in the liver (Wuarin & Schibler 1990, Lavery & Schibler 1993). The circadian control of transcription is significant because it provides an entry point to analyze the *cis*-acting regulatory elements and *trans*-acting factors through which the clock regulates gene expression. Only a few genes have been analyzed for *cis*-acting regulatory elements through which the clock might act (called circadian-clock responsive elements, or CCREs). In two cases, the CCREs are known elements: 1. A DBP DNA-binding site on the 7-α hydroxylase promoter appears to be important for binding an evening-specific complex of nuclear factors containing DBP in the liver (Lavery & Schibler 1993); and 2. a tandem array of four CRE sites in the P2 promoter of the CREM gene appears to mediate the rhythmic activation and repression of the CREM isoform ICER (Molina et al 1993, Stehle et al 1993). In addition, CCREs have been mapped to relatively small 5' upstream sequences in the wheat *Cab-1* gene (Fejes et al 1990) and the *Arabidopsis CAB2* gene (Millar & Kay 1991, Millar et al 1992, Anderson et al 1994). An intriguing prospect for future work is the possible existence of novel *trans*-acting factors whose primary function would be to regulate circadian transcription.

CLOCK GENES: LESSONS FROM *DROSOPHILA* AND *NEUROSPORA*

The genetic approach to circadian rhythms was pioneered by Konopka & Benzer (1971), who isolated single-gene mutations that altered circadian periodicity in *Drosophila*. In a chemical mutagenesis screen of the X chromosome, they found three mutants that either shortened (per^S), lengthened (per^L), or abolished (per^0) circadian rhythms of eclosion and adult locomotor activity. Surprisingly, the three mutants were allelic and defined a new gene, the *period* locus. This classic story has been told repeatedly and there are several excellent reviews on the subject (Konopka 1987; Hall & Rosbash 1988; Rosbash & Hall 1989; Dunlap 1990, 1993; Greenspan 1990; Hall 1990; Hall & Kyriacou 1990; Young 1992). In 1984, two groups at Brandeis and Rockefeller Universities independently cloned *per* in a series of ground-breaking experiments that were the first to show that germline transformation with DNA could rescue a complex behavioral program (reviewed in Rosbash & Hall 1989). Each of the mutant *per* alleles is caused by intragenic point mutations that produce missense mutations in per^S and per^L, and a nonsense mutation in per^0 (Baylies et al 1987, Yu et al 1987). Unfortunately, the sequence of the protein product of the *per* gene (PER) was not especially informative of its function, and a number of false starts and dead ends ensued during the following five years (reviewed in Kyriacou 1994). Only recently has the nature of PER become more clear.

The Drosophila Period *and* Timeless *Genes*

Clues to the nature of the *per* gene product appeared with the discovery of three genes that share sequence homology with PER. The *Drosophila* single-minded protein (SIM) (Nambu et al 1991), the human aryl hydrocarbon receptor nuclear translocator (ARNT) (Hoffman et al 1991), and the aryl hydrocarbon receptor (AHR) (Burbach et al 1992) all share a domain with PER called PAS (for PER, ARNT, SIM) (Nambu et al 1991). The PAS domain contains about 270 amino acids of sequence similarity with two 51–amino acid direct repeats. In SIM, ARNT, and AHR, the PAS domain lies adjacent to a basic helix-loop-helix (bHLH) domain that is involved in DNA binding and dimerization of transcriptional regulators. However, PER does not contain a bHLH motif or any other known DNA-binding motif. Recent work shows that the PAS domain can function as a dimerization domain and can mediate both homotypic and heterotypic protein interactions (Huang et al 1993). Interestingly, the ligand-binding domain of AHR (Burbach et al 1992) and the per^L mutation both map to the PAS domain. In addition, the per^L mutation abolishes PAS domain–mediated dimerization. Because SIM, ARNT, and AHR are transcriptional regulators and because PER can dimerize to these PAS-containing factors, PER could function as a transcriptional regulator either by working in concert with a partner that carries a DNA-binding domain or by acting as a dominant-negative regulator by competing with a transcriptional regulator for dimerization or DNA binding. Consistent with this role, PER is predominantly a nuclear protein in the adult central nervous system of *Drosophila* (Liu et al 1992).

The expression of PER itself is circadian, and both *per* mRNA and PER protein abundance levels oscillate. Hardin et al (1990) showed that *per* mRNA levels undergo a striking circadian oscillation. The *per* RNA rhythm persists in constant darkness, and the period of the RNA rhythm is ~24 h in *per*⁺ flies and is ~20 h in *per*ˢ flies. The RNA of *per*⁰ flies is present at a level ~50% of normal flies, but does not oscillate. In *per*⁰ flies that have been rescued by germline transformation with wild-type *per*⁺ DNA, both circadian behavior and *per* RNA cycles are restored. Importantly, in these transformed flies both the exogenous *per*⁺ RNA and the endogenous *per*⁰ RNA levels oscillate. In addition to a *per* RNA cycle, the PER protein also shows a circadian rhythm in abundance (Siwicki et al 1988, Zerr et al 1990, Edery et al 1994b). The rhythm in PER protein also depends on *per*, because *per*⁰ flies do not have a protein rhythm and because *per* mutants alter the PER rhythm (Zerr et al 1990). Therefore, the circadian expression of *per* mRNA and protein levels both depend on an active *per* gene. Because *per*ˢ shortens the period of the RNA cycle and because *per*⁺ DNA transformation rescues *per*⁰ RNA cycling, PER protein expression clearly regulates *per* RNA cycling. Hardin et al (1990) propose that feedback of the *per* gene product regulates its own mRNA levels. Support for such a model has been provided by showing that transient induction of PER from a heat-shock promoter driving a *per* cDNA transgene in a wild-type background can phase shift circadian activity rhythms in *Drosophila* (Edery et al 1994a).

The PER protein rhythm appears to be regulated at both transcriptional and posttranscriptional levels. Hardin et al (1992) have shown that levels of *per* precursor RNA cycle in concert with mature *per* transcripts. In addition, *per* promoter and CAT (chloramphenicol acetyltransferase) fusion gene constructs show that *per* 5' flanking sequences are sufficient to drive heterologous RNA cycles. These results suggest that circadian fluctuations in *per* mRNA abundance are controlled at the transcriptional level. In addition to a rhythm in *per* transcription and PER abundance, PER appears to undergo multiple phosphorylation events as it accumulates each cycle (Edery et al 1994b). The nature and functional significance of the PER phosphorylation sites, however, are not known at this time. Interestingly, the peak of the *per* RNA cycle precedes the peak of the PER protein cycle by about six hours. The reasons for the lag in PER accumulation are currently not understood. However, the recent isolation of a new clock mutant, named *timeless* (*tim*), may provide some insight here (Sehgal et al 1994). *Tim* mutants fail to express circadian rhythms in eclosion and locomotor activity, and more importantly, they also fail to express circadian rhythms in *per* mRNA abundance (Sehgal et al 1994). Furthermore, the nuclear localization of PER is blocked in *tim* mutants (Vosshall et al 1994). Interestingly, a domain of PER between amino acids 96 and 529, which includes the PAS domain, is necessary for the *tim*-sensitive block of nuclear localization. Deletion of these sequences results in nuclear localization of PER in a *tim* mutant background. This counterintuitive result suggests that these PER sequences inhibit nuclear localization of PER in the absence of *tim*. Thus, the *tim* gene product may interact with PER either directly or indirectly to facilitate nuclear entry of PER, or it may regulate factors that retain PER cytoplasmically. In either case, the identification of *tim* and its functional interaction with *per* is important because it suggests that elements of a transcription-translation nuclear-transport feedback loop may be central elements of the circadian mechanism in *Drosophila*.

The Neurospora Frequency *Gene*

In addition to the *Drosophila per* gene, significant progress has been made in elucidating the molecular nature of the *Neurospora frequency* (*frq*) gene (Dunlap 1993). Like *per,* the *frq* locus is defined by mutant alleles that either shorten, lengthen, or disrupt circadian rhythms (Feldman & Hoyle 1973, Feldman 1982, Dunlap 1993). Cloned in 1989, the sequence of FRQ shows little resemblance to PER (except for a region containing threonine-glycine repeats) (McClung et al 1989); however, recent molecular work shows striking functional similarities (Aronson et al 1994). The *frq* gene expresses a circadian oscillation of mRNA abundance whose period is altered by *frq* mutations (Aronson et al 1994). A null allele, *frq^9*, expresses elevated levels of *frq* transcript and does not show a rhythm in mRNA abundance (Aronson et al 1994). Interestingly, no level of constitutive expression of *frq$^+$* in a null background can rescue overt rhythmicity which suggests that the circadian rhythm of *frq* mRNA is a necessary component of the oscillator (Aronson et al 1994). However, overexpression of a *frq$^+$* transgene does negatively autoregulate expression of the endogenous *frq* gene (Aronson et al 1994). In addition, overexpression of a *frq$^+$* transgene in a wild-type background blocks overt expression of circadian rhythms (Aronson et al 1994). In the most definitive experiment of its kind yet performed, the phase of the overt circadian rhythm can be determined by a step reduction in FRQ protein expression (Aronson et al 1994). Taken together, these experiments strongly argue that *frq* is a central component of the *Neurospora* circadian oscillator and that a negative autoregulatory loop regulating *frq* transcription forms the basis of the oscillation (Aronson et al 1994).

Although there are remarkable functional similarities between *per* and *frq,* there are also distinct differences. The phases of the mRNA rhythms are different: *per* peaks at night (Hardin et al 1990), whereas, *frq* peaks during the day (Aronson et al 1994). While *per* overexpression shortens circadian period (Smith & Konopka 1982, Baylies et al 1987), *frq* overexpression does not change the period but rather abolishes overt rhythmicity (Aronson et al 1994). The null allele *per*[0] leads to a constant level of mRNA that is about 50% of the peak level of wild-type levels in *Drosophila* (Hardin et al 1990); while in *Neurospora, frq*[9] mRNA levels are significantly elevated relative to wild-type (Aronson et al 1994). These differences can be interpreted in three ways: (*a*) The differences could be quantitative and due to the nonlinear nature of oscillators, (*b*) *frq* and *per* could define different elements in a conserved pathway within the oscillator feedback loop, or (*c*) the *Drosophila* and *Neurospora* circadian clocks could be functionally analogous rather than phylogenetically homologous. Irrespective of the interpretation, however, a transcription-translation autoregulatory feedback loop is likely a common feature of circadian clocks.

GENETIC APPROACHES TO CIRCADIAN RHYTHMS IN MAMMALS

Given the substantial progress in our understanding of *Neurospora* and *Drosophila* clock genes, an obvious question is whether similar mechanisms may underlie the regulation of mammalian circadian rhythms. Unfortunately, searching for *per* homologues in mammals has not been very productive despite 10 years of effort by several laboratories. This is probably due to the relatively low level of sequence similarity of *per* even among the Drosophilids (Hall 1990). Putative *per* homologues in mammals have been reported in searches directed against the threonine-glycine (TG)-repeat region of PER (Shin et al 1985, Matsui et al 1993) and the region of the *per*[S] mutation (Siwicki et al 1992). However, the TG-repeat clones show no other sequence similarity to PER, and the antigens detected by antibodies to the *per*[S] region have not been characterized molecularly. New efforts targeted against the PAS dimerization domain (Huang et al 1993), which is moderately well conserved among insects (S Reppert, personal communication), using either PCR-based approaches or the yeast two-hybrid system (Fields & Song 1989) could eventually succeed as more bona fide *per* homologues are cloned in species more closely related to insects. Alternatively, as other *Drosophila* clock genes are cloned, some should have sequence conservation with mammals such as is found with genes regulating pattern formation (Krumlauf 1993) or signal transduction (Zipursky & Rubin 1994).

Quantitative Genetics and Spontaneous Variants

Very little information on the genetics of mammalian circadian rhythms is available. Most work in the field has used quantitative genetic approaches such as comparisons of circadian phenotype among inbred strains of mice and rats, recombinant inbred strain analysis, or selection of natural variants (for reviews see Hall 1990, Schwartz & Zimmerman 1990, Lynch & Lynch 1992). The most comprehensive analysis of inbred mouse strains was done by Schwartz & Zimmerman (1990), who compared 12 different strains and found that the most extreme strains (C57BL/6J and BALB/cByJ) had a period difference of about one hour in constant darkness. Reciprocal F_1 hybrid– and recombinant inbred–strain analysis provided no evidence of monogenic inheritance of the circadian period. Polygenic inheritance of circadian traits (or more strictly, failure to detect monogenic inheritance) has been the conclusion of every quantitative genetic analysis performed thus far.

A notable exception to the general finding of polygenic control of circadian phenotype is the spontaneous mutation *tau* found in the golden hamster (Ralph & Menaker 1988). *Tau* is a semidominant, autosomal mutation that shortens the circadian period by two hours in heterozygotes and by four hours in homozygotes. Its phenotype is remarkably similar to the *Drosophila per*^{S} allele; it is semidominant, changes period to the same extent, and increases the amplitude of the phase-response curve to light (Ralph & Menaker 1988, Ralph 1991). The *tau* mutation has been extremely useful for physiological analysis. For example, the circadian-pacemaker function of the SCN has been definitively demonstrated by transplantation of SCN tissue derived from *tau* mutant hamsters to establish that the genotype of the donor SCN determines the period of the restored rhythm (Ralph et al 1990). Furthermore, the effects of having both *tau* mutant and wild-type SCN tissue in the same animal show that mutant (~20 h) and wild-type (~24 h) periodicities can be expressed simultaneously, suggesting that very little interaction of the oscillators occurs under these conditions (Vogelbaum & Menaker 1992). Additional cellular interactions can also be studied by transplantation of dissociated SCN cells derived from *tau* mutant and wild-type animals (Ralph & Lehman 1991). Thus, several issues that could not be addressed previously have been resolved or approached by the use of the *tau* mutation.

Unfortunately, not much progress has been made on the genetic and molecular nature of *tau*. One paper reports differences as assessed by two-dimensional gel electrophoresis in two proteins in the *tau* mutant hamster (Joy et al 1992). Whether these proteins are tightly linked to *tau*, however, is unknown. Genetic mapping and molecular cloning of *tau* is extremely difficult because of the paucity of genetic information in the golden hamster. For example, out of the expected 22 linkage groups in the hamster, there are only 7 linkage groups defined by two mutations each. Clearly not an organism for genetic analysis! Thus far, the *tau* mutation has contributed substantially to physiological analysis, but it will remain difficult to elucidate the nature of the *tau* gene product unless candidate genes become apparent or the hamster is developed as a genetic system.

Forward-Genetic Approaches to Circadian Rhythms in the Mouse

Mammalian circadian rhythms are clearly under genetic control; however, the underlying mechanisms are largely unknown. In such situations, classical *forward genetics* has been a powerful approach to uncover unknown elements. Mutant screens make no assumptions concerning the mechanisms involved and require only a clear phenotype to be expressed. With efficient mutagenesis, coupled with careful screening procedures, informative mutants have been isolated in many systems, including the mouse. Given the widespread involvement of gene expression in circadian regulation, the circadian system is an ideal target for such procedures.

It is generally assumed that classical mutagenesis and screening procedures are not feasible in the mouse. However, the development of high-efficiency germline mutagenesis procedures with the alkylating agent, N-ethyl-N-nitrosourea (ENU) by Russell and colleagues (Russell et al 1979, Hitotsumachi et al 1985) have made it feasible to undertake large-scale mutant screens in the mouse (Rinchik 1991). Average forward-mutation frequencies of 0.0015 per locus per gamete can be achieved in the mouse with ENU (Rinchik 1991). This means that on average one has a 50% chance of finding a mutation in any single locus by screening about 655 gametes.

With these considerations in mind, we initiated a behavioral screen for ENU-induced mutations of the circadian system in the mouse. The laboratory mouse is an ideal organism for circadian research. Inbred strains, such as C57BL/6J, have robust and precise circadian rhythms of wheel-running activity that are well characterized (Pittendrigh & Daan 1976, Schwartz & Zimmerman 1990). Because the majority of clock mutants isolated in other organisms have been semidominant (Dunlap 1993), we screened heterozygotes directly in a search for dominant mutations with the procedure shown in Figure 1. In the case of a dominant screen, each G_1 mouse represents one mutagenized gamete. To approach saturation mutagenesis (i.e. to be able to mutagenize all loci at least once assuming Poisson statistics and a mutation frequency of 0.001 *per* locus per gamete), one would have to screen about 3000 gametes. Thus a circadian laboratory devoting 100 activity-recording channels to a screen could determine the phenotype of about 1000 mice per year.

In screening about 300 progeny (gametes) of ENU-treated mice, we identified one animal that expressed a circadian period that was more than one hour longer than normal (Vitaterna et al 1994). Although this one-hour difference sounds modest, this value was six standard deviations away from the

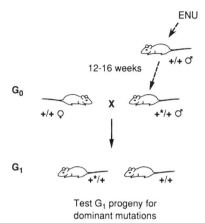

Figure 1 ENU mutagenesis screen in the mouse for dominant or semidominant mutations in first generation (G_1) offspring. Male mice are injected intraperitoneally with ENU to intervene at the stage of spermatogonia, and enter a period of sterility. Upon recovery of fertility (12–16 weeks after treatment) they may be bred with multiple females to produce G_1 progeny that will be heterozygous for any ENU-induced mutation. (From Takahashi et al 1994; Copyright © 1994 by the AAAS.)

mean. The long-period phenotype was inherited as a semidominant autosomal mutation, which we named *Clock* (Vitaterna et al 1994) (Figure 2). Homozygous *Clock* mice expressed extremely long periods of 27 to 28 h upon initial transfer to constant darkness, which was followed by a complete loss of circadian rhythmicity after about 2 weeks in constant darkness (Figure 2*C*). The *Clock* gene thus regulates at least two fundamental properties of the circadian clock system: the intrinsic circadian period and the persistence of circadian rhythmicity. Since a wild-type allele of *Clock* is necessary for sustained circadian rhythmicity, *Clock* defines an essential gene for this behavior. No anatomical defects in the SCN have been observed in association with the *Clock* mutation (Vitaterna et al 1994), which suggests that the loss of circadian rhythmicity in constant darkness cannot be attributed to a gross anatomical or developmental defect.

Because of the extensive genetic mapping information available in the mouse (Takahashi et al 1994), we were able to map *Clock* rapidly by linkage analysis, using intraspecific-mapping crosses and simple-sequence-length polymorphisms (SSLPs) from the MIT/Whitehead Institute genetic map (Vitaterna et al 1994). *Clock* mapped to the midportion of mouse chromosome 5 between two SSLP markers, D5Mit24 and D5Mit83, in a region that shows conserved synteny with human chromosome 4. The possibility of a human homologue of *Clock* on chromosome 4 is significant because it allows us to

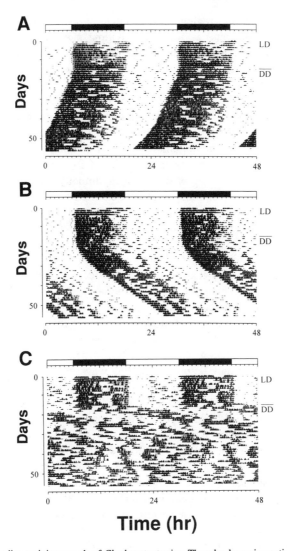

Figure 2 Circadian activity records of *Clock* mutant mice. The wheel-running activity records of three C57BL/6J F2 offspring are shown. All animals were kept on a light-dark cycle (LD12:12) for the first 10 days illustrated, then transferred to constant darkness (DD) on the day indicated by a line in the right margin. (*A*) Activity record of a wild-type F2 mouse. In DD, this animal's activity rhythm had a period of 23.6 h. (*B*) Activity record of a heterozygous *Clock/+* F2 mouse. In DD, this animal's activity rhythm had a period of 24.8 h. (*C*) Activity record of a homozygous *Clock/Clock* F2 mouse. This individual had a period of 27.1 h for the first 10 days in DD, which was followed by a complete loss of circadian rhythmicity thereafter. A residual ultradian periodicity emerges with the loss of the circadian rhythm in DD. (From Vitaterna et al 1994; Copyright © 1994 by the AAAS.)

focus attention upon this region of the genome for possible linkage to circadian traits in human subjects, as well as providing a candidate gene for other disorders associated with circadian-rhythm dysfunctions such as delayed-sleep-phase syndrome (Vignau et al 1993) and affective disorders (Wehr & Rosenthal 1989).

Aside from being the first clock mutation to be isolated in the mouse, which has obvious significance because of the feasibility of ultimately cloning the gene (Takahashi et al 1994), the phenotype of *Clock* is as robust as the best clock mutations in *Drosophila* and *Neurospora* (Dunlap 1993). By robust we mean that the period change is on the order of four to five hours, which is followed by a complete loss of circadian rhythmicity. The magnitude of the period change in *Clock* homozygotes is equivalent to that seen with the per^L and frq^7 alleles that also cause periods of 28–30 h in their respective organisms (Dunlap 1993). The loss of circadian rhythmicity seen in *Clock* homozygotes resembles that seen in per^0 and frq^9 alleles, which are null mutations in those respective genes (Dunlap 1993). The robustness of *Clock* is important for two reasons. First, mutations that have modest effects on period length (on the order of a one-hour change in period in homozygotes) could be due to secondary effects of mutations on the circadian clock system. Second, the most robust mutants in *Drosophila* and *Neurospora* are found at the *per* and *frq* loci, respectively, which are genes that appear to be critical and essential elements of the circadian mechanism in these organisms (discussed above).

Forward genetics may be one of the few ways in which to identify genes involved in the clock mechanism in mammals. In retrospect, it appears unlikely that *per* and *frq* could have been identified and their function studied by nongenetic means. It is also important to emphasize that in both *Drosophila* and mice, other genetic approaches such as selection of natural variants or comparison of different strains all yielded subtle differences that were polygenic and unmappable (Hall & Rosbash 1988, Schwartz & Zimmerman 1990). These results suggest that natural allelic variation is not as extreme as that attainable by induced mutagenesis, because in both *Drosophila* and mice, single genes that strongly influence circadian rhythmicity have been identified by forward genetics.

An important consideration in the use of forward genetics concerns the feasibility of approaching saturation mutagenesis. This is important for three reasons. First, one would like to identify and define all of the genes that are essential for circadian rhythmicity in order to have some measure of the complexity of the system. Will there be hundreds of genes involved or will only a handful be critical? Such assessments have only been achieved in organisms such as *Drosophila* for genes regulating pattern formation (Nüsslein-Volhard & Wieschaus 1980). Ultimately, the identification of all clock genes will be essential to understand the clock mechanism completely. The identification of multiple genes in a pathway will be a key method for trying to understand how clock elements interact and function. Second, the isolation of multiple mutations is crucial; many are likely to be uninformative because they are either nonspecific, pleiotropic, or secondary to a developmental effect. Because the recovery of such uninformative mutants is unavoidable, the only ways to insure that informative mutants will be found are to use stringent phenotyping methods and to test large numbers of mutagenized gametes. Finally, when attempting to clone genes by candidate gene or positional cloning methods (Collins 1992, Copeland et al 1993, Takahashi et al 1994), having multiple alleles at each locus makes it much easier to confirm that mutations are confined to a specific transcription unit in the final analysis. In addition, having multiple alleles with different phenotypes, as in the case of *per* and *frq*, reinforces the likelihood of the gene being an important element in the pathway.

To our knowledge, *Clock* represents the first mutation to be isolated in a behavioral screen of mutagenized mice. With the elegance and popularity of gene targeting in mice, molecular biologists have a tendency to avoid brute-force screening methods. This bias is especially true among those working with the mouse. One should consider, however, that it takes about the same length of time, one year, to isolate a mutation with ENU at a specific locus as it does to knock out a gene in the mouse. Obviously, gene targeting is enormously important, but it cannot yet replace mutant screens because the majority of genes have not been cloned. This is the major limitation of gene targeting — one needs a cloned gene in hand. In the case of circadian rhythms and many other fields of neuroscience and behavior, the genes are unknown and mutant screens are one of the few ways to uncover them. Because *Clock* was isolated in a behavioral screen, it represents an important demonstration that forward genetics can be applied to study a wide range of other behaviors. Like the use of gene targeting to study long-term potentiation (LTP) and spatial memory in mice (Grant et al 1992; Silva et al 1992a,b), the ENU mutagenesis approach provides a new perspective for behavior geneticists who have been primarily focused upon polygenic approaches.

CONCLUSIONS

While substantial progress has been achieved in our understanding of the neurobiology and physiology of circadian rhythms over the last two decades, much less is known about the cellular and molecular mechanisms that generate circadian rhythms in mammals. Continued application of reductionist approaches has shown that the circadian clock in mammals is likely to be cell autonomous and appears to require macromolecular synthesis as demonstrated more definitively in other organisms. However, it is not clear at this point in time whether continued analysis using perturbations with pharmacological and biochemical methods alone will ultimately reveal the mechanism of the circadian clock in mammals. In the study of other cellular functions of comparable complexity, such as the cell cycle, pattern formation, and signal transduction, the genetic approach has been extremely useful and perhaps essential in identifying the set of genes involved in regulating these processes. Similar success is also being achieved with the genetic analysis of circadian rhythms in *Drosophila* and *Neurospora*. Given the likelihood that the set of genes regulating circadian rhythms in mammals is largely undescribed and therefore unknown, a forward-genetic approach to mammalian circadian rhythms is clearly warranted. ENU mutagenesis and screening for clock mutants provides an efficient approach to identify previously unknown genes using phenotype-driven methods. The isolation of the *Clock* mutation in mice demonstrates that this approach is feasible and productive in mammals. Coupled together with the genetic and physical mapping resources available from the "new mouse genomics," mutations defined by phenotype alone can be molecularly identified by the method of positional cloning (Collins 1992, Copeland et al 1993, Takahashi et al 1994). Thus, the circadian biology of mammalian rhythms has entered a new era—the "genomics of circadian clocks" in the mouse are now tractable.

ACKNOWLEDGMENTS

I thank Dr. B Takahashi and members of my laboratory for helpful discussions. Research was supported by NIH grants R37 MH39592, R01 MH49241, R01 EY08467 to JST; a grant from the MacArthur Foundation Network on Depression and the NSF Center for Biological Timing; and a State of Illinois HECA award to JST, FW Turek, and LH Pinto.

Literature Cited

Abe H, Rusak B, Robertson HA. 1991. Photic induction of Fos protein in the suprachiasmatic nucleus is inhibited by the NMDA receptor antagonist MK-801. *Neurosci. Lett.* 127:9–12

Abe H, Rusak B, Robertson HA. 1992. NMDA and non-NMDA receptor antagonists inhibit photic induction of Fos protein in the hamster suprachiasmatic nucleus. *Brain Res. Bull.* 28: 831–35

Anderson SL, Teakle GR, Martino-Catt SJ, Kay SA. 1994. Circadian clock- and phytochrome-regulated transcription is conferred

by a 78-bp *cis*-acting domain of the *Arabidopsis CAB2* promoter. *Plant J.* 6:457–70

Aronin N, Sagar SM, Sharp FR, Schwartz WJ. 1990. Light regulates expression of a Fos-related protein in rat suprachiasmatic nuclei. *Proc. Natl. Acad. Sci. USA* 87:5959–62

Aronson BD, Johnson KA, Loros JJ, Dunlap JC. 1994. Negative feedback defining a circadian clock: autoregulation of the clock gene *frequency. Science* 263:1578–84

Baylies MK, Bargiello TA, Jackson FR, Young MW. 1987. Changes in the abundance and structure of the *per* gene product can alter periodicity of the *Drosophila* clock. *Nature* 326:390–92

Bell-Pedersen D, Dunlap JC, Loros JJ. 1992. The *Neurospora* circadian clock-controlled gene, *ccg-2*, is allelic to *eas* and encodes a fungal hydrophobin required for formation of the conidial rodlet layer. *Genes Dev.* 6:2382–94

Benzer S. 1973. Genetic dissection of behavior. *Sci. Am.* 229:24–37

Besharse JC, Iuvone PM. 1983. Circadian clock in *Xenopus* eye controlling retinal serotonin *N*-acetyltransferase. *Nature* 305:133–35

Block GD, Khalsa SBS, McMahon DG, Michel S, Geusz M. 1993. Biological clocks in the retina: cellular mechanisms of biological timekeeping. *Int. Rev. Cytol.* 146:83–144

Burbach KM, Poland A, Bradfield CA. 1992. Cloning of the Ah-receptor cDNA reveals a distinctive ligand-activated transcription factor. *Proc. Natl. Acad. Sci. USA* 89:8185–89

Cahill GM, Grace MS, Besharse JC. 1991. Rhythmic regulation of retinal melatonin: metabolic pathways, neurochemical mechanisms, and the ocular circadian clock. *Cell. Mol. Neurobiol.* 11:529–60

Chambille I, Doyle S, Serviére J. 1993. Photic induction and circadian expression of Fos-like protein: immunohistochemical study in the retina and suprachiasmatic nuclei of hamster. *Brain Res.* 612:138–50

Collins FS. 1992. Positional cloning: Let's not call it reverse anymore. *Nature Genet.* 1:3–6

Colwell CS, Foster RG. 1992. Photic regulation of Fos-like immunoreactivity in the suprachiasmatic nucleus of the mouse. *J. Comp. Neurol.* 324:135–42

Colwell CS, Foster RG, Menaker M. 1991. NMDA receptor antagonists block the effects of light on circadian behavior in the mouse. *Brain Res.* 554:105–10

Colwell CS, Ralph MR, Menaker M. 1990. Do NMDA receptors mediate the effects of light on circadian behavior? *Brain Res.* 523:117–20

Copeland NG, Jenkins NA, Gilbert DJ, Eppig JT, Maltais LJ, et al. 1993. A genetic linkage map of the mouse: current applications and future prospects. *Science* 262:57–66

Cutera RA, Kalsbeek A, Pévet P. 1993. No triazolam-induced expression of *fos* protein in raphe nuclei of the male Syrian hamster. *Brain Res.* 602:14–20

Deguchi T. 1979. A circadian oscillator in cultured cells of chicken pineal gland. *Nature* 282:94–96

Dunlap JC. 1990. Closely watched clocks: molecular analysis of circadian rhythms in *Neurospora* and *Drosophila*. *Trends Genet.* 6:159–65

Dunlap JC. 1993. Genetic analysis of circadian clocks. *Annu. Rev. Physiol.* 55:683–729

Earnest DJ, Iadarola M, Yeh HH, Olschowka

JA. 1990. Photic regulation of c-*fos* expression in neural components governing the entrainment of circadian rhythms. *Exp. Neurol.* 109:353–61

Earnest DJ, Sladek CD. 1986. Circadian rhythms of vasopressin release from individual rat suprachiasmatic explants in vitro. *Brain Res.* 382:129–33

Ebling FJP, Maywood ES, Staley K, Humby T, Hancock DC, et al. 1991. The role of *N*-methyl-D-aspartate-type glutamatergic neurotransmission in the photic induction of immediate-early gene expression in the suprachiasmatic nuclei of the Syrian hamster. *J. Neuroendocrinol.* 3:641–52

Edery I, Rutila JE, Rosbash M. 1994a. Phase shifting of the circadian clock by induction of the *Drosophila* Period protein. *Science* 263:237–40

Edery I, Zwiebel LJ, Dembinska ME, Rosbash M. 1994b. Temporal phosphorylation of the *Drosophila* Period protein. *Proc. Natl. Acad. Sci. USA* 91:2260–64

Edmunds LE. 1988. *Cellular and Molecular Bases of Biological Clocks*. New York: Springer-Verlag. 497 pp.

Falcón J, Marmillon JB, Claustrat B, Collin J-P. 1989. Regulation of melatonin secretion in a photoreceptive pineal organ: an in vitro study in the pike. *J. Neurosci.* 9:1943–50

Fejes E, Pay A, Kanevsky I, Szell M, Adam E, et al. 1990. A 268-bp upstream sequence mediates the circadian clock-regulated transcription of the wheat *Cab-1* gene in transgenic plants. *Plant Mol. Biol.* 15:921–32

Feldman JF. 1982. Genetic approaches to circadian clocks. *Annu. Rev. Plant Physiol.* 33:583–608

Feldman JF, Hoyle MN. 1973. Isolation of circadian clock mutants of *Neurospora crassa*. *Genetics* 75:605–13

Fields S, Song O-K. 1989. A novel genetic system to detect protein-protein interactions. *Nature* 340:245–46

Ginty DD, Kornhauser JM, Thompson MA, Bading H, Mayo KE, et al. 1993. Regulation of CREB phosphorylation in the suprachiasmatic nucleus by light and a circadian clock. *Science* 260:238–41

Grant SGN, O'Dell TJ, Karl KA, Stein PL, Soriano P, Kandel ER. 1992. Impaired long-term potentiation, spatial learning, and hippocampal development in *fyn* mutant mice. *Science* 258:1903–10

Green DJ, Gillette R. 1982. Circadian rhythm of firing rate recorded from single cells in the rat suprachiasmatic brain slice. *Brain Res.* 245:198–201

Greenspan RJ. 1990. The emergence of neurogenetics. *Semin. Neurosci.* 2:145–57

Hall JC. 1990. Genetics of circadian rhythms. *Annu. Rev. Genet.* 24:659–97

Hall JC, Kyriacou CP. 1990. Genetics of biological rhythms in *Drosophila. Adv. Insect Physiol.* 22:221–97

Hall JC, Rosbash M. 1988. Mutations and molecules influencing biological rhythms. *Annu. Rev. Neurosci.* 11:373–93

Handler AM, Konopka RJ. 1979. Transplantation of a circadian pacemaker in *Drosophila. Nature* 279:236–38

Hardin PE, Hall JC, Rosbash M. 1990. Feedback of the *Drosophila period* gene product on circadian cycling of its messenger RNA levels. *Nature* 343:536–40

Hardin PE, Hall JC, Rosbash M. 1992. Circadian oscillations in *period* gene mRNA levels are transcriptionally regulated. *Proc. Natl. Acad. Sci. USA* 89:11711–15

Hartwell LH, Culotti J, Pringle JR, Reid BJ. 1974. Genetic control of the cell division cycle in yeast. *Science* 183:46–51

Hartwell LH, Weinert TA. 1989. Checkpoints: controls that ensure the order of the cell cycle events. *Science* 246:629–34

Hastings JW, Rusak B, Boulos Z. 1991. Circadian rhythms: the physiology of biological timing. In *Neural and Integrative Animal Physiology,* ed. CL Prosser, pp. 435–546. New York: Wiley Interscience

Hitotsumachi S, Carpenter DA, Russell WL. 1985. Dose-repetition increases the mutagenic effectiveness of *N*-ethyl-*N*-nitrosourea in mouse spermatogonia. *Proc. Natl. Acad. Sci. USA* 82:6619–21

Hoffman EC, Reyes H, Chu F-F, Sader F, Conley LH, et al. 1991. Cloning of a factor required for activity of the Ah (dioxin) receptor. *Science* 252:954–58

Huang ZJ, Edery I, Rosbash M. 1993. PAS is a dimerization domain common to *Drosophila* Period and several transcription factors. *Nature* 364:259–62

Inouye ST, Takahashi JS, Wollnik F, Turek FW. 1988. Inhibitor of protein synthesis phase shifts a circadian pacemaker in mammalian SCN. *Am. J. Physiol.* 255:R1055–58

Jacklet JW. 1984. Neural organization and cellular mechanisms of circadian pacemakers. *Int. Rev. Cytol.* 89:250–95

Jacklet JW. 1989. Cellular neuronal oscillators. In *Neuronal and Cellular Oscillators,* ed. JW Jacklet, pp. 483–527. New York: Dekker

Janik D, Mrosovsky N. 1992. Gene expression in the geniculate induced by a nonphotic circadian phase stimulus. *NeuroReport* 3:575–78

Johnson CH, Nakashima H. 1990. Cycloheximide inhibits light-induced phase shifting of the circadian clock in *Neurospora. J. Biol. Rhythms* 5:159–67

Joy JE, Johnson GS, Lazar T, Ralph MR, Hochestrasser AC, et al. 1992. Protein differences in *tau* mutant hamsters: candidate clock proteins. *Mol. Brain Res.* 15:8–14

Kay SA. 1993. Shedding light on clock controlled *cab* gene transcription in higher plants. *Semin. Cell Biol.* 4:81–86

Khalsa SBS, Whitmore D, Block GD. 1992. Stopping the circadian pacemaker with inhibitors of protein synthesis. *Proc. Natl. Acad. Sci. USA* 89:10862–66

Klein DC, Moore RY, Reppert SM, eds. 1991. *Suprachiasmatic Nucleus: The Mind's Clock.* New York: Oxford Univ. Press. 467 pp.

Kondo T, Strayer CA, Kulkarni RD, Taylor W, Ishiura M, et al. 1993. Circadian rhythms in prokaryotes: luciferase as a reporter of circadian gene expression in cyanobacteria. *Proc. Natl. Acad. Sci. USA* 90:5672–76

Konopka RJ. 1987. Genetics of biological rhythms in *Drosophila. Annu. Rev. Genet.* 21:227–36

Konopka RJ, Benzer S. 1971. Clock mutants of *Drosophila melanogaster. Proc. Natl. Acad. Sci. USA* 68:2112–16

Kornhauser JM, Mayo KE, Takahashi JS. 1994. Light, immediate-early genes, and circadian rhythms. *Behav. Genet.* In press

Kornhauser JM, Nelson DE, Mayo KE, Takahashi JS. 1990. Photic and circadian regulation of c-*fos* gene expression in the hamster suprachiasmatic nucleus. *Neuron* 248:127–34

Kornhauser JM, Nelson DE, Mayo KE, Takahashi JS. 1992. Regulation of *jun-B* messenger RNA and AP-1 activity by light and a circadian clock. *Science* 255:1581–84

Koumenis C, Eskin A. 1992. The hunt for mechanisms of circadian timing in the eye of *Aplysia. Chronobiol. Int.* 3:201–21

Krumlauf R. 1993. *Hox* genes and pattern formation in the branchial region of the vertebrate head. *Trends Genet.* 9:106–12

Kyriacou CP. 1994. Clock research perring along: It's about time! *Trends Genet.* 10:69–71

Lavery DJ, Schibler U. 1993. Circadian transcription of the cholesterol 7α hydroxylase gene may involve the liver-enriched bZIP protein DBP. *Genes Dev.* 7:1871–84

Liu X, Zwiebel LJ, Hinton D, Benzer S, Hall JC, Rosbash M. 1992. The *period* gene encodes a predominantly nuclear protein in adult *Drosophila. J. Neurosci.* 12:2735–44

Loros JJ, Denome SS, Dunlap JC. 1989. Molecular cloning of genes under control of the circadian clock in *Neurospora. Science* 243:385–88

Loros JJ, Dunlap JC. 1991. *Neurospora crassa* clock-controlled genes are regulated at the level of transcription. *Mol. Cell Biol.* 11:558–63

Lynch GR, Lynch CB. 1992. Using quantitative genetic methods to understand mammalian circadian behavior and photoperiodism. In *Techniques for the Genetic Analysis of Brain* and Behavior, ed. D Goldowitz, D Wahlsten, RE Wimer, pp. 251–68. New York: Elsevier

Matsui M, Mitsui Y, Ishida N. 1993. Circadian regulation of per repeat mRNA in the suprachiasmatic nucleus of rat brain. Neurosci. Lett. 163:189–92

McClung CR, Fox BA, Dunlap JC. 1989. The Neurospora clock gene frequency shares a sequence element with the Drosophila clock gene period. Nature 339:558–62

Mead S, Ebling FJP, Maywood ES, Humby T, Herbert J, Hastings MH. 1992. A non-photic stimulus causes instantaneous phase advances of the light-entrainable circadian oscillator of the Syrian hamster but does not induce the expression of c-fos in the suprachiasmatic nuclei. J. Neurosci. 12: 2516–22

Meijer JH, Rietveld WJ. 1989. Neurophysiology of the suprachiasmatic circadian pacemaker in rodents. Physiol. Rev. 69:671–707

Michel S, Geusz ME, Zaritsky JJ, Block GD. 1993. Circadian rhythm in membrane conductance expressed in isolated neurons. Science 259:239–41

Millar AJ, Kay SA. 1991. Circadian control of cab gene transcription and mRNA accumulation in Arabidopsis. Plant Cell 3:541–50

Millar AJ, Short SR, Chua N-H, Kay S. 1992. A novel circadian phenotype based on firefly luciferase expression in transgenic plants. Plant Cell 4:1075–87

Miller JD. 1993. On the nature of the circadian clock in mammals. Am. J. Physiol. 264: R821–32

Molina CA, Foulkes NS, Lalli E, Sassone-Corsi P. 1993. Inducibility and negative autoregulation of CREM: an alternative promoter directs the expression of ICER, an early response repressor. Cell 75:875–86

Moore RY. 1992. The organization of the human circadian timing system. Prog. Brain Res. 93:101–17

Morgan JL, Curran T. 1991. Stimulus-transcription coupling in the nervous system: involvement of the inducible proto-oncogenes fos and jun. Annu. Rev. Neurosci. 14:421–51

Morse DS, Fritz L, Hastings JW. 1990. What is the clock? Translational regulation of circadian bioluminescence. Trends Biochem. Sci. 15:262–65

Murakami N, Takamure M, Takahashi K, Utunomiya K, Kuroda H, Etoh T. 1991. Long-term cultured neurons from rat suprachiasmatic nucleus retain the capacity for circadian oscillation of vasopressin release. Brain Res. 545:347–50

Nambu JR, Lewis JO, Wharton KA Jr., Crews ST. 1991. The Drosophila single-minded gene encodes a helix-loop-helix protein that acts as a master regulator of CNS midline development. Cell 67:1157–67

Norbury C, Nurse P. 1992. Animal cell cycles and their control. Annu. Rev. Biochem. 61: 441–70

Nüsslein-Volhard C, Wieschaus E. 1980. Mutations affecting segment number and polarity in Drosophila. Nature 287:795–80

Page TL. 1982. Transplantation of the cockroach circadian pacemaker. Science 216:73–75

Pardee AB. 1989. G1 events and regulation of cell proliferation. Science 246:603–8

Pierce ME, Sheshberadaran H, Zhang Z, Fox LE, Applebury ML, Takahashi JS. 1993. Circadian regulation of iodopsin gene expression in embryonic photoreceptors in retinal cell culture. Neuron 10:579–84

Pittendrigh CS. 1993. Temporal organization: reflections of a Darwinian clock-watcher. Annu. Rev. Physiol. 55:17–54

Pittendrigh CS, Daan S. 1976. A functional analysis of circadian pacemakers in nocturnal rodents: I. The stability and lability of spontaneous frequency. J. Comp. Physiol. 106:223–52

Raju U, Koumenis C, Nunez-Regueiro M, Eskin A. 1991. Alteration of the phase and period of a circadian oscillator by a reversible transcription inhibitor. Science 253:673–75

Raju U, Yeung SJ, Eskin A. 1990. Involvement of proteins in light resetting ocular circadian oscillators of Aplysia. Am. J. Physiol. 258: R256–62

Ralph MR. 1991. See Klein et al 1991, pp. 341–48

Ralph MR, Foster RG, Davis FC, Menaker M. 1990. Transplanted suprachiasmatic nucleus determines circadian period. Science 247: 975–78

Ralph MR, Lehman MN. 1991. Transplantation: a new tool in the analysis of the mammalian hypothalamic circadian pacemaker. Trends Neurosci. 14:362–66

Ralph MR, Menaker M. 1988. A mutation of the circadian system in golden hamsters. Science 241:1225–27

Rea MA. 1989. Light increases Fos-related protein immunoreactivity in the rat suprachiasmatic nuclei. Brain Res. Bull. 23: 577–81

Rea MA. 1992. Different populations of cells in the suprachiasmatic nuclei express c-fos in association with light-induced phase delays and advances of the free-running activity in hamsters. Brain Res. 579:107–12

Rea MA, Buckley B, Lutton LM. 1993. Local administration of EAA antagonists blocks light-induced phase shifts and c-fos expression in hamster SCN. Am. J. Physiol. 265: R1191–98

Reed SI. 1992. The role of p34 kinases in the G1 to S-phase transition. Annu. Rev. Cell Biol. 8:529–61

Rietveld WJ. 1992. Neurotransmitters and the

pharmacology of the suprachiasmatic nuclei. *Pharmacol. Ther.* 56:119–30

Rinchik EM. 1991. Chemical mutagenesis and fine-structure functional analysis of the mouse genome. *Trends Genet.* 7:15–21

Robertson LM, Takahashi JS. 1988a. Circadian clock in cell culture: I. Oscillation of melatonin release from dissociated chick pineal cells in flow-through microcarrier culture. *J. Neurosci.* 8:12–21

Robertson LM, Takahashi JS. 1988b. Circadian clock in cell culture: II. In vitro entrainment to light of melatonin oscillation from dissociated chick pineal cells. *J. Neurosci.* 8:22–30

Rosbash M, Hall JC. 1989. The molecular biology of circadian rhythms. *Neuron* 3:387–98

Rusak B. 1989. The mammalian circadian system: models and physiology. *J. Biol. Rhythms* 4:121–34

Rusak B, Bina KG. 1990. Neurotransmitters in the mammalian circadian system. *Annu. Rev. Neurosci.* 13:387–401

Rusak B, McNaughton L, Robertson HA, Hunt SP. 1992. Circadian variation in photic regulation of immediate-early gene mRNAs in rat suprachiasmatic nucleus cells. *Mol. Brain Res.* 14:124–30

Rusak B, Robertson HA, Wisden W, Hunt SP. 1990. Light pulses that shift rhythms induce gene expression in the suprachiasmatic nucleus. *Science* 248:1237–40

Russell WL, Kelly EM, Hunsicker PR, Bangham JW, Maddux SC, Phipps EL. 1979. Specific-locus test shows ethylnitrosourea to be the most potent mutagen in the mouse. *Proc. Natl. Acad. Sci. USA* 76:5818–19

Schwartz WJ. 1993. A clinician's primer on the circadian clock: its localization, function, and resetting. In *Advances in Internal Medicine*, ed. GH Stolleman, pp. 81–106. St. Louis, MO: Mosby-Year Book

Schwartz WJ, Takeuchi J, Shannon W, Davis EM, Aronin N. 1994. Temporal regulation of light-induced Fos and Fos-like protein expression in the ventrolateral subdivision of the rat suprachiasmatic nucleus. *Neuroscience* 58:573–83

Schwartz WJ, Zimmerman P. 1990. Circadian timekeeping in BALB/c and C57BL/6 inbred mouse strains. *J. Neurosci.* 10:3685–94

Sehgal A, Price JL, Man B, Young MW. 1994. Loss of circadian behavioral rhythms and *per* RNA oscillations in the *Drosophila* mutant *timeless. Science* 263:1603–6

Sheng M, Greenberg ME. 1990. The regulation and function of c-*fos* and other immediate early genes in the nervous system. *Neuron* 4:477–85

Shin H-S, Bargiello TA, Clark BT, Jackson FR, Young MW. 1985. An unusual coding sequence from a *Drosophila* clock gene is conserved in vertebrates. *Nature* 317:445–48

Silva AJ, Paylor R, Wehner JM, Tonegawa S. 1992a. Impaired spatial learning in α-calcium-calmodulin kinase II mutant mice. *Science* 257:206–11

Silva AJ, Stevens CF, Tonegawa S, Wang Y. 1992b. Deficient hippocampal long-term potentiation in α-calcium-calmodulin kinase II mutant mice. *Science* 257:201–6

Siwicki KK, Eastman C, Petersen G, Rosbash M, Hall JC. 1988. Antibodies to the *period* gene product of *Drosophila* reveal diverse tissue distribution and rhythmic changes in the visual system. *Neuron* 1:141–50

Siwicki KK, Schwartz WJ, Hall JC. 1992. An antibody to the *Drosophila period* protein labels antigens in the suprachiasmatic nucleus of the rat. *J. Neurogenet.* 8:33–42

Smith RF, Konopka RJ. 1982. Effects of dosage alterations at the *per* locus on the period of the circadian clock of *Drosophila. Mol. Gen. Genet.* 185:30–36

Stehle JH, Foulkes NS, Molina CA, Simonneaux V, Pévet P, Sassone-Corsi P. 1993. Adrenergic signals direct rhythmic expression of transcriptional repressor CREM in the pineal gland. *Nature* 365:314–20

Sutin EL, Kilduff TS. 1992. Circadian and light-induced expression of immediate early gene mRNAs in the rat suprachiasmatic nucleus. *Mol. Brain Res.* 15:281–90

Takahashi JS. 1991. Circadian rhythms: from gene expression to behavior. *Curr. Opin. Neurobiol.* 1:556–61

Takahashi JS. 1993. Circadian-clock regulation of gene expression. *Curr. Opin. Genet. Devel.* 3:301–9

Takahashi JS, Hamm H, Menaker M. 1980. Circadian rhythms of melatonin release from individual superfused chicken pineal glands in vitro. *Proc. Natl. Acad. Sci. USA* 77:2319–22

Takahashi JS, Kornhauser JM, Koumenis C, Eskin A. 1993. Molecular approaches to understanding circadian oscillations. *Annu. Rev. Physiol.* 55:729–53

Takahashi JS, Murakami N, Nikaido SS, Pratt BL, Robertson LM. 1989. The avian pineal, a vertebrate model system of the circadian oscillator: cellular regulation of circadian rhythms by light, second messengers, and macromolecular synthesis. *Rec. Prog. Horm. Res.* 45:279–352

Takahashi JS, Pinto LH, Vitaterna MH. 1994. Forward and reverse genetic approaches to behavior in the mouse. *Science* 264:1724–33

Takahashi JS, Turek FW. 1987. Anisomycin, an inhibitor of protein synthesis, perturbs the phase of a mammalian circadian pacemaker. *Brain Res.* 405:199–203

Takeuchi J, Shannon W, Aronin N, Schwartz WJ. 1993. Compositional changes of AP-1 DNA-binding proteins are regulated by light in a mammalian circadian clock. *Neuron* 11: 825–36

Truman JW, Riddiford LM. 1970. Neuroendocrine control of ecdysis in silkmoths. *Science* 167:1624–26

Turek FW. 1994. Circadian rhythms. *Rec. Prog. Horm. Res.* 49:43–90

Turek FW, Van Cauter E. 1994. Rhythms in Reproduction. In *The Physiology of Reproduction, Second Edition,* ed. E Knobil, JD Neill, pp. 487–540. New York: Raven

Underwood H. 1990. The pineal and melatonin: regulators of circadian function in lower vertebrates. *Experientia* 46:120–28

Underwood H, Barrett RK, Siopes T. 1990. The quail's eye: a biological clock. *J. Biol. Rhythms* 5:257–65

van den Pol AN, Dudek FE. 1993. Cellular communication in the circadian clock, the suprachiasmatic nucleus. *Neuroscience* 56: 793–811

Vignau J, Dahlitz M, Arendt J, English J, Parkes JD. 1993. Biological rhythms and sleep disorders in man: the delayed sleep phase syndrome. In *Light and Biological Rhythms in Man,* ed. L Wetterberg, pp. 261–71. Oxford: Pergamon

Vindlacheruvu RR, Ebling FJP, Maywood ES, Hastings MH. 1992. Blockade of glutamatergic neurotransmission in the suprachiasmatic nucleus prevents cellular and behavioral responses of the circadian system to light. *Eur. J. Neurosci.* 4:673–79

Vitaterna MH, King DP, Chang AM, Kornhauser JM, Lowrey JD, et al. 1994. Mutagenesis and mapping of a mouse gene, *Clock,* essential for circadian behavior. *Science* 264:719–25

Vogelbaum MA, Menaker M. 1992. Temporal chimeras produced by hypothalamic transplants. *J. Neurosci.* 12:3619–27

Vosshall LB, Price JL, Sehgal A, Saez L, Young MW. 1994. Block in nuclear localization of *period* protein by a second clock mutation, *timeless. Science* 263:1606–9

Watanabe K, Koibuchi N, Ohtake H, Yamaoka S. 1993. Circadian rhythms of vasopressin release in primary cultures of rat suprachiasmatic nucleus. *Brain Res.* 624:115–20

Wehr TA, Rosenthal NE. 1989. Seasonality and affective illness. *Am. J. Psychiatr.* 146:829–39

Wollnik F, Brysch W, Gillardon F, Bravo R, Zimmerman M, et al. 1994. Block of c-Fos and JunB expression by antisense-oligonucleotide inhibits light-induced phase shifts of the mammalian circadian clock. *Eur. J. Neurosci.* In press

Wollnik F, Turek FW, Majewski P, Takahashi JS. 1989. Phase shifting the circadian clock with cycloheximide: response of hamsters with an intact or split rhythm of locomotor activity. *Brain Res.* 496:82–88

Wuarin J, Falvey E, Lavery D, Talbot D, Schmidt E, et al. 1992. The role of the transcriptional activator protein DBP in circadian liver gene expression. *J. Cell Sci.* 16: 123–27

Wuarin J, Schibler U. 1990. Expression of the liver-enriched transcriptional activator protein DBP follows a stringent circadian rhythm. *Cell* 63:1257–66

Young MW, ed. 1993. *Molecular Genetics of Biological Rhythms.* New York: Dekker 319 pp.

Yu Q, Jacquier AC, Citri Y, Hamblen M, Hall JC, Rosbash M. 1987. Molecular mapping of point mutations in the *period* gene that stop or speed up biological clocks in *Drosophila melanogaster. Proc. Natl. Acad. Sci. USA* 84:784–88

Zatz M, Mullen DA, Moskal JR. 1988. Photoendocrine transduction in cultured chick pineal cells: effects of light, dark, and potassium on the melatonin rhythm. *Brain Res.* 438:199–215

Zerr DM, Hall JC, Rosbash M, Siwicki KK. 1990. Circadian fluctuations of Period protein immunoreactivity in the CNS and the visual system of *Drosophila. J. Neurosci.* 10: 2749–62

Zhang Y, Van Reeth O, Zee PC, Takahashi JS, Turek FW. 1993a. Fos protein expression in the circadian clock is not associated with phase shifts induced by a nonphotic stimulus, triazolam. *Neurosci. Lett.* 164:203–8

Zhang Y, Zee PC, Kirby JD, Takahashi JS, Turek FW. 1993b. A cholinergic antagonist, mecamylamine, blocks light-induced Fos immunoreactivity in specific regions of the hamster suprachiasmatic nucleus. *Brain Res.* 615:107–12

Zimmerman NH, Menaker M. 1979. The pineal gland: a pacemaker within the circadian system of the house sparrow. *Proc. Natl. Acad. Sci. USA* 76:999–1003

Zipursky SL, Rubin GM. 1994. Determination of neuronal cell fate: lessons from the R7 neuron of *Drosophila. Annu. Rev. Neurosci.* 17:373–97

Annu. Rev. Neurosci. 1995. 18:555–86

VISUAL FEATURE INTEGRATION AND THE TEMPORAL CORRELATION HYPOTHESIS

Wolf Singer

Max Planck Institute for Brain Research, Deutschordenstrasse 46, 6000 Frankfurt, Germany

Charles M. Gray

The Center for Neuroscience and Department of Neurobiology, Physiology, and Behavior, 1544 Newton Court, University of California, Davis, California 95616

KEYWORDS: visual cortex, physiology, pattern recognition, dynamics

INTRODUCTION

The Combinatorial Problem

The mammalian visual system is endowed with a nearly infinite capacity for the recognition of patterns and objects. To have acquired this capability the visual system must have solved what is a fundamentally combinatorial problem. Any given image consists of a collection of features, consisting of local contrast borders of luminance and wavelength, distributed across the visual field. For one to detect and recognize an object within a scene, the features comprising the object must be identified and segregated from those comprising other objects. This problem is inherently difficult to solve because of the combinatorial nature of visual images. To appreciate this point, consider a simple local feature such as a small vertically oriented line segment placed within a fixed location of the visual field. When combined with other line segments, this feature can form a nearly infinite number of geometrical objects. Any one of these objects may coexist with an equally large number of other

555

possible objects. The problem becomes daunting when we expand this scenario to account for the wide array of possible local features and the fact that objects may appear in different spatial locations and orientations. The possible combinations that confront the visual system are virtually unlimited. Yet when faced with any new scene, the visual system usually has no problem in segmenting the image into its component objects within a fraction of a second. This observation suggests that the visual system has adapted a very efficient mechanism for the flexible integration of featural information.

Population Coding and the Binding Problem

Considering what is presently known about the anatomical and functional organization of the mammalian visual system, we can clearly see that the requirement for flexible integration presents a fundamental problem to our understanding. The mammalian visual cortex consists of numerous interconnected areas (Rosenquist 1985, Felleman & van Essen 1991, Sereno & Allman 1991, Payne 1993). These different regions possess numerous feedforward, feedback, and intrinsic connections and are thought to be devoted to the analysis of different but often overlapping attributes of visual images (DeYoe & Van Essen 1988, Livingstone & Hubel 1988, Merigan & Maunsell 1993). At early stages of processing, cortical areas are retinotopically organized; neurons have relatively simple receptive field properties; and cells with similar functional properties are grouped together into functional streams (DeYoe & Van Essen 1988, Livingstone & Hubel 1988). This organization gradually gives way to a largely nonretinotopic mapping at higher levels in the hierarchy, and the receptive field properties of cortical neurons concurrently increase in size and complexity owing to convergence and divergence of connections from cells in lower areas. Although the functional streams are preserved, in the sense that separate areas tend to have distinct functional properties, extensive crosstalk occurs among areas at every processing stage.

It is apparent from this organization that the representation of a perceptual object, and its attributes such as location in space or direction of motion, are not likely to be processed at a single location but rather involve a large population of cells distributed over several different cortical areas. This raises the question as to how relations are established among the spatially distributed responses occurring within and between different levels of processing. Similar binding problems are also likely to arise at lower levels. A common assumption is that the representation of a particular local feature is achieved by the graded responses of a population of neurons. This notion is attractive because the relatively broad tuning of neuronal receptive fields suggests that single contours evoke simultaneous responses in many neurons. For the same reason, the response amplitude of any individual cell is an ambiguous descriptor of a particular feature because response vigor is equally influenced by the location of a contour and its orientation, contrast, and extent. Representation of a feature by a population of cells raises binding problems when nearby contours evoke graded responses in overlapping groups of neurons. Of the many simultaneous responses, those evoked by the same contour need to be distinguished and evaluated together to avoid interference with the responses elicited by neighboring contours. A similar need for response selection and binding arises in the context of perceptual grouping. Once the elementary features of a scene have been represented, some grouping operation must be performed to identify those neurons responding to the features of a particular object and to segregate the activity of neurons responding to the features of other objects or to the background. The implementation of units that receive converging inputs from cells whose responses require integration may allow this type of binding. The activity of such cells would then represent either elementary features or, at higher levels of processing, a particular constellation of elementary features. Finally, by iteration of this operation, units could be created that respond with high selectivity to single perceptual objects.

The analysis of single-cell receptive fields at different levels of visual processing suggests that the visual system does exploit the option of binding by convergence. Cells at higher levels of processing tend to have larger receptive fields and to respond selectively to rather complex constellations of elementary features, such as stereotypes of faces or patterns (Gross et al 1972, Baylis et al 1985, Desimone et al 1985, Perrett et al 1987, Sakai & Miyashita 1991, Fujita et al 1992, Gallant et al 1993). However, several observations indicate that binding is probably not achieved solely by the convergence of distributed signals onto specialized cells. Although such a mechanism could enable the rapid and unambiguous association of a limited set of key features, the number of units required to implement it as a universal mechanism scales very unfavorably with the number of possible patterns to be represented. Essentially, one cell would be required for every distinguishable feature, for each higher-order feature combination, and ultimately for every distinguishable perceptual object. Moreover, because of its inherent lack of flexibility, such a mechanism cannot easily cope with the representation of new or modified patterns. Experimental data also suggest that binding by convergence is probably not the only strategy for the association of distributed neuronal responses. The pattern-specific cells in the inferotemporal cortex are not selective for individual perceptual objects but for characteristic components of patterns (Fujita et al 1992). Hence, these cells respond to a whole family of related patterns, and conversely, a particular pattern is likely to activate many neurons simultaneously (Young & Yamane 1992).

The representation of features and objects by the joint activity of neuronal populations has several undisputed advantages (Hebb 1949; Braitenberg 1978; Ballard et al 1983; Singer 1985, 1990; von der Malsburg 1985; Edelman 1987;

Gerstein et al 1989; Grossberg 1980; Palm 1990; Abeles 1991). One essential feature of such assembly coding is that individual cells can participate at different times in the representation of different patterns. This strategy substantially reduces the number of cells required for the representation of different patterns and allows for greater flexibility in the generation of new representations. The assumption is that just as a particular feature can be present in many different patterns, the group of cells coding for this feature can be shared by many different representations in that they participate at different times in different assemblies of coactive neurons. The code is thus relational and the significance of an individual response depends entirely on the context set by the other members of the assembly.

To exploit the advantages of population coding, however, mechanisms are required that enable the flexible association of neuronal activity. Those active cells participating in a particular representation must be unambiguously identified as belonging together. One way to distinguish a given subset of cells is to enhance their saliency by increasing their relative firing rates (Olshausen et al 1993). The output of these cells then has a greater impact because of temporal summation. However, selecting neurons solely on the basis of enhanced discharge rates has two potential disadvantages: First, it limits the option of encoding information about stimulus properties by the graded activity of neuronal groups. Second, it limits the number of populations that can be enhanced simultaneously without becoming confounded. Only those populations that are clearly defined by a place code would remain segregated. However, although place codes can in principle reduce ambiguity, they are again expensive in terms of neuron numbers, and most importantly, they sacrifice flexibility. To maintain the position code, interactions between assemblies in different areas must be forbidden, because they would reintroduce the ambiguity that one wants to overcome. It has been proposed, therefore, that response selection could be achieved by the synchronization of activity among a distributed population of neurons rather than by solely increasing their discharge rate (Milner 1974; von der Malsburg 1981, 1985; von der Malsburg & Schneider 1986).

Two features of cortical connectivity suggest that synchronization may be a particularly effective way of enhancing response saliency. First, cortical cells contact each other with only a few synapses whose efficiency is usually low (Komatsu et al 1988, Mason et al 1991, Braitenberg & Schüz 1991, Nicoll & Blakemore 1993, Thomson & West 1993). Second, synaptic transmission among cortical neurons is characterized by pronounced frequency attenuation (Thomson & West 1993). Hence, increasing the summation of activity by synchronizing inputs may more effectively enhance transmission than raising discharge rates (see Abeles 1991). Another (and in this context) crucial advantage of synchronization is that it expresses unambiguous relations among neurons because it enhances selectively only the saliency of synchronous responses. Simulation studies by Softky & Koch (1993) suggest that the interval for effective summation of converging inputs is only a few milliseconds in cortical neurons. Thus, if synchronization of discharges can be achieved with a precision in the millisecond range, it can define relationships among neurons with very high precision. Moreover, if synchrony is established rapidly and maintained only over brief intervals, different assemblies can be organized in rapid temporal succession. In principle, a particular assembly can be defined by a single barrage of synchronous action potentials whereby each individual cell needs to contribute only a few spikes (Buzsaki et al 1992). Such synchronous events are likely to be very effective in eliciting responses in target populations, and because they are statistically improbable, their information content is high.

In summary, the hypothesis predicts that the discharges of neurons undergo a temporal patterning and become synchronous if they participate in the encoding of related information. This synchronization is thought to be based on a self-organizing process that is mediated by a selective network of corticocortical and corticothalamic connections. Thus, distributed groups of coactive neurons that code for a particular feature, or at higher levels, for constellations of features corresponding to a perceptual object, would be identifiable as members of an assembly because their responses would contain episodes during which their discharges are synchronous. These theoretical considerations yield several predictions regarding the organization of neuronal assemblies in cortical networks. In the following paragraphs we enumerate these predictions and review the experimental evidence for their validation.

PREDICTIONS

The first requirement of the temporal correlation hypothesis, and the one with the most supporting evidence, predicts that cells recorded simultaneously should, under appropriate conditions, exhibit synchronous firing on a millisecond time scale. Specifically, correlated firing should occur between cells recorded in (a) the same cortical column of a given area to enable the coding of local features, (b) different columns within an area to enable the linking of spatially disparate but related features, (c) different cortical areas to provide for the binding of information across different feature categories and different locations in space, (d) the two cerebral hemispheres to link information present in the two visual hemifields, and (e) different sensory and motor modalities to contribute to the processes of sensorimotor integration. Second, the probability for intra- and interareal response synchronization should reflect some of the Gestalt criteria for perceptual grouping (Koffka 1935). Third, individual cells should be able to rapidly change the partners with which they synchronize their responses if stimulus configurations change and require new associations. Fourth, if more than one object is present in a scene, several distinct assemblies should form. Cells belonging to the same assembly should exhibit synchronous response episodes, whereas no consistent temporal relationships should exist between the discharges of neurons belonging to different assemblies. This prediction, however, may apply differently to different levels in the cortical hierarchy. In retinotopically organized areas synchronous assemblies of active cells can be spatially separated. In nonretinotopic areas, such as inferotemporal cortex, spatial segregation is less likely. Thus, few assemblies or even a single assembly of synchronously active neurons may be present at any given time. This could contribute to the serial nature of visual attention (Crick & Koch 1990). Fifth, the connections determining synchronization probability should be specific and yet modifiable according to a correlation rule whereby synaptic connections should strengthen if pre- and postsynaptic activity is often correlated, and they should weaken when there is no correlation. This is required to enhance grouping of cells that code for features that often occur in consistent relations, as is the case for features constituting a particular object. Finally, the patterns of synchronized activity should bear some specific relation to visual discrimination behavior.

EXPERIMENTAL TESTING OF PREDICTIONS

Intracolumnar Interactions

Multielectrode recording experiments have revealed neuronal response synchronization over each of the predicted spatial scales. The bulk of these studies have focused on the occurrence and properties of local (<1 mm) intra- and intercolumnar interactions. The literature contains many examples of temporal synchrony between cells recorded within the same cortical column (<200 μm separation) in different areas of the cat or monkey visual cortex (Toyama et al 1981a,b; Michalski et al 1983; Ts'o et al 1986; Ts'o & Gilbert 1988; Aiple & Kruger 1988; Hata et al 1988, 1991; Gochin et al 1991; Schwarz & Bolz 1991; Gawne & Richmond 1993). These synchronous interactions occur among various cell types in different layers of cortex and are most often characterized by central peaks in cross-correlation histograms, which indicates a common excitatory or inhibitory input (Perkel et al 1967). Occasionally, the correlograms exhibit peaks or troughs at specific latencies indicative of direct excitatory or inhibitory synaptic interactions.

In our own studies, the systematic search for dynamic, stimulus-dependent interactions between cortical neurons was initiated by the finding that adjacent neurons in area 17 of the cat visual cortex often transiently engage in highly synchronous discharges when presented with their preferred stimulus (Gray &

Singer 1987, 1989). Groups of neurons recorded simultaneously with a single electrode discharge synchronously at intervals of 15–30 ms. These sequences of synchronous rhythmic firing occur preferentially when cells are activated with slowly moving contours of optimal orientation. They typically last a few hundred milliseconds and may occur several times during a single passage of a moving stimulus (Figure 1). Accordingly, autocorrelation histograms computed from such response epochs often exhibit a periodic modulation (Gray & Singer 1987, 1989; Eckhorn et al 1988; Gray et al 1990; Schwarz & Bolz 1991). During such episodes of synchronous firing, an oscillatory field potential can be recorded by the same electrode; here, the negative phase of the signal coincides with the cells' discharges (Figure 1). The occurrence of the local field response indicates that many cells in the vicinity of the electrode synchronize their discharges (Gray & Singer 1989). This conjecture is supported by recent multiple single-unit recordings employing new spike extraction techniques (Figure 2) (Gray & Viana Di Prisco 1993).

The locally synchronous firing has since been observed in recordings of multiple units and local field potentials (LFP) in several areas of the visual cortex of anesthetized cats [areas 17–19 and Posteromedial Lateral Suprasylvian (PMLS)] (Eckhorn et al 1988, 1992; Gray & Singer 1989; Gray et al 1990; Engel et al 1991c; Schwarz & Bolz 1991), in striate cortex of awake, behaving cats (Figure 2) (Raether et al 1989, Gray & Viana Di Prisco 1993) and monkeys (Eckhorn et al 1993), in the optic tectum of awake pigeons (Neuenschwander & Varela 1993), and in area 17 (Livingstone 1991) and area MT (Kreiter & Singer 1992, Engel et al 1992) of anesthetized and awake, behaving monkeys, respectively. In each instance the activity is characterized by properties similar to those observed in the cat striate cortex: 1. The spike trains consist of repetitive burst discharges at semiregular 15- to 30-ms intervals. 2. Neither the onset latency nor the phase of the synchronous episodes are precisely related to the position of the stimulus within the neuron's receptive field. When cross-correlation functions are computed between responses to identical stimuli, these shift predictors reveal no correlation (Gray & Singer 1989, Gray et al 1990, Jagadeesh et al 1992). This finding rules out the possibility that the synchronous firing is related to some fine spatial structure in the receptive fields of cortical neurons. 3. The locally synchronous firing often, but not always, results in the appearance of a correlated field potential signal at the same frequency (Eckhorn et al 1988, 1993; Gray & Singer 1989; Engel et al 1991c; Livingstone 1991; Kreiter & Singer 1992).

Although these properties appear to be general, several studies have found little or no evidence for rhythmic firing in single- and multiple-unit recordings from areas V1, MT, and the inferotemporal cortex of the macaque (Bair et al 1992, Tovee & Rolls 1992, Young et al 1992). The reasons for these conflicting results are not readily apparent. One possibility is suggested by data from the

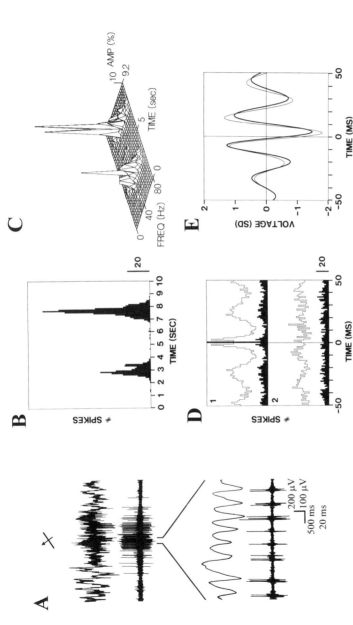

Figure 1 (A) Multiunit and local field-potential responses to the presentation of an optimally oriented light bar recorded from a single electrode in area 17 of an adult cat. Oscilloscope records show the response to the preferred direction of movement on a single trial. In the upper two traces, at a slow time scale, the onset of the response is associated with an increase in high-frequency activity in the local field potential. The lower two traces display the activity at an expanded time scale. Note the presence of oscillations in the local field potential correlated with the occurrence of the unit discharges. (*B–E*) Quantitative properties of multiunit and local field potential activity recorded on a single electrode in the striate cortex of a 6-week-old kitten in response to a moving light bar of optimal orientation. The light bar was first moved over the receptive field in one direction (1–4 s) and then in the opposite direction (6–9 s). (*B*) Post-stimulus-time histogram (PSTH) of the multiunit activity. The light bar was first moved over the receptive field in one direction (1–4 s) and then in the opposite direction (6–9 s). (*C*) Compressed spectral array of the LFP recorded on a single trial illustrating the time course and frequency content of the oscillatory response. The signal was bandpass filtered between 20 and 100 Hz. (*D*) Autocorrelation histograms and associated shift predictor histograms computed from the multiunit activity. The unfilled bars are for the second direction of stimulus movement (4–9 s). The shift predictors are flat, indicating that the oscillatory signals are not time locked to the visual stimulus. (*E*) Spike-triggered average of the LFP demonstrating that spike activity occurs with the highest probability during the negative phase of the field potential oscillations. The thick line represents the second direction

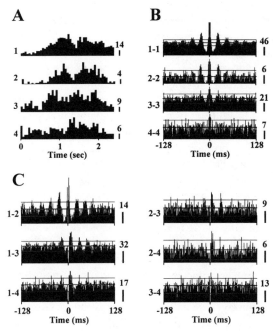

Figure 2 Rhythmic firing in area 17 of the alert cat is locally synchronous. Multiunit activity was recorded from a single electrode in area 17 while the cat maintained its gaze on a central fixation spot for a period of 2.4 s. Four single units were extracted from the multiunit recording using principal components analysis to identify separate spike waveforms. (*A*) PSTHs computed from each of the four spike trains (1–4). (*B*) Autocorrelation histograms computed from the activity of each single unit during the response to the visual stimulus. Note that oscillatory firing is readily apparent in the activity of units 1 and 2, less so for unit 3, and not at all for unit 4. (*C*) Cross-correlation histograms computed for each cell pair. Note that the discharges of cells 1 and 2 are in phase while cells 3 and 4 display a 4- to 5-ms lag relative to these cells. Also note that correlograms 1–4 and 2–4 exhibit a significant peak even though cell 4 shows little or no evidence of rhythmicity. The thick and thin horizontal lines passing through the histograms in *B* and *C* represent the mean and the 99% confidence interval computed from the shift predictors. (From Gray & Viana Di Prisco 1993.)

olfactory bulb (Freeman 1975, Gray & Skinner 1988) and the hippocampus (Alonso & Garcia-Austt 1987, Buzsaki et al 1992) where rhythmic activity is prominent at the level of field potentials but often not apparent in the autocorrelograms computed from single-unit activity. In these systems, however, the timing of spike discharges are often well correlated with the phase of the LFP. Thus, even though little evidence supports rhythmicity in the discharge of single units, the cells do participate as members of a population that exhibits oscillatory behavior. The lack of rhythmicity in the spike trains is thought to be a consequence of the low firing rates of the cells, the frequency variability of the rhythmic population events, and the variation of phase differences between the single-unit discharges and the population oscillations. These sources of variance combine to produce a spike train that appears Poisson when analyzed using relatively short epochs of data.

Figure 2 shows an example of this effect. The sampling problem is alleviated with multiunit recordings if several of the recorded cells are synchronized to the same rhythm. But variations in the frequency of the rhythm may yield autocorrelograms lacking any oscillatory modulation. Because the detectability of oscillatory activity in the gamma frequency range has become a major issue in the context of the correlation hypothesis, we emphasize that no conclusions regarding the formation of assemblies can be drawn from the presence or absence of oscillatory firing patterns in single-unit activity. First, oscillatory firing is in itself not thought to convey significant information regarding the stimulus. It is too variable in frequency and magnitude (Gray & Singer 1989; Engel et al 1990; Gray et al 1990, 1992; Ghose & Freeman 1992). Second, evidence for or against assembly formation can only be obtained with simultaneous recordings from two or more units or field-potential signals because the relevant parameter is the temporal correlation among the discharges of groups of neurons. Oscillations may provide an important mechanism for establishing synchrony, especially over large distances (König et al 1994), but the nervous system contains many examples of nonrhythmic synchronous firing.

Intercolumnar Interactions

Measurements of intercolumnar correlations are fewer in number but have revealed similar patterns of synchronous firing. The majority of these studies have been performed in area 17 and have sought to determine the relationship between receptive field properties and the occurrence of synchronous firing. Two general findings have emerged. First, the probability and strength of correlated firing falls off with distance in cortex (Michalski et al 1983, Ts'o et al 1986, Aiple & Kruger 1988, Ts'o & Gilbert 1988, Gray et al 1989, Engel et al 1990, Kruger 1990, Hata et al 1991, Schwarz & Bolz 1991). Cells that have overlapping receptive fields and that are separated by less than 2 mm are far more likely to exhibit synchronous firing than cells with nonoverlapping fields that have a separation greater than 2 mm. Nevertheless, numerous examples of long-range synchronous firing have been observed over distances spanning several hypercolumns in area 17 (Ts'o et al 1986, Ts'o & Gilbert 1988, Gray et al 1989, Engel et al 1990, Schwarz & Bolz 1991).

The second general finding obtained from these studies is that intercolumnar correlated firing occurs with greater probability for cells that have similar receptive field properties. Correlations tend to occur most often between cells with similar orientation preferences (Ts'o et al 1986, Ts'o & Gilbert 1988, Gray et al 1989, Hata et al 1991, Schwarz & Bolz 1991), similar ocular domninances (Ts'o et al 1986, Ts'o & Gilbert 1988), and similar color selectivities (Ts'o & Gilbert 1988). Moreover, cells with specific types of complex and simple receptive fields tend to interact with other cells of the same type (Ts'o et al 1986, Schwarz & Bolz 1991), although there are exceptions to this, as yet, poorly defined rule (Schwarz & Bolz 1991). These long-range correlations have generally been thought to reflect the specificity of intracortical horizontal axonal connections that preferentially link functional columns with similar feature specificities (Ts'o et al 1986; Ts'o & Gilbert 1988; Martin & Whitteridge 1984; Gilbert & Wiesel 1983, 1989; but see Matsubara et al 1987). In most instances, however, intercolumnar correlation measurements have been performed using two stimuli, one optimal for each of the two recorded cells. Cells that have different receptive field properties are thus activated by stimuli that have different features. Within the context of the theory presented here, such a featural difference in the stimuli introduces a potential cue for the segregation of assemblies and the consequent absence of correlated firing. This point is discussed in detail below.

When interactions occur over larger tangential distances in the cortex, the cross-correlograms exhibit two other prominent features. First, the peak in the histograms is usually centered around zero delay. The half width at half height of this peak is often on the order of 2–3 ms, indicating that most of the action potentials occur nearly simultaneously (Ts'o & Gilbert 1986, Gray et al 1989, Schwarz & Bolz 1991). Second, when cells engage in long-distance synchronization, the firing patterns of the local groups often exhibit synchronous repetitive discharges as described above. The central peak in the correlogram is thus often flanked on either side by troughs that result from refractory periods between the synchronous bursts. When the duration of these pauses is sufficiently constant throughout the episode of synchronization, the cross-correlograms show a periodic modulation with additional side peaks and troughs (Eckhorn et al 1988, Gray et al 1989, Engel et al 1990, Schwarz & Bolz 1991, Livingstone 1991). Synchronous oscillations are readily apparent if the LFP is also recorded from each electrode (Figure 3) (Engel et al 1990, Gray et al 1992). This pattern of rhythmic synchronization rarely, if ever, occurs in the absence of visual stimulation. Such intercolumnar synchronization has been observed over distances of up to 7 mm in the cat (Gray et al 1989) and up to 5 mm in area V1 of the squirrel monkey (M Livingstone, personal communication). In the cat, cells separated by more than 2 mm that have nonoverlapping receptive fields are more likely to exhibit synchronous firing if they have similar orientation preferences (Gray et al 1989, Engel et al 1990, Schwarz & Bolz 1991). If the cells are separated by less than 2 mm and have overlapping receptive fields, the synchronous firing occurs largely irrespective of the preferred orientation of the cells (Gray et al 1989, Engel et al 1990), provided that they can be activated with a single stimulus.

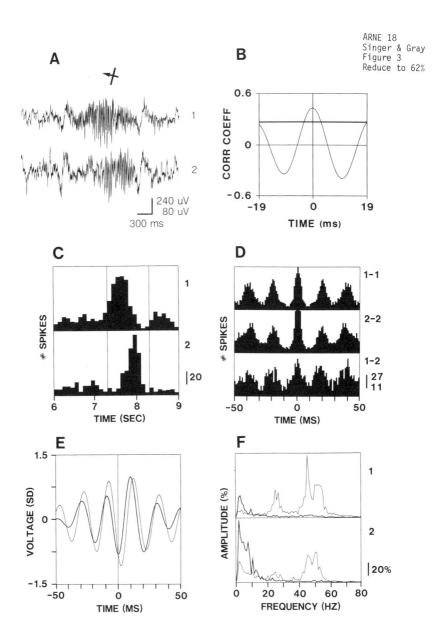

ARNE 18
Singer & Gray
Figure 3
Reduce to 62%

Interareal and Interhemispheric Interactions

In agreement with the predictions of the temporal correlation hypothesis, response synchronization has also been found between groups of cells located in different cortical areas. In the cat, stimulus-evoked synchronous firing has been observed between cells in areas 17 and 18 (Eckhorn et al 1988, 1992; Nelson et al 1992b); between cells in areas 17 and 19, and 18 and 19 (Eckhorn et al 1992); between cells in area 17 and area PLMS, an area specialized for motion processing (Figure 4) (Engel et al 1991c); and between cells in area 17 of the two hemispheres (Engel et al 1991a, Eckhorn et al 1992, Nelson et al 1992a). In the macaque, synchronous firing has been observed among neurons in areas V1 and V2 (Bullier et al 1992, Roe & Ts'o 1992, Nowak et al 1994).

In the studies of Nelson et al (1992b) interareal synchronous firing in cats occurs spontaneously and during the presentation of visual stimuli. The interactions span a wide tripartite range of temporal scales, which produces correlograms with central peaks of narrow, medium, and broad width. The narrow coupling is most often seen between cells that have overlapping receptive fields with similar properties. The broader coupling encompasses a much wider range of receptive-field separations and orientation-preference differences (Nelson et al 1992b). Synchronous interactions between cells in V1 and V2 in the monkey show similar broad peaks. Synchrony occurs between cells that have both overlapping and nonoverlapping receptive fields (Bullier et al 1992, Nowak et al 1994) and is most frequent between cells of similar color selectivity in the two areas (Roe & Ts'o 1992).

In the studies of Eckhorn et al (1988, 1992) and Engel et al (1991a,c), interareal and interhemispheric synchronous firing occurred primarily, if not exclusively, during coactivation of the cells by visual stimuli, and was partic-

---

Figure 3   The local field potential and multiunit activity recorded at two sites in area 17 separated by 7 mm show similar temporal properties and correlated interactions. (*A*) Plot of the LFP responses (1–100 Hz bandpass) recorded on a single trial to the presentation of two optimally oriented light bars passing over the receptive fields of the recorded neurons at each site. The peaks of the responses overlap in time but are not in precise register. (*B*) The average cross-correlogram computed between the two LFP signals (20–100 Hz bandpass) at a latency corresponding to the peak of the oscillatory responses. The thick horizontal line represents the 95% confidence limit for significant deviation from random correlation. (*C*) PSTHs of the multiunit activity recorded over 10 trials at the same two cortical sites as shown in panel A. Again the responses overlap but are not in precise register. (*D*) Auto- (1–1, 2–2) and cross- (1–2) correlograms of the multiunit activity recorded at each site. Note the presence of a clear periodicity in each correlogram, which indicates that the responses are oscillatory and that they show a consistent phase relationship. (*E*) Plots of the spike-triggered averages of the LFP signals at each site computed over all 10 trials. The thick and thin lines correspond to electrodes 1 and 2 respectively. Note that the peak negativity of the waveform is correlated with the occurrence of neuronal spikes at 0-ms latency. (*F*) Normalized average power spectrum of the LFP signals computed from periods of spontaneous (*thick line*) and stimulus-evoked (*thin line*) activity. The frequency of the activity is similar in both the autocorrelograms of the multiunit activity (MUA) and the power spectra of the LFPs. (From Gray et al 1992.)

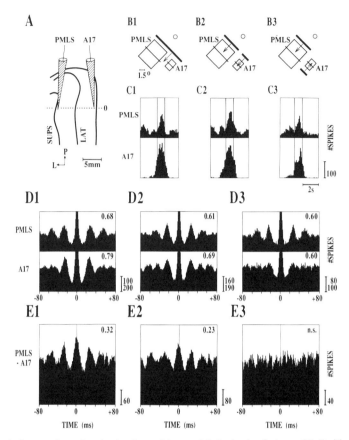

Figure 4   Interareal synchronization is sensitive to global stimulus features. (*A*) Position of the recording electrodes. A17, area 17; LAT, lateral sulcus; SUPS, suprasylvian sulcus; P, posterior; L, lateral. (*B1–B3*) Plots of the receptive fields of the PMLS and area 17 cells. The diagrams depict the three stimulus conditions tested. The circle indicates the visual field center. (*C1–C3*) PSTHs for the three stimulus conditions. The vertical lines indicate 1-s windows for which auto- and cross-correlation histograms were computed. (*D1–D3*) Comparison of the autocorrelation histograms computed for the three stimulus paradigms. (*E1–E3*) Cross-correlation histograms computed for the three stimulus conditions. Note that the response amplitudes are minimally affected by changes of stimulus configuration, whereas the synchronization of activity is largely absent for stimuli moving in opposite directions. (From Engel et al 1991c.)

ularly pronounced during periods of oscillatory firing. These data reveal several additional similarities to the intra- and intercolumnar oscillatory interactions. The occurrence of synchronous firing depends on, but is not locked to, the visual stimulus. The probability of occurrence and the magnitude of interareal temporal correlations are somewhat greater for cells that have overlapping receptive fields and similar orientation preferences. When receptive fields are nonoverlapping and cells are activated with two stimuli, synchronization probability is greatest when the stimuli move at the same speed in the same direction (Engel et al 1991a,c). Interhemispheric synchrony also requires the integrity of the corpus callosum. Surgical sectioning of this structure markedly reduces the probability of synchronous firing between the two hemispheres (Engel et al 1991a, Nelson et al 1992a). These studies provide the first results demonstrating that corticocortical connections are critical for the establishment of synchronous firing.

Evidence for Synchrony in Nonvisual Structures

In the preceding discussion, we have reviewed the evidence for synchronous activity revealed by multielectrode recordings from visual cortical areas. Comparable multicell recordings are still rare in nonvisual structures, but considerable data are available from field-potential studies. The data indicate that synchronous rhythmic activity occurs over a range of spatial scales in several different cortical and subcortical systems throughout the brains of mammals (Basar & Bullock 1992, Gray 1993, Singer 1993).

In the mammalian olfactory system, for example, 40- to 80-Hz oscillatory activity is evoked during inspiration in both the olfactory bulb and piriform cortex (Adrian 1950; Freeman 1975, 1978; Bressler 1984). This activity is synchronous over a scale of several millimeters both within and between the two structures (Bressler 1984, 1987; Freeman 1987). The patterns of activity that emerge during these coherent states correspond to specific odors, the animal's past experience with the odors, and their behavioral significance (Freeman 1987). The oscillatory activity by itself is not thought to convey any specific information; instead, it is viewed as a mechanism for establishing synchrony among large populations of coactive cells (Freeman 1987).

Similar rhythmic activities have been discovered in both the somatosensory and motor cortices of cats and monkeys. In the studies of Bouyer and colleagues, field potential oscillations in the beta and gamma ranges occur in the somatosensory cortex when animals are in a state of focused attention (Bouyer et al 1981). These rhythmic activities are synchronous over relatively large areas of cortex (Bouyer et al 1987), occur in phase with similar activity in the ventrobasal thalamus (Bouyer et al 1981), and are regulated by dopaminergic input from the ventral tegmentum (Montaron et al 1982). Oscillatory field potential and unit activities in the range of 20–40 Hz have also been observed in the motor cortex of alert monkeys (Murthy & Fetz 1992, Sanes & Donoghue 1993). These signals are synchronous over widespread areas of the motor cortical map within and between the two cerebral hemispheres, between the visual and motor cortices (E Fetz, personal communication), and between the motor and somatosensory cortices. They are also enhanced in amplitude when the animals are performing new and complicated motor acts (Murthy & Fetz 1992). The rhythms are suppressed, however, during the execution of trained movements (Sanes & Donoghue 1993).

The hippocampus exhibits several forms of synchronous rhythmic activity that are associated with particular behavioral states. Foremost among these is the theta rhythm, a 4- to 10-Hz oscillation of neuronal activity that occurs during active movement and alert immobility. Theta field potentials are often synchronous between the two hemispheres and over distances extending up to 8 mm along the longitudinal axis of the hippocampus (Bland et al 1975). Local populations of cells also exhibit a high degree of synchronous firing during theta activity (Kuperstein et al 1986). Furthermore, two other hippocampal neuronal rhythms have been discovered. One has a frequency of 30–90 Hz, occurs during a variety of behavioral states (Buzsaki et al 1983, Leung 1992, Bragin et al 1994), and is both locally and bilaterally synchronous. Another more recently discovered signal with a frequency around 200 Hz is associated with alert immobility and the presence of sharp waves in the hippocampal EEG (Buszaki et al 1992, Ylinen et al 1994). These events have been termed population oscillations because single cells do not exhibit high-frequency periodic firing. Rather they fire at low rates in synchrony with the surrounding population of cells, which in the composite yields a periodic signal that is synchronous over distances up to 2.1 mm (Buzsaki et al 1992, Ylinen et al 1994).

Oscillatory components in the beta- and gamma-frequency ranges have also been documented in humans. In several early studies, depth recordings of the local EEG from several cortical and subcortical sites revealed episodes of synchronous rhythmic activity during particular behavioral states (Sem-Jacobsen et al 1956, Chatrian et al 1960, Perez-Borja 1961). Surface EEG and MEG recordings have revealed gamma-frequency components in the auditory evoked potential (Galambos et al 1981, Basar 1988, Sheer 1989, Pantev et al 1991, Tiitinen et al 1993), the somatomotor cortex (Murphy et al 1994), and a broad distribution over the entire cerebral mantle (Ribary et al 1991). Recently, however, direct surface recordings of the EEG over the somatosensory cortex in humans have revealed that gamma-band activity does not predominate over activity in other frequency ranges (Menon et al 1994). In spite of this there is ample evidence from animals and humans that brain structures other than the visual cortex engage in synchronous rhythmic activity in the beta- and gamma-frequency ranges.

STIMULUS DEPENDENCE OF RESPONSE SYNCHRONIZATION

Another important prediction of the temporal correlation hypothesis is that the probability for distributed cells to join an assembly should reflect the Gestalt criteria, according to which, features in images tend to be grouped together into objects. Described early this century (see Koffka 1935), these criteria include the categories of proximity, similarity, continuity, and common fate, reflecting the fact that objects typically consist of features that exist in close proximity; that have common properties such as form, color, and depth; that are spatially contiguous; and that, when they move, tend to do so with a common or linked direction of motion. Evidence reviewed in the previous section supports the notion that response synchronization satisfies the criteria of proximity and similarity.

Additional evidence suggests that response synchronization fulfills the criteria of continuity and common fate. In one study, Gray et al (1989) recorded multiunit activity from two locations separated by 7 mm in striate cortex. The receptive fields of the cells were nonoverlapping, had nearly identical orientation preferences, and were spatially displaced along the axis of preferred orientation. This arrangement enabled stimulation of the cells with bars of the same orientation under three different conditions: two bars moving in opposite directions, two bars moving in the same direction, and one long bar moving across both fields coherently. Under these conditions, no significant correlation was found in the first case; a weak correlation was seen in the second; and a robust synchronization was found in the third case. This effect occurred in spite of the fact that the firing rates of the two cells and the oscillatory patterning of the responses were similar in the three conditions (Gray et al 1989). Further support for the notion that response synchronization is influenced by stimulus continuity was obtained in a study by Schwarz & Bolz (1991). They demonstrated that the probability of detecting intercolumnar synchronization between standard complex cells in layer 5 and simple or complex cells in layer 6 is highest not only when the cells have the same orientation preferences but when their receptive fields are aligned colinearly.

In a related experiment, Engel et al (1991c) obtained evidence that the criteria of continuity and common fate are also satisfied by interareal interactions. They made simultaneous recordings from cells in areas 17 and PMLS (a motion-sensitive area in the lateral suprasylvian sulcus of the cat) that had nonoverlapping receptive fields with similar orientation preference and colinear alignment. This technique allowed them to examine the effects of continuity and coherent motion on response synchronization using the paradigm described above. They found little or no correlation when the cells were activated by oppositely moving contours, a weak but significant correlation in response to two bars moving in the same direction, and a robust synchronization when the cells were coactivated by a single long bar moving over both fields (Figure 4) (Engel et al 1991c). Recently, Sillito et al (1994) demonstrated an analogous influence of stimulus continuity and coherent motion on the synchronization of activity in the lateral geniculate nucleus (LGN). They showed that cells having nonoverlapping receptive fields displayed little or no evidence of synchronization when activated by a pair of stationary spots. If, however, the cells were driven by a single long bar or an extended grating drifting over both receptive fields, the cells often engaged in synchronous firing. This effect was abolished by removal of the overlying visual cortex, suggesting that the thalamic synchronization is controlled by feedback from the visual cortex.

These findings, combined with the earlier results, suggest that the global properties of visual stimuli can influence the magnitude of synchronization between widely separated cells located within and between different cortical areas and thalamus. Single contours, but also spatially separate contours, which move coherently and therefore appear as parts of a single figure, are more efficient in inducing synchrony among the responding cell groups than oppositely moving contours that appear as parts of independent figures. This finding demonstrates that synchronization probability and magnitude depends not only on the spatial separation of cells and on their feature preferences, but also on the configuration of the stimuli.

A central prediction of the correlation hypothesis posits that individual cells must be able to change the partners with which they become synchronized in order to join different assemblies at different times. Thus, as the features in an image change, the relationships among the activity patterns of the cells responding to those features should change in a way that reflects the Gestalt properties of the image.

Tests of this prediction have been conducted by Engel et al (1991b) in cat area 17 and by Kreiter and colleagues in area MT of the anesthetized monkey (Kreiter et al 1992). In the cat, multiunit activity was recorded from up to four electrodes that had a spacing of approximately 0.5 mm. The proximity of the electrodes yielded recordings in which all the cells had overlapping receptive fields and covered a broad range of orientation preferences. This configuration enabled these investigators to compare the correlation of activity among cells coactivated by either one or two moving bars. In the majority of cases, cells with different orientation preferences fired synchronously when coactivated by a single bar of intermediate orientation. But when the same cells were activated by two independent bars of differing orientation, moving in different directions, the responses were not synchronized (Engel et al 1991b).

Kreiter & Singer (1994) recently demonstrated that this process is a general property of the visual cortex and is not confined to the visual cortex of anesthetized cats. Recordings of multiunit activity were made from two electrodes in area MT of an alert macaque. The electrode separation was less than 0.5 mm, which yielded cells with nearly completely overlapping receptive fields but often differing directional preferences. This setup enabled them to repeat the earlier experiment conducted in the cat (Engel et al 1991b) by coactivating the cells with one bar and then with two independently moving bars. The firing of the cells was synchronized when responses were evoked by a single bar moving over both fields but not when the responses were evoked by two

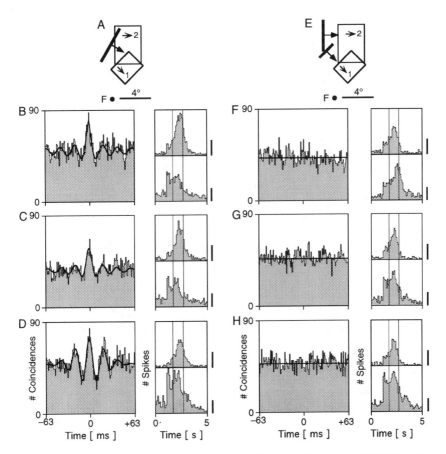

Figure 5 Stimulus-dependent synchronization in area MT of an awake macaque. (*A,E*) Plots of the receptive fields and the stimulus configurations. Panels *A–D* illustrate the single-bar configuration, while panels *E–H* show the results when two bars are moved simultaneously but in different directions over the two receptive fields. Both stimulus conditions were applied in 6 alternating blocks, each containing 10 trials. Cross-correlation histograms and PSTHs computed for the single-bar condition appear in panels *B, C,* and *D* and those for the dual bar condition in panels *F, G,* and *H.* The thin vertical lines in the PSTHs mark the time window in which the cross-correlation histograms were computed. The scale bars correspond to 40 spikes. Note that the single-bar stimulus results in a clear synchronization, as shown by the pronounced peaks in the cross-correlograms, while no synchronization occurs with the two-bar stimulus. (From Kreiter & Singer 1994.)

independent bars (Figure 5) (Kreiter & Singer 1994). Repeated measurements of the responses from the same cells under identical conditions revealed that the effect was stable.

Taken together, these data (Engel et al 1991b, Kreiter & Singer 1994) demonstrate that two overlapping, but independent, stimuli evoke simultaneous responses in a large array of spatially interleaved neurons. The active neuronal populations can become organized into two assemblies that are distinguished by the temporal coherence of activity within and the lack of coherence between the cell groups responding to the two different stimuli. Cells representing the same stimulus exhibit synchronized response epochs, while no consistent correlations occur between the responses of cells that are activated by different stimuli. This observation suggests that the pattern of temporal correlation among the responses of individual cells provides additional information to signal that two objects are present in nearby or overlapping regions of the visual field. More generally, these examples demonstrate that under appropriate conditions visual stimulus properties can influence the synchronization of activity of two or more groups of neurons, thus supporting the hypothesis that response synchrony can serve to establish relations between spatially distributed features in a visual image. It must be pointed out, however, that there do exist examples in which the correlated firing of neurons in visual cortex is not influenced by changes in the pattern of visual stimulation (Schwarz & Bolz 1991). Thus, further experiments are required to rigorously determine the conditions under which this type of dynamic correlation occurs.

The theoretical arguments put forth here with respect to visual processing also apply to tactile and auditory processing, motor control, and higher cognitive functions. Hence, transient synchronization of distributed neuronal responses should also occur within and across other regions of cortex. Moreover, these temporal correlations of neuronal firing should be flexible. Single cells or populations of cells should be able to participate in different assemblies at different times depending on stimulus properties or the context of a particular behavioral task. Supporting this prediction are recent studies in the alert monkey that were based on multineuron activities in the auditory and frontal cortex (Ahissar et al 1992, Vaadia & Aertsen 1992) and on field potential signals in sensory, motor, parietal, and frontal cortices (Bressler et al 1993). These investigators found instances in which changes in either stimulus properties or behavioral-task conditions led to marked changes in the correlated firing of two neurons without an appreciable change in their firing rate (Ahissar et al 1992, Vaadia & Aertsen 1992) or a change in the pattern of multiregional coherence of activity between different cortical areas (Bressler et al 1993). These findings suggest that dynamic, stimulus- and context-dependent modulations of response synchronization are a general property of cortical networks.

EXPERIENCE-DEPENDENT INFLUENCES ON INTRAAREAL SYNCHRONIZATION

The temporal correlation hypothesis implies that assemblies based on synchronous firing patterns form as a result of the synaptic interactions among the constituent cells. The experiments on interhemispheric synchronization have confirmed this supposition by identifying corticocortical connections as the anatomical substrate for synchrony. Therefore, the criteria by which particular features are grouped together must be built into the functional architecture of the corticocortical connections. If this architecture is specified entirely by genetic instructions, perceptual grouping criteria will have to be regarded as genetically determined. If the architecture is modifiable by activity and hence experience, some of these criteria could be acquired by learning. The numerous similarities in the layout of cortical connections indicate that some of the basic principles of cortical organization are determined genetically. But extensive evidence also supports epigenetic modifications. In mammals, corticocortical connections develop mainly postnatally (Innocenti & Frost 1979, Price & Blakemore 1985, Luhmann et al 1986, Callaway & Katz 1990) and attain their final specificity through an activity-dependent selection process (Innocenti & Frost 1979, Luhmann et al 1990, Callaway & Katz 1991).

Direct evidence for experience-dependent modifications of synchronizing connections comes from experiments with strabismic kittens (Löwel & Singer 1992). The primary visual cortex of kittens raised with artificially induced strabismus is split into two subpopulations of cells of about equal size, each responding rather selectively to stimulation of one eye only (Hubel & Wiesel 1965b). The changes in the thalamocortical connections elicited by squint suggest that input to cells driven by different eyes is only rarely correlated. A recent study showed that horizontal intercolumnar connections were reorganized in strabismic kittens, and unlike in normal animals, no longer connected neurons receiving input from different eyes (Löwel & Singer 1992). This finding suggests that experience-dependent selection of cortical connections occurs according to a correlation rule in a manner similar to other levels of the visual system (for review see Stryker 1990). Moreover, in strabismic cats response synchronization does not occur between cell groups connected to different eyes, whereas the incidence and magnitude of synchronization appears normal between cell groups connected to the same eye (König et al 1993). Because strabismic subjects are unable to bind signals conveyed by different eyes into coherent percepts (von Noorden 1990), this result also supports the hypothesis that response synchronization serves binding.

Further indications for a relation between response synchrony and visual function come from a recent study of strabismic cats that had developed amblyopia, a condition in which the spatial resolution and perception of patterns through one of the eyes is impaired. The identification of neuronal correlates of these deficits in animal models of amblyopia has remained inconclusive because the contrast sensitivity and spatial acuity of neurons in the retina and in the lateral geniculate nucleus were found to be normal, and evidence for correlates of the perceptual deficits in the visual cortex has

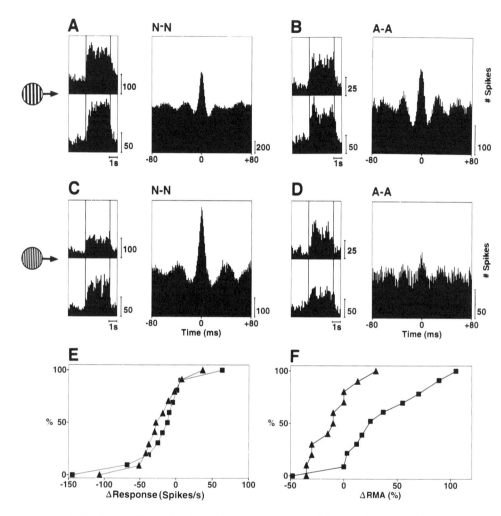

Figure 6 PSTHs (*left panels*) and associated cross-correlograms (*right panels*) computed from the activity recorded in response to the presentation of gratings of different spatial frequencies in cats with strabismic amblyopia. (*A–D*) Neuronal responses to low (*A, B*) and high (*C, D*) spatial frequency gratings, recorded simultaneously from two cell groups driven by the normal eye (N-sites) (*A, C*) and two groups driven by the amblyopic eye (A-sites) (*B, D*), respectively. Note that response amplitudes decrease at the higher spatial frequency in both cases, while the relative modulation amplitude of the cross-correlogram increases for the N-N pair but decreases for the A-A pair. (*E*) Cumulative distribution functions of the differences in response amplitudes to low and high spatial frequency gratings of optimal orientation. N-sites, squares ($n = 53$); A-sites, triangles ($n = 35$); abscissa, responses to high spatial-frequency minus responses to low spatial-frequency gratings. Note the similarity of the two distributions ($P > 0.1$). (*F*) Cumulative distribution functions of the differences between relative modulation amplitudes (DRMA) of cross-correlograms computed from responses to high and low spatial-frequency gratings of N-N pairs (*squares; n = 24*) and A-A pairs (*triangles; n = 11*). DRMA values (*abscissa*) were calculated by subtracting the relative modulation amplitude obtained with the low spatial frequency from that obtained with the high spatial frequency. The difference between the DRMA distributions of N-N pairs and A-A pairs is significant ($P < 0.001$). (From Roelfsema et al 1994.)

remained controversial (see Crewther & Crewther 1990, Blakemore & Vital Durand 1992). However, multielectrode recordings from the striate cortex of cats exhibiting behaviorally verified amblyopia have revealed significant differences in the synchronization behavior of cells driven by the normal and the amblyopic eye. The synchronization of activity among cells connected to the amblyopic eye was much less than that observed between cells responsive to the normal eye (Figure 6) (Roelfsema et al 1994). This difference was even more pronounced for responses elicited by gratings of high spatial frequency. Apart from these differences, the response properties of the cells appeared normal. Thus, cells connected to the amblyopic eye continued to respond vigorously to gratings whose spatial frequency had been too high to be discriminated with the amblyopic eye in the preceding behavioral tests. These results suggest that disturbed temporal coordination of responses may be one of the neuronal correlates of the amblyopic deficit.

MECHANISMS CONTROLLING OSCILLATIONS AND THEIR FUNCTIONAL IMPORTANCE

As reviewed above, synchronous activity is often associated with oscillatory firing patterns. This common finding raises the question of how oscillatory processes are generated and whether they contribute to synchronization. Our discussion of these mechanisms focuses on activity in the gamma-frequency band (30–80 Hz) because this is the range in which the majority of synchronous activity in the visual cortex has been described. There are three basic mechanisms likely to underlie the generation of oscillatory firing in the visual cortex. The first and most obvious mechanism to consider is oscillatory afferent input. The periodic temporal structure of cortical responses could result from rhythmic input from the lateral geniculate nucleus (LGN) or other thalamic nuclei. Support for this notion has come from studies demonstrating robust oscillatory activity in the frequency range of 30–80 Hz in both the retina and the LGN (Doty & Kimura 1963, Fuster et al 1965, Laufer & Verzeano 1967, Arnett 1975, Ariel et al 1983, Munemori et al 1984, Ghose & Freeman 1992). In both structures, the rhythmic activity is present in a subpopulation of roughly 10–20% of the cells, and it occurs spontaneously, in response to diffuse changes in illumination or specific stimulation of the receptive field (Laufer & Verzeano 1967, Arnett 1975, Robson & Troy 1987, Ghose & Freeman 1992). In some cases, however, neuronal oscillations in the LGN or perigeniculate nucleus are suppressed or not influenced by visual stimuli (Ghose & Freeman 1992, Pinault & Deschenes 1992). In spite of its prevalence, and surprising regularity, it is unlikely that oscillatory activity in the LGN can account for the cortical oscillations and their long-range synchronization.

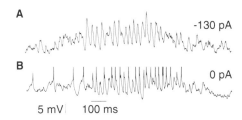

Figure 7 Intracellular recording of the response of a complex cell in area 17 of an anesthetized cat to the presentation of an optimally oriented light bar passed over the cell's receptive field. (*A*) Cell hyperpolarized by the injection of −130 pA of current. (*B*) The action potentials have been truncated. (From Jagadeesh et al 1992.)

Rather, the data of Sillito et al (1994) indicate that one form of thalamic synchronization depends on corticofugal influences.

A second mechanism likely to underlie the generation of cortical oscillations relies on intracortical network interactions. In this scenario, cyclical changes in firing probability are thought to arise through synaptic interactions of populations of excitatory and inhibitory neurons (Freeman 1968, 1975; Wilson & Cowan 1972; Bush & Douglas 1991; Wilson & Bower 1992, Bush et al 1993). Evidence supporting this hypothesis has come from intracellular recordings in cat striate cortex in vivo (Jagadeesh et al 1992, Bringuier et al 1992). These studies have revealed both high (30–60 Hz) and low (10–20 Hz) frequency oscillations of membrane potential closely resembling the properties of neuronal oscillations observed extracellularly (Gray & Singer 1989). The membrane potential oscillations occur during visual stimulation: They are orientation specific, and occur less often in cells receiving monosynaptic input from the LGN (Jagadeesh et al 1992). Moreover, spike discharges occur on the depolarizing phase of the oscillation, which mirrors the extracellularly observed negativity (Figure 7) (compare to Figure 1). When the cells exhibiting these fluctuations are hyperpolarized below the firing threshold by current injection, the visually evoked oscillations increase in amplitude but do not change in frequency (Figure 7) (Bringuier et al 1992, Jagadeesh et al 1992). These data suggest that the rhythmic fluctuations of membrane potential reflect a synchronous pattern of excitatory synaptic input onto the cells that arises in the cortex.

These data do not, however, rule out the possibility that cortical oscillations may arise as a consequence of activity in a subpopulation of cells that are intrinsically oscillatory (Llinas 1988). Support for this conjecture has been obtained from both in vitro and in vivo intracellular recordings. In the former case, a subpopulation of inhibitory interneurons in layer 4 of the rat frontal cortex exhibited subthreshold, voltage-dependent, 10- to 50-Hz oscillations in membrane potential in response to depolarizing current injection (Llinas et al 1991). These cells, if present in the visual cortex, would be expected to produce rhythmic hyperpolarizing potentials in their postsynaptic targets, a result for which the experimental evidence is currently limited (Ferster 1986). Recent intracellular recordings in vivo from cat striate cortex have revealed a subpopulation of cells that fire in regular repetitive bursts at 20–70 Hz in response to visual stimulation and intracellular depolarizing current injection (McCormick et al 1993). The firing properties of these cells are very similar to the burst-firing cells observed extracellularly in cat visual cortex (Hubel & Wiesel 1965a, Gray et al 1990, Gray & Viana Di Prisco 1993), and they are distinct from the previously defined class of intrinsic bursting cells known to be present in layer 5 (Connors 1982, McCormick et al 1985). Whether these cells are excitatory or inhibitory is unknown, but in either case their activation might serve to drive a local network of cells into a pattern of synchronous oscillation.

CONCLUSIONS

Visual scenes are inherently complex and can exist in a nearly infinite combinatorial variety. In spite of this complexity, mammalian species have evolved a virtually unlimited capacity for the rapid recognition of patterned visual information, which suggests that the visual system possesses effective mechanisms to enable the flexible integration of featural information. Anatomical and physiological evidence indicates that the featural attributes of visual images are processed in different parts of the visual cortex. Therefore, the representation of perceptual objects is not likely to take place at a single location but rather to involve the concerted action of large populations of cells distributed throughout the visual cortex. This raises the fundamental question, often referred to as the binding problem, of how the visual system establishes the appropriate relationships among the large number of neurons that respond to the many features in any given visual scene. Such relationships must be specific and yet flexible in order to cope with the combinatorial variety of features constituting visual images.

The experimental results reviewed in this chapter are compatible with the hypothesis that synchronization of neuronal activity on a millisecond time scale may be exploited to link featural information that is represented in different parts of the cortex (Milner 1974; von der Malsburg 1981, 1985). In this view the identification of related features, for example those belonging to the same perceptual object, is achieved by the temporal coincidence of the neuronal discharges evoked by those features. In a distributed network, such as the neocortex, where any given cell contacts any other cell with only a few synapses, but where individual cells receive converging input from many thousands of different cells (Braitenberg & Schüz 1991), synchronization may also be a particularly efficient mechanism to increase the saliency of activity. Thus, synchronization can, in principle, be used to select with high spatial and temporal resolution those activity patterns that belong together and to enhance the effect of this activity so it may be evaluated for further processing. The proposal presented here is that this selection is achieved in a distributed and parallel fashion by the system of corticocortical association fibers. In this context, the function of these connections consists largely of adjusting the timing of discharges rather than modulating discharge rates. Synchronization will of course affect discharge rates by enabling more effective summation than asynchronous inputs. Hence, modulation of response amplitude and synchronization of discharges can operate as complementary mechanisms for the selection and binding of responses.

In consideration of these functional issues, the fundamental question also arises as to how, at a cellular level, synchrony can be established on a millisecond time scale over the broad range of spatial scales that are required. One possibility is that during development those connections having a conduction velocity appropriate for the generation of synchrony get selectively stabilized. The fact that synchronizing connections are selected according to a correlation rule supports this possibility. Additional temporal patterning of discharges may further facilitate the establishment of synchronous firing. The evidence presented here suggests that this requirement may be met in many instances by discharge patterns in which bursts and pauses follow one another with a certain degree of rhythmicity. The refractory period following a burst will produce a brief period of decreased sensitivity, making synaptic inputs arriving at that time less likely to have an effective influence on the overall pattern of activity.

In this context, the question arises as to why the observed oscillations are so irregular and cover such a broad frequency range (Gray et al 1992). Functionally, such variation could facilitate the desynchronization of activity and thereby prevent the cortical network from entering global states of synchrony that would be inappropriate for information processing. Furthermore, the number of assemblies that can coexist in the same cortical region increases if the oscillation frequencies are variable, because spurious correlations arising from aliasing effects would be rare and only of short duration. Thus, a broad-banded oscillatory signal appears as a reasonable compromise between several opposing constraints. However, synchronous activity may arise in cortical networks without oscillatory firing, and the presence or absence of a regular oscillatory time structure in single-cell activity neither proves nor disproves that spatially segregated cells discharge in synchrony. Oscillations per se are thus of little diagnostic value for the testing of the temporal correlation hypothesis. The synchronization of activity is the relevant issue.

To directly test the hypothesis that correlated activity contributes to the integration of distributed featural information, experiments are needed in which the dependence of the occurrence and the magnitude of synchronized activity on stimulus configuration and behavioral performance can be assessed. This minimally requires the simultaneous recording of activity from several locations in the brain of awake, behaving animals and the evaluation of temporal correlations of activity at high resolution. With the techniques currently available, the number of recording sites that can be examined simultaneously is bound to remain small, which poses a severe sampling problem. For the brain, a brief sequence of correlated activity may be a highly significant event if it occurs in a large number of cells (Bressler et al 1993). However, for the experimenter, who can only look at a few cells (most often 2; occasionally as many as 100) (Wilson & McNaughton 1993), such brief synchronous events may pass undetected. They will only be recognized as significant if they recur. Thus, until new techniques become available, the relationships between synchronized neuronal activity and behavior are likely to be detectable only for conditions that maintain synchronization for longer periods of time. This limitation may confine the behavioral conditions suitable for such analyses to problem-solving tasks that are difficult, fraught with ambiguity, and require periods of sustained, focused attention.

ACKNOWLEDGMENTS

We thank Francis Crick and Christof Koch for helpful comments on an earlier version of the manuscript. This work was supported by the Max-Planck-Gesellschaft (WS) and grants from the National Science Foundation, the National Eye Institute, the Klingenstein Foundation, and the Sloan Foundation (CMG).

Literature Cited

Abeles M, ed. 1991. *Corticonics.* Cambridge: Cambridge Univ. Press

Adrian ED. 1950. The electrical activity of the mammalian olfactory bulb. *Electroencephalogr. Clin. Neurophysiol.* 2:377–88

Ahissar E, Vaadia E, Ahissar M, Bergman H, Arieli A, Abeles M. 1992. Dependence of cortical plasticity on correlated activity of single neurons and on behavioral context. *Science* 257:1412–15

Aiple F, Krüger J. 1988. Neuronal synchrony in monkey striate cortex: interocular signal flow and dependency on spike rates. *Exp. Brain Res.* 72:141–49

Alonso A, Garcia-Austt E. 1987. Neuronal

sources of theta rhythm in the entorhinal cortex of the rat. II. Phase relations between unit discharges and theta field potentials. *Exp. Brain Res.* 67:502–9

Ariel M, Daw NW, Rader RK. 1983. Rhythmicity in rabbit retinal ganglion cell responses. *Vision Res.* 23:1485–93

Arnett DW. 1975. Correlation analysis of units recorded in the cat dorsal lateral geniculate nucleus. *Exp. Brain Res.* 24:111–30

Bair W, Koch C, Newsome W, Britten K, Niebur E. 1994. Power spectrum analysis of bursting cells in area MT in the behaving monkey. *J. Neurosci.* 14:2870–92

Ballard DH, Hinton GE, Sejnowski TJ. 1983.

Parallel visual computation. *Nature* 306:21–26

Basar E. 1988. EEG—dynamics and evoked potentials in sensory and cognitive processing by the brain. In *Dynamics of Sensory and Cognitive Processing by the Brain. Springer Series in Brain Dynamics,* ed. E Basar, 1:30–55. Berlin/Heidelberg/New York: Springer

Basar E, Bullock TH. 1992. *Induced Rhythms in the Brain.* Boston: Birkhaeuser

Baylis GC, Rolls ET, Leonard CM. 1985. Selectivity between faces in the responses of a population of neurons in the cortex in the superior temporal sulcus of the monkey. *Brain Res.* 342:91–102

Blakemore C, Vital-Durand F. 1992. Different neural origins for 'blur' amblyopia and strabismic amblyopia. *Ophthalmic Physiol. Opt.* 12

Bland BH, Andersen P, Ganes T. 1975. Two generators of hippocampal theta activity in rabbits. *Brain Res.* 94:199–218

Bouyer JJ, Montaron MF, Rougeul A. 1981. Fast fronto-parietal rhythms during combined focused attentive behaviour and immobility in cat: cortical and thalamic localizations. *Electroencephalogr. Clin. Neurophysiol.* 51:244–52

Bouyer JJ, Montaron MF, Vahnee JM, Albert MP, Rougeul A. 1987. Anatomical localization of cortical beta rhythms in cat. *Neuroscience* 22:863–69

Bragin A, Jandó G, Nádasdy Z, Hetke J, Wise K. Busaki G. 1994. Gamma (40–100 Hz) oscillation in the hippocampus of the behaving rat. *J. Neurosci.* In press

Braitenberg V. 1978. Cell assemblies in the cerebral cortex. In *Lecture Notes in Biomathematics 21, Theoretical Approaches in Complex Systems,* ed. R Heim, G Palm, pp. 171–88. Berlin: Springer

Braitenberg V, Schüz A. 1991. *Anatomy of the Cortex: Statistics and Geometry.* Berlin/ Heidelberg/New York: Springer

Bressler SL. 1984. Spatial organization of EEGs from olfactory bulb and cortex. *Electroencephalgr. Clin. Neurophys.* 57:270–76

Bressler SL. 1987. Relation of olfactory bulb and cortex. I. Spatial variation of bulbocortical interdependence. *Brain Res.* 409:285–93

Bressler SL, Coppola R, Nakamura R. 1993. Episodic multiregional cortical coherence at multiple frequencies during visual task performance. *Nature* 366:153–56

Bringuier V, Fregnac Y, Debanne D, Shulz D, Baranyi A. 1992. Synaptic origin of rhythmic visually evoked activity in kitten area 17 neurons. *NeuroReport* 3:1065–68

Bullier J, Munk MHJ, Nowak LG. 1992. Synchronization of neuronal firing in areas V1 and V2 of the monkey. *Soc. Neurosci. Abstr.* 18:11.7

Bush P, Douglas RJ. 1991. Synchronization of

bursting action potential discharge in a model network of neocortical neurons. *Neural Comput.* 3:19–30

Bush P, Gray CM, Sejnowski T. 1993. Realistic simulations of synchronization in networks of layer V neurons in cat primary visual cortex. *Soc. Neurosci. Abstr.* 19:359.7

Buzsaki G, Horvath Z, Urioste R, Hetke J, Wise K. 1992. High-frequency network oscillation in the hippocampus. *Science* 256:1025–27

Buzsaki G, Leung LS, Vanderwolf CH. 1983. Cellular bases of hippocampal EEG in the behaving rat. *Brain Res. Rev.* 6:139–71

Callaway EM, Katz LC. 1990. Emergence and refinement of clustered horizontal connections in cat striate cortex. *J. Neurosci.* 10:1134–53

Callaway EM, Katz LC. 1991. Effects of binocular deprivation on the development of clustered horizontal connections in cat striate cortex. *Proc. Natl. Acad. Sci. USA* 88:745–49

Chatrian GE, Bickford RG, Uilein A. 1960. Depth electrographic study of a fast rhythm evoked from the human calcarine region by steady illumination. *Electroencephalgr. Clin. Neurophysiol.* 12:167–76

Connors BW, Gutnick MJ, Prince DA. 1982. Electrophysiological properties of neocortical neurons in vitro. *J. Neurophysiol.* 48:1302–20

Crewther DP, Crewther SG. 1990. Neural sites of strabismic amblyopia in cats: spatial frequency deficit in primary cortical neurons. *Exp. Brain Res.* 79:615–22

Crick F, Koch C. 1990. Towards a neurobiological theory of consciousness. *Semin. Neurosci.* 2:263–75

Desimone R, Schein SJ, Moran J, Ungerleider LG. 1985. Contour, color and shape analysis beyond the striate cortex. *Vision Res.* 24:441–52

DeYoe EA, Van Essen DC. 1988. Concurrent processing streams in monkey visual cortex. *Trends Neurosci.* 115:219–26

Doty RW, Kimura DS. 1963. Oscillatory potentials in the visual system of cats and monkeys. *J. Physiol.* 168:205–18

Eckhorn R, Bauer R, Jordan W, Brosch M, Kruse W, et al. 1988. Coherent oscillations: a mechanism for feature linking in the visual cortex? *Biol. Cybern.* 60:121–30

Eckhorn R, Frien A, Bauer R, Woelbern T, Kehr H. 1993. High Frequency 60–90 Hz oscillations in primary visual cortex of awake monkey. *Neuro Rep.* 4:243–46

Eckhorn R, Schanze T, Brosch M, Salem W, Bauer R. 1992. Stimulus-specific synchronizations in cat visual cortex: multiple microelectrode and correlation studies from several cortical areas. In *Induced Rhythms in the Brain,* ed. E Basar, TH Bullock, pp. 47–82. Boston: Birkhauser

Edelman GM. 1987. *Neural Darwinism: The*

Theory of Neuronal Group Selection. New York: Basic Books

Engel AK, König P, Gray CM, Singer W. 1990. Stimulus-dependent neuronal oscillations in cat visual cortex: inter-columnar interaction as determined by cross-correlation analysis. *Eur. J. Neurosci.* 2:588–606

Engel AK, König P, Kreiter AK, Singer W. 1991a. Interhemispheric synchronization of oscillatory neuronal responses in cat visual cortex. *Science* 252:1177–79

Engel AK, König P, Singer W. 1991b. Direct physiological evidence for scene segmentation by temporal coding. *Proc. Natl. Acad. Sci. USA* 88:9136–40

Engel AK, Kreiter AK, König P, Singer W. 1991c. Synchronization of oscillatory neuronal responses between striate and extrastriate visual cortical areas of the cat. *Proc. Natl. Acad. Sci. USA* 88:6048–52

Engel AK, Kreiter AK, Singer W. 1992. Oscillatory responses in the superior temporal sulcus of anesthetized macaque monkeys. *Soc. Neurosci. Abstr.* 18:11.10

Felleman DJ, van Essen DC. 1991. Distributed hierarchical processing in the primate cerebral cortex. *Cerebral Cortex* 1:1–47

Ferster D. 1986. Orientation selectivity of synaptic potentials in neurons of cat primary visual cortex. *J. Neurosci.* 6:1284–1301

Freeman WJ, ed. 1975. *Mass Action in the Nervous System.* New York: Academic

Freeman WJ. 1968. Analog simulation of prepiriform cortex in the cat. *Math. BioSci.* 2: 181–90

Freeman WJ. 1978. Spatial properties of an EEG event in the olfactory bulb and cortex. *Electroencephalogr. Clin. Neurophysiol.* 44: 586–605

Freeman WJ. 1987. Analytic techniques used in the search for the physiological basis for the EEG. In *Handbook of Electroencephalography and Clinical Neurophysiology,* ed. A Gevins, A Remond, 3A(Part 2):583–664. Amsterdam: Elsevier

Fujita I, Tanaka K, Ito M, Cheng K. 1992. Columns for visual features of objects in monkey inferotemporal cortex. *Nature* 360: 343–46

Fuster JM, Herz A, Creutzfeldt OD. 1965. Interval analysis of cell discharge in spontaneous and optically modulated activity in the visual system. *Arch. Ital. Biol.* 103: 159–77

Galambos R, Makeig S, Talmachoff PJ. 1981. A 40-Hz auditory potential recorded from the human scalp. *Proc. Natl. Acad. Sci. USA* 78: 2643–47

Gallant JL, Braun J, Van Essen DC. 1993. Selectivity for polar, hyperbolic, and Cartesian gratings in macaque visual cortex. *Science* 259:100–3

Gawne TJ, Richmond BJ. 1993. How independent are the messages carried by adjacent inferior temporal cortical neurons? *J. Neurosci.* 137:2758–71

Gerstein GL, Bedenbaugh P, Aertsen AMHJ. 1989. Neuronal assemblies. *IEEE Trans. Biomed. Eng.* 361:4–14

Ghose GM, Freeman RD. 1992. Oscillatory discharge in the visual system: does it have a functional role? *J. Neurophysiol.* 68:1558–74

Gilbert CD, Wiesel TN. 1983. Clustered intrinsic connections in cat visual cortex. *J. Neurosci.* 35:1116–33

Gilbert CD, Wiesel TN. 1989. Columnar specificity of intrinsic horizontal and corticocortical connections in cat visual cortex. *J. Neurosci.* 97:2432–42

Gochin PM, Miller EK, Gross CG, Gerstein GL. 1991. Functional interactions among neurones in inferior temporal cortex of the awake macaque. *Exp. Brain Res.* 84:505–16

Gray CM. 1993. Rhythmic activity in neuronal systems: insights into integrative function. In *Lectures in Complex Systems. Santa Fe Institute Studies in the Sciences of Complexity,* ed. L Nadel, D Stein, 5:89–161. Santa Fe, NM: Addison-Wesley

Gray CM, Engel AK, König P, Singer W. 1990. Stimulus-dependent neuronal oscillations in cat visual cortex: receptive field properties and feature dependence. *Eur. J. Neurosci.* 2:607–19

Gray CM, Engel AK, König P, Singer W. 1992. Synchronization of oscillatory neuronal responses in cat striate cortex: temporal properties. *Vis. Neurosci.* 8:337–47

Gray CM, König P, Engel AK, Singer W. 1989. Oscillatory responses in cat visual cortex exhibit inter-columnar synchronization which reflects global stimulus properties. *Nature* 338:334–37

Gray CM, Singer W. 1987. Stimulus-specific neuronal oscillations in the cat visual cortex: a cortical functional unit. *Soc. Neurosci. Abstr.* 13:404.3

Gray CM, Singer W. 1989. Stimulus-specific neuronal oscillations in orientation columns of cat visual cortex. *Proc. Natl. Acad. Sci. USA* 86:1698–1702

Gray CM, Skinner JE. 1988. Centrifugal regulation of neuronal activity in the olfactory bulb of the waking rabbit as revealed by reversible cryogenic blockade. *Exp. Brain Res.* 69:378–86

Gray CM, Viana Di Prisco G. 1993. Properties of stimulus-dependent rhythmic activity of visual cortical neurons in the alert cat. *Soc. Neurosci. Abstr.* 19:359.8

Gross CG, Rocha-Miranda EC, Bender DB. 1972. Visual properties of neurons in inferotemporal cortex of the macaque. *J. Neurophysiol.* 35:96–111

Grossberg S. 1980. How does the brain build a cognitive code? *Psychol. Rev.* 87:1–51

Hata Y, Tsumoto T, Sato H, Hagihara K, Tamura H. 1988. Inhibition contributes to orientation selectivity in visual cortex of cat. *Nature* 335:815–17

Hata Y, Tsumoto T, Sato H, Tamura H. 1991. Horizontal interactions between visual cortical neurones studied by cross-correlation analysis in the cat. *J. Physiol.* 441:593–614

Hebb DO. 1949. *The Organization of Behavior.* New York: Wiley

Hubel DH, Wiesel TN. 1965a. Receptive fields and functional architecture in two nonstriate visual areas 18 and 19 of the cat. *J. Neurophysiol.* 28:229–89

Hubel DH, Wiesel TN. 1965b. Binocular interaction in striate cortex of kittens reared with artificial squint. *J. Neurophysiol.* 28:1041–59

Innocenti GM, Frost DO. 1979. Effects of visual experience on the maturation of the efferent system to the corpus callosum. *Nature* 280:231–34

Jagadeesh B, Gray CM, Ferster D. 1992. Visually-evoked oscillations of membrane potential in neurons of cat striate cortex studied with in vivo whole cell patch recording. *Science* 257:552–54

König P, Engel AK, Löwell S, Singer W. 1993. Squint affects synchronization of oscillatory responses in cat visual cortex. *Eur. J. Neurosci.* 5:501–8

König P, Engel AK, Singer W. 1994. The relation between oscillatory activity and long-range synchronization in cat visual cortex. *Proc. Natl. Acad. Sci. USA* In press

Koffka K. 1935. *Principles of Gestalt Psychology.* New York: Harcourt

Komatsu Y, Nakajima S, Toyama K, Fetz E. 1988. Intracortical connectivity revealed by spike-triggered averaging in slice preparations of cat visual cortex. *Brain Res.* 442: 359–62

Kreiter AK, Engel AK, Singer W. 1992. Stimulus dependent synchronization in the caudal superior temporal sulcus of macaque monkeys. *Soc. Neurosci. Abstr.* 18:11.11

Kreiter AK, Singer W. 1992. Oscillatory neuronal responses in the visual cortex of the awake macaque monkey. *Eur. J. Neurosci.* 4:369–75

Kreiter AK, Singer W. 1994. Global stimulus arrangement determines synchronization of neuronal activity in the awake macaque monkey. *Suppl. Eur. J. Neurosci.* 7:153

Krüger J. 1990. Multi-electrode investigation of monkey striate cortex: link between correlational and neuronal properties in the infragranular layers. *Vis. Neurosci.* 5:135–42

Krüger J, Aiple F. 1988. Multimicroelectrode investigation of monkey striate cortex: spike train correlations in the infragranular layers. *J. Neurophysiol.* 60:798–828

Kuperstein M, Eichenbaum H, VanDeMark T. 1986. Neural group properties in the rat hippocampus during the theta rhythm. *Exp. Brain Res.* 61:438–42

Laufer M, Verzeano M. 1967. Periodic activity in the visual system of the cat. *Vision Res.* 7:215–29

Leung LS. 1992. Fast beta rhythms in the hippocampus: a review. *Hippocampus* 22: 93–98

Livingstone MS. 1991. Visually evoked oscillations in monkey striate cortex. *Soc. Neurosci. Abstr.* 17:73.3

Livingstone MS, Hubel DH. 1988. Segregation of form, color, movement, and depth: anatomy, physiology, and perception. *Science* 240:740–49

Llinas RR. 1988. The intrinsic electrophysiological properties of mammalian neurons: insights into central nervous system function. *Science* 242:1654–64

Llinas RR, Grace AA, Yarom Y. 1991. In vitro neurons in mammalian cortical layer 4 exhibit intrinsic oscillatory activity in the 10- to 50-Hz frequency range. *Proc. Natl. Acad. Sci. USA* 88:897–901

Löwel S, Singer W. 1992. Selection of intrinsic horizontal connections in the visual cortex by correlated neuronal activity. *Science* 255: 209–12

Luhmann HJ, Martinez-Millan L, Singer W. 1986. Development of horizontal intrinsic connections in cat striate cortex. *Exp. Brain Res.* 63:443–48

Luhmann HJ, Singer W, Martinez-Millan L. 1990. Horizontal interactions in cat striate cortex. I. Anatomical substrate and postnatal development. *Eur. J. Neurosci.* 2:344–57

Martin KAC, Whitteridge D. 1984. Form, function and intracortical projections of spiny neurons in the striate visual cortex of the cat. *J. Physiol.* 353:463–504

Mason A, Nicoll A, Stratford K. 1991. Synaptic transmission between individual pyramidal neurons of the rat visual cortex in vitro. *J. Neurosci.* 111:72–84

Matsubara JA, Cynader MS, Swindale NV. 1987. Anatomical properties and physiological correlates of the intrinsic connections in cat area 18. *J. Neurosci.* 75:1428–46

McCormick DA, Connors BW, Lighthall JW, Prince DA. 1985. Comparative electrophysiology of pyramidal and sparsely spiny stellate neurons of the neocortex. *J. Neurophysiol.* 54:782–806

McCormick DA, Gray CM, Wang Z. 1993. Chattering cells: a new physiological subtype which may contribute to 20–60 Hz oscillations in cat visual cortex. *Soc. Neurosci. Abstr.* 19:359.9

Menon V, Freeman WJ, Gevins AS. 1994. Bootstrap analysis of spatio-temporal correlations in human gamma band electroencephalograms. *Electroencephalogr. Clin. Neurophysiol.* In press

Merigan WH, Maunsell JH. 1993. How parallel are the primate visual pathways? *Annu. Rev. Neurosci.* 16:369–402

Michalski A, Gerstein GL, Czarkowska J, Tarnecki R. 1983. Interactions between cat striate cortex neurons. *Exp. Brain Res.* 51: 97–107

Milner P. 1974. A model for visual shape recognition. *Psychol. Rev.* 816:521–35

Montaron M, Bouyer J, Rougeul A, Buser P. 1982. Ventral mesencephalic tegmentum VMT controls electrocortical beta rhythms and associated attentive behaviour in the cat. *Behav. Brain Res.* 6:129–45

Munemori J, Hara K, Kimura M, Sato R. 1984. Statistical features of impulse trains in cat's lateral geniculate neurons. *Biol. Cybern.* 50: 167–72

Murthy VN, Aoki F, Fetz FE. 1994. Synchronous oscillations in sensorimotor cortex of awake monkeys and humans, See Pantev et al 1994, In press

Murthy VN, Fetz EE. 1992. Coherent 25- to 35-Hz oscillations in the sensorimotor cortex of awake behaving monkeys. *Proc. Natl. Acad. Sci. USA* 89:5670–74

Nelson JI, Nowak LG, Chouvet G, Munk MHJ, Bullier J. 1992a. Synchronization between cortical neurons depends on activity in remote areas. *Soc. Neurosci. Abstr.* 18: 11.8

Nelson JI, Salin PA, Munk MHJ, Arzi M, Bullier J. 1992b. Spatial and temporal coherence in cortico-cortical connections: a cross-correlation study in areas 17 and 18 in the cat. *Visual Neurosci.* 9:21–38

Neuenschwander S, Varela FJ. 1993. Visually-triggered neuronal oscillations in birds: an autocorrelation study of tectal activity. *Eur. J. Neurosci.* 5:870–81

Nicoll A, Blakemore C. 1993. Single-fiber EPSPs in layer 5 of rat visual cortex in vitro. *NeuroReport* 42:167–70

Nowak LG, Munk MHJ, Chounlamountri N, Bullier J. 1994. Temporal aspects of information processing in areas V1 and V2 of the macaque monkey. See Pantev et al 1994, In press

Olshausen BA, Anderson CH, Van Essen DC. 1993. A neurobiological model of visual attention and invariant pattern recognition based on dynamic routing of information. *J. Neurosci.* 1311:4700–19

Palm G. 1990. Cell assemblies as a guideline for brain research. *Concepts Neurosci.* 1: 133–37

Pantev C, Elbert T, Lutkenhoener B, eds. 1994. *Oscillatory Event-Related Brain Dynamics.* New York: Plenum. In press

Pantev C, Makeig S, Hoke M, Galambos R, Hampson S, Galen C. 1991. Human auditory evoked gamma-band magnetic fields. *Proc. Natl. Acad. Sci. USA* 88:8996–9000

Payne BR. 1993. Evidence for visual cortical area homologs in cat and macaque monkey. *Cerebral Cortex* 3:1–25

Perez-Borja C, Tyce FA, McDonald C, Uihlein A. 1961. Depth electrographic studies of a focal fast response to sensory stimulation in the human. *Electroencephalogr. Clin. Neurophysiol.* 13:695–702

Perkel DH, Gerstein GL, Moore GP. 1967. Neuronal spike trains and stochastic point processes. II. Simultaneous spike trains. *Biophys. J.* 7:419–40

Perrett DI, Mistlin AJ, Chitty AJ. 1987. Visual neurones responsive to faces. *Trends Neurosci.* 10:358–64

Pinault D, Deschenes M. 1992. Voltage-dependent 40 Hz oscillations in reticular thalamic neurons. *Neuroscience* 251:245–258

Price DJ, Blakemore C. 1985. The postnatal development of the association projection from visual cortical area 17 to area 18 in the cat. *J. Neurosci.* 5:2443–52

Raether A, Gray CM, Singer W. 1989. Intercolumnar interactions of oscillatoary neuronal responses in the visual cortex of alert cats. *Eur. Neurosci. Assoc.* 12:72.5

Ribary U, Joannides AA, Singh KD, Hasson R, Bolton JPR, et al. 1991. Magnetic field tomography of coherent thalamocortical 40 Hz oscillations in humans. *Proc. Natl. Acad. Sci. USA* 88:11037–41

Robson JG, Troy JB. 1987. Nature of the maintained discharge of Q, X, and Y retinal ganglion cells of the cat. *J. Opt. Soc. Am. Ser. A* 412:2301–7

Roe AW, Ts'o DY. 1992. Functional connectivity between V1 and V2 in the primate. *Soc. Neurosci. Abstr.* 18:11.4

Roelfsema PR, König P, Engel AK, Sireteanu R, Singer W. 1994. Reduced neuronal synchrony: a physiological correlate of strabismic ambylopia in cat visual cortex. *Eur. J. Neurosci.* In press

Rosenquist AC. 1985. Connections of visual cortical areas in the cat. In *Cerebral Cortex*, ed. A Peters, EG Jones, pp. 81–117. New York: Plenum

Sakai K, Miyashita Y. 1991. Neural organization for the long-term memory of paired associates. *Nature* 354:152–55

Sanes JN, Donoghue JP. 1993. Oscillations in local field potentials of the primate motor cortex during voluntary movement. *Proc. Natl. Acad. Sci. USA* 90:4470–74

Schwarz C, Bolz J. 1991. Functional specificity of the long-range horizontal connections in cat visual cortex: a cross-correlation study. *J. Neurosci.* 11:2995–3007

Sem-Jacobsen CW, Petersen MC, Dodge HW, Lazarte JA, Holman CB. 1956. Electroencephalographic rhythms from the depths of the parietal, occipital and temporal lobes in man. *Electroencephalogr. Clin. Neurophysiol.* 8:263–78

Sereno MI, Allman JM. 1991. Cortical visual areas in mammals. In *The Neural Basis of Visual Function,* ed. A Leventhal, pp. 160–72. New York: MacMillan

Sheer DE. 1989. Sensory and cognitive 40-Hz event-related potentials. In *Brain Dynamics,* ed. E Basar, TH Bullock, pp. 338–74. Berlin: Springer

Sillito AM, Jones HE, Gerstein GL, West DC. 1994. Feature-linked synchronization of thalamic relay cell firing induced by feedback from the visual cortex. *Nature* 369:479–82

Singer W. 1985. Activity-dependent self-organization of the mammalian visual cortex. In *Models of the Visual Cortex,* ed. D Rose, VG Dobson, pp. 123–36. Chichester: Wiley

Singer W. 1990. Search for coherence: a basic principle of cortical self-organization. *Concepts Neurosci.* 1:1–26

Singer W. 1993. Synchronization of cortical activity and its putative role in information processing and learning. *Annu. Rev. Physiol.* 55:349–74

Softky WR, Koch C. 1993. The highly irregular firing of cortical cells is inconsistent with temporal integration of random EPSPs. *J. Neurosci.* 13:334–50

Stryker MP. 1990. Activity-dependent reorganization of afferents in the developing mammalian visual system. In *Development of the Visual System,* ed. DMK Lam, CJ Shatz, pp. 267–87. Cambridge, MA: MIT Press

Thomson AM, West DC. 1993. Fluctuations in pyramidal-pyramidal excitatory postsynaptic potentials modified by presynaptic firing pattern and postsynaptic membrane potential using paired intracellular recordings in rat neocortex. *Neuroscience* 54:329–46

Tiitinen H, Sinkkonen J, Reinikainen K, Alho K, Lavikainen J, Naatanen R. 1993. Selective attention enhances the auditory 40-Hz transient response in humans. *Nature* 364:59–60

Tovee MJ, Rolls ET. 1992. Oscillatory activity is not evident in the primate temporal visual cortex with static stimuli. *NeuroReport* 3:369–72

Toyama K, Kimura M, Tanaka K. 1981a. Cross-correlation analysis of interneuronal connectivity in cat visual cortex. *J. Neurophysiol.* 46:191–201

Toyama K, Kimura M, Tanaka K. 1981b. Or-ganization of cat visual cortex as investigated by cross-correlation techniques. *J. Neurophysiol.* 46:202–14

Ts'o D, Gilbert C. 1988. The organization of chromatic and spatial interactions in the primate striate cortex. *J. Neurosci.* 8:1712–27

Ts'o D, Gilbert C, Wiesel TN. 1986. Relationship between horizontal interactions and functional architecture in cat striate cortex as revealed by cross-correlation analysis. *J. Neurosci.* 6:1160–70

Vaadia E, Aertsen A. 1992. Coding and computation in the cortex: single-neuron activity and cooperative phenomena. In *Information Processing in the Cortex. Experiments and Theory,* ed. A Aertsen, V Braitenberg, pp. 81–119. Berlin/New York: Springer-Verlag

von der Malsburg C. 1981. *The correlation theory of brain function.* Internal report. Max-Planck-Institute for Biophysical Chemistry, Gottingen, West Germany

von der Malsburg C. 1985. Nervous structures with dynamical links. *Ber. Bunsenges. Phys. Chem.* 89:703–10

von der Malsburg C, Schneider W. 1986. A neural cocktail-party processor. *Biol. Cybern.* 54:29–40

von Noorden GK. 1990. *Binocular Vision and Ocular Motility, Theory and Management of Strabismus.* St. Louis/Baltimore: Mosby

Wilson HR, Cowan JD. 1972. Excitatory and inhibitory interactions in localized populations of model neurons. *Biophys. J.* 12:1–24

Wilson MA, Bower JM. 1992. Cortical oscillations and temporal interactions in a computer simulation of piriform cortex. *J. Neurophysiol.* 674:981–95

Wilson MA, McNaughton BL. 1993. Dynamics of the hippocampal ensemble code for space. *Science* 261:1055–58

Ylinen A, Bragin A, Nádasdy Z, Jandó G, Szabó I, et al. 1994. Sharp wave-associated high frequency oscillation (200 Hz) in the intact hippocampus: network and intracellular mechanisms. *J. Neurosci.* In press

Young MP, Tanaka K, Yamane S. 1992. On oscillating neuronal responses in the visual cortex of the monkey. *J. Neurophysiol.* 67(6):1464–74

Young MP, Yamane S. 1992. Sparse population coding of faces in the inferotemporal cortex. *Science* 256:1327–31

SUBJECT INDEX

CUMULATIVE INDEXES

CONTRIBUTING AUTHORS, VOLUMES 11–18

CHAPTER TITLES, VOLUMES 11–18

ANNUAL REVIEWS

a nonprofit scientific publisher
**4139 El Camino Way
P.O. Box 10139
Palo Alto, CA 94303-0139 • USA**

Annual Reviews publications may be ordered directly from our office; through stockists, booksellers and subscription agents, worldwide; and through participating professional societies. **Prices are subject to change without notice. We do not ship on approval.**

- **Individuals:** Prepayment required on new accounts. in US dollars, checks drawn on a US bank.

- **Institutional Buyers:** Include purchase order. Calif. Corp. #161041 • ARI Fed. I.D. #94-1156476

- **Students / Recent Graduates:** $10.00 discount from retail price, per volume. *Requirements:* 1. be a degree candidate at, or a graduate within the past three years from, an accredited institution; 2. present proof of status (photocopy of your student I.D. or proof of date of graduation); 3. Order direct from Annual Reviews; 4. prepay. This discount **does not** apply to standing orders, *Index on Diskette*, Special Publications, ARPR, or institutional buyers.

- **Professional Society Members:** Many Societies offer *Annual Reviews* to members at reduced rates. Check with your society or contact our office for a list of participating societies.

- **California orders** add applicable sales tax. • **Canadian orders** add 7% GST. Registration #R 121 449-029.

- **Postage paid** by Annual Reviews (4th class bookrate/surface mail). UPS ground service is available at S2.00 extra per book within the contiguous 48 states only. UPS air service or US airmail is available to any location at actual cost. UPS requires a street address. P.O. Box, APO, FPO, not acceptable.

- **Standing Orders:** Set up a standing order and the new volume in series is sent automatically each year upon publication. Each year you can save 10% by prepayment of prerelease invoices sent 90 days prior to the publication date. Cancellation may be made at any time.

- **Prepublication Orders:** Advance orders may be placed for any volume and will be charged to your account upon receipt. Volumes not yet published will be shipped during month of publication indicated.

N O T E	For copies of individual articles from any *Annual Review*, or copies of any article cited in an *Annual Review*, call **Annual Reviews Preprints and Reprints (ARPR)** toll free 1-800-347-8007 (fax toll free 1-800-347-8008) from the USA or Canada. From elsewhere call 1-415-259-5017.

ANNUAL REVIEWS SERIES *Volumes not listed are no longer in print*	Prices, postpaid, per volume. USA/other countries	Regular Order Please send Volume(s):	Standing Order Begin with Volume:
Annual Review of ANTHROPOLOGY			
Vols. 1-20 (1972-91) ....$41 / $46			
Vols. 21-22 (1992-93) ....$44 / $49			
Vol. 23-24 (1994 and Oct. 1995) ....$47 / $52		Vol(s). _____	Vol. _____
Annual Review of ASTRONOMY AND ASTROPHYSICS			
Vols. 1, 5-14, 16-29 (1963, 67-76, 78-91) ....$53 / $58			
Vols. 30-31 (1992-93) ....$57 / $62			
Vol. 32-33 (1994 and Sept. 1995) ....$60 / $65		Vol(s). _____	Vol. _____
Annual Review of BIOCHEMISTRY			
Vols. 31-34, 36-60 (1962-65,67-91) ....$41 / $47			
Vols. 61-62 (1992-93) ....$46 / $52			
Vol. 63-64 (1994 and July 1995) ....$49 / $55		Vol(s). _____	Vol. _____
Annual Review of BIOPHYSICS AND BIOMOLECULAR STRUCTURE			
Vols. 1-20 (1972-91) ....$55 / $60			
Vols. 21-22 (1992-93) ....$59 / $64			
Vol. 23-24 (1994 and June 1995) ....$62 / $67		Vol(s). _____	Vol. _____

| ANNUAL REVIEWS SERIES
Volumes not listed are no longer in print | Prices, postpaid,
per volume.
USA/other countries | Regular
Order
Please send
Volume(s): | Standing
Order
Begin with
Volume: |

❏ *Annual Review of* **MICROBIOLOGY**
Vols. 21-24, 26-45   (1967-70, 72-91) ....$41 / $46
Vols. 46-47   (1992-93) ....$45 / $50
Vol.   48-49   (1994 and Oct. 1995) ....$48 / $53   Vol(s). _____ Vol. _____

❏ *Annual Review of* **NEUROSCIENCE**
Vols. 1-14   (1978-91) ....▸...........................$40 / $45
Vols. 15-16   (1992-93) ....$44 / $49
Vol.   17-18   (1994 and Mar. 1995) ....$47 / $52   Vol(s). _____ Vol. _____

❏ *Annual Review of* **NUCLEAR AND PARTICLE SCIENCE**
Vols. 12-41   (1962-91) ....$55 / $60
Vols. 42-43   (1992-93) ....$59 / $64
Vol.   44-45   (1994 and Dec. 1995) ....$62 / $67   Vol(s). _____ Vol. _____

❏ *Annual Review of* **NUTRITION**
Vols. 1-11   (1981-91) ....$43 / $48
Vols. 12-13   (1992-93)................................$45 /$50
Vol.   14-15   (1994 and July 1995) ....$48 / $53   Vol(s). _____ Vol. _____

❏ *Annual Review of* **PHARMACOLOGY AND TOXICOLOGY**
Vols. 2-3, 5-31   (1962-63, 65-91).....................$40 / $45
Vols. 32-33   (1992-93) ....$44 / $49
Vol.   34-35   (1994 and April 1995) ....$47 / $52   Vol(s). _____ Vol. _____

❏ *Annual Review of* **PHYSICAL CHEMISTRY**
Vols. 13-21, 23-27   (1962-70, 72-76)
       29-42   (1978-91) ....$44 / $49
Vols. 43-44   (1992-93) ....$48 / $53
Vol.   45-46   (1994 and Nov. 1995) ....$51 / $56   Vol(s). _____ Vol. ____/

❏ *Annual Review of* **PHYSIOLOGY**
Vols. 19-53   (1957-91) ....$42 / $47
Vols. 54-55   (1992-93) ....$46 / $51
Vol.   56-57   (1994 and Mar. 1995) ....$49 / $54   Vol(s). _____ Vol. _____

❏ *Annual Review of* **PHYTOPATHOLOGY**
Vols. 3-20, 22-29   (1965-82, 84-91) ....$42 / $47
Vols. 30-31   (1992-93) ....$46 / $51
Vol.   32-33   (1994 and Sept. 1995) ....$49 / $54   Vol(s). _____ Vol. _____

❏ *Annual Review of* **PLANT PHYSIOLOGY & PLANT MOLECULAR BIOLOGY**
Vols. 17-23, 26-29   (1966-72, 75-78)
       31-42   (1980-91) ....$40 / $45
Vols. 43-44   (1992-93) ....$44 / $49
Vol.   45-46   (1994 and June 1995) ....$47 / $52   Vol(s). _____ Vol. _____

❏ *Annual Review of* **PSYCHOLOGY**
Vols. 4, 5, 8, 10, 22-24 (1953, 54, 57, 59, 71-73)
       26-30, 33-37, 39-42 (75-79, 82-86, 88-91)....$40 / $45
Vols. 43-44   (1992-93) ....$43 / $48
Vol.   45-46   (1994 and Feb. 1995) ....$46 / $51   Vol(s). _____ Vol. _____

❏ *Annual Review of* **PUBLIC HEALTH**
Vols. 1-12   (1980-91) ....$45 / $50
Vols. 13-14   (1992-93) ....$49 / $54
Vol.   15-16   (1994 and May 1995) ....$52 / $57   Vol(s). _____ Vol. _____